ADVANCED ELECTROPORATION TECHNIQUES IN BIOLOGY AND MEDICINE

Biological Effects of Electromagnetics Series

Series Editors

Frank Barnes
University of Colorado
Boulder, Colorado, U.S.A.

Ben Greenbaum
University of Wisconsin–Parkside
Somers, Wisconsin, U.S.A.

Advanced Electroporation Techniques in Biology and Medicine, *edited by Andrei G. Pakhomov, Damijan Miklavčič, and Marko S. Markov*

Forthcoming Publications

The Physiology of Bioelectricity in Development, Tissue Regeneration, and Cancer, *edited by Christine Pullar*

Biological Effects of Electromagnetics Series

ADVANCED ELECTROPORATION TECHNIQUES IN BIOLOGY AND MEDICINE

Edited by
Andrei G. Pakhomov
Damijan Miklavčič
Marko S. Markov

CRC Press
Taylor & Francis Group
Boca Raton London New York

CRC Press is an imprint of the
Taylor & Francis Group, an **informa** business

CRC Press
Taylor & Francis Group
6000 Broken Sound Parkway NW, Suite 300
Boca Raton, FL 33487-2742

© 2010 by Taylor and Francis Group, LLC
CRC Press is an imprint of Taylor & Francis Group, an Informa business

No claim to original U.S. Government works

Printed in the United States of America on acid-free paper
10 9 8 7 6 5 4 3 2 1

International Standard Book Number: 978-1-4398-1906-7 (Hardback)

Visit the Taylor & Francis Web site at
http://www.taylorandfrancis.com

and the CRC Press Web site at
http://www.crcpress.com

Foreword

It is our pleasure to introduce this volume on electroporation as the first volume in a series of books on the effects of electric and magnetic fields on biological systems. This series of books is intended to both update and extend material that was reviewed in the third edition of the *Handbook of Biological Effects of Electric and Magnetic Fields*. The activity in this field includes both issues associated with low levels of exposures, including concerns about the possible health effects of extended exposures, and applications of electric and magnetic fields for therapeutic applications. Other volumes in various states of completion include detailed studies of the effects of electric fields on cell physiology, epidemiological studies of exposures to radio frequencies, and modeling of electric fields in biological systems.

Our hope is that this volume will be valuable both as an introduction and a guide to the current state of knowledge and techniques for those who wish to use these short electric field pulses for the clinical treatment of human diseases and to study subcellular structures. Future volumes in this series will have the same goals for other aspects of the biological effects of electric and magnetic fields.

Frank Barnes
Ben Freenebaum

Preface

It is well accepted today that intense, pulsed electric fields cause structural changes in biomembranes, which presumably involve a rearrangement of the phospholipid bilayer and a formation of aqueous pores. Consequently, the membrane becomes permeable to water-soluble molecules and ions that are otherwise deprived of membrane transport mechanisms. This phenomenon, widely known as membrane electroporation, integrates multiple effects predicted by theory and observed experimentally in biological and clinical settings. This book is intended to summarize the most recent experimental findings and theories related to the permeabilization of biomembranes by pulsed electric fields.

It was not until the 1970s that isolated biological experiments with electric pulses drew the attention of electrochemists and biophysicists. Within a short period, electroporation became the subject of special sessions and meetings under the framework of the Bioelectrochemical Society and Bioelectromagnetics Society. In the early 1990s, electroporation had already been recognized as a powerful tool for everyday laboratory research and had started showing promise as a therapeutic modality. As of today, the medical applications of electroporation include gene electrotransfer, transdermal drug delivery, tumor and tissue ablation, and the electrochemotherapy of tumors.

In general, the key parameter for electroporation is the induced transmembrane voltage generated by the external electric field due to the difference in the electric properties of the membrane and the surrounding medium, known as the Maxwell–Wagner polarization. Two main theoretical approaches have been developed to describe electroporation. The electromechanical approach considers membranes as viscoelastic bodies and applies principles of electrostatics and elasticity to predict membrane rupture above the critical membrane voltage. A conceptually different approach describing the formation and expansion of pores is based on energy consideration. It is assumed that the external electric field reduces the free energy barrier for the formation of hydrophilic pores due to the lower polarization energy of water in the pores as compared to the membrane. The molecular mechanisms of pore formation and stabilization during electroporation are not yet fully understood and rigorous experimental confirmation of different theories is still lacking.

For pulses of different durations, biological effects are to a great extent determined by the electrical charge that is transferred to the cell membrane during the treatment. During the last decade, experiments and theoretical approaches have been expanded to include nanosecond pulses. The electrical models of biological cells predict that reducing the duration of applied electrical pulses to values below the charging time of the outer cell membrane (which is on the order of 100 ns for mammalian cells) increases the probability of electric field interactions with intracellular structures.

This book is an interdisciplinary compilation intended mainly for engineers, physical and biomedical scientists, and clinicians. It can also be used as a textbook for students in advanced courses in biomedical engineering, molecular and cell biology, as well as in biophysics and clinical medicine.

Finally, we would like to thank all the authors who contributed their time and effort and delivered their chapters on time. It was their commitment and collaborative effort that ultimately made this project a success.

Andrei G. Pakhomov
Damijan Miklavčič
Marko S. Markov

Contents

PART I   Basics of Electroporation

PART II   Mechanisms of Electroporation in Lipid Systems

PART III   Mechanisms of Electroporation of Cells

PART IV   Mechanisms of Electroporation in Tissues

PART V   Technical Considerations

PART VI   Applications of Electroporation

Editors

Andrei G. Pakhomov is a research associate professor at Frank Reidy Research Center for Bioelectrics of Old Dominion University in Norfolk, Virginia. He received his MS in animal and human physiology from Moscow State University in Moscow, Russia in 1982 and his PhD in radiation biology/biophysics from the Medical Radiology Research Center (MRRC) in Obninsk, Russia, in 1989. Since 1982, he has been studying the bioeffects of electromagnetic fields, including millimeter waves, extremely high-peak power microwave pulses, and nanosecond pulsed electric fields. Dr. Pakhomov was a leading scientist at the MRRC until he received a Resident Research Associateship Award from the National Research Council. He moved to San Antonio, Texas, in 1994, and continued his research at Brooks AFB, Brooks City Base, Texas and UT Health Science Center, San Antonio, Texas. He is a member of the Bioelectromagnetics Society (BEMS), of the International Committee on Electromagnetic Safety (ICES), and of the Society for Neuroscience (SFN). He served on a number of grant review panels, including NIH study sections, and was an organizing committee member for several international meetings. In 2003–2004, he was a guest editor for *IEEE Transactions on Plasma Science*, and has served as an associate editor of *Bioelectromagnetics* since 2004. Dr. Pakhomov is the principal author of over 120 publications and meeting presentations, and has delivered keynote addresses and invited lectures at more than 20 meetings.

Damijan Miklavčič was born in Ljubljana, Slovenia, in 1963. He received his PhD in electrical engineering from the University of Ljubljana, Ljubljana, Slovenia. He is currently a professor in the Faculty of Electrical Engineering, University of Ljubljana, where he is the chair of the Department of Biomedical Engineering and the head of the Laboratory of Biocybernetics. He is active in the field of biomedical engineering. His current research interests include electroporation-assisted drug and gene delivery, including cancer treatment by means of electrochemotherapy, tissue oxygenation, and modeling.

Marko S. Markov is the president of Research International, Buffalo, New York. He graduated from Sofia University, Sofia, Bulgaria, with a PhD in biophysics and has been associated with the Department of Biophysics for 22 years. His research was on the biological and clinical effects of electromagnetic fields. In the second half of the 1980s, he became interested in studying electroporation. He is one of the founders of the European Bioelectromagnetic Association, the International Society of Bioelectricity, and the International Society of Magnetobiology and Magnetotherapy. As a member of the Bioelectromagnetics Society, he served a term as a member of the board of directors. Dr. Markov has published more than 160 papers in peer-reviewed journals; he is also the author and coeditor of seven books published by Plenum Press, Marcel Dekker, and Springer. He delivered keynote addresses and invited papers at more than 60 international meetings. He has also organized sessions, workshops, and meetings in Bulgaria, the United States, Brazil, Japan, Greece, Italy, and Armenia.

Contributors

Iskandar Barakat
Electrical Trauma Research
 Laboratory
Department of Surgery,
 Medicine, and Organismal
 Biology (Biomechanics)
Pritzker School of Medicine
The University of Chicago
Chicago, Illinois
ibarakat@surgery.bsd.
 uchicago.edu

Elisabeth Bellard
Institut de Pharmacologie et de
 Biologie Structurale
Unité Mixte de Recherches
Centre National de la Recherche
 Scientifique
Université Paul Sabatier
Université de Toulouse
Toulouse, France
Elisabeth.Bellard@ipbs.fr

Luca Giovanni Campana
Department of Oncological and
 Surgical Sciences
Sarcoma and Melanoma Unit
Istituto Oncologico Veneto
Padua, Italy
maximizing@hotmail.com

Maja Cemazar
Department of Experimental
 Oncology
Institute of Oncology Ljubljana
Ljubljana, Slovenia
mcemazar@onko-i.si

Hongfeng Chen
Electrical Trauma Research
 Laboratory
Department of Surgery,
 Medicine, and Organismal
 Biology (Biomechanics)
Pritzker School of Medicine
The University of Chicago
Chicago, Illinois
hchen@surgery.bsd.uchicago.edu

Lucie Delemotte
Equipe de Dynamique
 des Assemblages
 Membranaires
Centre National de la Recherche
 Scientifique
Unité Mixte de Recherche
Nancy Université
Nancy, France
Lucie.delemotte@srsmc.
 uhp-nancy.fr

Rumiana Dimova
Max Planck Institute of
 Colloids and Interfaces
Potsdam, Germany
dimova@mpikg.mpg.de

Witold Dyrka
Institute of Biomedical
 Engineering and
 Instrumentation
Wroclaw University of
 Technology
Wroclaw, Poland
witold.dyrka@pwr.wroc.pl

Jean-Michel Escoffre
Institut de Pharmacologie et de
 Biologie Structurale
Unité Mixte de Recherches
Centre National de la Recherche
 Scientifique
Université Paul Sabatier
Université de Toulouse
Toulouse, France
escoffre@ipbs.fr

Axel T. Esser
Division of Health Sciences
 and Technology
Harvard-Massachusetts
 Institute of Technology
Cambridge, Massachusetts
axel61@mit.edu

Patrick F. Forde
Leslie C Quick Jr. Laboratory
Cork Cancer Research Centre
BioSciences Institute
University College
Cork, Ireland
Pat@CCRC.ie

Jill Gallaher
Electrical Trauma Research
 Laboratory
Department of Surgery,
 Medicine, and Organismal
 Biology (Biomechanics)
Pritzker School of Medicine
The University of Chicago
Chicago, Illinois
jgallagher@surgery.bsd.
 uchicago.edu

Julie Gehl
Department of Oncology
Center for Experimental Drug
  and Gene Electrotransfer
Copenhagen University
  Hospital Herlev
Herlev, Denmark
JUGE@heh.regionh.dk

Muriel Golzio
Institut de Pharmacologie et de
  Biologie Structurale
Unité Mixte de Recherches
Centre National de la Recherche
  Scientifique
Université Paul Sabatier
Université de Toulouse
Toulouse, France
golzio@ipbs.fr

Thiruvallur R. Gowrishankar
Division of Health Sciences
  and Technology
Harvard-Massachusetts
  Institute of Technology
Cambridge, Massachusetts
tgowrish@mit.edu

Loree C. Heller
Frank Reidy Research Center
  for Bioelectrics
Old Dominion University
Norfolk, Virginia
LHeller@odu.edu

Richard Heller
Frank Reidy Research Center
  for Bioelectrics
Old Dominion University

and

Medical Laboratory and
  Radiation Sciences
Old Dominion University
Norfolk, Virginia
RHeller@odu.edu

Antoni Ivorra
Centre National de la Recherche
  Scientifique
Unité Mixte de Recherche

Institut Gustave-Roussy
Villejuif, France

and

Unité Mixte de Recherche
Univ Paris-Sud
Orsay, France
antoni.ivorra@gmail.com

Tomaz Jarm
Faculty of Electrical
  Engineering
University of Ljubljana
Ljubljana, Slovenia
tomaz.jarm@fe.uni-lj.si

Vanessa Joubert
Centre National de la Recherche
  Scientifique
Unité Mixte de Recherche
Institut Gustave-Roussy
Villejuif, France

and

Unité Mixte de Recherche
Univ Paris-Sud
Orsay, France
Vanessajoubert@aol.com

Sergej Kakorin
Faculty of Chemistry
University of Bielefeld
Bielefeld, Germany
sergej.kakorin@uni-bielefeld.de

Juergen F. Kolb
Frank Reidy Research Center
  for Bioelectrics
Old Dominion University
Norfolk, Virginia
kolb@odu.edu

Tadej Kotnik
Faculty of Electrical
  Engineering
University of Ljubljana
Ljubljana, Slovenia
tadej.kotnik@fe.uni-lj.si

Malgorzata Kotulska
Institute of Biomedical
  Engineering and
  Instrumentation

Wroclaw University of
  Technology
Wroclaw, Poland
malgorzata.kotubska@pwr.
  wroc.pl

Simona Kranjc
Department of Experimental
  Oncology
Institute of Oncology Ljubljana
Ljubljana, Slovenia
skranjc@onko-i.si

Jasna Krmelj
Faculty of Electrical
  Engineering
University of Ljubljana
Ljubljana, Slovenia
jasna.krmelj@gmail.com

Alenka Maček Lebar
Faculty of Electrical
  Engineering
University of Ljubljana
Ljubljana, Slovenia
alenka.maceklebar@fe.uni-lj.si

Nikolai Lebovka
Unité Transformations
  Intégrées de la Matière
  Renouvelable
Centre de Recherche de
  Royallieu
Université de Technologie de
  Compiègne
Compiègne, France

and

Ovcharenko Institute of
  Biocolloidal Chemistry
National Academy of Sciences
Kyiv, Ukraine
lebovka@roller.ukma.kiev.ua

Raphael C. Lee
Electrical Trauma Research
  Laboratory
Department of Surgery,
  Medicine, and Organismal
  Biology (Biomechanics)

Pritzker School of Medicine
The University of Chicago
Chicago, Illinois
rlee@surgery.bsd.uchicago.edu

Marko S. Markov
Research International
Buffalo, New York
msmarkov@aol.com

Damijan Miklavčič
Faculty of Electrical
    Engineering
University of Ljubljana
Ljubljana, Slovenia
damijan.miklavcic@
    fe.uni-lj.si

Lluis M. Mir
Centre National de la Recherche
    Scientifique
Unité Mixte de Recherche
Institut Gustave-Roussy
Villejuif, France

and

Unité Mixte de Recherche
Univ Paris-Sud
Orsay, France
luismir@igr.fr

Eberhard Neumann
Faculty of Chemistry
University of Bielefeld
Bielefeld, Germany
eberhard.neumann@
    uni-bielefeld.de

Gerald C. O'Sullivan
Leslie C. Quick Jr. Laboratory
Cork Cancer Research Centre
BioSciences Institute
University College
Cork, Ireland
Geraldc@ccrc.ie

Aurélie Paganin-Gioanni
Institut de Pharmacologie et de
    Biologie Structurale

Unité Mixte de Recherches
Centre National de la Recherche
    Scientifique
Université Paul Sabatier
Université de Toulouse
Toulouse, France
Aurelie.Paganin@ipbs.fr

Andrei G. Pakhomov
Frank Reidy Research Center
    for Bioelectrics
Old Dominion University
Norfolk, Virginia
andrei@pakhomov.net

Olga N. Pakhomova
Frank Reidy Research Center
    for Bioelectrics
Old Dominion University
Norfolk, Virginia
olga@pakhomova.net

Nataša Pavšelj
Faculty of Electrical
    Engineering
University of Ljubljana
Ljubljana, Slovenia
Natasa.pavselj@fe.uni-lj.si

Gorazd Pucihar
Faculty of Electrical
    Engineering
University of Ljubljana
Ljubljana, Slovenia
gorazd.pucihar@fe.uni-lj.si

Matej Reberšek
Faculty of Electrical
    Engineering
University of Ljubljana
Ljubljana, Slovenia
matej.rebersek@fe.uni-lj.si

Marie-Pierre Rols
Institut de Pharmacologie et de
    Biologie Structurale
Unité Mixte de Recherches
Centre National de la Recherche
    Scientifique
Université Paul Sabatier

Université de Toulouse
Toulouse, France

Carlo Ricardo Rossi
Department of Oncological and
    Surgical Sciences
Sarcoma and Melanoma Unit
Istituto Oncologico Veneto
Padua, Italy
carlor.rossi@unipd.it

Boris Rubinsky
School of Computer Science
    and Engineering
Hebrew University of Jerusalem
Jerusalem, Israel

and

Department of Mechanical
    Engineering
University of California at
    Berkeley
Berkeley, California
rubinsky@me.berkeley.edu

Przemyslaw Sadowski
Institute of Biomedical
    Engineering and
    Instrumentation
Wroclaw University of
    Technology
Wroclaw, Poland
przemyslaw.sadowski@pwr.
    wroc.pl

Gintautas Saulis
Department of Biology
Vytautas Magnus University
Kaunas, Lithuania
sg@kaunas@aol.com

Karl H. Schoenbach
Frank Reidy Research Center
    for Bioelectrics

and

Department of Electrical and
    Computer Engineering
Old Dominion University
Norfolk, Virginia
kschoenb@odu.edu

Gregor Sersa
Department of Experimental
    Oncology
Institute of Oncology Ljubljana
Ljubljana, Slovenia
gsersa@onko-i.si

Aude Silve
Centre National de la Recherche
    Scientifique
Unité Mixte de Recherche
Institut Gustave-Roussy
Villejuif, France

and

Unité Mixte de Recherche
Univ Paris-Sud
Orsay, France
aude.silve@ens-cachan.fr

Kyle C. Smith
Division of Health Sciences
    and Technology
Harvard-Massachusetts
    Institute of Technology

and

Department of Electrical
    Engineering and Computer
    Science
Massachusetts Institute of
    Technology
Cambridge, Massachusetts
smithkc@mit.edu

Declan M. Soden
Leslie C Quick Jr. Laboratory
Cork Cancer Research Centre
BioSciences Institute
University College
Cork, Ireland
d.soden@ucc.ie

Mounir Tarek
Equipe de Dynamique des
    Assemblages Membranaires
Centre National de la Recherche
    Scientifique

Unité Mixte de Recherche
Nancy Université
Nancy, France
Mounir.Tarek@edam.
    uhp-nancy.fr

Justin Teissié
Institut de Pharmacologie et de
    Biologie Structurale
Unité Mixte de Recherches
Centre National de la Recherche
    Scientifique
Université Paul Sabatier
Université de Toulouse
Toulouse, France
Justin.Teissie@ipbs.fr

Leila Towhidi
Department of Medical Physics
Tarbiat Modares University
Tehran, Iran
leilatowhidi@yahoo.com

Ciara Twomey
Leslie C Quick Jr. Laboratory
Cork Cancer Research Centre
BioSciences Institute
University College
Cork, Ireland
Ciara.Twomey@ucc.ie

P. Thomas Vernier
Department of Electrical
    Engineering-Electrophysics
Viterbi School of Engineering
University of Southern
    California

and

Viterbi School of Engineering
Information Sciences Institute
University of Southern
    California
Los Angeles, California
vernier@usc.edu

Julien Villemejane
Centre National de la Recherche
    Scientifique
Unité Mixte de Recherche
Institut Gustave-Roussy
Villejuif, France

and

Unité Mixte de Recherche
Univ Paris-Sud
Orsay, France

and

Centre National de la Recherche
    Scientifique
Systèms et Applications
    des Technologies de
    l'Information et de l'Energie
Institut d'Alembert
Ecole Normale Supérieure de
    Cachan
Cachan, France
julien.villemejane@ens-cachan.fr

Eugene Vorobiev
Département Génie Chimique
Laboratoire de Génie des
    Procédés Industriels
Université de Technologie de
    Compiègne
Compiègne, France
eugene.vorobiev@utc.fr

James C. Weaver
Division of Health Sciences
    and Technology
Harvard-Massachusetts
    Institute of Technology
Cambridge, Massachusetts
jcw@mit.edu

Anže Župani
Faculty of Electrical
    Engineering
University of Ljubljana
Ljubljana, Slovenia
anze.zupanic@fe.uni-lj.si

MATLAB® Disclaimer

I

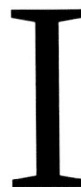

Basics of Electroporation

1

Physical Chemical Theory of Membrane Electroporation and Electrotransfer of Biogenic Agents

Eberhard Neumann

Sergej Kakorin

1.1 Introduction

The medical applications of membrane electroporation (MEP) in electrochemotherapy and gene electrotransfer are based on the principle of "functional electroporation." The concept dates back to 1982, where originally trains of electric pulses (with a time interval of 3 s between the individual pulses) have been applied to directly reprogram, that is, transform mouse lyoma cells with naked DNA. The DNA solution had been simply added to the dense cell suspension and, after a couple of minutes to allow for adsorption of the DNA to the cell surface, the high-voltage (HV) pulse trains had been applied

3

(Neumann et al. 1982, Wong and Neumann 1982). Previously, complementary to functional electro-uptake, the electric pulse technique had been used to achieve electro-release of cellular components, such as catecholamine, ATP, and proteins, from isolated chromaffin granules of bovine adrenal medullae (Neumann and Rosenheck 1972). These initial physical chemical data on electroporative uptake and release of molecules have been recently valued, among others, in *Nature Methods* (Eisenstein 2006) as seminal for the various biotechnological and medical applications of "functional electroporation," among them the clinical applications of voltage pulses combined with bioactive agents (Neumann 2007). The chemical electrodynamic concept of functional MEP also alludes to a molecular mechanism for the primary effect of the electric field forces (Neumann 2007) as directly affecting the hydrated polar head groups of lipids, leading to the formation of inverted or hydrophilic pores. Field-induced translational motions of the polar lipids in the curved pore wall also rationalize the huge acceleration of lipid flip-flop and other intra-wall motions such as the translocation of phosphatidylserine from the internal membrane monolayer to the outer monolayer. Electric pulses of low field intensity but longer pulse duration facilitate, apparently via electroporation, both the endocytotic uptake of external particles and the exocytotic release of intracellular components. The physical chemical theory of MEP for closed membrane shells, besides addressing the primary field-induced lipid processes, also rationalizes the observed longevity of the porous structures in terms of local cooperativity of the lipids in the highly curved pore walls of a hydrophilic pore. The theory-based technical developments have culminated in the clinical applications of electroporation: electrochemotherapy and gene electrotherapy (Okino and Mori 1987, Heller et al. 1996, Mir et al. 1996, 1991, Miklavcic et al. 1997). The following detailed digression is restricted to a critical appreciation of the contemporary physical chemical theory of MEP.

1.2 Physical Chemical Concepts of Membrane Electroporation

1.2.1 Physical Chemical Theory

The physical chemical theory, besides the pure physical aspects such as the electric polarization term of water entrance in pore, pore line tension and membrane surface tension, pore radius, and pore expansion (Chizmadzhev et al. 1997, Esser et al. 2007, Dimova et al. 2008, Smith and Weaver 2008), explicitly addresses the chemical free energy changes of pore formation and pore resealing. The structural changes cause the observed permeability changes, which are particularly apparent in the measured conductance relaxations of planar lipid bilayers and of cell suspensions. Changes in membrane structure are also indicated by electrooptical relaxation data. As to cell membranes, the closed membrane shells of vesicles are judged as a good model for the curved lipid parts of cell membranes. On the other hand, most planar lipid bilayers are under torus tension that enhances field-induced electromechanical stress finally leading to bilayer breakdown (rupture).

In summary of the relaxation kinetic data, the observed *exponential* time courses of both the conductometric signals and the electrooptical signals indicate entrance of water in the lipid phase and global shape changes. The single-current events of patch clamp measurements clearly indicate local transport sites, structurally specified as hydrophilic pores. In both the membranes of unilamellar lipid vesicles (Griese et al. 2002) and the cell membranes of densely packed CHO cells (Schmeer et al. 2004), respectively, there are at least two types of stable "electropores": (a) short-lived fluctuative hydrophilic pores (average pore radius of 0.6 ± 0.2 nm), after-field life time of 1–5 ms (Pliquett et al. 2002) specified below as type P_1 and (b) long-lived, larger fluctuative electropores (average pore radius ≥ 1 nm (Schmeer et al. 2004)), after-field life times of minutes, specified below as type P_2. The P_2 pores are the candidates to rationalize the structural longevity of the porous membrane states for the observed long-lived mass transport after the pulse. Usually, for cellular systems the pulse times should be short to prevent cell damage, thus mass transport within the field duration is very small as compared to after-field transport. This structural feature of pore longevity is also instrumental for rationalizing some of the voltage pulse data for pulse train combination modes of HV pulses and low voltage (LV) pulses, and the effects of a time interval between the pulses (Andre et al. 2008). Viewed afterward, the originally applied "exponential field pulses" (Neumann et al. 1982, Wong and Neumann 1982) with the longer RC-circuit discharge times combine the HV part of the initial time course with the LV part of the slower part.

1.2.2 Electric Field Is Force on Charge

In order to understand the field effects involved in the various medical electroporation treatments, it is essential to recall that the *electric field E* (of a voltage pulse) *acts as a force (simultaneously) on (all) polar, i.e., ionic and dipolar groups of the membrane components*. The induced membrane field, originating from the mobile ions near the two surface sides of the dielectric membrane, acts also over the membrane-adsorbed molecules. The field forces directly cause the structural reorganizations of the membrane lipids, finally leading to higher, field-stabilized order of the local permeation sites. The porous structures, once established, permit trans-membrane transport of small ions and of externally applied (and transiently adsorbed) substances into the inside, and transport of intracellular components to the outside.

1.2.3 Transport of Small Ions and Larger Molecules

Due to the small size of isolated fluctuating pores, the migration of substances, including small ions, through the porous parts is *always interactive*, involving transient electrostatic complexes between sites of the transported substance and sites of the structured pore wall. Patch clamp current data suggest that larger molecules like proteins and polyelectrolytes like DNA and RNA are in multiple contacts with the lipids (Hristova et al. 1997). It is thus essential that theory and analysis of transport also incorporate the thermodynamic interactions between the porous membrane patches and the adsorbed molecules, also during the transport phases. The amplified large membrane field may induce a kind of percolation of the small hydrophilic pores in the contact area of charged particles to form larger, but occluded pores. Lengthwise adsorbed DNA within the membrane may be viewed as occluding a long macro-pore. Similarly, larger globular molecules like proteins, and dyes and drugs appear to transiently stick in, or block or occlude, a crater-like large pore. It is recalled that DNA and RNA are polyanions. When exposed to an external electric field, they electro-migrate in the direction opposite to the direction of the applied field. So, at the cathodic cell hemisphere (facing the negative electrode), an external field draws the DNA toward, and once adsorbed, into the porous membrane. At the anodic hemisphere, these polyanions are drawn away from the membrane. Once within (i.e., occluding) a porous structure, the DNA is an interactive part of this local membrane patch. This important feature rationalizes that the small-ion leak currents, measurable in patch clamp configuration of lipid surface-adsorbed DNA, are linearly dependent on the length of the adsorbed parts of long DNA molecules (Hristova et al. 1997). Once a part of or within the membrane, after pulse termination the "membrane-bound particles" can thermally redistribute. Membrane-adherent DNA can dissociate after the pulses from the membrane into the cell interior, or alternatively, can return into the outside solution. The lipids of the pore wall of the formerly occluded large pore, freed from the macromolecule, now can reorganize back to the closed membrane configuration. In the absence of the ordering field forces, the lipids of the ordered pore walls will go through many mismatched lipid-water associations before the closed lipid structure is re-established. So, the random thermal nature of the pore resealing rationalizes the relatively slow return to the closed membrane structure. It is realized that the coupling of transport to preceding adsorption covering only the small contact part of the membrane surface, and the relatively small pore surface density, rationalizes the relatively small efficiency of interactive uptake of adsorbed molecules like DNA or proteins at lower concentrations into single cells in suspension. The higher cell density of cell aggregates as encountered in tissue, together with the higher local concentrations of externally applied molecules, all increase the uptake efficiency.

1.3 Chemical Schemes for Pore Formation

1.3.1 Data Basis

The data basis suggests that the structural transitions of pore formation and pore resealing involve a whole cascade of field-sensitive closed membrane states (C) and a sequence of porous states (P). Fundamental principles of chemical description and thermodynamic analysis of field effects had been previously outlined in terms of an overall two-state scheme $(C) \rightleftharpoons (P) \rightarrow X_{trs}$, where the field-sensitive membrane state transitions $(C) \rightleftharpoons (P)$ are coupled to specific transport processes X_{trs} (Neumann et al. 1982). The overall distribution function K_p had been defined as $K_p = [C]/[P]$ with reference to the total surface area A_0 and to the total state concentration $[C_0] = [C] + [P]$. The ratio $f_p = [(P)]/([P] + [C])$ represents an area ratio $f(A) = A(p)/A_0$, the pores covering the pore area $A(p)$. From experience, it is found that the limit is $f(A) = f_p \leq 10^{-2}$. However, for geometric and energetic reasons, the total membrane area can contain only a limited number N_p of aqueous pores. If this area for potential pore formation is denoted by C^*, where for lipid phases $[(C^*)] \leq 10^{-2}[C_0]$ holds true, the pore formation.

$$(C^*) \rightleftharpoons (P) \tag{1.1}$$

Scheme (1.1) may be viewed as representing pore formation in terms of field-induced cooperative rearrangements of n (dipolar head groups) lipids, to locally form hydrophilic pores L_n and $L_n(W)$ by water (W) entrance, according to

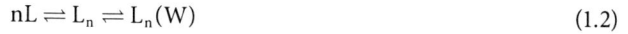

$$nL \rightleftharpoons L_n \rightleftharpoons L_n(W) \tag{1.2}$$

The water in the configuration $L_n(W)$ will be more polarized in the strong transmembrane field (E_m) as compared to bulk water. Therefore, water entrance contributes strongly to the thermodynamic stability of the aqueous hydrophilic pore (Abidor et al. 1979, Neumann et al. 1982). For comparison, $[L_n] = [(C^*)]$ and $[L_n(W)] = [(P)]$.

1.3.2 Hysteresis

Scheme (1.1) is also useful to feature another particular aspect of the electroporation-resealing cycle underlying the various transport phenomena. The data suggest that the very structural transitions of pore formation as well as those of pore closure may be viewed as a hysteresis cycle. The chemical processes as such are reversible. In presence of the field, net pore formation is unidirectional. Net pore resealing, in the absence of the external field, also proceeds unidirectionally, but is a purely thermal redistribution process. So, poration and resealing are different, involving different structural intermediates with different transition kinetics. The different branches of the structural transitions constitute the hysteresis loop (Neumann 1989). The directionality of the branches qualifies the field effect hysteresis as containing unidirectional, that is, irreversible pathways. If the resealed membrane state after pulse application is the same as that before pulsing, the hysteresis is called "overall reversible."

1.3.3 Equilibrium Distribution Constants

The distribution equilibrium constant for the overall Scheme (1.1) is defined as state density ratio or equilibrium constant:

$$K = \frac{[(P)]}{[(C^*)]} = \frac{f}{1-f} \tag{1.3}$$

The fraction f of pore states of the system (C*) and (P) is defined as

$$f = \frac{[(P)]}{[(P)] + [(C^*)]} \qquad (1.4)$$

Note that the numerical values of f are limited to the range $0 \leq f \leq 1$. In line with the actual data both K and f increase with increasing field. The analysis of the kinetic normal modes of conductometric and electrooptic relaxation spectrometric data (Griese et al. 2002, Schmeer et al. 2004) requires the formulations of at least two schemes in order to describe the two discernible kinetic phases: Scheme (P*) $\rightleftharpoons$ (P$_1$), for the rapid transition from a pre-pore state P* to the pore state P$_1$ associated with the fraction $f_1 = [P_1]/([P^*] + [P_1])$ and Scheme (P$_1$) $\rightleftharpoons$ (P$_2$), for which the distribution constant is given by $K_2 = [P_2]/[P_1] = f_2/(1 - f_2)$. The transition to larger pores of type P$_2$ is related to the fraction $f_2 = [(P_2)]/([(P_2)]+[(P_1)])$ in the range $0 \leq f_2 \leq 1$. Previously, the restricted assumption of $1 + K = K$, that is, $K \ll 1$, justified for the low-field range had been used for the analysis (Griese et al. 2002, Schmeer et al. 2004).

1.3.4 General Physical Chemical Energetics

The overall Scheme (1.1) is suited to describe the general effects of generalized thermodynamic forces on chemical processes in terms of a generalized van t' Hoff relationship (Neumann et al. 1982) covering, as a total differential d ln K, change dT in the Kelvin temperature T, change dp in the pressure p and change dE in the strength E of the "locally active" electric field, relative to the molar energy unit RT:

$$RT\, d\ln K = \Delta_r H^0_{p,E} dT - \Delta_r V^0_{T,E} dp + \Delta_r M^0_{p,T} dE \qquad (1.5)$$

where
   $R = k_B N_A$ is the gas constant
   k_B is the Boltzmann constant
   N_A is the Loschmidt–Avogadro constant

Equation 1.5 refers to the different poration phenomena: electroporation is associated with the standard value $\Delta_r M^0 = M^0(P) - M^0(C)$ of the reaction dipole moment, "sonoporation" with the standard value of the reaction volume $\Delta_r V^0$, "thermo-poration" and thermal aspects of laser "opto-poration" with the standard reaction enthalpy $\Delta_r H^0$. Note, for field effects, the total reaction energy at constant p,T and at given field strength E is the Legendre-transformed standard reaction enthalpy $\Delta_r \hat{H}^0 = \Delta_r \hat{G}^0 + T\Delta_r S^0$. The work potential is the Legendre-transformed Gibbs reaction energy in the field E: $\Delta_r \hat{G} = \Delta_r G - EM$, where M is the projection of the electric moment vector $\vec{M}$ onto the direction of $\vec{E}$, $\Delta_r G$ the ordinary Gibbs reaction energy, and $\Delta_r S$ the reaction entropy, all at constant pressure (Neumann and Kakorin 2000).

1.4 Field Amplification

1.4.1 Induced Membrane Potential (Difference)

The electric membrane current density vector j_m for the cross-membrane ionic flows (of both cations and the anions) is given by

$$\vec{j}_m = \lambda_m(-\nabla\varphi_m) = \lambda_m \vec{E}_m, \qquad (1.6)$$

where λ_m is the conductivity (or specific conductance) of the membrane (referring to all conductive pores). An externally applied electric field (E_{app}), of field strength $E = E_{app}$ induces the transmembrane electric potential difference $\Delta\varphi_m$. For spherical membrane shells in conductive media (after rapid buildup of the ionic Maxwell–Wagner polarization), the *stationary value* $\Delta\varphi_m(\theta, E)_{ss}$ at the polar angle θ to the direction of the (homogeneous, plate condenser geometry) external field E, see Figure 1.1, is given by (Neumann et al. 1999):

$$\Delta\varphi_m(\theta, E)_{ss} = -(3/2) \cdot a \cdot E \cdot f_\lambda \left|\cos\theta\right|. \tag{1.7}$$

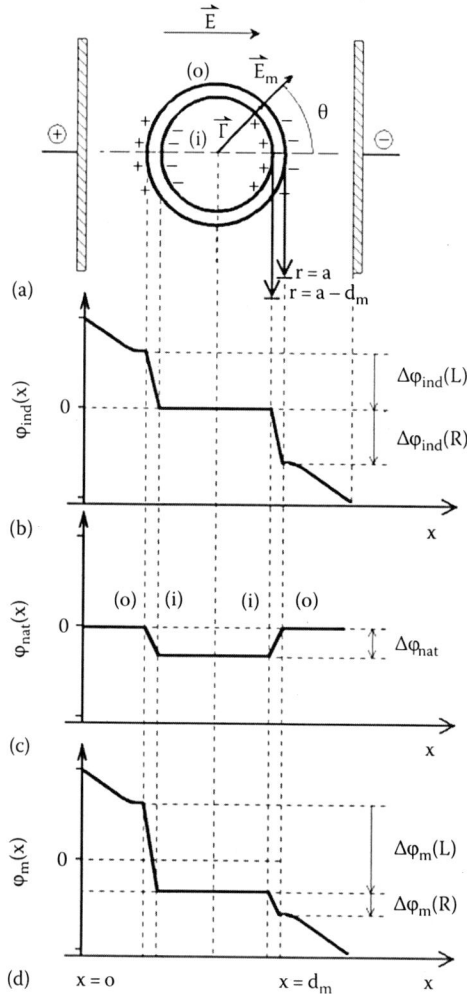

FIGURE 1.1 Interfacial (Maxwell–Wagner) polarization of a spherical membrane shell of outer radius $r = a$, inner radius $r = a - d_m$, exposed to an externally applied homogeneous field $\vec{E}$, where d_m is the membrane thickness, θ is the polar angle, and $\vec{r}$ is the radius vector. (a) Cross section through the center of the sphere. (b) The induced electrical potential profile $\varphi_{ind}(x)$ across the cell center in the direction of $\vec{E}$, the Cartesian x-axis; $\Delta\varphi_{ind}(R)$ is the drop in the induced membrane potential in the direction of $\vec{E}$ of the right hemisphere, $\Delta\varphi_{ind}(L)$ is the drop of the left hemisphere. (c) The natural (Nernstian) membrane potential $\varphi_{nat}(x)$ across the cell center, $\Delta\varphi_{nat} = \varphi_{nat}(i) - \varphi_{nat}(o)$, is the difference in the natural potential across the membrane (equal for all θ), where $\varphi_{nat}(o) = 0$ is taken as the reference potential. (d) The total electrical potential $\varphi_m(x) = \varphi_{ind}(x) + \varphi_{nat}(x)$ across the cell center in the direction of $\vec{E}$. The total membrane potential difference $\Delta\varphi_m$ across the cell center in the direction of $\vec{E}$ is $\Delta\varphi_m = \Delta\varphi_{ind} - \Delta\varphi_{nat} \cdot \text{sgn}(\cos\theta)$, see Equation 1.10.

Here, the term a is the outer radius (i.e., r = a) of the spherical membrane shell of a cell. The inner radius is $r = a - d_m$ and d_m is the membrane thickness. Note that the polar angle for spherical geometry is confined to the range $0° \leq \theta \leq 180°$ (or $0 \leq \theta \leq \pi$), thus $1 \geq |\cos\theta| \geq 0$.

The practical cases of larger cells of very low membrane conductivity are qualified by $d_m \ll a$ and by $\lambda_m \ll \lambda_{ex}, \lambda_{in}$, where λ_{ex} is the external medium conductivity and λ_{in} the conductivity of the cell interior. For this case, the conductivity factor for spherical shells in ionic media reduces to

$$f_\lambda = 1 - \lambda_m \frac{2\lambda_{ex} + \lambda_{in}}{2\lambda_{ex}\lambda_{in}d_m/a} \tag{1.8}$$

Note, Equation 1.7 is consistent with the Maxwell definition of the vector $\vec{E}$ as the negative gradient $(\vec{E} = -\nabla\varphi)$ of the electric potential φ. The induced membrane field at the polar angle θ in the direction of $\vec{E}$ (Cartesian coordinate system) is given by

$$E_m(\theta) = \frac{U_m(\theta)}{d_m} = \frac{-\Delta\varphi_m(\theta)}{d_m} = E_m^{cap}|\cos\theta| \tag{1.9}$$

Here, E_m^{cap} is the (maximum) membrane field at the pole caps. Note, because of the fundamental relationship $(\vec{E} = -\nabla\varphi)$, $\vec{E}_m^{cap}$ has the same direction as $\vec{E}_{app}$. In the notation of the absolute $|\cos\theta|$, there is no change in sign when going from the "cathodic" pole cap, in the range $1 \leq \cos\theta \leq 0$, to the anodic one with the angular range $0 \geq \cos\theta \geq (-1)$. Hence, Equation 1.7 is unrestrictive, that is, generally applicable for the description of current flows, see Equation 1.6, *in the direction* of the respective E vector, through the two electroporated hemispheres of a spherical membrane shell in an external field.

Living cell membranes have a finite natural membrane potential (difference), $\Delta\varphi_{nat}$, defined as $\Delta\varphi_{nat} = \varphi(i) - \varphi(o)$, where the outside potential is taken as the reference $\varphi(o) = 0$. In many cells, the membrane potential is dominated by the Nernst potential $\Delta\varphi(K^+)$ for K^+-ions. Due to the dominant permselectivity for K^+, we may use the approximation $\Delta\varphi_{nat} = \Delta\varphi(K^+)$. Typically, $\Delta\varphi_{nat} = -70\,mV$. The contribution to $\Delta\varphi_m$ of the natural membrane potential (difference) $\Delta\varphi_{nat}$ is readily incorporated. Since the membrane potential difference is given by $\Delta\varphi_m = \Delta\varphi_m(\theta, E)_{ss} - \Delta\varphi_{nat} \cdot \text{sgn}(\cos\theta)$, where θ is the polar angle restricted to the range $0° < \theta < 180°$, $\Delta\varphi_{nat}$ is independent of E and of θ, and $\text{sgn}(\cos\theta) = \cos\theta/|\cos\theta|$ is the sign of $\cos\theta$. Insertion of Equation 1.7 yields (Neumann et al. 1999)

$$\Delta\varphi_m = -\frac{3}{2} \cdot a \cdot E \cdot f_\lambda |\cos\theta| - \Delta\varphi_{nat} \cdot \text{sgn}(\cos\theta) \tag{1.10}$$

As required, the total membrane potential (difference) for E = 0 in the direction of the external field vector E is given by $\Delta\varphi_m = -\Delta\varphi_{nat} \cdot \text{sgn}(\cos)$.

1.4.2 Potential and Ionic Current Flow Direction

Figure 1.1 graphically "visualizes" the direction of the E vector as the direction of current of the positive ions along and "down" the negative gradient of the potential from left to right, accounting for the minus sign in Equations 1.6 and 1.7). The absolute sign in $|\cos\theta|$ is also required for correct relations of other practical quantities. In experiments with spherical cells, the measured quantity, say for instance membrane current or conductance, has contributions from all positions (i.e., all angles θ) on the spherical shell. The $\cos\theta$ average within the boundaries $0 \leq \theta \leq \pi$ of the polar angle is $\langle|\cos\theta|\rangle = (1/2)\int|\cos\theta|\sin\theta\,d\theta = 1/2$. Hence, the $\cos\theta$ average of the stationary membrane potential of a spherical membrane shell is given by

$$\langle\Delta\varphi_m(\theta)\rangle = -\frac{3}{4}E \cdot a \tag{1.11}$$

The average membrane field amplification is then

$$\left\langle E_m(\theta) \right\rangle = \frac{3}{4} E \frac{a}{d_m}$$ (1.12)

The geometrical amplification factor a/d_m of a larger spherical shell, say, with radius $a = 5\,\mu m$ and $d_m = 5\,nm$, for example, is as large as $a/d_m = 10^3$. It is this geometrical amplification which rationalizes that comparatively small external fields in the range of $E = 1\,kV/cm$ are amplified to yield the large field strength $E_m = 10^3\,kV/cm$, which then has such high (electroporative) power on membrane structure. Equations 1.11 and 1.12 provide the correct tools for the respective estimates of the membrane field E_m and membrane voltage. When an experimental (visibility) threshold field strength E_{th} is used, the effectively porated area, A_p, may be expressed by $A_p = A_o(1 - |\cos\theta| = A_o(1 - E_{th}/E)$, where $A_o = (4/3)\pi a^3$ is the area of the spherical shell.

1.5 Molecular Field Effect Analysis

1.5.1 Experimental Signal

Often the experimental signal (S), for instance optical absorbance, current, or conductance, can be cast into a fractional "effect," for example, $f = S/S_{max}$, where S_{max} is the maximum signal. If we can justify that f is proportional to S, it is the fractional extent f that connects experiment to thermodynamic distribution constant, $K = f/(1 - f)$, and thus to the field of strength E. For electroporation field effects, Equation 1.5 is rewritten as

$$\left(\frac{\partial \ln K}{\partial E}\right)_{p,T} = \frac{\Delta_r M}{RT} = dX(E)$$ (1.13)

Integration leads to

$$K(E) = K_0 \exp X(E),$$ (1.14)

where K_0 is the state distribution constant at zero applied field, shown below to be related to the half point field strength E_{half}, that is, E at $f = 0.5$. The field effect term X(E) is specified in terms of the reaction moment $\Delta_r M$ as

$$X(E_{loc}) = \frac{\int \Delta_r M dE_{loc}}{RT}$$ (1.15)

1.5.2 Reaction Moment for Permanent Dipoles

If permanent dipoles are involved, the average molecular reaction moment is defined by $\left\langle \Delta_r m \right\rangle = m(P) - m(C) = \Delta_r M/N_A$. Insertion into Equation 1.15 leads to

$$X(E_{loc}) = \frac{\int \left\langle \Delta_r m \right\rangle dE_{dir}}{k_B T}$$ (1.16)

Note in this context, $E = E_{loc}$ is the local field at the (molecular) sites of field action. For permanent dipoles, $E_{loc} = E_{dir}$, the directing field calculated from the induced membrane field E_m, see Equations 1.9 and 1.12.

1.5.3 Reaction Moments for Induced Polarizations

For continuum polarization processes, the reaction dipole moment is expressed as $\Delta_r M = V_p \Delta_r P$, where $V_p = N_A \langle v_p \rangle$ is the polarization volume. Here, $\langle v_p \rangle = d_m \pi \langle r_p^2 \rangle$ is the mean pore volume and $<r_p>$ the mean pore radius (of the cylindrical pore model). The reaction polarization for the entrance of water in the lipid phase forming an aqueous pore is specified by $\Delta_r P = \varepsilon_0 (\varepsilon_W - \varepsilon_L) E_m$, where ε_0 is the dielectric permittivity of the vacuum, ε_W and ε_L the dielectric constants of water and the lipid phase, respectively (Abidor et al. 1979):

$$X(E_{loc}) = \frac{\langle v_p \rangle \varepsilon_0 (\varepsilon_W - \varepsilon_L) E_m^2}{2 k_B T} \tag{1.17}$$

1.5.4 Field Dependence of the Fractional Extent

Substitution of Equation 1.14 into Equation 1.3, $K = f/(1 - f)$, yields

$$f(E) = \frac{K_0 \exp X(E)}{1 + K_0 \exp X(E)} \tag{1.18}$$

For each special case of application, the respective field has to be used. For instance, if $E = E_m$, the polarization field factor $X(E)$ is specified as

$$X(E_m) = b E_m^2 \tag{1.19}$$

where the b term is defined by

$$b = \frac{\langle v_p \rangle \varepsilon_0 (\varepsilon_W - \varepsilon_L)}{2 k_B T} \tag{1.20}$$

$$X(E_{app}) = b * E_{app}^2 \tag{1.21}$$

Since $b^* = b \, (3 \, a/4 \, d_m)^2$, the b^* factor is given by

$$b^* = \frac{9 \langle v_p \rangle \varepsilon_0 (\varepsilon_W - \varepsilon_L) a^2}{32 k_B T d_m^2} \tag{1.22}$$

1.5.5 Half-Point Field Strength

Many experimental data correlations with respect to the applied field strength E_{app} permit the estimation of a half-point field strength $E_{half} = E_{app,0.5}$, defined for the fractional extent $f = 0.5$, $E_{app}(f = 0.5) = E_{app,0.5}$. So, the half signal $S(E_{0.5})$ is defined via $f = S(E_{0.5})/S_{max} = 0.5$. Insertion of $f = 0.5$ into Equation 1.18 and specifying for $E = E_{app}$ leads to

$$K_0 = \exp\left[-X(E_{app,0.5})\right] \tag{1.23}$$

Insertion of Equation 1.23 leads to those practical expressions, which can be applied to the signal fraction $f(S) = S/S_{max}$ as $f(S) = f(E_{app})$:

$$f(E_{app}) = \frac{\exp[b*(E_{app}^2 - E_{app,0.5}^2)]}{1 + \exp[b*(E_{app}^2 - E_{app,0.5}^2)]} = \frac{1}{1 + \exp[-b*(E_{app}^2 - E_{app,0.5}^2)]} \tag{1.24}$$

Previously, current relaxations of densely packed cell suspensions have been analyzed using only the nonlinear onset part of the amplitude values as a function of E, in terms of the simple first order approximation $f(E) = K_0 \exp(X)$ (Schmeer et al. 2004) readily derived from Equation 1.18 for $K_0 \ll 1$ and $K_0 \exp(X) \ll 1$. For practical purposes, the fraction $f(D)$ of uptake of dye or DNA, properly defined by experimentally accessible quantities, can be rationalized to reflect effective poration and coupled transport. So, if $f(D) = f(E_{app})$ is justifiable, Equation 1.24 can be used for quantitative analysis to yield physical parameters for judging the credibility of data and analysis. In particular, the transport quantities are specific for the transported species. Permeabilization as a general term is incomplete. To be meaningful, it has to be specified with respect to the permeating species and the membrane; for instance, permeability of the plasma membrane for a dye molecule or DNA.

1.5.6 Field Effects and Threshold of Detection

In MEP, like any other non-linear dependence of experimental signals on the applied force parameter, data indicate an experimental "threshold field strength E_{th}" (Neumann 2000). In particular, at short pulse times, the threshold signal S_{thr} at E_{th} is dependent on pulse length, quantified in the known experimental strength–duration correlation. Experimental experience approximates E_{th} as that estimate of the field strength where effects become visible. Thus, E_{th} may be qualified as "visibility" threshold. In physical chemical theory, it is not customary to incorporate such an experimental detection threshold. Applying an external field, there is always some finite field effect, provided the reaction dipole moment or reaction susceptibility is finite, no matter how small the acting field force is. The relevant question is how large is the effect, as compared to the randomizing thermal motion. In electroporation theory of lipid systems, the experimental threshold field strength, whether dependent on pulse length or not, had been used to formulate the energetic balance of electric polarization and surface tension on the one hand and of the (counter-acting) line tension energy on the other hand.

1.6 Electroporation Kinetics

1.6.1 Pore Type and Size from Relaxation Kinetics

Experimental current relaxations of single cells as well as those of densely packed cell suspensions had been analyzed in terms of amplitudes and relaxation time constants of the single normal modes. At least two types of pores are apparent. The first type is like permselective "Nernst-Planck" pores, permitting flow of either of cations or of anions going separately through different pores of electroporated parts of the membrane shell. So, on average, half of the pores transport cations and, parallel to it, the other half of the pores transport anions. A transport of this type is a kind of overall ion exchange, e.g., cations go into the cell interior on one hemisphere and go out of the cell on the other hemisphere. Hence, there is no net transport of ions, for instance, out of the cell. The data further suggest that these ion exchange pores (of radius $r_p = 0.6 \pm 0.2$ nm) can, at a higher pore density, occasionally develop to larger pores at the expense of the smaller ones. The larger pores ($r_p \geq 1$ nm) permit the net transport, for instance, net antiparallel, single file outflow of both cations and anions, mechanically driven by Maxwell stress (on equatorial regions of a cell) causing cell elongation.

1.6.2 Pore Expansion Redefined

The kinetic feature of exponential relaxation modes indicates that it is the *number of pores* of defined size (r_p) that increases with time and field strength (Griese et al. 2002, Schmeer et al. 2004). Therefore, the transport quantities of ionic current, conductance, resistance indicate that the transport cross section for flow is structurally controlled. These kinetic features, indicating single pores (of given average size), require a re-evaluation of the concept of pore expansion. As a diffusive process, expansion *sensu strictu* is associated with a square root dependence on time. The time dependence of electroporation events is however, exponential, consistent with defined types of pores.

1.6.3 Lipid Rearrangements

The previously presumed lipid rearrangements during pore formation have been specified as directed rotations of the dipolar lipid head groups to form a specific pore wall like that in hydrophilic or inverted pore. In 1982, the dipolar head groups were drawn as aligned parallel to the external field direction (Neumann et al. 1982). This presumption is now supported by both, relaxation kinetic data obtained with small lipid bilayer vesicles, and by molecular dynamics simulations of the molecular rearrangements of lipid and water molecules involved in electric pore formation. The technique of cell electroporation has been recently extended to ultra-short pulses with nominally very high external electric field strengths. The applied field rapidly increases and acts internally primarily on the intracellular organelles. Due to the short pulse duration, the field only marginally affects the slowly responding plasma membrane. Ultra-electroporation is expected to have powerful clinical potentials, for instance, for inducing cell apoptosis in malignant tissue (Esser et al. 2007). Analytically, if the experimental electroporation times are expressed as a function of the respective directing field, (calculated from the external field using the dielectric constant of the polar environment), the entire field strength range, from moderate field strengths and short pulse times up to the very high external fields of the ultra-short pulses, can be consistently described with one and the same permanent dipole moment. The thermodynamic analysis along the physical chemical theory yields the mean dipole moment associated with the elementary unit, involved in a defined dipolar rotation process. This value compares well with the dipole moment of the zwitterionic phosphatidylcholine head group of $(41 \pm 5)\ 10^{-30}$ Cm $= (12 \pm 2)$ D. The hydrated ionic and dipolar lipid head groups, where the water molecules in the asymmetric hydration shells of the ionic groups contribute to higher dipole moments, are the molecular receptors for the interaction of the local field with the membranes. The analysis of the relaxation data and the molecular dynamics calculations quantify the alignment processes of these dipolar field receptors into field-parallel positions in the wall of the hydrophilic (or inverted) electropore (Boeckmann et al. 2008).

1.7 Nonequilibrium Thermodynamics

1.7.1 Unidirectionality of Hysteresis Processes

For the analysis reaction flows, it is recalled that MEP is recognized as a hysteresis cycle of the (rapid) electroporation processes to produce long-lived porous structures and the (slow) resealing processes, which couple in, and are reflected in, the observed material transport through the electroporated structures. It is recalled that the longevity of the porous structures is the structural reason for the large after-field material transport, such as release of small ions, net uptake (or release) of DNA, RNA, or proteins, larger-sized dye or drug molecules, through the slowly resealing MEP-structures.

1.7.2 Conventional Flows

The concept of the overall hysteresis cycle intrinsically implies that the structural (equilibrium) transitions, along each branch of the hysteresis loop, occur net unidirectionally. This important feature justifies the introduction of (unidirectional) structural reaction flows modifying (controlling) ordinary particle flow, by nature qualified as net unidirectional (irreversible). Thus, physical chemical electroporation theory also comprises, besides conventional kinetics, flow analysis aiming at numerical values of the flow coefficients. The flow coefficient includes the permeability coefficient (diffusion coefficient, thickness of transport area, Nernst–Planck distribution coefficient for the transport compartment and the outer compartments), the surface volume ratio, and the fraction of actually transporting area or pore fraction, both for the rapid in-field processes as well as for the slower after-field processes.

1.7.3 Nonlinear Flow Analysis

Recently, the newly introduced concept of time-dependent flow coefficients has turned out to be instrumental for proper flow analysis (Neumann 2000). It is recalled that in particular the after-field conductance relaxations (resealing curves) reflect transport through decreasing transport cross sections, that is, current modulated by the decreasing number of pores. The actual time course of the fraction $Y(t) = (g(t) - g(0))/g(0) = (I(t) - I(0))/I(0)$ of the after-field currents, I, or conductances, g, relative to the zero time current $I(0)$, $g(0)$ conductance, that is, before pulse application, in the simplest case is given by (Kakorin et al. 1998, Schmeer et al. 2004)

$$Y(t) = Y_{max}\{1 - \exp[-k_0 \tau_R (1 - \exp[-t/\tau_R])]\} \qquad (1.25)$$

where Y_{max} refers to the maximum value for the case of complete equilibration between the intracellular ion concentration and that of the external medium. The experimental stationary value $Y_\infty = Y(t \to \infty) \leq Y_{ss}$ refers to complete resealing before equilibration and is given by

$$Y_\infty = Y_{max}(1 - \exp[-k_0 \tau_R]) \qquad (1.26)$$

In Equation 1.26, k_0 is the flow coefficient at a given field at the time point $t = 0$ of switching off the single pulse or the pulse train, and τ_R the (field-independent zero field) time constant of resealing of the fractional transporting area $f_p = (A_{trp}/A_0)$, see Equation 1.28, or the fraction of conductive pores $f(N_p) = N_p/N_{p(max)}$, N_p the number of pores. Practically, the approximation $f_p = f(N_p)$ is applied. The time course of $Y(t)$ indicates that the pore resealing starts with the value $f(E)$, at the given field at the pulse end, denoted here as after-field time zero, $t_0 = 0$. In the simplest case, the resealing of the porous area in the absence of the field is exponential resealing according to

$$f_p(t) = f_p(E) \exp[-t/\tau_R] \qquad (1.27)$$

This procedure of analysis has, for instance, been successfully applied for the resealing phase of densely packed CHO cells. As rationalized with Equation 1.25, the measured transport curves are therefore exponentials of exponentials. (Note that the actually measured curve types can deceive smeared exponentials of the Kohlrausch-type.) As seen in the case of exponential decrease in the pore fraction, the after-field kinetics provides mechanistic details of the long-lived electroporative membrane states. This analytical framework has been used to obtain the values of k_0 (E, t_E), dependent on E and on the pulse duration t_E, and the time constant τ_R of the resealing process (Schmeer et al. 2004).

1.7.4 Mole Flow and Mole Flux

Any analysis in terms of flow coefficients is suggested to start with the Nernst–Planck *mole flow* equation. For the unidirectional case (1-D), the mole flow is defined as $\partial n_i/\partial t$, where n_i is the amount of substance of species or ion of type i. If the transport is *orthogonal* across a slab of thickness d_m, the *actually transporting area* A_{trp} lies within the membrane range $0 \leq x \leq d_m$ of the flow pathway. Explicitly, the mole flow equation for species i reads

$$\left(\frac{\partial n_i}{\partial t}\right)_x = A_{trp}\left(D_i\left(-|\nabla c_i|\right) + u_i c_i\left(-|\nabla \varphi|\right)\right) \tag{1.28}$$

where

$c_i = n_i/V$ is the molar concentration

volume $V = 1\,dm^3$

$u_i = z_i e_0 D_i/k_B T$ the "signed" electric mobility; u_i having the same sign as the ionic charge number z_i

∇ the nabla vector (of concentration and electric potential, respectively)

D_i the diffusion coefficient

The absolute signs refer to the scalar values of the gradients. For membranes, the actual transport area is $A_{trp} = N_p \pi r_p^2 = f_p A_0$, recalling $f_p = A_{trp}/A_0$ the fraction of pores, A_0 the total area, N_p the number of pores and r_p the mean pore radius of the assumed cylindrical pore. The *mole flux* (or flow density) vector $\vec{J}_i$ here refers to, and is expressed as, the flow *perpendicular* across A_0. The scalar value of the mole flux is given by the (modified) Nernst–Planck equation:

$$J_i = \left(\frac{\partial n_i}{\partial t}\right)_x \frac{1}{A_0} = f_p \cdot \left(D_i\left(-|\nabla c_i|\right) + u_i c_i\left(-|\nabla \varphi|\right)\right) \tag{1.29}$$

Note that electro-diffusion in free solution corresponds to $f = A_{trp}/A_0 = 1$. The flux vector, $\vec{J}_i$, involved in the expression for the electric current (besides the small diffusion potentials of ions with different mobilities, reducing the actual field in the solution) is solely given by the potential term, and is expressed in vector notation by

$$\vec{J}_i = \vec{v}_i c_i = u_i c_i \vec{E} \tag{1.30}$$

where $\vec{v}_i = u_i \vec{E}$ is the drift velocity. The mobility u_i is given by

$$u_i = z_i D_i \frac{F}{RT} = z_i D_i \frac{e_0}{k_B T} \tag{1.31}$$

In this context, note that the (ionic) Poisson equation for the potential φ is given by

$$\nabla^2 \varphi = -\nabla \cdot (-\nabla \varphi) = -\rho/\varepsilon_0 \varepsilon_M \tag{1.32}$$

where

$\rho = F \sum_i z_i c_i$ is the ionic charge density

F the Faraday

ε_M the dielectric constant of the medium

The term $\vec{J}$ and the ionic solution conductivity λ are "ion-specified" as

$$\vec{j} = \vec{E}\lambda = F\left(\sum_i z_i c_i u_i\right)\vec{E} \tag{1.33}$$

$$\lambda = F\left(\sum_i z_i c_i u_i\right) \tag{1.34}$$

The electric current due to ion flows, cations in the direction of $\vec{E}$, anions opposite to cations, where $z_i u_i$ is always positive, is then given by

$$I = A \; F \sum_i z_i \left| \vec{J}_i \right| = A \; F \left(\sum_i z_i c_i u_i \right) |\vec{E}|, \tag{1.35}$$

Divergence (nabla scalar) operation, $\nabla \cdot$, on the current density $\vec{j}$ yields

$$\nabla \cdot \vec{j} = -F \sum_i z_i \left(\frac{\partial c_i}{\partial t} \right) \tag{1.36}$$

The "conservative continuity equation," meaning inflow = outflow at the site of the concentration change with time is expressed as:

$$\left(\frac{\partial c_i}{\partial t} \right)_{x,y,z} = -\nabla \cdot \vec{J}_i \tag{1.37}$$

In Equation 1.35, it is readily seen that the drift-diffusion contribution, cations and anions flowing in the same direction, does not contribute as such to electric current I. This framework of "ion equations" is necessary in a molecular interpretation of electroporation currents and of transport flows of charged and uncharged particles.

1.7.5 General Integral Flow Equations

Inspecting Equation 1.25, the time courses reflect, in a folded form, the change in the fraction of resealing pores because the underlying mole flow is proportional to the flow area, that is, decreasing number of pores. Therefore, proper analysis preferentially starts with the mole flow, and not with the mole flow density, in order to rationalize time-dependent flow coefficients. In addition, the flow coefficient at the time point of the end of the applied pulse yields the kinetic parameters for the rate limiting structural transitions, preceding the actual transport processes. The flow coefficient k comprises the (free) diffusion coefficient D, the Nernst distribution constant γ for the distribution of the particle in the pore relative to a bulk compartment, and the thickness of the passage d_m. For instance, $k = (D\gamma/d_m) \, A_{trp}/V_0 = P \, f_p \, A_0/V_0$, where $P = D\gamma/d_m$ is the permeability coefficient. It is the area A_{trp}, and thus the pore area fraction f_p that are time dependent. For instance, $f_p(t) = f_p(E) \exp[-t/\tau_R]$ yields the time-dependent flow coefficient $k(t) = P \, f_p(t) \, A_0/V_0$ (Neumann et al. 1999). Obviously, for transport of larger molecules, the coefficients for permeation are specific for both the transported species and the membrane. So, any permeation quantity such as electropermeabilization or permeability coefficient must be specified for species and membrane.

1.7.6 Fractional Signal Change

For proper flow analysis, for instance, the measured signals S(t) are expressed as the fraction $y(t) = S(t)/S_{ss}$, where the stationary signal S_{ss} is the amplitude, $S_{ss} = S(t \to \infty)$). Next, the proper differential equation is selected, formally analogous to the linear form of dy(t)/d(t):

$$\frac{dy(t)}{dt} = -k(t)(y(t) - y_{ss}) \tag{1.38}$$

Here, $y_{ss} = S(t \rightarrow \infty)/S_{ss} = 1$. The integrated form is called integral flow equation (Kakorin et al. 1998):

$$y(t) = y_{ss} \exp\left[-\int k(t)dt\right] \tag{1.39}$$

If we use the explicit flow coefficients $k(t) = k_0 \exp[-t/\tau_R]$ or $k(t) = k_0 (1 - \exp[-t/\tau_R])$, expressions of the type of Equation 1.25 are obtained. It is stressed, that in each case, it must be carefully checked, which equation can be applied and whether existing equations have to be modified or expanded, as dictated by proper physical chemical reasoning along the fundamental laws of thermodynamics, in particular those of nonequilibrium (or flow) thermodynamics. This applies, too, to the claims of small electromagnetic field (EMF) effects, where, at a first glance, the data appear "unbelievable for physical chemical reasons" (Neumann 2000, 2004).

References

Abidor, I.G., Arakelyan, V.B., Chernomordik, L.V., Chizmadzhev, Y.A., Pastuchenko, V.P., Tarasevich, M.R. 1979. Electric breakdown of bilayer lipid membrane. I. The main experimental facts and their theoretical discussion, *Bioelectrochem. Bioenerg.* 6: 37–52.

Andre, F.M., Gehl, J., Sersa, G., Preat, V., Hojman, P., Eriksen, J., Golzio, M. et al. 2008. Efficiency of high- and low-voltage pulse combinations for gene electrotransfer in muscle liver, tumor, and skin, *Human Gene Ther.* 19: 1261–1271.

Boeckmann, R.A., de Groot, B.L., Kakorin, S., Neumann, E., Grubmüller, H. 2008. Kinetics, statistics, and energetics of lipid membrane electroporation, studied by molecular dynamics simulations, *Biophys. J.* 95: 1837–1850.

Chizmadzhev, Y.A., Zarnitsin, V.G., Weaver, J.C., Potts, R.O. 1997. Mechanism of electro-induced ionic species transport through a multilamellar lipid system, *Biophys. J.* 68: 749–756.

Dimova, R., Riske, K.A., Aranda, S., Bezlyepkina, N., Knorr, R.L., Lipkowsky, R. 2008. Giant vesicles in electric fields, *Soft Matter* 3: 828–836.

Eisenstein, M. 2006. A look back: A shock to the system, *Nat. Methods* 3: 66.

Esser, A.T., Smith, K.C., Gowrishankar, T.R., Weaver, J.C. 2007. Towards solid tumor treatment by irreversible electroporation intrinsic redistribution of fields and currents in tissue, *Technol. Cancer Res. Treat.* 6: 261–273.

Griese, T., Kakorin, S., Neumann, E. 2002. Conductometric and electrooptic relaxation spectrometry of lipid vesicle electroporation at high fields, *Phys. Chem. Chem. Phys.* 4: 1217–1227.

Heller, R., Jaroszeski, M., Glass, L., Messina, J., Rapaport, D., DeConti, R., Fenske, N., Gilbert, R., Mir, L.M., Reintgen, D. 1996. Phase I/II trial for treatment of cutaneous and subcutaneous tumors using electrochemotherapy, *Cancer* 77: 964–971.

Hristova, N., Tsoneva, I., Neumann, E. 1997. Sphingosine-mediated electroporative DNA transfer through lipid bilayers, *FEBS Lett.* 415: 81–86.

Kakorin, S., Redeker, E., Neumann, E. 1998. Electroporative deformation of salt filled vesicles, *Eur. Biophys. J.* 27: 43–53.

Miklavcic, D., Jarm, T., Cemazar, M., Sersa, G., An, D.J., Beleradek, J., Jr., Mir, L.M. 1997. Tumor treatment by direct electric current, Tumor perfusion changes. *Bioelectrochem. Bioenerg.* 43: 253–256.

Mir, L.M., Orlowski, S., Belehradek, J., Jr., Paoletti, C. 1991. Electrochemotherapy: Potentiation of antitumor effect of bleomycin by local electric pulses. *Eur. J. Cancer* 27: 68–72.

Mir, L.M., Tounekti, O., Orlowski, S. 1996. Review. Bleomycin: Revival of an old drug, *Gen. Pharamacol.* 27: 745–748.

Neumann, E. 1989. The relaxation hysteresis of membrane electroporation. In *Electroporation and Electrofusion in Cell Biology.* E. Neumann, A.E. Sowers, and C.A. Jordan (Eds.). Plenum Press, New York, pp. 61–82.

Neumann, E. 2000. Digression on chemical electromagnetic field effects in membrane signal transduction the experimental paradigm of the acetylcholine receptor, *Bioelectrochemistry* 52: 43–49.

Neumann, E. 2004. Electric and magnetic field reception. In *Encyclopedia of Molecular Cell Biology and Molecular Medicine*, Vol. 4, R.A. Meyers (Ed.). Wiley-VCH, New York, pp. 1–20.

Neumann, E. 2007. Systemic electroporation–combining electric pulses with bioactive agents, *Int. Fed. Med. Biol. Eng. Proc.* 16: 40–43.

Neumann, E., Kakorin, S. 2000. Electroporation of curved lipid membranes in ionic strength gradients, *Biophys. Chem.* 85: 249–271.

Neumann, E., Rosenheck, K. 1972. Permeability changes induced by electric impulses in vesicular membranes, *J. Membr. Biol.* 10: 279–290.

Neumann, E., Schaefer-Ridder, M., Wang, Y., Hofschneider, P.H. 1982. Gene transfer into mouse lyoma cells by electroporaton in high electric fields, *EMBO J.* 1: 841–845.

Neumann, E., Tönsing, K., Kakorin, S., Budde, P., Frey, J. 1998. Mechanism of electroporative dye uptake by mouse B cells, *Biophys. J.* 74: 98–108.

Neumann, E., Kakorin, S., Toensing, K. 1999. Fundamentals of electroporative delivery of drugs and genes, *Bioelectrochem. Bioenerg.* 48: 3–16.

Okino, M., Mori, H. 1987. Effects of a high voltage impulse and an anticancer drug on *in vivo* growing tumors, *Jpn. J. Cancer Res.* 78: 1319–1321.

Pliquett, U., Schmeer, M., Seipp, T., Neumann, E. 2002. Fast recovery process after electroporation, *Int. Fed. Med. Biol. Eng. Proc.* 3: 98–99.

Pliquett, U., Gallo, S., Hui, S.W., Gusbeth, Ch., Neumann, E. 2005. Local and transient structural changes in stratum corneum at high electric fields: Contribution of Joule heating, *Bioelectrochemistry* 67: 37–46.

Schmeer, M., Seipp, T., Pliquett, U., Kakorin, S., Neumann, E. 2004. Mechanism for the conductivity changes caused by membrane electroporation of CHO cell-pellets, *Phys. Chem. Chem. Phys.* 6: 5564–5574.

Smith, K.C., Weaver, J.C. 2008. Active mechanisms are needed to describe cell responses to submicrosecond, megavolt-per-meter pulses: cell models for ultrashort pulses, *Biophysical J.* 95: 1547–1563.

Wong, T.K., Neumann, E. 1982. Electric field mediated gene transfer, *Biophys. Biochem. Res. Commun.* 107: 584–587.

2

Bioelectric Effect of Intense Nanosecond Pulses

Karl H. Schoenbach

2.1 Introduction

The effect of intense pulsed electric fields on biological cells and tissues has been the topic of research since the late 1950s. Intense means that the electric field is of a sufficient magnitude to cause nonlinear changes in cell membranes. The first paper that reported the reversible breakdown of cell membranes when electric fields are applied was published in 1958 (Staempfli, 1958). The first report on the increase in the permeability of the plasma membrane of a biological cell, an effect subsequently named "electroporation," appeared in 1972 (Neumann and Rosenheck, 1972). The electric fields that are required to achieve electroporation depend on the duration of the applied pulse. Typical pulses range from tens of milliseconds with amplitudes of several hundred V/cm to pulses of a few microseconds and several kV/cm.

More recently, the pulse duration has been shortened into the nanosecond range. The effects of such short pulses have been shown to reach into the cell interior (Schoenbach et al., 2001). Pulse durations are as brief as several nanoseconds, with pulse amplitudes as high as 300 kV/cm for short pulses. A new field of research opens when the pulse duration is decreased even further into the sub-nanosecond range. By applying such ultrashort pulses, it will also become possible to use wideband antennas, rather than direct contact electrodes, to deliver the pulses to tissues.

In order to understand the effect of ultrashort pulses, often only labeled as nanosecond pulses, we can use a simple analytical, passive, and linear model of the cell. By passive and linear, we mean that changes in the properties of the cell structures, such as electroporation, are not considered. The assumption holds, therefore, only for electric field amplitudes below those required for electroporation or nanoporation. However, it provides useful information on the threshold for the onset of nonlinear effects and the temporal range for which subcellular effects can be expected.

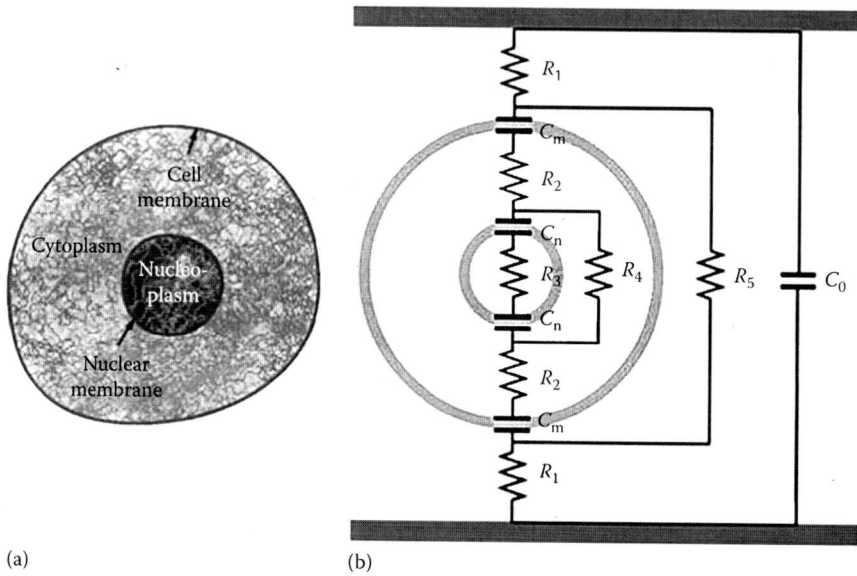

FIGURE 2.1   (a) Cross section of a cell with nucleus as would be observed with a light microscope. The typical dimension of a mammalian cell (diameter) is on the order of 10 μm. (b) Double shell model of a biological cell, and superimposed equivalent circuit of the cell between two electrodes. The cell membrane and the nuclear membrane are described by the capacitances C_m and C_n, the cytoplasm and nucleoplasm by the resistances R_2, R_4, and $R_3 \cdot R_1$, R_5 and C_0 are dependent on the electrical properties of the medium in which the cell is embedded, and the geometry of the system. (From Schoenbach, K.H. et al., *Proc. IEEE*, 92, 1122, 2004. With permission.)

Figure 2.1a shows a cross section of a mammalian cell, with the only membrane-bound substructure shown being the nucleus. The cytoplasm, which fills much of the cell, contains dissolved proteins, electrolytes, and forms of glucose and is moderately conductive, as are the nucleoplasm and the contents of other organelles. On the other hand, the membranes that surround the cell and subcellular structures have a low conductivity. We can, therefore, think of the cell as a conductor surrounded by a lossy, insulating envelope containing substructures with similar properties. Data on the dielectric constants and conductivities of cell membranes and cytoplasm, as well as nuclear membranes and nucleoplasm, have been obtained using dielectric spectroscopy of cells (Ermolina et al., 2001; Feldman et al., 2003). Typical values for the plasma membrane of mammalian cells, e.g., B- or T-cells, are relative permittivities on the order of 10 and conductivities of approximately 10^{-5} S/m. For the cytoplasm, the relative permittivity is approximately that of water, 80, and the conductivity is typically one-fifth that of seawater, 1 S/m.

Based on the simple equivalent circuit (Figure 2.1), the voltages across membranes (the plasma membrane as well as membranes of subcellular structures) can be expressed in terms of the frequency, ω, of the applied voltage between the two electrodes. In the case of a spherically symmetric cell, the voltage across the plasma membrane, V_m, at the poles of the cell (assuming a negligible membrane conductance) is

$$V_m = 1.5 \frac{aE}{1 + j\omega\tau_m} \qquad (2.1)$$

where
   E is the applied electric field
   a is the radius of the spherical cell
   τ_m is the charging time constant of the plasma membrane

τ_m for cells in suspensions is given as (Cole, 1937)

$$\tau_m = aC_m\left[\rho_i + \rho_e \frac{1+2V}{2(1-V)}\right] \tag{2.2}$$

with
   C_m being the capacitance per unit area of the membrane
   V being the volume fraction of the cells in suspension
   ρ_i and ρ_e being the internal and external resistivity, respectively

The amplitude of the voltage across the plasma membrane is consequently

$$V_m = 1.5\frac{aE}{\sqrt{\left(1+(\omega\tau_m)^2\right)}} \tag{2.3}$$

The equation for the polar voltage across the membrane of a concentric, spherical, subcellular organelle is, again assuming negligible conductance of the membrane,

$$V_n = 1.5^2 bE\rho_i\left[\frac{j\omega C_n}{(1+j\omega\tau_m)(1+j\omega\tau_n)}\right] \tag{2.4}$$

where
   b is the radius of the spherical organelle
   C_n is the capacitance of the organelle membrane per unit area
   τ_n is the charging time constant of the organelle (assuming that the terms with V can be neglected)

$$\tau_n = bC_n\left(\rho_n + \frac{\rho_i}{2}\right) \tag{2.5}$$

with ρ_n being the resistivity of the interior of the organelle. If the resistivities inside and outside the organelle are identical, τ_n reduces to

$$\tau_n = 1.5\rho_i bC_m \tag{2.6}$$

The amplitude of the voltage across the organelle membrane is then

$$V_n = 1.5\frac{Eb\omega\tau_n}{\sqrt{\left(1+(\omega\tau_m)^2\right)\left(1+(\omega\tau_n)^2\right)}} \tag{2.7}$$

Plotting the normalized membrane voltages, $2V_{n,m}/3Ea$, versus the normalized frequency, ω/ω_c, where

$$\omega_c = \frac{1}{\tau_m} \tag{2.8}$$

is known as the β relaxation frequency leads to the diagrams in Figure 2.2 (Foster and Schwan, 1995). For the organelles in this double shell model, two values for b were chosen: one with $a/b = 2$ and one with $a/b = 4$. C_n was assumed to be identical to C_m and ρ_i was set to the same value as ρ_e.

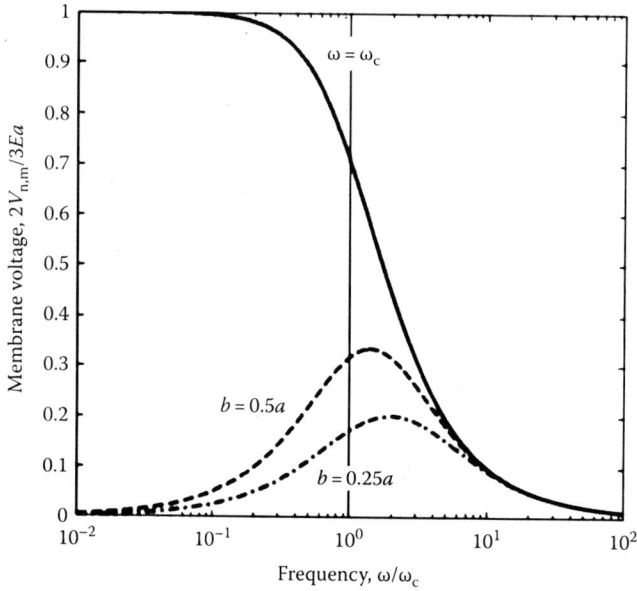

FIGURE 2.2 Spectral distribution of the voltage induced in an alternating electric field across the plasma membrane of a spherical cell (solid line) and the membranes of spherical, concentric organelles with a radius of half and of a quarter of the cell radius (dashed and dashed-dotted lines, respectively).

For frequencies exceeding the β-relaxation frequency, the voltage across the plasma membrane decreases with a slope of ω/ω_c, whereas the voltage across internal membranes increases, reaching a peak at

$$\omega = \sqrt{(\tau_m \tau_n)^{-1}} \tag{2.9}$$

and then decreases with the same slope as the voltage across the plasma membrane. The smaller the organelle, the higher the frequency where one can expect a full effect on the internal membranes. In general, the frequency should exceed the β-relaxation frequency to obtain a reasonable voltage across subcellular organelle membranes.

Under the conditions mentioned above (assuming that all membranes are equal and the conductivity of organelle interiors and cytosol are the same), the voltage across the organelle membranes never exceeds the voltage across the plasma membrane. However, if we deviate from these oversimplified assumptions and take into account that cell and organelle membranes differ electrically as do their interiors, it is possible to construct scenarios where the voltage across the organelle membranes exceeds that of the plasma membrane in a certain spectral range. This has been shown by using the simple equivalent circuit (Figure 2.1) but assuming a lower capacitance of subcellular membranes (thicker membranes) than that of the plasma membrane (Schoenbach et al., 2001, 2002). Similar results have been reported using a more elaborate continuum model by Kotnik and Miclavcic (2006). A diagram from their paper, illustrating such an effect, is shown in Figure 2.3. Here the conductivity of the organelle interior was assumed to be higher than that of the cytosol, and the membrane permittivity lower than that of the plasma membrane. In such a situation, it can be expected that at high frequencies the poration of subcellular membranes will be more likely than plasma membrane poration.

In the time domain, this corresponds to a higher likelihood for ultrashort pulses—with a Fourier spectrum that extends into the high (critical) frequency range as shown in Figure 2.3—to permeate membranes of cellular organelles before porating the plasma membrane. This has been confirmed through modeling in the time domain. The temporal response of the membrane voltages to an applied

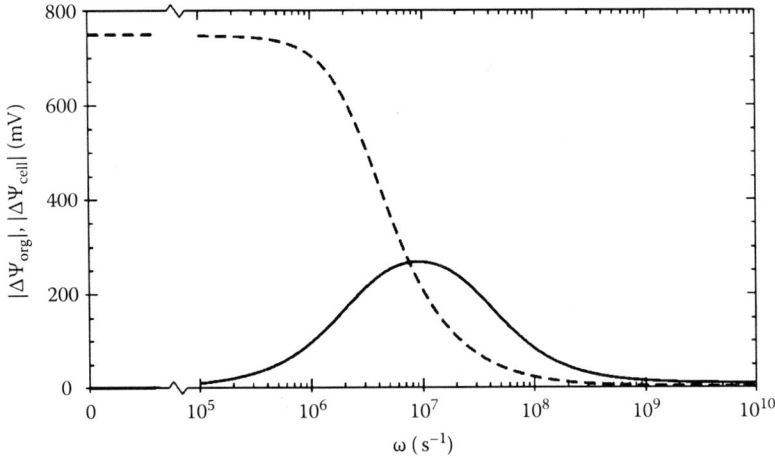

FIGURE 2.3 The frequency dependence of the voltages induced across the cell membrane (dashed line) and an organelle membrane (solid line) in an alternating field for a cell where the conductivity of the organelle interior was increased and the capacitance of the organelle membrane was decreased with respect to their default values. (From Kotnik, T. and Miclavcic, D., *Biophys. J.*, 90, 480, 2006. With permission.)

fast-rising (1 ns) electric field was calculated by Kotnik and Miclavcic (2006) (with the same cell parameters as those used to compute the frequency domain response). The results showed that for the first 117 ns, the voltage across a spherical organelle with a 3 μm diameter exceeded that of the voltage across the plasma membrane of the 10 μm cell. Similar results were obtained by Joshi et al. (Joshi et al., 2004a,b; Schoenbach et al., 2004). In a study where a distributed electrical double-shell model for current flow was coupled with the Smoluchowski equation, the transmembrane voltages across the cell membrane and the membrane of a 2 μm diameter organelle in a 10 μm diameter cell were computed. The cell parameter values were mainly based on measured electrical cell parameters (Ermolina et al., 2001).

Figure 2.4a and b show the temporal development of membrane voltage across the outer and inner membranes for a trapezoidal 300 ns pulse and an 11 ns pulse. For the 300 ns pulse, the membrane voltage across the inner organelle membrane exceeds that of the outer one during the first 75 ns, for the 11 ns pulse, for the entire pulse duration.

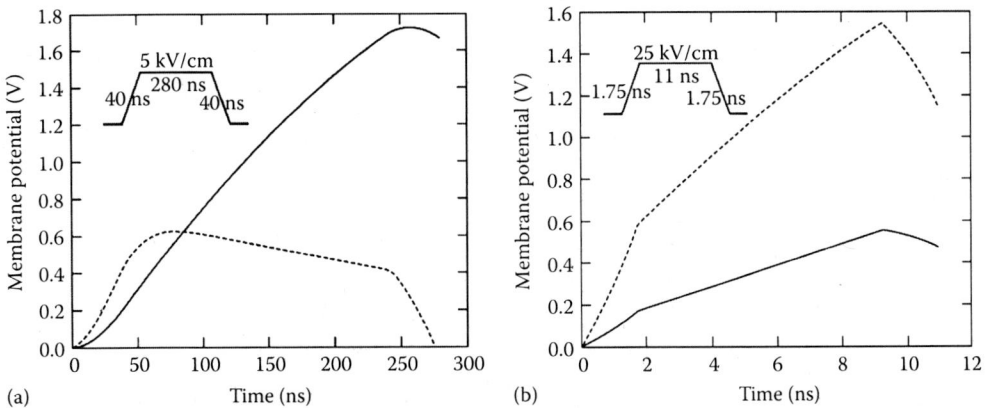

FIGURE 2.4 (a) Membrane voltage across plasma membrane (solid line) and inner (mitochondrial) membrane (dotted line) for a 300 ns trapezoidal pulse. (b) The same for an 11 ns long pulse. The thickness of the inner membrane was assumed to be twice that of the plasma membrane. (From Schoenbach, K.H. et al., *Proc. IEEE*, 92, 1122, 2004. With permission.)

2.2 The Temporal Range of "Nanosecond" Pulse Effects

With few exceptions, studies on the effects of high intensity electric field pulses on subcellular structures are performed with wideband (ultrashort) pulses. To our knowledge, only the research group at Kumamoto University is focusing their research on the effect of radio- and microwave-frequency modulated intense electric field pulses on cells and tissue (Nomura et al., 2009). Modeling efforts have therefore been focused on the modeling of the cell response to electrical pulses, particularly nanosecond pulses, and studies that deal with the effects of intense, pulsed electric fields on subcellular structures are often referred to as nanosecond pulsed electric field (nsPEF), nanosecond electrical pulse (nsEP), or simply nanopulse studies. However, these expressions don't provide information on the science of "intracellular electromanipulation," another term that was used to describe these bioelectric intracellular effects.

A systematic approach for determining the limits of such intracellular effects, beyond the generic term "nano," requires studies in the time domain rather than in the frequency domain. However, the frequency response of cell membranes based on the simple equivalent circuit still allows us to estimate not only the upper limit in pulse duration, but also the effects of pulse rise time (pulse shape) for intracellular pulse effects. A modification of this circuit, including the previously neglected electrical parameters of the membrane and cytosol, also allows us to define a lower limit in pulse duration for "nanosecond" pulse effects.

In order to obtain a quantitative criterion for the upper limit in pulse duration and pulse rise time, let us consider the spectral distribution of a pulse and compare it with that of the cell membrane. Obviously, a condition for intracellular effects would be that the spectrum of the applied electric pulse reaches into the spectral range where subcellular effects approach those of plasma membrane effects (Figure 2.2). For a monopolar pulse, this means that the corner frequency of the pulse should be on the order of, or higher than, the β-relaxation frequency of the plasma membrane. In addition, it is desirable to have a pulse shape such that the decrease in the frequency spectrum above the corner frequency is a minimum, in order to extend the range of interaction with subcellular structures as far as possible into the high frequency range.

The applied pulses in bioelectrics are ideally square-wave, but realistically have a trapezoidal shape. For a trapezoidal pulse with a pulse duration of τ_p, and a rise and fall time of τ_r (as shown in Figure 2.5), the frequency spectrum is given as

$$X(\omega) = \frac{2A}{\omega^2 \tau_r}\left[\cos\left(\frac{\omega \tau_p}{2}\right) - \cos\left(\omega\left(\frac{\tau_p}{2} + \tau_r\right)\right)\right] \tag{2.10}$$

where A is a measure for the amplitude of the pulse. The envelope of this spectrum is shown in Figure 2.5. The first corner frequency for such a trapezoidal pulse is at

$$\omega_{c1} = \frac{2}{\tau_p} \tag{2.11}$$

The spectral envelope then decreases by $1/\omega$ up to the second corner frequency

$$\omega_{c2} = \frac{2}{\tau_r} \tag{2.12}$$

For frequencies beyond the second corner frequency, the spectrum decreases by $1/\omega^2$.

The condition that, for intracellular effects the corner frequency of a trapezoidal pulse should be equal or greater than the β-relaxation frequency of the plasma membrane leads to a condition for the

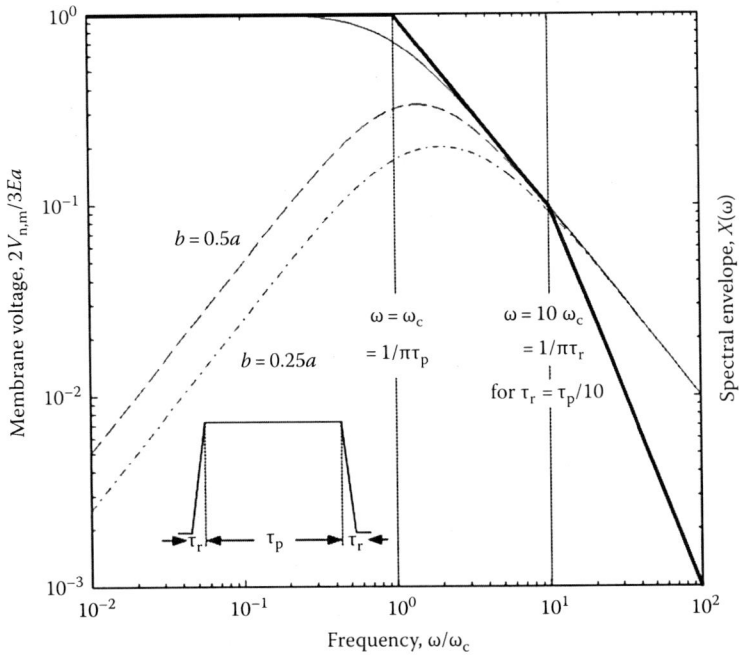

FIGURE 2.5 Envelope of the spectrum (bold, solid line) of a trapezoidal pulse as shown in the insert, having a rise and fall time of 10% of the pulse duration. The first corner frequency ($\omega_{c1} = 1/\tau_p$) is chosen such that it is located at the maximum in the membrane voltage spectrum of the ($b = 0.5a$) organelle. It represents the minimum conditions for efficient membrane manipulation for such an organelle. Longer pulses and longer rise times will have diminishing intracellular effects.

maximum pulse length where we still could expect intracellular effects. According to Equations 2.8 and 2.11, this condition reads

$$\tau_p < 2\tau_m \qquad (2.13)$$

For a spherical cell with a 5 μm radius, a plasma membrane capacitance of 1 μF/cm², and internal and external resistivities of 100 Ω-cm in a dilute solution, the charging time constant is 75 ns. Consequently, for intracellular effects to be stimulated by a trapezoidal pulse, the pulse duration should be less than 150 ns.

Almost equally important is the rise time of the pulse. As shown in Figure 2.5, where the frequency spectrum of a trapezoidal pulse (with $\tau_p < 2\tau_m$) is superimposed to the frequency spectrum of membrane voltages, it is important to have the rise (and fall) time as short as possible to cover the spectral range above the β-relaxation frequency effectively. For long rise times, the spectrum falls off by $1/\omega^2$ just above the first corner frequency, and consequently, has a diminishing effect on the organelle membranes.

According to these considerations, which rely on the simple equivalent circuit shown in Figure 2.1, there shouldn't be a lower limit in pulse duration for these kinds of intracellular effects. Reducing the duration further and further will just shift the corner frequency of the trapezoidal pulse to higher and higher frequencies and, consequently, will allow better access to subcellular structures. It will come at a cost, however, keeping the spectral amplitude the same at higher frequencies would require a linear increase in pulse amplitude with the inverse of pulse duration, a fact that is discussed in the following paragraph of this chapter. What also needs to be considered when we reduce the pulse duration is the change in the physics of the membrane effects, something that requires a closer look at, and a modification of, the cell model shown in Figure 2.1.

In the equivalent circuit shown in Figure 2.1, the conductance of the plasma membrane is assumed to be zero, and the capacitive components of cytoplasm (the interior of the cell) are neglected. However, these assumptions limit the applicability of the circuit model to times that are short relative to the dielectric relaxation time of the membrane, and to times that are long relative to the dielectric relaxation time of the cytoplasm. The dielectric relaxation time, τ_d, provides information on the importance of the resistive or capacitive component of the membrane and cytoplasm, respectively, with respect to the duration of an applied electric field, τ_d

$$\tau_d = \frac{\varepsilon}{\sigma} \tag{2.14}$$

where
   ε is the permittivity
   σ is the conductivity

For a pulse duration, τ_p, long compared with τ_d, the resistive component dominates; for the opposite case, it is the capacitive component.

In order to describe the effects of electrical pulses on cells over a wide range of pulse duration, we need to consider the equivalent circuit of a cell where these additional circuit elements are taken into account (Schoenbach et al., 2007). For simplicity, in the following discussion we will focus on one part of the equivalent circuit only, which includes a section of the plasma membrane and cytoplasm. The equivalent circuit for this case is shown in Figure 2.6. Although we focus here only on the plasma membrane, the conclusions that are drawn from the discussion of this simple model can easily be extended to predict the electrical effects on the inner cell structures.

For long pulses, long meaning that the capacitive term in the membrane impedance can be neglected compared with the resistive term, the equivalent circuit is reduced to two resistors in series. This condition holds for $\varepsilon_m/\sigma_m > \tau_m$. With the relative membrane permittivity, ε_{mr}, being approximately 10 and the membrane conductivity, σ_m, being on the order of 10^{-5} A/Vm, this condition is satisfied for pulse durations large compared with $10\,\mu$s. Consequently, for pulses long compared with $10\,\mu$s duration, the capacitive effects play a diminishing role, but resistive effects (thermal effects) in the plasma membrane need to be considered.

For very short pulses, the dielectric properties of the m, rather than its resistive characteristics, determine the electric field distribution (Schoenbach et al., 2008). The condition that the resistive term in the cytoplasm impedance can be neglected compared with the capacitive term requires that the pulse duration is short compared with the dielectric relaxation time of the cytoplasm ($\varepsilon_{cp}/\sigma_{cp}$). Assuming that the relative permittivity of the cytoplasm is 80, and the conductivity is 1 S/m, this is only true if the

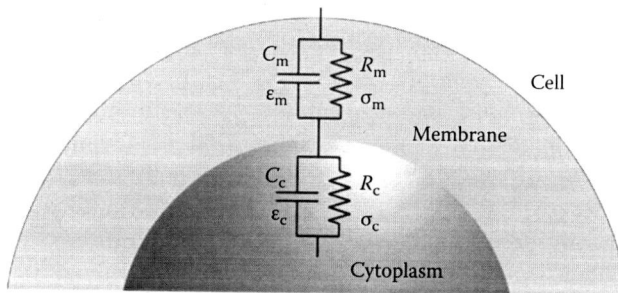

FIGURE 2.6 Equivalent circuit of the membrane and cytoplasm section of a biological cell. It includes the membrane resistance and the capacitance (permittivity) of the cytoplasm.

pulse duration is less than 1 ns. The electric fields in the various parts of the cell are then defined by the continuity of the electric flux density:

$$\varepsilon_m E_m = \varepsilon_c E_c \tag{2.15}$$

For a membrane with a relative dielectric constant of 10 (Ermolina et al., 2001; Feldman et al., 2003), the electric field in the membrane is eight times higher than the electric field in the adjacent cytoplasm, which has a dielectric constant of 80. For very high applied electric fields of several hundred kV/cm, it might be possible to reach voltage levels across the membrane that cause direct and instant conformational changes of membrane proteins. This range of operation, where the cell is defined by only its permittivity, opens a new temporal domain for cell responses to pulsed electric fields.

In summary, the range for nanopulse effects as defined by the charging of subcellular membranes reaches from pulses with durations comparable to the charging time, τ_m, of the plasma membrane (and short rise times compared with the pulse duration) to pulses with durations on the order of the dielectric relaxation time of the cytoplasm:

$$2\tau_m < \tau_p(\tau_r \ll \tau_p) < \tau_{ds} \tag{2.16}$$

This condition should not only hold for cells in dilute solution but also for cells that are densely packed. The difference between the two cases is the difference in the characteristic plasma membrane charging time, τ_m. For dilute suspensions, the volume fraction term in Equation 2.2 can be neglected, and the characteristic charging time, e.g., of a cell with 5 μm radius, a capacitance of 1 μF/cm^2, and internal and external resistivities of 100 Ω-cm, is 75 ns. For a dense suspension (and even more so for tissue), the charging time constant and, consequently, the β-relaxation frequency change considerably, e.g., for a volume fraction of V = 0.95, assuming that the internal and external resistivities, ρ_i and ρ_e, are the same, the charging time constant is 14 times longer than for cells in a dilute suspension, where the V-term can be neglected. In our example of cells with a 5 μm radius and external and internal resistivities of 100 Ω-cm, the plasma membrane charging time constant, and consequently the maximum pulse duration for intracellular effects, would increase from 75 ns to 1.05 μs. However, what is gained by relaxing the condition of the pulse duration is lost due to the reduced sensitivity of the membranes of cell organelles at longer pulses (Figure 2.7).

2.3 Modeling–Nanoporation

After the onset of poration, this simple linear, passive element approach is no longer applicable to describe electric field-cell interactions. The membrane then becomes an "active" cell element with variable resistivity and variable permeability. The modeling of cells with "active" membranes has been a topic of publications by the team at Old Dominion University with R.P. Joshi as the principal investigator and at MIT with J.C. Weaver as the principal investigator. Weaver's group has focused on a lattice model (Smith et al., 2006; Stewart et al., 2006) and Joshi's group has used a distributed circuit model with current continuity (Joshi et al., 2004b; Schoenbach et al., 2007). The Smoluchowski equation was applied for the voltage-dependent description of the nonlinear membrane resistance and pore development. Details of the Smoluchowski equation, its application, and the dynamical effects of electroporation with ultrashort pulses have been addressed in a number of papers by the Joshi group (Joshi et al., 2001, 2002).

Figure 2.8 shows the results of a continuum model (Gowrishankar et al., 2006; Weaver, 2006). Here, the poration of the plasma membrane and subcellular membranes was compared for pulses of 7 μs and 60 ns duration. The electric field of the pulses was adjusted such that the energy density in both cases was identical. For 60 ns long pulses (Figure 2.8a), the cell membranes, both plasma and internal, are fully exposed to the applied 60 kV/cm pulse, clearly demonstrating that nanosecond pulses allow us to

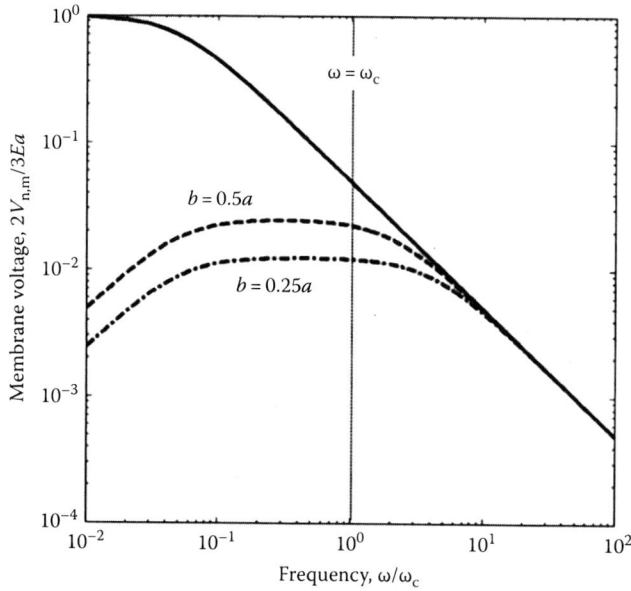

FIGURE 2.7 Spectrum of the voltage across the plasma membrane (solid line) and subcellular membranes (dashed and dashed-dotted line) for the case where the volume fraction is high ($V = 0.95$). The curves are plotted relative to the corner frequency (β-relaxation frequency) of the spectra for dilute suspensions. There is a shift to lower frequencies, but also a reduction in sensitivity of the subcellular membranes, compared to that of dilute suspensions.

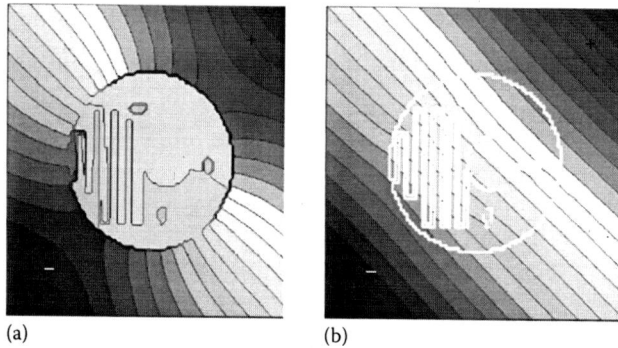

FIGURE 2.8 The effect of 7 μs long pulses with 1.1 kV/cm field amplitude (a) and that with 60 ns at 60 kV/cm amplitude (b) on cells. The electrical parameters were chosen such that the electrical energy for both cases is identical. (From Weaver, J.C., Harvard-MIT Division of Health Sciences and Technology, Cambridge, MA, private communication, 2006. With permission.)

affect subcellular membrane potentials. When the cell was exposed to a long pulse (7 μs) at 1.1 kV/cm, the effects were only seen on the plasma membrane. The cell interior was shielded.

Whereas long pulses cause the creation of membrane pores large enough for large molecules to pass (electroporation), using a Smoluchowski equation-based model, the application of nanosecond pulses was found to cause the creation of a high density of "nanopores" (Joshi et al., 2004b; Gowrishankar et al., 2006; Vasilkoski et al., 2006; Schoenbach et al., 2007). These are pores of such small diameters that they become permeable only for small ions, but not large molecules such as propidium iodide. Molecular dynamics (MD) simulation is a method that is particularly suitable to study these nanopores. The part of the cell that is to be modeled is considered to be a collection of interacting particles. In the case of a cell membrane, which is a lipid bilayer, these particles are dioleoyl-phosphatidyl-choline

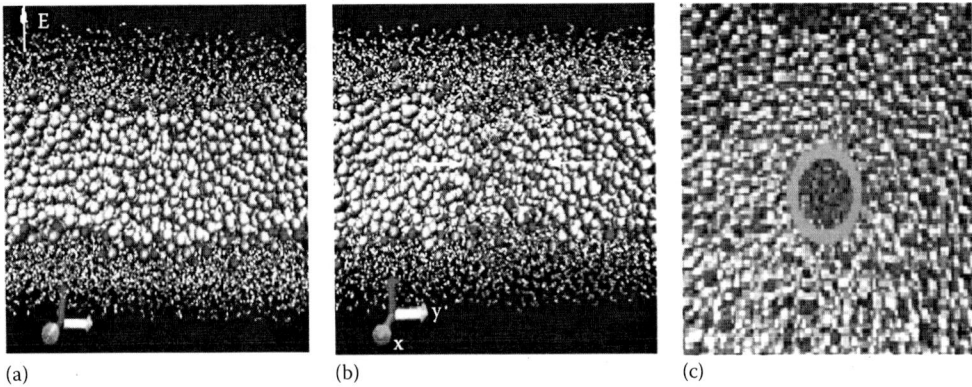

(a)   (b)   (c)

FIGURE 2.9   (See color insert following page 268.) Molecular dynamics results showing nano-pore formation at a membrane by an external 0.5 V/nm pulsed electric field. (a) Before application of electric field, (b) 3.3 ns after application of pulsed field (cross section), and (c) same, but end-on view. (From Schoenbach, K.H. et al., *IEEE Trans. Dielect. Electr. Insul.*, 14, 1088, 2007. With permission.)

(DPPC) molecules that are characterized by charged subgroups. For each of the molecules, the equation of motion (Newton's equation) is solved, with the force on each charge and dipole within the molecular structure generated by the surrounding charges and dipoles.

The use of the MD codes is a computationally intensive approach requiring a large number of particles to be considered as well as small time steps. Therefore, MD is currently restricted to modeling small parts of a cell and pulses on the order of nanoseconds. The importance of this method is in the visualization of the membrane effects on this timescale, the inclusion of complex underlying physics, and the determination of critical electrical fields for membrane changes on the nanoscale, e.g., for pore formation (Tieleman et al., 2003; Hu et al., 2005a,b; Vernier et al., 2005, 2006b).

The results of an MD simulation, shown in Figure 2.9 for a fixed field of 0.5 V/nm, demonstrate the nano-pore formation (Schoenbach et al., 2007). Figure 2.9a shows the initial configuration of the lipid bilayer with aqueous media on either side. Figure 2.9c shows the top view of the final membrane configuration obtained from MD simulations after about 3.3 ns at the constant field of 0.5 V/nm. The cross-sectional view appears in Figure 2.9b and clearly shows the head groups lining the pore. The pore is predicted to be about 1.5 nm in diameter, a typical value for "nanopores." The short duration of nsPEF prevents the expansion of these nano-pores to sizes that allow the passage of large molecules at moderate electric fields. Of course, multiple pulsing and/or high electric fields will eventually lead to the formation of large pores (Schoenbach et al., 2008).

2.4 Pulse Generators for Nanosecond Bioelectric Studies

The experimental study of intracellular electro-effects and applications of such effects requires electrical pulse generators that provide well-defined high voltage pulses with a fast current rise to the load. The load could be either cells in suspension or tissue. The applied voltage is determined by the field threshold required for intracellular effects and the distance between the electrodes. The value depends strongly on pulse duration: shorter pulses require higher amplitudes for the same effect (see Section 2.9). The onset field for any effect is also strongly dependent on the cell type. For single pulses or pulse trains with only a few pulses, generally up to five, typical electric fields required for inducing apoptosis range from tens to hundreds of kV/cm. The highest values for 10 ns pulses were 300 kV/cm. These fields are lower for pulse trains with an extended number of pulses.

In order to study the intracellular electro-effects, two types of pulse generators have been used. One was developed for experiments that require large numbers of cells, such as those in which flow cytometry is used or where studies are performed on tissues. The pulse generator is based on the Blumlein concept and can deliver pulses ranging from 10 to 600 ns and voltages of up to 40 kV into a low resistance (generally 10 Ω) load. A second class of pulse generators was designed for observations of individual cells under a microscope. Consequently, the gap distance can be reduced to the 100 μm range, and simultaneously, voltage requirements can be relaxed. Instead of typical pulsed power components, low voltage, high frequency cables can be used in the design together with fast semiconductor switches. Such types of pulse generators, micropulsers, operate at voltages of less than 1 kV. However, because of the small electrode gap, electric field intensities of up to 100 kV/cm are possible and can be applied to cells in suspension that are placed in the electrode gap. Details of the pulse generator and pulse delivery devices are described in a paper by Kolb et al. (2006) and in Chapter 17.

2.5 Effects on Plasma Membrane

The charging of the outer membrane is the predominant primary response of a mammalian cell to an externally applied electric field. With the accumulation of ions along the cell membrane, the potential difference across the membrane increases and the electric field inside the cell is simultaneously reduced. Membrane lipids are perturbed, and when a potential difference above a certain threshold is sustained long enough, pores will begin to open and will allow molecular transport. If neither field strength nor exposure time is extensive, the pores will reseal again within several seconds after the electric field has dropped (Benz and Zimmermann, 1981; Neumann, 1989; Shirakashi et al., 2004). The poration effect is related to an increase in membrane conductance, and consequently, a change in transmembrane voltage and current.

Using laser stroboscopy, we were able to measure the change in transmembrane voltage with a temporal resolution determined by the laser pulse duration of 5 ns (Frey et al., 2006). The transmembrane voltage was quantified by staining the cell membrane with a fluorescent voltage-sensitive dye. Changes in fluorescence intensity quantify the change in transmembrane voltage. In order to utilize the short illumination time of the laser, the response of the dye needs to be faster than the laser pulse. The voltage-sensitive dye that was used, Annine-6, has a voltage-regulated fluorescence response that depends only on the shift of energy levels by Stark-effect (Kuhn et al., 2004). This fast electronic process theoretically enables a sub-nanosecond voltage sensitive response.

To investigate the membrane charging in response to an ultrashort pulsed electric field, we stained Jurkat cells (a T-cell leukemia cell line) with Annine-6 and exposed them for 60 ns to an electric field of 100 kV/cm. At different times during the exposure, the cells were illuminated with a laser pulse of 5 ns duration full-width-at-half maximum (FWHM). The recorded changes in fluorescence intensity allow us to estimate the transmembrane voltages based on calibration curves (which were developed for transmembrane voltages of up to 0.3 V and extrapolated for our results) and monitor their temporal development. The results of measurements at the cathodic pole of a cell with a 60 ns pulse of 100 kV/cm amplitude applied to Jurkat cells are shown in Figure 2.10.

The results of this study show that an increase in plasma membrane conductivity occurs after only a few nanoseconds (less than the resolution of our method), indicated by the deviation of the measured voltage from the ideal temporal voltage development that would be expected for a passive membrane (solid curve). After 20 ns, a second increase in membrane conductance is seen, which might be due to the formation of nanopores. The membrane voltage decrease after the pulse is determined by the discharge of the membrane, but also by the residual applied voltage.

The observed change in membrane conductance in the single pulse experiment (Frey et al., 2006) is assumed to be due to the formation of a large concentration of nanopores, as discussed in the previous section. Using a whole cell patch clamp to measure plasma membrane conductance changes resulting from nsPEF exposure, such nanopores have been observed following nsPEF exposure (Pakhomov et al., 2007). It was found that the nanopores exhibit inward rectification along with the inhibition of voltage-gated outward K^+ current at positive membrane potentials. These behaviors clearly distinguish the permeability increase resulting from nsPEF from that resulting from the much longer pulses used for classical electroporation. Two other important characteristics of the nsPEF-induced permeability increase are that it is long-lasting and increases with field strength. The membrane resistance (R_m) after

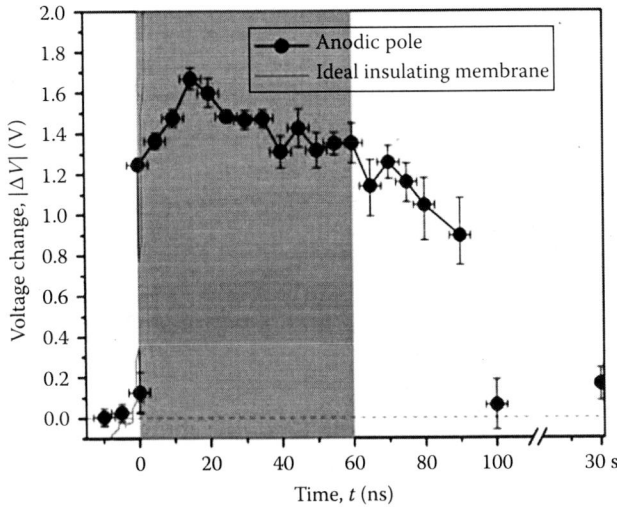

FIGURE 2.10 Temporal development of the transmembrane voltage at the anode pole of a Jurkat cell when a 60 ns long 100 kV/cm pulse was applied. (From Frey, W. et al., *Biophys. J.*, 90, 3608, 2006. With permission.)

exposure to 60 ns pulses at 15 kV/cm decreased threefold or more after exposure to a single pulse in GH3, PC12, and Jurkat cells, but not in HeLa cells, which were some 5- to 10-fold more resistant to nsPEF exposure. The permeabialized state is maintained for at least 100 s after nsPEF exposure. Furthermore, subsequent experiments established that R_m recovers rather gradually and remains below the control levels even at 10 and 15 min after the exposure.

Whereas these studies were performed at relatively low electric fields, the effects of high intensity pulses (or a large number of pulses) on the plasma membrane demonstrated an uptake of large molecules. Three of the dyes most often used to study changes in plasma membrane permeability are trypan blue, propidium iodide (PI), and ethidium homodimer. Figure 2.11 shows the uptake of PI by

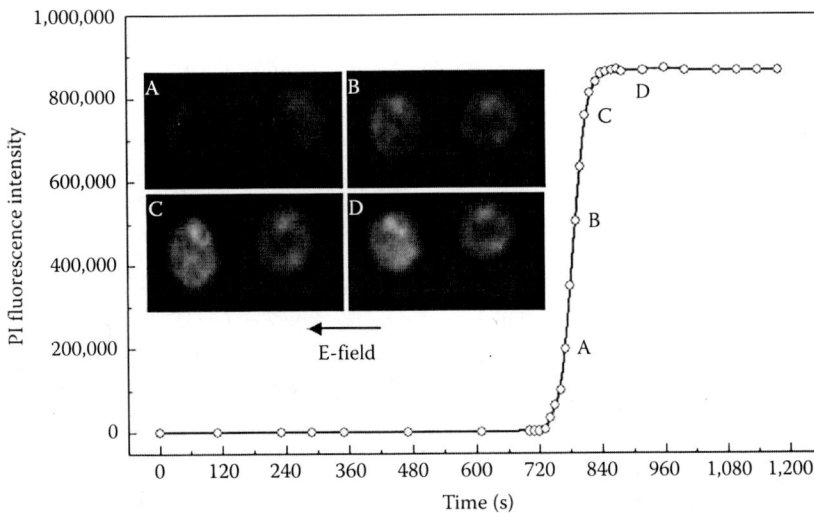

FIGURE 2.11 (**See color insert following page 268.**) The temporal development of the PI uptake and microscopic real-time images of typical HL-60 cells undergoing PI uptake (A) 770 s, (B) 790 s, (C) 810 s, and (D) 920 s, following a 60 ns, 26 kV/cm electric pulse. The electric field orientation is marked. (From Chen, N. et al., *Biochem. Biophys. Res. Commun.*, 317, 421, 2004. With permission.)

HL-60 cells following a single 60 ns, 65 kV/cm pulse (Chen et al., 2004). It is interesting that the increase in fluorescence is observed only after approximately 12 min and then completed in 1 min. This indicates that the formation of pores large enough to allow the passage of PI is a secondary effect, following the formation of nanopores, which occurs on a nanosecond timescale. PI uptake has been observed 3–15 min after exposure. A slow diffusion cannot be ruled out, but a rapid increase (within several seconds) indicates a greatly increased membrane permeability spanning the dynamic range of the PI response (i.e., the values are well above simple threshold values). It is likely that the nanopores are too small to allow PI uptake immediately and the secondary PI uptake is due to membrane failure as *in vitro* necrosis, secondary to apoptosis.

Usually, the uptake of these membrane-integrity dyes is considered an indication of cell death. However, experiments with nanosecond pulses (60 ns) (Beebe et al., 2003a; Hall et al., 2007) as well as 800 ps pulses (Schoenbach et al., 2008) indicate that these effects can be transient and are not necessarily harbingers of cell death. When Jurkat cells were exposed to 60 ns pulses in the presence of ethidium homodimer and analyzed by flow cytometry, the cells exhibited increased fluorescence, indicating that the membrane had been breached to allow uptake of the dye. However, when added 5 min after the pulse, no ethidium fluorescence was observed, indicating that the membrane had "resealed" (Beebe et al., 2003b). Likewise, when HCT116 colon carcinoma cells were exposed to three 60 ns pulses at 60 kV/cm in the presence of ethidium homodimer, the cells took up the dye up to and even after 1 h beyond the pulse. However, 5 h after the pulse, the cells were no longer permeable to the dye. Under these conditions, only 10%–20% of the cells died, so the reversible effects on the plasma membrane were not due to the elimination of permeable cells. Similar results were observed for Annexin-V-FITC binding, which is a marker for the externalization of phosphatidylserine and apoptosis. Under minimally lethal conditions, these pulses had reversible effects on the plasma membrane integrity and phospholipid orientation. Similar results were observed when B16-F10 cells were exposed to 800 ps pulses at electric fields between 350 and 1000 kV/cm with energy densities up to 2 kJ/cc using trypan blue (Schoenbach et al., 2008). Initially, the cells took up the dye. After a certain time, generally about 1 h, the membranes of most of the cells seemed to have lost their permeability and the dye was excluded.

These results indicate that nsPEFs have significant effects on the structure of the plasma membrane, but under certain conditions the structural changes are reversible. The time required to return to the control state can be considered as a recovery time, presumably to allow for "repair" of the structure and function. Most importantly, when using nsPEFs, these membrane integrity dyes should not be considered to be cell death markers and Annexin-V-FITC binding may not represent a valid apoptosis marker.

2.6 Effect on Subcellular Membranes

In order to prove the validity of the hypothesis that shorter pulses increasingly affect membranes of subcellular structures, the effect of 60 ns pulses on human eosinophils has been studied (Schoenbach et al., 2001). Eosinophils are one of the five different types of white blood cells and are characterized by large red (i.e., eosinophilic) cytoplasmic granules when the cells are fixed and stained. The cells were loaded with calcein-AM (calcein-acetoxymethylester), an anionic fluorochrome that enters cells freely and becomes trapped in the cytoplasm by an intact cell surface membrane following removal of the AM group. The granules in the eosinophils stay unlabeled because the cytosolic calcein is impermeant to the granular membrane.

When ultrashort pulses with an electric field amplitude of 50 kV/cm and higher were applied to the eosinophils suspended in Hanks Balanced Salt Solution, the granules, which were dark (nonfluorescent) before pulsing, began to fluoresce brightly (Figure 2.12). This is strong evidence for the breaching of the granule membranes and ionic binding of free calcein from the cytosol to the cationic granule components. On the other hand, the retention of the cytoplasmic calcein staining indicated that the surface (outer membrane) was still intact after the pulsing, or, more accurately, was not electroporated such that

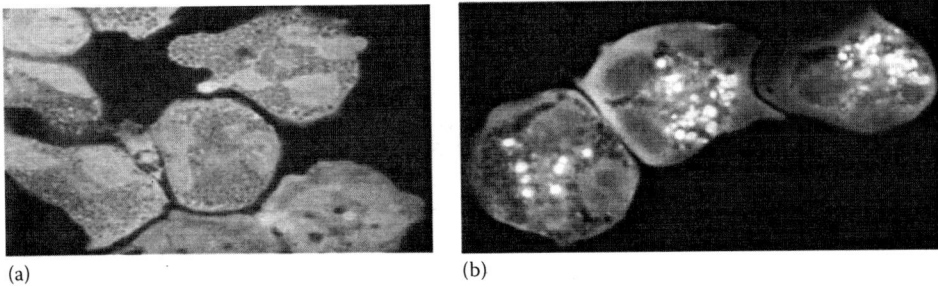

(a)   (b)

FIGURE 2.12   Eosinophils (white blood cells) before (a) and after (b) the application of 60 ns pulses with electric fields of 50 kV/cm—photograph on left shows that inner structures have opened and taken up dyes—shown as "sparklers." (From Schoenbach, K.H. et al., *J. Bioelectromagn.*, 22, 440, 2001. With permission.)

it became permeable for these ions. This does not exclude that it was "nanoporated." The experiments with 60 ns pulses showed, for the first time, that poration/disruption of intracellular membranes can be achieved with ultrashort electrical pulses without the loss of the surface membrane integrity. Experiments with shorter pulses (10 ns) have confirmed the earlier results (Buescher and Schoenbach, 2003).

Similar experimental results have been reported by Tekle et al. (2005). A mixed population of phospholipid vesicles and single COS-7 cells, in which vacuoles were induced by stimulated endocytosis, were exposed to nanosecond pulses. Under appropriate conditions, the preferential permeabilization of one vesicle population in a mixed preparation of vesicles of similar size was obtained. It was also shown that vacuoles in COS-7 cells could be selectively permeabilized with little effect on the integrity of the plasma membrane.

The effect of short electrical pulses on the cell nucleus has been explored using acridine orange (AO), a vital fluorescent dye (Chen et al., 2004). It is able to permeate the plasma membrane, nuclear membrane, and other organelle membranes of living cells and interacts with DNA and RNA by intercalation or electrostatic attraction, respectively. Experiments with HL-60 cells, where 10 ns pulses of 65 kV/cm field amplitude were applied, showed an exponential decrease in the average fluorescence intensity in the nucleus with a time constant of approximately 3 min. These results showed that nanosecond pulses cause an increase in the permeability of the nuclear membrane, with a subsequent outflow of labeled DNA.

2.7 Secondary Effects

Whereas the studies described in the previous sections dealt with primary effects, particularly the effect of ultrashort pulses on transport through membranes, in the following, we will focus on changes in cell functions in response to ultrashort pulses. Dependent on pulse duration, pulse amplitude, and on the number of pulses in a pulse train, various effects have been observed. For high electric fields, which for 10 ns pulses may need to be higher than 200 kV/cm and lower for longer pulses or pulse trains, apoptosis has been observed (Beebe et al., 2002, 2003a,b; Vernier et al., 2003a). When the pulse amplitude was lowered, calcium release from intracellular stores and subsequent calcium influx through store-operated channels in the plasma membrane was observed (Buescher and Schoenbach, 2003; Beebe et al., 2003b; Vernier et al., 2003b; White et al., 2004).

2.7.1 Intracellular Calcium Release

Ultrashort pulses with electric field amplitudes less than required to induce apoptosis in cells have been found to trigger physiological responses in cells. Most of the research in the lower range of pulse amplitudes has focused on its effect on the release of intracellular free calcium. Calcium is known as a ubiquitous second messenger molecule that regulates a number of responses in cell signaling, including enzyme activation, gene transcription, neurotransmitter release, secretion, muscle contraction, and apoptosis, among others. Intracellular calcium is stored in the endoplasmic reticulum (ER) compartments and mitochondria (and for platelets in α-granules), and calcium release from mitochondria is considered to be an initiation event for apoptosis (Berridge et al., 1998; Susin et al., 1998).

When 300 ns long pulses were applied to polymorphonuclear leukocytes, immediate, but transient, rises in the intracellular calcium concentration could be obtained with pulse amplitudes as low as 12 kV/cm (Buescher and Schoenbach, 2003). In experiments where the cells were actively crawling over a slide surface, with associated fluctuations in $[Ca^{2+}]$ prior to pulse application (observed by using Fluo-3, a calcium indicator), pulsing caused an abrupt loss of mobility that correlated to a rise in intracellular calcium. The immobilization phase of the cells was found to be dependent on the amplitude of the field. Lowering the electric field allowed the cells to recover more quickly, an effect that has previously been observed when aquatic organisms, such as brine shrimp (Schoenbach et al., 1997) or hydrozoans (Ghazala and Schoenbach, 2000), were subjected to pulsed electric fields.

Pulse-induced calcium release was studied in experiments with HL-60 cells, which were loaded with Fura-2, a calcium indicator, and exposed to pulses with 60 ns durations and electric fields that were below those used to induce apoptosis (Beebe et al., 2003b). To compare the pulse effect to chemically induced calcium release, the purinergic agonist UTP was used, which is known to release calcium from the ER. In the absence of extracellular calcium, both UTP and a single 60 ns pulse with 10 kV/cm amplitude induced calcium release from intracellular stores. The kinetics of the responses were similar; both were rapid and transient. When calcium was subsequently added to the extracellular medium, an influx of calcium was observed in response to both agonists with similar kinetics. Thus, ultrashort pulses and UTP cause intracellular calcium release followed by influx through calcium channels in the plasma membrane with similar kinetics.

Fluorescence microscopy with a temporal resolution of milliseconds confirmed the ER as the likely source of calcium following pulse application (Scarlett et al. 2009). Other internal calcium stores, in particular mitochondria, might be affected in a similar way by these nanosecond pulses. However, the much smaller size of mitochondria, when compared with the volume filled by the ER, has prevented the temporal and spatial resolution of calcium release from these sources so far. The release of calcium was observed within the first 18 ms after exposure to a 60 ns pulse and reaches a maximum in less than a second, before it decreases in an exponential manner over 2–3 min (Figure 2.13). With 25 and 50 kV/cm field strengths, the response is independent of the presence of extracellular calcium. For 100 kV/cm field strength, calcium concentrations increase after the initial peak for another 10–25 s, which suggests the influx of calcium.

nsPEFs also release calcium from human platelets (Zhang et al., 2008). The results of studies where nanosecond pulse effects were compared with that of thrombin suggested that the nsPEF calcium release site in platelets were α-granules. That these granules are affected by nanosecond pulses is also indicated by the observed release of platelet-derived growth factors (PDGF) known to be stored in α-granules.

Based on what is known about ligand-induced, physiological calcium mobilization and electric field effects on membranes, it is possible that the pulsed field-induced calcium release could occur by at least one of two mechanisms. One mechanism is an effect on the plasma membrane that mimics the ligand-induced, physiological calcium mobilization response acting through G-protein-coupled receptors or elements downstream of the receptor. These receptors generate IP3, which binds to specific IP3 receptor channels in the ER for calcium release into the cytoplasm. This is followed by capacitative calcium influx through store-operated channels in the plasma membrane (Taylor, 2006). UTP, a purinergic ligand agonist, and nsPEFs, a nonligand agonist, achieved the same calcium mobilization responses indicated above (Beebe et al., 2003b). This could occur by a conformational change and an activation of a plasma membrane receptor, or an effect on some other step in the pathway that leads to ER IP3 receptor channel activation. The other mechanism for nsPEF-induced calcium release is suggested by our understanding of electric field effects on membranes, especially those with shorter pulses affecting intracellular membranes, which are not readily affected by conventional plasma membrane electroporation. Thus, it

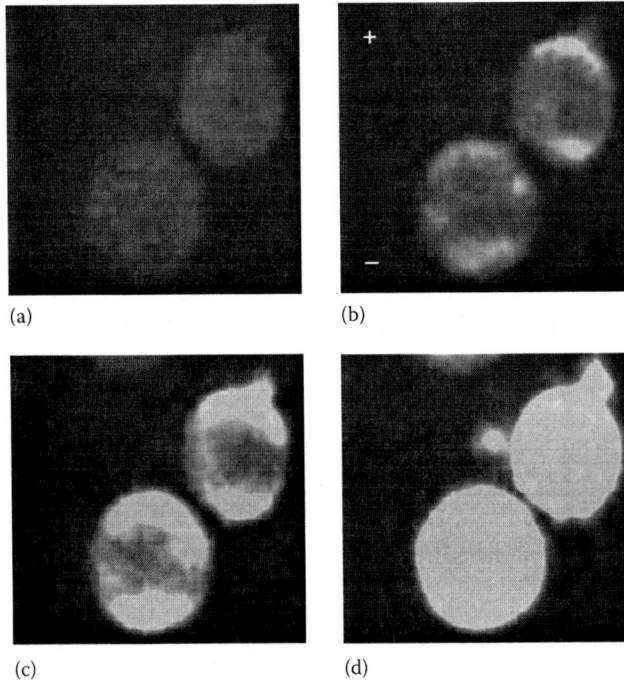

FIGURE 2.13 Jurkat cells loaded with Fluo-4 were imaged every 18 ms with a 5 ms exposure after delivery of a 60 ns pulse of 100 kV/cm. (a) Before pulse, (b) at 18 ms, (c) at 36 ms, and (d) at 54 ms. (From Scarlett, S.S. et al., *Biochim. Biophys. Acta Biomembr.*, 1788, 1168, 2009. With permission.)

is possible that electric field effects have direct supra-electroporation effects on the ER, or other intracellular storage sites, releasing intracellular calcium that mimics the purinergic ligand- and thrombin-induced responses. Likewise, a longer pulse duration, higher electric fields, and/or more pulses will have proportionally greater effects on the plasma membrane, which can result in an influx of calcium through pores or aqueous channels.

2.7.2 Apoptosis

Mammalian cells use programmed cell death (PCD) to efficiently eliminate themselves from a tissue system when they are no longer functional or necessary (for more information on apoptosis, see e.g., Afford and Randhawa (2000)). This process is different from necrosis in that it does not cause inflammation or gross damage to surrounding healthy cells. One of the most detailed forms of PCD is termed apoptosis, and following an apoptosis-inducing stimulus, several signaling pathways are activated in order to bring about cell death. These pathways include specific enzymes and signaling molecules that cause distinct morphological changes to the dying cell. The physical changes in the cell reflect the degradation that is occurring to its DNA and proteins and allow it to be recognized by other cells as components that are meant to be scavenged by the immune system. Although the cell is being systematically broken down on the inside, the plasma membrane integrity is generally maintained until a much later time, so that the enzymes responsible for degradation are not released into the surrounding healthy tissue areas, thus avoiding gross damage to the area.

During apoptosis, the plasma membrane becomes marked for disposal by phosphatidyl serine (PS) molecules on its surface and, *in vivo*, the task of the final removal of this apoptotic body is given to macrophages that engulf the dying cell and finish the breakdown. However, when monitoring apoptosis progression in cells in culture, and thus in the absence of macrophages, the loss of membrane integrity is considered to be a sign of late apoptosis and is termed secondary necrosis. Phosphatidyl serine externalization following the application of ultrashort pulses has been studied extensively by Vernier et al., using pulses as short as 3 ns (Vernier et al., 2003a, 2004, 2006a,b). Experiments with Jurkat cells loaded with Calcium Green, a calcium indicator, showed that calcium release is coupled to phosphatidylserine externalization (Vernier et al., 2003a). The pulses applied to the cells had a duration of 30 ns and an amplitude of 25 kV/cm. Although phosphatidylserine externalization is a generally accepted apoptosis marker, it does not necessarily, specifically with this type of stimulation, indicate or guarantee cell death.

In order to study apoptosis induction through nanosecond pulses, experiments on Jurkat cells and HL-60 cells subjected to one to five pulses of 10, 60, and 300 ns were performed with electric fields ranging from 300 kV/cm (for 10 ns) to several tens of kV/cm for long pulses (Beebe et al., 2002). Apoptosis was induced by ultrashort pulses when the electric field amplitude exceeded a threshold value. For 10 ns pulses, almost 300 kV/cm was required to induce apoptosis in Jurkat cells. For 300 kV/cm, increased Annexin-V-FITC fluorescence is seen, which indicates the externalization of phosphatidylserine, a marker for apoptosis. Annexin-V-FITC specifically binds to the cell membrane lipid, phosphatidylserine, which is on the inside of the plasma membrane of nonapoptotic cells and flips to the outside of the plasma membrane when the cell becomes apoptotic. For longer pulses, e.g., 60 ns, phosphatidylserine externalization was observed at electric fields as low as 40 kV/cm. Whereas Jurkat cells showed signs of apoptosis at relatively low electric fields, HCT116 colon carcinoma cells required higher electric fields to see the same effect (Beebe et al., 2003a; Hall et al., 2007).

Ultrashort pulse-induced apoptosis markers appear in tens of minutes following treatment. Following the pulse application, cells bind Annexin-V-FITC and, only later, take up ethidium homodimer. Caspases were activated in 5–20 min. The Annexin-V-FITC binding occurred rapidly and permanently and 30% of the cells had proceeded to exhibit membrane rupture by 30 min, a typical characteristic of, secondary necrosis. This rapid progression of apoptosis is quite different from that obtained with other apoptotic stimuli, such as UV light and toxic chemicals, which require hours for apoptosis markers to appear. However, the kinetics of ultrashort pulse-induced apoptosis depends on the pulse duration. Shorter pulses result in slower apoptosis progression than longer pulses for the same electrical energy density. Cytochrome c release (another apoptosis marker) was also measured for the presence of caspase activation, when ultrashort pulses were applied to Jurkat cells (Beebe et al., 2002), indicating mitochondria-dependent apoptosis mechanisms. Instead of using single pulses or small numbers of subsequent pulses at high electric fields (>100 kV/cm at 10 ns), apoptosis can also be induced by applying large numbers of pulses at lower electric field intensity. Vernier et al. (2003a) reported phosphatidylserine translocation and caspase activation in Jurkat cells with 10 ns pulses of 25 kV/cm amplitude. Unlike the Jurkat cells, rat glioma C6 cells, which normally grow attached to a surface, were found to be highly resistant to the same pulses and pulse sequences. In other experiments where cell survival was explored after ultrashort pulse application, the results also indicated that nonadherent cells seem to be more sensitive to this application than adherent cells (Stacey et al., 2003).

Since the initial discovery that nsPEFs trigger an apoptosis signaling pathway, there have been significant efforts made to determine the exact mechanism of this process. One of the possible reasons for apoptosis induction was assumed to be extensive calcium release, caused by nanosecond pulses (Vernier et al., 2003b). Calcium at high concentrations is known to induce apoptosis. However, in experiments on Jurkat and HL-60 cells (Beebe et al., 2003b), it was found that chelation of calcium in the extracellular media with ethylene glycol tetraacetic acid (EGTA) and in the cytoplasm with BAPTA-AM had little or no effect on caspase activation, suggesting that calcium was not required for nsPEF-induced apoptosis.

Another possible mechanism for apoptosis induction with ultrashort electrical pulses was suggested by Weaver (2003). Ultrashort (nanosecond) pulses are known to affect subcellular structures, either through the displacement current flowing during the early part of the pulse, or through the conduction current after the conductance of the plasma membrane is increased through poration. Consequently, the electric field correlated to the current density in the cytoplasm will affect subcellular membranes,

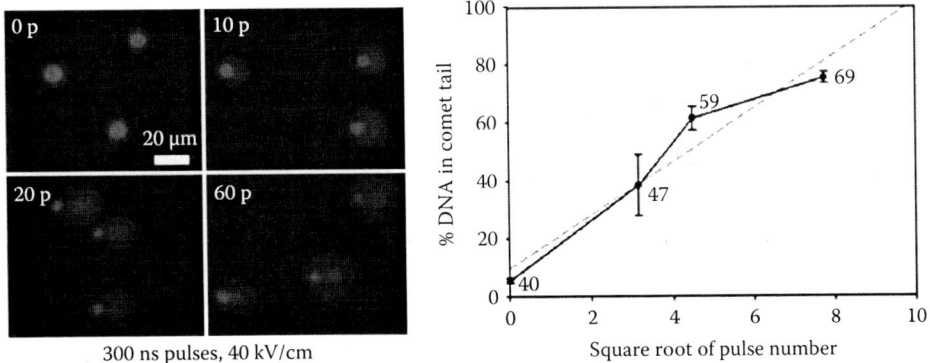

300 ns pulses, 40 kV/cm

FIGURE 2.14   (See color insert following page 268.) Comet assay using B16 cells *in vitro*. 40 kV/cm pulses 300 ns long were used and the pulse number is indicated on each figure. Quantification of propidium iodide fluorescence allows us to estimate the percentage of total DNA in the comet tail. When plotted against the square root of the pulse number, a linear dependence is revealed that predicts 100% DNA fragmentation when cells are exposed to 100 pulses. The straight line is a least squares fit to the four data points and the error bars represent the SEM with the number of cells averaged in each point written next to it. (From Nuccitelli, R. et al., *Int. J. Cancer*, 125, 438, 2009. With permission.)

e.g., the mitochondrial membranes. It is known that biochemically induced apoptosis involves the mito-chondrial permeability transition pore complex (MPTP) (Halestrap et al., 2002) and/or mitochondrial membrane voltage-dependent anion channels (Savill and Haslett, 1995; Tsujimot and Shimizu, 2002). Weaver's hypothesis is that ultrashort pulses change the transmembrane voltage at mitochondrial mem-brane sites, which leads to an opening of the MPTP, inducing apoptosis (Weaver, 2003).

The production of reactive oxygen species and subsequent DNA damage has also been proposed as a possible effect causing apoptosis (Walker et al., 2006). There is evidence that nsPEF stimulation with multiple, intense pulses causes damage to DNA or other critical proteins. Studies by comet assay have shown that significant DNA damage does occur very quickly after treatment (90 s post-treatment) with nsPEFs (Nuccitelli et al., 2009) (Figure 2.14). Some DNA repair was observed when the permeabiliza-tion of the cells for the comet assay was delayed for an hour. The degree of fragmentation was found to be linearly proportional to the square root of the pulse number. At this time, it is not known if DNA damage is a primary (direct) effect of intense nanosecond electric fields or a secondary effect caused by membrane permeabilization.

2.8 Applications

Applications for nanosecond pulse effects cover a wide range: from rather mundane uses, such as bio-fouling protection for cooling water systems (Schoenbach et al., 1997), to medical applications such as wound healing (Zhang et al., 2008) and cancer treatment (Nuccitelli et al., 2006, 2009; Schoenbach et al., 2006).

2.8.1 Wound Healing

It has been shown that nanosecond pulses have an effect on platelets similar to thrombin, an agonist that promotes aggregation (Beebe et al., 2004). Results of studies on the effect are shown in Figure 2.15. This process of aggregation is initiated by calcium release from internal stores. This is consistent with the well-known fact that aggregation of platelets by known agonists, such as thrombin, requires an increase in intracellular free calcium. The data obtained with nsPEF on calcium mobilization and calcium influx is reminiscent of the well-known capacitive or store-operated calcium entry process

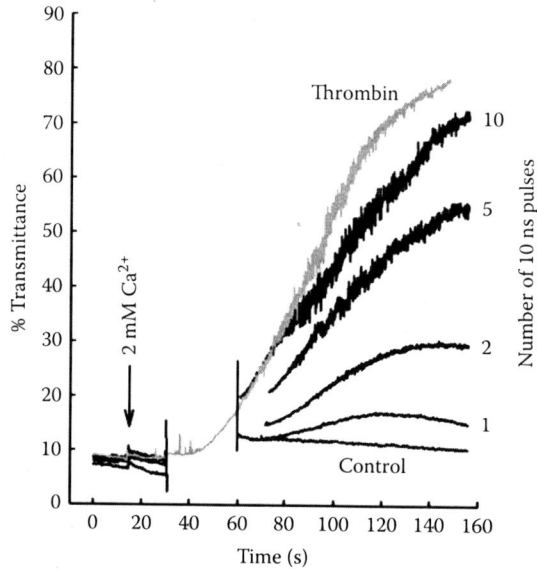

FIGURE 2.15 Influence of 10 ns electrical pulses with electric field amplitudes of 125 kV/cm on the aggregation of platelets. The results obtained with various pulse numbers are compared to the effect of thrombin. Platelet aggregation was measured in a Chrono-Log Whole Blood Aggregometer (model 560-VS). The amount of transmitted light that passed through the sample cuvette (containing washed human platelets) was compared to the reference cuvette that contained buffer only. As platelets aggregated, there was an increased amount of light transmitted through the sample cuvette, which reflected increased platelet aggregation. Experimentally, a baseline light reading was performed, platelets were removed from the aggregometer into a pulser cuvette, pulsed, then returned to the aggregometer where light transmission was again measured. (From Schoenbach, K.H. et al., *IEEE Trans. Dielect. Electr. Insul.*, 14, 1088, 2007. With permission.)

induced by hormones (e.g., thrombin) in many nonexcitable cells (Taylor, 2006). In this process, when intracellular calcium is mobilized from the ER, there is an activation of an influx process in the plasma membrane, i.e., the two processes are coupled. The platelet aggregation correlates to internal calcium mobilization and calcium influx (Zhang et al., 2008).

In addition to aggregation, an increase in PDGF release has been observed after washed platelets were pulsed with nanosecond pulses (Zhang et al., 2008). This release is most likely due to subcellular electromanipulation, similar to that reported in reference (Schoenbach et al., 2001). Platelets are rich in alpha-granule growth factors (i.e., PDGF and transforming growth factor beta [TGF-β]). The release of this growth factor is essential for wound healing. In addition to growth factor release, recent studies have also shown that nsPEFs have an antibacterial effect. The concentration of *Staphylococcus aureus* decreased considerably after treatment with nsPEF-activated platelet gel and decreased significantly compared with bacteria treated with thrombin.

2.8.2 Treating Skin Cancer

A wide range of stimuli with either plasma membrane or intracellular membrane targets can be responsible for the induction of apoptosis. This includes pharmacological agents, toxins, and radiation; or simply the removal of serum from the culture medium to trigger an apoptotic cascade. As discussed in one of the previous sections, intense nsPEFs joined the rank of apoptosis-inducing agents in 2002 (Beebe et al., 2002). This observation was exciting in that the application of nsPEFs can be considered to be a way of treating tumors without the additional requirement of chemotherapeutic agents, with no residual side effects following treatment and with limited inflammation at the treatment site.

Nuccitelli et al. (2006) confirmed this hypothesis. It was found that electric pulses (300 ns in duration and 40 kV/cm) can cause a total remission of skin cancer in mice. One million B16-F10 melanoma cells injected into SKH-1 mice generate melanoma tumors that are 3–5 mm in diameter within 4 days following injection. When these tumors are treated, they exhibit three rapid changes: (1) the nuclei of the tumor cells begin to shrink rapidly and the average nuclear volume shrinks to 50% within an hour; (2) vasculature providing blood flow to the tumor is disrupted within a few minutes and is not restored for a week or two; and (3) the tumor begins to shrink and within two weeks, has shrunk by 90%. Most of the roughly 200 tumors treated begin to grow again at this point, but if treated a second time, will completely disappear.

The studies on tumor treatment continued with 300 ns pulses applied to murine melanomas to explore the long-term effects on tumors. The studies have shown that nanosecond pulses trigger both necrosis and apoptosis resulting in complete tumor regression within an average of 47 days in the 17 animals treated (Figure 2.16). None of these melanomas recurred during the 4 month period after the initial melanoma had disappeared.

Immunohistochemistry studies on the tumors clearly showed the apoptotic effect of nanosecond pulses (Chen et al., 2009; Nuccitelli et al., 2009). Apoptosis studies were performed using Bcl-2 as a marker. Bcl-2 is a family of proteins involved in the response to apoptosis. Some of these proteins (such as Bcl-2 and Bcl-XL) are anti-apoptotic, while others (such as Bad, Bax, or Bid) are pro-apoptotic. The sensitivity of cells to apoptotic stimuli can depend on the balance of pro- and anti-apoptotic Bcl-2 proteins. The involvement of Bcl-2 and Bad using immunohistochemistry was investigated on sections from both nsPEF-treated and untreated tumors. An average decrease of 74% in the anti-apoptotic Bcl-2 labeling was observed when comparing nine nsPEF-treated tumors with nine untreated tumors. In contrast, the apoptotic label, Bad, increased by an average of 320% (N = eight treated and eight untreated). Both of these changes suggest that nsPEF initiates apoptosis in the tumor cells. This is also addressed in a modeling paper by the Weaver group, where apoptosis induction by nanosecond pulses is discussed in terms of nanoporation of the plasma membrane (Esser et al., 2009).

In addition, the changes in microvessel density caused by nanosecond pulses were explored, again using immunohistochemistry. Tumor growth depends on the availability of nutrients which, after the disruption of vasculature, requires the formation of new capillaries in a process known as angiogenesis.

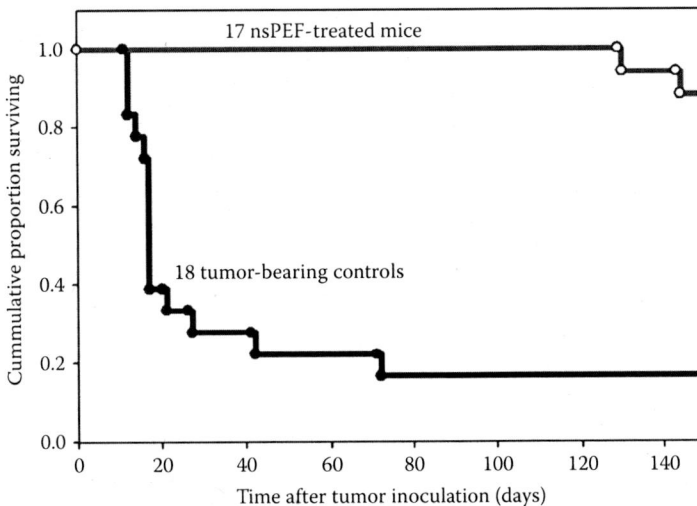

FIGURE 2.16 Survival curve for 17 nsPEF-treated mice and 18 untreated controls with one melanoma each. All 17 treated mice exhibited complete tumor regression without recurrence during the 150 days prior to euthanizing. One of these mice was euthanized on day 130 due to a 20% weight loss in a week but did not exhibit metastasis to the lungs or liver. A second treated mouse was euthanized on day 144 due to an eye infection. Controls were euthanized when tumors ulcerated. (From Nuccitelli, R. et al., *Int. J. Cancer*, 125, 438, 2009. With permission.)

Endothelial cells forming capillaries can be detected by using antibodies to the endothelial cell marker, CD31 (Folpe and Cooper, 2007). The endothelial cell density in both nsPEF-treated and untreated sections from five different melanomas was recorded and an average reduction of more than 90% in CD31 expression was found in nsPEF-treated tumors. This suggests that the microcirculation to the treated tumors is severely reduced and this should lead to necrosis and tumor shrinkage.

This nsPEF therapy has also been used to treat skin tumors by another group at the University of Southern California (Garon et al., 2007). They have found it to be effective against pancreatic tumors developing from cells injected beneath mouse skin as well as for a single case of a human basal cell carcinoma that exhibited complete remission and very little scarring after one treatment with nsPEF (200 pulses, 20 ns long, 43 kV/cm). This suggests that this new therapy, which has proven very effective for treating mouse skin cancer, might be equally as effective in humans.

2.9 Scaling of Nanosecond Pulse Effects

Experimental results of bioelectric studies with intense (high electric field), square wave electrical pulses of nanosecond duration indicate a scaling law for membrane polarization and related bioelectric effects of the form (Schoenbach et al., 2009a)

$$S = S(E\tau\sqrt{N}) \qquad (2.17)$$

where

E is the electrical field intensity
τ is the pulse duration
N is the number of pulses
S denotes the intensity of an observable effect

Identical results can be expected for experiments in which the product of electric field amplitude and pulse duration (electrical impulse) times the square root of the number of pulses is kept constant.

An example for the validity of this scaling law is shown in Figure 2.17 for platelet aggregation caused by Ca^{++} release (Zhang et al., 2008). Another example is shown in Figure 2.18, where the trypan blue exclusion in B16 melanoma cells is plotted versus the scaling parameter, $E\tau\sqrt{N}$, and for comparison, versus the electrical energy density, $\sigma E^2\tau N$ (Schoenbach et al., 2008). There are also other observations where only two of the three variables in the scaling parameter were varied, which supports this scaling law for intense nanosecond pulse effects with emphasis on the word intense (Beebe et al., 2004; Schoenbach et al., 2008; Nuccitelli et al., 2009). This observable effect is either directly related to membrane permeabilization or can be traced back to the membrane permeabilization of either the plasma membrane or internal membranes.

This empirical law for single pulse effects is consistent with the results of a simple model that is based on the hypothesis that the intensity of bioelectric effects is determined by the charge per unit area carried across cellular membranes (current density) above the threshold for electroporation (permeabilization) or nanoporation, with the transmembrane voltage assumed to be constant (Schoenbach et al., 2009a). In the case of electroporation, this constant membrane voltage is generally assumed to be $V_c = 1$ V (Neumann, 1989), but it can be less (Melikov et al., 2001) or, for very short pulses, more than 1 V (Frey et al., 2006).

These assumptions lead to an equation which allows us to calculate the charge carried across the membrane dependent on electric field intensity and pulse duration for single pulses ($N = 1$)

$$Q_{no} = \left[E_{no} - 1 \right] \left[\tau_{no} - \ln\left\{ \frac{fE_{no}}{fE_{no} - 1} \right\} \right] \qquad (2.18)$$

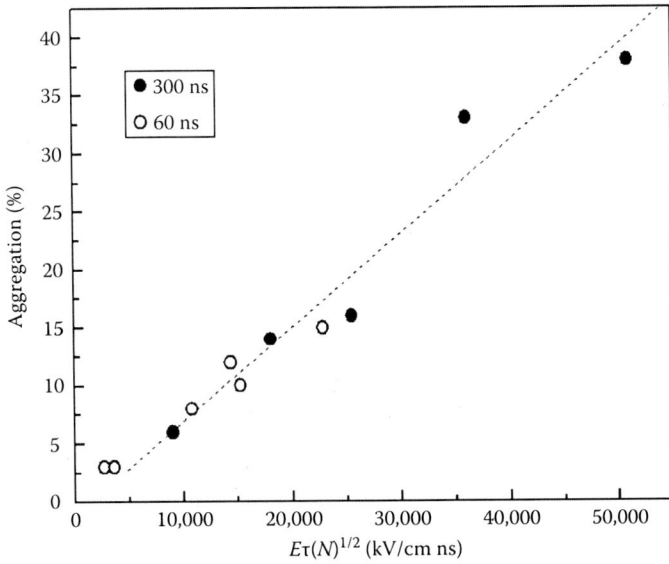

FIGURE 2.17 Aggregation of platelets plotted versus the similarity parameter, $E\tau\sqrt{N}$ (electric field intensity times pulse duration time square root of the number of pulses) for pulses with duration of 60 and 300 ns. (From Schoenbach, K.H. et al., *IEEE Trans. Dielect. Electr. Insul.*, 14, 1088, 2007. With permission.)

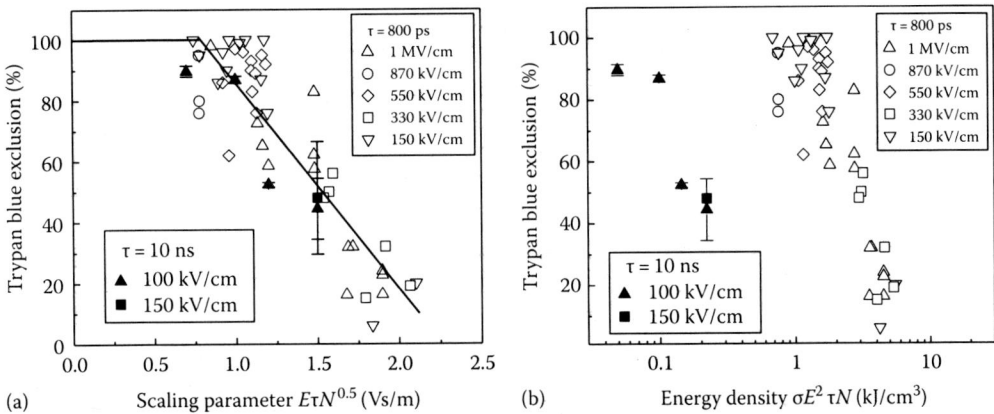

FIGURE 2.18 (a) Trypan blue exclusion for B16 cells versus the scaling parameter $E\tau\sqrt{N}$. The number of pulses in the 800 ps experiment was varied between 1 and 20,000, the number in the 10 ns experiment, between 50 and 200. (b) Trypan blue exclusion for B16 cells versus energy density (same data as plotted in Figure 2.18a). The results show that trypan blue exclusion is not a dose effect. (From Schoenbach, K.H. et al., 2008, *IEEE Trans. Plasma Sci.*, 36, 414, 2008. With permission.)

where f is a factor determined by the shape of the cell (1.5 for spherical cells)

$$Q_{no} = \frac{Q\rho_i a}{V_c \tau_m} \tag{2.19}$$

is the charge density normalized with respect to the charge density, Q_c, at the membrane poles required for the onset of electroporation at $V = 2V_c$

E_{no} is the electric field, E, normalized with respect to the electric field just at (above) the threshold of the bioelectric effect (e.g., electroporation at $V_a = 2V_c$)

$$E_{no} = \frac{Ea}{V_c} \qquad (2.20)$$

An E_{no} of 1 for a 10 μm-diameter, spherical cell with $V_c = 1$ V corresponds to an electric field of 2 kV/cm. The normalized pulse duration (with respect to the charging time constant of the plasma membrane) is given as

$$\tau_{no} = \frac{\tau_p}{\tau_m} \tau \qquad (2.21)$$

The normalized electric field is plotted in Figure 2.19 versus the normalized pulse duration, with the normalized charge density as a parameter for single pulses ($N = 1$). The curves of the constant Q_n represent, according to our hypothesis, curves where identical bioelectric effects can be expected. The curve

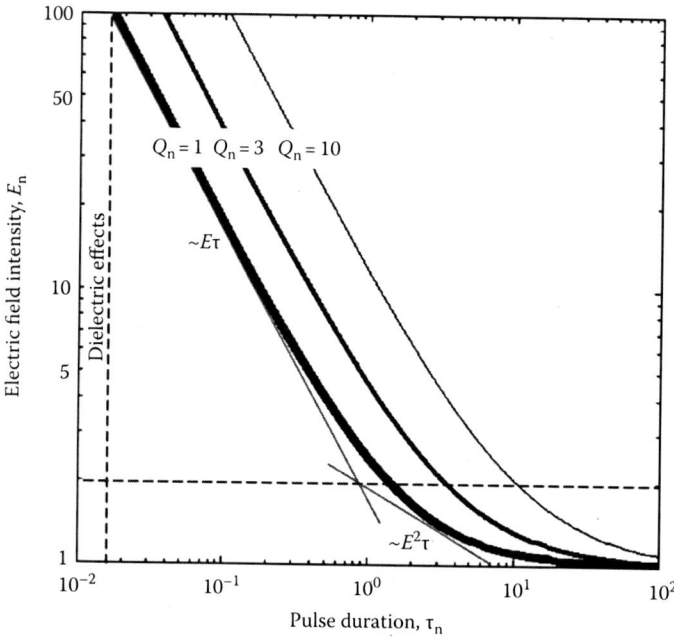

FIGURE 2.19 Normalized electric field intensity–normalized pulse duration range for single pulse electroporation of the plasma membrane of a spherical cell ($f = 1.5$). The graph shows curves for constant values of membrane permeabilization. Parameter is the ratio of total charge to critical charge required to initiate the effect. The validity of these analytic results is limited for short pulse durations to the dielectric relaxation time, ε/σ, of the medium or cytoplasm, whichever is larger. It is also limited to relatively small values of Q_n above $Q_n = 1$, due to the assumption that the poration is limited to the area of the poles. This limitation is indicated by a reduced line thickness for the $Q_n > 1$ curves. For electric fields greater than approximately twice the critical field for membrane permeabilization ($E_n = 1$), the curves can be approximated by a scaling law, $S = S(E\tau)$. For lower electric fields, the electric field changes nonlinearly with the pulse duration (due to the logarithmic term in Equation 2.18 which can be expanded as a power series in E_n). The possibility of approximating the curve for low electric fields, but still above the threshold for electroporation, by an $E^2\tau$ dependence (as shown in this figure) could lead to the wrong conclusion that bioelectric effects at lower electrical fields are dose effects. (From Schoenbach, K.H. et al., *IEEE Trans. Dielect. Electr. Insul.*, 16, 1224, 2009a. With permission.)

for $Q_n = 1$ can be considered to be a "strength-duration" relationship between the threshold electric field intensity of a square wave pulse and its duration (Reilly, 1998). What is obvious from these curves is that for E_n large compared to one, which is the case in most of the bioelectric studies with pulse durations below 100 ns, the product of E and τ is constant. That means that electropermeabilization at high electric fields can be considered an electric impulse effect, rather than a dose effect. This is expected from Equation 2.18: for high E_n, the logarithmic term approaches zero, and the equation leads to a scaling law of the form

$$Q_{no} \sim S = S(E\tau) \tag{2.22}$$

a result that is consistent with the empirical results for $N = 1$ (Equation 2.17).

There is no assumption in the model that leads to the scaling law (Equation 2.22) that would limit its validity to nanosecond pulses only. It should be valid for long pulses, e.g., multimicrosecond pulses, as long as the condition that the pulse rise time is very short is also satisfied. Indeed, in experiments with 10 pulses varying in duration, τ, from approximately 20–100 μs (Rols and Teissie, 1990), the threshold for permeabilization (onset of trypan blue uptake) was found to follow a scaling law similar to Equation 2.22

$$E_p[\text{kV/cm}] = \frac{1.5}{\tau_p}[(\mu s)^{-1}] + 0.3 \tag{2.23}$$

E_p is the threshold value in electric field intensity for electropermeabilization (Rols and Teissie, 1990). The slight offset in the electric field (+0.3) would be negligible for submicrosecond pulses.

For multiple exposures to intense electric field pulses, the intensity of observable bioelectric effects in a well-defined cell suspension (one cell type, no variation in the properties of the surrounding medium) scales with the square root of the number of pulses, as expressed in the general scaling law (Equation 2.17). This square root–dependence on the pulse number points to a statistical motion of cells between pulses with respect to the applied electric field and can be explained using an extension of the random walk statistical results for random rotations. For multiple pulses (*N* pulses), and keeping the cell as the frame of reference, we can add the randomly distributed electric field vectors (each of them with the same magnitude, E_0) and describe the observed effect as the result of a single electric field vector with an amplitude, E. It is known that the resulting field vector, E, for identical electric field vectors with amplitude E_0, but randomly varying field directions, is given as (Sommerfeld, 1959)

$$E = E_0\sqrt{N} \tag{2.24}$$

The $\sqrt{N}$ dependence in the scaling law (Equation 2.17) is treated in detail in a paper by Schoenbach et al. (2009a).

The random rotation concept becomes less valid for electric fields and pulse durations far above the values that define the onset of electroporation ($Q_n = 1$). With increasing electric field and/or pulse duration, the permeabilized surface of the cell will expand from the poles to the equator, and random motion will lose its importance for the intensity of the observable effect. The validity of the "random rotation" law also requires that the recovery time of the cell membrane be long compared with the time between pulses. It also requires that the motion of the cells between shots (rotation with respect to the electric field direction) is completely random.

The scaling law not only holds for the primary effects on the membrane, such as electroporation, but also for secondary effects and as long as they are intensity-related (not necessarily linearly) to membrane charging, that means they are stimulated through membrane charging effects. With platelet aggregation, for example, this connection between effect and cause is as follows: nanosecond pulses cause the release of calcium from the ER through the charging of the ER membrane (White et al., 2004).

The increase in intracellular calcium causes an influx of extracellular calcium, which serves as an agent for aggregation. Consequently, aggregation has its cause in the charging of the ER membrane (Zhang et al., 2008).

2.10 From Nanosecond to Picosecond Pulses

By reducing the duration of electrical pulses from microseconds into the nanosecond range, the electric field-cell interactions shift increasingly from the plasma (cell) membrane to subcellular structures. Yet another domain of pulsed electric field interactions with cell structures and functions opens when the pulse duration is reduced to values such that membrane charging becomes negligible, and direct electric field-molecular effects determine the biological mechanisms. The condition for the dominance of such effects is that the pulse duration needs to be less than the dielectric relaxation time of the cytoplasm (see Section 2.2). For mammalian cells, this holds for a pulse duration of less than 1 ns.

Besides entering a new field of bioelectrics by moving into the subnanosecond temporal range, there is a practical reason for entering this new field. It is the possible use of antennas as pulse delivery systems. The use of needle or plate electrodes in therapeutic applications that rely on electroporation (Hofmann, 2000) or nsPEF (Nuccitelli et al., 2006) requires that the electrodes are brought into close contact with the treated tissue. This limits the application to treatments of tissue close to the surface of the body. The use of antennas, on the other hand, would allow one to apply such electric fields to tissues (tumors) that are not easily accessible with needles. Also, the focusing of electrical energy on the target would reduce the damage to the tissue layers surrounding the target and the skin. The spatial resolution of an electric field generated in tissue depends on the pulse duration and the permittivity of the tissue.

Besides using ultrashort electrical pulses for medical therapies, the use of focusing antennas may lead to medical imaging methods with centimeter spatial resolution (Xiao et al., 2008, Schoenbach et al., 2009b). Imaging is based on the measurement of changes in the complex permittivity of tissue and may complement methods based on the measurement of other physical parameters, such as X-ray computed tomography (CT), magnetic resonance imaging (MRI), and ultrasound. For example, breast tumors have an almost one order of magnitude higher electrical conductivity and permittivity than normal breast tissue over a wide frequency range (10^7–10^{10} Hz) (Fear et al., 2002) and can therefore be easily differentiated from normal tissue. In such an imaging system, the electromagnetic waves are focused on a target inside the body. The scattered signal contains information on the dielectric properties and the geometry of the target and allows its identification through an inverse scattering method. 3D scanning allows us to obtain the dielectric profile of the tissue and to detect any abnormality in a uniform background.

An "antenna," or better, a pulse-delivery system that is used for bioelectrics applications would need to operate in the antenna's near-field. There are several concepts for such a delivery system. These are mainly based on the use of a prolate spheroidal reflector, where the pulse is launched from one focal point and reflected into a second focal point (Baum, 2007). This second focal point would, for our application, be the location of a tumor inside the body (Figure 2.20).

At Old Dominion University, we have concentrated our efforts on a near-field antenna where the electric field generated in the target is focused by means of a prolate spheroidal reflector, in connection with a conical wave launcher. The dielectric in the reflector volume is a high permittivity liquid, e.g., water. For a 100 ps rise time, step-function input pulse, the focusing spot size (FWHM) in tissue is 2–3 cm along the z-axis and <1 cm along the x-axis. For a 100 kV input pulse, the peak E-field at the second focal point was calculated to be 250 kV/cm.

In order to explore the biological effects of wideband pulses emitted by an antenna, we have begun to study the effects of 120–180 ps long pulses with electric fields of up to 150 kV/cm (Camp et al., 2008). The electric field value is anticipated to be achievable by using antennas with focusing reflectors as shown in Figure 2.20 or with a novel antenna design that includes lenses in addition to a reflector

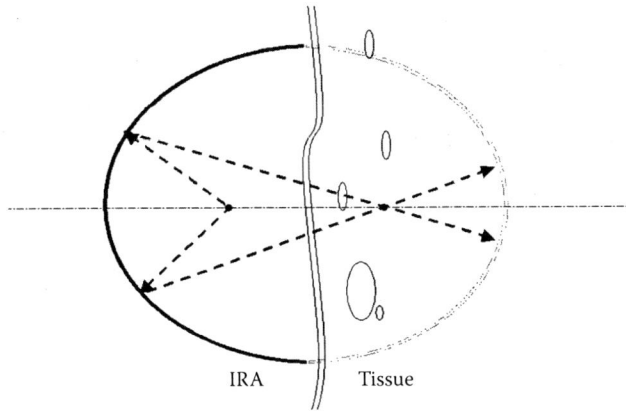

FIGURE 2.20 Schematics of an antenna with a prolate spheriodal reflector used in bioelectric applications. (From Schoenbach, K.H. et al., Wideband, high-amplitude, pulsed antennas for medical therapies and medical imaging, in *Proceedings 2009 International Conference on Electromagnetics in Advanced Applications*, September 14–18, 2009, Torino, Italy, 2009b, pp. 580–583, IEEE catalog number: CFP0968B-CDR. With permission.)

(Altunc et al., 2009). For the biological studies, we have used the pulse generator described in reference (Camp et al., 2008) with a coaxial exposure chamber as the load.

The lethality of such wideband pulses was studied by measuring the trypan blue uptake in pancreatic cancer cells (Panc-1). With 1000 pulses with a duration of 180 ps and electric fields of 150 kV/cm applied to Panc-1 cells in cell culture media, trypan blue uptake of 10%–15% was measured, almost independent of temperature, in the range from 23°C to 35°C. It must be noted that this is a preliminary result and further experiments are required to verify these results. However, even if confirmed, the removal of cancer cells would, according to these results, require a high number of pulses and, consequently, relatively long exposure times in therapeutic applications. By utilizing effects synergistic to the wideband pulse effects, it may be possible to reduce the threshold for induced cell death, apoptosis, considerably. A promising approach, according to our latest studies, is that of pulsed heating of tissue in addition to wideband pulse exposure. This can be achieved either by using the wideband pulses themselves, when operated in a burst mode at a high repetition rate, or by using high power microwaves emitted into the tissue with an antenna similar to the one used for wideband pulse exposure.

References

Afford, S. and Randhawa, S., 2000, Apoptosis, *J. Clin. Pathol. Mol. Path.*, 53, 55–63.

Altunc, S., Baum, C.E., Buchenauer, C.J., Christodoulou, C.G., and Schamiloglu, E., 2009, Design of a special dielectric lens for concentrating a subnanosecond electromagnetic pulse on a biological target (Special issue on Bioelectrics), *IEEE Trans. Dielect. Electr. Insul.*, 16, 1364–1375.

Baum, C.E., 2007, Focal waveform of a prolate-spheroidal impulse-radiating antenna, *Radio Sci.*, 42, RS6S27.

Beebe, S.J., Fox, P.M., Rec, L.C., Somers, K., Stark, R.H., and Schoenbach, K.H., 2002, Nanosecond pulsed electric field (nsPEF) effects on cells and tissues: Apoptosis induction and tumor growth inhibition, *IEEE Trans. Plasma Sci.*, 30, 286–292.

Beebe, S.J., Fox, P.M., Rec, L.J., Willis, L.K., and Schoenbach, K.H., 2003a, Nanosecond, high intensity pulsed electric fields induce apoptosis in human cells, *FASEB J.*, 17, 1493.

Beebe, S.J., White, J.A., Blackmore, P.F., Deng, Y., Somers, K., and Schoenbach, K.H., 2003b, Diverse effects of nanosecond pulsed electric fields on cells and tissues, *DNA Cell Biol.*, 22, 785–796.

Beebe, S.J., Blackmore, P.F., White, J., Joshi, R.P., and Schoenbach, K.H., 2004, Nanosecond pulsed electric fields modulate cell function through intracellular signal transduction mechanisms, *Physiol. Meas.*, 25, 1077–1093.

Benz, R. and Zimmermann, U., 1981, The resealing process of lipid bilayers after reversible electrical breakdown, *Biochim. Biophys. Acta*, 640, 169–178.

Berridge, M.J., Bootman, M.D., and Lipp, P., 1998, Calcium—A life and death signal, *Nature*, 395, 645–648.

Buescher, E.S. and Schoenbach, K.H., 2003, Effects of submicrosecond, high intensity pulsed electric fields on living cells—Intracellular electromanipulation, *IEEE Trans. Dielect. Electr. Insul.*, 10, 788–794.

Camp, J.T., Xiao, S., and Schoenbach, K.H., 2008, Development of a high voltage, 150 ps pulse generator for biological applications, in *Proceedings of the 2008 Power Modulator Conference*, Las Vegas, NV, pp. 338–341.

Chen, N., Schoenbach, K.H., Kolb, J.F., Swanson, R.J., Garner, A.L., Yang, J., Joshi, R.P., and Beebe, S.J., 2004, Leukemic cell intracellular responses to nanosecond electric fields, *Biochem. Biophys. Res. Commun.*, 317, 421–427.

Chen, X., Swanson, R.J., Kolb, J.F., Nuccitelli, R., and Schoenbach, K.H., 2009, Histopathology of normal skin and melanomas after nanosecond pulsed electric field treatment, *Melanoma Res*, 19, 361–371.

Cole, K.S., 1937, Electrical impedance of marine egg membranes, *Trans. Faraday Soc.*, 23, 966–972.

Ermolina, I., Polevaya, Y., Feldman, Y., Ginzburg, B., and Schlesinger, M., 2001, Study of normal and malignant white blood cells by time domain dielectric spectroscopy, *IEEE Trans. Dielect. Electr. Insul.*, 8, 253–261.

Esser, A.T., Smith, K.C., Gowrishankar, T.R., and Weaver, J.C., 2009, Towards solid tumor treatment by nanosecond pulsed electric fields, *Technol. Cancer Res. Treat.*, 8, 289–306.

Fear, E.C., Hagness, S.C., Meaney, P.M., Okoniwski, M., and Stuchly, M.A., 2002, Enhancing breast tumor detection with near-field imaging, *IEEE Microwave Magazine*, March 2002, pp. 48–56.

Feldman, Y., Ermolina, I., and Hayashi, Y., 2003, Time domain dielectric spectroscopy study of biological systems, *IEEE Trans. Dielect. Electr. Insul.*, 10, 728–753.

Folpe, A.L. and Cooper, K., 2007, Best practices in diagnostic immunohistochemistry: Pleomorphic cutaneous spindle cell tumors, *Arch. Pathol. Lab. Med.*, 131, 1517–1524.

Foster K.R. and Schwan, H.P., 1995, Dielectric properties of tissues, in *Handbook of Biological Effects of Electromagnetic Fields*, Polk, C. and Postow, E., Eds., CRC Press, Boca Raton, FL, p. 90, Fig. 28.

Frey, W., White, J.A., Price, R.O., Blackmore, P.F., Joshi, R.P., Nuccitelli, R., Beebe, S.J., Schoenbach, K.H., and Kolb, J.F., 2006, Plasma membrane voltage changes during nanosecond pulsed electric field exposure, *Biophys. J.*, 90, 3608–3615.

Garon, E.B., Sawcer, D., Vernier, P.T., Tang, T., Sun, Y., Marcu, L., Gundersen, M.A., and Koeffler, H.P., 2007, In vitro and in vivo evaluation and a case report of intense nanosecond pulsed electric field as a local therapy for human malignancies, *Int. J. Cancer*, 121, 675–682.

Ghazala, A. and Schoenbach, K.H., 2000, Biofouling prevention with pulsed electric fields, *IEEE Trans. Plasma Sci.*, 28, 115–121.

Gowrishankar, T.R., Esser, A.T., Vasilkoski, Z., Smith, K.C., and Weaver, J.C., 2006, Microdosimetry for conventional and supra-electroporation in cells with organelles, *Biochem. Biophys. Res. Commun.*, 341, 1266–1276.

Halestrap, A.P., McStay, G.P., and Clarce, S.J., 2002, The permeability transition pore complex: Another view, *Biochimie*, 84, 153–166.

Hall, E., Schoenbach, K.H., and Beebe, S.J., 2007, Nanosecond pulsed electric fields have differential effects on cells in S-phase, *DNA Cell Biol.*, 26(3), 160–171.

Hofmann, G.A., 2000, Instrumentation and electrodes for in vivo electroporation, in *Electrochemotherapy, Electrogenetherapy, and Transdermal Drug Delivery*, Jaroszewski, M.J., Heller, R., and Gilbert, R., Eds., Humana Press, Totowa, NJ, pp. 37–61.

Hu, Q., Viswanadham, S., Joshi, R.P., Schoenbach, K.H., Beebe, S.J., and Blackmore, P.F., 2005a, Simulations of transient membrane behavior in cells subjected to a high-intensity ultrashort electrical pulse, *Phys. Rev. E*, 71, 031914-1–031914-9.

Hu, Q., Joshi, R.P., and Schoenbach, K.H., 2005b, Simulations of nanopore formation and phosphatidyl-serine externalization in lipid membranes subjected to a high-intensity, ultrashort electric pulse, *Phys. Rev. E*, 72, 031902-1–031902-10.

Joshi, R.P., Hu, Q., Aly, R., Schoenbach, K.H., and Hjalmarson, H.P., 2001, Self-consistent simulations of electroporation dynamics in biological cells subjected to ultrafast electrical pulses, *Phys. Rev. E*, 64, 11913-01–11913-03.

Joshi, R.P., Hu, Q., Schoenbach, K.H., and Hjalmarson, H.P., 2002, Improved energy model for membrane electroporation in biological cells subjected to electrical pulses, *Phys. Rev. E*, 65, 041920-01–041920-07.

Joshi, R.P., Hu, Q., Schoenbach, K.H., and Beebe, S.J., 2004a, Energy-landscape-model analysis for irreversibility and its pulse-width dependence in cells subjected to a high-intensity ultrashort electrical pulse, *Phys. Rev. E*, 69, 051901-1–051901-10.

Joshi, R.P., Hu Q., and Schoenbach, K.H., 2004b, Modeling studies of cell response to ultrashort, high-intensity electric fields—Implications for intracellular manipulation, *IEEE Trans. Plasma Sci.*, 32, 1677–1686.

Kolb, J.F., Kono, S., and Schoenbach, K.H., 2006, Nanosecond pulsed electric field generators for the study of subcellular effects, *Bioelectromagn. J.*, 27, 172–187.

Kotnik, T. and Miclavcic, D., 2006, Theoretical evaluation of voltage inducement on internal membranes of biological cells exposed to electric fields, *Biophys. J.*, 90, 480–491.

Kuhn, B., Fromherz, P., and Denk, W., 2004, High sensitivity of Stark-shift voltage-sensing dyes by one- or two-photon excitation near the red spectral edge, *Biophys J.*, 87, 631–639.

Melikov, K.C., Frolov, V.A., Shcherbakov, A., Samsonov, A.V., Chizmadzhev, Y.A., and Chernomordik, L.V., 2001, Voltage-induce nonconductive pre-pores and metastable single pores in unmodified planar lipid bilayer, *Biophys. J.*, 80, 1829–1836.

Neumann, E., 1989, The relaxation hysteresis of membrane electroporation, in *Electroporation and Electrofusion in Cell Biology*, Neumann, E., Sowers, A.E., and Jordan, C.A., Eds., Plenum Press, New York, pp. 61–82.

Neumann, E. and Rosenheck, K., 1972, Permeability changes induced by electrical impulses in vesicular membranes, *J. Membr. Biol.*, 10, 279–290.

Nomura, N., Yano, M., Katsuki, S., Akiyama, H., Abe, K., and Abe, S.-I., 2009, Intracellular DNA damage induced by non-thermal, intense narrowband electric fields, *IEEE Trans. Dielect. Electr. Insul.*, 16, 1288–1293.

Nuccitelli, R., Pliquett, U., Chen, X., Ford, W., Swanson, R.J., Beebe, S.J., Kolb, J.F., and Schoenbach, K.H., 2006, Nanosecond pulsed electric fields cause melanomas to self-destruct, *Biochem. Biophys. Res. Commun.*, 343, 351–360.

Nuccitelli, R., Chen, X., Pakhomov, A.G., Baldwin, W.H., Sheikh, S., Pomicter, J.L., Ren, W. et al., 2009, A new pulsed electric field therapy for melanoma disrupts the tumor's blood supply and causes complete remission without recurrence, *Int. J. Cancer*, 125, 438–445.

Pakhomov, A.G., Shevin, R., White, J., Kolb, J.F., Pakhomova, O.N., Joshi, R.P., and Schoenbach, K.H., 2007, Membrane permeabilization and cell damage by ultrashort electric field shocks, *Arch. Biochem. Biophys.*, 465(1), 109–118.

Reilly, J.P., 1998, *Applied Bioelectricity*, Springer, New York, Chap. 6.3.

Rols, M.P. and Teissie, J., 1990, Electropermeabilization of mammalian cells: Quantitative analysis of the phenomenon, *Biophys. J.*, 58, 1089–1098.

Savill, J. and Haslett, C., 1995, Granulocyte clearance by apoptosis in the resolution of inflammation, *Semin. Cell Biol.*, 6, 385–360.

Scarlett, S.S., White, J.A., Blackmore, P.F., Schoenbach, K.H., and Kolb, J.F., 2009, Regulation of intracellular calcium concentration by nanosecond pulsed electric fields, *Biochim. Biophys. Acta Biomembr.*, 1788, 1168–1175.

Schoenbach, K.H., Peterkin, F.E., Alden, R.W., and Beebe, S.J., 1997, The effect of pulsed electric fields on biological cells: Experiments and applications, *Trans. Plasma Sci.*, 25, 284–292.

Schoenbach, K.H., Beebe, S.J., and Buescher, E.S., 2001, Intracellular effect of ultrashort electrical pulses, *J. Bioelectromagn.*, 22, 440–448.

Schoenbach, K.H., Katsuki, S., Stark, R.H., Buescher, E.S., and Beebe, S.J., 2002, Bioelectrics—New applications for pulsed power technology, *IEEE Trans. Plasma Sci.*, 30, 293–300.

Schoenbach, K.H., Joshi, R.P., Kolb, J.F., Chen, N., Stacey, M., Blackmore, P.F., Buescher, E.S., and Beebe, S.J., 2004, Ultrashort electrical pulses open a new gateway into biological cells, *Proc. IEEE*, 92, 1122–1137.

Schoenbach, K.H., Nuccitelli, R.L., and Beebe, S.J., 2006, ZAP: Extreme voltage could be a surprisingly delicate tool in the fight against cancer, *IEEE Spectr.*, 43(8), 20–26.

Schoenbach, K.H., Hargrave, B., Joshi, R.P., Kolb, J.F., Osgood, C., Nuccitelli, R., Pakhomov, A. et al., 2007, Bioelectric effects of nanosecond pulses, *IEEE Trans. Dielect. Electr. Insul.*, 14, 1088–1119.

Schoenbach, K.H., Xiao, S., Joshi, R.P., Camp, J.T., Heeren, T., Kolb, J.F., and Beebe, S.J., 2008, The effect of intense subnanosecond electrical pulses on biological cells, *IEEE Trans. Plasma Sci.*, 36, 414–424.

Schoenbach, K.H., Baum, C.E., Joshi, R.P., and Beebe, S.J., 2009a, A scaling law for membrane permeabilization with nanopulses, *IEEE Trans. Dielect. Electr. Insul.*, 16, 1224–1235.

Schoenbach, K.H., Xiao, S., Camp, J.T., Migliaccio, M., Beebe, S.J., and Baum, C.E., 2009b, Wideband, high-amplitude, pulsed antennas for medical therapies and medical imaging, in *Proceedings 2009 International Conference on Electromagnetics in Advanced Applications*, September 14–18, 2009, Torino, Italy, IEEE catalog number: CFP0968B-CDR, pp. 580–583.

Shirakashi, R., Sukhorukov, V.L., Tanasawa, I., and Zimmermann, U., 2004, Measurement of the permeability and resealing time constant of the electroporated mammalian cell membranes, *Int. J. Heat Mass Transfer*, 47, 4517–4524.

Smith, K.C., Gowrishankar, T.R., Esser, A.T., Stewart, D.A., and Weaver, J.C., 2006, The spatially distributed dynamic transmembrane voltage of cells and organelles due to 10 ns pulses: Meshed transport networks, *IEEE Trans. Plasma Sci.*, 34, 1394–1404.

Sommerfeld, A., 1959, Optik, paragraph 33 in Vorlesungen über Theoretische Physik, Vol. IV. A kademische verlagsgesellschaft, Leipzig, Germany.

Stacey, M., Stickley, J., Fox, P., Statler, V., Schoenbach, K.H., Beebe, S.J., and Buescher, S., 2003, Differential effects in cells exposed to ultra-short, high intensity electric fields: Cell survival, DNA damage, and cell cycle analysis, *Mutat. Res.*, 542, 65–75.

Staempfli, R., 1958, Reversible breakdown of the excitable membrane of a Ranvier node, *An. Acad. Brasil. Ciens*, 30, 57.

Stewart, D.A., Gowrishankar, T.R., and Weaver, J.C., 2006, Three dimensional transport lattice model for describing action potentials in axons stimulated by external electrodes, *Bioelectrochemistry*, 69, 88–93.

Susin, S.A., Zamzami, N., and Kroemer, G., 1998, Mitochondria as regulators of apoptosis: Doubt no more, *Biochim. Biophys. Acta*, 1366, 151–165.

Taylor, C.W., 2006, Store-operated Ca^{2+} entry: A STIMulating stOrai, *Trends Biochem. Sci.*, 31, 597–601.

Tekle, E., Oubrahim, H., Dzekunov, S.M., Kolb, J.F., Schoenbach, K.H., and Chock, B.P., 2005, Selective field effects on intracellular vacuoles and vesicle membranes with nano-second electric pulses, *Biophys. J.*, 89, 274–284.

Tieleman, D., Leontiadou, H., Mark, A.E., and Marrink, S.J., 2003, Simulation of pore formation in lipid bilayers by mechanical stress and electric fields, *J. Am. Chem. Soc.*, 125, 6382–6383.

Tsujimot, Y. and Shimizu, S., 2002, The voltage-dependent anion channel: An essential player in apoptosis, *Biochimie*, 84, 187–193.

Vasilkoski, Z., Esser, A.T., Gowrishankar, T.R., and Weaver, J.C., 2006, Membrane electroporation: The absolute rate equation and nanosecond time scale pore creation, *Phys. Rev.*, E74, 021904-1–021904-12.

Vernier, P.T., Li, A., Marcu, L., Craft, C.M., and Gundersen, M.A., 2003a, Ultrashort pulsed electric fields induce membrane phospholipid translocation and caspase activation: Differential sensitivities of Jurkat T lymphoblasts and rat glioma C6 cells, *IEEE Trans. Dielect. Electr. Insul.*, 10, 795–809.

Vernier, P.T., Sun, Y., Marcu, L., Salemi, S., Craft, C.M., and Gundersen, M.A., 2003b, Calcium bursts induced by nanosecond electrical pulses, *Biochem. Biophys. Res. Commun.*, 310, 286–295.

Vernier, P.T., Sun, Y., Marcu, L., Craft, C.M., and Gundersen, M.A., 2004, Nanoelectropulse-induced phosphatidylserine translocation, *Biophys. J.*, 86, 4040–4048.

Vernier, P.T., Ziegler, M.J., Sun, Y., Gundersen, M.A., and Tieleman, T.P., 2005, Nanopore-facilitated, voltage-driven translocation in lipid bilayers—In cells and in silico, *Phys. Biol.*, 3, 233–247.

Vernier, P.T., Sun, Y., and Gundersen, M.A., 2006a, Nanoelectropulse-driven membrane perturbation and small molecule permeabilization, *BMC Cell Biol.*, 7, 37.

Vernier, P.T., Ziegler, M.J., Sun, Y., Chang, W.V., Gundersen, M.A., and Tieleman, D.P., 2006b, Nanopore formation and phosphatidylserine externalization in a phospholipid bilayer at high transmembrane potential, *J. Am. Chem. Soc.*, 128, 6288–6289.

Walker, K. III, Pakhomova, O.N., Kolb, J.F., Schoenbach, K.H., Stuck, B.E., Murphy, M.R., and Pakhomov, A.G., 2006, Oxygen enhances lethal effect of high-intensity, ultrashort electrical pulses, *Bioelectromagn. J.*, 27, 221–225.

Weaver, J.C., 2003, Electroporation of biological membranes from multicellular to nanoscales, *IEEE Trans. Dielect. Electr. Insul.*, 10, 754–768.

Weaver, J.C., 2006, Harvard-MIT Division of Health Sciences and Technology, Cambridge, MA, private communication.

White, J.A., Blackmore, P.F., Schoenbach, K.H., and Beebe, S.J., 2004, Stimulation of capacitive calcium entry in HL-60 cells by nanosecond pulsed electric fields (nsPEF), *J. Biol. Chem.*, 279, 22964–22972.

Xiao, S., Schoenbach, K.H., and Baum, C.E., 2008, Time-domain focusing radar for medical imaging, XXIX Gen. Assembly of URSI, Chicago, IL, 2008, http://ursi-france.institut-telecom.fr/pages/pages_ursi/URSIGA08/papers/E02p4.pdf

Zhang, J., Blackmore, P.F., Hargrave, B.Y., Xiao, S., Beebe, S.J., and Schoenbach, K.H., 2008, The characteristics of nanosecond pulsed electrical field stimulation on platelet aggregation in vitro, *Arch. Biochem. Biophys.*, 471, 240–248.

3

Induced Transmembrane Voltage—Theory, Modeling, and Experiments

Tadej Kotnik

Gorazd Pucihar

3.1 The Cell and the Induced Transmembrane Voltage

From the electrical point of view, the cell can roughly be described as an electrolyte (the cytoplasm) surrounded by an electrically insulating shell (the plasma membrane). Physiologically, the surroundings of the cell also resemble an electrolyte quite closely. Under such conditions, when a cell is exposed to an external electric field, the electric field in its very vicinity concentrates within the membrane, which thus shields the cytoplasm from the exposure (this is the reason why the internal structure of the cell is not too important, except for very short pulses and very high field frequencies discussed in Section 3.2.4). The concentration of the electric field inside the membrane results in an electric potential difference across it, termed the induced transmembrane voltage, which superimposes onto the resting transmembrane voltage typically present under physiological conditions. As the electric field vanishes, so does the induced component of transmembrane voltage. This voltage affects the functioning of voltage-gated membrane channels, initiates the action potentials, stimulates cardiac cells, and when sufficiently large, it can also lead to cell membrane electroporation (Bedlack et al. 1994, Cheng et al. 1999, Neumann et al. 1999, Teissié et al. 1999, Burnett et al. 2003, Sharma and Tung 2004, Huang et al. 2006).

With rapidly time-varying electric fields, such as waves with frequencies in the megahertz range or higher, or electric pulses with durations in the submicrosecond range, both the membrane and its surroundings have to be treated as materials with both a nonzero electric conductivity and a nonzero dielectric permittivity.

From the geometrical point of view, the cell can be characterized as a geometric body (the cytoplasm) surrounded by a shell of uniform thickness (the membrane). For a suspended cell, the simplest model is a sphere surrounded by a spherical shell. For augmented generality, the sphere can be replaced by a spheroid (or an ellipsoid), but in this case, the requirement of uniform thickness complicates the description of the shell substantially. If its inner surface is a spheroid or an ellipsoid, its outer surface lacks a simple geometrical characterization, and vice versa. Still, in the steady state, this does not affect the induced transmembrane voltage, which can still be determined analytically.

Spheres, spheroids, and ellipsoids may be reasonable models for suspended cells, but not for cells in tissues. No simple geometrical body can model a typical cell in a tissue, and furthermore every cell generally differs in its shape from the rest. With irregular geometries and/or with cells close to each other, the induced voltage cannot be determined analytically, and thus cannot be formulated as an explicit function. This deprives us of some of the insight available from explicit expressions, but using modern computers and numerical methods, the voltage induced on each particular irregular cell can still be determined quite accurately.

An alternative to both analytical and numerical determination of the induced transmembrane voltage is the experimental approach, which can be performed invasively using microelectrodes, or noninvasively by loading the cells with a potentiometric dye and measuring its fluorescence.

In Sections 3.2 through 3.4 we focus separately on the analytical derivation of the induced transmembrane voltage for the cells with simple shapes, numerical computation for the cells for which the analytical approach fails, and noninvasive experimental determination by means of potentiometric dyes.

3.2 Analytical Derivation

In this section, we present the course of derivation of the transmembrane voltage induced on a spherical cell placed into a homogeneous electric field. The reasons for including a detailed derivation, and not only a sketch of the essential steps, are the frequent requests (from both researchers and students) for a handout containing such a derivation, and the lack of readily available sources containing such a derivation (which might also explain these requests). Readers who are not interested in the technical details can skip to the final result, which is given by Equation 3.18, and continue reading from that point on.

3.2.1 Laplace's Equation

Transmembrane voltage is defined as the difference between the values of the electric potential on both sides of the membrane. The derivation of the induced component of this voltage is based on solving the equation

$$\nabla\left(\left(\sigma+\varepsilon\frac{\partial}{\partial t}\right)\nabla\Psi(x,y,z,t)\right)=0 \tag{3.1}$$

which describes the spatial and temporal distribution of the electric potential. For the steady-state situation, in which the time derivatives are zero, Equation 3.1 simplifies into the well-known Laplace's equation

$$\nabla\cdot\nabla\Psi(x,y,z)=0 \tag{3.2}$$

Solving this equation in a particular coordinate system gives the mathematical solution for the steady-state spatial distribution of Ψ in systems of objects that can be described in such a coordinate system. The mathematical solution is typically a rather large set of functions containing a number of arbitrary constants. By applying physically realistic boundary conditions, the number of functions in this set is reduced, and the values of the constants are determined, yielding the physical solution that describes the actual spatial distribution of Ψ in the given system. The induced transmembrane voltage is then calculated as the difference between the electric potentials on both sides of the membrane.

In the following section, we illustrate this principle by solving Laplace's equation in spherical coordinates, thereby obtaining the description of Ψ in and around a spherical cell.

3.2.2 Spherical Cells

Although biological cells are not perfect spheres, in theoretical treatments they are often considered as such: a spherical interior (the cytoplasm) surrounded by a concentric spherical shell of uniform thickness (the membrane). For certain types of cells, and particularly for cells in suspensions, this is also a reasonable approximation.

To determine the spatial distribution of the electric potential in and around a spherical cell placed into a homogeneous electric field, we write Laplace's equation in the spherical coordinate system (Figure 3.1):

$$\nabla \cdot \nabla \Psi(r,\theta,\varphi) = \frac{1}{r^2}\left(\frac{\partial}{\partial r}\left(r^2\frac{\partial \Psi(r,\theta,\varphi)}{\partial r}\right) + \frac{1}{\sin\theta}\frac{\partial}{\partial\theta}\left(\sin\theta\frac{\partial \Psi(r,\theta,\varphi)}{\partial\theta}\right) + \frac{1}{\sin^2\theta}\frac{\partial^2\Psi(r,\theta,\varphi)}{\partial\varphi^2}\right) = 0 \quad (3.3)$$

We align the center of the cell with the origin of the system and orient its coordinates so that the direction of the external field is parallel to the coordinate line traced by r for $\theta = 0°$, $\varphi = 0°$, as shown in Figure 3.1. This yields a symmetry with respect to φ, in the sense that any coordinate circle traced by φ for (r = constant, θ = constant) is everywhere perpendicular to the field, and consequently $\partial\Psi/\partial\varphi = 0$.

Thus, for the treated case, we have $\Psi(r, \theta, \varphi) = \Psi(r, \theta)$, and Equation 3.3 simplifies into

$$\frac{1}{r^2}\left(\frac{\partial}{\partial r}\left(r^2\frac{\partial\Psi(r,\theta)}{\partial r}\right) + \frac{1}{\sin\theta}\frac{\partial}{\partial\theta}\left(\sin\theta\frac{\partial\Psi(r,\theta)}{\partial\theta}\right)\right)$$

$$= \frac{\partial^2\Psi(r,\theta)}{\partial r^2} + \frac{2}{r}\frac{\partial\Psi(r,\theta)}{\partial r} + \frac{1}{r^2}\frac{\partial^2\Psi(r,\theta)}{\partial\theta^2} + \frac{\text{ctg}\,\theta}{r^2}\frac{\partial\Psi(r,\theta)}{\partial\theta} = 0 \quad (3.4)$$

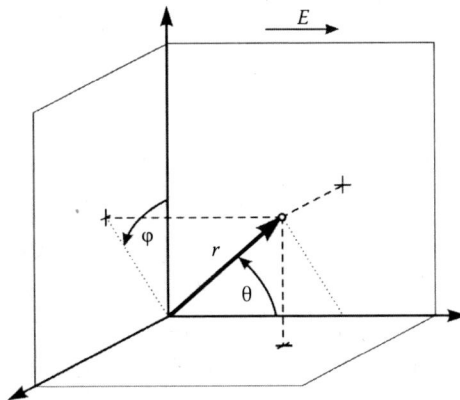

FIGURE 3.1 The spherical coordinate system and the orientation of the external field with respect to this system as used here in solving Laplace's equation for a spherical cell. The center of the cell is aligned with the origin of the system.

where we have applied the chain rule for derivatives. We now perform the separation of variables: we write the function Ψ as a product of two functions of a single variable

$$\Psi(r,\theta) = G(r)H(\theta) \tag{3.5}$$

and insert this form into Equation 3.4, obtaining

$$H(\theta)\left(\frac{\partial^2 G(r)}{\partial r^2} + \frac{2}{r}\frac{\partial(G(r))}{\partial r}\right) + \frac{G(r)}{r^2}\left(\frac{\partial^2 H(\theta)}{\partial \theta^2} + \mathrm{ctg}\,\theta\frac{\partial H(\theta)}{\partial \theta}\right) = 0 \tag{3.6}$$

Dividing both sides by $G(r)H(\theta)/r^2$, writing $\partial G/\partial r = G'(r)$, $\partial H/\partial\theta = H'(\theta)$, and transferring the functions of θ to the right-hand side of the equation, we get

$$\frac{r^2 G''(r) + 2rG'(r)}{G(r)} = -\frac{H''(\theta) + \mathrm{ctg}(\theta)H'(\theta)}{H(\theta)} \tag{3.7}$$

The left-hand side contains only functions of r, and is thus independent of the value of θ. Similarly, the right-hand side contains only functions of θ, and is thus independent of the value of r. Since according to Equation 3.7 the two sides are equal, it follows that they are independent of both r and θ, and must thus be equal to a certain constant. Denoting this constant by K, we thus get a system of two ordinary differential equations of second order

$$\begin{cases} r^2 G''(r) + 2rG'(r) - KG(r) = 0 \\ H''(\theta) + \mathrm{ctg}(\theta)H'(\theta) + KH(\theta) = 0 \end{cases} \tag{3.8}$$

The general solution of the first equation in Equation 3.8 is

$$G(r) = \begin{cases} C_1 r^{-1/2}\sin\left(\sqrt{-\frac{1}{4}-K}\,\log r\right) + C_2 r^{-1/2}\cos\left(\sqrt{-\frac{1}{4}-K}\,\log r\right), & K < -\frac{1}{4} \\ \dfrac{C_1}{r} + C_2, & K = -\frac{1}{4} \\ C_1 r^{-1/2\left(1-\sqrt{1+4K}\right)} + C_2 r^{-1/2\left(1+\sqrt{1+4K}\right)}, & K > -\frac{1}{4} \end{cases} \tag{3.9}$$

where C_1 and C_2 are arbitrary constants.

We now apply the first physically realistic boundary condition: as the distance from the cell increases, the distortion of the field by the cell decreases, and the electric field asymptotically approaches homogeneity, i.e., the state in which $\Psi(r, 0)$ is directly proportional to r. Since $H(\theta)$ is not a function of r, this implies that $G(r)$ is directly proportional to r, and in Equation 3.9 this only occurs for $K = 2$, where

$$G(r) = C_1 r + \frac{C_2}{r^2} \tag{3.10}$$

Since the value of K is the same in both differential equations in Equation 3.8, we can insert $K = 2$ into the second equation and solve it in this more specific form, obtaining

$$H(\theta) = C_3 \cos\theta + C_4 \left(1 - \cos\theta \log\sqrt{\frac{1+\cos\theta}{1-\cos\theta}} \right) \tag{3.11}$$

where C_3 and C_4 are arbitrary constants.

Accounting for the physically realistic boundary condition that $\Psi(r, \theta)$ is finite for all finite r and θ, it follows that $C_4 = 0$, since the term inside the parentheses in Equation 3.11 has singularities at $\theta = 0°$ and $\theta = 180°$. Therefore,

$$H(\theta) = C_3 \cos\theta \tag{3.12}$$

Joining Equation 3.10 and Equation 3.12 according to Equation 3.5, we get

$$\Psi(r,\theta) = \left(Ar + \frac{B}{r^2} \right) \cos\theta \tag{3.13a}$$

where A and B are arbitrary constants. This is the general physical solution for $\Psi(r, \theta)$ in a system consisting of a sphere, an arbitrary number of concentric spherical shells surrounding it, and the (infinite) space surrounding them, provided that the sphere and its shells are the only objects distorting the homogeneity of the electric field.

To come from the general physical solution to the specific one, in which also the constants A and B are determined, and thus the electric potential is fully described, we must now account for the particular geometrical and electrical properties of the system under our consideration. Our system consists of three regions: the cell interior (cytoplasm), the cell membrane, and the exterior, which differ in these properties, and hence also the values of A and B are in general different for each of these regions. Thus, we write

$$\Psi(r,\theta) = \begin{cases} \Psi_i(r,\theta) = \left(A_i r + B_i r^{-2} \right)\cos\theta, & 0 \le r \le (R-d) \\ \Psi_m(r,\theta) = \left(A_m r + B_m r^{-2} \right)\cos\theta, & (R-d) \le r \le R \\ \Psi_e(r,\theta) = \left(A_e r + B_e r^{-2} \right)\cos\theta, & r \ge R \end{cases} \tag{3.13b}$$

where R is the cell radius and d is the membrane thickness.

We now proceed with applying the boundary conditions to determine the six constants in (Equation 3.13b). Since the actual Ψ is finite at $r = 0$, it follows that $B_i = 0$. Requiring once again the field homogeneity far from the cell, this time writing explicitly the electric potential in a homogeneous field E,

$$\Psi(r, \theta) = -Er\cos\theta \tag{3.14}$$

we see that $A_e = -E$.

The remaining four constants are determined by applying the continuity of the electric potential and the electric current density at the two interfaces between the regions.

$$\Psi_i(R-d,\theta) = \Psi_m(R-d,\theta) \tag{3.15a}$$

$$\Psi_m(R,\theta) = \Psi_e(R,\theta) \tag{3.15b}$$

$$\sigma_i \left.\frac{\partial \Psi_i(r,\theta)}{\partial r}\right|_{r=R-d} = \sigma_m \left.\frac{\partial \Psi_m(r,\theta)}{\partial r}\right|_{r=R-d} \tag{3.15c}$$

$$\sigma_m \left.\frac{\partial \Psi_m(r,\theta)}{\partial r}\right|_{r=R} = \sigma_e \left.\frac{\partial \Psi_e(r,\theta)}{\partial r}\right|_{r=R} \tag{3.15d}$$

where σ_i, σ_m, and σ_e denote the electric conductivities of the cytoplasm, the membrane, and the exterior, respectively. Inserting the explicit forms of Ψ_i, Ψ_m, and Ψ_e as given by (Equation 3.13b) into (Equations 3.15a through d), applying the already determined values $B_i = 0$ and $A_e = -E$, and treating the three conductivities as known constants, we obtain a system of four equations with four unknown constants (A_i, A_m, B_m, and B_e). Upon solving this system, we get

$$A_i = \frac{-9ER^3\sigma_e\sigma_m}{2R^3\left(\sigma_m + 2\sigma_e\right)\left(\sigma_m + \frac{1}{2}\sigma_i\right) - 2\left(R-d\right)^3\left(\sigma_e - \sigma_m\right)\left(\sigma_i - \sigma_m\right)} \tag{3.16a}$$

$$A_m = \frac{-3ER^3\sigma_e\left(\sigma_i + 2\sigma_m\right)}{2R^3\left(\sigma_m + 2\sigma_e\right)\left(\sigma_m + \frac{1}{2}\sigma_i\right) - 2\left(R-d\right)^3\left(\sigma_e - \sigma_m\right)\left(\sigma_i - \sigma_m\right)} \tag{3.16b}$$

$$B_m = \frac{3ER^3\left(R-d\right)^3\sigma_e\left(\sigma_i - \sigma_m\right)}{2R^3\left(\sigma_m + 2\sigma_e\right)\left(\sigma_m + \frac{1}{2}\sigma_i\right) - 2\left(R-d\right)^3\left(\sigma_e - \sigma_m\right)\left(\sigma_i - \sigma_m\right)} \tag{3.16c}$$

$$B_e = \frac{ER^3\left[R^3\left(\sigma_m - \sigma_e\right)\left(2\sigma_m + \sigma_i\right) - \left(R-d\right)^3\left(\sigma_m - \sigma_i\right)\left(2\sigma_m + \sigma_e\right)\right]}{2R^3\left(\sigma_m + 2\sigma_e\right)\left(\sigma_m + \frac{1}{2}\sigma_i\right) - 2\left(R-d\right)^3\left(\sigma_e - \sigma_m\right)\left(\sigma_i - \sigma_m\right)} \tag{3.16d}$$

Inserting the expressions for A_m and B_m into Ψ_m as given by (Equation 3.13b), the induced transmembrane voltage can now be expressed as

$$\Delta\Psi_m = \Psi_m(R-d,\theta) - \Psi_m(R,\theta) = f_S ER\cos\theta \tag{3.17a}$$

where

$$f_S = \frac{3\sigma_e\left[3dR^2\sigma_i + \left(3d^2R - d^3\right)\left(\sigma_m - \sigma_i\right)\right]}{2R^3\left(\sigma_m + 2\sigma_e\right)\left(\sigma_m + \frac{1}{2}\sigma_i\right) - 2\left(R-d\right)^3\left(\sigma_e - \sigma_m\right)\left(\sigma_i - \sigma_m\right)} \tag{3.17b}$$

By applying a simplifying assumption that the membrane is a pure insulator, $\sigma_m = 0$, the function f_S turns into a constant, $f_S = 3/2$, and we obtain the well-known formula often referred to as the (steady-state) Schwan's equation:

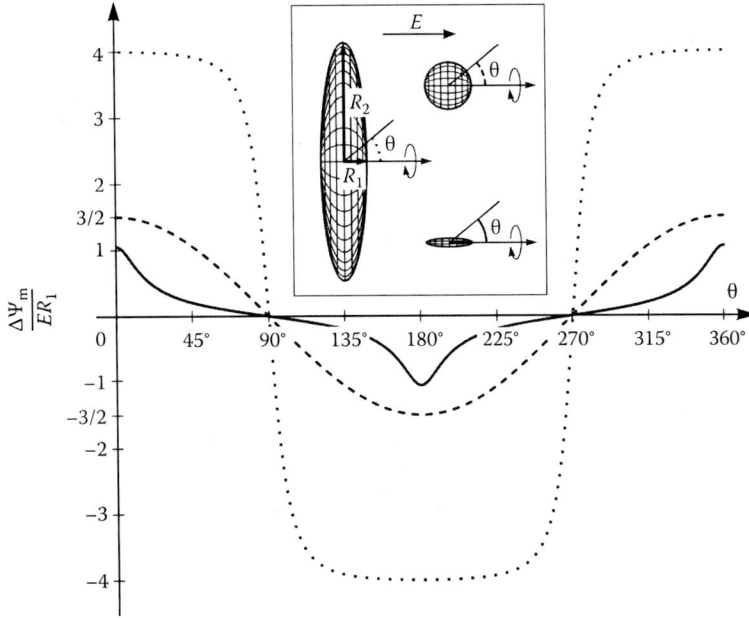

FIGURE 3.2 Normalized steady-state $\Delta\Psi_m$ as a function of the polar angle θ for spheroidal cells with the axis of rotational symmetry (ARS) aligned with the direction of the field. Solid: a prolate spheroidal cell with $R_2 = 0.2 \times R_1$. Dashed: a spherical cell, $R_2 = R_1 = R$. Dotted: an oblate spheroidal cell with $R_2 = 5 \times R_1$.

$$\Delta\Psi_m = \frac{3}{2} ER \cos\theta \tag{3.18}$$

This formula tells that the induced transmembrane voltage is proportional to the applied electric field and to the cell radius. Furthermore, it has extremal values at the points where the field is perpendicular to the membrane, i.e., at $\theta = 0°$ and $\theta = 180°$ (the "poles" of the cell), and in-between these poles it varies proportionally to the cosine of θ (see Figure 3.2, dashed).

Equation 3.18 describes the steady-state situation, which is typically established several microseconds after the onset of the electric field. With exposures to a DC field lasting hundreds of microseconds or more, this formula can safely be applied to yield the maximal, steady-state value of the induced transmembrane voltage. To describe the transient behavior during the initial microseconds, in addition to the electric conductivities one also has to account for the dielectric permittivity of the membrane, ε_m. Such a derivation, which is a slight extension of the derivation presented above (Pauly and Schwan 1959), yields the first-order Schwan's equation that reads

$$\Delta\Psi_m = \frac{3}{2} ER \cos\theta \left(1 - e^{-t/\tau_m}\right) \tag{3.19a}$$

where τ_m is the time constant of the membrane charging,

$$\tau_m = \frac{R\varepsilon_m}{2d\left(\sigma_i\sigma_e/(\sigma_i + 2\sigma_e)\right) + R\sigma_m} \tag{3.19b}$$

In certain experiments *in vitro*, where artificial extracellular media with conductivities substantially lower than physiological are used, the factor 3/2 is an oversimplification, and the more precise form given by Equation 3.17 must be used, as discussed in detail in Kotnik et al. (1997). But generally, the formulae

Equation 3.18 and Equation 3.19 are applicable to exposures to sine (AC) electric fields with frequencies below 1 MHz, and to rectangular electric pulses longer than 1 μs.

To determine the voltage induced by even higher field frequencies or even shorter pulses, the dielectric permittivities of the electrolytes also have to be accounted for. This leads to a further generalization of Equation 3.17 and/or Equation 3.19 to a second-order model (Grosse and Schwan 1992, Kotnik et al. 1998, Kotnik and Miklavčič 2000a), and the results it yields will be outlined in Section 3.2.4.

3.2.3 Spheroidal, Ellipsoidal, and Cylindrical Cells

Another direction of generalization is to assume a cell shape more general than that of a sphere. The most straightforward generalization is to a spheroid (a geometrical body obtained by rotating an ellipse around one of its radii, so that one of its orthogonal projections is a sphere, and the other two are the same ellipse) and further to an ellipsoid (a geometrical body in which each of its three orthogonal projections is a different ellipse). To obtain the analogues of Schwan's equation for such cells, one solves Laplace's equation in spheroidal and ellipsoidal coordinates, performing the same steps as in the solution in spherical coordinates described in detail in Section 3.2.2. A detailed description of the derivation in prolate and oblate spheroidal coordinates is given in Kotnik and Miklavčič (2000b), Gimsa and Wachner (2001), and Valič et al. (2003), and in analogy to Equation 3.18, for an oblate spheroid with the ARS aligned with the field it yields

$$\Delta\Psi_m = E \frac{R_2^2 - R_1^2}{\frac{R_2^2}{\sqrt{R_2^2 - R_1^2}} \operatorname{arcctg} \frac{R_1}{\sqrt{R_2^2 - R_1^2}} - R_1} \frac{R_2 \cos\theta}{\sqrt{R_1^2 \sin^2\theta + R_2^2 \cos^2\theta}} \tag{3.20}$$

and for a prolate spheroid with the ARS aligned with the field

$$\Delta\Psi_m = E \frac{R_1^2 - R_2^2}{R_1 - \frac{R_2^2}{\sqrt{R_1^2 - R_2^2}} \ln\frac{R_1 + \sqrt{R_1^2 - R_2^2}}{R_2}} \frac{R_2 \cos\theta}{\sqrt{R_1^2 \sin^2\theta + R_2^2 \cos^2\theta}} \tag{3.21}$$

where R_1 and R_2 are the radii of the spheroid in the directions parallel and perpendicular to the field, respectively (see also Figure 3.2).

Besides the fact that the expressions obtained for Ψ are somewhat more intricate than the one in spherical coordinates, the generalization of the shape from spherical to spheroidal invokes two additional complications outlined in the next two paragraphs.

A description of a cell is geometrically realistic if the thickness of its membrane is uniform. This is the case if the membrane represents the space between two concentric spheres, but not two confocal spheroids or ellipsoids. As a result, the thickness of the membrane modeled in spheroidal or ellipsoidal coordinates is necessarily nonuniform. By solving Laplace's equation in these coordinates, we thus obtain the spatial distribution of the electric potential in a nonrealistic setting. However, under the assumption that the membrane conductivity is zero, the induced transmembrane voltage obtained in this manner is still realistic. Namely, the shielding of the cytoplasm is then complete, and hence the electric potential everywhere inside the cytoplasm is constant. Therefore, the geometry of the inner surface of the membrane does not affect the potential distribution outside the cell, which is the same as if the cell would be a homogeneous nonconductive body of the same shape. A more rigorous discussion of the validity of this approach can be found in Kotnik and Miklavčič (2000b). Figure 3.2 compares the transmembrane voltage induced on two spheroids with the ARS aligned with the direction of the field, and that induced on a sphere.

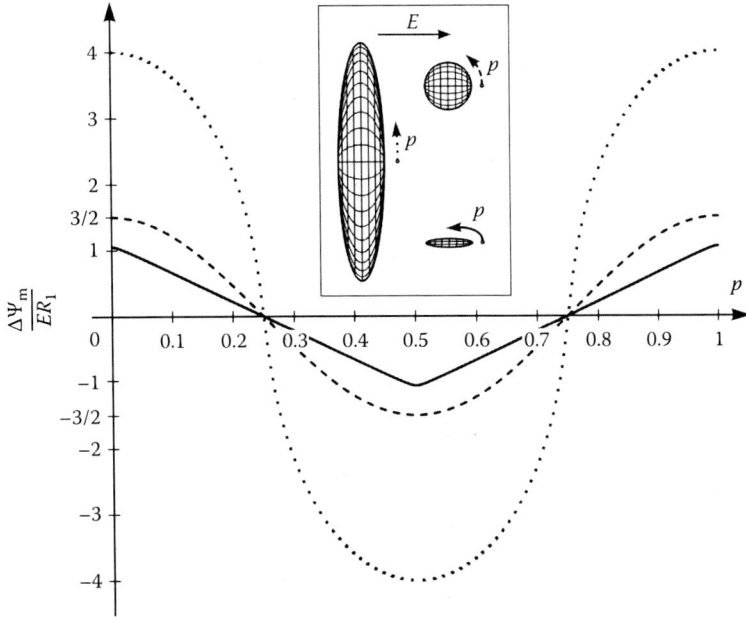

FIGURE 3.3 Normalized steady-state $\Delta\Psi_m$ as a function of the normalized arc length p for spheroidal cells with the ARS aligned with the direction of the field. Solid: a prolate spheroidal cell with $R_2 = 0.2 \times R_1$. Dashed: a spherical cell, $R_2 = R_1 = R$. Dotted: an oblate spheroidal cell with $R_2 = 5 \times R_1$.

For nonspherical cells, it is generally more revealing to express $\Delta\Psi_m$ as a function of the arc length than as a function of the angle θ (for a sphere, the two quantities are directly proportional). For uniformity, the normalized version of the arc length is used, denoted by p and increasing from 0 to 1 equidistantly along the arc of the membrane. This is illustrated in Figure 3.3 for the cells for which $\Delta\Psi_m(\theta)$ is shown in Figure 3.2, and all the plots of $\Delta\Psi_m$ henceforth will be presented in this manner.

Typically, $\theta(p)$ cannot be expressed by an elementary function, and $\Delta\Psi_m(p)$ has to be determined by numerical mapping. In analytical treatment of $\Delta\Psi_m$ in regular cell shapes for which such a treatment is possible, this mapping can be performed to arbitrary accuracy, but the process is somewhat tedious. In contrast, when $\Delta\Psi_m$ is computed numerically (see Section 3.3), the accuracy is limited by the size of the mesh employed, but the arc length is readily determined by the software, and the plots of $\Delta\Psi_m(p)$ are easy to generate.

Another complication caused by generalizing the cell shape from a sphere to a spheroid or an ellipsoid is that the induced voltage now also becomes dependent on the orientation of the cell with respect to the electric field. To deal with this, one decomposes the field vector into the components parallel to the axes of the spheroid or the ellipsoid, and writes the induced voltage as a corresponding linear combination of the voltages induced for each of the three coaxial orientations (Gimsa and Wachner 2001, Valič et al. 2003). Figures 3.4 and 3.5 show the effect of rotation of two different spheroids with respect to the direction of the field.

An analytical solution for $\Delta\Psi_m$ is also attainable for circular cylinders with the axis oriented perpendicularly to the external field, and is given by

$$\Delta\Psi_m = 2ER\cos\theta \tag{3.22}$$

This is a suitable approximation for elongated cellular structures such as muscle cells and axons of nerve cells, provided that the field is roughly perpendicular to their direction.

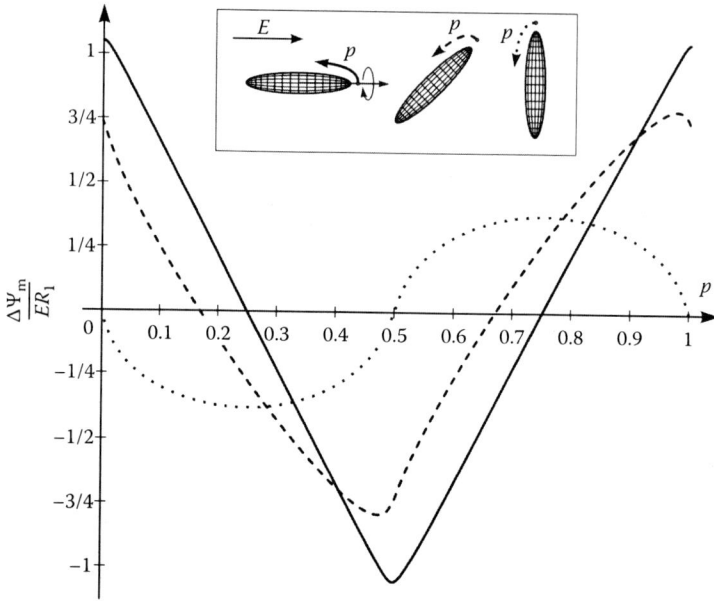

FIGURE 3.4 Normalized steady-state $\Delta\Psi_m(p)$ for a prolate spheroidal cell with $R_2 = 0.2 \times R_1$. Solid: ARS aligned with the field. Dashed: ARS at 45° with respect to the field. Dotted: ARS perpendicular to the field.

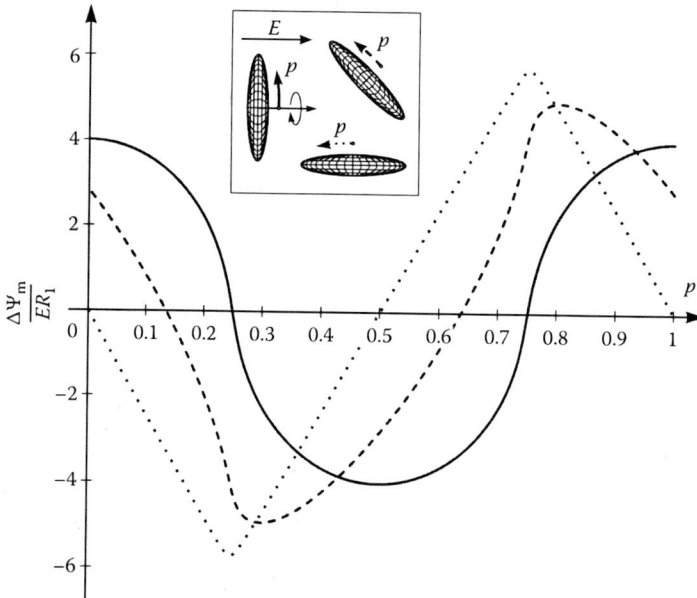

FIGURE 3.5 Normalized steady-state $\Delta\Psi_m(p)$ for an oblate spheroidal cell with $R_2 = 5 \times R_1$. Solid: ARS aligned with the field. Dashed: ARS at 45° with respect to the field. Dotted: ARS perpendicular to the field.

3.2.4 High Frequencies and Very Short Pulses

The time constant of the membrane charging (τ_m) given by (Equation 3.19b) implies that there is a delay between the time courses of the external field and the voltage induced by this field. As mentioned above, τ_m (and thus the delay) is somewhat below a microsecond under physiological conditions, but can be larger when cells are suspended in a low-conductivity medium. For alternating (AC) fields with the oscillation period much longer than τ_m, as well as with rectangular pulses much longer than τ_m, the amplitude of the induced voltage is very close to the steady-state value given by Equation 3.18. However, for AC fields with the period comparable or shorter than τ_m, as well as for rectangular pulses shorter than τ_m, the amplitude of the induced voltage starts to decrease.

To illustrate how the amplitude of the induced transmembrane voltage gets attenuated as the frequency of the AC field increases, we plot the normalized amplitude of the induced voltage as a function of the field frequency. For a spherical cell, the plot obtained is shown in Figure 3.6. The low-frequency plateau and the downward slope that follows are both described by the first-order Schwan's equation, but the high-frequency plateau is only described by the second-order model (Grosse and Schwan 1992, Kotnik et al. 1998, Kotnik and Miklavčič 2000a), in which all electric conductivities and dielectric permittivities have nonzero values.

With field frequencies approaching the GHz range, or with pulse durations in the nanosecond range, the attenuation of the voltage induced on the cell plasma membrane becomes so pronounced that this voltage becomes comparable to the voltage induced on organelle membranes in the cell interior. In certain circumstances, particularly if the organelle interior is electrically more conductive than the cytosol, or if the organelle membrane has a lower dielectric permittivity than the cell membrane, the voltage induced on the membrane of this organelle can temporarily even exceed the voltage induced on the plasma membrane (Kotnik and Miklavčič 2006). In principle, this could provide a theoretical explanation for a number of recent reports that very short and intense electric pulses (tens of ns, millions or tens of millions of V/m) can also induce electroporation of organelle membranes (Schoenbach et al. 2001, Beebe et al. 2003, Tekle et al. 2005).

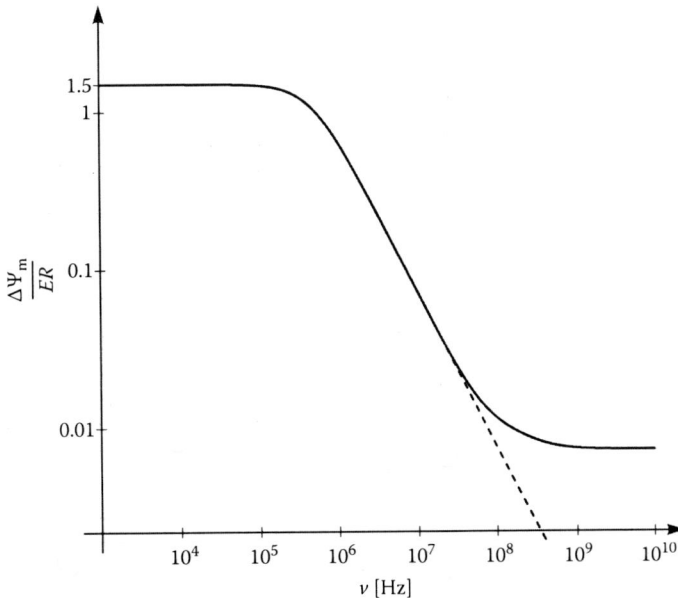

FIGURE 3.6 The amplitude of normalized $\Delta\Psi_m$ as a function of the frequency of the AC field. The dashed curve shows the first-order Schwan's equation, and the solid one the second-order Schwan's equation. Note that both axes are logarithmic.

3.3 Numerical Computation

Realistic cells can deviate considerably from regular shapes considered in Section 3.2. Moreover, in actual situations, cells are rarely isolated, and when sufficiently close to each other, the mutual distortion of the field they cause cannot be neglected. Often, the cells are even in direct contact, forming two-dimensional (monolayers attached to the bottom of a dish) or three-dimensional structures (tissues), and they can even be interconnected by structures such as gap junctions. None of these cases allows for analytical derivation of the induced transmembrane voltage ($\Delta\Psi_m$), while employing analytical solutions for spherical, spheroidal, or cylindrical cells as approximations can lead to rather large errors. In practice, there are two approaches for obtaining accurate estimates of $\Delta\Psi_m$ on irregularly shaped cells: numerical computation and experimental determination. Here, we focus on the numerical methods, while the experimental approach will be the subject of Section 3.4.

3.3.1 Computational Methods

Numerical computation of $\Delta\Psi_m$ is generally performed in several steps. First, the continuous geometry of the model and/or the differential equations describing the electric or electromagnetic field are transformed into their discrete counterparts. Next, these equations are solved either directly or iteratively until adequate convergence is reached. Finally, the electric potential on both sides of the membrane is extracted from the computed data, and $\Delta\Psi_m$ is computed as their difference. Elementary methods, such as solving a linear system of equations, are mostly inadequate for this purpose, while advanced methods, such as the finite difference method and particularly the finite element method, are well suited for this task.

3.3.1.1 Finite Difference Method

The finite difference method is a method for solving differential and integral equations, or systems of such equations. In this method, the continuous geometry of the model is replaced by a grid, which is restricted in the basic form of the method to rectangular shapes and simple alterations thereof. At each grid point, the differential terms of the equation are replaced by the difference terms, and the obtained difference equations are then solved to yield the electric potential in the points of the grid. If this method is used to solve time-dependent partial differential equations (so that time also proceeds in discrete steps), the method is termed the finite-difference time-domain method. The attractive feature of this method is its straightforward implementation, but due to the rectangular mesh it is generally inaccurate with complicated object shapes, and particularly close to the curved boundaries.

3.3.1.2 Finite Element Method

The finite element method is another method for finding approximate solutions of differential and integral equations. The method is based on discretization of the geometry (meshing) into subregions, which are referred to as the finite elements (Reddy 2005). These elements can be of different shapes and sizes, which allow to model intricately shaped objects and to focus on the regions of interest (by making the mesh locally more dense). Similar to the finite difference method, the finite element method can also be extended to time-dependent problems. However, unlike with finite differences, the solution obtained for each finite element is a function varying smoothly between the nodes of the element and preserving the continuity also between the elements. The main advantage of the finite element method is its ability to handle complicated geometries and boundaries with relative ease, and while some attempts were made to model irregularly shaped cells or clusters of such cells with finite differences, these were confined to a 2-D space (Gowrishankar and Weaver 2003, Esser et al. 2007, Joshi et al. 2008). In contrast, the finite element approach can handle realistic 3-D shapes quite easily (Miller and Henriquez 1988, Pucihar et al. 2006, 2009a), and in the next section we illustrate this in more detail.

3.3.2 Irregularly Shaped Cells

Before discretization, a realistic model of the cell under consideration must be constructed and stored in the computer. While this process is straightforward for cells of simple geometric shapes (e.g., cells in suspension can be modeled as spheres), it becomes problematic when irregularly shaped cells are modeled (an extreme example of this is a nerve cell). During discretization, an even more compelling problem occurs, namely, the meshing of the cell membrane, which is over 1000-fold thinner than the dimensions of a typical cell. Even the modern adaptive-size meshing methods generally fail when such disproportions are involved. Another condition related to the membrane meshing that is difficult to meet is the uniform membrane thickness. Once all these difficulties are overcome, the final step is the computation of the electric potential and hence of the induced transmembrane voltage.

3.3.2.1 Constructing a 3-D Model of the Cell

The simplest approach in modeling an irregularly shaped cell is to compose it from several simple geometrical objects (e.g., hemispheres, circular or elliptic cylinders) (Fear and Stuchly 1998, Buitenweg et al. 2003, Valič et al. 2003, Huang et al. 2004). However, typical cells growing in a dish or in a tissue have markedly irregular shapes, and this approach can only yield a rough approximation of the actual situation.

A more realistic three-dimensional model of an irregularly shaped cell can be constructed from a sequence of cross sections of the cell under consideration, as sketched in Figure 3.7. The cross sections are obtained by staining the cell with a membrane marker and acquiring images in different focal

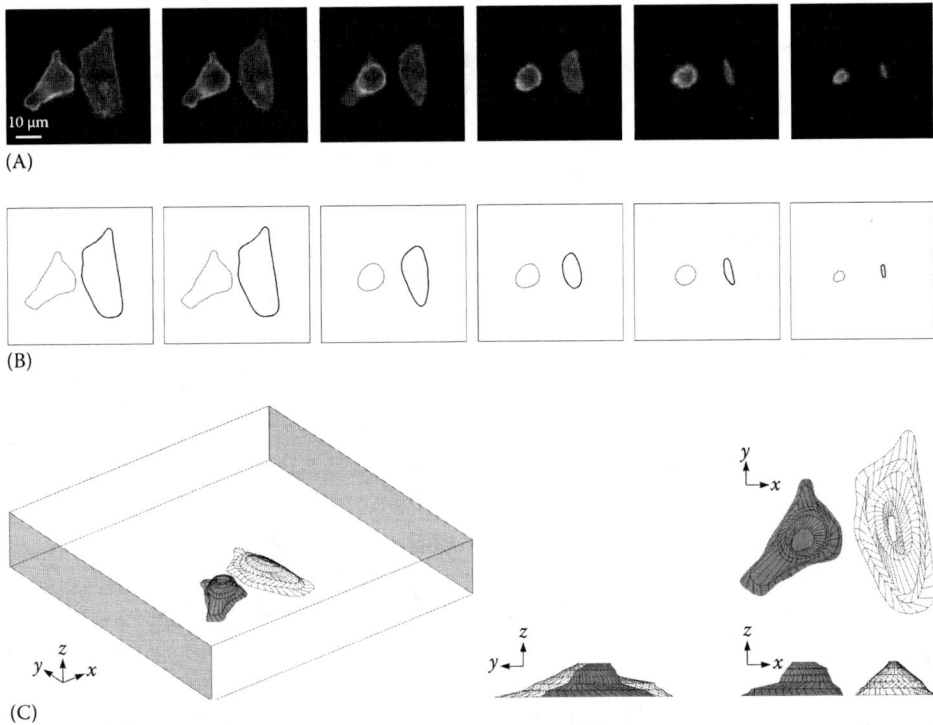

FIGURE 3.7 Construction of a 3-D model of two irregularly shaped CHO cells. (A) Fluorescence cross-section images of the cells stained with di-8-ANEPPS, acquired from the bottom to the top of the cells in 1 μm steps. (B) The contours. (C) The 3-D model in COMSOL Multiphysics 3.4. The interior of the rectangular block represents the extracellular medium, the gray-shaded faces are the electrodes, and the other four faces are insulating. (Adapted from Pucihar G. et al., *Ann. Biomed. Eng.*, 34, 642, 2006.)

planes. In these images, the contours of the cell are then detected, transformed into solid planes, combined into a 3-D object, and imported into the workspace of a finite element software, such as COMSOL Multiphysics (COMSOL Inc., Burlington, MA).

3.3.2.2 Modeling the Cell Membrane

The normal component of the current density in the membrane, J, is given by

$$J(t) = \frac{\sigma_m\left(\Psi_i(t) - \Psi_e(t)\right)}{d} + \frac{\varepsilon_m}{d}\frac{\partial\left(\Psi_i(t) - \Psi_e(t)\right)}{\partial t} \tag{3.23}$$

with σ_m, ε_m, d, Ψ_i, and Ψ_e having the same meaning as in Equations 3.13, 3.15, and 3.19. The first term on the right-hand side represents the conductive component, and the second term the capacitive component of the electric current flowing through the membrane.

When constructing a finite element model of the cell, direct incorporation of a realistic cell membrane (i.e., a very thin layer of uniform thickness enclosing the cell) would require the model to consist locally of an extremely large number of finite elements. Even with the modern adaptive-size mesh generation algorithms, this is often prohibitively time consuming and demanding on computer memory. However, unless the spatial distribution of the electric potential inside the membrane is of interest, this can be avoided. Namely, as far as the electric potentials in the cytoplasm and the cell exterior are concerned, the effect of the membrane with thickness d, electric conductivity σ_m, and dielectric permittivity ε_m is equivalent to the effect of an interface with thickness 0 (i.e., a mathematical surface) separating these two regions and characterized by surface electric conductivity $\kappa_m = \sigma_m/d$, and surface dielectric permittivity $\beta_m = \varepsilon_m/d$. Thus, we can rewrite Equation 3.23 as

$$J(t) = \kappa_m\left(\Psi_i(t) - \Psi_e(t)\right) + \beta_m\frac{\partial\left(\Psi_i(t) - \Psi_e(t)\right)}{\partial t} \tag{3.24}$$

Despite the membrane as such being absent from the model, the drop of electric potential at such an interface is equivalent to the $\Delta\Psi_m$ induced on the membrane characterized by corresponding values of d, σ_m, and ε_m. In models constructed in this way, the mesh of finite elements is generated without difficulty, as disproportionally small elements corresponding to the membrane interior are avoided (Pucihar et al. 2006, 2009a). By assuming that κ_m is a function of $\Delta\Psi_m$, this approach can be extended further, e.g., to simulate the course of electroporation (Pucihar et al. 2009a).

3.3.2.3 Computation of $\Delta\Psi_m$

In COMSOL Multiphysics, the electric potential Ψ is computed numerically by solving the discretized form of Equation 3.1, where σ and ε are the electric conductivity and dielectric permittivity of each region under consideration. In general, this gives Ψ as a function of both space and time, and thus describes both the transient and the steady state. The induced transmembrane voltage, $\Delta\Psi_m(t)$, is then calculated as the difference between electric potentials on both sides of the membrane:

$$\Delta\Psi_m = \Psi_i(t) - \Psi_e(t) \tag{3.25}$$

While the results obtained in this manner are quite accurate, they are only applicable to the particular cell shape for which they were computed. Figure 3.8 displays the steady-state $\Delta\Psi_m(p)$ computed for the two irregularly shaped cells shown in Figure 3.7.

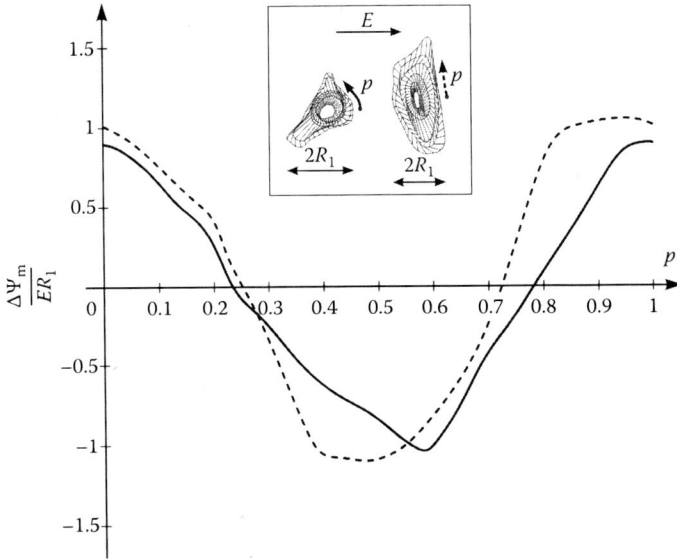

FIGURE 3.8 Normalized steady-state $\Delta\Psi_m(p)$ for two irregularly shaped cells from Figure 3.7. (Adapted from Pucihar G. et al., *Ann. Biomed. Eng.*, 34, 642, 2006.)

3.3.3 Cells in Dense Suspensions and Tissues

In dilute cell suspensions, the distance between the cells is much larger than the cell sizes themselves, and the local field outside each cell is practically unaffected by the presence of other cells. Thus, for cells representing less than 1% of the suspension volume (for a spherical cell with a radius of 10 μm, this means up to 2 million cells/mL), the deviation of the induced transmembrane voltage from the prediction given by Schwan's equation (3.18) is negligible. However, for larger volume fractions occupied by the cells, the distortion of the local field around each cell by the adjacent cells becomes more pronounced, and the deviation from Schwan's equation is also larger (Figure 3.9). For suspensions with cell volume fractions over 10%, as well as for cells in clusters and lattices, a reliable determination of $\Delta\Psi_m$ requires numerical computation (Susil et al. 1998, Pavlin et al. 2002, Pucihar et al. 2007). Regardless of the volume fraction they occupy, as long as the cells are suspended, they float freely, and their arrangement is rather uniform. Asymptotically, this would correspond to a face-centered lattice, and this lattice is also the most appropriate for the analysis of the transmembrane voltage induced on cells in suspension.

For even larger volume fractions, the electrical properties of the suspension start to resemble that of a tissue, but only to a certain extent. The arrangement of cells in tissues does not necessarily resemble a face-centered lattice, since cells can form specific structures (e.g., layers). In addition, cells in tissues can be directly electrically coupled (e.g., through gap junctions). Numerical modeling and computation of electric fields and currents in tissues is discussed in detail in Chapter 15.

3.4 Experimental Determination

An alternative to the analytical and numerical methods for determining the induced transmembrane voltage ($\Delta\Psi_m$) are the experimental techniques. These include the measurements of $\Delta\Psi_m$ with microelectrodes and with potentiometric fluorescent dyes. Microelectrodes (either conventional or patch clamp) were used in pioneering measurements of the action potential propagation (Ling and Gerard

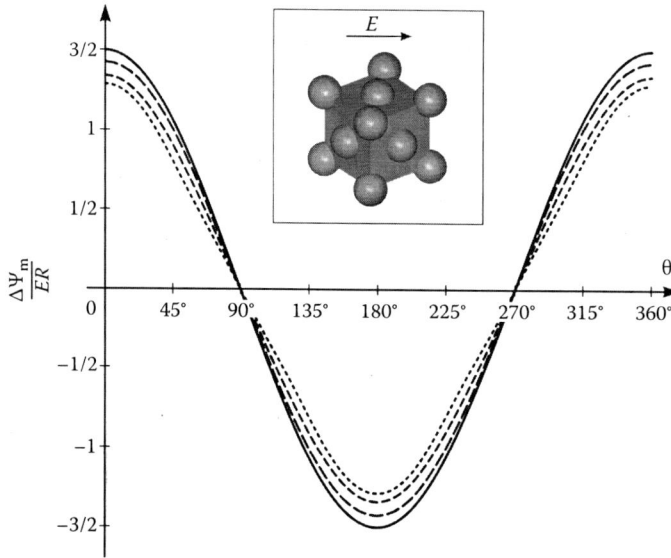

FIGURE 3.9 Normalized steady-state $\Delta\Psi_m(\theta)$ for spherical cells in suspensions of various densities (intercellular distances). Solid: The analytical result for a single cell as given by Equation 3.18. Dashed: numerical results for cells arranged in a face-centered cubic lattice and occupying (with decreasing dash size) 10%, 30%, and 50% of the total suspension volume.

1949, Neher and Sakmann 1976) and were preferred for their simple use and high temporal resolution. However, the invasive nature of measurements and low spatial resolution are considerable shortcomings of this approach. Moreover, the physical presence of the electrodes during the measurement affects the distribution of the electric field around them, and thus also the value of $\Delta\Psi_m$. In contrast, measurement by means of potentiometric dyes is noninvasive, offers higher spatial resolution, does not distort the field and thus $\Delta\Psi_m$. Moreover, it can be performed simultaneously on a number of cells. For these reasons, during the last decades, the potentiometric dyes have become the preferred tool in experimental studies and measurements of $\Delta\Psi_m$.

3.4.1 Potentiometric Dyes

Based on their response mechanism, potentiometric dyes are divided into two classes (Invitrogen Corp. 2009): (1) slow potentiometric dyes that are translocated across the membrane by an electrophoretic mechanism, which is accompanied by a fluorescence change and (2) fast potentiometric dyes that incorporate into the membrane, with their electronic structure and consequently their fluorescence properties dependent on transmembrane voltage.

Electric pulses used in electrophysiological and electroporation-based applications usually have durations in the range of microseconds to milliseconds. In order to measure $\Delta\Psi_m$ induced by such pulses, fast potentiometric dyes have to be used. These dyes respond to changes in $\Delta\Psi_m$ within microseconds or less, which makes them suitable even for measurements of the transient effects. Slow dyes, on the other hand, need several seconds to respond to a change of $\Delta\Psi_m$.

One of the fast potentiometric dyes widely used for measuring $\Delta\Psi_m$ is di-8-ANEPPS (di-8-butyl-amino-naphthyl-ethylene-pyridinium-propyl-sulfonate), developed by Leslie Loew and colleagues at the University of Connecticut (Fluhler et al. 1985, Gross et al. 1986, Loew 1992). This dye is nonfluorescent in water, but becomes strongly fluorescent when incorporated into the lipid bilayer of the cell membrane, thereby making the membrane highly visible. This enables the construction of numerical models of cells from microscopic fluorescence images, as described in Section 3.3, and thereby provides a possibility to compute $\Delta\Psi_m$ on the same cells on which an experiment was carried out.

The fluorescence intensity of di-8-ANEPPS varies proportionally to the change of $\Delta\Psi_m$; the response of the dye is linear for voltages ranging from −280 to +250 mV (Lojewska et al. 1989, Cheng et al. 1999). Relatively small changes in fluorescence of the dye, uneven membrane staining, and dye internalization make di-8-ANEPPS less suitable for absolute measurements of membrane voltage, such as the resting membrane voltage, although such efforts were also reported (Zhang et al. 1998). It is, however, well suited for measuring larger changes in membrane voltage, such as the onset of induced transmembrane voltage in nonexcitable cells exposed to external electric fields (Gross et al. 1986, Montana et al. 1989), or action potentials in excitable cells (Bedlack et al. 1994, Cheng et al. 1999). di-8-ANEPPS also allows for determination of $\Delta\Psi_m$ by ratiometric measurements of fluorescence excitation (Montana et al. 1989, Hayashi et al. 1996) or emission (Knisley et al. 2000), which increases the sensitivity of the response.

3.4.2 Image Acquisition and Data Processing

Since the sensitivity of fast potentiometric dyes to the changes of $\Delta\Psi_m$ is low (typically, a change of $\Delta\Psi_m$ by 100 mV results in the change of fluorescence intensity by 2%–12%), the fluorescence changes are hardly discernible by the naked eye and become apparent only after image processing and analysis.

This procedure is performed in several steps (Pucihar et al. 2009b). The first step is acquiring a pair of images: a control image (immediately before the exposure to the electric field) and the pulse image (during the exposure). To get a more reliable measurement, a sequence of pulses can be applied, with both the control and the pulse image acquired for each pulse. The background fluorescence is then subtracted from both images. For the cell under investigation, the region of interest corresponding to the membrane is determined, and the fluorescence intensities along this region in the control and pulse image are measured. For each pulse, the control data are subtracted from the pulse data, and the result divided by the control data to obtain the relative fluorescence changes. If a sequence of pulses is applied, the values of relative fluorescence changes determined for each pulse can be averaged. The relative fluorescence changes are then transformed into values of $\Delta\Psi_m$ using a calibration curve. A rough estimation of this curve can be obtained from the literature, but for higher accuracy, it has to be measured for each particular setup, as shown in Figure 3.10. Calibration is performed with either (1) potassium

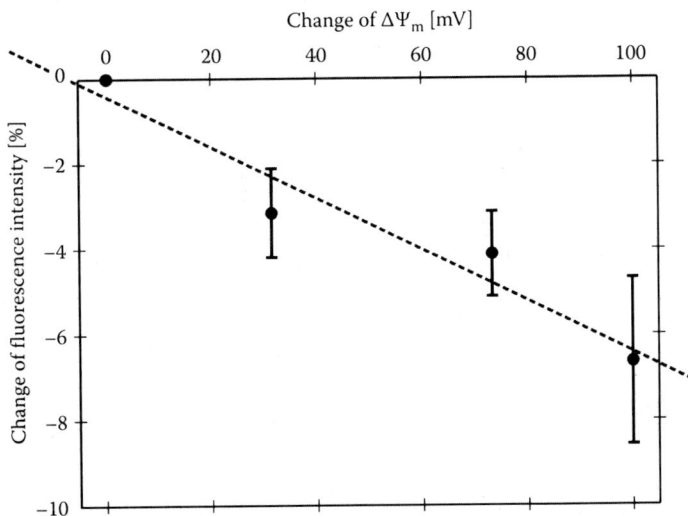

FIGURE 3.10 The calibration curve for measurements of $\Delta\Psi_m$ using di-8-ANEPPS.

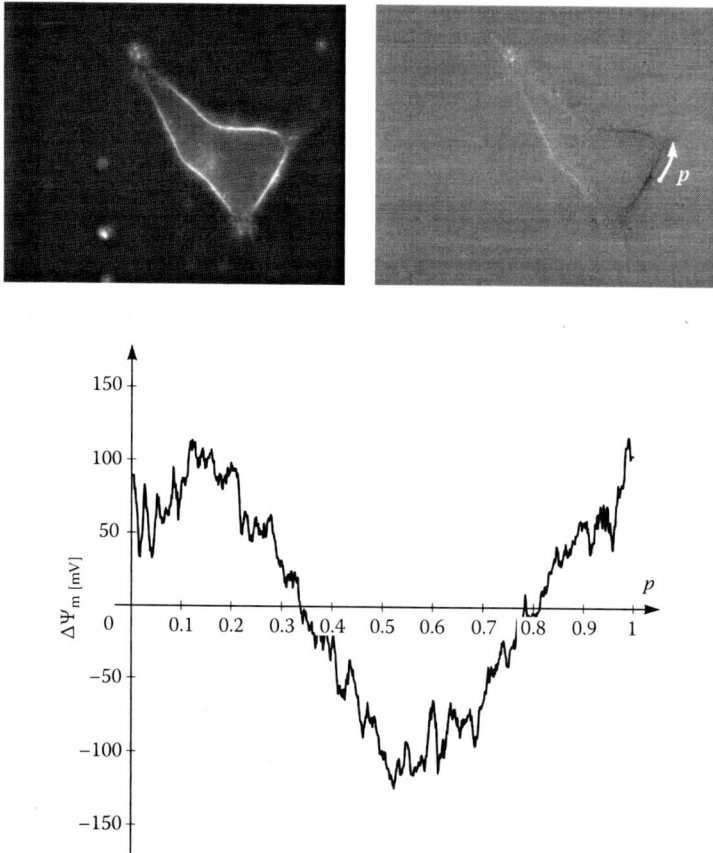

FIGURE 3.11   Determination of steady-state $\Delta\Psi_m$ using di-8-ANEPPS. Top left: Raw fluorescence image of a cell stained with di-8-ANEPPS. Top right: Processed image. Bottom: $\Delta\Psi_m(p)$ determined from the image using the calibration curve shown in Figure 3.10.

ionophore valinomycin and a set of different potassium concentrations in external medium (Montana et al. 1989, Pucihar et al. 2006), or (2) patch clamp in voltage clamp mode (Zhang et al. 1998; Pakhomov and Pakhomova 2010). Finally, the voltage is plotted as a function of the relative arc length. To remove some of the noise inherent to potentiometric measurements, the curve can be smoothed using a suitable filter (e.g., the moving average). Figure 3.11 shows a cell stained with di-8-ANEPPS, the processed image reflecting $\Delta\Psi_m$, and the plot of $\Delta\Psi_m$ along the cell membrane.

Acknowledgment

This work was supported by the Slovenian Research Agency (Programme P2-0249 and Project Z2-9229).

References

Bedlack RS, Wei M, Fox SH, Gross E, Loew LM. 1994. Distinct electric potentials in soma and neurite membranes. *Neuron* 13: 1187–1193.

Beebe SJ, Fox PM, Rec LJ, Willis EL, Schoenbach KH. 2003. Nanosecond, high-intensity pulsed electric fields induce apoptosis in human cells. *FASEB J* 17: 1493–1495.

Buitenweg JR, Rutten WL, Marani E. 2003. Geometry-based finite-element modeling of the electrical contact between a cultured neuron and a microelectrode. *IEEE Trans Biomed Eng* 50: 501–509.

Burnett P, Robertson JK, Palmer JM, Ryan RR, Dubin AE, Zivin RA. 2003. Fluorescence imaging of electrically stimulated cells. *J Biomol Screen* 8: 660–667.

Cheng DK, Tung L, Sobie EA. 1999. Nonuniform responses of transmembrane potential during electric field stimulation of single cardiac cells. *Am J Physiol* 277: H351–H362.

Esser AT, Smith KC, Gowrishankar TR, Weaver JC. 2007. Towards solid tumor treatment by irreversible electroporation: Intrinsic redistribution of fields and currents in tissue. *Technol Cancer Res Treat* 6: 261–273.

Fear EC, Stuchly MA. 1998. Modeling assemblies of biological cells exposed to electric fields. *IEEE Trans Biomed Eng* 45: 1259–1271.

Fluhler E, Burnham VG, Loew LM. 1985. Spectra, membrane binding, and potentiometric responses of new charge shift probes. *Biochemistry* 24: 5749–5755.

Gimsa J, Wachner D. 2001. Analytical description of the transmembrane voltage induced on arbitrarily oriented ellipsoidal and cylindrical cells. *Biophys J* 81: 1888–1896.

Gowrishankar TR, Weaver JC. 2003. An approach to electrical modeling of single and multiple cells. *Proc Natl Acad Sci U S A* 100: 3203–3208.

Gross D, Loew LM, Webb W. 1986. Optical imaging of cell membrane potential changes induced by applied electric fields. *Biophys J* 50: 339–348.

Grosse C, Schwan HP. 1992. Cellular membrane potentials induced by alternating fields. *Biophys J* 63: 1632–1642.

Hayashi Y, Zviman MM, Brand JG, Teeter JH, Restrepo D. 1996. Measurement of membrane potential and $[Ca^{2+}]_{(i)}$ in cell ensembles: Application to the study of glutamate taste in mice. *Biophys J* 71: 1057–1070.

Huang X, Nguyen D, Greve DW, Domach MM. 2004. Simulation of microelectrode impedance changes due to cell growth. *IEEE Sensors J* 4: 576–583.

Huang CJ, Harootunian A, Maher MP, Quan C, Raj CD, McCormack K, Numann R, Negulescu PA, Gonzalez JE. 2006. Characterization of voltage-gated sodium-channel blockers by electrical stimulation and fluorescence detection of membrane potential. *Nat Biotechnol* 24: 439–446.

Invitrogen Corp. 2009. *Molecular Probes—The Handbook*. Carlsbad, CA: Invitrogen Corp. (available electronically at http://www.invitrogen.com/site/us/en/home/References/Molecular-Probes-The-Handbook.html).

Joshi RP, Mishra A, Schoenbach KH. 2008. Model assessment of cell membrane breakdown in clusters and tissues under high-intensity electric pulsing. *IEEE Trans Plasma Sci* 36: 1680–1688.

Knisley SB, Justice RK, Kong W, Johnson PL. 2000. Ratiometry of transmembrane voltage-sensitive fluorescent dye emission in hearts. *Am J Physiol Heart Circ Physiol* 279: H1421–H1433.

Kotnik T, Miklavčič D. 2000a. Second-order model of membrane electric field induced by alternating external electric fields. *IEEE Trans Biomed Eng* 47: 1074–1081.

Kotnik T, Miklavčič D. 2000b. Analytical description of transmembrane voltage induced by electric fields on spheroidal cells. *Biophys J* 79: 670–679.

Kotnik T, Miklavčič D. 2006. Theoretical evaluation of voltage inducement on internal membranes of biological cells exposed to electric fields. *Biophys J* 90: 480–491.

Kotnik T, Bobanović F, Miklavčič D. 1997. Sensitivity of transmembrane voltage induced by applied electric fields—A theoretical analysis. *Bioelectrochem Bioenerg* 43: 285–291.

Kotnik T, Miklavčič D, Slivnik T. 1998. Time course of transmembrane voltage induced by time-varying electric fields—A method for theoretical analysis and its application. *Bioelectrochem Bioenerg* 45: 3–16.

Ling G, Gerard RW. 1949. The normal membrane potential of frog sartorius fibers. *J Cell Comp Physiol* 34: 383–396.

Loew LM. 1992. Voltage sensitive dyes: Measurement of membrane potentials induced by DC and AC electric fields. *Bioelectromagnetics* Suppl. 1: 179–189.

Lojewska Z, Farkas DL, Ehrenberg B, Loew LM. 1989. Analysis of the effect of medium and membrane conductance on the amplitude and kinetics of membrane potentials induced by externally applied electric fields. *Biophys J* 56: 121–128.

Miller CE, Henriquez CS. 1988. Three-dimensional finite element solution for biopotentials: Erythrocyte in an applied field. *IEEE Trans Biomed Eng* 35: 712–718.

Montana V, Farkas DL, Loew LM. 1989. Dual-wavelength ratiometric fluorescence measurements of membrane-potential. *Biochemistry* 28: 4536–4539.

Neher E, Sakmann B. 1976. Single-channel currents recorded from membrane of denervated frog muscle fibres. *Nature* 260: 779–802.

Neumann E, Kakorin S, Toensing K. 1999. Fundamentals of electroporative delivery of drugs and genes. *Bioelectrochem Bioenerg* 48: 3–16.

Pakhomov AG, Pakhomova ON. 2010. Nanopores: A distinct transmembrane passageway in electroporated cells. In: *Advanced Electroporation Techniques in Biology and Medicine*. Boca Raton, FL: Taylor & Francis.

Pauly H, Schwan HP. 1959. Über die Impedanz einer Suspension von kugelförmigen Teilchen mit einer Schale. *Z Naturforsch* 14B: 125–131.

Pavlin M, Pavšelj N, Miklavčič D. 2002. Dependence of induced transmembrane potential on cell density, arrangement, and cell position inside a cell system. *IEEE Trans Biomed Eng* 49: 605–612.

Pucihar G, Kotnik T, Valič B, Miklavčič D. 2006. Numerical determination of the transmembrane voltage induced on irregularly shaped cells. *Ann Biomed Eng* 34: 642–652.

Pucihar G, Kotnik T, Teissié J, Miklavčič D. 2007. Electroporation of dense cell suspensions. *Eur Biophys J* 36: 173–185.

Pucihar G, Miklavčič D, Kotnik T. 2009a. A time-dependent numerical model of transmembrane voltage inducement and electroporation of irregularly shaped cells. *IEEE Trans Biomed Eng* 56: 1491–1501.

Pucihar G, Kotnik T, Miklavčič D. 2009b. Measuring the induced membrane voltage with di-8-ANEPPS. *J Vis Exp* 33: 1659 (available as video at http://www.jove.com/index/details.stp?ID=1659).

Reddy JN. 2005. *An Introduction to the Finite Element Method*, 3rd edn. New York: McGraw-Hill.

Schoenbach KH, Beebe SJ, Buescher ES. 2001. Intracellular effect of ultrashort electrical pulses. *Bioelectromagnetics* 22: 440–448.

Sharma V, Tung L. 2004. Ionic currents involved in shock-induced nonlinear changes in transmembrane potential responses of single cardiac cells. *Pflugers Arch* 449: 248–256.

Susil R, Šemrov D, Miklavčič D. 1998. Electric field induced transmembrane potential depends on cell density and organization. *Electro Magnetobiol* 17: 391–399.

Teissié J, Eynard N, Gabriel B, Rols MP. 1999. Electropermeabilization of cell membranes. *Adv Drug Deliv Rev* 35: 3–19.

Tekle E, Oubrahim H, Dzekunov SM, Kolb JM, Schoenbach KH, Chock PB. 2005. Selective field effects on intracellular vacuoles and vesicle membranes with nanosecond electric pulses. *Biophys J* 89: 274–284.

Valič B, Golzio M, Pavlin M, Schatz A, Faurie C, Gabriel B, Teissié J, Rols MP, Miklavčič D. 2003. Effect of electric field induced transmembrane potential on spheroidal cells: Theory and experiment. *Eur Biophys J* 32: 519–528.

Zhang J, Davidson RM, Wei MD, Loew LM. 1998. Membrane electric properties by combined patch clamp and fluorescence ratio imaging in single neurons. *Biophys J* 74: 48–53.

4

Electroporation: A Review of Basic Problems in Theory and Experiment

Marko S. Markov

4.1 Introduction

It has long been accepted that biological membranes play a critical role in the functioning of all living organisms. These soft condensed matter structures envelope the cells and their inner organelles, maintaining relevant concentration gradients by acting as selective filters toward ion and molecule transport. In general, the biological membrane is not only a separator of the cell interior from the surrounding media, but also a "transporter" of material, energy, and information. Therefore, the biological membrane represents a powerful amplifier in the signal transduction. Besides their passive role, cell membranes also host a number of metabolic and biosynthetic activities related to their important physiologic functions.

The interaction of electric fields with biological membranes and pure phospholipid bilayers has been extensively studied in recent decades. It has been shown, both theoretically and experimentally, that a strong external electric field can destabilize membranes causing charge redistribution over the membrane bilayer and inducing changes in their structure. The key parameter is the induced-transmembrane voltage generated by an external electric field due to the difference in the electric properties of the membrane and the external medium. The observed effects could be reversible or irreversible depending on the parameters of the applied electric field and the specificity of the targeted membrane.

The reversible "electrical breakdown" of the membrane was first reported by Stampfli (1958), but for some time this report remained unnoticed. Nearly a decade later, Sale and Hamilton (1967) reported the nonthermal electrical destruction of microorganisms under strong electric pulses. In 1972, Neumann and Rosenheck showed that electric pulses induce a large increase of membrane permeability in vesicles (Neumann and Rosenheck, 1972). Following these pioneering studies, three important works have motivated a series of further investigations. First, Neumann and coworkers showed in 1982 that genes could be transferred into the cells by using exponentially decaying electric pulses. Later, in 1987, Okino and Mohri and in 1988 Mir and coworkers showed that certain cytotoxic molecules could be introduced into the cells by using electric pulses in either *in vivo* or *in vitro* conditions (Okino and Mohri, 1987; Mir et al., 1988).

Electroporation has been observed and studied in many different systems, such as artificial planar lipid bilayers, lipid vesicles, cells (*in vitro* and *in vivo*), as well as tissues. Cell membranes are obviously much more complicated than artificial lipid structures (bilayers or vesicles) with respect to geometry, composition, and the presence of active processes. The problems associated with the complexity of natural cell membranes can be avoided by investigating synthetic liposomes or vesicles that mimic the geometry and the size of cells and cell membranes, but do not possess ion channels and multitudes of other embedded components, mainly proteins. The artificial planar lipid bilayer is the simplest model of the lipid system that also has the geometric advantage of providing electrical and chemical access to both sides of a membrane (see also Kramar et al., 2009).

Electric fields/pulses with selected parameters cause a rearrangement of the membrane lipids to form aqueous pores, which increases the conductivity of the membrane and its permeability to water-soluble molecules otherwise deprived of membrane transport mechanisms. The advantage of using lipid bilayers to study the effects of electric fields also includes the homogeneity of lipid structures from a chemical and electrical point of view.

By applying an electric field of adequate strength and duration, the membrane can undergo some temporary perturbations and return to its normal state when the exposure to the electric field ends, making the electroporation reversible. However, if the exposure to the electric field is too long or the strength of the electric field is too high, the membrane does not reseal after the end of the exposure and electroporation becomes irreversible. Depending on the type of electroporation (i.e., reversible or irreversible), two groups of applications exist: functional, where the functionality of the cells or tissues must be sustained; and destructive, where the electric fields are used to destroy the plasma membranes of cells or microorganisms (Miklavcic and Puc, 2006). The irreversible electroporation can be used for nonthermal food and water preservation, mainly causing permanent destruction of microorganisms (Haas and Aturaliye, 1999; Rowan et al., 2000; Teissie et al., 2002; Lebovka and Vorobiev, 2010).

It was also shown that by electroporation small and/or large molecules can be introduced into cells or extracted from cells, proteins can be inserted into the membrane, and cells can be fused. As a result of its efficiency, electroporation has rapidly found applications in many fields of biochemistry, molecular biology, and medicine. Probably, the most important functional application of electroporation is that it creates conditions for the transport of small or large molecules to the cytoplasm through the cellular membrane (Mir, 2001; Gehl, 2003; Escoffre et al., 2010; Heller and Heller, 2010). Electrochemotherapy became a plausible therapeutic approach in cancer treatment modalities that combines chemotherapy and electroporation (Rossi and Campana, 2010; Sersa et al., 2010). The application of electric pulses at the time a chemotherapeutic drug reaches its highest extracellular concentration considerably increases the transport through the membrane toward the intracellular targets and the cytotoxicity of a drug is enhanced (Mir et al., 1995; Sersa et al., 1998; Heller et al., 1999; Heller and Heller, 2010, Jarm et al., 2010). The application of electroporation for the transfer of DNA molecules into the cell is referred to as gene electrotransfection (Gehl, 2003; Escoffre et al., 2010).

Under certain experimental conditions, a delivery of electric pulses can lead to the fusion of membranes of adjacent cells. Electrofusion has been observed between suspended cells (Zimmermann, 1982; Abidor and Sowers, 1992; Sowers, 1993) and even between cells in tissue (Mekid and Mir, 2000). For successful electrofusion in suspension, the cells must previously be brought into close contact, for example, by dielectrophoresis (Abidor and Sowers, 1992). Electrofusion has proved to be a successful approach in the production of vaccines (Scott-Taylor et al., 2000) and antibodies (Schmidt et al., 2001).

The molecular mechanisms of the electropermeabilization of membranes are still not fully understood and there are often contradictions between experimental data and theoretical descriptions of pore formation (Zimmermann, 1982; Neumann et al., 1989; Rols and Teissie, 1990; Tsong, 1991; Weaver and Chizmadzhev, 1996; Esser et al., 2010; Kotnik and Pucihar, 2010). It was shown that pore formation and the effectiveness of cell electroporation depend on the parameters of electric pulses like number, duration, frequency, and electric field strength. It was also shown that neither electrical energy nor charge of the electric pulses alone determine the extent of electroporation consequences and that the dependency on voltage, duration, and number of pulses is more complex (Maček-Lebar and Miklavčič, 2001; Miklavcic and Kotnik, 2004).

4.2 Electroporation of Planar Lipid Bilayers

The planar lipid bilayer is often considered to be an approximation of a small fraction of the total cell membrane. The simple geometry and homogeneous composition of the lipid bilayers have often been used to investigate the basic aspects of electroporation especially because of its geometric advantage allowing chemical and electrical access to both sides of a membrane.

The contribution of factors, such as lipid composition, organic solvent, temperature, and electrolyte composition, on the electroporation has been intensively studied.

The incorporation of nonphospholipid substances into planar lipid bilayer changes the intensity and duration of the electrical stimulus needed for electroporation to occur. Such an effect depends on the surfactant molecular shape altering the spontaneous charge distribution (in the case of the planar bilayer) or the membrane curvature, which is especially important during the defects formation process. In the 1980s, a plausible theory of electroporation related to the flexoelectrical properties of membranes was proposed (Petrov 1984, 2002; Petrov and Bivas 1984).

4.3 Theoretical Models

A number of theoretical models have been proposed for the explanation of electroporation. These models attempted to address the following issues:

- To create a physically realistic picture of both the nonpermeabilized and the permeabilized membrane
- To investigate the dependence of the electropermeabilization yield and rate on the pulse parameters
- To establish a realistic value of the minimal transmembrane voltage at which permeabilization occurs
- To define the conditions at which electroporation becomes irreversible

Due to space constrains, I will only briefly mark the theoretical models discussed in the literature.

4.3.1 Hydrodynamic Model

The hydrodynamic model, developed in the early 1970s (Michael and O'Neill, 1970; Taylor and Michael, 1973), describes the membrane as a charged bilayer of a nonconductive and noncompressible material separating two conductive liquids. Above a threshold voltage, an instability occurs in the membrane; the compressive pressure prevails, causing a breakdown of the membrane.

There are two basic problems with the hydrodynamic model. First, it assumes that liquids have isotropic fluidity, which is not true for lipid bilayers, where transverse movement of molecules is very restricted. Second, it does not describe the permeabilized membrane.

4.3.2 Elastic Model

In contrast with the hydrodynamic model, which assumes a membrane with a constant volume and a variable surface, the elastic model (Crowley, 1973) assumes a variable volume and a constant surface. In the elastic model, the pressure exerted by the transmembrane voltage leads to a decrease in the volume of the membrane and an increase of elastic pressure opposing the compression.

4.3.3 Hydroelastic Model

The assumptions on which the hydrodynamic and the elastic models are built are mutually excluding and both are unrealistic. Similar to the two models described previously, the hydroelastic model predicts a compressive instability. Therefore, the assumption of constant value of the elasticity module is much more reasonable than in the elastic model, in which the membrane breakdown is associated with a volume reduction of almost 40%.

While the description of the compressive instability provided by the hydroelastic model is more realistic than those of the hydrodynamic and the elastic models, it has the same general drawback—it gives a nonrealistic picture of the electropermeabilized membrane: a uniform, infinitely thin layer, rippled at some locations (Pavlin et al., 2008).

4.3.4 Viscohydroelastic Model

The viscohydroelastic model expands the hydroelastic model by adding to it the membrane viscosity (Maldarelli et al., 1980; Steinchen et al., 1982). This model considers the charged membrane as both compressed and rippled. However, the compression is not instantaneous, but follows the onset of the transmembrane voltage gradually. Up to a certain voltage, the membrane is still in equilibrium, but at higher voltages, the integrity of the membrane cannot be sustained. The analysis of such instability is elaborate and is described in more detail elsewhere (Dimitrov and Jain, 1984).

The viscohydroelastic model offers a principal advantage over all of the previous models, as the requirement of an above-critical voltage is linked to that of above-critical duration, providing a possible explanation of the dependence of permeabilization on the duration of electric pulses.

4.3.5 Phase Transition Model

The above-mentioned models consider the electropermeabilization as a modification of the supramolecular membrane structure. In contrast, the phase transition model describes this phenomenon as conformational changes of the membrane proteins (Sugar, 1979). On the molecular scale, the pressure is replaced by molecular energies, and the pressure equilibrium corresponds to the state of minimum free energy. With several minima of the free energy, several stable states are possible, each corresponding to a distinct phase. In lipid bilayers, there are in general two such phases: solid (gel) and liquid. Regrettably, the model contains several parameters with unknown values and only by adapting these values does one get an arbitrarily agreement with experimental data.

To estimate a realistic value of the critical transmembrane voltage, the parameters of the model (which currently are arbitrarily chosen) need to be determined experimentally. Still, the phase transition model meets several requirements of which all the previous models fail. The permeabilized state is the minimum of free energy, and a return to the nonpermeable state requires a sufficient amount of energy, which offers a possible explanation of the observed durability of the permeabilized state.

4.3.6 Domain-Interface Breakdown Model

The domain-interface breakdown model takes into account the fact that cell membranes consist of distinct lipid domains that differ in their structure, and embedded proteins, involved in a complex interaction and charge allocation with the lipid environment. According to this model, electroperme-abilization occurs at the boundaries between the domains. Similar to the viscohydroelastic model, this model considers the increased permeability as a result of fractures, with the difference that in the former model they occur along the ripples, while in the latter they form along the domain interfaces. In addition, while the model describes permeabilization as localized to the domain interfaces, the phenomenon is also observed experimentally in bilayers and vesicles with homogeneous lipid structures. Thus, the domain-interface breakdown can only serve as an additional mechanism, eventually enhancing permeabilization in cell membranes with respect to that in artificial bilayers and vesicles.

4.3.7 Aqueous Pore Formation Model

The hydrodynamic, the elastic, the viscoelastic, and the viscohydroelastic models consider electroporation as a large scale phenomenon, with no direct role attributed to the molecular structure of the membrane. The phase transition model and the domain-interface breakdown models represent the other extreme, attempting to explain the electroporation by the properties of individual lipid molecules and the interactions between them.

The aqueous pore formation offers a compromise between these two approaches considering the electropermeabilization as a result of the formation of transient aqueous pores in the lipid bilayer. Each pore is formed by a large number of lipid molecules, while the shape, size, and stability of the pore are strongly influenced by the nature of these molecules and their local electrochemical interactions. As pointed out by Neumann and Kakorin (2010), an applied electric field is capable of affecting the hydrated polar head groups of lipids, which leads to the formation of hydrophilic pores. The field-induced translational motions of the polar lipids in the curved pore wall also rationalize the huge acceleration of lipid flip-flop and the eventual redistribution of the lipids and charges from the internal membrane monolayer to the outer monolayer and vice versa.

The possibility of spontaneous pore formation in lipid bilayers was first analyzed in 1975, independently by two groups (Litster, 1975; Taupin et al., 1975). If a larger pore is artificially created (e.g., by mechanical pressure), the energy barrier is overcome, and since no stable state exists at larger pore radii, the membrane breaks down. In the presence of a transmembrane voltage, pore formation affects the capacitive energy of the system, while the transmembrane voltage reduces both the critical radius of the pore and the energy barrier.

The hydrophilic structure of the pore is reached through a transition from an initial, hydrophobic state in which the lipids still have their original orientation. The expressions do not deal with this transition and at all radii they treat the pore as fully formed, i.e., hydrophilic.

One deficiency of the model is the lack of consideration of the hydration of proteins and lipid polar headgroups as well as the hydrated ions that are subject of transport through the pores. It should be taken into account that the lipids adjacent to the aqueous molecules inside of the pore are reoriented in a manner that their hydrophilic heads are facing the pore, while their hydrophobic tails are hidden inside the membrane. Some authors argued that such a reorientation requires an input of energy, which is larger for smaller pores, correspondingly increasing the free energy of the membrane at small pore radii.

Today, the aqueous pore formation model is considered to be the most convincing explanation of electroporation. Further research, especially measurements of the relevant physical quantities, should gradually result in the improvement of the current aqueous pore model. It is also reasonable to expect that insights obtained by molecular dynamics (MD) simulations, and perhaps by an advanced method of visualization or another type of detection, should also yield a clearer picture of the electropermeabilized membrane on a nanometer scale, thereby providing the final verdict on the validity of the concept of electroporation (Neumann and Kakorin, 2010).

4.4 Molecular Dynamics for Modeling the Interaction of Pulsed Electric Fields with Cells

One useful approach to understanding the pore formation was offered by MD simulation. Modeling of pulsed electric field effects on the charge distribution in biological cells ranges from simple analytical models to those generated using MD. MD simulations provide the most basic and fundamental approaches for modeling the effect of electric fields on cells. Considering the cell membrane as a lipid bilayer, the interacting particles are characterized by charged polar heads of the lipids. Thus, the structural details on the nanoscale can be included into the simulation for maximum accuracy and relevant physics. This method has only recently been used to model cellular membranes under the electrical stress (Tieleman et al., 2003; Hu et al., 2005a,b; Vernier et al., 2006; Tarek and Delemotte, 2010; Vernier, 2010).

The numerical modeling consistently predicts that the pores have a distribution of radii in the nanometer range. At longer times and upon higher repetitive pulsing, the pore distribution is seen to broaden due to diffusion. The total area of pores created dynamically by the electric pulsing can also be computed at any given membrane by the distributed circuit approach. As might be expected, the total area of pores across the entire surface of the plasma membrane increases monotonically with the number of pulses.

A saturating effect is predicted as the interplay between the increasing local conductivity of a porated membrane and the associated transmembrane potential that drives the growth of the pores. With an increase in the number of pulses, the pore density grows, leading to larger increases in membrane conductivity. This in turn reduces the transmembrane potentials. Since the pore formation and growth rates are dependent on the transmembrane potential, the reduction effectively leads to progressive decreases in new pore formation. Physically, any cell membrane can only support a finite amount of pores. Beyond the critical area, the electroporation becomes irreversible.

MD appears to be the most plausible approach to understanding spatial and temporal electric field–membrane interactions on a nanoscale. MD is a time-dependent kinetic scheme that follows the trajectories of N-interaction bodies subject to chosen external fields. It is a microscopic approach that specifically treats every atom within the chosen simulation region. MD relies on the application of classical Newtonian mechanics for the dynamical movement of ions and neutral atoms, taking into account the multiple interactions within a realistic molecular representation of the biosystem. The results of MD calculations shed light on the physics of such interactions and provide information on critical electric fields for field–membrane interactions. Thus, for example, a segment of the lipid bilayer membrane or a channel protein is first constructed taking into account the initial geometric arrangement of all the atoms and their bonding angles. Regions of water containing user-specified ion densities are then defined on either side of the membrane to form the total simulation space.

4.5 Some Considerations in Regards to Electroporation of Cellular Organelles

In a recent review, Schoenbach et al. (2007) discussed a simple analytical, passive, and linear model of the cell that does not consider the changes in the properties of the cell structures. The assumption holds only for electric field amplitudes below those required for electroporation or nanoporation. The model includes the cell nucleus as the only subcellular structure that has membrane properties. However, it is known that the cytoplasm contains dissolved proteins, electrolytes, and forms of glucose and is moderately conductive, as are the nucleoplasm and the contents of other organelles. On the other hand, the cellular membrane has a low conductivity and therefore the cell can be considered as a conductor surrounded by a lossy, insulating envelope containing substructures with similar properties. The dielectric constants and conductivities of cell membranes and cytoplasm, as well as nuclear membranes and nucleoplasm, have been obtained using a dielectric spectroscopy of cells (Ermolina et al., 2001; Feldman et al., 2003).

The assumptions used in most models for membrane charging and electroporation as well as intracellular electromanipulation are that the membranes are perfect insulators and that the permittivity of the liquids in and outside the cell can always be neglected. It is limited, however, to times that are short relative to the dielectric relaxation time of the membrane and to times that are long relative to the dielectric relaxation time of the cytoplasm. The electric field in the membrane at the poles of the cell can be estimated from the condition that the current density at the interface of the cytoplasm and membrane is continuous. The electric fields in the various parts of the cell are then defined by the continuity of the electric flux density (Hu et al., 2005a). In this analytic approach, the substructures of the cell are neglected.

Taking into account cellular organelles, the modeling of cells would require more complicated numerical methods. As a first step towards a more sophisticated analytical model, it would be useful to introduce a two-shell model: a cell that contains only one substructure, e.g., the nucleus. In this case, it is assumed that the plasma membrane and the membrane of the organelle (e.g., the nucleus) are perfect insulators and that the cytoplasm and the nucleoplasm can be fully described as resistive media. The equivalent circuit provides two time constants: the charging time constant for the plasma membrane and one for the substructure membrane, which is almost always less than that of the plasma membrane.

The charging time constant of the plasma membrane defines the range of intracellular electromanipulation, a research area that becomes the focus of increasing experimental studies. The concept of intracellular electromanipulation is easy to understand. During the charging time of the outer membrane, the potential differences generated across subcellular membranes strongly depend on the pulse duration: the effect will be stronger with a shorter pulse rise time (Schoenbach et al., 2001).

The subcellular effect will diminish for long rise times compared with the characteristic charging time of the membranes. Such charging times are in the 100 ns range for human cells and are longer for tissue. If direct current electric fields or pulses of long duration (compared with the charging time of the capacitor formed by the outer membrane) are applied, eventually only the outer membrane will be charged; the electric field generated across subcellular membranes during the charging will be zero for an ideal, fully insulating outer membrane.

The membrane then becomes an "active" cell element with variable resistivity and, to a lesser extent, variable permeability. The modeling of cells with "active" membranes has been discussed within a lattice model (Stewart et al., 2006) and a coupled Smoluchowski equation for pore development. Joshi et al. (2004) used a spatially distributed, time-dependent cell model.

It is important to consider the fluctuations in the "effective" radius in response to water molecules and other species constantly entering and leaving the pores (Markov, 1991).

The transmembrane potential initially is the highest at the poles. However, the value at longer times is guided by the interplay between the external electric field, the poration event, and the resulting localized conductivity modulation at the membrane. For instance, a large external electric field value would lead to the creation of a high pore density at the poles. This, in turn, would work to increase the conductivity and reduce the local voltage drop across the porated membrane. The poles are predicted to have the largest transmembrane potential at early times.

In the late 1980s, Markov and coworkers (Markov, 1988; Goltzev et al., 1989; Markov and Guttler, 1989) demonstrated that the electroporation of erythrocytes and chloroplasts membranes occurred at approximately the same signal parameters, mainly at the polar area of these elongated cells. Surprisingly, it was found that the long axis of these two cells is exactly equal, which could be a reason for the observed similarity. These studies bridge the observations that the electroporation of the plasma membrane (in the case of erythrocytes) and organelle membrane (in the case of chloroplasts) occurs at polar regions at approximately similar field parameters.

4.6 Experimental Observations of Cell Electroporation

The function of the plasma membrane is not only to protect the cell interior, but also to facilitate the flow of selected types of ions and molecules from and to its surroundings. At the same time, the plasma membrane is a powerful amplifier for signal transduction processes. Having in mind the fact that the membrane possesses electrical charges, any external electric field could cause charging of the cell membrane and/or charge redistribution over the membrane surface. During this charging, membrane lipids could be perturbed and membrane proteins could undergo conformational changes. When the potential difference above a certain threshold is sustained long enough, pores will be formed and will allow ion and molecular transport. The poration is therefore related to an increase in the membrane conductance, and consequently, to a change in transmembrane voltage and current.

Still the exact molecular mechanisms of the formation, structure, and stability of these permeable structures are not completely understood. Theoretical descriptions that were developed for lipid bilayers do not include proteins and cell structures such as the cytoskeleton and organelles. At the same time, the increased permeability after the pulses, which enables the delivery of molecules (drugs, DNA), is crucial for the application of cell electroporation in biotechnology and in medicine.

4.6.1 Induced Transmembrane Voltage and Forces on the Cell Membrane

Even though the exact physical–chemical mechanisms of cell electroporation are not clear and several theoretical models exist, it is generally accepted that one of the key parameters for successful membrane permeabilization is the induced transmembrane voltage. This voltage is generated by an external electric field due to the difference in the electric properties of the membrane and the external medium. If the induced transmembrane voltage is large enough, and above the critical value, electroporation could be observed (Neumann and Rosenheck, 1982; Neumann et al., 1982, 1989; Zimmermann, 1982; Weaver and Chizmadzhev, 1996; Jaroszeski et al., 1999).

Most of the authors agree that the induced transmembrane voltage is superimposed on the resting transmembrane voltage, resulting in an asymmetric transmembrane voltage (Zimmermann, 1982, 1996; Tekle et al., 1990; Teissie and Rols, 1993). For a typical biological cell, the time constant is directly related to the beta-dispersion range in the impedance spectra and is around a microsecond, which represents the time needed for the induced transmembrane voltage to build up on the cell membrane and for electroporation to occur. It was shown that the transport of ions and molecules during electroporation occurs only through the permeabilized area (Schwister and Deuticke, 1985; Hibino et al., 1991; Gabriel and Teissie, 1998, 1999).

For times shorter than a microsecond, the cell interior is also exposed to an electric field, resulting in the induced transmembrane voltage across the membrane of cell organelles. Thus, for very short high-voltage pulses, cell organelles can be permeabilized (Schoenbach et al., 2001, 2007; Kotnik and Miklavcic, 2006).

In the case of dense cell systems, such as dense suspensions and tissues with high cell volume fraction, the local electric field due to the effect of neighboring cells is lower than the applied electric field (Susil et al., 1998; Pavlin et al., 2002).

4.6.2 *In Vitro* Cell Electroporation

Different studies analyzed cell electroporation *in vitro* on single cells, attached cells, cells in suspensions, or multicell layers. A common observation is that the electroporation is a threshold phenomenon that is governed by pulse parameters (duration, number, and repetition frequency). Sometimes the critical voltage above which the transport is observed is defined as the reversible threshold since the changes are reversible and the cell membrane reseals after a given time lasting from minutes to hours. The value of the critical (threshold) transmembrane voltage at room temperature was reported to be between 0.2 and 1 V (Zimmermann, 1982; Teissie and Rols, 1993) depending on the pulse parameters and experimental conditions (Miklavcic et al., 2000).

4.6.3 Experimental Studies of Molecular Transport

The transport, which governs the uptake of ions and the leakage of cytoplasm contents, depends on both the status of test cells and pulse parameters. The extensive studies of Teissie, Rols, and colleagues (Rols and Teissie 1989, 1992, 1993, 1998) examined the effect of electric field strength, number of pulses, and pulse duration on the extent of the permeabilization uptake of exogenous molecules, cell survival, release of intracellular ATP, and resealing. The authors define the electropermeabilization threshold below, in which no transport is observed for the given pulse parameters. They also define threshold (Rols and Teissie, 1990) as the field strength value below which no permeabilization occurs no matter how long the pulses are or how many are used. Two other extensive studies of electroporation *in vitro* were reported in which the authors studied the uptake and viability for different electric field strength, duration, number of pulses, and also different cell volume fractions (Canatella et al., 2001; Maček-Lebar and Miklavčič, 2001).

4.6.4 Dependence of Calcium Release from Nanoporation Parameters

Various effects have been observed dependent on pulse duration, pulse amplitude, and the number of pulses in a pulse train. When the pulse amplitude was lowered, calcium release from intracellular stores and subsequent calcium influx through store-operated channels in the plasma membrane was observed (Buescher and Schoenbach, 2003; Vernier et al., 2003a,b; Beebe et al., 2004; White et al., 2004). Calcium is known as a ubiquitous second messenger molecule that regulates a number of responses in cell signaling, including enzyme activation, gene transcription, neurotransmitter release, secretion, muscle contraction, and apoptosis, among others. One known role for nanosecond pulsed electric fields (nsPEFs)-induced calcium mobilization is platelet activation, which is important for hemostasis and wound healing. It is known that the intracellular calcium is stored in the endoplasmic reticulum and mitochondria, and calcium release from the mitochondria is considered an initiation event of apoptosis (Susin et al., 1998).

It is possible that an applied electric field causes direct supraelectroporation effects on the endoplasmic reticulum, or other intracellular storage sites releasing intracellular calcium. Likewise, a longer pulse duration, higher electric fields, and/or more pulses will have proportionally greater effects on the plasma membrane, which can result in an influx of calcium through pores or aqueous channels.

The viability of human lymphoma cells (U937) after applying 10 ns long pulses has been studied by Pakhomov et al. (2004). Trypan blue was added 24 h after pulsing to ensure that cell recovery, which occurs on a much faster timescale, is not an issue. The survival curves closely resemble well-known types of dose responses for bioeffects of ionizing radiation and can be conveniently described in the same terms. The results showed that cell survival could be described as a dose effect: the viability was determined by the square of the electric field intensity, times the number of pulses. Particularly interesting is the weak response to a low number of pulses followed by an exponential decline with a further increase in pulse number. The same type of survival curve is also typical for sparsely ionizing radiations (x-rays, γ-rays, electrons), which kill cells primarily by the generation of free radicals that damage biological macromolecules, e.g., DNA. Although pulsed electric fields are to be considered nonionizing radiation, one cannot exclude that, at extreme voltages, they might be able to cause water ionization and/or dissociation, leading to free radical formation and cell death. This hypothesis gained strong support in a recent study that established the reduction of cell killing by nsPEFs in oxygen-deprived cells (Walker et al., 2006), similar to the oxygen radiomodification effect.

It was shown that any increase in membrane voltage could be expressed in terms of the product of electric field intensity and pulse duration. Accordingly, the increase in the transmembrane potential of organelles is dependent on the product of pulse duration and an electric field. Based on the hypothesis that any nanosecond effect requires an increase in membrane voltage to a critical level (e.g., for poration of the outer membrane this increase would be on the order of 1 V) consequently leads to a second similarity law for nanosecond pulse effects. That means that identical effects can be expected when the applied electric field (E), the pulse duration (τ), and the number of pulses (n) are varied in such a way that the product of ($E^2\tau^2n$) stays constant (Pakhomov et al., 2007).

The transient increase of membrane conductivity indicates the occurrence of short-lived membrane structures that enable ion permeation. In the aqueous pores formation model of electroporation, this corresponds to conductive hydrophilic pores (Weaver and Chizmadzhev, 1996). The membrane conductivity drops to the initial level in a range of a millisecond after the pulses. This could be explained only with the existence of many small pores, transient during the electric pulses, which close very rapidly (milliseconds) after the pulse. The number of these short-lived pores does not depend on the number of applied pulses but solely on the electric field strength and pulse duration.

The state of increased permeability can last tens of minutes after pulse application. Therefore, it is clear that in contrast to transient pores that reseal in milliseconds some pores are stabilized enabling transport across the membrane for minutes after pulses. In contrast to transient short-lived pores, these stable pores are governed both by electric field strength as well as the number of pulses. The fraction of long-lived stable pores increases with the electric field due to a larger area of the cell membrane being exposed to above critical voltage and due to higher energy available for pores formation. Moreover, each pulse further increases the probability for the formation of the stable pores.

The resealing of the cell membrane strongly depends on the temperature and lasts from minutes to hours after pulse application. This clearly shows that long-lived pores are thermodynamically stable. It appears that the nature of long-lived "transport" pores is different from that of transient pores, which are present only during the pulses.

Hydrophilic pores are not stable after pulse application; therefore, some additional process must be involved in the formation of stable pores. There are inter-relations between structural changes, conductivity changes, and permeabilization (increased transport of molecules). The nature of long-lived pores and the relation between short-lived and long-lived structural changes is still not completely understood.

Sugar and Neumann (1984) suggested that larger pores are formed by coalescence of smaller pores/defects, which travel within the membrane. Some authors suggested that these pores/defects migrate along the membrane surface and are grouped around inclusions.

There is general agreement that proteins are involved in the stabilization of larger pores (Glaser et al., 1988; Weaver and Chizmadzhev, 1996). These authors speculated that the cytoskeleton structure could act similarly to the macroscopic aperture of planar membrane experiments leading to the rupture of limited portions of a cell membrane but not of the entire membrane. According to Rols and Teissie (1992), the disintegration of the cytoskeleton network could affect electroporation making some specific sites in the membrane more susceptible to pore formation.

4.7 Comparison of Electroporation of Planar Lipid Bilayers and Cells

It should be pointed out that most of theoretical models were developed for planar bilayer membranes, which differ from cell membranes where the membrane proteins and cytoskeleton are present. In addition, the membrane curvature should be considered. A number of studies have demonstrated that the structural changes probably occur in the lipid region of the cell membrane (Zimmermann, 1982; Chernomordik et al., 1987; Weaver and Chizmadzhev, 1996). In both, planar bilayer membranes and cell membranes, the authors obtained a gradual increase of conductivity in high electric fields. The greatest observed difference is that the reversible electroporation in cells is much more common than in planar bilayer membranes and that resealing of artificial bilayer membranes takes milliseconds whereas the resealing of cells can last for several minutes.

The long-lasting increased permeability of the cell membrane is crucial for biotechnological and biomedical applications. For a complete description of cell electroporation, the role of the curvature, colloid osmotic swelling, and especially cell structures, such as cytoskeleton, domains, and membrane proteins have to be discussed and examined.

The theoretical model of aqueous pores describes the experimental observations in lipid bilayers relatively well: discussing critical transmembrane voltage and stochastic nature of the process. However, there is still no theoretical description that completely describes all observable phenomena during cell electroporation and the underlying physical mechanism: the occurrence of structural changes in the membrane during the electric pulse, the stochastic nature of electroporation, the observed dependence of molecular uptake on pulse duration and the number of pulses, field strength, repetition frequency, and the stability of "pores" after pulsation as well as the resealing dynamics.

Weaver and Chizmadzhev (1996) suggested that some additional processes/structures have to be considered to obtain a realistic theoretical model of stable pores capable of explaining the long-lived permeability of the cell membrane after electroporation. The more realistic theoretical description should incorporate membrane proteins and the cytoskeleton as crucial factors that enable pore stabilization and prevent the breakdown of the cell membrane, thus enabling the most important applications of electroporation: electrogene transfer and electrochemotherapy.

The next approximation of lipid bilayers is seen in lipid made vesicles. The vesicle behavior under electric pulses is discussed in detail by Dimova (2010).

4.8 Membrane Fluidity

It has been shown that short, high intensity electrical pulses are capable of causing reversible cell membrane permeabilization. The efficiency of electroporation strongly depends on the membrane composition and its fluidity. The membrane fluidity is a function of the temperature of the medium. Kanduser et al. (2008) reported that the chilling of cell suspension from physiological temperature (of 37°C) to 4°C has a significant effect on the cell membrane electropermeabilization, leading to a lower fraction of permeabilized cell membrane. At the same time, with the decreasing temperature, the cell membranes become less fluid with higher order parameters in all three types of domains and higher proportion of domain with highest order parameter. Since the temperature affects both cell membrane fluidity and cell membrane electropermeabilization, it makes sense to search for a potential correlation between both effects. It was determined that a temperature decrease from the physiological temperature of 37°C to 4°C decreases the membrane fluidity and the membrane domain structure.

Some studies were focused on the effect of chilling on the electropermeabilization of erythrocytes, algae, and porcine skin (Coster and Zimmermann, 1975; Kinosita and Tsong, 1979; Zimmermann, 1982; Gallo et al., 2002). It has been shown that the chilling of the stratum corneum had a significant effect on electropermeabilization. Decreasing the temperature from 25°C to 4°C required higher voltages for electropermeabilization, suggesting that higher temperatures facilitated the formation of electro pores (Gallo et al., 2002). The authors further proposed that lipid fluidity, which is controlled by the temperature, is the major factor affecting the electropermeabilization of the stratum corneum.

More detailed information about the membrane domain alterations was obtained by Kanduser et al. (2008) using a computer simulation of the EPR spectra. This simulation shows that the reduced membrane fluidity, due to the chilling of cells to 4°C, reflected in the higher order parameters in all the domain types that compose cell membrane and higher proportion of domain types with the highest order parameter when compared with cell suspension maintained at 37°C.

Several studies on lecithin cholesterol model membranes indicate that not only average membrane fluidity but rather membrane domain structure is important for the membrane permeabilization (Raffy and Teissie, 1997; Koronkiewicz and Kalinowski, 2004; Kanduser et al., 2008).

The temperature effect on electropermeabilization could be explained by thermal effects during pore formation (Weaver and Chizmadzhev, 1996; Kotnik and Miklavčič 2000; Pavlin et al., 2008). The cell membrane fluidity and domain structure eventually play a significant role in that process. According to the electroporation theory, hydrophobic pores in the cell membrane are formed spontaneously by lateral thermal fluctuations of the lipid molecules.

The structural rearrangements of hydrophobic to hydrophilic pores during electroporation occur when the radius of the hydrophobic pore exceeds critical value and when the reorientation of the lipid molecules becomes energetically favorable (Weaver and Chizmadzhev, 1996). The energy needed for reorientation of lipid molecules to form hydrophilic pores is expected to be lower in more fluid cell membranes or membrane domains with lower order parameters.

It needs to be taken into account, however, that temperature affects the physiology and metabolism of the cell. Therefore, the optimal temperature for electropermeabilization is the physiological temperature and it is not surprising that chilling significantly reduces electropermeabilization effectiveness.

4.9 Short-Lived versus Long-Lived Pores

Several experimental studies showed a complex dependence of the membrane transport on pulse parameters. In only a few studies, however, the actual transport across the membrane was quantified. Theoretical studies describe pore formation in artificial lipid membranes but still cannot explain the mechanisms of formation and properties of long-lived pores formed during cell electroporation. By analyzing a transient increase in conductivity during the pulses in parallel with ion efflux after the pulses, the relation between short-lived and long-lived pores could be investigated.

Pavlin and Miklavčič (2008) proposed a simple model that interprets an increase of the fraction of long-lived pores under higher electric fields due to larger areas of the cell membrane being exposed to above-critical voltage and due to higher energy being available for pore formation. It was also shown that each consecutive pulse increases the probability for the formation of long-lived pores.

It is well accepted that the cell membrane has low permeability for most molecules and ions due to the hydrophobic nature of the lipid bilayer. Transport through the membrane occurs only for certain molecules and ions through membrane channels by means of diffusion or by active transport. This becomes a very important issue because recent developments in contemporary medicine slowly move into delivering medication through the cell membrane. This plausible method requires an increased permeability of the membrane in order to introduce certain drugs into the cell.

The principle question is: what is the amplitude of the applied voltage or duration and frequency of pulses in order to distinguish the electric field rupture of the membrane from the reversible pore formation? It has long been known that the cell membrane becomes permeable for transmembrane voltages above a certain threshold, which was estimated to be between 200 mV and 1 V (Zimmermann, 1982; Tsong, 1991; Teissie and Rols, 1993).

After the pulse application and pore opening, the process of resealing takes several minutes allowing for the transport of molecules and ions across the cell membrane. The electroporation has been proven to be reversible only for a given range of parameters of electric pulses. When an electric field is increased above the threshold or for longer pulses or for a larger number of pulses, the changes in the cell structure become irreversible finally leading to the cell death.

Although there is a lot of indirect experimental evidence of the increased membrane permeability, these observations alone do not explain the underlying physical–chemical mechanisms that cause structural changes of the membrane and finally lead to its increased permeability long after the electric pulses are switched off (Weaver and Chizmadzhev, 1996; Teissie et al., 2002).

As of today, none of the existing experimental methods enable the direct visualization of pores. MD studies showed the formation of aqueous pores in the lipid bilayer under very strong nanoseconds electric field pulses (Tieleman et al., 2003), which is in agreement with the experimental observation of the electroporation and bioelectric effects observed with nanoseconds pulses (Schoenbach et al., 2007). However, current MD simulations are limited to short nanosecond time–scales, while "classical" electroporation occurs in microsecond time–scales.

Pavlin and Miklavčič (2008) reported a study of the effect of cell electroporation on ion permeability during the electric pulses and ion diffusion after electric pulses on dense cell suspension of B16F1 mouse melanoma cells with the objective of understanding the relation between short-lived pores and long-lived pores in cell electroporation. Based on measured conductivity, ion diffusion, and molecular transport and using theoretical models, the authors quantified the fraction of short-lived and long-lived pores and further analyzed which parameters affected the stabilization of the long-lived pores. They also reported transient conductivity changes as well as the changes of conductivity between the pulses. These relatively slow changes in conductivity could be attributed to the ion efflux, which occurs between the pulses and after the pulsing was completed. The efflux of ions increases for higher electric fields as well as with a larger number of pulses. Since the increased ion diffusion is observed for several seconds after the pulses, it can be attributed to the existence of long-lived pores, which enable the diffusion of ions and molecules. The diffusion of ions is a slow process compared with the duration of the electric pulses, and thus we can assume that the major contribution to the efflux of ions occurs in the absence of the electric field.

The electric field dependent permeability is, in general, a function of the number and the length of pulses, which govern pore size and number. For a complete description, a model should provide a formal description on a molecular level as well as incorporate other parameters such as pulse length and temperature. A remarkably good coincidence of the theoretical and experimental data was demonstrated recently (Pavlin and Miklavčič, 2008).

The fraction of long-lived pores also increases approximately linearly with the number of pulses, which is in agreement with the measurements of molecular uptake (Rols and Teissie, 1990; Kotnik et al., 2000; Maček-Lebar and Miklavčič, 2001; Pavlin et al., 2002).

To determine the relation between short-lived and long-lived pores during cell electroporation, one has to analyze the quantities related to transient changes and the quantities related to the long-lived increased permeability of the cell membrane. It can be seen that transient conductivity changes and consequently the fraction of short-lived pores are almost identical for all pulses, whereas the fraction of long-lived pores and the percentage of permeabilized cells gradually increase with the number of pulses.

Since the relaxation of short-lived transient pores is fast, the membrane conductivity decreases to an initial level after each pulse, and when the next pulse is applied, the cell membrane again behaves as an insulator. Therefore, the value of the induced transmembrane voltage is the same for the second, third, and even eighth pulse (Pavlin and Miklavčič, 2008).

However, the nature of long-lived pores must be different than that of the transient pores. Pavlin and Miklavčič (2008) suggest that during each pulse a given fraction of transient pores is "stabilized," forming long-lived pores. However, since the fraction of these pores is much smaller compared with transient pores, it is difficult to detect them performing conductivity measurements. The quantity that approximately correlates with the molecular uptake is the conductivity change due to ion efflux and consequently the fraction of long-lived pores.

The increase in the fraction of long-lived pores with the number of pulses suggests that each individual pulse (independently of the previous pulse) adds to the formation/stabilization of new long-lived pores. Therefore, the ion diffusion can be used to determine the fraction of the long-lived pores and long-lasting permeability of the cell membrane. Furthermore, the transient changes are much larger than the fraction of long-lived pores, clearly indicating that both quantities have to be analyzed separately.

Pavlin and Miklavčič (2008) suggested that the number and/or size of short-lived pores is at least an order of magnitude larger compared with long-lived pores, which is in agreement with previous studies (Kinosita and Tsong, 1977, 1979; Schwister and Deuticke, 1985; Hibino et al., 1991, 1993; Schmeer et al., 2004; Pavlin et al., 2005). However, a short lifetime after pulses indicates that these pores are not stable in the absence of the electric field and that only a few of these pores remain present after the pulse. Possible explanations of pore stabilization could be a coalescence of smaller pores into larger ones

(Sugar and Neumann, 1984), stabilization due to membrane proteins and cytoskeleton (Rols and Teissie, 1992; Teissie and Rols, 1994; Weaver and Chizmadzhev, 1996; Fosnaric et al., 2003) or the formation of long-lived pores due to structural discontinuities at domain interfaces (Kotulska et al., 2007).

It has been shown experimentally that the membrane conductivity drops to the initial level in a range of a millisecond after pulses. This could be explained only by assuming the existence of many small pores transient during the electric pulses, which close very rapidly after the pulse. The number of these short-lived pores does not depend on the number of applied pulses but solely on the electric field strength and pulse duration. Therefore, by measuring transient conductivity changes during the electric pulses, one can detect the permeabilization threshold but not the level of permeabilization.

The state of increased permeability can last for several minutes after pulse application. Therefore, it is clear that in contrast to transient pores with fast resealing in milliseconds, some pores are stabilized enabling transport across the membrane in minutes after the pulses. The fraction of long-lived pores increases with higher electric fields due to a larger area of the cell membrane being exposed to above-critical voltage and due to higher energy, which is available for pores formation. Moreover, each consecutive pulse increases the probability for the formation of the long-lived pores (Pavlin and Miklavčič, 2008; Miklavcic and Towhidi, 2010).

4.10 Applications of Electroporation

The high interest of basic science in laboratory research (both theoretical and experimental) quickly opened the avenues for practical application in clinical medicine and biotechnology. Not surprisingly, the first applications were for the treatment of superficial tumors. There are seven chapters in this book, allocated in Part VI "Applications of Electroporation," that are directly or indirectly related to the treatment of cancer. One chapter (Lebovka and Vorobiev, 2010) is on electroporation for biotechnology purposes.

4.10.1 Skin Electropermeabilization

Having a large size and easy accessibility, skin is an attractive target tissue for the applications of electric fields for transdermal drug delivery and *in vivo* gene delivery.

The skin epidermis contains different layers, but the one that largely defines its electrical properties is the outermost layer, the stratum corneum. Although very thin (typically about 20 μm), it overwhelms the electrical properties of skin. Its protective function and low permeability can be temporarily breached by electroporation, creating aqueous pathways across otherwise nonpermeable lipid-based structures.

The four layers of skin, the stratum corneum, epidermis, dermis, and the underlying layer of fat and connective tissue, were included in the models. When the electric field is applied to the skin fold, almost the entire voltage drop is on the outermost layer of the skin, stratum corneum, due to its lowest conductivity. Applying electric pulses on such a voltage divider causes the voltage to be distributed between the resistors proportionally to their resistivity.

However, once the stratum corneum is permeabilized, the electric field "penetrates" to the layers underneath it. This process was modeled as an irreversible phase transition, taking into account the increase in tissue conductivity due to cell membrane electropermeabilization. When the electric field exceeds the predefined threshold, tissue conductivity increases. This change subsequently causes a change in the electric field distribution and of the corresponding current (Pavselj and Miklavcic, 2008b).

The electropermeabilization process in skin was described theoretically by numerical modeling, based on data derived from previously published *in vivo* experimental data. The numerical models took into account the layered structure of skin, macroscopical changes of its bulk electrical properties during electroporation, as well as the presence of localized sites of increased molecular transport termed local transport regions. In addition, permeabilizing voltage amplitudes suggested by the model are also well in the range of the voltage amplitudes reported.

The transdermal drug delivery is used to insert drugs into the skin for therapeutic purposes by chemical and/or physical enhancers (Prausnitz, 1997) as an alternative to oral, intravascular, subcutaneous, and transmucosal routes. Skin is also an attractive target tissue for *in vivo* gene delivery (Drabick et al., 1994). After pulsing, the cell membrane reseals, provided that the applied voltage was not too high to cause permanent cell membrane damage. Pavselj and Miklavcic (2008a,c) developed a model of the electropermeabilization process in skin by means of numerical modeling, based on data derived from *in vivo* experiments (Prausnitz et al., 1993; Pliquett et al., 1995; Gabriel et al., 1996a,b; Pliquett and Weaver, 1996; Gallo et al., 1997; Jadoul et al., 1999; Pliquett, 1999; Pavselj and Préat, 2005).

Successful DNA delivery to the dermis and the viable epidermis by means of electroporation has been shown in *in vivo* experiments (Pavselj and Préat, 2005). An increase in tissue conductivity due to cell membrane electroporation was observed (Prausnitz et al., 1993; Pliquett et al., 1995; Jadoul et al., 1999). Furthermore, it has been shown that the electropermeabilization and, consequently, the increase in the conductivity of the stratum corneum are not homogeneous throughout the electroporated area. Molecular and ionic transport across the skin subjected to high-voltage pulses is highly localized in the so-called local transport regions. It has been found that longer electric pulses produce larger local transport areas, as the alteration of the stratum corneum structure is caused by a synergistic effect between electroporation and Joule heating, while a higher pulse amplitude means higher pore density (Pliquett et al., 2004).

The dynamics of the change of tissue conductivities with the electric field is yet another unknown of tissue electropermeabilization. Due to the nonuniformity of the cell size and shape in the tissue, not all the cells are permeabilized at the same time once the threshold electric field is reached. The electric field threshold value needed for skin electropermeabilization was found to be approximately 400–600 V/cm.

It has been shown that molecular and ionic transport across skin subjected to high-voltage pulses is highly localized (Vanbever et al., 1994; Pliquett, 1999; Pliquett and Gusbeth, 2004). The size of these localized places depends on pulse duration, while pulse amplitude dictates the density. In most cases, they are formed in the sites of the so-called stratum corneum "defects" and are further expanded by Joule heating caused by the high local current density due to the drop in the resistivity of the stratum corneum inside them.

4.10.2 Electroporation in Cancer Treatment

During the last two decades, the achievements in theoretical and experimental electroporation have been introduced in clinical medicine for the treatment of various problems, including superficial cancer and wound healing. The ability of electroporation to "drill" passages for molecules that cannot otherwise penetrate the cell membrane has been used for drug delivery. Electroporation is currently an established method for the delivery of molecules into cells *in vitro* and *in vivo*, as well as an integral part of the electrochemotherapy of tumors (Okino and Mohri, 1987; Serša et al., 2000, 2010; Mir, 2001) and gene electrotransfer (Neumann et al., 1982; Wong and Neumann, 1982; Jaroszeski et al., 1999; Heller and Heller, 2010). Several recent *in vivo* studies (Liu et al., 2006; Prud'homme et al., 2006) have suggested that gene electrotransfer could become an important method for DNA transfer in the gene therapy of various illnesses as an alternative method to viral transfection, while electrochemotherapy is starting to be used in clinics as a possible therapy for cutaneous and subcutaneous tumors (Jaroszeski et al., 1999a,b; Marty et al., 2006; Moller et al., 2009).

In another application, Nuccitelli et al. (2006) have shown that 30 ns, 40 kV/cm electric pulses could cause total remission of skin cancer in mice. When tumors are treated with 300 pulses, they exhibit three rapid changes: (1) the nuclei of the tumor cells begin to shrink rapidly and the average nuclear area falls by 50% within an hour; (2) blood flow to the tumor stops within a few minutes and does not recover for a week or two; and (3) the tumor begins to shrink and within two weeks is shrunk by 90%. Most of the roughly 200 tumors treated began to regrow at this point, but if treated a second time, they completely disappeared. All animals in the treated group exhibited complete remission with no recurrence after 4 months.

While not yet tested on humans, it is likely that this technique may have advantages over the surgical removal of skin lesions because nsPEFs kill the tumor without disrupting the dermis so that scarring is less likely.

It should also be effective on other tumor types located deeper in the body if a catheter electrode can be guided to the tumor. Among its most intriguing characteristics is the incredibly short time that these cells have been exposed to the electric field. All of the tumor regression shown here resulted from a total electric field exposure time of $120\,\mu s$ or less. A second important characteristic is the low energy delivered to the tissue. Each 300 ns pulse of 40 kV/cm delivers only 0.18 J to the tissue between the 5 mm plates. Based on the specific heat capacity of water, this would increase in tissue temperature by only $2^\circ C$–$4^\circ C$ (Nuccitelli et al., 2006).

By reducing the duration of electrical pulses from microseconds into the nanosecond range, the electric field–cell interactions shift increasingly from the plasma (cell) membrane to subcellular structures. Yet another domain of pulsed electric field interactions with cell structures and functions opens when the pulse duration is reduced to values such that membrane charging becomes negligible and direct electric field-molecular effects determine the biological mechanisms.

4.10.3 Electroporation for Wound Healing

The concept of using nsPEFs for wound healing is exciting for several reasons. First, the activation of platelets using nsPEFs will provide the medical community with an alternative to the use of thrombin (which has been associated with allergic reactions in some patients). Secondly, nsPEFs provide a focused, localized stimulus for platelet activation, whereas, other platelet activators such as thrombin, adenosine diphosphate (ADP), thromboxane, or collagen can all enter the circulation and have systemic effects. It has been shown that nanosecond pulses have a similar effect on platelets as thrombin, an agonist that promotes aggregation (Beebe et al., 2004). This process of aggregation is initiated by a calcium release from internal stores (Beebe et al., 2004; White et al., 2004). This is consistent with the well-known fact that the aggregation of platelets by known agonists, such as thrombin, requires an increase in intracellular free calcium. The effects of nsPEFs on calcium mobilization and calcium influx are reminiscent of the well-known capacitative or store-operated calcium entry process induced by hormones (e.g., thrombin) in many nonexcitable cells. In this process, when intracellular calcium is mobilized from the endoplasmic reticulum, there is an activation of an influx process in the plasma membrane, i.e., the two processes are coupled. Therefore, if the degree of platelet aggregation is correlated with internal calcium mobilization and calcium influx, platelet aggregation should also follow the similarity law. In addition to aggregation, an increase in platelet derived growth factor (PDGF) released from washed platelets has been observed after pulsing with nanosecond pulses. Platelets are rich in alpha granule growth factors (i.e., PDGF and transforming growth factor beta (TGF-β)). The release of this growth factor is essential for wound healing.

4.10.4 Biotechnology

As it has been pointed out in other chapters of this book (Dimova, 2010; Lebovka and Vorobiev, 2010), electric fields induce various electrokinetic phenomena, which have found wide application in biotechnology as well as in micro- and nano-technologies. Electro-neutral particles (droplets, bubbles, lipid vesicles, solid beads) suspended in a medium of different polarizability acquire a charge at their surfaces when exposed to electric fields. These kinetic phenomena are widely used in biology and biotechnology as methods for cell or membrane characterization and manipulation (Vorobiev and Lebovka, 2008).

By studying the AC field-induced flows in giant lipid vesicles, we have learned that there are possible applications in microfluidic technologies. Giant vesicles in inhomogeneous AC fields or in hydrodynamic flows mimicking the situation of red blood cells in capillaries may be used as nano-reactors for fluid manipulation, i.e., displacing, mixing, trapping, etc. These AC field-induced flows in giant vesicles have possible applications in microfluidic technologies.

References

Abidor IG, Sowers AE. 1992. Kinetics and mechanism of cell membrane electrofusion. *Biophys J* 61: 1557–1569.

Beebe SJ, Blackmore PF, White J, Joshi RP, Schoenbach KH. 2004. Nanosecond pulsed electric fields modulate cell function through intracellular signal transduction mechanisms. *Physiol Meas* 25: 1077–1093.

Buescher ES, Schoenbach KH. 2003. Effects of submicrosecond, high intensity pulsed electric fields on living cells—Intracellular electromanipulation. *IEEE Trans Dielectr Electr Insul* 10: 788–794.

Canatella PJ, Karr JF, Petros JA, Prausnitz MR. 2001. Quantitative study of electroporation-mediated molecular uptake and cell viability. *Biophys J* 80: 755–764.

Chernomordik LV, Sukharev SI, Popov SV, Pastushenko VF, Sokirko AV, Abidor IG, Chizmadzev YuA. 1987. The electrical breakdown of cell and lipid membranes: The similarity of phenomenologies. *Biochim Biophys Acta* 902: 360–373.

Coster HGL, Zimmermann U. 1975. The mechanism of electrical breakdown in the membranes of *Valonia utricularis*. *J Membr Biol* 22: 73–90.

Crowley JM. 1973. Electrical breakdown of bimolecular lipid membranes as an electro-mechanical instability. *Biophys J* 13: 711–724.

Dimitrov DS, Jain RK. 1984. Membrane stability. *Biochim Biophys Acta* 779: 437–468.

Dimova R. 2010. Electrodeformation, electroporation and electrofusion of cell-sized lipid vesicles. In Pakhomv A, Miklavcic D, Markov M (eds.) *Advanced Electroporation Techniques in Biology and Medicine*, Taylor & Francis, Boca Raton, FL.

Drabick JJ, Glasspool-Malone J, King A, Malone RW. 1994. Cutaneous transfection and immune responses to intradermal nucleic acid vaccination are significantly enhanced by in vivo electropermeabilization. *Mol Ther* 3(2): 249–255.

Ermolina I, Polevaya Y, Feldman Y, Ginzburg B, Schlesinger M. 2001. Study of normal and malignant white blood cells by time domain dielectric spectroscopy. *IEEE Trans Dielectr Electr Insul* 8: 253–261.

Escoffre JM, Gioanni AP, Bellard E, Golzio M, Ross MP, Teissie J. 2010. Gene transfection: From basic processes to preclinical applications. In Pakhomv A, Miklavcic D, Markov M (eds.) *Advanced Electroporation Techniques in Biology and Medicine*, Taylor & Francis, Boca Raton, FL.

Esser AT, Smith KC, Gowrishankar TR, Weaver JC. 2010. Drug-free, solid tumor ablation by electroporating pulses. In Pakhomv A, Miklavcic D, Markov M (eds.) *Advanced Electroporation Techniques in Biology and Medicine*, Taylor & Francis, Boca Raton, FL.

Feldman Y, Ermolina I, Hayashi Y. 2003. Time domain dielectric spectroscopy study of biological systems. *IEEE Trans Dielectr Electr Insul* 10: 728–753.

Fošnarič M, Kralj-Iglič V, Bohinc K, Iglič A, May 2003. Stabilization of pores in lipid bilayers by anisotropic inclusions. *J Phys Chem B* 107: 12519–12526.

Gabriel C, Gabriel S, Corthout E. 1996a. The dielectric properties of biological tissues: I. Literature survey. *Phys Med Biol* 41(11): 2231–2249.

Gabriel S, Lau RW, Gabriel C. 1996b. The dielectric properties of biological tissues: II. Measurements in the frequency range 10 Hz to 20 GHz. *Phys Med Biol* 41(11): 2251–2269.

Gabriel B, Teissie J. 1998. Fluorescence imaging in the millisecond time range of membrane electropermeabilization of single cells using a rapid ultra-low-light intensifying detection system. *Eur Biophys J* 27: 291–298.

Gabriel B, Teissie J. 1999. Time courses of mammalian cell electropermeabilization observed by millisecond imaging of membrane property changes during the pulse. *Biophys J* 76: 2158–2165.

Gallo SA, Oseroff AR, Johnson PG, Hui SW. 1997. Characterization of electric-pulse-induced permeabilization of porcine skin using surface electrodes. *Biophys J* 72: 2805–2811.

Gallo SA, Sen A, Hensen ML, Hui SW. 2002. Temperature dependent electrical and ultrastructural characterizations of porcine skin upon electroporation. *Biophys J* 82: 109–119.

Gehl J. 2003. Review: Electroporation: Theory and methods, perspectives for drug delivery, gene therapy and research. *Acta Physiol Scand* 177: 437–447.

Glaser RW, Leikin SL, Chernomordik LV, Pastushenko VF, Sokirko AI. 1988. Reversible electrical breakdown of lipid bilayers: Formation and evolution of pores. *Biochim Biophys Acta* 940: 275–287.

Goltzev VN, Markov MS, Doltchinkova VR, Michailova D. 1989. Modification of the chloroplast delayed fluorescence by electroporation. *Semin Biophys* 5: 71–78.

Haas CN, Aturaliye DN. 1999. Kinetics of electroporation assisted chlorination of *Giardia muris*. *Water Res* 33: 1761–1766.

Heller LC, Heller R. 2010. Translation of electrically mediated DNA delivery to the clinic. In Pakhomv A, Miklavcic D, Markov M (eds.) *Advanced Electroporation Techniques in Biology and Medicine*, Taylor & Francis, Boca Raton, FL.

Heller R, Gilbert R, Jaroszeski MJ. 1999. Clinical application of electrochemotherapy. *Adv Drug Deliv Rev* 35: 119–129.

Hibino M, Shigemori M, Itoh H, Nagayama K, Kinosita K. 1991. Membrane conductance of an electroporated cell analyzed by sub-microsecond imaging of transmembrane potential. *Biophys J* 59: 209–220.

Hibino M, Itoh H, Kinosita K Jr. 1993. Time courses of cell electroporation as revealed by submicrosecond imaging of transmembrane potential. *Biophys J* 64: 1789–1800.

Hu Q, Viswanadham S, Joshi RP, Schoenbach KH, Beebe SJ, Blackmore PF. 2005a. Simulations of transient membrane behavior in cells subjected to a high-intensity ultrashort electrical pulse. *Phys Rev E* 71: 031914-1–031914-9.

Hu Q, Joshi RP, Schoenbach KH. 2005b. Simulations of nanopore formation and phosphatidylserine externalization in lipid membranes subjected to a high intensity, ultrashort electric pulse. *Phys Rev E* 72: 031902-1–031902-10.

Jadoul A, Bouwstra J, Préat V. 1999. Effects of iontophoresis and electroporation on the stratum corneum. Review of the biophysical studies. *Adv Drug Deliv Rev* 35(1): 89–105.

Jarm T, Gemazar M, Sersa G. 2010. Tumor blood flow-modifying effects of electroporation and electrochemotherapy—Experimental evidence and implication for the therapy. In Pakhomv A, Miklavcic D, Markov M (eds.) *Advanced Electroporation Techniques in Biology and Medicine*, Taylor & Francis, Boca Raton, FL.

Jaroszeski MJ, Gilbert R, Nicolau C, Heller R. 1999a. In vivo gene delivery by electroporation. *Adv Drug Deliv Rev* 35: 131–137.

Jaroszeski MJ, Heller R, Gilbert R. 1999b. *Electrochemotherapy, Electrogenetherapy and Transdermal Drug Delivery: Electrically Mediated Delivery of Molecules to Cells*, Humana Press, Totowa, NJ.

Joshi RP, Hu Q, Schoenbach KH. 2004. Modeling studies of cell response to ultrashort, high-intensity electric fields—Implications for intracellular manipulation. *IEEE Trans Plasma Sci* 32: 1677–1686.

Kanduser M, Sentjurc M, Miklavcic D. 2008. The temperature effect during pulse application on cell membrane fluidity and permeabilization. *Bioelectrochemistry* 74: 52–57.

Kinosita K, Tsong TY. 1977. Voltage-induced pore formation and hemolysis of human erythrocytes. *Biochim Biophys Acta* 471: 227–242.

Kinosita K, Tsong TY. 1979. Voltage-induced conductance in human erythrocyte. *Biochim Biophys Acta* 554: 479–497.

Koronkiewicz S, Kalinowski S. 2004. Influence of cholesterol on electroporation of bilayer lipid membranes: Chronopotentiometric studies. *Biochim Biophys Acta Biomembr* 1661: 196–203.

Kotnik T, Maček-Lebar A, Miklavčič D, Mir LM. 2000. Evaluation of cell membrane electropermeabilization by means of non-permeant cytotoxic agent. *BioTechniques* 28: 921–926.

Kotnik T, Miklavčič D. 2000. Theoretical evaluation of the distributed power dissipation in biological cells exposed to electric fields. *Bioelectromagnetics* 21: 385–394.

Kotnik T, Miklavcic D. 2006. Theoretical evaluation of voltage inducement on internal membranes of biological cells exposed to electric fields. *Biophys J* 90: 480–491.

Kotnik T, Pucihar G. 2010. Induced transmembrane voltage—Theory, modeling, and experiments. In Pakhomv A, Miklavcic D, Markov M (eds.) *Advanced Electroporation Techniques in Biology and Medicine*, Taylor & Francis, Boca Raton, FL.

Kotulska M, Kubica K, Koronkiewicz S, Kalinowski S. 2007. Modeling the induction of lipid membrane electropermeabilization. *Bioelectrochemistry* 70: 64–70.

Kramar P, Miklavcic D, Macek-Lebar A. 2009. A system for the determination of planar lipid bilayer breakdown voltage and its applications. *IEEE Trans Nanobiosci* 8: 132–138.

Lebovka N, Vorobiev E. 2010. Food and biomaterials processing assisted by electroporation. In Pakhomv A, Miklavcic D, Markov M (eds.) *Advanced Electroporation Techniques in Biology and Medicine*, Taylor & Francis, Boca Raton, FL.

Litster JD. 1975. Stability of lipid bilayers and red blood cell membranes. *Phys Lett* 53A: 193–194.

Liu F, Heston S, Shollenberger LM, Sun B, Mickle M, Lovell M, Huang L. 2006. Mechanism of in vivo DNA transport into cells by electroporation: Electrophoresis across the plasma membrane may not be involved. *J Gene Med* 8: 353–361.

Maček-Lebar A, Miklavčič D. 2001. Cell electropermeabilization to small molecules in vitro: Control by pulse parameters. *Radiol Oncol* 35: 193–202.

Maldarelli C, Jain RK, Ivanov LB, Rckenstein E. 1980. Stability of symmetric and unsymmetric, thin liquid films to short and long wavelength perturbations. *J Colloid Interface Sci* 78: 118–143.

Markov MS. 1988. Electromanipulation in biotechnology. In Ottova A (ed.) *Mathematical Modeling in Biotechnology*, Elsevier, Amsterdam, the Netherlands, pp. 85–92.

Markov MS. 1991. The contribution of water in the stabilization of biological membranes. In Bender M (ed.) *Interfacial Phenomena in Biological Systems*, Marcel Dekker, New York, pp. 153–170.

Markov MS, Guttler JP. 1989. Electromagnetic field influence on electroporation of erythrocytes. *Studia Biophys* 130: 211–214.

Marty M, Serša G, Garbay JA, Gehl J, Collins CG, Snoj M, Billard V et al. 2006. Electrochemotherapy—An easy, highly effective and safe treatment of cutaneous and subcutaneous metastases: Results of ESOPE (European Standard Operating Procedures of Electrochemotherapy) study. *Eur J Cancer Suppl* 4: 3–13.

Mekid H, Mir LM. 2000. In vivo cell electrofusion. *Biochim Biophys Acta* 1524: 118–130.

Michael DH, O'Neill ME. 1970. Electrohydrodynamic instability in plane layers of fluid. *J Fluid Mech* 41: 571–580.

Miklavcic D, Kotnik T. 2004. Electroporation for electrochemotherapy and gene therapy. In Rosch PJ, Markov MS (eds.) *Bioelectromagnetic Medicine*, Marcel Dekker, New York, pp. 637–656.

Miklavcic D, Puc M. 2006. *Electroporation. Wiley Encyclopedia of Biomedical Engineering*, Wiley, New York.

Miklavcic D, Towhidi L. 2010. Model of cell membrane electroporation and transmembrane molecular transport. In Pakhomv A, Miklavcic D, Markov M (eds.) *Advanced Electroporation Techniques in Biology and Medicine*, Taylor & Francis, Boca Raton, FL.

Miklavcic D, Semrov D, Mekid H, Mir LM. 2000. A validated model of in vivo electric field distribution in tissues for electrochemotheapy and for DNA electrotransfer for gene therapy. *Biochim Biophys Acta* 1519: 73–83.

Mir LM. 2001. Therapeutic perspectives of in vivo cell electropermeabilization. *Bioelectrochemistry* 53: 1–10.

Mir LM, Banoun H, Paoletti C. 1988. Introduction of definite amounts of nonpermeant molecules into living cells after electropermeabilization: Direct access to the cytosol. *Exp Cell Res* 175: 15–25.

Mir LM, Orlowski S, Belehradek J, Teissie J, Rols MP, Sersa G, Miklavcic D, Gilbert R, Heller R. 1995. Biomedical applications of electric pulses with special emphases on antitumor electrochemotherapy. *Bioelectrochem Bioenerg* 38: 203–207.

Moller MG, Salwa S, Soden DM, O'Sullivan GC. 2009. Electrochemotherapy as an adjunct or alternative to other treatments for unresectable or in-transit melanoma. *Expert Rev Anticancer Ther* 9: 1611–1630.

Neumann E, Kakorin S. 2010. Physical chemical theory of membrane electroporation and electrotransfer of biogenic agents. In Pakhomv A, Miklavcic D, Markov M (eds.) *Advanced Electroporation Techniques in Biology and Medicine*, Taylor & Francis, Boca Raton, FL.

Neumann E, Rosenheck K. 1972. Permeability changes induced by electric impulses in vesicular membranes. *J Membr Biol* 10: 279–290.

Neumann E, Schaefer-Ridder M, Wang Y, Hofschneider PH. 1982. Gene transfer into mouse lyoma cells by electroporation in high electric fields. *EMBO J* 1: 841–845.

Neumann E, Sowers A, Jordan C. 1989. *Electroporation and Electrofusion in Cell Biology*, Plenum Press, New York.

Nuccitelli RL, Pliquett U, Chen X, Ford W, Swanson RJ, Beebe SJ, Kolb JP, Schoenbach KH. 2006. Nanosecond pulsed electric fields cause melanomas to self-destruct. *Biochem Biophys Res Commun* 343: 351–360.

Okino M, Mohri H. 1987. Effects of high-voltage electrical impulse and an anticancer drug on in vivo growing tumors. *Jpn J Cancer Res* 78: 1319–1321.

Pakhomov AG, Phinney A, Ashmore J, Walker KJK, Kono S, Schoenbach KS, Murphy MR. 2004. Characterization of the cytotoxic effect of high-intensity, 10-ns duration electrical pulses. *IEEE Transactions on Plasma Science* 32(4): 1579–1585.

Pakhomov AG, Shevin R, White JA, Kolb JA, Pakhomova ON, Joshi RP, Schoenbach KH. 2007. Membrane permeabilization and cell damage by ultrashort electric field shocks. *Arch Biochem Biophys* 465: 109–118.

Pavlin M, Miklavčič D. 2008. Theoretical and experimental analysis of conductivity, ion diffusion and molecular transport during cell electroporation—Relation between short-lived and long-lived pores. *Bioelectrochemistry* 74: 38–46.

Pavlin M, Pavselj N, Miklavcic D. 2002. Dependence of induced transmembrane potential on cell density arrangement and cell position inside a cell system. *IEEE Trans Biomed Eng* 49: 605–612.

Pavlin M, Kotnik T, Miklavcic D, Kramar P, Macek-Lebar A. 2008. Electroporation of planar lipid bilayers and membranes. In Leitmanova Liu A (ed.) *Advances in Planar Lipid Bilayers and Liposomes*, Vol 6, Elsevier, Amsterdam, the Netherlands, pp. 165–226.

Pavlin M, Kandušer M, Reberšek M, Pucihar G, Hart FX, Magjarević R, Miklavčič D. 2005. Effect of cell electroporation on the conductivity of a cell suspension. *Biophys J* 88: 4378–4390.

Pavselj N, Miklavcic D. 2008a. Numerical models of skin electropermeabilization taking into account conductivity changes and the presence of local transport regions. *IEEE Trans Plasma Sci* 36: 1650–1658.

Pavselj N, Miklavcic D. 2008b. Numerical modeling in electroporation-based biomedical applications. *Radiol Oncol* 42: 159–168.

Pavselj N, Miklavcic D. 2008c. A numerical model of permeabilized skin with local transport regions. *IEEE Trans Biomed Eng* 55: 1927–1930.

Pavselj N, Préat V. 2005. DNA electrotransfer into the skin using a combination of one high- and one low-voltage pulse. *J Control Release* 106(3): 407–415.

Petrov AG. 1984. Flexoelectricity of lyotropics and biomembranes. *Nuovo Cimento Soc Ital Fis D* 3: 174–192.

Petrov AG. 2002. Flexoelectricity of model and living membranes. *Biochim Biophys Acta* 1561: 1–25.

Petrov AG, Bivas I. 1984. Elastic and flexoelectric aspects of out-of-plane fluctuations in biological and model membranes. *Prog Surf Sci* 16: 389–512.

Pliquett U. 1999. Mechanistic studies of molecular transdermal transport due to skin electroporation. *Adv Drug Deliv Rev* 35: 41–60.

Pliquett U, Gusbeth C. 2004. Surface area involved in transdermal transport of charged species due to skin electroporation. *Bioelectrochemistry* 65: 27–32.

Pliquett U, Weaver JC. 1996. Electroporation of human skin: Simultaneous measurement of changes in the transport of two fluorescent molecules and in the passive electrical properties. *Bioelectrochem Bioenerg* 39: 1–12.

Pliquett U, Langer R, Weaver JC. 1995. Changes in the passive electrical properties of human stratum corneum due to electroporation. *Biochim Biophys Acta* 1239(2): 111–121.

Pliquett U, Elez R, Piiper A, Neumann E. 2004. Electroporation of subcutaneous mouse tumors by rectangular and trapezium high voltage pulses. *Bioelectrochemistry* 62: 83–93.

Prausnitz MR. 1997. Reversible skin permeabilization for transdermal delivery of macromolecules. *Crit Rev Ther Drug Carr Syst* 14(4): 455–483.

Prausnitz MR, Bose VG, Langer R, Weaver JC. 1993. Electroporation of mammalian skin: A mechanism to enhance transdermal drug delivery. *Proc Natl Acad Sci USA* 90: 10504–10508.

Prud'homme GL, Glinka Y, Khan AS, Draghia-Akli R. 2006. Electroporation-enhanced nonviral gene transfer for the prevention or treatment of immunological, endocrine and neoplastic diseases. *Curr Gene Ther* 6: 243–273.

Raffy S, Teissie J. 1997. Electroinsertion of glycophorin A in interdigitation-fusion giant unilamellar lipid vesicles. *J Biol Chem* 272: 25524–25530.

Rols MP, Teissie J. 1989. Ionic-strength modulation of electrically induced permeabilization and associates fusion of mammalian cells. *Eur J Biochem* 179: 109–115.

Rols MP, Teissie J. 1990. Electropermeabilization of mammalian cells. Quantitative analysis of the phenomenon. *Biophys J* 58: 1089–1098.

Rols MP, Teissie J. 1992. Experimental evidence for involvement of the cytoskeleton in mammalian cell electropermeabilization. *Biochim Biophys Acta* 1111: 45–50.

Rols MP, Teissie J. 1993. The time course of electropermeabilization. In Blank M (ed.) *Electricity and Magnetism in Biology and Medicine*, San Francisco Press, San Francisco, CA, pp. 151–154.

Rols MP, Teissie J. 1998. Electropermeabilization of mammalian cells to macromolecules: Control by pulse duration. *Biophys J* 75: 1415–1423.

Rossi CR, Campana LG. 2010. Clinical electrochemotherapy: Italian experience. In Pakhomv A, Miklavcic D, Markov M (eds.) *Advanced Electroporation Techniques in Biology and Medicine*, Taylor & Francis, Boca Raton, FL.

Rowan NJ, MacGregor SJ, Anderson JG, Fouracre RA, Farish O. 2000. Pulsed electric field inactivation of diarrhoeagenic *Bacillus cereus* through irreversible electroporation. *Lett Appl Microbiol* 31: 110–114.

Sale AJH, Hamilton A. 1967. Effects of high electric fields on microorganisms: I. Killing of bacteria and yeasts. *Biochem Biophys Acta* 148: 781–788.

Schmeer M, Seipp T, Pliquett U, Kakorin S, Neumann E. 2004. Mechanism for the conductivity changes caused by membrane electroporation of CHO cell-pellets. *Phys Chem Chem Phys* 6: 5564–5574.

Schmidt EU, Leinfelder P, Gessner D, Zillikens EB, Brocker U, Zimmermann. (2001). CD19+ B lymphocytes are the major source of human antibody-secreting hybridomas generated by electrofusion. *J Immunol Methods* 255: 93–102.

Schoenbach KH, Beebe SJ, Buescher ES. 2001. Intracellular effect of ultrashort electrical pulses. *Bioelectromagnetics* 22: 440–448.

Schoenbach KH, Hargrave B, Joshi RP, Kolb JF, Nuccitelli R, Osgood S, Pakhomov A et al. 2007. Bioelectric effects of intense nanosecond pulses. *IEEE Trans Dielectr Electr Insul* 14: 1088–1109.

Schwister K, Deuticke B. 1985. Formation and properties of aqueous leaks induced in human erythrocytes by electrical breakdown. *Biochim Biophys Acta* 816: 332–348.

Scott-Taylor TH, Pettengell R, Clarke I, Stuhler G, La Barthe MC, Walden P, Dalgleish AG. 2000. Human tumour and dendritic cell hybrids generated by electrofusion: Potential for cancer vaccines. *Biochim Biophys Acta* 1500: 265–267.

Sersa G, Stabuc B, Cemazar M, Jancar B, Miklavcic D, Rudolf Z. 1998. Electrochemotherapy with cisplatin: Potentiation of local cisplatin antitumor effectiveness by application of electric pulses in cancer patients. *Eur J Cancer* 34: 1213–1218.

Serša G, Kranjc S, Čemažar M. 2000. Improvement of combined modality therapy with cisplatin and radiation using electroporation of tumors. *Int J Radiat Oncol Biol Phys* 46: 1037–1041.

Sersa G et al. 2010. Combined modality therapy: electrochemotherapy with tumor irradiation. In Pakhomv A, Miklavcic D, Markov M (eds.) *Advanced Electroporation Techniques in Biology and Medicine*, Taylor & Francis, Boca Raton, FL.

Sowers AE. 1993. Membrane electrofusion: A paradigm for study of membrane fusion mechanisms. *Methods Enzymol* 220: 196–211.

Stampfli R. 1958. Reversible electrical breakdown of the excitable membrane of a Ranvier node. *Ann Acad Brasil Ciens* 30: 57–63.

Steinchen A, Gallez D, Sanfeld A. 1982. A viscoelastic approach to the hydrodynamic stability of membranes. *J Colloid Interface Sci* 85: 5–15.

Stewart DA, Gowrishankar TR, Weaver JC. 2006. Three dimensional transport lattice model for describing action potentials in axons stimulated by external electrodes. *Bioelectrochemistry* 69: 88–93.

Sugar IP. 1979. A theory of the electric field-induced phase transition of phospholipid bilayers. *Biochim Biophys Acta* 556: 72–85.

Sugar IP, Neumann E. 1984. Stochastic model for electric field-induced membrane pores electroporation. *Biophys Chem* 19: 211–225.

Susil R, Semrov D, Miklavcic D. 1998. Electric field-induced transmembrane potential depends on cell density and organization. *Electro Magnetobiol* 17: 391–399.

Susin SA, Zamzami N, Kroemer G. 1998. Mitochondria as regulators of apoptosis: Doubt no more. *Biochim Biophys Acta* 1366: 151–165.

Tarek M, Delemotte L. 2010. Electroporation of lipid membranes: Insights from molecular dynamic simulations. In Pakhomv A, Miklavcic D, Markov M (eds.) *Advanced Electroporation Techniques in Biology and Medicine*, Taylor & Francis, Boca Raton, FL.

Taupin C, Dvolaitzky M, Sauterey C. 1975. Osmotic pressure induced pores in phospholipid vesicles. *Biochemistry* 14: 4771–4775.

Taylor GI, Michael DH. 1973. On making holes in a sheet of fluid. *J Fluid Mech* 58: 625–639.

Teissie J, Rols MP. 1993. An experimental evaluation of the critical potential difference inducing cell membrane electropermeabilization. *Biophys J* 65: 409–413.

Teissie J, Rols MP. 1994. Manipulation of cell cytoskeleton affects the lifetime of cell membrane electropermeabilization. *Ann NY Acad Sci* 720: 98–110.

Teissie J, Eynard N, Vernhes MC, Benichou A, Ganeva V, Galutzov B, Cabanes PA. 2002. Recent biotechnological developments of electropulsation. A prospective review. *Bioelectrochemistry* 55: 107–112.

Tekle E, Astumian RD, Chock PB. 1990. Electro-permeabilization of cell membranes: Effect of the resting membrane potential. *Biochem Biophys Res Commun* 172: 282–287.

Tieleman DP, Leontiadou H, Mark AE, Marrink SJ. 2003. Simulation of pore formation in lipid bilayers by mechanical stress and electric fields. *J Am Chem Soc* 125: 6282–6383.

Tsong TY. 1991. Electroporation of cell membranes. *Biophys J* 60: 297–306.

Vanbever R, Lecouturier N, Preat V. 1994. Transdermal delivery of metoprolol by electroporation. *Pharmacol Res* 11: 1657–1662.

Vernier T. 2010. Nanoscale restructuring of lipid bilayers in nanosecond electric fields. In Pakhomv A, Miklavcic D, Markov M (eds.) *Advanced Electroporation Techniques in Biology and Medicine*, Taylor & Francis, Boca Raton, FL.

Vernier PT, Li A, Marcu L, Craft CM, Gundersen MA. 2003a. Ultrashort pulsed electric fields induce membrane phospholipid translocation and caspase activation: Differential sensitivities of Jurkat T lymphoblasts and rat glioma C6 cells. *IEEE Trans Dielectr Electr Insul* 10: 795–809.

Vernier PT, Sun Y, Marcu L, Salemi S, Craft CM, Gundersen MA. 2003b. Calcium bursts induced by nanosecond electrical pulses. *BBRC* 310: 286–295.

Vernier PT, Sun Y, Gundersen MA. 2006. Nanoelectropulse driven membrane perturbation and small molecule permeabilization. *BMC Cell Biol* 7: 37.

Vorobiev E, Lebovka N. 2008. *Electrotechnologies for Extraction from Food Plants and Biomaterials*, Springer Science, New York.

Walker III K, Pakhomova ON, Kolb JF, Schoenbach KH, Stuck BE, Murphy MR, Pakhomov AG. 2006. Oxygen enhances lethal effect of high-intensity, ultrashort electrical pulses. *Bioelectromagnetics* 27: 221–225.

Weaver JC, Chizmadzhev YA. 1996. Theory of electroporation: A review. *Bioelectrochem Bioenerg* 41: 135–160.

White JA, Blackmore PF, Schoenbach KH, Beebe SJ. 2004. Stimulation of capacitive calcium entry in HL-60 cells by nanosecond pulsed electric fields (nsPEF). *J Biol Chem* 279: 22964–22972.

Wong TK, Neumann E. 1982. Electric field mediated gene transfer. *Biochem Biophys Res Commun* 107: 584–587.

Zimmermann U. 1982. Electric field-mediated fusion and related electrical phenomena. *Biochim Biophys Acta* 694: 227–277.

Zimmermann U. 1996. The effect of high-intensity electric pulses on eukaryotic cell membranes. In Zimmermann U, Neil GA (eds.) *Electromanipulation of Cells*, CRC Press, London, U.K., pp. 1–105.

II

Mechanisms of Electroporation in Lipid Systems

5

Electrodeformation, Electroporation, and Electrofusion of Cell-Sized Lipid Vesicles

Rumiana Dimova

5.1 Introduction

Cells, being the basic building units of most living creatures, are the obvious object of interest when the properties and functions of organisms are to be deduced. The autonomy of a cell is ensured by a bounding membrane. The links between membrane physics and the other areas of research are becoming increasingly evident and important. As a result, the field of membrane structure and characteristics is attracting the attention of a growing number of researchers. Membrane biophysics builds upon studies performed predominantly on model membranes, among which lipid monolayers at the air–water interface, solid-supported bilayers, black lipid membranes, vesicles, and bilayer stacks.

5.1.1 Giant Vesicles as Biomimetic Compartments

Vesicles are membrane "bubbles" formed by bending and closing of a lipid bilayer. Various experimental techniques have been developed for preparing liposomes of different sizes (from nanometers to tens of microns). The largest (several tens of microns) are called "giant vesicles" and are an extraordinarily convenient system for studying the membrane behavior (Luisi and Walde 2000, Dimova et al. 2006). They are well visible under an optical microscope using various enhancing techniques like phase contrast, differential interference contrast, or confocal and standard fluorescence microscopy (see Figure 5.1). Thus, giant vesicles allow for a direct manipulation and observation of membrane interactions and responses to external perturbations. On the contrary, working with conventional vesicles (few hundreds of nanometers) usually involves the application of indirect methods and techniques for observation. In addition, their small sizes often raise questions about the effects due to high membrane curvature when molecular interactions are considered. In contrast, giant vesicles, which have sizes in the micrometer range, i.e., cell size, and therefore have nearly zero membrane curvature, reflect the properties and behavior of cell plasma membranes.

There are two widely used techniques for the formation of giant vesicles: (1) spontaneous swelling, introduced by Reeves and Dowben (1969) and (2) electroswelling, introduced by Angelova and Dimitrov (1986); for a brief description of the two preparation protocols, see Dimova et al. (2006). Interestingly enough, as discussed in Section 5.2.1, the underlying mechanism of the electroformation protocol, which is based on exposing lipid layers to alternating electric fields, is still poorly understood even though widely used. Both protocols yield giant vesicles with sizes in the range of a few tens of micrometers.

Giant unilamellar vesicles (GUVs) are increasingly employed for the quantitative characterization of membrane-related processes like cell adhesion, phase separation and domain formation, protein sorting in lipid rafts, membrane fusion, the effects of anchored molecules, protein mobility, to mention just a few of the studied fields. They are also a very practical tool to study the response

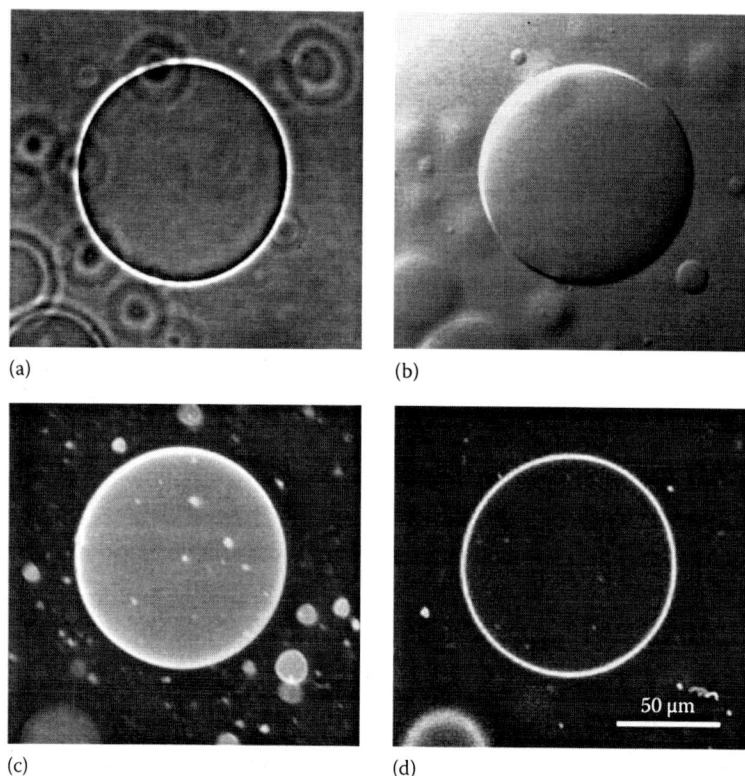

(a)   (b)

(c)   (d)

50 μm

FIGURE 5.1 Snapshots of one giant vesicle observed under different microscopy modes: (a) phase contrast; (b) differential interference contrast; (c, d) confocal microscopy, where (c) is a projection-averaged image and (d) is an equatorial section image. (Adapted from Dimova, R. et al., *J. Phys. Condens. Matter*, 18, S1151, 2006.)

of membranes to external perturbations. This chapter is dedicated to GUVs as a suitable biomimetic system in which the effects induced by electric fields can be directly visualized and interpreted with the general concepts of membrane biophysics.

5.1.2 Mechanical and Rheological Properties of Lipid Bilayers

To understand the membrane response to external perturbations like electromagnetic fields, it is necessary to know the physical properties of the lipid bilayer. How easy is it to bend this membrane sheet? How easy is it to stretch it? How much can one stretch it before it ruptures? If subjected to shear, how easy would the lipids flow relative to each other? These are some of the questions addressed in this section. Having described the mechanical and rheological properties of lipid membranes, we will be prepared to tackle problems related to stress induced in the bilayer by electric fields and the phenomena that it triggers, for example, the dynamics of vesicle and cell deformation, lipid flows along the membrane, bilayer instability and electroporation, and electrofusion. All of these processes will be discussed in detail and their nature will be examined and illustrated with experimental evidence collected on giant vesicles. In the following sections, we will find out that all these processes are strongly governed by the mechanical and rheological properties of the membrane.

Simplistically, the lipid bilayer may be viewed as a single film or slab, which may be bent, compressed or dilated, and sheared. In the equilibrium state, the membrane responses to various constraints are characterized by the following constants: bending rigidity, κ, Gaussian curvature modulus, κ_G, area compressibility modulus, K_a, and shear elastic modulus, μ. At physiological temperatures, most natural lipid membranes are fluid, therefore $\mu = 0$. Below the lipid phase transition temperature, single-component membranes crystallize. Mechanically, the bilayer acquires nonzero shear elasticity ($\mu \neq 0$). In this so-called gel phase (discussed in Section 5.3.3), the relative motion of membrane inclusions is principally hindered. The fluidity of the membrane at higher temperatures is essential for cellular membranes because it permits the displacement of membrane-anchored macromolecules or inclusions, for example, transmembrane proteins. The membrane, on the one hand, and the surrounding fluid, on the other, impose a hydrodynamic drag to the motion of an inclusion. The resistance or shear in the plane of the film is characterized by the shear viscosity, η_S. One may equivalently define a viscosity η_D related to the dilation and compression of the membrane. The fluid membranes are also very flexible as characterized by the bending elasticity modulus, κ. It is because κ is only on the order of several k_BT (where k_BT is the thermal energy) that the membrane fluctuates due to thermal noise. These fluctuations can be directly observed on tensionless giant vesicles under the microscope. When tension is applied to the membrane and gradually increased, one first pulls out the bilayer fluctuations and then, at high tension stretches it, changing the area per lipid molecule. The stretching elasticity of lipid bilayers, K_a, is on the order of that of a rubber sheet with the same thickness (≈ 4 nm). Upon stretching, the lipid bilayer can sustain tensions up to about several milli-Newton per meter (mN/m). At certain critical tension, also known as the lysis tension, σ_{lys}, the membrane ruptures. Note that the membrane tensile strength depends on the tension loading rate (Evans et al. 2003).

In Table 5.1 we give some typical values of the constants discussed above for both fluid-phase and gel-phase membranes. In the gel phase, the membrane viscosity diverges and the bending rigidity drastically increases.

5.1.3 Some Basic Relations about Membranes in Electric Fields

5.1.3.1 Characteristic Times in Vesicle Response

The response of membranes to electric fields involves dynamic physical processes occurring on different timescales. Free charges accumulate on boundaries separating media with different electric properties. A spherical vesicle polarizes on the Maxwell–Wagner timescale (Jones 1995):

TABLE 5.1 Typical Values for the Characteristic Properties of Lipid Membranes in Fluid and Gel Phase

Bilayer Property	Fluid-Phase Lipid Membranes	Gel-Phase Lipid Membranes
Bending rigidity, κ	0.9×10^{-19} J ($\sim 20\ k_B T$) (for κ-values of different lipid bilayers, see, e.g., Seifert and Lipowsky 1995, Rawicz et al. 2000)	14×10^{-19} J ($\sim 350\ k_B T$)[a] (Dimova et al. 2000, Mecke et al. 2003)
Stretching elasticity, K_a	240 mN/m (see, e.g., Rawicz et al. 2000)	~ 850 mN/m (Needham and Evans 1988)
Shear surface viscosity, η_S	$\sim 5 \times 10^{-9}$ N s/m (Dimova et al. 1999)	Diverges (Dimova et al. 2000)
Dilational surface viscosity, η_D	3.5×10^{-7} N s/m (Brochard-Wyart et al. 2000)	—
Lysis tension, σ_{lys}	5–10 mN/m (Needham and Hochmuth 1989, Olbrich et al. 2000)	≥ 15 mN/m (Evans and Needham 1987)

[a] This value corresponds to the bending rigidity of membranes at temperatures about 5°C below the main phase transition temperature of the lipid.

$$t_{MW} = \frac{\varepsilon_{in} + \varepsilon_{ex}}{\lambda_{in} + 2\lambda_{ex}} \tag{5.1}$$

where

ε_{in} and ε_{ex} are the dielectric constants

λ_{in} and λ_{ex} are the conductivities of the solutions inside and outside the vesicle, respectively

The lipid bilayer is impermeable to ions and free charges accumulate on both membrane surfaces. Hence, the vesicle membrane acts as a capacitor, which charges on a timescale (Schwan 1985, Grosse and Schwan 1992):

$$t_c = RC_m \left(\frac{1}{\lambda_{in}} + \frac{1}{2\lambda_{ex}} \right) \tag{5.2}$$

where

R is the vesicle radius

C_m is the membrane capacitance

The capacitor-charging time t_c is typically much longer than the Maxwell–Wagner time t_{MW}. For example, we can estimate $t_c \sim 10\,\mu s$ and $t_{MW} \sim 0.01\,\mu s$ for conditions corresponding to experiments on vesicles in 1 mM NaCl, namely $\varepsilon_{in} \sim \varepsilon_{ex} = 80\ \varepsilon_0$, where ε_0 is the vacuum permittivity, $\lambda_{in} \sim \lambda_{ex} \sim 10\,mS/m$, $C_m \sim 0.01\,F/m^2$, and $R \sim 10\,\mu m$.

These timescales are a key to understanding the dynamic response of vesicles subjected to short electric pulses discussed in Section 5.3, as well as the frequency dependence of vesicle deformation discussed in Section 5.2. Note that characteristic angular frequencies are defined as the inverse of the timescales in Equations 5.1 and 5.2, e.g., $\omega_{MW} = 1/t_{MW}$. The experimental frequency, ν, is related to the angular one via $\nu = \omega/2\pi$.

5.1.3.2 Vesicles in Alternating Electric (AC) Fields

When exposed to AC fields, spherical vesicles assume elliptical shapes. The type and degree of deformation depends on several parameters such as the field strength and frequency and media conductivity. The tension imposed by the AC field is usually weak and results in pulling out the membrane undulations. Only relatively high AC fields (>3 kV/m) can lead to the electroporation of giant vesicles as visualized by large pore formation on both vesicle poles facing the electrodes (Harbich and Helfrich 1979).

In the regime of low field strengths, where no poration is achieved, the tension acting on the membrane can be obtained from (Kummrow and Helfrich 1991)

$$\sigma = g\varepsilon_w E^2 \frac{1}{\left(c_1 + c_2\right)_{\text{pol}} - \left(c_1 + c_2\right)_{\text{equ}}} \tag{5.3}$$

where

ε_w is the dielectric constant of water

E is the field strength far away from the vesicle

c_1 and c_2 are the principal curvatures taken either at the equator (equ) or the pole (pol), and therefore measurable from the geometry of the vesicle

g is a dimensionless parameter, which is a function of the field frequency, the dielectric constants of the membrane, and the solutions inside and outside the vesicle

Using the above expression, the analyses of vesicle shape deformations induced by AC field have been used to measure the bending rigidity of membranes (Kummrow and Helfrich 1991, Niggemann et al. 1995, Dimova et al. 2009, Gracia et al. 2010). The protocol of such measurements consists of subjecting a vesicle to an AC electric field of increasing strength and recording the induced deformation. For the conductivity conditions and frequency range (between 1 and 300 kHz) in such experiments, the vesicles adopt prolate deformation as discussed in Section 5.2.1.

5.1.3.3 Transmembrane Potential and Conditions for Vesicle Electroporation

As already mentioned, lipid membranes are essentially impermeable to ions. Thus, in the presence of an electric field, the charges accumulate on both sides of the bilayer and give rise to a transmembrane potential (Kinosita et al. 1988)

$$\Psi_m\left(t\right) = 1.5R\left|\cos\theta\right|E\left[1 - \exp\left(-\frac{t}{t_c}\right)\right] \tag{5.4}$$

where

R is the radius of a spherical vesicle as before

θ is the tilt angle between the electric field and the surface normal

t is time

t_c is the charging time as defined in Equation 5.2

Equations 5.2 and 5.4 are valid only for a nonconductive membrane. Above some electroporation threshold, the transmembrane potential Ψ_m cannot be further increased, and can even decrease due to the transport of ions across the membrane (Kinosita et al. 1988, Hibino et al. 1991).

The electroporation phenomenon can also be understood in the terms of a stress in the bilayer created by the electric field. The transmembrane potential, Ψ_m, induces an effective electrical tension σ_{el}, as defined by the Maxwell stress tensor (Abidor et al. 1979, Needham and Hochmuth 1989, Riske and Dimova 2005a). This tension is given by

$$\sigma_{el} = \varepsilon_m \frac{h}{2h_e^2}\Psi_m^2 \tag{5.5}$$

where

> h is the total bilayer thickness, $h \approx 4\,\text{nm}$
>
> h_e the dielectric thickness, $h_e \approx 2.8\,\text{nm}$ for lecithin bilayers (Simon and McIntosh 1986, Nagle and Tristram-Nagle 2000)
>
> ε_m is the membrane permittivity, $\varepsilon_m \approx 2\,\varepsilon_0$

For vesicles with some initial tension σ_0, the total tension reached during the pulse is

$$\sigma = \sigma_0 + \sigma_{el} \tag{5.6}$$

If the total membrane tension exceeds the lysis tension σ_{lys}, the vesicle ruptures. This corresponds to building up a certain critical transmembrane potential, $\Psi_m = \Psi_c$. According to Equations 5.5 and 5.6, this poration potential Ψ_c depends on the initial membrane tension σ_0 as previously reported (Needham and Hochmuth 1989, Akinlaja and Sachs 1998, Riske and Dimova 2005a). The critical transmembrane potential for cell membranes is $\Psi_c \approx 1$ V (see, e.g., Tsong 1991, Weaver and Chizmadzhev 1996). Similarly, for tension-free vesicles, the critical potential was measured to be similar (Needham and Hochmuth 1989). Its value decreases when the initial membrane tension increases. In agreement with this expectation, nonfluctuating vesicles that exhibit an appreciable tension, readily porate at $\Psi_c < 1$ V (Riske and Dimova 2005b, Portet et al. 2009), i.e., they porate at fields weaker than those needed to porate deflated vesicles with visibly undulating membranes.

5.2 Vesicle Response to AC Fields

The question "how cells respond to AC fields?" has been addressed in a number of studies. In some of them, cells have been found to orient parallel or perpendicular to the field direction (Griffin 1970, Iglesias et al. 1985); others report that cells deform in electric fields (Engelhardt et al. 1984, Iglesias et al. 1985, Engelhardt and Sackmann 1988). This difference is probably due to the membrane elasticity and coupling to the cytoskeleton as well as to the excess membrane area. The orientation of the cells in the field was found to depend on the solution conductivity (Griffin 1970, Iglesias et al. 1985). Similarly, the direction of cell elongation is influenced by the external conductivity (Zimmermann 1982, Engelhardt et al. 1984). In this section, we will aim at drawing an overall picture of the spectrum of the responses of model membranes to alternating electric fields at various field frequency and conductivity conditions. Vesicle deformations have been treated both experimentally and theoretically, but a comprehensive description reconciling observations and calculations is still to emerge.

5.2.1 Vesicle Deformation in AC Fields

In AC fields, spherical giant vesicles deform adopting ellipsoidal shapes. Initiated by the seminal work of Winterhalter and Helfrich (1988), this effect has been considered theoretically by several groups (Hyuga et al. 1991c, 1993, Mitov et al. 1993, Peterlin et al. 2007). Experimental studies have shown that in AC fields of intermediate frequencies (~2 kHz), vesicles in pure water assume prolate deformation with the longer axis oriented along the field direction (Niggemann et al. 1995). At higher frequencies, a prolate-to-oblate transition was reported (Mitov et al. 1993, Peterlin et al. 2000).

Thus, analogously to cells, the deformation of vesicles subjected to AC fields depends on the field frequency ν (or ω) and the conductivity conditions. The latter can be described by the ratio between the internal and the external conductivities: $x = \lambda_{in}/\lambda_{ex}$. Systematically varying the field frequency and solution conductivities allows one to construct a morphological diagram of the shape transitions observed in neutral phosphatidylcholine vesicles (Aranda et al. 2008, Dimova et al. 2009); see Figure 5.2. At high frequencies, the vesicles are spherical independently of x. As the frequency decreases, vesicles with $x > 1$,

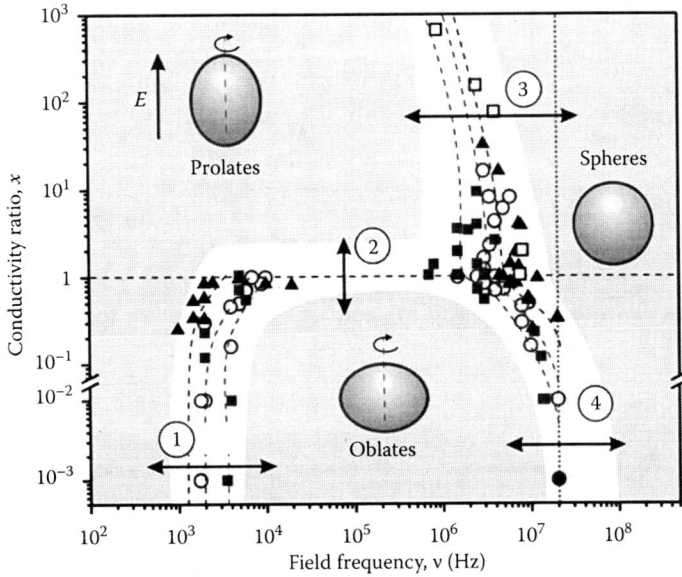

FIGURE 5.2 Morphological diagram of the shapes of vesicles at different conductivity conditions ($x = \lambda_{in}/\lambda_{ex}$) and various field frequencies as determined experimentally. The symbols correspond to different internal conductivity, λ_{in}, in units mS/m: 1.5 (solid squares), 6.5 (open circles), 13 (solid triangles), 1000 (open squares). The dashed lines are guides to the eye and the shaded areas indicate zones of specific morphology. The four types of morphological transitions are discussed in the text. The dotted vertical line shows the experimentally accessible frequency limit ($\nu = 2 \times 10^7$ Hz). Schematic views of the vesicle shapes are included as insets, and the electric field is indicated by an arrow. (Adapted from Dimova, R. et al., *Soft Matter*, 5, 3201, 2009. With permission.)

i.e., with internal conductivity higher than the external one, become prolate ellipsoids corresponding to transition 3 in Figure 5.2, while vesicles with $x < 1$ adopt oblate shapes after undergoing transition 4. Further decrease in frequency changes the vesicle shape at transition 1 from oblate to prolate for $x < 1$. For intermediate frequencies, an oblate vesicle can become prolate at transition 2. At the border conductivity range, $x \cong 1$, a coexistence of the two shapes, prolate or oblate, can be observed (Dimova et al. 2007).

The physical mechanism responsible for the vesicle electrodeformation is the interplay between the electric field partitioning in normal and tangential components, and the charging of the membrane interfaces. The lipid bilayer is an insulator, and acts as a capacitor. At low frequencies, $\omega \ll 1/t_c$ (see Equation 5.2 for the definition of t_c), the large membrane impedance blocks current from flowing into the vesicle interior and the electric field lines are tangent to the membrane. The vesicle is squeezed at the equator and pulled at the poles by the radial Maxwell stress or pressure arising from the tangential electric field. As a result, the vesicle adopts a prolate shape.

At intermediate frequencies, $1/t_c < \omega < 1/t_{MW}$ (see Equation 5.1 for the definition of t_{MW}), the membrane is capacitively short-circuited and displacement currents flow through it. The electric field lines penetrate the vesicle interior and the electric field acquires a component normal to the membrane. When $x \neq 1$, i.e., because of the asymmetry of the internal and external conductivities, the charge densities on the inner and outer membrane interfaces become imbalanced. These charges arise from the discontinuity of the permittivities across the interfaces and represent the local accumulation of cations and anions at these interfaces. The sign of the net free-charge density at the membrane is determined mainly by the conductivity ratio: schematic snapshots for $x > 1$ and $x < 1$ are sketched in Figure 5.3b and c. The interaction of the tangential and normal electric fields with the free charges produces lateral and normal forces, f_t and f_n, respectively. Depending on the polarity of the net free charge, f_t is directed

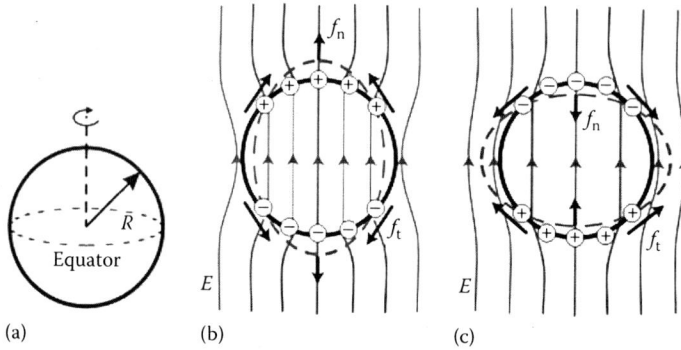

FIGURE 5.3 Origin of the vesicle deformation in AC fields at intermediate frequency (see transition 2 in Figure 5.2): (a) the vesicle geometry in the absence of field and (b, c) snapshots of the net-charge distribution at the vesicle interfaces at intermediate frequencies. Due to the difference in the conductivity conditions, the net charges across the membrane, illustrated with pluses and minuses, differ depending on the value of the conductivity ratio $x = \lambda_{in}/\lambda_{ex}$. The forces ($f_n$ and f_t) applied to the charges by the normal and the tangential electric fields deform the vesicles into prolates for higher internal conductivity, $x > 1$ (b), and oblates for higher external conductivity $x < 1$ (c). (Adapted from Dimova, R. et al., *Soft Matter*, 5, 3201, 2009. With permission.)

either toward the poles or the equator, and f_n is directed inward or outward (Vlahovska et al. 2009), leading to prolate or oblate vesicle shapes as indicated by transition 2 in Figure 5.2 and sketched in Figure 5.3b and c.

In the high-frequency regime, $\omega > 1/t_{MW}$, the electric charges cannot follow the oscillations of the electric fields. As a result, the net charge density decreases with the field frequency. This relaxes the shape of the vesicle from prolate ($x > 1$) or oblate ($x < 1$) to spherical (transitions 3 and 4 in Figure 5.2).

Theoretical studies of vesicle deformation in AC fields have been limited to rather simple systems. For example, these studies omit the asymmetry in the media conductivities (Winterhalter and Helfrich 1988, Mitov et al. 1993, Peterlin et al. 2007), and their theoretical predictions are at odds with experiment [see, e.g., the supplementary material in Aranda et al. (2008)]. Recently, another approach based on the balance of all forces exerted on the membrane was reported (Vlahovska et al. 2009). This model, accounting for variable membrane tension and hydrodynamic forces, correctly describes the vesicle deformation in AC fields.

A detailed understanding of the membrane behavior in AC fields is important for various electromanipulation techniques, as well as for the traditional protocol for giant vesicle preparation based on electroformation (Angelova and Dimitrov 1986, Angelova et al. 1992) and some of its recent improvements (Taylor et al. 2003a,b, Estes and Mayer 2005, Montes et al. 2007, Pott et al. 2008). Even though vesicle electroformation is widely used, the underlying mechanism is not well understood (Angelova 2000, Sens and Isambert 2002). This motivates further studies on the effects of AC fields on membranes.

5.2.2 Lipid Flows Triggered by Inhomogeneous AC Fields

Electric fields induce various electrokinetic phenomena, which have found wide application in micro- and nanotechnologies in the past decade. Electro-neutral particles (droplets, bubbles, lipid vesicles, solid beads) suspended in a medium of different polarizability acquire charge at their surfaces when exposed to electric fields. The interaction of the fields with the surface charges may result in particle kinetics, electroosmotic fluid flow, or a combination of both (Squires and Bazant 2004). These kinetic phenomena are widely used in biology and biotechnology as methods for cell or membrane characterization (Schwan 1983, Yang et al. 1999) and micro-manipulation (Zimmermann and Neil 1996, Voldman 2006, Lecuyer et al. 2008).

As discussed in Section 5.2.1, electric fields induce forces at the vesicle interface. At intermediate frequencies, $1/t_c < \omega < 1/t_{MW}$, the lateral force f_t is involved in the vesicle deformation, as shown in Figure 5.3. In addition, this force may also lead to fluid flows, analogous to the flows induced in liquid droplets. However, there is a fundamental difference between droplets and vesicles, which arises from the properties of the lipid bilayer (Taylor 1966). The membrane behaves as a two-dimensional nearly incompressible fluid. Under stress, it develops tension to keep its surface area constant. In uniform AC fields, membrane flow in the vesicle is not expected because the lateral electric stress is counterbalanced by the resulting axially symmetric membrane tension. In inhomogeneous fields, however, this force balance is broken and a flow of lipids set off in order to restore it.

To visualize such flows along the membrane, one can employ giant vesicles with fluorescent domains as markers for the movement of the membrane (Staykova et al. 2008). Such fluid domains can be obtained if the membranes are made of a lipid mixture, which, at room temperature, separates in two fluid phases (see, e.g., Veatch and Keller 2002, Lipowsky and Dimova 2003). The two phases show as domains on the vesicles if a fluorescent dye is employed, which preferentially partitions in one of the phases. The membrane flow pattern can be resolved by following the motion of the domains with confocal microscopy (Staykova et al. 2008). To generate the inhomogeneous AC field, the vesicles can be prepared to be heavy, for example, by loading them with sucrose solution and placing them in glucose media. Due to the density difference of the two solutions, the vesicles sediment at the bottom of the observation chamber. The proximity of the bottom glass to the vesicle, as shown in Figure 5.4a, leads to an asymmetric field distribution at the membrane surface. The field strength is much higher at the lower vesicle part, facing the glass, than at the top part (Staykova et al. 2008).

Such asymmetric field distribution leads to special membrane flow patterns, consisting of concentric closed trajectories organized in four symmetric quadrants, each extending from the bottom to the top of the vesicle; see Figure 5.4a through e. The flow is the fastest at the periphery of the quadrant and at the bottom of the vesicle. The top and the bottom of the vesicle are stagnation points. The velocity of the domains (in the order of micrometers per second) can be altered by the field strength and frequency, and by the conductivity of the external solution. The calculations of the lateral electric stress or surface force density on the membrane suggest that the vesicle experiences significant shear stress in the vicinity of the solid substrate (Staykova et al. 2008). As a result, a nonuniform and nonsymmetric membrane tension builds up. It triggers lipid flow toward the regions of highest tension, in analogy to Marangoni flows in monolayers (Marangoni 1871, Edwards et al. 1991, Darhuber and Troian 2005).

These AC field-induced flows in giant vesicles have possible applications in microfluidic technologies. Giant vesicles in inhomogeneous AC fields or in hydrodynamic flows mimicking, e.g., the situation of red blood cells in capillaries may be used as nano-reactors for fluid manipulation, i.e., displacing, mixing, trapping, etc. Lipid mixing is demonstrated in Figure 5.4f through h. If we expose a vesicle with only two domains to an AC field for a certain period of time, the domains undergo fission, leading to the appearance of a large number of smaller domains. For sufficiently strong membrane flows, the number of domains grows with the time of exposure. The growing number of domains, on the other hand, increases the probability of domain encounter and fusion. Domain fusion counterbalances the fission and, therefore, the domains will reach a stationary state characterized by a certain size distribution after a certain time.

5.3 Vesicle Response to DC Pulses

While the discussion of vesicles exposed to AC fields was limited to stationary shapes, DC pulses induce short-lived shape deformations. The pulse duration is usually varied from several microseconds to milliseconds, while studies on cells have investigated a much wider range of pulse durations—from tens of nanoseconds to milliseconds and even seconds (Neumann et al. 1989), as discussed in various chapters of this book (see e.g. Chapters 2, 7 through 10, 12, 17 and 18). Because the application of both AC fields

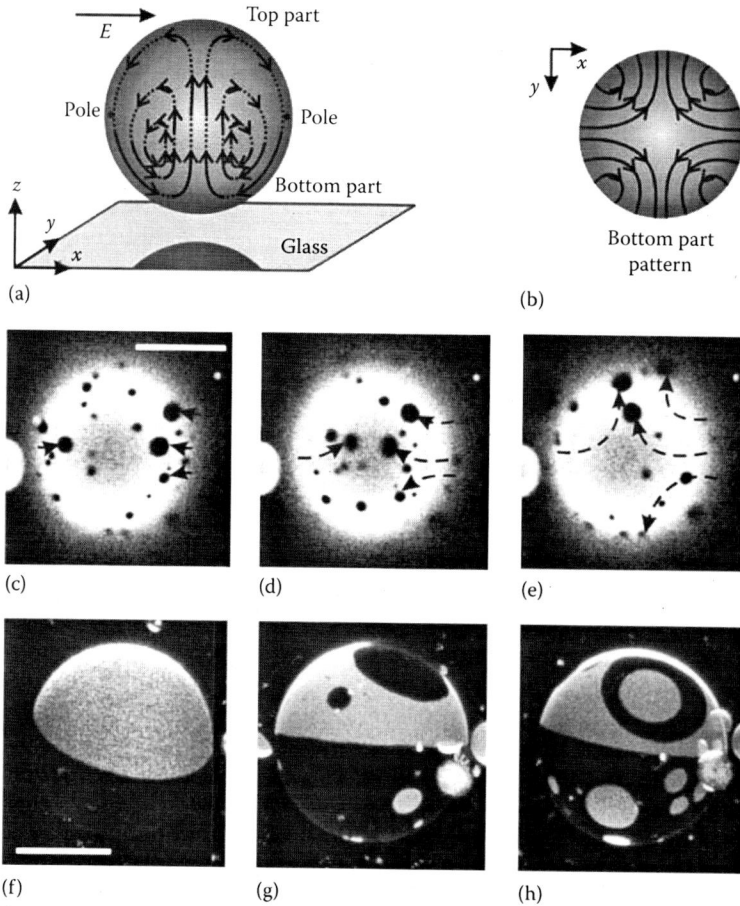

FIGURE 5.4 Membrane flow triggered on a giant vesicle exposed to an inhomogeneous AC field. The vesicle is located close to the bottom of the observation chamber as illustrated in (a), where the vesicle top and bottom parts, the poles, and the field directions are indicated. The side and the bottom views of the flow lines on the vesicle surface are sketched in (a) and (b), respectively. The length of the arrows in (a) roughly corresponds to the amplitude of the flow velocity. The lipid flow induced by an AC field (36 kV/m, 80 kHz) on a giant vesicle with a diameter of about 150 μm, at external and internal conductivities of 25 and 0.3 mS/m, respectively, is shown in micrographs (c–e). The time between the consecutive snapshots is approximately 1.3 s. The dashed arrows indicate the trajectories of selected domains in the consecutive snapshots. Lipid mixing is demonstrated on another vesicle (~95 μm diameter), which initially had two domains as seen with three-dimensional confocal scans of the lower vesicle hemisphere (f). The domains break apart after continuous field exposure (80 kHz, 50 kV/m) of 2 (g) and 3 min (h). The scale bars corresponds to 50 μm. (Adapted from Staykova, M. et al., *Soft Matter*, 4, 2168, 2008. With permission.)

and DC pulses creates a transmembrane potential, the vesicle deformations of a similar nature are to be expected in both cases. Indeed, the AC field frequency, ν, can be compared to the inverse duration of the DC pulse, $1/t_p$. This correspondence should be valid for a certain shape of the AC field signal (whether sinusoidal, rectangular, or any other shape) and of the DC pulse (square, triangular, etc.). For example, applying a square wave AC field is analogous to applying a sequence of square-wave DC pulses with alternating electrode polarity. At equivalent conductivity conditions, a DC pulse of duration $t_p = 100\,\mu s$ should induce a deformation similar to the one obtained for a square-wave AC field of frequency ν = 5 kHz, i.e., a correspondence between ν and $1/(2t_p)$ is to be expected. However, the working field strength for the DC pulses is usually higher by several orders of magnitude. Thus, the degree of deformation can be different.

As mentioned in Section 5.1.3.1, in theoretical studies typically the angular field frequency, ω, is used ($\omega = 2\pi\nu$). The latter is preferred because it simplifies the mathematical treatment of the problem.

Vesicle *deformation* induced by DC pulses has been studied theoretically (Hyuga et al. 1991a,b, Sokirko et al. 1994). The majority of experimental studies were preformed on small vesicles with hundreds of nanometers in size (Neumann et al. 1998, Griese et al. 2002, Kakorin and Neumann 2002). The *poration* of small vesicles induced by DC pulses has attracted even stronger interest (Teissie and Tsong 1981, Glaser et al. 1988, Kakorin and Neumann 2002). Because of the small size of the vesicles, the direct observation of the deformation and poration is not feasible. Thus, experiments on giant vesicles are of special relevance because their size allows for direct observation using optical microscopy (Kinosita et al. 1992, Zhelev and Needham 1993, Sandre et al. 1999, Tekle et al. 2001, Rodriguez et al. 2006). However, the microscopy observation of effects caused by electric pulses on lipid giant vesicles is difficult because of the short duration of the pulse. A typical video frequency of 30 frames per second (fps) would provide a snapshot every 33 ms. Having in mind that the pulse duration can be about three orders of magnitude shorter, the immediate dynamics of the vesicle response would be indiscernible at such acquisition speeds. Recently, the direct experimental visualization of the vesicle response was achieved via fast digital imaging with temporal resolution of up to 30,000 fps, i.e., acquiring one image every 33 µs (Riske and Dimova 2005a, 2006, Haluska et al. 2006). In this way, the dynamics of the vesicle response during and after the applied DC pulse was resolved. The following sections introduce some characteristic features of the deformation of spherical vesicles when exposed to DC pulses.

5.3.1 Dynamic of Vesicle Deformation in Salt-Free Solutions

5.3.1.1 Response of Vesicles Composed of Zwitterionic Lipids

Spherical vesicles subjected to electric pulses adopt ellipsoidal shapes, which relax back to the initial vesicle shapes after the end of the pulse. The degree of deformation of an ellipsoidal vesicle can be characterized by the aspect ratio of the two principal radii, a and b (see sketch in Figure 5.5a). For $a/b = 1$ the vesicle is a sphere. The relaxation dynamics of this aspect ratio depends on whether the vesicle has been

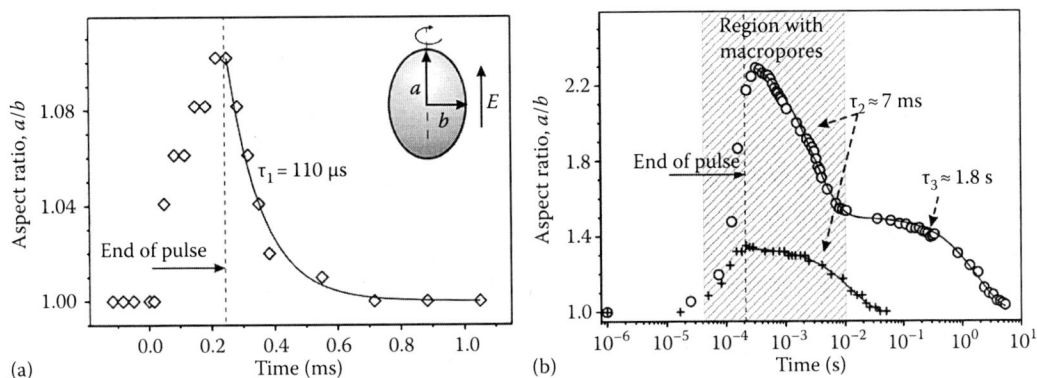

FIGURE 5.5 Deformation of vesicles exposed to square-wave DC pulses. (a) Response of a vesicle subjected to a pulse below the poration threshold: field strength $E = 100\,\text{kV/m}$, pulse duration $t_p = 250\,\mu\text{s}$; $\Psi_m(t = t_p) < \Psi_c$. The solid curve is an exponential fit with a decay time τ_1 as given in the figure. (b) Data from the response and relaxation of two vesicles, in which macropores were observed: $E = 200\,\text{kV/m}$, $t_p = 200\,\mu\text{s}$, $\Psi_m(t = t_p) > \Psi_c$. One of the vesicles (crosses) did not have excess area and the relaxation is described by a single exponential fit (solid curve) with a decay time τ_2. The other vesicle (open circles) had excess area and its relaxation is described by a double exponential fit (solid curve) with decay times τ_2 and τ_3 as described in the text. The shaded area indicates the time interval when macropores were optically detected. In both plots, time $t = 0$ was set as the beginning of the pulse. The dashed lines indicate the end of the pulse. (Adapted from Dimova, R. et al., *Soft Matter*, 3, 817, 2007. With permission.)

porated or not. In the absence of poration, the relaxation can be described by a single exponential with a characteristic decay time, τ_1. Figure 5.5a gives one example of the response of a giant vesicle, which is initially spherical. The pulse conditions in this case build up a transmembrane potential, which is below the poration limit. The maximum deformation of this vesicle corresponds to about 10% change in the vesicle aspect ratio. The degree of vesicle deformation depends on the initial tension of the vesicle as well as on the excess area. The latter represents an excess compared to the area of a spherical vesicle of the same volume.

The typical decay time for the relaxation of nonporated vesicles, τ_1, is on the order of 100 μs. It is defined by the relaxation of the total membrane tension achieved at the end of the pulse, which is the sum of the electrotension σ_{el} and the initial tension σ_0; see Equation 5.6. Thus, τ_1 relates mainly to the relaxation of membrane stretching: $\tau_1 \sim \eta_D/\sigma$, where η_D is the surface dilatational viscosity of the membrane (see Section 5.1.2). For membrane tensions of the order of 5 mN/m (which should be around the maximum tension before the membrane ruptures, $\sigma \cong \sigma_{lys}$) and for typical values of η_D (see Table 5.1), one obtains $\tau_1 \sim 100$ μs, which corresponds to the value experimentally measured; see Figure 5.5a.

Porated vesicles exhibit more complex dynamics. The pores can reach various sizes depending on the location on the vesicle or cell surface (see, e.g., Weaver and Chizmadzhev 1996, Krassowska and Filev 2007 and work cited therein). For the case of plane parallel electrodes, the poration occurs predominantly in the area at the poles of the vesicle facing the electrodes. This is because the transmembrane potential attains its maximal value at the two poles as expressed by the angular dependence in Equation 5.4. With optical microscopy, only pores that are of diameter larger than about half a micron can be resolved. We refer to them as macropores since they are much larger than the average ones. With fast camera observation on giant vesicles, macropores can be visualized in the following way. If the vesicles are prepared in sucrose solution and then diluted in isotonic glucose solution, the refractive indices of the internal and external vesicle media are different. Thus, with phase contrast microscopy, the vesicles appear as dark objects on a light gray background. When macropores are formed, there is an efflux of the darker sucrose solution. Figure 5.6 shows an example of an electroporated vesicle.

The lifetime of macropores, τ_{pore}, observed in vesicles in the fluid state, varies with pore radius, r_{pore} (Sandre et al. 1999), and depends on the membrane edge tension, γ, and the membrane dilatational viscosity: $\tau_{pore} \sim 2r_{pore}\eta_D/\gamma$. For vesicles with low tension, the lifetime τ_{pore} is typically shorter than 30 ms (Riske and Dimova 2005a).

Two different types of dynamics can be distinguished for porated vesicles. The relaxation of vesicles with no excess area is described by a single exponential decay, while vesicles with excess area exhibit two characteristic decay times; see Figure 5.5b. The maximum deformation achieved in both cases is much higher than the one observed for nonporated vesicles (compare with Figure 5.5a). Naturally, vesicles with excess area deform much more than those without (compare the two curves in Figure 5.5b). For vesicles with no excess area, the relaxation time is $\tau_2 \cong 7 \pm 3$ ms. When the vesicles have some excess area,

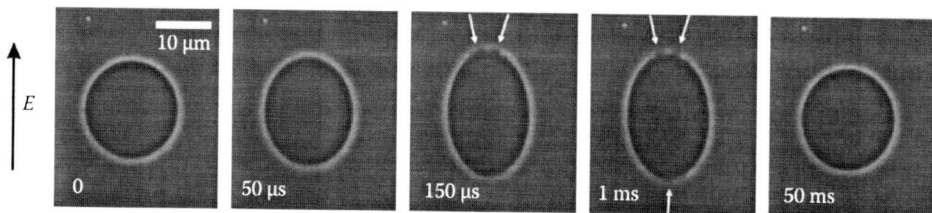

FIGURE 5.6 Poration of a vesicle (phase contrast microscopy; the internal solution is sucrose and the external one is glucose). The DC pulse duration is 200 μs and the field strength is 140 kV/m. The direction of the field is indicated by the arrow on the left. The time period in the lower left corner of each snapshot is the time after the beginning of the pulse. The arrows in the third and fourth snapshots indicate pores at the vesicle poles visualized by dark eruptions of sucrose solution leaking out of the vesicle. The pores reach a diameter up to 4 μm. (Adapted from Dimova, R. et al., *Soft Matter*, 3, 817, 2007. With permission.)

the relaxation proceeds in two steps, fast relaxation characterized by τ_2, and a second, longer, relaxation with decay time τ_3: $0.5\,s < \tau_3 < 3\,s$.

The relaxation process associated with τ_2 takes place during the time interval when the pores are present (see shaded region in Figure 5.5b). Thus, τ_2 is determined by the closing of the pores: $\tau_2 \sim \eta_D r_{pore}/(2\gamma)$. The line energy per unit length as above is $\gamma \sim \kappa/2h$. Taking some typical values for the bending stiffness κ (see Table 5.1) and the membrane thickness, yields for γ a value on the order of 10^{-11} N (Harbich and Helfrich 1979). Then, for a typical pore radius of $1\,\mu m$ one obtains $\tau_2 \sim 10\,ms$.

The relaxation time τ_3 is related to the presence of some excess area available for shape changes. The latter can be characterized by a dimensionless volume-to-area ratio $\upsilon = (3V/4\pi)(4\pi/A)^{3/2}$, where V and A are the vesicle volume and area, respectively. This reduced volume υ is 1 when the vesicle is a sphere, and smaller than 1 in the rest of the cases. The relaxation described by τ_3 is associated with the process of pushing away the volume of fluid involved in the ellipsoidal vesicle deformation compared to a relaxed spherical shape. The restoring force is related to the bending elasticity of the lipid bilayer. Then, the decay time can be presented as $\tau_3 \sim (4\pi\eta R^3/3\kappa)(1/\upsilon - 1)$, where η is the bulk viscosity of the immersing media and R is the vesicle radius. For typical values of υ between 0.99 and 0.94, one obtains τ_3 between approximately 0.5 and 3 s, which corresponds excellently to the measured data (Riske and Dimova 2005a); see Figure 5.5b. All characteristic decay times for vesicle relaxation and the macropore lifetime are summarized in Table 5.2.

5.3.1.2 Unusual Behavior of Charged Membranes Exposed to DC Pulses

As discussed in Section 5.3.1.1, strong electric pulses applied to single-component giant vesicles made of zwitterionic lipids like phosphatidylcholines (PC) induce the formation of pores, which reseal within milliseconds. When negatively charged lipids, like phosphatidylglycerol (PG) or phosphatidylserine (PS), are present in the membrane, a very different response of the vesicles can be observed, partially influenced by the medium conditions (Riske et al. 2009).

In buffered solutions containing EDTA, PC:PG vesicles with molar ratios 9:1, 4:1, and 1:1 behave in the same way as pure PC vesicles, i.e., the pulses induce opening of macropores with a diameter up to about $10\,\mu m$, which reseal within tens of milliseconds. In non-buffered solutions, membranes with low molar fractions of charged lipids (9:1 and 4:1) retain this behavior, but very surprising response can observed for 1:1 PC:PG vesicles—they generally disintegrate after electroporation (Riske et al. 2009); see Figure 5.7. Typically, one macropore forms and expands in the first 50–100 ms at a very high speed of approximately 1 mm/s. The entire vesicle content is released seen as darker fluid in Figure 5.7. The bursting is followed by the restructuring of the membrane into what seem to be interconnected bilayer fragments in the first seconds, and a tether-like structure in the first minute. Then the membrane stabilizes

TABLE 5.2   Characteristic Times of Vesicle Relaxation, Poration, and Fusion

Characteristic Time	Specification	Typical Values	Dependence on Membrane Material Properties
τ_1	Decay time for conditions of no poration	$\sim 100\,\mu s$	$\tau_1 \sim \eta_D/\sigma$
τ_2	Decay time for conditions of poration	$3\text{--}10\,ms$	$\tau_2 \sim \eta_D\, r_{pore}/(2\gamma)$
τ_3	Relaxation time of vesicles with excess area	$0.5\text{--}3\,s$	$\tau_3 \sim (4\pi\eta R^3/3\kappa)(1/\upsilon - 1)$
τ_{pore}	Lifetime of macropores	$<30\,ms$	$\tau_{pore} \sim 2r_{pore}\eta_D/\gamma$
τ_{early}	Early stage of fusion neck expansion	$\sim 100\,\mu s$	$\tau_{early} \sim \tau_1 \sim \eta_D/\sigma$
τ_{late}	Late stage of fusion neck expansion	$\sim 100\,s$	$\tau_{late} \sim \eta R^3/\kappa$

Note: See text for details.

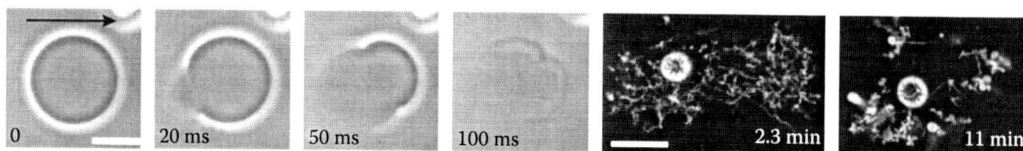

FIGURE 5.7 Bursting of charged (PC:PG 1:1) vesicles subjected to electric pulses. The time after the beginning of the pulse is marked on each image. The first four images are phase contrast microscopy snapshots from fast camera observation of a vesicle in salt solution subjected to a pulse with field strength 120 kV/m and duration 200 μs. The field direction is indicated in the first snapshot with an arrow. The vesicle bursts and disintegrates. The last two images show confocal cross-sections of a vesicle, which has been subjected to an electric pulse and has burst and rearranged into a network of tubes and smaller vesicles.

into interconnected micron-sized tubules and small vesicles. Similar behavior is observed on vesicles prepared from lipid extracts from the plasma membrane of the red blood cells, which also contain a fraction of charged lipids (Riske et al. 2009). These observations suggest that the vesicle bursting and membrane instability is related to the amount of the charged lipid in the bilayer and to the medium. No vesicle disintegration is observed in buffered solution containing EDTA and for a low content of the charged lipid. Indeed, instabilities in membranes are related to the charged state of the bilayer (Isambert 1998, Betterton and Brenner 1999, Kumaran 2000).

The amount of charged lipid in the bilayer is not the only factor triggering a burst of the synthetic membranes. In particular, as mentioned above, vesicles with a high content of PG (50 mol%) do not burst in solutions containing EDTA, which is a chelating agent generally added in solutions to bind possible multivalent ions present as impurities in the solution, like calcium (Riske et al. 2003). The observation that EDTA suppresses bursting suggests that divalent ions present as impurities can act as destabilizing agent in the process of vesicle bursting.

Plasma membranes should exhibit similar bursting behavior as that of the vesicles made of lipid extracts, because their lipid composition is similar. However, cell membranes are subjected to internal mechanical constraints imposed by the cytoskeleton, which prevents their disintegration even if their membranes are prone to disruption when subjected to pulses. Instead, the pores in the cell membrane are stable for a long time (Schwister and Deuticke 1985), and can either lead to cell death by lysis or reseal depending on the media (Kinosita and Tsong 1977a, Tekle et al. 1994). The latter is the key to efficient electroporation-based protocols for drug or gene transfer in cells. The results discussed in this section suggest that membrane charge as well as minute amounts of molecules such as EDTA might be important but not yet well-understood regulating agents in these protocols.

5.3.2 Deformation Dynamics of Vesicles in Salt Solutions

In the presence of salt in the vesicle exterior (e.g., NaCl solution with concentration above 0.1 mM), unusual shape changes are observed (Riske and Dimova 2006). The vesicles assume cylindrical shapes during the pulse. These deformations are short-lived (their lifetime is about 1 ms) and occur only in the presence of salt outside the vesicles, irrespective of their inner content. When the solutions conductivities inside and outside are the same, $\lambda_{in} \approx \lambda_{ex}$ ($x \approx 1$), vesicles with square cross-section are observed. For the case $\lambda_{in} < \lambda_{ex}$ ($x < 1$), the vesicles assume disc-like shapes, while in the opposite case, $\lambda_{in} > \lambda_{ex}$ ($x > 1$), they deform into long cylinders with rounded caps; see Figure 5.8. The transition from tubes to discs is analogous to transition 2 (from prolates to oblates) in Figure 5.2 observed in AC fields.

The detected cylindrical deformations are nonequilibrium shapes and have a very short lifetime, which is why they have not been observed at standard video acquisition speed. The flattening of the vesicle walls starts during the applied pulse and is observed throughout a period of about one millisecond. The formation of these shapes is not well understood. They are observed also in the presence of small negatively charged nanoparticles in the vesicle exterior (Dimova et al. 2009). As discussed in a

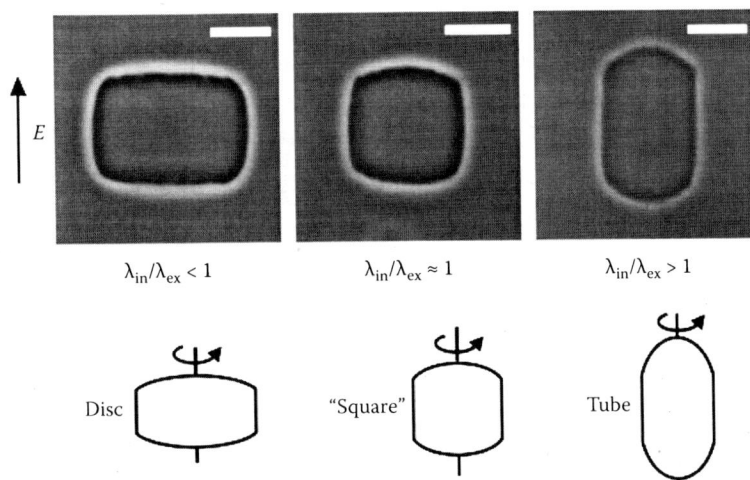

FIGURE 5.8 Deformation of vesicles at different conductivity conditions subjected to DC pulses. Schematic illustrations of the cross sections of the vesicles are given below every snapshot. The field direction is indicated with an arrow on the left. The presence of salt in the vesicle exterior causes flattening of the vesicle membrane into disc-like, "square"-like, and tube-like shapes, whereby the overall vesicle shape depends on the conductivity ratio $x = \lambda_{in}/\lambda_{ex}$. The scale bars correspond to 15 μm. (Adapted from Dimova, R. et al., *Soft Matter*, 3, 817, 2007. With permission.)

previous report (Riske and Dimova 2006), one possible explanation could be that ions or particles flatten the equatorial zone of the deformed vesicle. During the pulse, there is an inhomogeneity in the membrane tension due to the fact that the electric field is the strongest at the poles of the vesicle, and almost zero close to the equator. The kinetic energy of the accelerated ions or particles hitting the equatorial (tensionless) region of the vesicle is higher than the energy needed to bend the membrane, thus leading to the observed deformation. In addition, particle-driven flows may be inducing membrane instability, giving rise to higher-order modes of the vesicle shape (Kantsler et al. 2007). Yet another possible explanation may be related to a change in the spontaneous curvature of the bilayer due to the particle or ion asymmetry across the membrane (Lipowsky and Dobereiner 1998). During the pulse, the local and transient accumulation of particles or ions in the membrane vicinity can occur. The mechanism driving the cylindrical deformations might be a combination of nanoparticle electrophoresis and the changes in the membrane spontaneous curvature. Furthermore, another influencing factor might be an electrohydrodynamic instability caused by electric fields interacting with flat membranes, which was predicted to increase the membrane roughness (Sens and Isambert 2002). Finally, the flexoelectric properties of the lipid membrane, first postulated by Petrov (1984, 2002) and Petrov and Bivas (1984), may also be involved as recently proposed (Gao et al. 2008, 2009). The interplay between interface ion concentration gradients combined with the overall ionic strength and the bilayer material properties and tension could be expected to produce the cylindrical deformation we observe.

5.3.3 Behavior of Gel-Phase Vesicles Exposed to DC Pulses

In the previous Sections (5.2, 5.3.1 and 5.3.2), we discussed the response of membranes in the fluid state. The mechanical and rheological properties of such membranes differ significantly from those of membranes in the gel phase. For example, the bending stiffness and the shear surface viscosity of gel-phase membranes are of higher orders of magnitude than those of membranes in the fluid phase (see Table 5.1) (Heimburg 1998, Dimova et al. 2000, Lee et al. 2001, Mecke et al. 2003), and membranes in the gel phase are thicker (Nagle and Tristram-Nagle 2000). These differences introduce new features in the response of gel-phase membranes to electric fields, which we discuss next.

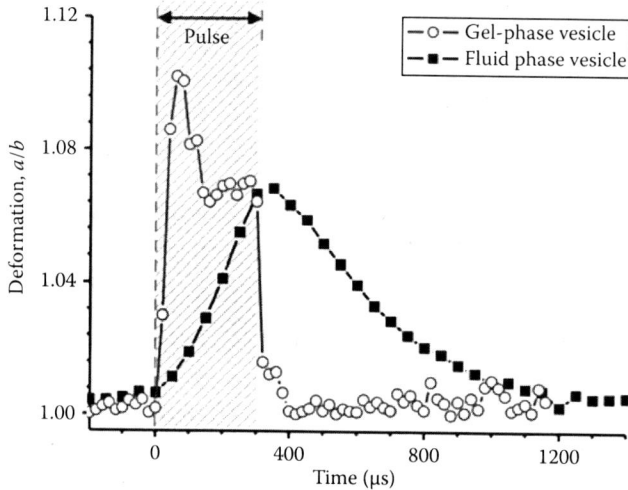

FIGURE 5.9 Deformation response of a gel-phase vesicle made of dipalmitoylphosphatidylcholine with a radius of 22 μm, and a fluid-phase vesicle made of palmitoyloleoylphosphatidylcholine with a radius of 20 μm to DC pulses with duration of 300 μs. The pulse length is indicated by the shaded zone. The field strength of the pulses was 500 and 80 kV/m for the gel and the fluid vesicle, respectively. (Adapted from Dimova, R. et al., *Soft Matter*, 5, 3201, 2009. With permission.)

We compared the response to DC pulses of a vesicle in the fluid phase, with that of a vesicle in the gel phase. The applied DC pulses were weak enough not to induce the formation of macropores in the membranes, and no leakage of the internal sucrose solution outside the vesicle was observed. Figure 5.9 shows the deformation of the two vesicle types in response to DC pulses 300 μs long. To achieve similar maximal degree of deformation (on the order of 10%), stronger pulses had to be applied to the gel-phase vesicle as compared to the fluid one. Pulses with field strength about 100 kV/m do not produce optically detectable deformations in gel-phase vesicles, while strong pulses about 500 kV/m applied to the fluid-phase vesicles cause poration.

The responses of the two vesicles differ significantly. The fluid vesicle gradually deforms and reaches maximal deformation (i.e., maximal aspect ratio a/b) at the end of the pulse. The gel-phase vesicle responds significantly faster, and exhibits a relaxation with a decay time of about 30 μs during the pulse. To our knowledge, such intra-pulse relaxation has not been previously reported and is related to membrane wrinkling (Knorr et al. 2010). The vesicles had similar size and were both in salt-free solutions. Gel-phase membranes are thicker, and thus have lower membrane capacitance (Stuchly et al. 1988), leading to shorter charging times. The faster response of the gel-phase vesicle, as shown in Figure 5.9, correlates with the shorter charging time as compared to the fluid vesicle.

After the end of the pulse, the relaxation of the gel vesicle is also much faster than that of the fluid membrane. The relaxation behavior depends on whether the membrane is porated or not; see Section 5.3.1.1. In the example given in Figure 5.9, no macropores were detected, but it is plausible that in the gel-phase, vesicle pores with sizes in the sub-optical range were formed during the pulse. The formation of such pores and deviations from ellipticity as characterized by the vesicle aspect ratio a/b may explain the intra-pulse relaxation in the vesicle deformation.

If DC pulses of field strength larger than the discussed above are applied, the gel-phase vesicles rupture: the pores resemble micrometer-sized cracks on a solid shell (Dimova et al. 2007). The electroporation studies on small unilamellar vesicles (about 100 nm in diameter) in the gel phase suggested the formation of transient pores (Teissie and Tsong 1981). These pores were presumably much smaller than 100 nm, and the poration was reported not to induce a global damage of the vesicles. It was believed that the electroporation of small vesicles in the gel phase is a reversible process (Teissie and

Tsong 1981). However, the direct observations on giant vesicles suggest that the formation of cracks in the gel-phase membrane appears to be an irreversible process in laboratory timescale. Within a period of more than 10 min, one does not observe resealing of these cracks. In some cases, the vesicles collapse at a later stage.

Compared to the macropores in fluid membranes, which have a lifetime shorter than about 30 ms, the pore resealing in vesicles in the gel phase seems to be arrested. This is to be expected, considering the dependence of the pore lifetime, τ_{pore}, on the membrane viscosity; see Table 5.2. The dilatational surface viscosity is expected to show similar behavior to that of the shear surface viscosity, which diverges when the membrane crosses the fluid-to-gel transition (see Table 5.1). Thus, the resealing of pores formed on membranes in the gel phase is hampered.

Understanding the response of the gel-phase membranes will require a thorough consideration of the membrane mechanical and rheological properties as well as the interaction of electric fields with such membranes. Both the intra- and after-pulse relaxations of the vesicles in gel phase (see Figure 5.9) are poorly understood and should be subjected to further investigation.

5.4 Vesicle Electrofusion

When a DC pulse is applied to a couple of fluid-phase vesicles, which are in contact and oriented in the direction of the field, electrofusion can be observed. Vesicle orientation and alignment into pearl chains can be achieved by application of AC field to a vesicle suspension. This phenomenon is also observed with cells (Zimmermann 1986, Zimmermann and Neil 1996) and is due to the dielectric screening of the field. When the suspension is dilute, two vesicles can be brought together via the AC field and aligned. A subsequent application of a DC pulse to such a vesicle couple can lead to fusion. The necessary condition is that poration is induced in the contact area between the two vesicles. The possible steps of the electrofusion of two membranes are schematically illustrated in Figure 5.10.

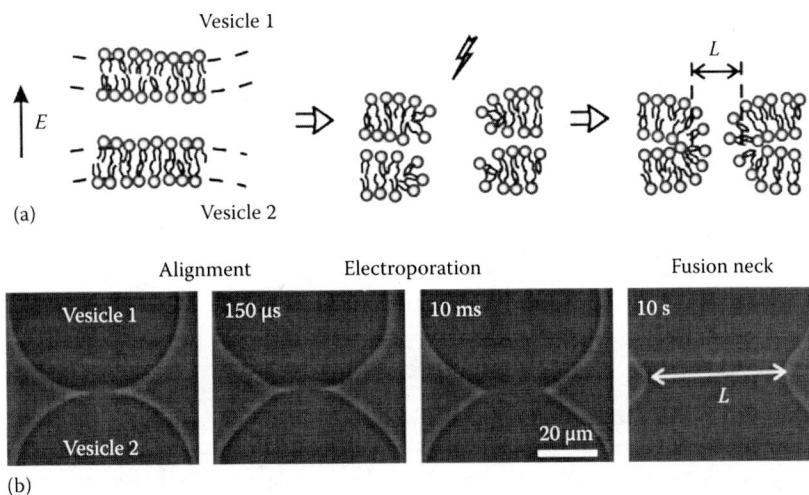

FIGURE 5.10 (a) Schematic illustration of the possible steps of the electrofusion process: two lipid vesicles are brought into contact (only the membranes in the contact zone of the vesicles are sketched), followed by electroporation and formation of a fusion neck of diameter L. (b) Micrographs from electrofusion of a vesicle couple. Only segments of the vesicles are visible. The external solution contains 1 mM NaCl, which causes flattening of the vesicle walls in the second snapshot (see Section 5.3.2). The amplitude of the DC pulse was 240 kV/m, and its duration was 120 μs. The time after the beginning of the pulse is indicated on the snapshots. (Adapted from Dimova, R. et al., *Soft Matter*, 3, 817, 2007. With permission.)

5.4.1 Dynamics of the Opening of the Fusion Neck

Membrane fusion is a fast process. The time needed for the formation of a fusion neck can be rather short as follows from electrophysiological methods applied to the fusion of small vesicles with cell membranes (Llinas et al. 1981, Lindau and de Toledo 2003, Hafez et al. 2005, Shillcock and Lipowsky 2005). The time evolution of the observed membrane capacitance indicates that the formation of the fusion neck is presumably faster than 100 μs.

Previously, the limits of direct observation with optical microscopy were in the range of milliseconds. The usage of fast digital imaging shifted the resolution limit by about two orders of magnitude and allowed studying the fusion dynamics with microsecond resolution (Haluska et al. 2006, Riske et al. 2006). An example of a few snapshots taken from the electrofusion of two vesicles in the presence of salt is given in Figure 5.10. Note that in this case, the overall deformation of each vesicle during and shortly after the pulse corresponds to the cylindrical shapes as observed with individual vesicles in the presence of salt (see Section 5.3.2).

From such micrographs, one can measure the fusion neck diameter, denoted by L in Figure 5.10, and follow the dynamics of its expansion. In Figure 5.11, the time evolution of the fusion neck of one vesicle couple is given. The data spans more than five orders in magnitude in time. A first inspection of the fusion neck expansion data shows that two stages of the fusion process can be distinguished: an early one, which is very fast and with an average expansion velocity of about 2×10^4 μm/s, followed by a later slower one.

Intuitively, one would relate the early stage with a fast relaxation of the membrane tension built during the pulse, whereby the dissipation occurs in the bilayer. Essentially, the driving forces here are the same as those responsible for the relaxation dynamics of nonporated vesicles (as characterized by τ_1 in Section 5.3.1.1). Thus, the characteristic time for this early stage of fusion, τ_{early}, can be expressed as $\tau_{early} \sim \eta_D / \sigma$. The membrane tension σ should be close to the tension of rupture σ_{lys}; see Table 5.1. Thus for τ_{early}, one obtains 100 μs, which is in agreement with the experimental observations for the time needed to complete the early stage of fusion.

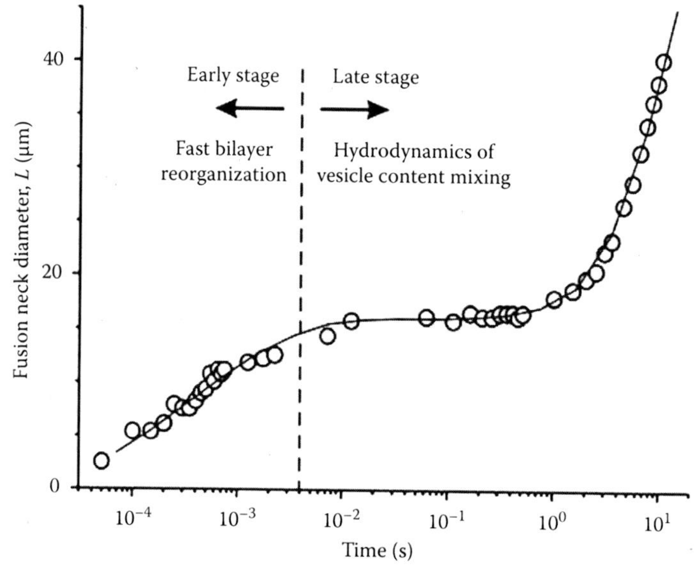

FIGURE 5.11 Time evolution of the fusion neck, L, formed between two vesicles about 15 μm in radius. The solid curve is a guide to the eye. The vertical dashed line indicates the border between the two stages in the fusion dynamics. The early stage of fusion is characterized by an average expansion velocity of 2×10^4 μm/s. The later stage of fusion is characterized by slow opening of the fusion neck with the velocity of 2.4 μm/s. (Adapted from Dimova, R. et al., *Soft Matter*, 3, 817, 2007. With permission.)

The linear extrapolation of the data in the early stage predicts that the formation of the fusion neck with a diameter of about 10 nm should occur within a time period of about 250 ns (Haluska et al. 2006). It is quite remarkable that this timescale of the order of 200 ns was also obtained from computer simulations of a vesicle fusing with a tense membrane segment (Shillcock and Lipowsky 2005).

In the later stage of fusion, the neck expansion velocity slows down by more than two orders of magnitude. Here the dynamics is mainly governed by the displacement of the volume ΔV of fluid around the fusion neck between the fused vesicles. The restoring force is related to the bending elasticity of the lipid bilayer. The parameters that determine this late relaxation are those that govern the relaxation of porated vesicles with excess area, as characterized by τ_3; see Table 5.2 (note that after the two vesicles fuse, the resulting vesicle has a relatively small reduced volume υ and, thus, some excess area). The corresponding decay time in this later stage can be presented as $\tau_{late} \sim \eta \Delta V / \kappa$, where η is the bulk viscosity of the media as before, $\Delta V \sim R^3$, and κ is the bending elasticity modulus of the membrane. Thus, for a typical vesicle size of $R = 20\,\mu m$, we obtain $\tau_{late} \sim 100\,s$, which is the timescale that we measure for a complete opening of the fusion neck. The characteristic fusion times are summarized in Table 5.2.

5.4.2 Some Applications of Vesicle Electrofusion

The phenomenon of electrofusion is of particular interest, because of its vast use in cell biology and biotechnology (see, e.g., Kinosita and Tsong 1977b, Chang et al. 1992, Zimmermann and Neil 1996 and the studies cited therein). The application of electrofusion to cells can lead to the creation of multinucleated viable cells with new properties (this phenomenon is also known as hybridization; see, e.g., Zimmermann and Neil 1996). In addition, electroporation and electro-fusion are often used to introduce molecules like proteins, foreign genes (plasmids), antibodies, and drugs into cells.

Out of the biological context, vesicle fusion can be employed to scale down the interaction volumes of chemical reactions and reduce it to a few picoliters or less. Thus, the fusion of two vesicles of different contents is an illustration for the realization of a tiny microreactor (Chiu et al. 1999, Fischer et al. 2002, Kulin et al. 2003, Noireaux and Libchaber 2004, Yang et al. 2009). The principle of fusion-mediated synthesis is simple: the starting reagents are separately loaded into different vesicles, and then the reaction is triggered by the fusion of these vesicles, which allows the mixing of their contents. The success of this approach is guaranteed by two important factors. First, the lipid membrane is impermeable to the reactants such as ions or macromolecules. Second, fusion can be initiated by a variety of fusogens such as membrane stress (Cohen et al. 1982, Shillcock and Lipowsky 2005, Grafmuller et al. 2007), ions or synthetic fusogenic molecules (Estes et al. 2006, Haluska et al. 2006, Kunishima et al. 2006, Lentz 2007), fusion proteins (Jahn et al. 2003), or electric fields (Riske et al. 2006, Yang et al. 2009). Among the fusion methods listed above, electrofusion becomes increasingly important because of its reliable, fast, and easy handling. An immediate benefit of this strategy is that the precise temporal and spatial control on the synthesis process can be easily achieved.

Electrofusion can be employed as a way to mix solutions enclosed by giant vesicles as membrane compartments, but fusing two vesicles, differing not in their internal content, but in the composition of their membranes can provide a promising tool for studying raft-like domains in membranes (Dietrich et al. 2001, Baumgart et al. 2003, Kahya et al. 2003, Lipowsky and Dimova 2003, Veatch and Keller 2003). Thus, vesicle electrofusion is a very attractive experimental approach for producing multicomponent vesicles of well-defined composition (Riske et al. 2006).

One example for the fusion of two vesicles with different membrane composition is given in Figure 5.12. To distinguish the vesicles according to their composition, different fluorescent markers have been used. In this particular example, one of the vesicles (vesicle 1) is composed of sphingomyelin and cholesterol in 7:3 molar ratio. The other vesicle (vesicle 2) is composed of dioleoylphosphatidylcholine and cholesterol in 8:2 molar ratio (with fluorescence microscopy, the two vesicle can be distinguished by their color; in Figure 5.12 only difference in the intensity can be seen). Thus, the membrane of the fused vesicle is a three-component one. At room temperature, this mixture separates into two phases, liquid ordered (rich in sphingomyelin and cholesterol) and liquid disordered (rich in dioleoylphosphatidylcholine),

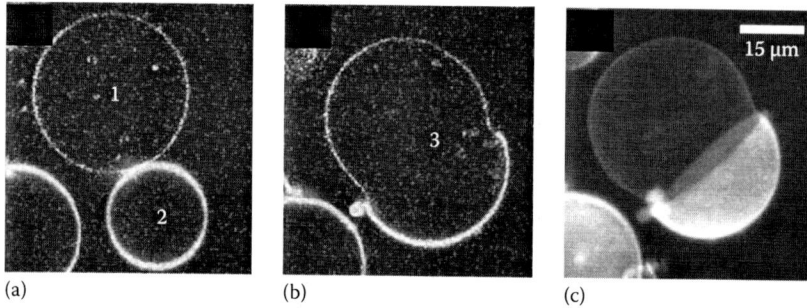

FIGURE 5.12 Creating a multidomain vesicle by the electrofusion of two vesicles of different composition as observed with fluorescence microscopy. The images (a, b) are acquired with a confocal microscopy scans nearly at the equatorial plane of the fusing vesicles. (a) Vesicle 1 is made of sphingomyelin and cholesterol (7:3) and labeled with one fluorescent dye. Vesicle 2 is composed of dioleoylphosphatidylcholine and cholesterol (8:2) and labeled with another fluorescent dye (brighter). (b) The two vesicles were subjected to an electric pulse (220 kV/m, duration 300 μs) and fused to form vesicle 3. (c) A three-dimensional image projection of vesicle 3 with the two domains formed from vesicles 1 and 2. (Adapted from Dimova, R. et al., *Soft Matter*, 3, 817, 2007. With permission.)

which is why the final vesicle exhibits immiscible fluid domains. The exact composition of each of these domains is not well known. However, from the domain area and the area of the initial vesicles before fusion, one can judge whether there is a redistribution of cholesterol between the domains, and eventually calculate the actual domain composition.

5.5 Concluding Remarks

The issues addressed in this chapter demonstrate that cell-sized giant vesicles provide a very useful model for resolving the effect of electric fields on lipid membranes because vesicle dynamics can be directly observed with optical microscopy. We have examined the behavior of giant vesicles exposed to AC fields of various frequencies and elucidated the underlying physical mechanism for the vesicle deformations as well as stress-induced lipid flows in inhomogeneous AC fields. We have shown that the vesicle response to electric fields can be interpreted and understood considering the basic mechanical properties of the membrane.

Until recently, the dynamics of vesicle relaxation and poration, which occur on microsecond timescales, has eluded direct observation because the temporal resolution of optical microscopy acquisition with analog video technology is in the range of milliseconds. The recent use of fast digital imaging has helped discover new features in the membrane response arising from the presence of charged lipids in the membrane, nanoparticles in the surrounding media, and allowed us to compare the response of gel-phase membranes with that of fluid ones. Due to this high temporal resolution, new shape deformations, such as cylindrical ones with square cross-section have been detected. The observations on vesicle fusion revealed the presence of two stages of the fusion process. Finally, we introduced a novel application of membrane electrofusion, which allowed us to construct vesicles with fluid domains.

In conclusion, the reported observations demonstrate that giant vesicles as biomimetic membrane compartments can be of significant help to advance fundamental knowledge about the complex behavior of cells and membranes in electric fields and can inspire novel practical applications.

Acknowledgments

In preparing this chapter, I have profited from numerous experimental contributions from the skillful members of my group and current and former collaborators, among whom Karin A. Riske, Roland L. Knorr, Margarita Staykova, Natalya Bezlyepkina, Peng Yang, Said Aranda, Marie Domange, and

D. Duda. I also wish to acknowledge the valuable insight I got from the enlightening discussions with theoreticians like Reinhard Lipowsky, Petia M. Vlahovska, Tetsuya Yamamoto, and Rubèn S. Gracià.

References

Abidor IG, Arakelyan VB, Chernomordik LV, Chizmadzhev YA, Pastushenko VF, Tarasevich MR. 1979. Electrical breakdown of bilayer lipid-membranes. 1. Main experimental facts and their qualitative discussion. *Bioelectrochem Bioenerg* 6:37–52.

Akinlaja J, Sachs F. 1998. The breakdown of cell membranes by electrical and mechanical stress. *Biophys J* 75:247–254.

Angelova MI. 2000. Liposome electroformation. In *Giant Vesicles*, P. L. Luisi and P. Walde (eds.). Chichester, U.K.: John Wiley & Sons Ltd., pp. 27–36.

Angelova MI, Dimitrov DS. 1986. Liposome electroformation. *Faraday Discuss* 81:303–311.

Angelova MI, Soleau S, Meleard P, Faucon JF, Bothorel P. 1992. Preparation of giant vesicles by external ac electric-fields—Kinetics and applications. *Trends Colloid Interface Sci* 89:127–131.

Aranda S, Riske KA, Lipowsky R, Dimova R. 2008. Morphological transitions of vesicles induced by alternating electric fields. *Biophys J* 95:L19–L21.

Baumgart T, Hess ST, Webb WW. 2003. Imaging coexisting fluid domains in biomembrane models coupling curvature and line tension. *Nature* 425:821–824.

Betterton MD, Brenner MP. 1999. Electrostatic edge instability of lipid membranes. *Phys Rev Lett* 82:1598–1601.

Brochard-Wyart F, de Gennes PG, Sandre O. 2000. Transient pores in stretched vesicles: Role of leak-out. *Physica A* 278:32–51.

Chang DC, Chassy BM, Saunders JA, Sowers AE. 1992. *Guide to Electroporation and Electrofusion*. New York: Academic Press.

Chiu DT, Wilson CF, Ryttsen F, Stromberg A, Farre C, Karlsson A, Nordholm S et al. 1999. Chemical transformations in individual ultrasmall biomimetic containers. *Science* 283:1892–1895.

Cohen FS, Akabas MH, Finkelstein A. 1982. Osmotic swelling of phospholipid-vesicles causes them to fuse with a planar phospholipid-bilayer membrane. *Science* 217:458–460.

Darhuber AA, Troian SM. 2005. Principles of microfluidic actuation by modulation of surface stresses. *Annu Rev Fluid Mech* 37:425–455.

Dietrich C, Bagatolli LA, Volovyk ZN, Thompson NL, Levi M, Jacobson K, Gratton E. 2001. Lipid rafts reconstituted in model membranes. *Biophys J* 80:1417–1428.

Dimova R, Aranda S, Bezlyepkina N, Nikolov V, Riske KA, Lipowsky R. 2006. A practical guide to giant vesicles. Probing the membrane nanoregime via optical microscopy. *J Phys Condens Matter* 18:S1151–S1176.

Dimova R, Bezlyepkina N, Jordo MD, Knorr RL, Riske KA, Staykova M, Vlahovska PM, Yamamoto T, Yang P, Lipowsky R. 2009. Vesicles in electric fields: Some novel aspects of membrane behavior. *Soft Matter* 5:3201–3212.

Dimova R, Dietrich C, Hadjiisky A, Danov K, Pouligny B. 1999. Falling ball viscosimetry of giant vesicle membranes: Finite-size effects. *Eur Phys J B* 12:589.

Dimova R, Pouligny B, Dietrich C. 2000. Pretransitional effects in dimyristoylphosphatidylcholine vesicle membranes: Optical dynamometry study. *Biophys J* 79:340–356.

Dimova R, Riske KA, Aranda S, Bezlyepkina N, Knorr RL, Lipowsky R. 2007. Giant vesicles in electric fields. *Soft Matter* 3:817–827.

Edwards DA, Brenner H, Wasan DT. 1991. *Interfacial Transport Processes and Rheology*. Boston, MA: Butterworth-Heinemann.

Engelhardt H, Gaub H, Sackmann E. 1984. Viscoelastic properties of erythrocyte-membranes in high-frequency electric-fields. *Nature* 307:378–380.

Engelhardt H, Sackmann E. 1988. On the measurement of shear elastic-moduli and viscosities of erythrocyte plasma-membranes by transient deformation in high-frequency electric-fields. *Biophys J* 54:495–508.

Estes DJ, Lopez SR, Fuller AO, Mayer M. 2006. Triggering and visualizing the aggregation and fusion of lipid membranes in microfluidic chambers. *Biophys J* 91:233–243.

Estes DJ, Mayer M. 2005. Giant liposomes in physiological buffer using electroformation in a flow chamber. *Biochim Biophys Acta* 1712:152–160.

Evans E, Heinrich V, Ludwig F, Rawicz W. 2003. Dynamic tension spectroscopy and strength of biomembranes. *Biophys J* 85:2342–2350.

Evans E, Needham D. 1987. Physical-properties of surfactant bilayer-membranes—Thermal transitions, elasticity, rigidity, cohesion, and colloidal interactions. *J Phys Chem* 91:4219–4228.

Fischer A, Franco A, Oberholzer T. 2002. Giant vesicles as microreactors for enzymatic mRNA synthesis. *ChemBioChem* 3:409–417.

Gao LT, Feng XQ, Gao HJ. 2009. A phase field method for simulating morphological evolution of vesicles in electric fields. *J Comput Phys* 228:4162–4181.

Gao LT, Feng XQ, Yin YJ, Gao HJ. 2008. An electromechanical liquid crystal model of vesicles. *J Mech Phys Solids* 56:2844–2862.

Glaser RW, Leikin SL, Chernomordik LV, Pastushenko VF, Sokirko AI. 1988. Reversible electrical breakdown of lipid bilayers—Formation and evolution of pores. *Biochim Biophys Acta* 940:275–287.

Gracia RS, Bezlyepkina N, Knorr RL, Lipowsky R, Dimova R. 2010. Effect of cholesterol on the rigidity of saturated and unsaturated membranes: Fluctuation and electrodeformation analysis of giant vesicles. *Soft Matter*, DOI:10.1039/B920629A (in press).

Grafmuller A, Shillcock J, Lipowsky R. 2007. Pathway of membrane fusion with two tension-dependent energy barriers. *Phys Rev Lett* 98:218101.

Griese T, Kakorin S, Neumann E. 2002. Conductometric and electrooptic relaxation spectrometry of lipid vesicle electroporation at high fields. *Phys Chem Chem Phys* 4:1217–1227.

Griffin JL. 1970. Orientation of human and avian erythrocytes in radio-frequency fields. *Exp Cell Res* 61:113–120.

Grosse C, Schwan HP. 1992. Cellular membrane-potentials induced by alternating-fields. *Biophys J* 63:1632–1642.

Hafez I, Kisler K, Berberian K, Dernick G, Valero V, Yong MG, Craighead HG, Lindau M. 2005. Electrochemical imaging of fusion pore openings by electrochemical detector arrays. *Proc Natl Acad Sci U S A* 102:13879–13884.

Haluska CK, Riske KA, Marchi-Artzner V, Lehn JM, Lipowsky R, Dimova R. 2006. Time scales of membrane fusion revealed by direct imaging of vesicle fusion with high temporal resolution. *Proc Natl Acad Sci U S A* 103:15841–15846.

Harbich W, Helfrich W. 1979. Alignment and opening of giant lecithin vesicles by electric-fields. *Z Naturforsch A Phys Sci* 34:1063–1065.

Heimburg T. 1998. Mechanical aspects of membrane thermodynamics. Estimation of the mechanical properties of lipid membranes close to the chain melting transition from calorimetry. *Biochim Biophys Acta* 1415:147–162.

Hibino M, Shigemori M, Itoh H, Nagayama K, Kinosita K. 1991. Membrane conductance of an electroporated cell analyzed by submicrosecond imaging of transmembrane potential. *Biophys J* 59:209–220.

Hyuga H, Kinosita K, Wakabayashi N. 1991a. Deformation of vesicles under the influence of strong electric-fields. *Jpn J Appl Phys*, Part 1 30:1141–1148.

Hyuga H, Kinosita K, Wakabayashi N. 1991b. Deformation of vesicles under the influence of strong electric-fields. 2. *Jpn J Appl Phys*, Part 1 30:1333–1335.

Hyuga H, Kinosita K, Wakabayashi N. 1991c. Transient and steady-state deformations of a vesicle with an insulating membrane in response to step-function or alternating electric-fields. *Jpn J Appl Phys*, Part 1 30:2649–2656.

Hyuga H, Kinosita K, Wakabayashi N. 1993. Steady-state deformation of a vesicle in alternating electric-fields. *Bioelectrochem Bioenerg* 32:15–25.

Iglesias FJ, Lopez MC, Santamaria C, Dominguez A. 1985. Orientation of *Schizosaccharomyces pombe* nonliving cells under alternating uniform and nonuniform electric-fields. *Biophys J* 48:721–726.

Isambert H. 1998. Understanding the electroporation of cells and artificial bilayer membranes. *Phys Rev Lett* 80:3404–3407.

Jahn R, Lang T, Sudhof TC. 2003. Membrane fusion. *Cell* 112:519–533.

Jones TB. 1995. *Electromechanics of Particles*. New York: Cambridge University Press.

Kahya N, Scherfeld D, Bacia K, Poolman B, Schwille P. 2003. Probing lipid mobility of raft-exhibiting model membranes by fluorescence correlation spectroscopy. *J Biol Chem* 278:28109–28115.

Kakorin S, Neumann E. 2002. Electrooptical relaxation spectrometry of membrane electroporation in lipid vesicles. *Colloids Surf A* 209:147–165.

Kantsler V, Segre E, Steinberg V. 2007. Vesicle dynamics in time-dependent elongation flow: Wrinkling instability. *Phys Rev Lett* 99:178102.

Kinosita K, Ashikawa I, Saita N, Yoshimura H, Itoh H, Nagayama K, Ikegami A. 1988. Electroporation of cell-membrane visualized under a pulsed-laser fluorescence microscope. *Biophys J* 53:1015–1019.

Kinosita K, Tsong TY. 1977a. Formation and resealing of pores of controlled sizes in human erythrocyte-membrane. *Nature* 268:438–441.

Kinosita K, Tsong TY. 1977b. Voltage-induced pore formation and hemolysis of human erythrocytes. *Biochim Biophys Acta* 471:227–242.

Kinosita KJ, Hibino M, Itoh H, Shigemori M, Hirano K, Kirino Y, Hayakawa T. 1992. Events of membrane elecroporation visualized on a time scale from microseconds to seconds. In *Guide to Electroporation and Electrofusion*, D. C. Chang, B. M. Chassy, J. A. Saunders and A. E. Sowers (eds.). New York: Academic Press, pp. 29–46.

Knorr RL, Staykova M, Gracia RS, Dimova R. 2010. Wrinkling and electroporation of giant vesicles in the gel phase. *Soft Matter*, DOI: 10.1039/B925929E (in press).

Krassowska W, Filev PD. 2007. Modeling electroporation in a single cell. *Biophys J* 92:404–417.

Kulin S, Kishore R, Helmerson K, Locascio L. 2003. Optical manipulation and fusion of liposomes as microreactors. *Langmuir* 19:8206–8210.

Kumaran V. 2000. Instabilities due to charge-density-curvature coupling in charged membranes. *Phys Rev Lett* 85:4996–4999.

Kummrow M, Helfrich W. 1991. Deformation of giant lipid vesicles by electric-fields. *Phys Rev A* 44:8356–8360.

Kunishima M, Tokaji M, Matsuoka K, Nishida J, Kanamori M, Hioki K, Tani S. 2006. Spontaneous membrane fusion induced by chemical formation of ceramides in a lipid bilayer. *J Am Chem Soc* 128:14452–14453.

Lecuyer S, Ristenpart WD, Vincent O, Stone HA. 2008. Electrohydrodynamic size stratification and flow separation of giant vesicles. *Appl Phys Lett* 92:104105.

Lee CH, Lin WC, Wang JP. 2001. All-optical measurements of the bending rigidity of lipid-vesicle membranes across structural phase transitions. *Phys Rev E* 6402:020901.

Lentz BR. 2007. PEG as a tool to gain insight into membrane fusion. *Eur Biophys J* 36:315–326.

Lindau M, de Toledo GA. 2003. The fusion pore. *Biochim Biophys Acta* 1641:167–173.

Lipowsky R, Dimova R. 2003. Domains in membranes and vesicles. *J Phys Condens Matter* 15:S31–S45.

Lipowsky R, Dobereiner HG. 1998. Vesicles in contact with nanoparticles and colloids. *Europhys Lett* 43:219–225.

Llinas R, Steinberg IZ, Walton K. 1981. Relationship between presynaptic calcium current and postsynaptic potential in squid giant synapse. *Biophys J* 33:323–351.

Luisi PL, Walde P. 2000. *Giant Vesicles*. Chichester, U.K.: John Wiley & Sons, Ltd.

Marangoni C. 1871. Ueber die Ausbreitung der Tropfen einer Flüssigkeit auf der Oberfläche einer anderen. *Ann Phys Chem* 219:337–354.

Mecke KR, Charitat T, Graner F. 2003. Fluctuating lipid bilayer in an arbitrary potential: Theory and experimental determination of bending rigidity. *Langmuir* 19:2080–2087.

Mitov MD, Meleard P, Winterhalter M, Angelova MI, Bothorel P. 1993. Electric-field-dependent thermal fluctuations of giant vesicles. *Phys Rev E* 48:628–631.

Montes LR, Alonso A, Goni FM, Bagatolli LA. 2007. Giant unilamellar vesicles electroformed from native membranes and organic lipid mixtures under physiological conditions. *Biophys J* 93:3548–3554.

Nagle JF, Tristram-Nagle S. 2000. Structure of lipid bilayers. *Biochim Biophys Acta* 1469:159–195.

Needham D, Evans E. 1988. Structure and mechanical-properties of giant lipid (DMPC) vesicle bilayers from 20-degrees-C below to 10-degrees-C above the liquid-crystal crystalline phase-transition at 24-degrees-C. *Biochemistry* 27:8261–8269.

Needham D, Hochmuth RM. 1989. Electro-mechanical permeabilization of lipid vesicles—Role of membrane tension and compressibility. *Biophys J* 55:1001–1009.

Neumann E, Kakorin S, Toensing K. 1998. Membrane electroporation and electromechanical deformation of vesicles and cells. *Faraday Discuss* 111:111–125.

Neumann E, Sowers AE, Jordan C. 1989. *Electroporation and Electrofusion in Cell Biology.* New York: Plenum.

Niggemann G, Kummrow M, Helfrich W. 1995. The bending rigidity of phosphatidylcholine bilayers—dependences on experimental-method, sample cell sealing and temperature. *J Phys II* 5:413–425.

Noireaux V, Libchaber A. 2004. A vesicle bioreactor as a step toward an artificial cell assembly. *Proc Natl Acad Sci U S A* 101:17669–17674.

Olbrich K, Rawicz W, Needham D, Evans E. 2000. Water permeability and mechanical strength of polyunsaturated lipid bilayers. *Biophys J* 79:321–327.

Peterlin P, Svetina S, Zeks B. 2000. The frequency dependence of phospholipid vesicle shapes in an external electric field. *Pfluegers Arch/Eur J Physiol* R139–R140.

Peterlin P, Svetina S, Zeks B. 2007. The prolate-to-oblate shape transition of phospholipid vesicles in response to frequency variation of an AC electric field can be explained by the dielectric anisotropy of a phospholipid bilayer. *J Phys Condens Matter* 19:136220.

Petrov AG. 1984. Flexoelectricity of lyotropics and biomembranes. *Nuovo Cimento Soc Ital Fis D* 3:174–192.

Petrov AG. 2002. Flexoelectricity of model and living membranes. *Biochim Biophys Acta* 1561:1–25.

Petrov AG, Bivas I. 1984. Elastic and flexoelectric aspects of out-of-plane fluctuations in biological and model membranes. *Prog Surf Sci* 16:389–512.

Portet T, Febrer FCI, Escoffre JM, Favard C, Rols MP, Dean DS. 2009. Visualization of membrane loss during the shrinkage of giant vesicles under electropulsation. *Biophys J* 96:4109–4121.

Pott T, Bouvrais H, Meleard P. 2008. Giant unilamellar vesicle formation under physiologically relevant conditions. *Chem Phys Lipids* 154:115–119.

Rawicz W, Olbrich KC, McIntosh T, Needham D, Evans E. 2000. Effect of chain length and unsaturation on elasticity of lipid bilayers. *Biophys J* 79:328–339.

Reeves JP, Dowben RM. 1969. Formation and properties of thin-walled phospholipid vesicles. *J Cell Physiol* 73:49–60.

Riske KA, Bezlyepkina N, Lipowsky R, Dimova R. 2006. Electrofusion of model lipid membranes viewed with high temporal resolution. *Biophys Rev Lett* 1:387–400.

Riske KA, Dimova R. 2005a. Electro-deformation and poration of giant vesicles viewed with high temporal resolution. *Biophys J* 88:1143–1155.

Riske KA, Dimova R. 2005b. Timescales involved in electro-deformation, poration and fusion of giant vesicles resolved with fast digital imaging. *Biophys J* 88:241A–241A.

Riske KA, Dimova R. 2006. Electric pulses induce cylindrical deformations on giant vesicles in salt solutions. *Biophys J* 91:1778–1786.

Riske KA, Dobereiner HG, Lamy-Freund MT. 2003. Comment on "Gel-Fluid transition in dilute versus concentrated DMPG aqueous dispersions." *J Phys Chem B* 107:5391–5392.

Riske KA, Knorr RL, Dimova R. 2009. Bursting of charged multicomponent vesicles subjected to electric pulses. *Soft Matter* 5:1983–1986.

Rodriguez N, Cribier S, Pincet F. 2006. Transition from long- to short-lived transient pores in giant vesicles in an aqueous medium. *Phys Rev E* 74:061902.

Sandre O, Moreaux L, Brochard-Wyart F. 1999. Dynamics of transient pores in stretched vesicles. *Proc Natl Acad Sci U S A* 96:10591–10596.

Schwan HP. 1983. Electrical-properties of blood and its constituents—Alternating-current spectroscopy. *Blut* 46:185–197.

Schwan HP. 1985. Dielectric properties of cells and tissues. In *Interactions between Electromagnetic Fields and Cells*, A. Chiabrera, C. Nicolini and H. P. Schwan (eds.). New York: Plenum Press, pp. 75–97.

Schwister K, Deuticke B. 1985. Formation and properties of aqueous leaks induced in human-erythrocytes by electrical breakdown. *Biochim Biophys Acta* 816:332–348.

Seifert U, Lipowsky R. 1995. Morphology of vesicles. In *Structure and Dynamics of Membranes (Handbook of Biological Physics)*, R. Lipowsky and E. Sackmann (eds.). Amsterdam, the Netherlands: Elsevier, pp. 403–463.

Sens P, Isambert H. 2002. Undulation instability of lipid membranes under an electric field. *Phys Rev Lett* 88:128102.

Shillcock JC, Lipowsky R. 2005. Tension-induced fusion of bilayer membranes and vesicles. *Nat Mater* 4:225–228.

Simon SA, McIntosh TJ. 1986. Depth of water penetration into lipid bilayers. *Methods Enzymol* 127:511–521.

Sokirko A, Pastushenko V, Svetina S, Zeks B. 1994. Deformation of a lipid vesicle in an electric-field—A theoretical study. *Bioelectrochem Bioenerg* 34:101–107.

Squires TM, Bazant MZ. 2004. Induced-charge electro-osmosis. *J Fluid Mech* 509:217–252.

Staykova M, Lipowsky R, Dimova R. 2008. Membrane flow patterns in multicomponent giant vesicles induced by alternating electric fields. *Soft Matter* 4:2168–2171.

Stuchly MA, Stuchly SS, Liburdy RP, Rousseau DA. 1988. Dielectric-properties of liposome vesicles at the phase-transition. *Phys Med Biol* 33:1309–1324.

Taylor G. 1966. Studies in electrohydrodynamics. I. The circulation produced in a drop by an electric field. *Proc R Soc Lond Ser A* 291:159–166.

Taylor P, Xu C, Fletcher PDI, Paunov VN. 2003a. Fabrication of 2D arrays of giant liposomes on solid substrates by microcontact printing. *Phys Chem Chem Phys* 5:4918–4922.

Taylor P, Xu C, Fletcher PDI, Paunov VN. 2003b. A novel technique for preparation of monodisperse giant liposomes. *Chem Commun* 14:1732–1733.

Teissie J, Tsong TY. 1981. Electric-field induced transient pores in phospholipid-bilayer vesicles. *Biochemistry* 20:1548–1554.

Tekle E, Astumian RD, Chock PB. 1994. Selective and asymmetric molecular-transport across electroporated cell-membranes. *Proc Natl Acad Sci U S A* 91:11512–11516.

Tekle E, Astumian RD, Friauf WA, Chock PB. 2001. Asymmetric pore distribution and loss of membrane lipid in electroporated DOPC vesicles. *Biophys J* 81:960–968.

Tsong TY. 1991. Electroporation of cell-membranes. *Biophys J* 60:297–306.

Veatch SL, Keller SL. 2002. Organization in lipid membranes containing cholesterol. *Phys Rev Lett* 89:4.

Veatch SL, Keller SL. 2003. Separation of liquid phases in giant vesicles of ternary mixtures of phospholipids and cholesterol. *Biophys J* 85:3074–3083.

Vlahovska PM, Gracia RS, Aranda-Espinoza S, Dimova R. 2009. Electrohydrodynamic model of vesicle deformation in alternating electric fields. *Biophys J* 96:4789–4803.

Voldman J. 2006. Electrical forces for microscale cell manipulation. *Annu Rev Biomed Eng* 8:425–454.

Weaver JC, Chizmadzhev YA. 1996. Theory of electroporation: A review. *Bioelectrochem Bioenerg* 41:135–160.

Winterhalter M, Helfrich W. 1988. Deformation of spherical vesicles by electric-fields. *J Colloid Interface Sci* 122:583–586.

Yang J, Huang Y, Wang XJ, Wang XB, Becker FF, Gascoyne PRC. 1999. Dielectric properties of human leukocyte subpopulations determined by electrorotation as a cell separation criterion. *Biophys J* 76:3307–3314.

Yang P, Lipowsky R, Dimova R. 2009. Nanoparticle formation in giant vesicles: Synthesis in biomimetic compartments. *Small* 5:2033–2037.

Zhelev DV, Needham D. 1993. Tension-stabilized pores in giant vesicles—Determination of pore-size and pore line tension. *Biochim Biophys Acta* 1147:89–104.

Zimmermann U. 1982. Electric field-mediated fusion and related electrical phenomena. *Biochim Biophys Acta* 694:227–277.

Zimmermann U. 1986. Electrical breakdown, electropermeabilization and electrofusion. *Rev Physiol Biochem Pharmacol* 105:175–256.

Zimmermann U, Neil GA. 1996. *Electromanipulation of Cells.* Boca Raton, FL: CRC Press.

6

Fluorescent Methods in Evaluation of Nanopore Conductivity and Their Computational Validation

Malgorzata
Kotulska

Witold Dyrka

Przemyslaw
Sadowski

6.1 Introduction

The transport of ions and molecules that do not readily permeate the plasma membrane is significantly enhanced by transient hydrophilic nanopores induced by the application of an electric field, which is known as membrane electroporation. The electroporation allows the conduction of ions and molecules across the plasma membrane barrier. This phenomenon has been applied to amplify the insertion of nucleic acid molecules in genetic modifications (Neumann et al. 1982), to enhance drug transport in cancer treatment—named electrochemotherapy (Mir et al. 2006), and for immune stimulation of cells under treatment (Daud et al. 2008). Research on other potential applications is ongoing. A choice of optimal electroporation protocol depends on the application and must take into account both the membrane composition and physicochemical properties of transported molecules. Inadequate exposure to the electric field will not permit the creation of sufficiently permeable pores, but an overly intense field application will result either in an excessive opening of the cell membrane and detrimental contact of the cytoplasm with the extracellular medium or in irreversible electroporation and membrane rupture. The electric field is determined by the pulse duration, frequency, and amplitude. Extensive experiments have shown that the most effective electroporation for transport of small molecules is obtained using short pulses with high intensity ($\sim 100\,\mu s$, $\sim 10^5$ V/m), while large and charged molecules, e.g., DNA, are best transported by long electrical pulses of lower intensity ($\sim 10\,ms$, ~ 1 V/m), with notable electrophoretic component. Recent studies demonstrate interesting electroporation effects for pulse conditions outside typical. At one extreme are nanosecond pulses of very high intensity ($\sim 1\,ns$, $\sim 10^6$ V/m) (Vernier et al. 2006b) inducing very fast appearing small pores, while at the other are current-controlled methods ($\sim 0.1\,nA$) producing a single nanopore lasting a few hours (Kalinowski et al. 1998).

In this chapter, we review the application of dyes and fluorescent probes to test the efficiency and selectivity of the molecular transport through electropores. These methods can be used to optimize the experimental conditions for electroporation-supported drug delivery. They can also be used for estimation of electropore characteristics, such as the pore density and median radius. These evaluations, however, can be very imprecise if applied without a regard for the nanometer size range of the pore, or assuming an overly simplified model of its shape. The nanometer scale of the markers and pores greatly influence the diffusion of molecules, which interact with the pore walls and other transported molecules and ions. In this chapter, we present computational results showing the sensitivity of the diffusion to the ratio of the pore and probe radii, and to the pore shape. Computer simulations based on the three-dimensional Poisson–Nernst–Planck (PNP) model show how the free diffusion assumption and cylindrical representation of the pore affect the error scale in modeling molecular probe transport.

6.2 Fluorescent Probes and Their Application in Characterizing Membrane Electropermeabilization

The main objective of conductance experiments combined with the electroporation is determining the transport efficiency of the electropores under different conditions. The studies concern transport characteristics in all the stages of electroporation, detection of distinctive time points, such as the onset of molecular transport and its termination, and abrupt changes of the conductance that are related to different pore geometries, densities, and distributions. The particles transported through the electropores differ in their size and physicochemical properties, depending on the potential applications of the electroporation. The area of interest includes small drug molecules used in electrochemotherapy, such as cisplatin and bleomycin with molecular masses of 300 Da and 1.4 kDa, respectively, and larger molecules like nucleic acid plasmids with molecular mass over 2 MDa. The experiments on cytotoxic drugs evaluating cell survival rate, or nucleic acids with tests on gene expression levels, are significantly extended by studies on the fluxes of fluorescent probes and dyes that are not cytotoxic. The direct flux measurements of fluorescent probes or dyes can be applied to quantify the transport rate. The molecular transportation rate across the membrane can be evaluated by the fluorescence intensity at the transport destination site. Another approach is transport evaluation of molecules that modify fluorescent indicators nested in the transport destination. Probe molecules are usually charged and their flux is driven by both the electric potential and concentration gradients.

The onset of electropermeabilization in the cellular membrane was studied by the uptake of calcium ions (Hibino et al. 1993, Tekle et al. 1994, Gabriel and Teissie 1999, Tekle et al. 2001, Pucihar et al. 2008) and thallium ions (Pakhomov et al. 2009). Ions with sizes in the angstrom range are the most sensitive indicators of the electroporation since they are conducted by any hydrophilic electropore that may form, and their diffusion through the electropores indicates that a hydrophilic pore has been created. A change of free Ca^{2+} concentration can be evaluated by the fluorescence level of the calcium indicator dye Fluo-3 (Pucihar et al. 2008). Tl^+ ions were detected with a Tl^+-sensitive fluorophore (Pakhomov et al. 2009). Another method for identification of pore occurrence is phosphatidylserine (PS) externalization determined by florescent methods, such as with Annexin-V-FITC (Vernier et al. 2004a,b, 2006a,b, Pakhomov et al. 2009). PS-externalization experiments are based on the asymmetric lipid composition of the cell membrane, applying the fact that PS is located inside the cell. Following electropore appearance, the PS is translocated to the outer membrane leaflet.

Larger molecules can be used to estimate the upper limit of electropore diameter. The most common test molecule is propidium iodide ($M_r = 668.4$ Da), which enhances its fluorescence intensity 20–30 times upon binding to DNA inside the cell (Tekle et al. 1994, Djuzenova et al. 1996, Gabriel and Teissie 1997, 1999, Golzio et al. 2002, Shirakashi et al. 2004, Kennedy et al. 2008, Pucihar et al. 2008, Pakhomov et al. 2009). Other probes used to study electropore conductance are lucifer yellow (Puc et al. 2003), calcein (Prausnitz et al. 1993, Gift and Weaver 2000, Canatella et al. 2001), ethidium bromide (Sixou and Teissie 1993, Tekle et al. 1994, Gabriel and Teissie 1997), ethidium homodimer (Tekle et al. 1994), YO-PRO-1 (Vernier et al. 2008), and TOTO cyanine fluorochromes (Golzio et al. 2002, Rols 2006). The luciferin–luciferase reaction was used to quantify the efflux of ATP (Rols et al. 1998). Electropore conductance was also studied by nonfluorescent dyes, such as trypan blue (Wolf et al. 1994, Gabriel and Teissie 1995, Teissie et al. 1999, Zhen et al. 2006) and SERVA blue G (Neumann et al. 1998). The probes listed above range from a few hundred daltons (ethidium bromide $M_r = 394$ Da) to a few kilodaltons in their molecular mass (TOTO, $M_r = 1.3$ kDa) and have Stokes radii below 1 nm. A whole variety of fluorescein-labeled molecules with selected specific properties can also be employed. For example, fluorescein-labeled bovine serum albumin (BSA), an 18-residue signal peptide ($M_r = 69$ kDa), was used to study protein transport (Prausnitz et al. 1994). In particular, the use of fluorescein-labeled dextrans has a very high potential since they are available in a wide range of sizes. Fluorescein-labeled dextrans are used for modeling moderate-sized drug molecule transport, e.g., FD-4 with a molecular mass of 4 kDa (Sixou and Teissie 1993, Zaharoff et al. 2008). The transport of plasmids used in genetic transfers is most efficiently studied by dyes of very large molecules, typically with molecular masses around 1 MDa. A large dye molecule, blue fluorescein isothiocyanate-labeled dextran FD-2000 with a molecular mass of 2 MDa and Stokes radius 28 nm, was used for modeling the transport characteristics of plasmid DNA and RNA. These studies showed temporal behavior and time constants of very large electropores (Zaharoff et al. 2008). However, due to the negligibly small charge of FD-2000, when compared to plasmids, it is not the optimal model of plasmid transport.

Some fluorophores are inherently fluorescent, but several dyes used in experiments with living cells fluoresce only when attached to intracellular components, such as specific proteins, ATP, or nucleic acids in the nucleus. This approach excludes artifacts coming from dye molecules that could remain in the extracellular medium. Ethidium bromide, ethidium homodimer, and propidium iodide, for example, exhibit significantly enhanced fluorescence when bound to DNA.

A combination of different indicators of the membrane electropermeabilization in one study can define both the size limits of the molecular transport under specific experimental conditions and also the predominant mechanisms responsible for ionic and molecular transport of larger molecules. Two driving forces are responsible for the transmembrane flux: the electrophoretic movement related to the electric field gradient and the concentration gradient-associated diffusion. The time delay between the application of the electric field and the transport onset shows which of these components is dominant for particular probe size and charge.

When high-intensity pulses are used in nanosecond electroporation experiments, the pores open within less than 2 ns (Vernier et al. 2008) and they remain open for more than a minute (Pakhomov et al. 2009). These pores are less than 1.5 nm in diameter, as judged by the absence of propidium uptake concurrent with the increased ionic permeability of the membrane. However, in about a minute range they can form larger pores, and will then conduct dye molecules. These experiments show rectification properties in the current–voltage relationship, which resembles the characteristics of other nanopores such as protein ion channels and artificial nanopores induced in polymer foil (Siwy et al. 2005), and may indicate an asymmetry in the electropore shape.

As pulse lengths increase and field intensities decrease, larger and larger molecules are transported through the electropores. Membrane electropermeabilization at moderate pulse intensity is observed a few microseconds after the onset of the electric field. The transport of larger molecules at these conditions, monitored by the fluorescence increase, starts milliseconds after the field exposure (Tekle et al. 1990, 1994, Gabriel and Teissie 1997, Pucihar et al. 2008). However, the transport is not homogeneous in time and three distinct stages are observed depending on the pore dynamics. The ensemble of pores created during an electric pulse consists of small pores, with diameters within single nanometer, and larger pores that are capable of conducting large molecules. As pulses that are not very long do not maintain the pore stability, the majority of large pores disappear within a few milliseconds after the pulse ends and they are completely absent after 1 s. Small electropores, in contrast, survive a few minutes, allowing for transport of small molecules (Neumann et al. 1998, Rols and Teissie 1998, Pavlin and Miklavcic 2008, Zaharoff et al. 2008). Therefore, the transport of small molecules is governed by the concentration gradient to much greater extent than is the case with large particles. The probe charge also plays an important role, enhancing transport through electropores under applied electric field.

A further elongation of the pulse decreases alterations in the flux characteristic. The measurements of electropore conductivity using current-clamp methods (Kalinowski et al. 1998, Kotulska et al. 2004) or in voltage-clamp conditions on previously formed electropore (Kotulska 2007) indicate the creation of a single pore with a fairly stabilized geometry, time-independent conductance, and self-scaling dynamical properties. Such pores act similar to ion channels or artificial nanopores with well-defined static geometry (Siwy and Fulinski 2002). When an electropore is fairly stable over a period of time, its diffusion behavior is similar to molecular transport through a fixed-shape nanopore. Modeling approaches similar to those used for protein-ion channels can be applied for stabilized electropores or for transient electropores over short time intervals.

6.3 Modeling Molecular Transport through Nanopores

Electropore properties can be derived from the temporal fluorescence or conductance characteristics combined with the pore radius and diffusion capability. The transmembrane transport of charged molecules through electropores is typically modeled by one-dimensional Nernst–Planck equation (Neumann et al. 1998, Pavlin and Miklavcic 2008, Pucihar et al. 2008, Zaharoff et al. 2008), which combines molecular diffusion due to the concentration gradient with the electrophoretic component of a charged molecule. The flux density J through a singular pore is defined in the Nernst–Planck equation in the following:

$$J = -D\left(\nabla n + n \cdot \nabla \Phi \cdot \frac{ze}{kT} \right), \tag{6.1}$$

where
D is a diffusion constant
k the Boltzmann constant
n the ionic density
∇n the ionic density gradient
$\nabla \Phi$ the gradient of the electric potential
z represents the ionic charge
e the elementary charge
T is the temperature in Kelvin

Unlike constant-current electroporation, under a pulsating electric field, pores appear and disappear, changing their dimensions and shapes throughout the experiment. For calculating and modeling the conductance, the ensemble of pores conductive at a particular time instant should be taken into account and the total transport represented as a sum of all contributing fluxes. Integrating the total flux over the time provides the fluorescent probe concentration.

The main problem in modeling the phenomenon is the diversity in electropore geometry and the instability associated with the temporal changes of the electric field. Unlike constant-current electroporation, where a single pore of a fairly stable radius is formed, a typical pulse-field electroporation creates pores that change their shape in phase with the pulse. To resolve this problem, it was proposed to model electropores as objects of three (Neumann et al. 1998) to five different conductivity states (Teissie et al. 2005). In the first approach, only conductive states are considered; two states represent pores active during the pulse time; and a third state appears after the pulse termination. A chemical-kinetic analysis with appropriate rate constants was used to mimic the pore state transitions during their opening and resealing (Neumann et al. 1998). The model was employed for an experiment with dye SERVA blue G molecules. A similar approach was adopted by Pucihar et al. (2008), where the uptake of propidium into electropermeabilized Chinese hamster ovary cells was modeled using the three-state pore representation with an exponential time-course evolution of the pores.

During modeling a fluorescent probe flux in electroporation, the fact that this is not free diffusion is often neglected. Although small molecules are transported differently than large plasmids, diffusion is always affected when the molecular dimension approaches the size of the nanopore. Ignoring this fact leads to imprecise results or severe errors. The transport of ions and particles through nanosized pores is not a trivial problem (e.g., Kosinska and Fulinski 2005). A confined conductance pathway of a very narrow opening poses an entropic barrier for traveling ions and molecules, which occurs due to the reduction of the space available for the diffusing particles (Jacobs 1967, Zwanzig 1992, Reguera and Rubí 2001). The transport is hindered due to the interactions between transported molecule and the pore walls as well as the interactions between the transported particles and ions. As a result, the effective diffusion coefficient is always below its free value. Based on the Fick–Jacobs equation (Jacobs 1967), Zwanzig proposed the following relationship for the position dependent effective diffusion coefficient $D(x)$ (Zwanzig 1992):

$$D(x) \cong \frac{D_0}{1 + \gamma w'(x)^2}, \tag{6.2}$$

where

D_0 is diffusion coefficient in bulk
$w(x)$ is the pore radius at x of the axis normal to the pore cross section

The coefficient γ is determined by the potential confining diffusion to the vicinity of the center line; for a three-dimensional tube, it was derived that $\gamma = 1/2$, and $\gamma = 1/3$ for a two-dimensional channel.

In electroporation experiments, the diffusion suffers not only from space confinement, but also from changes in the dipole charge distribution in the lipid molecules, related to the reorientation during electroporation process, and binding of water molecules to the lipids (Joshi et al. 2006). Hence, the diffusion constant of a nanometer-scale probe transported through a nanopore is different from its bulk value, which is often mistakenly used in modeling the transport through electropores. Some authors introduce a correction that accounts for the impeded diffusion. For example, the effective diffusion constant for SERVA blue inside an electropore ($D = 10^{-10}$ m²/s) was assumed five times lower than its bulk value, $D_0 = 5 \times 10^{-10}$ m²/s (Neumann et al. 1998). The molecular dynamics (MD) simulations of electroporation experiments indicate that the value of the diffusion constant depends on the field intensity and frequency (Joshi et al. 2006), showing the peak diffusion coefficient in bulk water at around 2 THz ($D_0 \cong 7 \times 10^{-9}$ m²/s) and monotonically decreasing for higher frequencies to $D_0 \cong 5 \times 10^{-10}$ m²/s at 10 THz. The value inside the pore was always below its bulk equivalent. In these simulations, the static diffusion coefficient $D_0 = 4.5 \times 10^{-9}$ m²/s was reduced threefold inside the electropore to the value $D = 1.3 \times 10^{-9}$ m²/s.

The correct value of the diffusion constant should take into account the probe and the pore geometry. It is therefore reasonable that the diffusion constant takes into consideration the ratio between the probe and pore radii. If the pore size is unknown, it can be estimated based on the membrane electrical conductance for very small ions, where the free diffusion representation is more acceptable, although not entirely correct. A model by Paine and Scherr (1975), which has been successfully used for protein ionic channels (Noskov et al. 2004), can also be applied in electroporation. Using continuum hydrodynamics, an approximate dependence was formulated of the pore radius on the diffusion coefficient of a spherical particle conducted through a pore (Paine and Scherr 1975), extending the Stokes–Einstein relation for the diffusion coefficient D_{ion} of a molecule to the following form:

$$D_{\text{ion}} = \frac{k_{\text{B}}T}{6\pi\eta R_{\text{ion}}k_1},$$ (6.3)

where

R_{ion} is the particle radius
k_{B} is the Boltzmann constant
T is the temperature
η is the viscosity of media

The novel component k_1 represents the friction. The friction value depends on R_{ion}, R_{pore}, and the radial distance d of the particle from the central axis of the pore. Values of k_1 were tabulated based on experimental results. Equation 6.3 has been reformulated by Noskov and coworkers (Noskov et al. 2004) into the analytical expression dependent on the radii ratio $\beta = R_{\text{ion}}/R_{\text{pore}}$:

$$D_{\text{ion}}(r) = \frac{D_{\text{ion}}}{A + B\exp(\beta/C) + D\exp(\beta/E)}.$$ (6.4)

The best fit with tabulated values from Paine and Scherr (1975) was obtained with the following parameters (Noskov et al. 2004): $A = 0.64309$, $B = 0.00044$, $C = 0.06894$, $D = 0.35647$, and $E = 0.19409$. Figure 6.1 presents the function showing, for example, that if a particle occupies half of the pore cross section, then its effective diffusion constant should be reduced five times when compared to its bulk value.

The other issues are the pore shape and possible charge distribution on the pore walls. Although the Nernst–Planck relation may be a good approximate of the transport characteristic in some situations, it does not take into account the pore shape, electrostatic interactions between probes traversing the channel, ions of the solute, and the channel walls. The cylindrical shape, assumed in most calculations, is very imprecise. Not only does the cylindrical pore not seem a plausible physical model, it has not been confirmed by any computational or experimental study. A study presented in Pakhomov et al. (2009) indicates possible asymmetry occurrence in the pore shape. MD simulations show pores with a very irregular rim (Tieleman 2004, Tarek 2005).

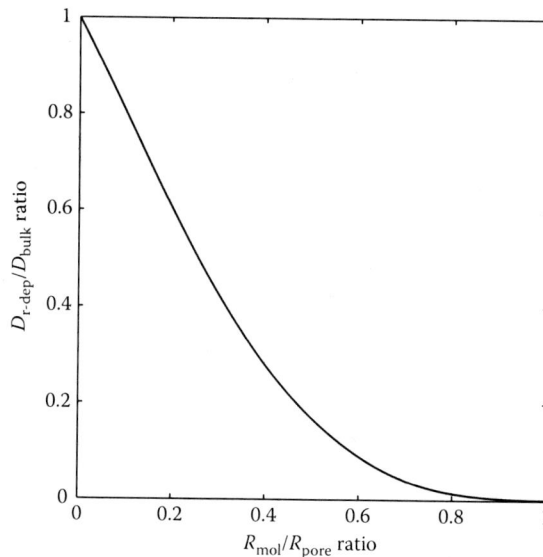

FIGURE 6.1 Effective diffusion coefficient. Dependence of the ratio of the effective diffusion coefficient to its bulk value on the probe radius related to the pore dimension. (According to Noskov, S.Y. et al., *Biophys. J.*, 87, 2299, 2004.)

Obtaining the full characteristics of the molecular transport is possible through more advanced models, either discrete or continuous. Discrete modeling by MD simulations is the most computationally demanding, but also the most accurate in the case of irregular channels with electrostatic charges. Semi-discrete Brownian dynamics (BD) modeling (Corry et al. 2000a) provides a more economical although less accurate technique. The continuous PNP model is the fastest method; however, due to its accuracy it is the most appropriate for a regular pore (Corry et al. 2000b) at its selected (or averaged) static conformation. Although the shapes of the electropores have not yet been resolved, pores seem regular enough to qualify for three-dimensional PNP simulations that would represent short time intervals if the pulsating field is applied. Despite these limitations, the model is still much more accurate than the Nernst–Planck approximation.

The PNP provide a basis for computing charge fluxes through narrow pores, and are thus especially applicable to biological ion channels. The PNP equations offer an averaged theory, applicable under steady-state conditions for a large number of atoms over a long timescale relative to atomic and molecular processes. The theory is based entirely on the continuum approach to charge transport, which permits the inference of analytical laws relating the flux to pore structure. Importantly, it provides the fastest computational approach to transport phenomena on the nanometer and sub-nanometer scale. The PNP model consists of two equations, defining all intensive (size-independent) and extensive quantities. In addition to the Nernst–Plank Equation (6.1), the electric potential in the pore, dependent on the accumulated charge, is defined by the Poisson equation:

$$\varepsilon_0 \nabla \cdot \left[\varepsilon(r) \nabla \Phi(r) \right] = -e \sum_v z_v n_v(r) - \rho_{ex}, \tag{6.5}$$

where

ε_0 is the permittivity of free space
$\varepsilon(r)$ is permittivity of the medium
ρ_{ex} represents the external charges embedded in the pore walls

The summation of v takes into account a variety of ion species flowing through the pore.

The PNP model is nonlinear and therefore has no general analytical solution, but under specific conditions some special cases have been solved. An example is the Goldman–Hodgkin–Katz (GHK) equation, derived in the limited case of a constant electric field applied across a planar membrane, which reduces the PNP model effectively to a single-dimensional dependence (Hodgkin and Katz 1949, Schultz 1980). The same solution arises in the limit of low ionic concentration or for a sufficiently short channel in an arbitrary electric field. The general PNP model can only be solved numerically, typically by finite difference or finite element methods (see, e.g., Kurnikova et al. 1999).

As described by Corry and coworkers (Corry et al. 2000b), the standard computational PNP algorithm is based on the discrete versions of Equations 6.1 and 6.5, calculated over a system discretized into a three-dimensional rectangular grid of points over the pore, membrane, and reservoirs. The grid defines cells of dimensions: $h_x \times h_y \times h_z$. The Nernst–Planck equation in its discrete form reads as

$$J_j = -D \left(\frac{n_j - n_i}{h_j} + \frac{n_j + n_i}{2} \cdot \frac{\Phi_j - \Phi_i}{h_j} \cdot \frac{ze}{kT} \right), \tag{6.6}$$

where J_i, $i = 1, \ldots, 6$ denotes the flux through each of the six cell walls. Equation 6.3 under steady-state conditions, i.e., $\nabla \cdot J = 0$, which takes into account a position-dependent diffusion constant (Noskov 2004), gives a formula for the density n_i at the central grid point:

$$n_i = \frac{\sum_{j=1}^{6} n_j \left((D_i + D_j)/h_j \right)\left[1 + (ez/2kT)(\Phi_j - \Phi_i) \right]}{\sum_{j=1}^{6} \left((D_i + D_j)/h_j \right)\left[1 - (ez/2kT)(\Phi_j - \Phi_i) \right]}.$$ (6.7)

Since there is no flux through the boundary, only the cells adjoining the pore are included in the calculations. The discrete Poisson equation has the following form:

$$\varepsilon_0 \sum_{j=1}^{6} \varepsilon_j \frac{\Phi\left(r_i + h_j \vec{j} \right) - \Phi(r_i)}{h_j} \cdot \frac{V}{h_j} = -Ve \sum_v z_v n_v(r_i) - \rho_{ex},$$ (6.8)

where

r_i is a central grid point of the cell
$\vec{j}$ is a unit vector normal to the plane dividing cells i and j
v represents an ion species
V denotes each cell volume

The potential Φ_i ($=\Phi(r_i)$) can be derived from Equation 6.8, giving the following formula:

$$\Phi_i = \frac{\sum_j \varepsilon_j \Phi_j / h_j^2 + \sum_v z_v e n_{vi}/\varepsilon_0 + q_i/(\varepsilon_0 V)}{\sum_j \varepsilon_j / h_j^2}.$$ (6.9)

Equations 6.7 and 6.9 are solved simultaneously, according to an interchangeable iterative scheme. The iteration starts with some initial guesses concerning the values of concentration and potential, and proceeds until both variables converge to stable values. In each iteration step, both variables are updated in the same manner:

$$n_{iNEW} = w_n \cdot n_{iOLD} + \left(1 - w_n \right) \cdot n_{iCALC}$$ (6.10a)

$$\Phi_{iNEW} = w_\Phi \cdot \Phi_{iOLD} + \left(1 - w_\Phi \right) \cdot \Phi_{iCALC},$$ (6.10b)

where

n_i denotes the concentration
Φ_i denotes the potential in the ith cell
n_{iOLD} and Φ_{iNEW} are values from the previous iteration step
n_{iCALC} and Φ_{iCALC} are values derived from Equations 6.7 and 6.9
n_{iNEW} and Φ_{iNEW} are values finally accepted for further calculation

A value of the convergence coefficient w_n is typically assumed as a constant value from the range $(0, 1)$ or $(-1, 1)$ if over-relaxation is acceptable. The iteration stops if both differences, n_{diff} and Φ_{diff}, between subsequent values of both variables, i.e., n_{iOLD}, n_{iCALC} (and Φ_{iOLD}, Φ_{iCALC} respectively), change below the assumed maximal limit $n_{diffMAX}$ ($\Phi_{diffMAX}$):

$$n_{diff} = \max_i \left(n_{iCALC} - n_{iOLD} \right) < n_{diffMAX}$$ (6.11)

$$\Phi_{\text{diff}} = \max_i \left(\Phi_{i\text{CALC}} - \Phi_{i\text{OLD}} \right) < \Phi_{\text{diff MAX}}. \qquad (6.12)$$

The PNP model, although most efficient for modeling the ionic or molecular transport through nanopores, is nonetheless still computationally demanding. A method to reduce its complexity was proposed in Dyrka et al. (2008).

6.4 Effect of the Nanopore Diameter and Shape in the Three-Dimensional Transport Model

Using the three-dimensional PNP model, we tested the extent of the correction of the diffusion constant introduced in Noskov et al. (2004) could influence the predicted transport characteristics, with regard to possible shapes of the pore and properties of the fluorescent probe. An electropore reminiscent of the irregular hole, observed in MD simulations, was represented by a pore with an irregular cross section (Figure 6.2). Other more regular symmetrical and asymmetrical pores were represented by cones, which also seem more physical models of an electropore than a classical cylinder. Two cones were tested: a two-sided symmetric cone (Figure 6.3) and a one-sided asymmetric cone (Figure 6.4). The minimal pore diameter ranged from 1.2 to 4 nm. The simulations were run for a fluorescent or dye marker of 1 nm in diameter, at the marker concentration of 5 μM on one side of the pore. The membrane 4.2 nm thick was surrounded by 0.1 M solution of NaCl. The probe molecules were modeled as spheres with net charges of −1 (Figures 6.5a,b, 6.6a,b, 6.7, and 6.8) or −2 (Figures 6.5c,d, 6.6c,d, and 6.9). They were located in the left-hand-side reservoir (Figure 6.8b) or in the right-hand-side reservoir at concentration of 5 μM (all other figures). The probe molecule diffusion coefficient in bulk was set to 0.5×10^{-9} m^2/s and the external voltage ranged from −300 to 300 mV in several steps. The convention with electric potential zero at the left-hand-side system boundary was assumed. The calculations were carried out in the cube of 8.2 nm in the edge length that included 68,921 cells of dimensions 2 Å × 2 Å × 2 Å. The transport ratio was calculated as a ratio of transport with bulk-diffusion (D_{bulk}) coefficient to transport with radius-ratio-dependent diffusion coefficient ($D_{\text{r-dep}}$). The modeling results presented here were obtained using

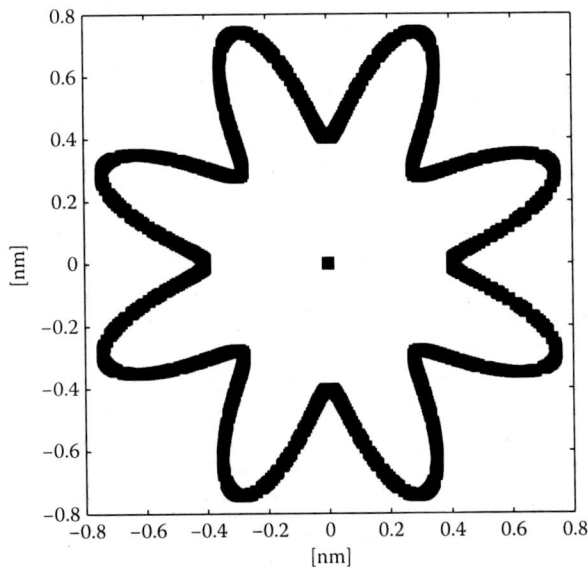

FIGURE 6.2 Irregular channel cross section. Idealized view of the irregular channel cross-section in the *XY* plane. The minimal diameter is $d_{\text{min}} = 0.8$ nm and maximal diameter is $d_{\text{max}} = 1.6$ nm. The average diameter is ca. 1.2 nm.

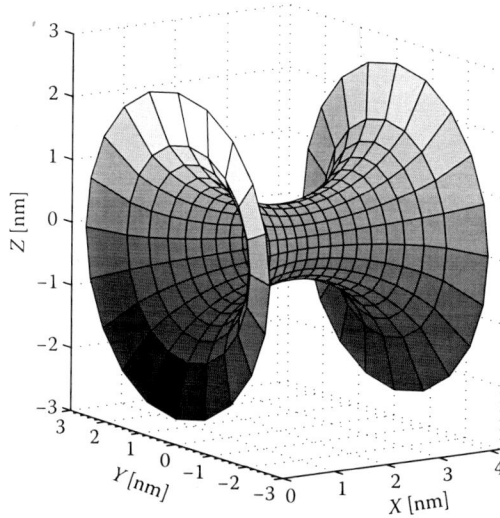

FIGURE 6.3 Two-sided conical nanopore. Schematic view of the two-sided conical nanopore. The minimal diameter is $d = 1.2$ nm, the maximal diameter $D = 5.2$ nm.

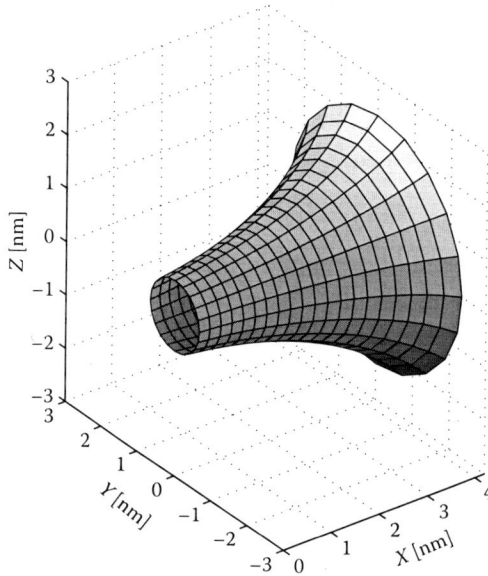

FIGURE 6.4 Asymmetric conical nanopore. Schematic view of the asymmetric conical nanopore. The minimal diameter $d = 1.2$ nm, the maximal diameter $D = 4.8$ nm.

our algorithm of improved efficiency (Dyrka et al. 2008). The software used for modeling is available at www.3dpnp.biolab24.com.

The results prove a very significant impact of the diffusion-coefficient model on the calculated marker transport through a nanopore. The magnitude of this influence depends on several factors. More specifically, Figure 6.5a and c shows the transport magnitude for a marker with the charge of −1 and −2, respectively. Figure 6.5b and d shows the calculated transport ratio between bulk, i.e., radius-ratio-independent (RRI) and radius-ratio-dependent (RRD) models. It can be observed that if the cylindrical pore diameter is only 1.2 times larger than the marker diameter (assumed as 1 nm), the transport ratio implied by the RRI model is from 10 (at 300 mV of the external voltage) to almost 80 (50 mV) times

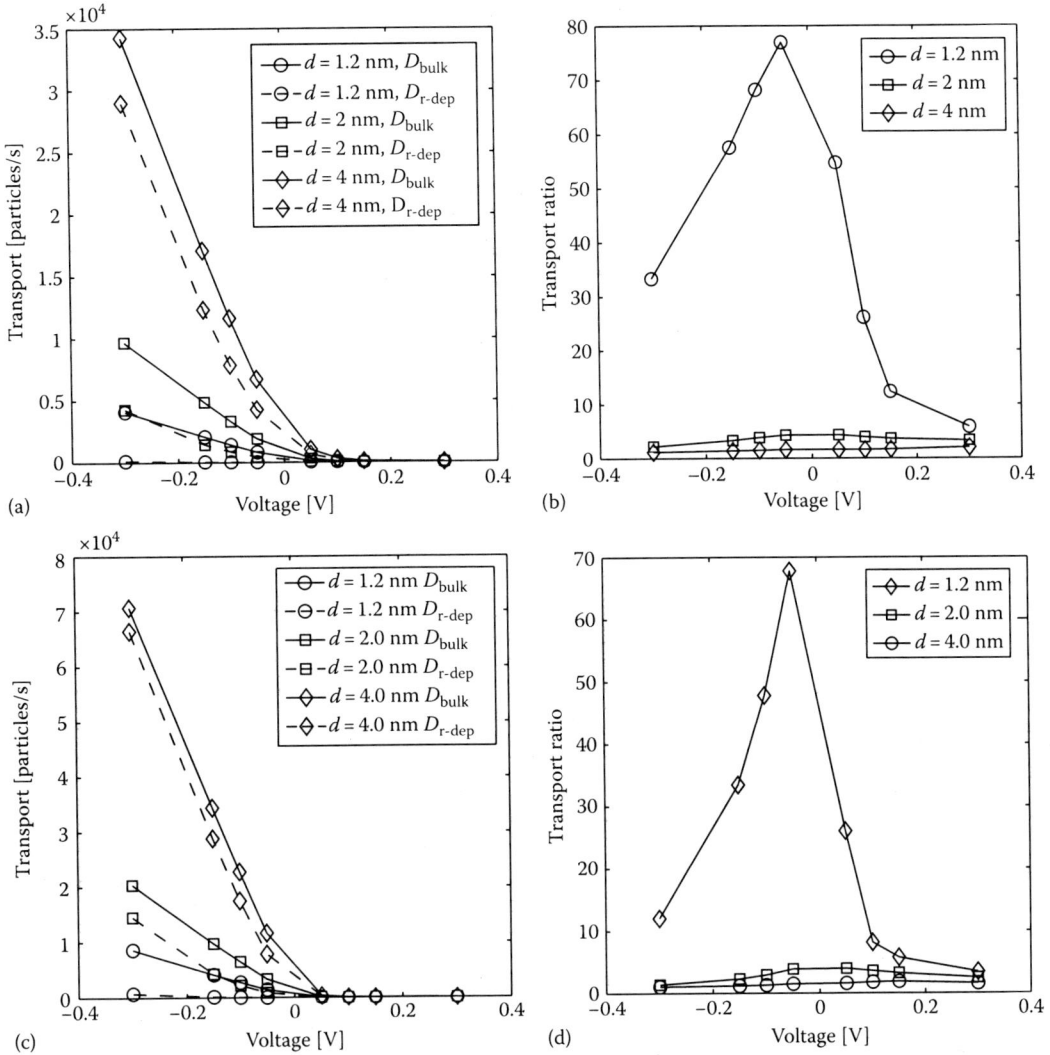

FIGURE 6.5 Probe molecule transport in symmetric cylindrical pores. Pore shape is a regular cylinder with a constant diameter. Comparison of the transport rates (a,c) and impact made by the diffusion-coefficient model (b,d) for pores of different diameters. The probe molecule net charge is −1 (a,b) or −2 (c,d).

greater in comparison to the RRD diffusion-coefficient model for the same dimensions of the nanopore. For cylinders wider than 2 nm in diameter, the impact of the diffusion-coefficient model is an order of magnitude weaker than in the previous case. However, the drop in the transport rate between RRD and bulk diffusion-coefficient systems is still from 1.5 to 5 times. This shows that RRI model applied to nanometer-scale pores gives too large transport rate. Therefore, the nanopore diameter, estimated on the basis of the experimentally obtained transport magnitude and the RRI transport model, will be significantly smaller than in reality. For example, in the bulk RRI model, a cylindrical pore of 2 nm in diameter can be mistakenly recognized as a 1.2 nm pore (see Figure 6.5a). In the case of 4 nm channels, RRD diffusion coefficient does not fall below 50% of its bulk value and these systems are less affected by the diffusion-coefficient definition, especially when the marker carries the higher charge (Figure 6.5c and d).

Overestimation of the transport rate, observed for cylindrical channels in the RRI model, was also recorded for the symmetric conical pores (Figure 6.6a through d), and even stronger for the irregular

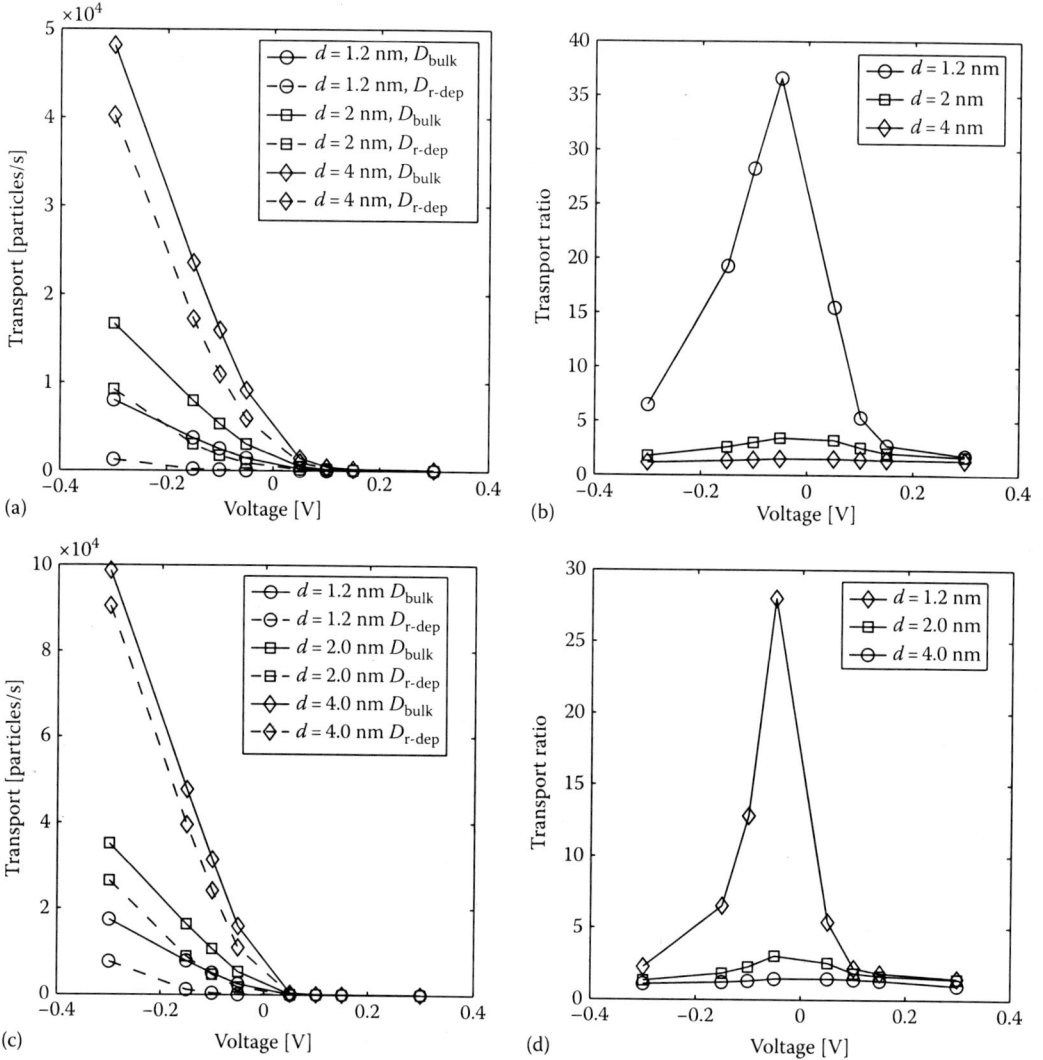

FIGURE 6.6 Probe molecule transport in symmetric conical pores. Pore shape is a two-sided symmetrical cone. Comparison of the transport rates (a,c) and impact made by the diffusion coefficient model (b,d) for pores of different minimal diameters. The probe molecule net charge is −1 (a,b) or −2 (c,d).

channel (Figure 6.7), although the RRD diffusion coefficient used in modeling of the irregular pore was only based on the averaged radius of the pore and the obstruction created by narrow parts of the channel, irregular cross section was not taken into account. In conical pores, the lesser impact of the diffusion coefficient model can be attributed to larger average radius of the pore than in the case of a cylinder.

A qualitative analysis of the transport–voltage curves, obtained from calculations with RRD diffusion coefficient (Figures 6.5a,c, 6.6a,c, and 6.7a), shows a nonlinear dependence. This effect is stronger when the pore diameter is smaller. The transport–voltage curves for wide 4 nm channels and for all the systems with the bulk diffusion model are near linear in the part where the electric field and concentration gradients induce flows in the same direction. Hence, it can be claimed that the bulk diffusion constant introduces significant qualitative and quantitative inaccuracies to the transport model when the probe molecule radius is half of the pore radius or larger.

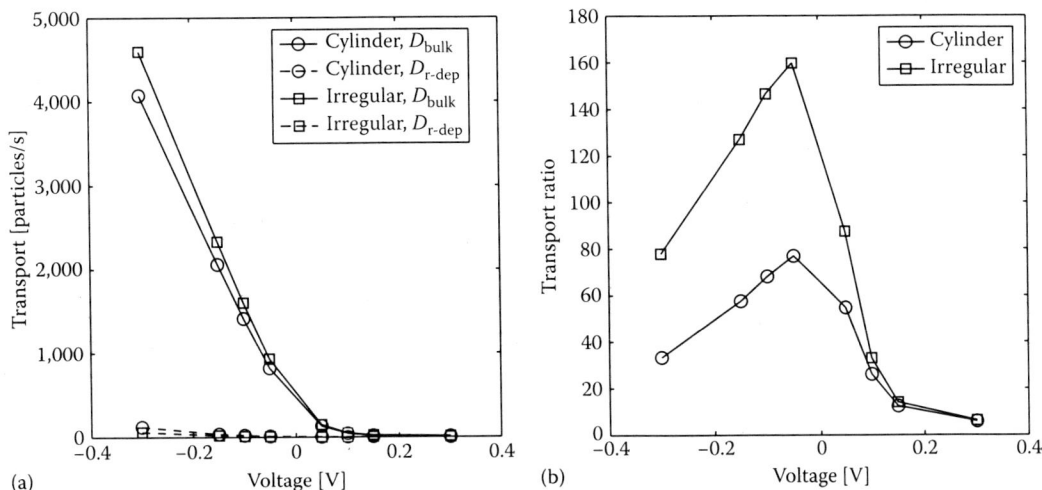

FIGURE 6.7 Probe molecule transport in symmetric cylindrical and irregular pores. Regular cylindrical pore (diameter of 1.2 nm) and irregular pore (effective diameter of 1.2 nm, see cross section in Figure 6.2) are compared in terms of the transport rates (a) and impact made by the diffusion coefficient model (b). The probe molecule net charge is −1.

Molecular transport in asymmetric pores is shown in Figure 6.8a,b, which compares marker transport rates in the 1.2 nm regular cylinder and asymmetric cones (1.2 × 2.4 nm and 1.2 × 4.8 nm). The probe molecules were located on the wide mouth of the pore (Figure 6.8a) or on the narrow entrance to the channel (Figure 6.8b). If the pore is symmetrical, its transport–voltage characteristics show mirror symmetry; however, the transport symmetry is broken for an asymmetric channel. In case of the asymmetry, the transport rate from the wide to the narrow side of the pore is approximately 20% higher when the bulk diffusion model is adopted. This effect becomes even more evident in the system with the RRD

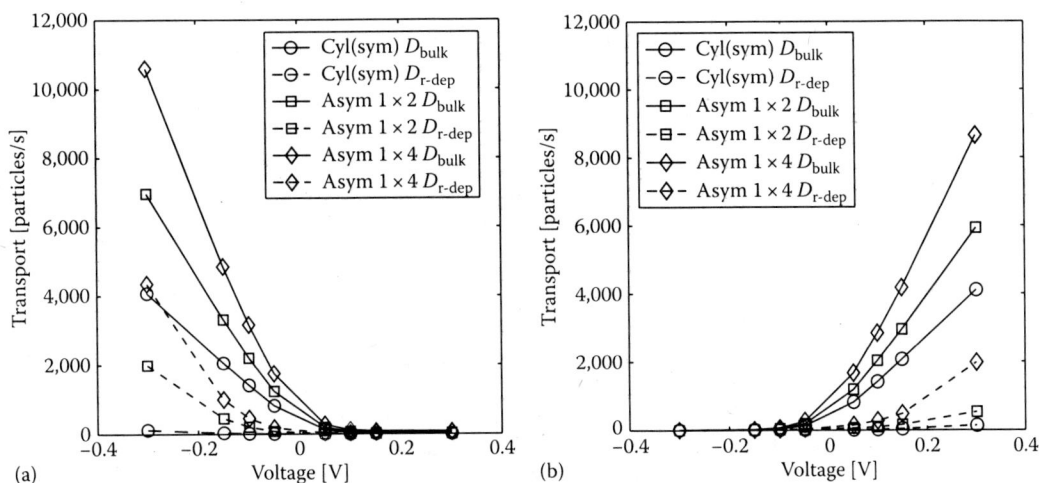

FIGURE 6.8 Probe molecule transport in symmetric and asymmetric pores. Symmetric regular cylindrical pore and asymmetric (one-sided, asymmetry 1:2 and 1:4) conical pore are compared in terms of the transport rates. (a) Probe molecules located on the wide end of the pore—right-hand side. (b) Probe molecules located on the narrow end of the pore—left-hand side. Electric potential is zero on the left-hand side of the pore in both cases. The probe molecules net charge is −1.

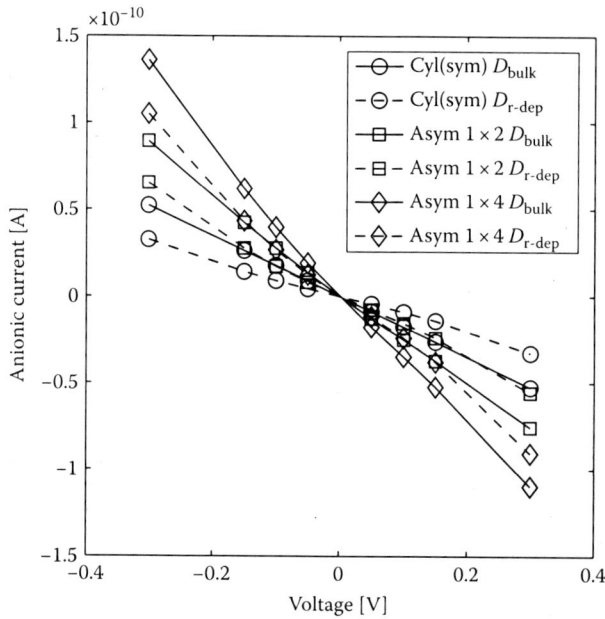

FIGURE 6.9 Anionic current in symmetric and asymmetric pores. Comparison of the anionic (chloride) current in symmetric regular cylindrical pore and asymmetric (one-sided, asymmetry 1:2 and 1:4) conical pore.

diffusion coefficient, where transport from wide to narrow end of the pore is several times faster than transport in the other direction.

The transport characteristic for a single chloride ion in both diffusion models is presented in Figure 6.9. Despite the small size of the anion (radius below 2 Å) in comparison to the pore diameter (1.2 nm at least), the effect of the RRD diffusion coefficient on the anionic current is considerable, even when the hydration shell is not taken into account. For example, the *I–V* curve for the 1.2 × 4.8 nm asymmetric pore obtained with the RRD diffusion-coefficient model is quantitatively similar to the *I–V* curve for the 1.2 × 2.4 nm asymmetric pore obtained with the RRI model. These results suggest that using an improper diffusion model can potentially lead to failed predictions from current–voltage experimental data.

6.5 Conclusions

This chapter reviews fluorescent probes, dyes, and methods used to discover conductance properties of electropores. The probe dimensions range from angstroms, in the case of ions, to tens of nanometers for the particles modeling transport of plasmids. The combinations of different probe species in a single experiment indicate temporal changes in the electropore geometry, pronounced mainly at short-pulse fields. Fluorescence intensity, related to the total number of molecules that cross plasma membrane through an electropore, quantifies the transportation efficiency and may be used to evaluate pore dimensions. To accomplish this task with a high accuracy, an appropriate theoretical model should be implemented. In the case of electropores, there are two main difficulties to be addressed. The first is high pore dynamics and distinct pore species appearing at the subsequent stages of the pulse application. The other problem occurs when the diameters of the probe molecules and electropore are comparable, both within single nanometer scale. The molecular diffusion in such systems is affected by hindrances posed by the entropic barrier and by molecular interactions. As shown in this chapter, disregard of these effects causes significant underestimation of the nanopore dimensions. Moreover, a simple cylindrical shape does not seem a good model for an electropore and, as a result, a classical one-dimensional Nernst–Planck equation with a bulk value of the diffusion constant is very imprecise. Advanced three-dimensional modeling is necessary, which takes into account more physical representation of the nanopore shape, charge distribution, and interactions between molecules, ions, and pore walls. When the sizes of the pores and probes are both of single nanometer scale, this should be reflected by assuming an effective diffusion constant value. The impact of such improvements on the results, which can be applied to a quasi-static, possibly long-lived (or time-frozen) nanopore, was demonstrated by means of the three-dimensional PNP model with adjustable diffusion coefficient. The results showed significant sensitivity to the pore shape and pore-marker diameter ratio, which are neglected in widely used simplified models.

Acknowledgments

This work was partially supported by Lower Silesian Grant II/37/2009 and the grant of Ministry of Science and Higher Education of Poland N N519 401537.

References

Canatella PJ, Karr JF, Petros JA, Prausnitz MR. 2001. Quantitative study of electroporation-mediated molecular uptake and cell viability. *Biophys. J.* 80(2):755–764.

Corry B, Kuyucak S, Chung SH. 2000a. Tests of continuum theories as models of ion channels. I. Poisson-Boltzmann theory versus Brownian dynamics. *Biophys. J.* 78:2349–2363.

Corry B, Kuyucak S, Chung SH. 2000b. Tests of continuum theories as models of ion channels. II. Poisson-Nernst-Planck theory versus Brownian dynamics. *Biophys. J.* 78:2364–2381.

Daud AI, DeConti RC, Andrews S et al. 2008. Phase I trial of interleukin-12 plasmid electroporation in patients with metastatic melanoma. *J. Clin. Oncol.* 26(36):5896–5903.

Djuzenova CS, Zimmermann U, Frank H, Sukhorukov VL, Richter E, Fuhr G. 1996. Effect of medium conductivity and composition on the uptake of propidium iodide into electropermeabilized myeloma cells. *Biochim Biophys Acta* 1284(2):143–152.

Dyrka W, Augousti AT, Kotulska M. 2008. Ion flux through membrane channels—An enhanced algorithm for the Poisson-Nernst-Planck model. *J. Comput. Chem.* 29(12):1876–1888.

Gabriel B, Teissie J. 1995. Control by electrical parameters of short- and long-term cell death resulting from electropermeabilization of Chinese hamster ovary cells. *Biochim Biophys Acta* 1266:171–178.

Gabriel B, Teissie J. 1997. Direct observation in the millisecond time range of fluorescent molecule asymmetrical interaction with the electropermeabilized cell membrane. *Biophys. J.* 73(5):2630–2637.

Gabriel B, Teissie J. 1999. Time courses of mammalian cell electropermeabilization observed by millisecond imaging of membrane property changes during the pulse. *Biophys. J.* 76:2158–2165.

Gift EA, Weaver JC. 2000. Simultaneous quantitative determination of electroporative molecular uptake and subsequent cell survival using gel microdrops and flow cytometry. *Cytometry* 39(4):243–249.

Golzio M, Teissié J, Rols MP. 2002. Cell synchronization effect on mammalian cell permeabilization and gene delivery by electric field. *Biochim Biophys Acta* 1563(1–2):23–28.

Hibino M, Itoh H, Kinosita K. 1993. Time courses of cell electroporation as revealed by submicrosecond imaging of transmembrane potential. *Biophys. J.* 64:1789–1800.

Hodgkin AL, Katz B. 1949. The effect of sodium ions on the electrical activity of giant axon of the squid. *J. Physiol.* 108:37–77.

Jacobs MH. 1967. *Diffusion Processes.* Springer, New York.

Joshi RP, Sridhara V, Schoenbach KH. 2006. Microscopic calculations of local lipid membrane permittivities and diffusion coefficients for application to electroporation analyses. *Biochem Biophys Res Commun.* 348(2):643–648.

Kalinowski S, Ibron G, Bryl K, Figaszewski Z. 1998. Chronopotentiometric studies of electroporation of bilayer lipid membranes. *Biochim Biophys Acta* 1369(2):204–212.

Kennedy SM, Ji Z, Hedstrom JC, Booske JH, Hagness SC. 2008. Quantification of electroporative uptake kinetics and electric field heterogeneity effects in cells. *Biophys J* 94(12):5018–5027.

Kosinska ID, Fulinski A. 2005. Asymmetric nanodiffusion. *Phys Rev E* 72:011201.

Kotulska M. 2007. Natural fluctuations of an electropore show fractional Lévy stable motion. *Biophys J* 92:2412–2421.

Kotulska M, Koronkiewicz S, Kalinowski S. 2004. Self-similar processes and flicker noise from a fluctuating nanopore in a lipid membrane. *Phys Rev E* 69:031920.

Kurnikova MG, Coalson RD, Graf P, Nitzan A. 1999. A lattice relaxation algorithm for three-dimensional Poisson-Nernst-Planck theory with application to ion transport through the gramicidin A channel. *Biophys J* 76:642–656.

Mir LM, Gehl J, Sersa G, Collins CG, Garbay JR, Billard V, Geertsen PF, Rudolf Z, O'Sullivan GC, Marty M. 2006. Standard operating procedures of the electrochemotherapy: Instructions for the use of bleomycin or cisplatin administered either systemically or locally and electric pulses delivered by the CliniporatorTM by means of invasive or non-invasive electrodes. *Eur J Cancer* Suppl. S4:14–25.

Neumann E, Schaefer-Ridder M, Wang Y, Hofschneider PH. 1982. Gene transfer into mouse lyoma cells by electroporation in high electric fields. *EMBO J* 1:841–845.

Neumann E, Toensing K, Kakorin S, Budde P, Frey J. 1998. Mechanism of electroporative dye uptake by mouse B cells. *Biophys J* 74:98–108.

Noskov SY, Im W, Roux B. 2004. Ion permeation through the alpha-hemolysin channel: Theoretical studies based on Brownian dynamics and Poisson-Nernst-Plank electrodiffusion theory. *Biophys J* 87:2299–2309.

Paine PL, Scherr P. 1975. Drag coefficients for the movement of rigid spheres through liquid-filled cylindrical pores. *Biophys J* 15(10):1087–1091.

Pakhomov AG, Bowman AM, Ibey BL, Andre FM, Pakhomova ON, Schoenbach KH. 2009. Lipid nanopores can form a stable, ion channel-like conduction pathway in cell membrane. *Biochem Biophys Res Commun* 385(2):181–186.

Pavlin M, Miklavcic D. 2008. Theoretical and experimental analysis of conductivity, ion diffusion and molecular transport during cell electroporation—Relation between short-lived and long-lived pores. *Bioelectrochemistry* 74:38–46.

Prausnitz MR, Lau BS, Milano CD, Conner S, Langer R, Weaver JC. 1993. A quantitative study of electroporation showing a plateau in net molecular transport. *Biophys J* 65(1):414–422.

Prausnitz MR, Milano CD, Gimm JA, Langer R, Langer R, Weaver JC. 1994. Quantitative study of molecular transport due to electroporation: Uptake of bovine serum albumin by erythrocyte ghosts. *Biophys J* 66:1522–1530.

Puc M, Kotnik T, Mir LM, Miklavcic D. 2003. Quantitative model of small molecules uptake after in vitro cell electropermeabilization. *Bioelectrochemistry* 60:1–10.

Pucihar G, Kotnik T, Miklavcic D, Teissie J. 2008. Kinetics of transmembrane transport of small molecules into electropermeabilized cells. *Biophys J* 95(6):2837–2848.

Reguera D, Rubí JM. 2001. Kinetic equations for diffusion in the presence of entropic barriers. *Phys Rev E* 64(6):061106.

Rols MP. 2006. Electropermeabilization, a physical method for the delivery of therapeutic molecules into cells. *Biochim Biophys Acta* 1758:423–428.

Rols MP, Teissie J. 1998. Electropermeabilization of mammalian cells to macromolecules: Control by pulse duration. *Biophys J* 75:1415–1423.

Rols MP, Delteil C, Golzio M, Teissie J. 1998. Control by ATP and ADP of voltage-induced mammalian-cell-membrane permeabilization, gene transfer and resulting expression. *Eur J Biochem* 254:382–388.

Schultz SG. 1980. Chapter 2. In *Basic Principles of Membrane Transport*. Cambridge University Press: Cambridge, U.K.

Shirakashi R, Sukhorukov VL, Tanasawa I, Zimmermann U. 2004. Measurement of the permeability and resealing time constant of the electroporated mammalian cell membranes. *Int J Heat Mass Transf* 47:4517–4524.

Siwy Z, Fulinski A. 2002. Origin of $1/f\alpha$ noise in membrane channel currents. *Phys Rev Lett* 89(15):158101.

Siwy Z, Kosinska ID, Fulinski A, Martin CR. 2005. Asymmetric diffusion through synthetic nanopores. *Phys Rev Lett* 94(4):048102.

Sixou S, Teissie J. 1993. Exogenous uptake and release of molecules by electroloaded cells: A digitized videomicroscopy study. *Bioelectrochem Bioenerget* 31:237–257.

Tarek M. 2005. Membrane electroporation: A molecular dynamics simulation. *Biophys J* 88(6):4045–4053.

Teissie J, Eynard N, Gabriel B, Rols MP. 1999. Electropermeabilization of cell membranes. *Adv Drug Deliv Rev* 35:3–19.

Teissie J, Golzio M, Rols MP. 2005. Mechanisms of cell membrane electropermeabilization: A mini review of our present (lack of ?) knowledge. *Biochim Biophys Acta* 1724:270–280.

Tekle E, Astumian RD, Chock PB. 1990. Electro-permeabilization of cell membranes: Effect of the resting membrane potential. *Biochem Biophys Res Commun* 172(1):282–287.

Tekle E, Astumiant R, Chock PB. 1994. Selective and asymmetric molecular transport across electroporated cell membranes. *Biochemistry* 91:11512–11516.

Tekle E, Astumian RD, Friauf WA, Chock PB. 2001. Asymmetric pore distribution and loss of membrane lipid in electroporated DOPC vesicles. *Biophys J* 81(2):960–968.

Tieleman DP. 2004. The molecular basis of electroporation. *BMC Biochem* 5:10.

Vernier PT, Sun YH, Marcu L, Craft CM, Gundersen MA. 2004a. Nanoelectropulse-induced phosphatidylserine translocation. *Biophys J* 86:4040–4048.

Vernier PT, Sun YH, Marcu L, Craft CM, Gundersen MA. 2004b. Nanosecond pulsed electric fields perturb membrane phospholipids in T lymphoblasts. *FEBS Lett* 572:103–108.

Vernier PT, Ziegler MJ, Sun YH, Chang WV, Gundersen MA, Tieleman DP. 2006a. Nanopore formation and phosphatidylserine externalization in a phospholipid bilayer at high transmembrane potential. *J Am Chem Soc* 128:6288–6289.

Vernier PT, Sun Y, Gundersen MA. 2006b. Nanoelectropulse-driven membrane perturbation and small molecule permeabilization. *BMC Cell Biol* 7:37.

Vernier PT, Sun Y, Chen MT, Gundersen MA, Craviso GL. 2008. Nanosecond electric pulse-induced calcium entry into chromaffin cells. *Bioelectrochemistry* 73:1–4.

Wolf H, Rols MP, Boldt E, Neumann E, Teissie J. 1994. Control by pulse parameters of electric field-mediated gene transfer in mammalian cells. *Biophys J* 66:524–531.

Zaharoff DA, Henshaw JW, Mossop B, Yuan F. 2008. Mechanistic analysis of electroporation-induced cellular uptake of macromolecules. *Exp Biol Med* 233:94–105.

Zhen J, Kennedy SM, Booske JH, Hagness SC. 2006. Experimental studies of persistent poration dynamics of cell membranes induced by electric pulses. *IEEE Trans Plasma Sci* 34:1416–1424.

Zwanzig R. 1992. Diffusion past an entropy barrier. *J Phys Chem* 96(10):3926–3930.

7

Electroporation of Lipid Membranes: Insights from Molecular Dynamics Simulations

Mounir Tarek

Lucie Delemotte

7.1 Introduction

Membranes consist of an assembly of a wide variety of lipids, proteins, and carbohydrates self-organized into a thin barrier that separates the interior of the cell from the outside environment, preventing the free diffusion of small molecules (Gennis 1989). The main lipid constituents of natural membranes are phospholipids: amphiphilic molecules that have a hydrophilic head and two hydrophobic tails. When exposed to water, lipids arrange themselves into a two-layered sheet (a bilayer) with all their tails pointing toward the center of the sheet. Although they are only a few nanometers thick, lipid bilayers are impermeable to most water-soluble (hydrophilic) molecules.

To assume a lot of the biological functions necessary for the cell machinery, e.g., the passive and active transport of matter, the capture and storage of energy, the control of the ionic balance, or the cellular recognition and signaling, cells require the assistance of membrane proteins. In 1972, Singer and Nicolson (1972) introduced a model to describe the molecular organization of biological membranes, called the "fluid mosaic" model, according to which the lipid bilayer forms the membrane matrix in which proteins are embedded and free to diffuse at physiological temperatures in the plane of the membrane. Membranes were since shown to be more mosaic than fluid (Engelman 2005). They are formed by several heterogeneous regions (micro-domains) with variable thicknesses, some of which have high protein concentrations, while others are rich in cholesterol and sphingolipids (Brown 1998; Brown and London 2000; Simons and Ikonen 1997), resulting in more compact and thicker regions (Anderson and Jacobson 2002; Edidin 2003).

Electroporation is a phenomenon that affects the fundamental behavior of cells since it disturbs transiently or permanently the integrity of their membrane (Eberhard et al. 1989; Li 2008; Nickoloff 1995). This process relates to the cascade of events that follows the application of high electric fields and leads to the cell membrane permeabilization (Glaser et al. 1988; Needham and Hochmuth 1989; Neumann and Rosenheck 1972; Teissié et al. 1999; Tsong 1987, 1991; Weaver 1995; Zimmerman 1996; Zimmerman et al. 1976). It finds numerous applications today since, under certain conditions, it is reversible, and hence permits an efficient transmembrane transfer of small molecules. Electroporation is routinely used in molecular biology, biotechnology, and has found applications in medicine (Golzio et al. 2002; Harrison et al. 1998; Lee et al. 1992; Lundqvist et al. 1998; Neumann et al. 1982; Nishi et al. 1996). The method is efficient for transdermal drug delivery, the transport of drugs, oligonucleotides, antibodies, and plasmids across cell membranes (Prausnitz et al. 1993; Suzuki et al. 1998; Tsong 1983, 1987). Recent reviews of its application may be found in Bodles-Brakhop et al. (2009), Campana et al. (2009), Cemazar and Sersa (2007), Cemazar et al. (2008), Teissié et al. (2008), Testori et al. (2008), Villemejane and Mir (2009), and the reader is referred to the other chapters of the book for further reading.

Experimental evidence suggests that the direct effect of the external stress is to produce aqueous pores in the lipid bilayer, showing that the site of interaction of the electric field with the cell membrane is the lipid component (Abidor et al. 1979; Benz et al. 1979; Chen et al. 2006; Weaver 2003; Weaver and Chizmadzhev 1996). The sequence of events describing the phenomenon is gathered mainly from the measurements of electrical currents through planar bilayer membranes (BLM) under the influence of strong electric fields, and from the characterization of the molecular transport of molecules into (or out of) cells subjected to electric field pulses. It takes place as follows: The application of millisecond to nanosecond electrical pulses triggers the reorganization of ions that leads to the production of a transient, elevated, transmembrane voltage ΔV and creates a local electric field (Pucihar et al. 2006). This field induces a rearrangement of the membrane components (mainly water and lipids) that ultimately leads to the formation of aqueous (hydrophilic) pores, whose presence increases substantially the ionic and molecular transport through the cell membrane (Pucihar et al. 2008). Under appropriate circumstances, turning off the electric field can lead to the recovery of the membrane.

Due to the complexity and heterogeneity of membrane systems, it is difficult to interpret electroporation data in terms of atomically resolved structural and dynamical processes, even if the samples studied are as simple as planar lipid membranes (Kramar et al. 2009). A method that could provide such a level of detail that is inaccessible to conventional experimental techniques would therefore be extremely valuable to improve our understanding of how membranes behave under the influence of high electric fields. Atomic simulations in general, and molecular dynamics (MD) simulations in particular, have proven to be an effective approach for this purpose, providing new insights into both the structure and the dynamics of these membrane systems (see below). Here we report on the progress made so far using such a technique to model the electroporation of lipid bilayers. We will start with a description of the commonly used methodologies and protocols in the MD simulations of lipid bilayers, with an emphasis on their limitations and their strengths in reproducing with satisfactory agreement many experimental observables. We then present an account of early models of the *in silico* electroporation of lipid membranes resulting from the direct application of an electric field, and discuss their connection with experiments. Section 7.4 of the chapter will briefly describe a method recently proposed that better "mimics" the experimental conditions, avoiding therefore several shortcomings of the electric field method. The preliminary results of the electroporation of phospholipid bilayers using this new method will also be presented and discussed.

7.2 MD Simulations of Membranes

We briefly review in this section some basic concepts and terminology concerning the MD simulation techniques, and recall the actual state of their application to study simple membranes. Comprehensive reviews on modeling lipid bilayers can be found in Anézo et al. (2003), Berkowitz et al. (2006), Edholm (2008), Feller (2000), Feller (2008), Forrest and Sansom (2000), Lindahl and Sansom (2008), Marrink et al. (2009), Mashl et al. (2001), Saiz and Klein (2002a), Tieleman et al. (1997), Tobias (2001), and Tobias et al. (1997).

7.2.1 MD Simulations: Basics

MD simulations involve solving the equations of motion for all the molecules in the considered system (lipids, water, etc.), using forces derived from a potential energy function: U (Allen and Tildesley 1987; Frenkel and Smit 1996; Leach 1996; Yip 2005). Force fields most commonly used in chemistry and biophysics, e.g., GROMOS (Schuler et al. 2001), CHARMM (MacKerell et al. 1998), and AMBER (Case et al. 1999), are based on a classical treatment of particle–particle interactions that preclude bond dissociation: the function U is represented in terms of interactions between pairs of atoms and is typically divided into "bonded" and "nonbonded" terms. The bonded interactions consist of intramolecular harmonic bond stretching, angle bending, and the dihedral angle deformation of individual molecules. Nonbonded interactions include intermolecular electrostatic and van der Waals terms. The former are calculated from partial charges assigned to each atom, whereas the latter describe short-range repulsion and long-range attraction between pairs of atoms. A typical potential function is of the form:

$$U = \sum_{i<j} \frac{q_i q_j}{4\pi\varepsilon r_{ij}} + \sum_{i<j} 4\varepsilon_{ij}\left(\frac{\sigma_{ij}^{12}}{r_{ij}^{12}} - \frac{\sigma_{ij}^{6}}{r_{ij}^{6}}\right)$$

$$+ \sum_{bonds} \frac{1}{2}k_{ij}^{b}\left(r_{ij} - b_{ij}^{0}\right) + \sum_{angels} \frac{1}{2}ki_{jk}^{\theta}\left(\theta_{ijk} - \theta_{ijk}^{0}\right)^2$$

$$+ \sum_{dihedrals} k^{\varphi}\left[1 + \cos\left(n\left(\varphi - \varphi^{0}\right)\right)\right]$$

where
   r_{ij} is the distance between atoms i and j
   q_i is the partial charge on atom i
   ε_{ij} and σ_{ij} are Lennard-Jones parameters for van der Waals interactions
   k_{ij}, k_{ijk}, and k^{φ} on one hand and b_{ij}^{0}, θ_{ijk}^{0}, and φ^{0} on the other are the force constants and equilibrium
      values for the bond stretch, angle bend, and dihedral torsion deformations

MD simulations generate a set of atomic positions and velocities as a function of time that evolve deterministically from an initial configuration according to the interaction potential U. MD simulations use information (positions, velocities or momenta, and forces) at a given instant in time, t, to predict the positions and momenta at a later time, $t + \Delta t$, by integrating the classical equations of motion. Δt is the time step, of the order of a femtosecond (10^{-15} s), and remains constant throughout the simulation. The numerical solutions of the equations of motion are thus obtained by an iteration of this elementary step. Using statistical mechanics, the simulation results may be gathered to calculate observable quantities if the trajectory is long enough to yield satisfactory time averages. Computer simulations are usually performed on a small number of molecules (few tens to few hundred-thousand atoms), the system size being limited by the speed of the execution of the programs and the availability of computer power. In practice, the trajectories for complex systems such as lipid membranes span a few nanoseconds (10^{-9} s). This in general is long enough to permit a relaxation of many degrees of freedom of the system and allows the determination of local structural properties. The quality of the proper sampling of the trajectory may be assessed, in part, by comparing calculated quantities to experiment (see below) (Anézo et al. 2003; Chipot et al. 2005; Tobias 2001; Tobias et al. 1997). For the computation to be tractable, the number of lipids in the system is typically less than 100. In order to reduce the effects caused by such a relatively small system size, the sample is placed in a central cell, called the simulations box, which is replicated infinitely in three dimensions using periodic boundary conditions (PBCs) (Figure 7.1B). Every molecule in the box interacts therefore with its neighbors or their replicas. The long-range van der Waals interactions are typically turned off beyond a certain cutoff distance (generally, 10–15 Å), which is smaller than half the box size to avoid interactions between a molecule and its own image. Hence, for

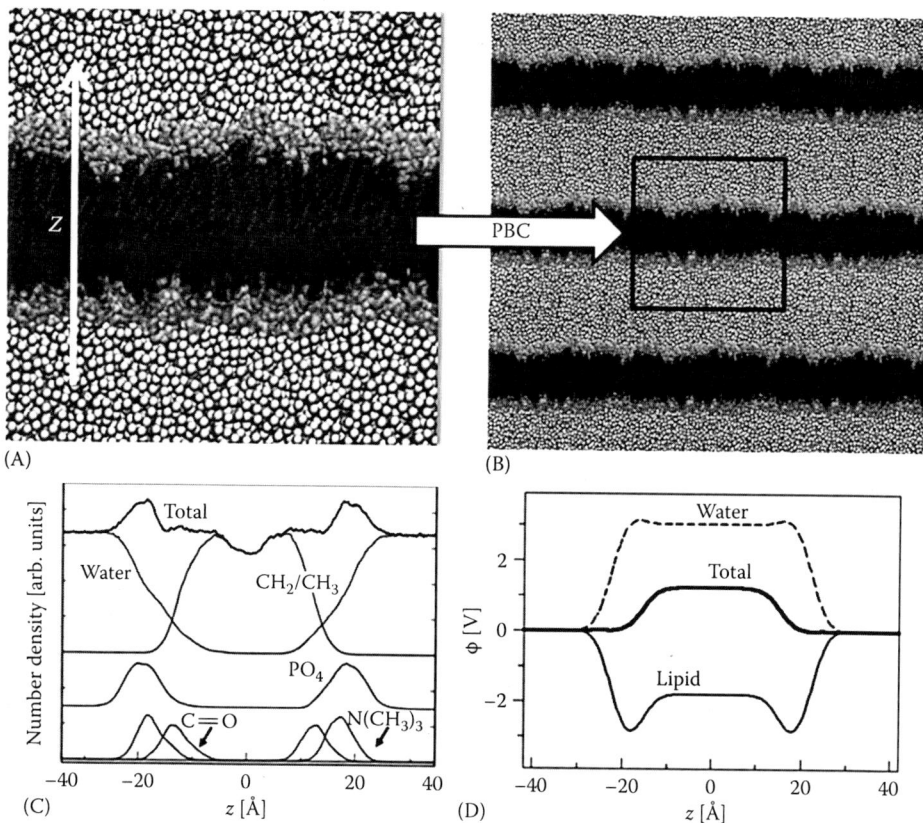

FIGURE 7.1 General features of lipid bilayers gathered from MD simulations. (A) Configuration of a POPC hydrated bilayer system from a well-equilibrated constant pressure MD simulation performed at 300 K. Only the water molecules (as gray solid van der Waals spheres), the phosphate and nitrogen atoms (gray transparent spheres) of the lipid head groups and the acyl chains (black) in the simulation cell are shown. The axis normal to the membrane (Z) is also represented. (B) Illustration of the use of periodic boundary conditions (PBCs) in MD simulations: The system from (A) is replicated infinitely in the three directions of space, resulting in the simulation of a multilamellar system. (C) Number density profiles (arbitrary units) along the bilayer normal averaged over 2 ns of the MD trajectory. The total density, water, and hydrocarbon chain contributions are indicated, along with those from the POPC head group moieties. The bilayer center is located at $z = 0$. (D) Corresponding electrostatic potential profile calculated using Equation 7.1 and symmetrized over the two sides of the bilayer. Contributions of water and lipids are shown separately.

membranes, for instance, the simulated system would correspond to a small fragment of either a planar lipid membrane, a liposome, or a multilamellar-oriented lipid stack deposited on a substrate.

7.2.2 MD Simulations of Lipid Bilayers

Nowadays, academic MD packages, such as AMBER (Case et al. 2008), CHARMM (Brooks et al. 2009), GROMACS (Hess et al. 2008), or NAMD (Phillips et al. 2005), have benefited from several recent methodological developments on the algorithmic front, and incorporate more or less all the key features that are necessary to investigate membrane assemblies rigorously. From both a theoretical and an experimental perspective, the bilayers formed of zwitterionic phosphatidylcholine (PC) lipids constitute the best-characterized systems. Among these, hydrated dimyristoyl-phosphatidyl-choline (DMPC) (Chiu et al. 1995; Damodaran and Merz 1994) and dipalmitoyl-phosphatidyl-choline (DPPC) (Berger et al. 1997; Essman and Berkowitz 1999; Feller et al. 1997; Tieleman and Berendsen 1996; Tu et al. 1995; Venable et al. 1993) bilayers have been so far probably the most extensively surveyed membrane mimics. Furthermore, at the exception of a handful of simulations of lipid bilayers in the gel phase (Essmann et al. 1995; Leekumjorn and Sum 2007; Qin et al. 2009; Tu et al. 1996; Venable et al. 1993), most investigations have focused on the biologically relevant, so-called liquid crystal (L_α) phase.

Since the early simulations, several authors have turned to lipids that are more relevant to biosystems like palmitoyl-oleyl-phosphatidylcholine (POPC) (Chiu et al. 1999; Rög et al. 2002) for examining membrane proteins in a realistic environment, and lipids based on the mixtures of saturated and polyunsaturated alkyl chains stearoyl-docosahexaenoyl-glycero-phosphocholine (SDPC) (Feller et al. 2002; Saiz and Klein 2001). Lipids featuring different head groups, e.g., phosphatidyl-ethanolamin (Berkowitz and Raghavan 1991; Damodaran and Merz 1994), phosphatidyl-serine (Cascales et al. 1996; Mukhopadhyay et al. 2004), or sphingomyelin (Chiu et al. 2003), and more recently mixed bilayer compositions (Dahlberg and Maliniak 2008; Gurtovenko and Vattulainen 2008; Li et al. 2009; Pandit et al. 2003a; Patel and Balaji 2008; Rog et al. 2009; Vacha et al. 2009) have also been explored. In most of these studies, however, the modelers lack experimental data with which the results of atomic simulations can be compared.

Simple bilayers built from PC lipids represent remarkable test systems not only to probe the simulations methodology, but also to gain additional insight into the physical properties of membranes. The MD simulations of such lipids are often shown to reproduce and predict the structural and dynamical properties of lipid membranes. Hence, the information provided by the density distributions of the bilayer components can be confronted directly to x-ray and/or neutron diffraction measurements (Armen et al. 1998; Benz et al. 2006; Klauda et al. 2006; Wiener and White 1992). Atomistic modeling was also shown to provide valuable insight into bilayer structure by the mutual refinement of neutron reflectivity data (Krueger et al. 2001; Majkrzak et al. 2000; Tarek et al. 1999). A better refinement of nuclear magnetic resonance (NMR) data has been targeted at obtaining quantities such as the average lipid chain length, the surface area per molecule (Petrache et al. 1999), or the acyl chain order parameters (Vermeer et al. 2007) and infrared (IR) data have been reinterpreted to estimate the concentrations of *gauche–gauche*, *trans–gauche* and *trans–trans* conformational sequences in a DPPC bilayer (Snyder et al. 2002).

Similar efforts were undertaken to probe the dynamics of lipids in bilayer assemblies. Hence, MD simulations were used to analyze NOESY cross-relaxation rates in lipid bilayers (Feller et al. 1999), to calculate relaxation rates, and to assign them to various motions of the lipid molecules. The dynamics of individual lipid molecules have also been proposed based on a thorough comparison to the ^{13}C NMR T_1 relaxation rates (Pastor et al. 2002) and P-31-NMR spin-lattice (R-1) relaxation rates (Klauda et al. 2008) of DPPC alkyl chains. Detailed models for lipid dynamics in bilayers on the 100 ps timescale were proposed based on the comparison of MD simulation results to inelastic neutron scattering (INS) data (Tobias 2001). MD simulations have also been used to complement the inelastic x-ray data of lipid bilayers both in the gel and the liquid crystal phases (Hub et al. 2007; Tarek et al. 2001). The results supported the applicability of generalized hydrodynamics to describe the motion of carbon atoms in the hydrophobic core, and to extract key parameters, such as sound-mode propagation velocity, thermal diffusivity, and kinematic longitudinal viscosity. Large systems (over 1000 lipids) studied over 10 ns led to the direct observation of bilayer undulations and thickness fluctuations of mesoscopic nature (Lindahl and Edholm 2000). Continuum properties such as bending modulus, surface compressibility, and mode relaxation times were calculated and confronted with experimental data.

7.2.3 Membranes Electrostatic Properties

Before proceeding with an in-depth account of the simulations of membrane electroporation results, we shall describe in the following some aspects of the electrostatic properties of lipid bilayers at rest.

On the atomic scale, the lipid bilayer appears as a broad hydrophilic interface, with only a thin slab of pure hydrocarbon fluid in the middle (Figure 7.1C). The density distributions of different atom types along the bilayer normal, which can be measured by neutron and x-ray diffraction techniques (Hristova and White 1998), or calculated from MD simulations, indicate that the membrane–water interface is characterized by a rough lipid head group area, across which water density decays smoothly from the bulk value and penetrates deeply into the bilayer. To this heterogeneous atomic distributions are associated charge and molecular dipole distributions that are at the origin of an intrinsic electrostatic profile (EP) across the bilayer. In their pioneering work, Liberman and Topaly (1969), from the measurements of the partition coefficients of fat-soluble tetraphenylboron anions and fat-soluble triphenylphosphonium cations between the membrane and aqueous phases, hypothesized that the inner part of the bilayer membrane must be initially positively charged. The existence of a positive potential difference between the membrane interior and the adjacent aqueous phase implied that, for the purpose of describing the electrostatics, the membrane can be thought of as a planar array of dipoles whose negative ends point toward the water. The absolute value of this "dipole potential" has been very difficult to measure or predict (see Clarke 2001; Demchenko and Yesylevskyy 2009; Mc Laughlin 1989; Wang et al. 2006), and estimates obtained from various methods and various lipids range from ~200 to 1000 mV. More recent and direct measurement based on Cryo-EM imaging (Wang et al. 2006) and atomic force microscopy (AFM) (Yang et al. 2008) techniques show that the dipole potential can be "measured" in a noninvasive manner and estimate its value to few 100 mV.

In *in silico* investigations, the EP profile along the lipid bilayer normal Z may be estimated from MD simulations using Poisson's equation, and can be derived directly as the double integral of $\rho(Z)$, the molecular-charge density distributions:

$$\Delta\phi(Z) = \phi(Z) - \phi(0) = \frac{1}{\varepsilon_0} \int_0^Z \int_0^{Z'} \rho(Z'') dZ'' dZ' \tag{7.1}$$

neglecting therefore the explicit electronic polarization (Tieleman et al. 1997; Tobias et al. 1997). Indeed, conventional force fields only include point charges and pair-additive Coulomb potentials, which prevent them from describing realistic collective electrostatic effects, such as charge transfer, electronic excitations, or electronic polarization, which is often considered as a major limitation. Note that constant efforts are undertaken on the development of potential functions that explicitly treat electronic polarizability in empirical force fields (Halgren and Damm 2001; Lopes et al. 2009; Warshel et al. 2007), but none of these "polarizable" force fields is routinely used in large-scale simulations for now, the main reasons for that being the dramatic increase of the computational time and additional complications with their parameterization. A recent MD simulation of a solvated DMPC bilayer using a polarizable force field reported a dipole potential 950 mV compared to the value of 800 mV based on nonpolarizable force-field calculations (Davis et al. 2009). In contrast, for DPPC monolayers, the dipole potential is shown to decrease from 800 to 350 mV when many-body polarization effects are included (Harder et al. 2009).

The large body of results from the simulations of fully hydrated lipid bilayers using nonpolarizable force fields are found in qualitative agreement with experiments, showing that the EP profile monotonically increases across the membrane–water interface. Given the diversity of the lipids studied and more significantly of the force fields used, the values of the dipole potential, ranging from 500 to 1200 mV, have been reported (Berkowitz et al. 2006; Demchenko and Yesylevskyy 2009; Mashl et al. 2001; Sachs et al. 2004; Smondyrev and Berkowitz 1999; Wang et al. 2006).

Hence, both experimental and theoretical investigations clearly show that, for the purpose of describing membrane electrostatics, each lipid–water interface can be thought of as a planar array of dipoles whose negative ends point toward the water. Detailed analysis from MD simulations allows the further characterization of the specific contribution of each lipid component to this total dipole potential. The MD simulations of fully hydrated PC bilayers show that the phospholipid head groups adopt in general a preferential orientation: on average, the P–N dipoles point 30° away from the membrane normal (Saiz and Klein 2002b; Tobias 2001). The organization of the phosphate (PO_4^-) and the choline ($N(CH_3)_3^+$) groups on one hand, and of the carbonyl (C=O) groups on the other hand give rise to a permanent dipole. The solvent molecules bound to the lipid head-group moieties tend therefore to orient their dipoles to compensate the latter. As a result, the typical contributions of each membrane component to the total EP profile (cf. Figure 7.1D) consist of a negative lipid contribution on the hydrocarbon side of the interface balanced by a positive contribution of the water side, the net positive potential being primarily due to an excess of water molecules orientation so as their dipoles point toward the membrane surface (Gawrisch et al. 1992).

Similar findings were reported for charged lipid bilayers, and bilayers with added-salt concentrations (see Berkowitz et al. 2006; Bockmann et al. 2003; Demchenko and Yesylevskyy 2009; Gurtovenko and Vattulainen 2008; Lee et al. 2008 and references therein). Note that in such systems, the presence of the uncompensated charges of the lipid head groups, adsorbed ions, and the counter ions that reside in the vicinity of the head groups on the both sides of the membrane contribute to the total electrostatic potential (Pandit et al. 2003a,b). This surface potential that can be reasonably equated to the zeta potential is much smaller than the intrinsic dipole potential (Mc Laughlin 1989; Xu and Loew 2003).

As one may gather from the previous paragraphs, exquisite experimental results have led to the implementation of refined MD techniques that have been used to elucidate precise microscopic events taking place in the lipid bilayer at a level of detail not always attainable by experiment alone. Most often, the degree of cooperation between experiment and MD simulations has provided, to a certain level, a validation of the theoretical approach. Additionally, MD allows the construction of systems in a selective manner in such a way that specific atomic details can be isolated or modified in order to determine the governing influences on the structure and dynamics of the bilayer. This can be done while maintaining consistency and agreement with experimentally determined observables. It is therefore with greater confidence that one expects computational studies to provide a unique and direct measure of microscopic properties that govern the mechanisms of lipid membrane electroporation. The following sections report the first attempts to do so and clearly show that indeed the picture emerging from modeling the process is consistent with several hypotheses based on experimental observations.

7.3 Membrane Electroporation: Early MD Studies

In the MD simulations of membranes, the often-used strategy to induce a transmembrane (TM) voltage ΔV is to apply a constant electric field $\vec{E}$, perpendicular to the membrane plane (Crozier et al. 2001; Roux 2008; Tieleman et al. 2001; Yang et al. 2002; Zhong et al. 1998). MD simulations adopting such an approach have been used so far to study membrane electroporation (Bockmann et al. 2008; Hu et al. 2005; Tarek 2005; Tieleman 2004; Ziegler and Vernier 2008) and lipid externalization (Vernier et al. 2006), to activate voltage-gated K+ channels (Treptow et al. 2004), and to determine the transport properties of ion channels (Aksimentiev and Schulten 2005; Chimerel et al. 2008; Khalili-Araghi et al. 2006; Sotomayor et al. 2007).

In practice, the electric field is implemented by applying a force $\vec{F} = q_i\vec{E}$ to all the atoms bearing a charge q_i. For lipid bilayers, the field acts primarily on the interfacial water dipoles (small polarization

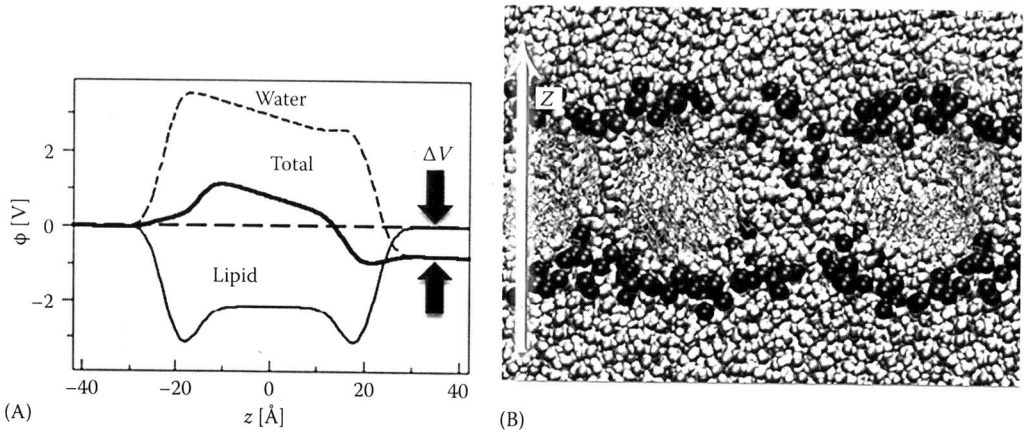

FIGURE 7.2 Applying a transmembrane potential: the electric field method. (A) Electrostatic potential profile estimated from the MD simulations of a hydrated POPC lipid bilayer subject to a constant electric field perpendicular to the membrane plan pointing upward and creating a transmembrane potential $\Delta V \sim 1$ V. (B) Configuration taken from an MD simulation of a large POPC bilayer, subject to an electric field generating a transmembrane potential of ~3 V, after a 2 ns run. Note the simultaneous presence of water (gray van der Waals spheres) wires and of large water pores stabilized by lipid head groups (black spheres). The lipid tails are represented as gray sticks.

of bulk water molecules). The reorientation of the lipid head groups appears not to be affected at short timescales (Tarek 2005; Vernier and Ziegler 2007), and not exceeding few degrees toward the field direction at longer timescales (Bockmann et al. 2008). The electrostatic potential profile, estimated as above from the charge distribution, shows that the electric field within a very short timescale—typically few picoseconds (Tarek 2005)—induces an overall transmembrane potential difference $\Delta V \approx |\vec{E}| \cdot L_z$ over the whole system, L_z being the size of the simulation box in the field direction (Figure 7.2). Interestingly, the simulations of a hydrated octane slab as membranes mimic confirms the main role played by the reorientation of interfacial water since even without the presence of charged lipid head groups the field $\vec{E}$ induces a transmembrane voltage $\Delta V \approx |\vec{E}| \cdot L_z$.

For the electric fields of high-enough magnitude, all simulations performed under this protocol show more or less a common poration sequence: the electric field creates a force on the interfacial water dipoles proportional to the field gradient, favoring therefore the formation of water wires deeper toward the hydrophobic core (Tieleman 2004). It turns out that the forces on both water–lipid interfaces are symmetric, which explains why the "fingers" of water molecules begin to penetrate the interior of the bilayer from both sides. Ultimately these fingers join up to form water channels (often termed pre-pores or hydrophobic pores) that span the membrane. Within a few nanoseconds, some lipid head groups start to migrate from the membrane–water interface to the interior of the bilayer, stabilizing hydrophilic pores (~5 nm radius) that increase in size as the electric field is maintained (Figure 7.2).

The timescales associated with this sequence of events depend on the strength of the applied electric field, and are within the nanosecond timescale for moderate to highly moderate applied fields (corresponding to $\Delta V \sim 2.5$–5 V). Several very interesting details could be gathered from the different investigations cited above.

7.3.1 Time Scales and Voltage Threshold

The simulations of the same system performed at various field strengths show that the electroporation process takes place much more rapidly under higher fields, without a major change in the pore formation characteristics. For voltages above ~2.5–3 V, pore formation occurs in a reasonable time from the point of view of a modeler (few nanoseconds). The lowest voltages reported to electroporate a lipid bilayer were above 2 V (estimates from the electric field and the size of the system reported by Bockmann et al. (2008) and Vernier and Ziegler (2007)). Ziegler and Vernier (2008) reported minimum poration external field strengths for four different PC lipids with different chain lengths and composition (the number of unsaturations). The authors find a direct correlation between the minimum porating field (ranging from 0.26 to 0.38 V nm^{-1}) and the membrane thickness (ranging from 2.92 to 3.92 nm). The size L_z of the systems was not reported, but all the lipid systems contained the same amount of water molecules. A rough estimate from the system characteristics indicates that the lower voltage threshold (~1.7 V) was found for the shorter lipid (dilauroyl PC; DLPC). Note that the estimates of the electroporation threshold from simulations may be considered only as indicative since it is related to the timescale the pore formation may take. In general, a field strength threshold was "assumed" to be reached when no pores were observed for the extended simulation lengths of 100–200 ns.

7.3.2 Pore Sizes and Pore Density

For a typical MD system size (128 lipids; 6 × 6 nm section), most of the simulations reported a single pore formation at high field strength. In two instances, for much larger systems, multiple pore formations were witnessed (Tarek 2005; Tieleman 2004), with diameters ranging from few to 10 nm. All the studies reported pore expansion as the electric field was maintained. Simulations have also shown, as predicted by Lewis (2003), that upon pore creation, the electric field induces a significant lateral stress, of the order of 1 mN m^{-1}. It is unclear how the induced lateral tension relaxes in a macroscopic system when a voltage pulse is applied, which may have an incidence on the density of pores that could nucleate. Furthermore, regardless of the topology of the bilayer, i.e., in planar lipid membranes or in a liposome, one expects a strong correlation between the size of the defect created, the density of pores, and the maintained electrical stress. Interestingly enough, it was shown in one instance (Tarek 2005) that a hydrophilic pore could reseal when the applied field was switched off within few nanoseconds. Membrane complete recovery, i.e., the migration of the lipid head group forming the hydrophilic pore toward the lipid–water interface, being a much longer process, was on the other hand not observed.

7.3.3 Electroporation of Complex Systems

Because of their interactions with lipids, some integral membrane proteins, e.g., gramicidin, are shown to increase the membrane electroporation threshold as it has been demonstrated by Troiano et al. (1999). The studies of a peptide nanotube (ion channel) (Tarek 2005) and of gramicidin A (GA) (Siu and Bockmann 2007) embedded in membranes both showed that no pore formation in the vicinity of the ion channels was observed, and that the time needed for pore formation was increased by a factor of two for bilayers containing GA. It was suggested that lipid molecules and therefore the membrane are stabilized due to their strong hydrogen bonding with the peptides: at high membrane protein concentrations, higher TM voltages are needed in order to break these hydrogen bonds.

Finally, one notes that a set of simulations (Tarek 2005) has allowed the modeling of the formation of pre-pores, large hydrophilic pores, and a subsequent partial transport of a large molecule (DNA plasmid) across the membrane. The strand considered diffused toward the interior of the bilayer when a pore was created beneath it and formed a stable DNA–lipid complex in which the lipid head groups encapsulate the strand. The process provides support to the gene delivery model proposed by Golzio et al. (2002) in which an "anchoring step" connecting the plasmid to permeabilized cells membranes that takes place during DNA transfer assisted by electric pulses.

Despite the satisfactory insight provided, the electric method suffers from many shortcomings. First, each charged particle in the system "feels" both the electric field generated by the induced transmembrane voltage ($\Delta V \approx |\vec{E}| \cdot L_z$) and the applied electric field ($\vec{F} = q_i \vec{E}$). It is very difficult to estimate the incidence of such supplementary force on the electroporation process. One intriguing phenomenon that is witnessed with the electric field method is the pore expansion as the field is maintained. Indeed, one expects a different scenario since the transmembrane potential collapses when hydrophilic pores form in the bilayer after electroporation (Hibino et al. 1993, 1991; Pavlin and Miklavcic 2003; Tarek 2005).

Moreover, despite the apparent modeling of the effect of an electric field on a membrane, the method is far from "reproducing" a realistic experiment. Indeed, when an electric field is applied to a cell (by placing the cell in a conductive medium between two electrodes and applying a voltage pulse to the electrodes), the resulting current causes the accumulation of electrical charges (ions) at the cell membrane and consequently a voltage difference across it. Considering ions in the electrolytes and a different approach not implying an external field may help one to model more realistically the collapse of the transmembrane potential when hydrophilic pores are created, and investigate the incidence it may have on the electroporation process performed under controlled current or voltage clamp conditions.

7.4 Transmembrane Voltage Induced by Ionic Salt Concentration Gradients

Regardless of how an electric field is applied, the ultimate step is the charging of the membrane due to ion flow. The resulting ionic-charge imbalance between both the sides of the lipid bilayer induces a TM potential. In a classical setup of membrane simulations, due to the use of three-dimensional periodic boundary conditions, the TM voltage cannot be controlled by imposing a charge imbalance Q across the bilayer, even when ions are present in the solution.

Recently, a method allowing the simulations of realistic TM potential gradients across bilayers has been proposed (Sachs et al. 2004). A unit cell consisting of three saltwater baths separated by two bilayers and full three-dimensional periodicity is used. A net charge imbalance between the two water baths induces TM voltage by explicit ion dynamics. For large-enough TM voltages, it was shown (Gurtovenko and Vattulainen 2005; Kandasamy and Larson 2006) that such protocol allows the electroporation of one of the bilayers, with subsequent ionic conduction and pore resealing when the transmembrane voltage collapses to lower values.

We have since introduced a variant of this method where the double layer is not needed, avoiding therefore the over-cost of a large and asymmetrical system (Delemotte et al. 2008). The method consists in simulating a unique bilayer surrounded by electrolyte baths, each of them terminated by an air–water interface. The system is set up as follows (Figure 7.3): First, we consider well-equilibrated bilayer at a given salt concentration using three-dimensional periodic boundary conditions. Air–water interfaces are then created on both the sides of the membrane, and further equilibration is then undertaken at constant volume, maintaining therefore a separation between the upper and lower electrolyte. In Figure 7.3 we report the electrostatic potential profiles along the normal to the membrane generated from MD simulations a POPC bilayer in contact with 1 M NaCl saltwater baths. In practice, charge imbalances (Qs)—varying here from 0 to 8e—are generated by simply displacing at time $t = 0$ an adequate number of ions from one slab to the other. For all simulations, the profiles show plateau values in the aqueous regions and an increasing potential difference between the two electrolytes indicative of a TM potential ΔV. Note that the electrostatic potentials are perturbed when approaching the air–water interface both due to water dipoles and to the nonuniform distribution of ions at the interface. The effect of the latter can be neglected as far as this interface is at more than 25–30 Å from the lipid. The bilayer "feels" as if it is embedded in infinite baths, whose characteristics are those of the modeled finite solution.

7.4.1 Membrane Capacitance

The linear variation of ΔV as a function of Qs reported in Figure 7.3 shows that, as expected, the lipid bilayer behaves as a capacitor. Its capacitance C, estimated considering $\Delta V = Qs/C$, amounts here to $0.85\,\mu\text{F cm}^{-2}$. The capacitance values extracted from similar simulations would depend on the lipid

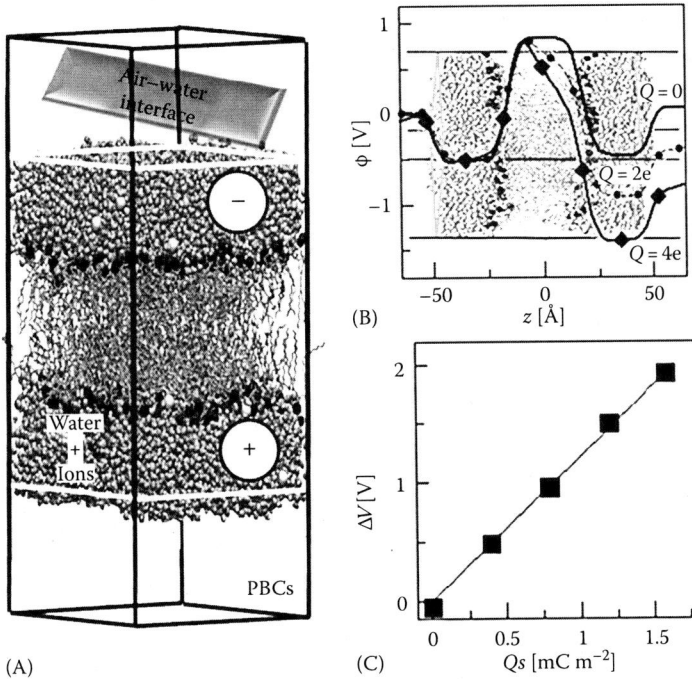

FIGURE 7.3   Applying a transmembrane potential: the charge imbalance method. (A) Configuration of a hydrated POPC lipid bilayer, embedded in a 1 M NaCl electrolyte solution in the setup allowing to induce a TM voltage by imposing a net charge imbalance across the membrane. The white box represents the simulation box used to generate the three-dimensional periodic boundary conditions, and the black box the one where air–water interfaces are considered. (B) Electrostatic potential across a POPC lipid bilayer for different net charge imbalances (Q) between the upper and lower electrolytes from MD simulations considering the setup in (A). $\phi(z)$, is estimated as an in-plane average of the EP distributions (Equation 7.1). As a reference it was set to zero in the lower electrolyte. (C) TM potential ΔV as a function of the charge imbalance Q_s par unit area. The capacitance of the bilayer can be derived from the slope of the curve.

composition (charged or not) and on the force-field parameters used. Interestingly enough, in our case, it is close to the value usually assumed in the literature, e.g., $1.0\,\mu F\,cm^{-2}$ (Roux 1997; Sachs et al. 2004) and to recent measurements for POPC lipid bilayer membranes ($0.5\,\mu F\,cm^{-2}$) (Kramar et al. 2007). Note that this original setup considering separated baths, by allowing a direct measurement of membrane capacitance, constitutes a supplementary way of checking the accuracy of lipid force-field parameters used in the simulation.

7.4.2 Membrane Electroporation

For TM voltages above a certain threshold, the following processes take place in the simulations using either protocols: double bilayer (Gurtovenko and Vattulainen 2005; Kandasamy and Larson 2006) or air–water interfaces (unpublished data, Figure 7.4). The lowest voltages reported to electroporate the membrane were between 1.5 and 2.5 V.

1. Within few nanoseconds, water fingers start to form inside the hydrophobic core.
2. Later on, water wires connecting both the sides of the membrane appear within 4 ns.
3. Lipid head groups start to migrate along one wire and form a connected pathway large enough to conduct ions.

4. Both cations and anions exchange between the two baths; the overall flux of charges within the hydrophilic pathway having a tendency to compensate for the remaining charge imbalance. Ion-translocation time ranges from few tens to few hundred picoseconds.

5. In some instances, when the charge imbalance reaches a level where the TM potential is down to a couple of hundred mV, the hydrophilic pores collapse (closed), and no more ionic translocation is witnessed.

6. The final topology of the pores toward the end of the simulations remain stable for time spans exceeding the 10 ns timescale, probably because, as reported in previous simulations (Tarek 2005), the complete recovery of the membrane requires a much longer timescale.

Clearly, despite the seemingly overall similar response of the membrane as in the case of electric field method, the second protocol presents many advantages over the electric field method, and is believed to describe the processes taking place during the electroporation of systems alike, e.g., black lipid membranes (BLMs). Exquisite details about the ionic transport, selectivity of the hydrophilic pores, and salt concentration effects may be gathered using the method from thorough comparative studies (Kandasamy and Larson 2006).

The single bilayer setup using the air–water interfaces offers few advantages. On one hand it avoids the over-cost of a large system. As a result, the affordable treatment of a larger single bilayer allowed the "capture" of multiple pore formation. Interestingly, as ions started migrating through the hydrophilic

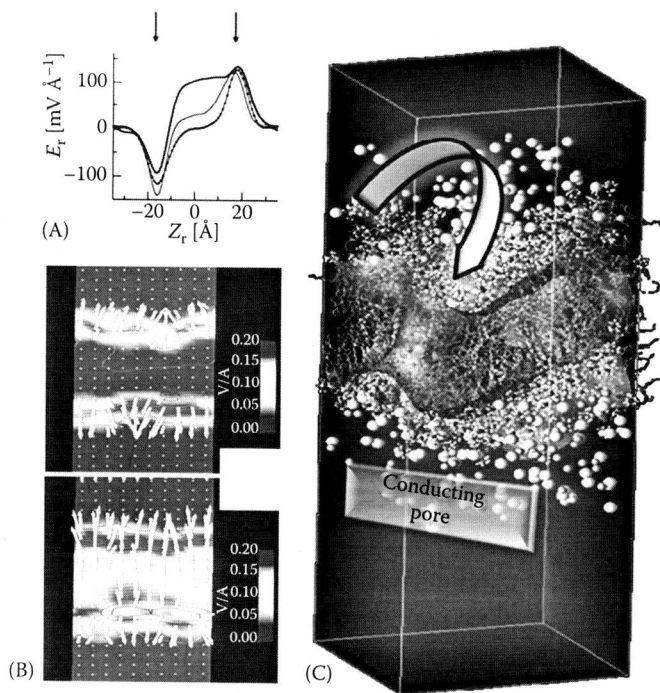

FIGURE 7.4 Electroporation of a lipid membrane using the charge imbalance method. (A) Electric field (E_r) profile along the bilayer normal for increasing TM voltages imposed by a net charge imbalance across the membrane, derived from the electrostatic potential shown in Figure 7.3. The arrows indicate the approximate position of the water–membrane interface. Note that the electric field is oriented at each interface toward the electrolytes. (B) Corresponding cross-sectional view of the three-dimensional map of the electric field derived as the gradient of the electrostatic potential obtained by solving the Poisson equation. The arrows indicate the direction and strength of the field. Top: under no TM potential, Bottom: subject to a TM potential (C) three-dimensional snapshot of a conducting pore in a membrane undergoing electroporation triggered by a net charge imbalance. Ions are represented in light colors.

stable pore, water molecules forming the other wires migrated back to the bulk solution, leaving the membrane with one single ionic conduction pathway.

One of the interesting features revealed by the charge imbalance method is the hydrophilic pore closure following the collapse of the TM potential. This is expected, given the simulation conditions, but can be prevented if ΔV is maintained constant. In order to do so, the modeler needs to maintain the initial charge imbalance by "injecting" charges (ions) in the electrolytes at a paste equivalent to the rate of ions translocation through the hydrophilic pore. The protocols are, and in particular in the single bilayer setup, hence shown to allow us to perform simulations under constant voltage or constant current conditions and to compare the obtained results with appropriate experiments.

7.5 Conclusion

Large computer simulations, in particular MD simulations, are now able to provide a novel insight into membrane electroporation processes, thereby serving as an additional, complementary source of information to the current arsenal of experimental tools. At rest, i.e., before the membrane breakdown, many characteristics of the bilayer (hydrophobic core thickness, area per lipid, intrinsic dipole potential, capacitance, etc.) are in satisfactory agreement with experiment, which indicate that the force fields used are rather well optimized. The membrane breakdown voltage (electroporation threshold), in the order of 1–2 V, are also within the bulk value of measurements. Of course, much more sampling through repeated simulations and under various voltage conditions is necessary before reaching to a quantitative agreement with experiments.

There are several points that need further investigation before fully characterizing the phenomena at the molecular level. Perhaps the most important of these in our opinion is the interplay between pore densities (number of hydrophilic pores per unit area that can form) and ionic transport rate that can be maintained at a given imposed voltage or current condition. Combined experimental/theoretical (MD) studies on the same systems should allow one to check and tune, in a self-consistent manner, these parameters that can hardly be controlled independently. Only then can we be confident in determining with accuracy the length and timescales involved in membrane electroporation, and start investigating the cascade of events involved in much more complex events such as electroporation and transport of molecules across the membranes.

Acknowledgments

The simulations presented in this work benefited from access to the HPC resources of the Centre Informatique National de l'Enseignement Supérieur (CINES) FRANCE. The authors would like to acknowledge the very fruitful and insightful discussion with Damijan Miklavčič and Peter Kramar from the University of Ljubljana.

References

Abidor IG, Arakelyan VB, Chernomordink LV, Chizmadzhev YA, Pastushenko VF, Tarasevich MR. 1979. Electrical breakdown of BLM: Main experimental facts and their qualitative discussion. *Bioelectrochem. Bioenerg.* 6:37–52.

Aksimentiev A, Schulten K. 2005. Imaging α-hemolysin with molecular dynamics: Ionic conductance, osmotic permeability, and the electrostatic potential map. *Biophys. J.* 88:3745–3761.

Allen MP, Tildesley DJ. 1987. *Computer Simulation of Liquids.* Oxford, U.K.: Clarendon Press.

Anderson RG, Jacobson K. 2002. A role for lipid shells in targeting proteins to caveolae, rafts, and other lipid domains. *Science* 296:1821–1825.

Anézo C, de Vries AH, Höltje HD, Tieleman DP, Marrink SJ. 2003. Methodological issues in lipid bilayer simulations. *J. Phys. Chem. B* 107:9424–9433.

Armen RS, Uitto OD, Feller SE. 1998. Phospholipid component volumes: Determination and application to bilayer structure calculations. *Biophys. J.* 75:734–744.

Benz R, Beckers F, Zimmerman U. 1979. Reversible electrical breakdown of lipid bilayer membranes—Charge-pulse relaxation study. *J. Membr. Biol.* 48:181–204.

Benz RW, Nanda H, Castro-Roman F, White SH, Tobias DJ. 2006. Diffraction-based density restraints for membrane and membrane-peptide molecular dynamics simulations. *Biophys. J.* 91:3617–3629.

Berger O, Edholm O, Jahnig F. 1997. Molecular dynamics simulations of a fluid bilayer of dipalmitoylphosphatidylcholine at full hydration, constant pressure, and constant temperature. *Biophys. J.* 72:2002–2013.

Berkowitz ML, Bostick DL, Pandit S. 2006. Aqueous solutions next to phospholipid membrane surfaces: Insights from simulations. *Chem. Rev.* 106(4):1527–1539.

Berkowitz ML, Raghavan MJ. 1991. Computer simulation of a water/membrane interface. *Langmuir* 7:1042–1044.

Bockmann RA, de Groot BL, Kakorin S, Neumann E, Grubmuller H. 2008. Kinetics, statistics, and energetics of lipid membrane electroporation studied by molecular dynamics simulations. *Biophys. J.* 95:1837–1850.

Bockmann RA, Hac A, Heimburg T, Grubmuller H. 2003. Effect of sodium chloride on a lipid bilayer. *Biophys. J.* 85:1647–1655.

Bodles-Brakhop AM, Heller R, Draghia-Akli R. 2009. Electroporation for the delivery of DNA-based vaccines and immunotherapeutics: Current clinical developments. *Mol. Ther.* 17:585–592.

Brooks BR, Brooks III CL, Mackerell AD, Nilsson L, Petrella RJ, Roux B, Won Y et al. 2009. CHARMM: The biomolecular simulation Program. *J. Comput. Chem.* 30:1545–1615.

Brown DA, London E. 2000. Structure and function of sphingolipid- and cholesterol-rich membrane rafts. *J. Biol. Chem.* 275:17221–17224.

Brown RE. 1998. Sphingolipid organization in biomembranes: What physical studies of model membranes reveal. *J. Cell Sci.* 111:1–9.

Campana LG, Mocellin S, Basso M, Puccetti O, De Salvo GL, Chiarion-Sileni V, Vecchiato A, Corti L, Rossi CR, Nitti D. 2009. Bleomycin-based electrochemotherapy: Clinical outcome from a single institution's experience with 52 Patients. *Ann. Surg. Oncol.* 16:191–199.

Cascales JJL, Berendsen HJC, Garcia de la Torre J. 1996. Molecular dynamics simulation of water between two charged layers of dipalmitoylphosphatidylserine. *J. Phys. Chem.* 100:8621–8627.

Case DA, Darden TA, Cheatham III TE, Simmerling CL, Wang J, Duke RE, Luo R et al. 2008. *AMBER 10.* San Francisco, CA: University of California.

Case DA, Pearlman DA, Caldwell JW, Cheatham III TE, Ross WS, Simmerling CL, Darden TA et al. 1999. *AMBER 6.* San Francisco, CA: University of California.

Cemazar M, Sersa G. 2007. Electrotransfer of therapeutic molecules into tissues. *Curr. Opin. Mol. Ther.* 9:554–562.

Cemazar M, Tamzali Y, Sersa G, Tozon N, Mir LM, Miklavcic D, Lowe R, Teissie J. 2008. Electrochemotherapy in veterinary oncology. *J. Vet. Int. Med.* 22:826–831.

Chen C, Smye SW, Robinson MP, Evans JA. 2006. Membrane electroporation theories: A review. *Med. Biol. Eng. Comput.* 44:5–14.

Chimerel C, Movileanu L, Pezeshki S, Winterhalter M, Kleinekathofer U. 2008. Transport at the nanoscale: Temperature dependence of ion conductance. *Eur. Biophys. J. Biophys. Lett.* 38:121–125.

Chipot C, Klein ML, Tarek M. 2005. Modeling lipid membranes. In: Yip S (ed.) *Handbook of Materials Modeling.* Dordrecht, the Netherlands: Springer, pp. 929–958.

Chiu SW, Clark M, Balaji V, Subramaniam S, Scott HL, Jakobsson E. 1995. Incorporation of surface tension into molecular dynamics simulation of an interface: A fluid phase lipid bilayer membrane. *Biophys. J.* 69:1230–1245.

Chiu SW, Clark M, Jakobsson E, Subramaniam S, Scott HL. 1999. Optimization of hydrocarbon chain interaction parameters: Application to the simulation of fluid phase lipid bilayers. *J. Phys. Chem. B* 103:6323–6327.

Chiu SW, Vasudevan S, Jakobsson E, Mashl RJ, Scott HL. 2003. Structure of sphingomyelin bilayers: A simulation study. *Biophys. J.* 85:3624–3635.

Clarke RJ. 2001. The dipole potential of phospholipid membranes and methods for its detection. *Adv. Colloid Interface Sci.* 89–90:263–281.

Crozier PS, Henderson D, Rowley RL, Busath DD. 2001. Model channel ion currents in NaCl extended simple point charge water solution with applied-field molecular dynamics. *Biophys. J.* 81(6):3077–3089.

Dahlberg M, Maliniak A. 2008. Molecular dynamics simulations of cardiolipin bilayers. *J. Phys. Chem. B* 112:11655–11663.

Damodaran KV, Merz KM. 1994. A comparison of DMPC and DLPE-based lipid bilayers. *Biophys. J.* 66:1076–1087.

Davis JE, Rahaman O, Patel S. 2009. Molecular dynamics simulations of a DMPC bilayer using nonadditive interaction models. *Biophys. J.* 96:385–402.

Delemotte L, Dehez F, Treptow W, Tarek M. 2008. Modeling membranes under a transmembrane potential. *J. Phys. Chem. B* 112:5547–5550.

Demchenko AP, Yesylevskyy SE. 2009. Nanoscopic description of biomembrane electrostatics: Results of molecular dynamics simulations and fluorescence probing. *Chem. Phys. Lipids* 160:63–84.

Eberhard N, Sowers AE, Jordan CA. 1989. *Electroporation and Electrofusion in Cell Biology*. New York: Plenum Press.

Edholm O. 2008. Time and length scales in lipid bilayer simulations. In: Feller SE (ed.) *Computational Modeling of Membrane Bilayers*. San Diego, CA: Elsevier, pp. 91–110.

Edidin M. 2003. The state of lipid rafts: From model membranes to cells. *Annu. Rev. Biophys. Biomol. Struct.* 32:257–283.

Engelman DM. 2005. Membranes are more mosaic than fluid. *Nature* 438:578–580.

Essman U, Berkowitz M. 1999. Dynamical properties of phospholipid bilayers from computer simulations. *Biophys. J.* 76:2081–2089.

Essmann U, Perera L, Berkowitz ML. 1995. The origin of the hydration interaction of lipid bilayers from MD simulation of dipalmitoylphosphatidylcholine membranes in gel and liquid crystalline phases. *Langmuir* 11:4519–4531.

Feller SE. 2000. Molecular dynamics simulations of lipid bilayers. *Curr. Opin. Colloid Inerface Sci.* 5:217–223.

Feller SE. 2008. Computational modeling of membrane bilayers. In: Benos DJ, Simon SA (eds.) *Current Topics in Membranes*. San Diego, CA: Elsevier.

Feller SE, Gawrisch K, MacKerell AD. 2002. Polyunsaturated fatty acids in lipid bilayers: Intrinsic and environmental contributions to their unique physical properties. *J. Am. Chem. Soc.* 124:318–326.

Feller SE, Huster D, Gawrisch K. 1999. Interpretation of NOESY cross-relaxation rates from molecular dynamics simulation of a lipid bilayer. *J. Am. Chem. Soc.* 121:8963–8964.

Feller SE, Venable RM, Pastor RW. 1997. Computer simulation of a DPPC phospholipid bilayer: Structural changes as a function of molecular surface area. *Langmuir* 13:6555–6561.

Forrest LR, Sansom MSP. 2000. Membrane simulations: Bigger and better. *Curr. Opin. Struct. Biol.* 10:174–181.

Frenkel D, Smit B. 1996. *Understanding Molecular Simulations: From Algorithms to Applications*. San Diego, CA: Academic Press.

Gawrisch K, Ruston D, Zimmerberg J, Parsegian V, Rand R, Fuller N. 1992. Membrane dipole potentials, hydration forces, and the ordering of water at membrane surfaces. *Biophys. J.* 61:1213–1223.

Gennis RB. 1989. *Biomembranes: Molecular Structure and Function*. Heidelberg, Germany: Springer-Verlag.

Glaser RW, Leiken SL, Chernomordik LV, Pastushenko VF, Sokirko AI. 1988. Reversible electrical breakdown of lipid bilayers: Formation and evolution of pores. *Biochim. Biophys. Acta* 940:275–287.

Golzio M, Teissie J, Rols M-P. 2002. Direct visualization at the single-cell level of electrically mediated gene delivery. *Proc. Natl. Acad. Sci. U.S.A.* 99:1292–1297.

Gurtovenko AA, Vattulainen I. 2005. Pore formation coupled to ion transport through lipid membranes as induced by transmembrane ionic charge imbalance: Atomistic molecular dynamics study. *J. Am. Chem. Soc.* 127:17570–17571.

Gurtovenko AA, Vattulainen I. 2008. Effect of NaCl and KCl on phosphatidylcholine and phosphatidylethanolamine lipid membranes: Insight from atomic-scale simulations for understanding salt-induced effects in the plasma membrane. *J. Phys. Chem. B* 112:1953–1962.

Halgren TA, Damm W. 2001. Polarizable force fields. *Curr. Opin. Struct. Biol.* 11:236–242.

Harder E, MacKerell AD, Roux B. 2009. Many-body polarization effects and the membrane dipole potential. *J. Am. Chem. Soc.* 131:27760–27761.

Harrison RL, Byrne BJ, Tung L. 1998. Electroporation-mediated gene transfer in cardiac tissue. *FEBS Lett.* 435:1–5.

Hess B, Kutzner C, van der Spoel D, Lindahl E. 2008. GROMACS 4: Algorithms for highly efficient, load-balanced, and scalable molecular simulation. *J. Chem. Theory Comput.* 4:435–447.

Hibino M, Itoh H, Kinosita Jr. K. 1993. Time courses of cell electroporation as revealed by submicrosecond imaging of transmembrane potential. *Biophys. J.* 64:1789–1800.

Hibino M, Shigemori M, Itoh H, Nagayama K, Kinosita Jr. K. 1991. Membrane conductance of an electroporated cell analyzed by submicrosecond imaging of transmembrane potential. *Biophys. J.* 59:209–220.

Hristova K, White SH. 1998. Determination of the hydrocarbon core structure of fluid DOPC bilayers by x-ray diffraction using specific bromination of the double-bonds: Effect of hydration. *Biophys. J.* 74:2419–2433.

Hu Q, Viswanadham S, Joshi RP, Schoenbach KH, Beebe SJ, Blackmore PF. 2005. Simulations of transient membrane behavior in cells subjected to a high-intensity ultrashort electric pulse. *Phys. Rev. E* 71:031914.

Hub JS, Salditt T, Rheinstader MC, de Groot BL. 2007. Short-range order and collective dynamics of DMPC bilayers: A comparison between molecular dynamics simulations, X-ray, and neutron scattering experiments. *Biophys. J.* 93:3156–3168.

Kandasamy SK, Larson RG. 2006. Cation and anion transport through hydrophilic pores in lipid bilayers. *J. Chem. Phys.* 125:074901.

Khalili-Araghi F, Tajkhorshid E, Schulten K. 2006. Dynamics of K+ ion conduction through Kv1.2. *Biophys. J.* 91:L72–L74.

Klauda JB, Kucerka N, Brooks BR, Pastor RW, Nagle JF. 2006. Simulation-based methods for interpreting x-ray data from lipid bilayers. *Biophys. J.* 90:2796–2807.

Klauda JB, Roberts MF, Redfield AG, Brooks BR, Pastor RW. 2008. Rotation of lipids in membranes: Molecular dynamics simulation, P-31 spin-lattice relaxation, and rigid-body dynamics. *Biophys. J.* 94:3074–3083.

Kramar P, Miklavcic D, Lebar AM. 2007. Determination of the lipid bilayer breakdown voltage by means of linear rising signal. *Bioelectrochemistry* 70:23–27.

Kramar P, Miklavcic D, Macek-Lebar A. 2009. A system for the determination of planar lipid bilayer breakdown voltage and its applications. *IEEE Trans. Nanobiosci.* 8:132–138.

Krueger S, Meuse CW, Majkrzak CF, Dura JA, Berk NF, Tarek M, Plant AL. 2001. Investigation of hybrid membranes with neutron reflectometry: Probing the interactions of melittin. *Langmuir* 17:511–521.

Leach AR. 1996. *Molecular Modelling: Principles and Applications*. Harlow, U.K.: Addison Wesley Longman Limited.

Lee RC, River LP, Pan FS, Li J, Wollman RF. 1992. Surfactant-induced sealing of electropermeabilized skeletal muscle membrane *in vivo*. *Proc. Natl. Acad. Sci. U.S.A.* 89:4524–4528.

Lee SJ, Song Y, Baker NA. 2008. Molecular dynamics simulations of asymmetric NaCl and KCl solutions separated by phosphatidylcholine bilayers: Potential drops and structural changes induced by strong Na+ lipid interactions and finite size effects. *Biophys. J.* 94:3565–3576.

Leekumjorn S, Sum AK. 2007. Molecular studies of the gel to liquid-crystalline phase transition for fully hydrated DPPC and DPPE bilayers. *Biochim. Biophys. Acta Biomembr.* 1768:354–365.

Lewis TJ. 2003. A model for bilayer membrane electroporation based on resultant electromechanical stress. *IEEE Trans. Dielectr. Electr. Insul.* 10:769–777.

Li S (ed.) 2008. *Electroporation Protocols: Preclinical and Clinical Gene Medicine.* Totowa, NJ: Humana Press.

Li Z, Venable RM, Rogers LA, Murray D, Pastor RW. 2009. Molecular dynamics simulations of PIP2 and PIP3 in lipid bilayers: Determination of ring orientation, and the effects of surface roughness on a Poisson-Boltzmann description. *Biophys. J.* 97:155–163.

Liberman YA, Topaly VP. 1969. Permeability of biomolecular phospholipid membranes for fat-soluble ions. *Biophys. USSR* 14:477.

Lindahl E, Edholm O. 2000. Mesoscopic undulations and thickness fluctuations in lipid bilayers from molecular dynamics simulations. *Biophys. J.* 79:426–433.

Lindahl E, Sansom MSP. 2008. Membrane proteins: Molecular dynamics simulations. *Curr. Opin. Struct. Biol.* 18:425–431.

Lopes PEM, Roux B, MacKerell AD. 2009. Molecular modeling and dynamics studies with explicit inclusion of electronic polarizability: Theory and applications. *Theoret. Chem. Acc.* 124:11–28.

Lundqvist JA, Sahlin F, Aberg MA, Stromberg A, Eriksson PS, Orwar O. 1998. Altering the biochemical state of individual cultured cells and organelles with ultra microelectrodes. *Proc. Natl. Acad. Sci. U.S.A.* 95:10356–10360.

MacKerell Jr. AD, Bashford D, Bellott M, Dunbrack Jr. RL, Evanseck J, Field MJ, Fischer S et al. 1998. All-atom empirical potential for molecular modeling and dynamics studies of proteins. *J. Phys. Chem. B* 102:3586–3616.

Majkrzak CF, Berk NF, Krueger S, Dura JA, Tarek M, Tobias DJ, Silin V, Meuse CW, Woodward J, Plant AL. 2000. First principle determination of hybrid bilayer membrane structure by phase-sensitive neutron reflectometry. *Biophys. J.* 79:3330–3340.

Marrink SJ, de Vries AH, Tieleman DP. 2009. Lipids on the move: Simulations of membrane pores, domains, stalks and curves. *Biochim. Biophys. Acta. Biomembr.* 1788:149–168.

Mashl RJ, Scott HL, Subramaniam S, Jakobsson E. 2001. Molecular simulation of dioleylphosphatidylcholine bilayers at differing levels of hydration. *Biophys. J.* 81:3005–3015.

Mc Laughlin S. 1989. The electrostatic properties of membranes. *Annu. Rev. Biophys. Biophys. Chem.* 18:113–136.

Mukhopadhyay P, Monticelli L, Tieleman DP. 2004. Molecular dynamics simulation of a palmitoyl-oleoyl phosphatidylserine bilayer with Na+ counterions and NaCl. *Biophys. J.* 86:1601–1609.

Needham D, Hochmuth RM. 1989. Electro-mechanical permeabilization of lipid vesicles. *Biophys. J.* 55:1001–1009.

Neumann E, Rosenheck K. 1972. Permeability changes induced by electric impulses in vesicular membranes. *J. Membr. Biol.* 10:279–290.

Neumann E, Schaefer-Ridder M, Wang Y, Hofschneider PH. 1982. Gene transfer into mouse lyoma cells by electroporation in high electric fields. *EMBO J.* 1:841–845.

Nickoloff JA (ed.) 1995. *Animal Cell Electroporation and Electrofusion Protocols.* Totowa, NJ: Humana Press.

Nishi T, Yoshizato K, Yamashiro S, Takeshima H, Sato K, Hamada K, Kitamura I et al. 1996. High-efficiency *in vivo* gene transfer using intraarterial plasmid DNA injection following *in vivo* electroporation. *Cancer Res.* 56:1050–1055.

Pandit SA, Bostick D, Berkowitz ML. 2003a. Mixed bilayer containing dipalmitoylphosphatidylcholine and dipalmitoylphosphatidylserine: Lipid complexation, ion binding, and electrostatics. *Biophys. J.* 85:3120–3131.

Pandit SA, Bostick D, Berkowitz ML. 2003b. Molecular dynamics simulation of a dipalmitoylphosphatidylcholine bilayer with NaCl. *Biophys. J.* 84:3743–3750.

Pastor RW, Venable RM, Feller SE. 2002. Lipid bilayers, NMR relaxation, and computer simulations. *Acc. Chem. Res.* 35:438–446.

Patel RY, Balaji PV. 2008. Characterization of symmetric and asymmetric lipid bilayers composed of varying concentrations of ganglioside GM1 and DPPC. *J. Phys. Chem. B* 112:3346–3356.

Pavlin M, Miklavcic D. 2003. Effective conductivity of a suspension of permeabilized cells: A theoretical analysis. *Biophys. J.* 85:719–729.

Petrache HI, Tu K, Nagle JF. 1999. Analysis of simulated NMR order parameters for lipid bilayer structure determination. *Biophys. J.* 76:2479–2487.

Phillips JC, Braun R, Wang W, Gumbart J, Tajkhorshid E, Villa E, Chipot C, Skeel RD, Kale L, Schulten K. 2005. Scalable molecular dynamics with NAMD. *J. Comput. Chem.* 26:1781–1802.

Prausnitz MR, Bose VG, Langer R, Weaver JC. 1993. Electroporation of mammalian skin: A mechanism to enhance transdermal drug delivery. *Proc. Natl. Acad. Sci. U.S.A.* 90:10504–10508.

Pucihar G, Kotnik T, Miklavcic D, Teissie J. 2008. Kinetics of transmembrane transport of small molecules into electropermeabilized cells. *Biophys. J.* 95:2837–2848.

Pucihar G, Kotnik T, Valic B, Miklavcic D. 2006. Numerical determination of transmembrane voltage induced on irregularly shaped cells. *Ann. Biomed. Eng.* 34:642–652.

Qin SS, Yu ZW, Yu YX. 2009. Structural characterization of the gel to liquid-crystal phase transition of fully hydrated DSPC and DSPE bilayers. *J. Phys. Chem. B* 113:8114–8123.

Rog T, Martinez-Seara H, Munck N, Oresic M, Karttunen M, Vattulainen I. 2009. Role of cardiolipins in the inner mitochondrial membrane: Insight gained through atom-scale simulations. *J. Phys. Chem. B* 113:3413–3422.

Rög T, Murzyn K, Pasenkiewicz-Gierula M. 2002. The dynamics of water at the phospholipid bilayer: A molecular dynamics study. *Chem. Phys. Lett.* 352:323–327.

Roux B. 1997. Influence of the membrane potential on the free energy of an intrinsic protein. *Biophys. J.* 73:2980–2989.

Roux B. 2008. The membrane potential and its representation by a constant electric field in computer simulations. *Biophys. J.* 95:4205–4216.

Sachs JN, Crozier PS, Woolf TB. 2004. Atomistic simulations of biologically realistic transmembrane potential gradients. *J. Chem. Phys.* 121:10847–10851.

Saiz L, Klein ML. 2001. Structural properties of a highly polyunsaturated lipid bilayer from molecular dynamics simulations. *Biophys. J.* 81:204–216.

Saiz L, Klein ML. 2002a. Computer simulation studies of model biological membranes. *Acc. Chem. Res.* 35:482–489.

Saiz L, Klein ML. 2002b. Electrostatic interactions in a neutral model phospholipid bilayer by molecular dynamics simulations. *J. Chem. Phys.* 116:3052–3057.

Schuler LD, Daura X, van Gunsteren WF. 2001. An improved GROMOS96 force field for aliphatic hydrocarbons in the condensed phase. *J. Comput. Chem.* 22:1205–1218.

Simons K, Ikonen E. 1997. Functional rafts in cell membranes. *Nature* 387:569–572.

Singer SJ, Nicolson GL. 1972. The fluid mosaic model of the structure of cell membranes. *Science* 175:720–731.

Siu SWI, Bockmann RA. 2007. Electric field effects on membranes: Gramicidin A as a test ground. *J. Struct. Biol.* 157:545–556.

Smondyrev AM, Berkowitz ML. 1999. Structure of dipalmitoylphosphatidylcholine/cholesterol bilayer at low and high cholesterol concentrations: Molecular dynamics simulation. *Biophys. J.* 77:2075–2089.

Snyder RG, Tu K, Klein ML, Mendelssohn R, Strauss HL, Sun W. 2002. Acyl chain conformation and packing in dipalmitoylphosphatidylcholine bilayers from MD simulations and IR spectroscopy. *J. Chem. Phys. B* 106:6273–6288.

Sotomayor M, Vasquez V, Perozo E, Schulten K. 2007. Ion conduction through MscS as determined by electrophysiology and simulation. *Biophys. J.* 92:886–902.

Suzuki T, Shin BC, Fujikura K, Matsuzaki T, Takata K. 1998. Direct gene transfer into rat liver cells by in vivo electroporation. *FEBS Lett.* 425:436–440.

Tarek M. 2005. Membrane electroporation: A molecular dynamics simulation. *Biophys. J.* 88:4045–4053.

Tarek M, Tobias DJ, Chen SH, Klein ML. 2001. Short wavelength collective dynamics in phospholipid bilayers: A molecular dynamics study. *Phys. Rev. Lett.* 87:238101.

Tarek M, Tu K, Klein ML, Tobias DJ. 1999. Molecular dynamics simulations of supported phospholipid/alkanethiol bilayers on a gold(111) surface. *Biophys. J.* 77:464–472.

Teissié J, Escoffre JM, Rols MP, Golzio M. 2008. Time dependence of electric field effects on cell membranes. A review for a critical selection of pulse duration for therapeutical applications. *Radiol. Oncol.* 42:196–206.

Teissié J, Eynard N, Gabriel B, Rols MP. 1999. Electropermeabilization of cell membranes. *Adv. Drug Deliv. Rev.* 35:3–19.

Testori A, Soteldo J, Di Pietro A, Verrecchia F, Rastrelli M, Zonta M, Spadola G. 2008. The treatment of cutaneous and subcutaneous lesions with electrochemotherapy with bleomycin. *Eur. Dermatol.* 3:1–3.

Tieleman DP. 2004. The molecular basis of electroporation. *BMC Biochem.* 5:10.

Tieleman DP, Berendsen HJC. 1996. Molecular dynamics simulations of a fully hydrated dipalmitoylphosphatidylcholine bilayer with different macroscopic boundary conditions and parameters. *J. Chem. Phys.* 105:4871–4880.

Tieleman DP, Berendsen JHC, Sansom MSP. 2001. Voltage-dependent insertion of alamethicin at phospholipid/water and octane water interfaces. *Biophys. J.* 80:331–346.

Tieleman DP, Marrink SJ, Berendsen HJC. 1997. A computer perspective of membranes: Molecular dynamics studies of lipid bilayer systems. *Biochim. Biophys. Acta* 1331:235–270.

Tobias DJ. 2001. Membrane simulations. In: Becker OH, Mackerell AD, Roux B, Watanabe M (eds.) *Computational Biochemistry and Biophysics*. New York: Marcel Dekker.

Tobias DJ, Tu K, Klein ML. 1997. Atomic-scale molecular dynamics simulations of lipid membranes. *Curr. Opin. Colloid Interface Sci.* 2:15–26.

Treptow W, Maigret B, Chipot C, Tarek M. 2004. Coupled motions between pore and voltage-sensor domains: A model for *Shaker* B, a voltage-gated potassium channel. *Biophys. J.* 87:2365–2379.

Troiano GC, Stebe KJ, Raphael RM, Tung L. 1999. The effects of Gramicidin on the electroporation of lipid bilayers. *Biophys. J.* 76:3150–3157.

Tsong TY. 1983. Voltage modulation of membrane permeability and energy utilization in cells. *Biosci. Rep.* 3:487–505.

Tsong TY. 1987. Electric modification of membrane permeability for drug loading into living cells. *Methods Enzymol.* 149:248–259.

Tsong TY. 1991. Electroporation of cell membrane. *Biophys. J.* 60:297–306.

Tu K, Tobias DJ, Klein ML. 1995. Constant pressure and temperature molecular dynamics simulation of a fully hydrated liquid crystal phase DPPC bilayer. *Biophys. J.* 69:2558–2562.

Tu K, Tobias DJ, Klein ML. 1996. Molecular dynamics investigation of the structure of a fully hydrated gel-phase dipalmitoylphosphatidylcholine bilayer. *Biophys. J.* 70:595–608.

Vacha R, Berkowitz ML, Jungwirth P. 2009. Molecular model of a cell plasma membrane with an asymmetric multicomponent composition: Water permeation and ion effects. *Biophys. J.* 96:4493–4501.

Venable RM, Zhang Y, Hardy BJ, Pastor RW. 1993. Molecular dynamics simulations of a lipid bilayer and of hexadecane: An investigation of membrane fluidity. *Science* 262:223–226.

Vermeer LS, de Groot BL, Reat V, Milon A, Czaplicki J. 2007. Acyl chain order parameter profiles in phospholipid bilayers: Computation from molecular dynamics simulations and comparison with H-2 NMR experiments. *Eur. Biophys. J. Biophys.* 36:919–931.

Vernier PT, Ziegler MJ. 2007. Nanosecond field alignment of head group and water dipoles in electroporating phospholipid bilayers. *J. Phys. Chem. B* 111:12993–12996.

Vernier PT, Ziegler MJ, Sun Y, Chang WV, Gundersen MA, Tieleman DP. 2006. Nanopore formation and phosphatidylserine externalization in a phospholipid Bilayer at high transmembrane potential. *J. Am. Chem. Soc.* 128:6288–6289.

Villemejane J, Mir LM. 2009. Physical methods of nucleic acid transfer: General concepts and applications. *Br. J. Pharmacol.* 157:207–219.

Wang L, Bose PS, Sigworth FJ. 2006. Using cryo-EM to measure the dipole potential of a lipid membrane. *Proc. Natl. Acad. Sci. U.S.A.* 103:18528–18533.

Warshel A, Kato M, Pisliakov AV. 2007. Polarizable force fields: History, test cases, and prospects. *J. Chem. Theory Comput.* 3:2034–2045.

Weaver JC. 1995. Electroporation theory. In: Nickoloff JA (ed.) *Methods in Molecular Biology*, Vol. 55. Totowa, NJ: Humana Press, Inc., pp. 3–29.

Weaver JC. 2003. Electroporation of biological membranes from multicellular to nano scales. *IEEE Trans. Dielectr. Electr. Insul.* 10:754–768.

Weaver JC, Chizmadzhev YA. 1996. Theory of electroporation. A review. *Bioelectrochem. Bioenerg.* 41:135–160.

Wiener MC, White SH. 1992. Structure of fluid dioleylphosphatidylcholine bilayer determined by joint refinement of x-ray and neutron diffraction data. III. Complete structure. *Biophys. J.* 61:434–447.

Xu C, Loew LM. 2003. The effect of asymmetric surface potentials on the intramembrane electric field measured with voltage-sensitive dyes. *Biophys. J.* 84:2768–2780.

Yang Y, Henderson D, Crozier P, Rowley RL, Busath DD. 2002. Permeation of ions through a model biological channel: Effect of periodic boundary condition and cell size. *Mol. Phys.* 100:3011–3019.

Yang Y, Mayer KM, Wickremasinghe NS, Hafner JH. 2008. Probing the lipid membrane dipole potential by atomic force microscopy. *Biophys. J.* 95:5193–5199.

Yip S. (ed.) 2005. *Handbook of Materials Modeling*. Dordrecht, the Netherlands: Springer.

Zhong Q, Moore PB, Newns DM, Klein ML. 1998. Molecular dynamics study of the LS3 voltage-gated ion channel. *FEBS Lett.* 427:267–270.

Ziegler MJ, Vernier PT. 2008. Interface water dynamics and porating electric fields for phospholipid bilayers. *J. Phys. Chem. B* 112:13588–13596.

Zimmerman U. 1996. The effect of high intensity electric field pulses on eukaryotic cell membranes: Fundamentals and applications. In: *Electromanipulation of Cells*. Boca Raton, FL: CRC Press, pp. 1–106.

Zimmerman U, Pilwat G, Beckers F, Riemann F. 1976. Effects of external electrical fields on cell membranes. *Bioelectrochem. Bioenerg.* 3:58–83.

8

Nanoscale Restructuring of Lipid Bilayers in Nanosecond Electric Fields

P. Thomas Vernier

... and now remains
That we find out the cause of this effect,
Or rather say, the cause of this defect,
For this effect defective comes by cause.

Polonius,
Hamlet, Act ii. Sc. 2.

8.1 Electropermeabilization, Charge Migration, and the Dielectric Shell Model

8.1.1 Conventional ("Long Pulse") Electroporation

A long-standing guiding model for the analysis of the effects of pulsed and AC electric fields on biological cells is the dielectric shell, in which the cell is a membrane-bound ellipsoidal volume of a conductive fluid suspended in an external conductive medium (Plonsey and Altman, 1988). To represent the properties of real cell suspensions, the membranes are assigned a low permittivity and a low conductivity, and the media have relatively much higher permittivities and conductivities. Ions and other charged species migrate in the conductive intracellular and extracellular media in the presence of an electric field, accumulating at the essentially impermeable membrane–medium interfaces, where the resulting charge separation across the membrane produces a transmembrane potential. Because the membrane is so thin (about 5 nm), this charge accumulation effectively magnifies the externally applied field a thousand-fold or more in the membrane interior, with the maximum value occurring where the applied field is normal to the membrane surface. When the transmembrane potential exceeds a critical value (400 mV–1 V), the membrane becomes electrically conductive and permeable to ions and small molecules (Kinosita and Tsong, 1977; Zimmermann, 1982; Teissie and Rols, 1993). The influx of fluorescent dyes like propidium iodide (PI), a salt of the divalent propidium ion (the solubilized cation is the actual fluorochrome), which do not pass through intact membranes, is a common diagnostic for this electric field–driven restructuring of the membrane, which results in permeabilization.

In a popular hypothetical mechanism for this process, the transmembrane electric field causes the formation of pores in the membrane through which normally blocked ions and molecules can pass (Zimmermann et al., 1974; Abidor et al., 1979; Chizmadzhev and Abidor, 1980). The diameter of these pores depends on the duration of the porating electric pulse and ranges from 1 nm or less to 100 nm or more. Because of the widespread acceptance of this mechanism, this process is commonly called electroporation. In the absence of direct evidence for the existence of such pores, the term electropermeabilization may be preferable, but since both terms are widely used, we will consider them interchangeable here.

8.1.2 Biological Cells in High-Frequency Electric Fields

It has been known for a long time that electric-field effects on biological systems are highly dependent on the duration of the applied pulses, or the corresponding frequency of a continuous or burst-mode AC signal (Sher et al., 1970; Drago et al., 1984; Schoenbach et al., 2004). The electric properties of the various components of a cell or a tissue (extracellular and intracellular aqueous electrolytes, cell membranes, organelles, and other biomolecular structures) are not constant but change at higher frequencies (or with shorter pulses), a phenomenon known as dielectric dispersion, and there is considerable variability of these properties among cellular constituents, cell types, and tissues (Schwan and Foster, 1980; Stuchly et al., 1981; Pethig and Kell, 1987; Gabriel et al., 1996). This adds a substantial complexity to the simplest pictures of electroporation, but also opens the door to the possibility of the production of a range of electroporative effects—long-lasting or short-lived permeabilization, large- or small-diameter pores, selectively permeabilizing one cell type or size, minimizing or enhancing electrochemical effects—and even combining porating and non-porating fields, territories which have not been extensively explored.

8.1.3 Nanosecond Bioelectrics

In what we call conventional electroporation, relatively long pulses (µs or longer) and relatively low fields (kV/m) are delivered to cells suspended in an osmotically balanced medium. Under these conditions, the cell membrane charges very early in the pulse period so that the cell interior is shielded from the external field. The idea, built on the theoretical work described above, that very short pulses, in the nanosecond regime, (or very high frequencies) could "bypass" the plasma membrane (which can be conceptualized as a shell capacitor) and deposit most of their energy inside the cell, was investigated experimentally beginning in the late 1990s (Hofmann et al., 1999; Schoenbach et al., 2001), and a number of interesting and unique effects (intracellular calcium release, apoptosis, etc.) have been discovered that are still today the subject of ongoing studies.

The notion that effects on the plasma membrane would be minimized if the pulses were short enough was a consequence of the dielectric shell model, in which the permeabilizing transmembrane voltage arises from charge migration, a process which has a characteristic time constant that is a function of the conductivities and permittivities of the membranes and media in the system. It was expected that pulses with a duration significantly less than the charging time constant of the cells (>50 ns for typical systems) would not charge the membrane to the critical porating voltage long enough to cause damage, and experimental efforts seemed to confirm this idea. Although one early report indicated that the electric field–driven conductive breakdown of membranes can occur in as little as 10 ns (Benz and Zimmerman, 1980), and a well grounded model predicted "poration everywhere" in the nanosecond pulse regime (Gowrishankar et al., 2006), procedures used to detect conventional electroporation of the plasma membrane (and the loss of membrane integrity in general), for example, detection of propidium influx using fluorescence microscopy, fluorometry, or flow cytometry produced negative results for pulses with durations less than or equal to 30 ns.

A more careful theoretical analysis, still using the simple dielectric shell model but taking into account the very high fields (MV/m) required to produce observable biological effects with nanosecond pulses, demonstrated that, in general, pulses longer than 2 ns and having field amplitudes as large as those found to be effective in the laboratory (MV/m) will produce porating transmembrane potentials (Vernier et al., 2004a). The failure to demonstrate nanoelectropermeabilization experimentally was reconciled to this prediction from the model with the explanation that nanosecond pulse exposures did not result in "conventional electroporation."

In addition to highlighting the limitations of the traditional experimental methods for observing membrane permeabilization, this apparent discrepancy between model and observation pointed also to the need to address inadequacies in the dielectric shell model itself at timescales below the membrane (cell) charging time constant. Some investigators have pointed to the higher frequency effects associated with the different dielectric properties of high-permittivity aqueous media and low-permittivity biological membranes (Grosse and Schwan, 1992; Kotnik and Miklavcic, 2000; Gowrishankar and Weaver, 2003; Timoshkin et al., 2006), but this has until recently not received much attention since these are secondary and minor factors for the electropermeabilizing conditions that are most commonly studied (μs, kV/m pulses). We discuss this "dielectric amplification" below.

8.2 Nanosecond Pulses Restructure the Plasma Membrane

Several lines of experimental evidence, direct and indirect, indicate that nanosecond electric pulses do in fact cause changes in the integrity and organization of the cell membrane.

8.2.1 Trypan Blue Permeabilization

Early in our own studies of nanoelectropulse-induced apoptosis, we observed symptoms of "unconventional electroporation" in Jurkat T lymphoblasts exposed to a series of 50 pulses, 20 ns in duration, with an applied field of 4 MV/m (Figure 8.1). Even though these cells remained propidium-negative in flow cytometry and direct fluorescence microscopy observations, the cell volume increased slightly, and they became somewhat permeable to Trypan blue (TB), a normally impermeant dye commonly used to separate viable (unstained) cells from dead cells. When we added TB after nanoelectropulse exposure, we observed a few densely stained cells but also an unusual population comprising 50% or more of the

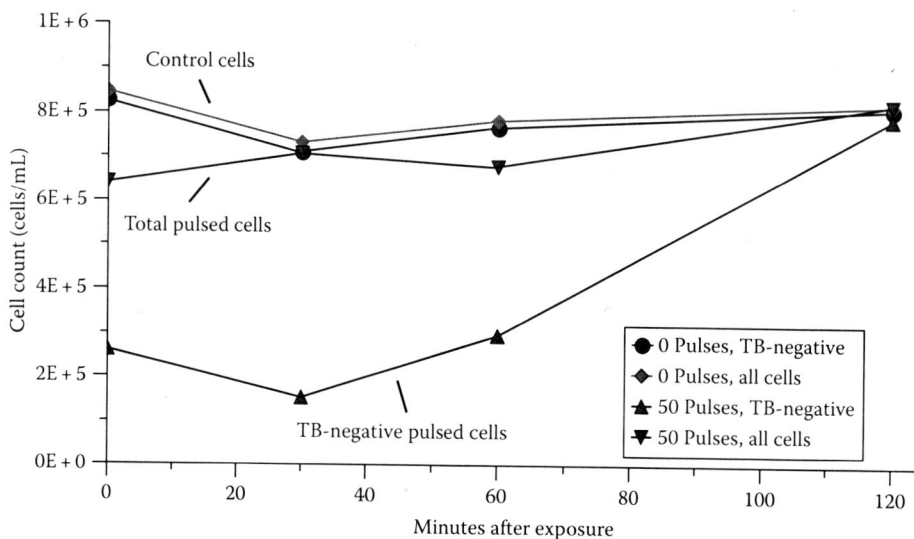

FIGURE 8.1 Nanoelectropulsed cells recover from initial Trypan blue permeabilization. About 70% of Jurkat T lymphoblasts exposed to 50 ns, 20 ns, and 4 MV/m pulses at 20 Hz in RPMI 1640 growth medium containing 10% fetal bovine serum take up Trypan blue immediately after treatment. Over a period of 2 h after exposure, the percentage of cells (incubated at 37°C in a humidified, 5% CO_2 atmosphere) which exclude Trypan blue (TB-negative) increases from 30% to nearly 100% of the total population. TB was not present in the medium at the time of pulse exposure, but was added to aliquots of cell suspension at the indicated sampling times (immediately after pulsing for the zero time samples).

total cells that were lightly stained with TB. A technician probably would not have scored them "dead," but they had clearly taken up some of the dye. When we monitored these cells over time after pulse exposure, we found that essentially all of these initially lightly stained, weakly TB-positive cells recovered their ability to exclude TB. Cell volumes also returned to normal, indicating that control over osmotic balance had been regained.

Schoenbach and colleagues have recently published similar results for B16 murine melanoma cells exposed to sub-nanosecond (800 ps) pulses at very high fields (55 MV/m) (Schoenbach et al., 2008).

8.2.2 Nanoelectropulse-Induced Porating Transmembrane Potentials

In another set of observations, Frey, Kolb, Schoenbach, and colleagues, using the membrane potential–sensitive dye annine-6, demonstrated directly with fluorescence microscopic imaging that porating transmembrane potentials are generated during nanoelectropulse exposure (Frey et al., 2006). Although they did not show that the potentials generated during high-field, nanosecond pulse events are associated with permeabilization, these data provide an important substantiating link between conventional electroporation models and experiments and the nanosecond pulse regime.

8.2.3 Nanoelectropulse-Induced Phospatidylserine Externalization

Studies of nanoelectropulse-induced apoptosis in which phosphatidylserine (PS) externalization was used as a diagnostic indicator of apoptotic cells revealed that the loss of asymmetry in membrane phospholipid distribution occurs immediately after pulse exposure (Vernier et al., 2004b), much earlier than would be expected from the normal progression of symptoms after induction of apoptosis. This is consistent with a reorganization of membrane constituents driven directly by nanosecond-duration electric fields. In the simplest mechanism proposed for this nanoelectropulse-induced phospholipid flip-flop

(Vernier et al., 2004a), nanometer-diameter, hydrophilic membrane pores appear in response to the electric field, providing a low-energy path for electrophoretically facilitated diffusion of PS from its normal location on the cytoplasmic leaflet of the plasma membrane to the external face of the cell.

8.2.4 MD Simulations Link PS Externalization and Pore Formation in Lipid Bilayers

Although nanoelectropore-facilitated electrophoretic translocation of PS has been observed only indirectly, through the responses of PS-specific ligands and fluorescent dyes, molecular dynamics (MD) simulations of phospholipid bilayers in electric fields provide further support for this model. In simulations of the electroporation of simple membranes, water bridges and then hydrophilic pores appear within a few nanoseconds (Tieleman, 2004). When PS is incorporated into one of the leaflets of the bilayer, the anionic phospholipid migrates electrophoretically along the walls of these pores to the anode-facing side of the membrane, and then diffuses away from the pore (Hu et al., 2005a; Vernier et al., 2006b), replicating—in silico—experimental observations in living cells showing that nanoelectropulse-induced PS externalization occurs primarily at the anode-facing pole of the cells (Vernier et al., 2006c).

8.2.5 Nanoelectropermeabilization

All of the experimental and modeling results above are consistent with immediate permeabilization of cell membranes by pulsed electric fields on a nanosecond timescale, but the evidence until recently for pulses with durations under 30 ns was indirect or overlooked, despite strong indications that even sub-10 ns pulses must be permeabilizing the cells. Eventually, a way was found to demonstrate it.

The first direct evidence for nanoelectropermeabilization was obtained by monitoring the pulse-induced influx of YO-PRO-1 (YP), a more sensitive indicator of membrane permeabilization than the traditional PI (Idziorek et al., 1995). YP influx after exposure of Jurkat cells to 4 ns pulses is dependent on pulse count, pulse amplitude, and pulse repetition rate (Vernier et al., 2006a).

Corroborating these findings, as mentioned above, Karl Schoenbach and members of his group at Old Dominion University (ODU), who are leading in efforts to develop picosecond pulse generators for bioelectrics applications, have recently reported TB uptake in cells exposed to large numbers (625) of 800 ps, 55 MV/m pulses (Schoenbach et al., 2008).

Although their pulses are longer than 4 ns, Pucihar, Kotnik, Miklavcic, and Teissie have developed a sensitive, non-imaging method for detecting early influx of dye after pulse exposure using a photomultiplier tube and very high concentrations of propidium (Pucihar et al., 2008). It is likely that earlier failures to detect nanoelectropermeabilization with propidium resulted from a lack of sensitivity in the detection system, and the fact that YP has a smaller steric cross-sectional area than propidium (Figure 8.2) is probably at least part of the reason that YP influx is easier to detect. More sensitive instrumentation than has been used in previous studies will almost certainly detect propidium influx after nanosecond pulse exposure. We discuss this further below.

8.2.6 Nanoelectroporation

Additional direct evidence for nanoelectropermeabilization comes from the work of Andrei Pakhomov at ODU. Patch clamp experiments on living cells reveal long-lasting increases in membrane conductance following exposure to single, 60 ns (and 600 ns) pulses (Pakhomov et al., 2007a,b, 2009). These are long pulses relative to the 4 ns pulses shown to permeabilize cells to YP, and it will be interesting to see in future work whether the 60 ns pulse–induced conductance changes arise from membrane structural modifications similar to those that account for the YP permeabilization observed after 4 ns pulse exposure, perhaps at a higher density, or whether the 60 ns pulses, because they are so much longer, cause more extensive alterations in membrane configuration.

YO-PRO-1   Propidium   Trypan blue

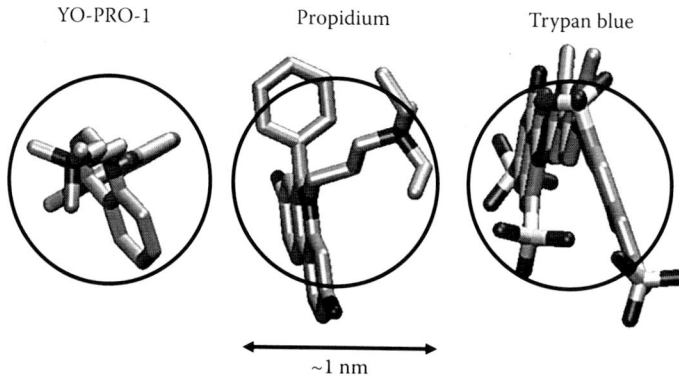

~1 nm

FIGURE 8.2 Minimum areal cross-section views of YO-PRO-1, propidium, and Trypan blue, superimposed on 1 nm diameter circles. YO-PRO-1 is the most sensitive indicator of permeabilization after exposure of cells to nanosecond electric pulses. Propidium and Trypan blue influx have also been observed, but at higher doses. Molecular dynamics and continuum models and experimental observations are consistent with a minimum electropore diameter of about 1 nm.

8.2.7 Nanosecond Activation of Electrically Excitable Cells

All of the experimental observations mentioned above involve cell types that are not electrically excitable. Nerve, neuroendocrine, and muscle cells, which respond functionally to small changes in transmembrane potential provide a distinct and more sensitive environment for the investigation of the effects of nanosecond electric pulses on biological systems. Adrenal chromaffin cells (Vernier et al., 2008) and cardiomyocytes (Wang et al., 2009) respond strongly to a single 4 ns pulse, and muscle fiber has been shown to respond to a 1 ns stimulus (Rogers et al., 2004). Although these investigations are in their early stages, the nanoelectropulse-induced responses of these electrically excitable cells seem to result from a combination of the direct activation of voltage-sensitive membrane channel proteins and the permeabilization of the cell membrane (Craviso et al., 2009).

8.3 Nanosecond Bioelectrics and the Dielectric Stack Model

8.3.1 Temporal Sequence of Electrical Polarization of a Biological System

To properly characterize what is happening to cells exposed to electric pulses in the sub-microsecond regime, and to understand better why a 60 ns pulse might produce very different effects from a 4 ns pulse, and why a 4 ns pulse might produce very different effects from a 1 ns pulse, we consider briefly the time course of events in an aqueous suspension of living cells and electrolytes between two electrodes after an electric pulse is applied (Figure 8.3). In a real system, the switching time of the pulse may be an important variable. A 10 ns pulse with 2 ns rise and fall times may affect cells differently from a 10 ns pulse with 100 ps rise and fall times, even if we adjust the actual pulse durations so that the total integrated energy delivered is the same in each case, as a consequence of the higher frequency spectral components present in the faster rising pulse. For simplicity, we ignore the rise time of the signal here and treat it as a step function.

8.3.2 The First Nanosecond

When the electrodes are charged, an electromagnetic wave propagates through the system. For small samples like commercial electroporation cuvettes or for the application of pulses to tissues through needle electrode arrays spaced by a few millimeters, the time involved for this is negligible relative to the duration of the pulse (the speed of light in physiological saline is about 225 mm/ns).

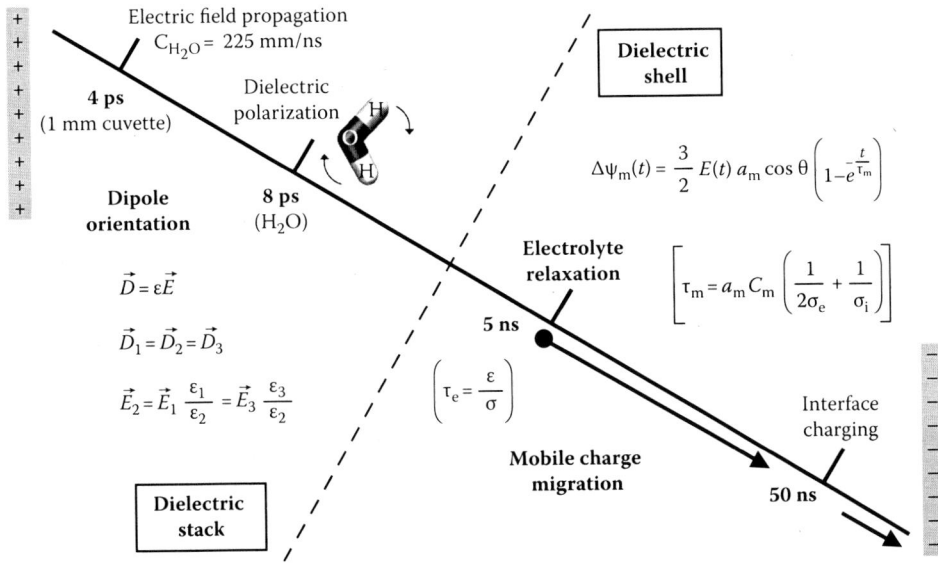

FIGURE 8.3 Timeline representing the sequence of events following electrical polarization of a biological tissue or aqueous suspension of cells. The shorter, sub-nanosecond regime, dominated by the reorientation of dipoles, can be modeled by the dielectric properties of the system components. For times longer than a few nanoseconds, the evolution of the distribution of electric fields and potentials is affected much more strongly by the migration of mobile charged species. Dielectric stack and dielectric shell models can be used to describe and analyze these two regimes, as discussed in the text. The diagram, which is nonlinear, includes simple expressions for the electric field E in the dielectric stack model, for the electrolyte relaxation time (τ_e), for the transmembrane potential ($\Delta\psi_m$), and the membrane interface charging (Maxwell–Wagner) time constant (τ_m) in the dielectric shell model, and is framed by a symbolic anode and cathode.

In response to the electric field now present in the system, water (and other) dipoles reorient. This is also a fast process. The water dipole relaxation time at 25°C is about 8 ps (Buchner et al., 1999), again very short relative to the width of the bioelectric pulses that are technologically feasible today. This dielectric rearrangement is a dominant component of the electric-field landscape that develops in a cell suspension during the first nanosecond of the application of an external electric field, discussed in more detail below.

8.3.3 Electrolyte Equilibrium

The electric field also disturbs the equilibrium established among charged species in the system (ions and counterions) and their hydrating water molecules. The time constant for the restoration of this electro-diffusive equilibrium, neglecting hydrodynamic factors, is a function primarily of the ratio of the conductivity and permittivity of the medium (ε/τ) (Hubbard and Onsager, 1977; Lopez-Garcia et al., 2000; Shilov et al., 2000; Martinsen et al., 2002); its magnitude in biological systems *in vitro* and *in vivo* ranges roughly from 0.5 to 7 ns. This establishes a boundary for the applicability of the dielectric shell model to the analysis of cell suspensions in electric fields. As the pulse duration is reduced below the electrolyte relaxation time to smaller and smaller values, the amount of charge accumulation at membrane interfaces becomes insufficient to produce porating transmembrane potentials. In this perspective, a 10 ns pulse with a high enough field may drive a porating quantity of charge to the plasma membrane. A 1 ns pulse may not be long enough. Even if one counters the decreased pulse width with a higher pulse amplitude, at some point the combined electrostatic and hydrodynamic inertia will dominate, and membrane charging will be negligible. But nanoelectropermeabilization may involve a different mechanism.

8.3.4 Dielectric Stack

For electric-field durations of a few nanoseconds or less, depending on the system, before significant charge migration occurs, the dielectric shell model can be replaced with a simpler model, a dielectric stack (Figure 8.4). The local electric field in this model depends only on the external (applied) electric field (the field that would be present if the space between the electrodes were filled only with vacuum) and the dielectric permittivity of each component of the system.

Note that if we fill the space between two electrodes with a suspension of cells and measure the potential between the electrodes with an oscilloscope or other instrument we are recording the effective field—the applied field (produced by the separation of charges at the electrodes) reduced by the polarization of the medium. If the system is homogeneous, the local field will equal the effective field everywhere in the system. In a heterogeneous system like a suspension of cells, variations in the local permittivity such as those encountered in the vicinity of a membrane-aqueous interface will determine the value of the local field.

For the simple case represented in Figure 8.4, where the electric field is normal to the plane of the membrane, and where no free charge has accumulated at the interface, we can consider the slabs of medium, membrane, and cytoplasm (water–lipid–water) to be an ideal dielectric stack. Because the electric displacement ($D = \varepsilon_0 \varepsilon_r E$) is continuous across the interfaces, the effective electric fields in each region are related inversely by their relative permittivities. Consider a system in which sufficient charge is transferred to two electrodes to

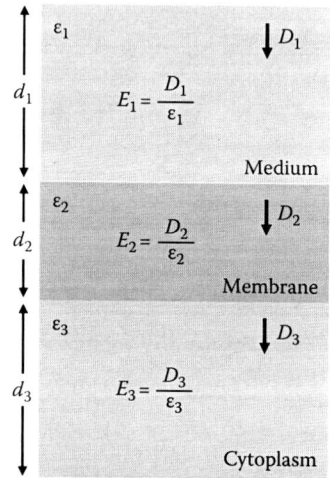

FIGURE 8.4 An idealized dielectric stack composed of slabs of extracellular medium ($\varepsilon_r = 80$), cell membrane ($\varepsilon_r = 4$), and cytoplasm ($\varepsilon_r = 80$). To achieve a nominal porating transmembrane potential of 0.8 V (200 MV/m across a 4 nm membrane) before mobile charge migration, the associated field in the suspending medium must be at least 10 MV/m.

create a field of 800 MV/m in the vacuum separating the electrodes. When the stack in Figure 8.4 is placed in this gap, the effective field in the regions labeled medium and cytoplasm, assuming a relative permittivity of 80, will be 10 MV/m ($E_{net} = E_{free}/\varepsilon_r = 800$ MV/m/80), equivalent to the fields found experimentally to be required in cell suspensions for nanosecond electropermeabilization (Vernier et al., 2006a). The corresponding effective field in the membrane, assuming a relative permittivity of 4, is 200 MV/m (800/4), equivalent to a transmembrane potential of approximately 0.8 V, again in the range observed experimentally for porating transmembrane potentials (Kinosita and Tsong, 1977; Zimmermann, 1982; Teissie and Rols, 1993). An introduction to a more complete analysis of this dielectric regime as it applies to biological systems can be found in Timoshkin et al. (2006).

From the point of view of a nanoelectropulse experimentalist, who must adjust the pulse generator to deliver enough charge to create an effective 10 MV/m field between the electrodes, the field is "amplified" by the relative dielectric permittivities in the system from 10 MV/m in the bulk suspension (which is less than 10% cells by volume at normal cell concentrations) to a value of 200 MV/m in the membrane (a gain factor of 20). More accurately, the dielectric polarization term arising from the hydrated phospholipid head groups and the hydrocarbon tails in the membrane interior is smaller than the dielectric polarization arising from water molecules in the aqueous media, and so the net field in the membrane (the vector sum of the field from the free charge at the electrodes and the dipole field in the membrane) is larger, by a factor equivalent to the ratio of the permittivities of medium and membrane, than the net field in the cytoplasm and external medium.

Within the limits of this dielectric stack model, we can say that for any pulses longer than the dielectric relaxation time of water (8 ps), porating transmembrane potentials (0.4–1 V) will be produced if the effective field in the aqueous suspending medium is on the order of 10 MV/m or greater. (Pulses in the microsecond and millisecond regimes produce porating potentials with much lower fields, but as a result primarily of charge migration rather than direct dielectric effects.) Two simplifying assumptions we have made in this discussion should be mentioned here. First, we have ignored the time required for the dielectric relaxation of the hydrocarbon interior of the lipid bilayer and of the phospholipid head group dipoles, which in the latter case is on the order of several nanoseconds (Shepherd and Buldt, 1978; Svanberg et al., 2009). This would result in an even larger dielectric "amplification" of the field in the membrane during a nanosecond pulse exposure, while the lipid dipoles are responding to the external electric field. Second, for pulse durations approaching the electrolyte relaxation time (which can vary over an order of magnitude, from less than 0.5 ns to greater than 5 ns, depending on the electrical properties of the medium and the cells), there will be an accumulation of free charge at the interface, further increasing the transmembrane potential, as the dielectric stack model yields to the dielectric shell.

8.3.5 Nanoelectropore Formation Time

Now that it has been established that single pulses as short as 3 ns can permeabilize and restructure cell membranes (Benz and Zimmermann, 1980; Rogers et al., 2004; Tieleman, 2004; Vernier et al., 2006a, 2008; Wang et al., 2009), and that the dielectric properties of biological systems are consistent with mechanisms that can produce porating transmembrane fields even with pulse durations deep in the sub-nanosecond regime, what biological responses can be expected from exposure of cells and tissues to pulses 1 ns and shorter?

Because the TB permeabilization reported already for pulses of 800 ps duration involves pulse counts in the hundreds at extremely high fields (Schoenbach et al., 2008), we might wonder if the TB uptake in that case is not a consequence of an accumulation of electromechanical assaults rather than the result of the formation of discrete, nanoscale pores. On the other hand, we have as yet no direct experimental evidence for the formation of physical pores (cylindrical or quasi-cylindrical openings in the bilayer) with 4, 60 ns, or even the longer pulses used for conventional electroporation. All of the electrical conductance and small molecule influx and efflux data are consistent with a permeabilization mechanism involving the formation of statistical populations of nanometer-diameter hydrophilic pores (Sugar and Neumann, 1984; Popescu et al., 1991; Weaver and Chizmadzhev, 1996), but except for one isolated observation (Chang and Reese, 1990), no one has actually seen a pore.

We do have MD simulations of phospholipid bilayers, and these show us a mechanism, consistent with stochastic and asymptotic pore formation models (DeBruin and Krassowska, 1998; Neu and Krassowska, 1999; Weaver, 2003; Vasilkoski et al., 2006), in which the electric field–facilitated construction of water bridges across the low-dielectric membrane interior is followed by head group migration along the water column and the establishment of a quasi-stable, nanometer-diameter hydrophilic pore (Tieleman et al., 2003; Hu et al., 2005b; Tarek, 2005; Vernier et al., 2006b). In these simulations, the time to initiation of individual pore formation is random, but strongly dependent on the applied electric field. Once initial water intrusion occurs, the hydrophilic pore is fully formed within about 2 ns, even at the lowest porating field magnitudes.

The MD models suggest that pulses of 1 ns duration or less may be ineffective at causing electropore formation, regardless of the field amplitude, since the field will not be present long enough to support the entire sequence of water and lipid reorganization necessary to form a stable pore. The recent results from Schoenbach and colleagues describing TB permeabilization with 800 ps pulses (but only with very high pulse counts and electric fields) (Schoenbach et al., 2008) may be at the threshold of this new short-pulse regime, which presents the possibility of manipulating biomolecular structures with high electric fields without forming hydrophilic pores in membranes.

8.3.6 Nanoelectropermeabilization and Continuum Electroporation Models

MD simulations at present provide the only available molecular-scale windows on electropore formation in lipid bilayers. We have confidence in them to the extent that they reproduce key physical

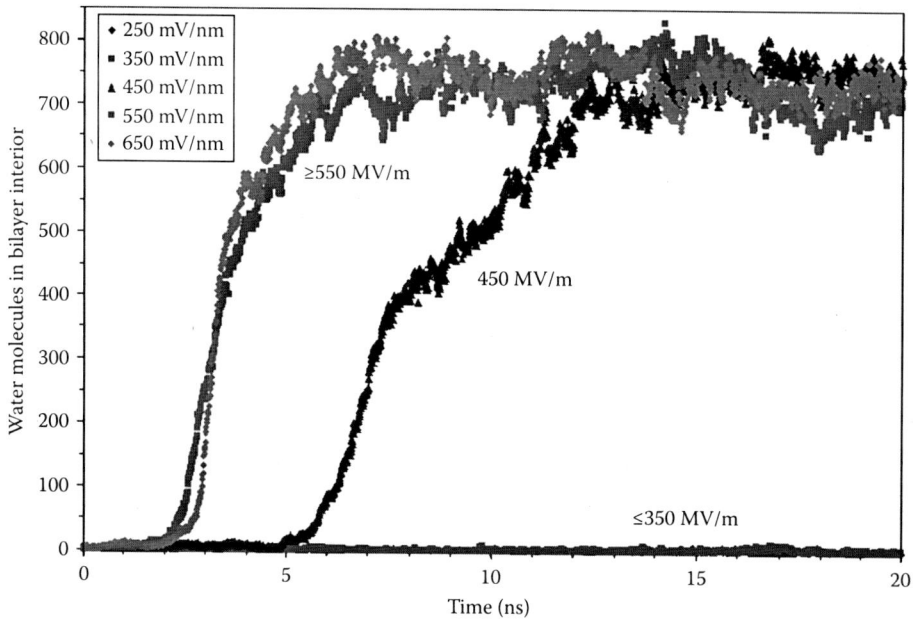

FIGURE 8.5 Electric field–driven intrusion of water into a simulated lipid bilayer. Intruding waters are within 0.5 nm of the bilayer mid-plane (midway between the mean phosphorus planes of the two leaflets), averaged over each 10 ps window. Water intrusion, the first step in nanoelectropore formation, has a strong, nonlinear dependence on the electric field, consistent with an Arrhenius pore formation rate (DeBruin and Krassowska, 1998). The MD system consists of an asymmetric DOPC:DOPS bilayer (68:4, with 36 DOPC in one leaflet, 32 DOPC and 4 DOPS in the other), 2873 H_2O, 4 Na^+, as described in Vernier et al., (2006b). (Vernier, P. T., et al., *J. Am. Chem. Soc.*, 128, 6288, 2006b. With permission.)

characteristics of lipid membranes that can be measured in the laboratory—area per lipid, order parameters, phase transitions, ion binding, etc. Current models perform reasonably well by these metrics, but there are still many assumptions and simplifications built into MD simulations of electroporation. To validate the models and to extend them eventually to enable accurate simulations of the electropermeabilization of large areas of cell membrane and beyond that to cells and groups of cells, it will be useful to find intersections, or to build bridges where necessary, between all-atom molecular assemblies, continuum representations of cell suspensions and tissues, and experimental observations of cells and whole organisms. Calibrations of the various models at these intersections will be essential to ensure accuracy and consistency at all levels. For example, the stochastic asymptotic pore model assumes an exponential relation between the transmembrane potential and several indices of electropore formation (DeBruin and Krassowska, 1998). The MD results in Figure 8.5, showing water intrusion into the membrane interior as a function of applied electric field, qualitatively demonstrate a comparable nonlinear relation between field and poration. The challenge is to carefully define the simulations and the analysis, and then to allocate the computing power necessary to generate statistically significant data sets.

8.4 Nanosecond Restructuring of Lipid Bilayers and Cell Membranes

8.4.1 Simulations, Simplicity, Observations, Complexity

Experiments and simulations tell us that electric pulses as short as a few nanoseconds, if they are of sufficient magnitude, can nondestructively restructure cell membranes so that they become permeable to water and other substances than normally cannot readily pass through the membrane. MD simulations

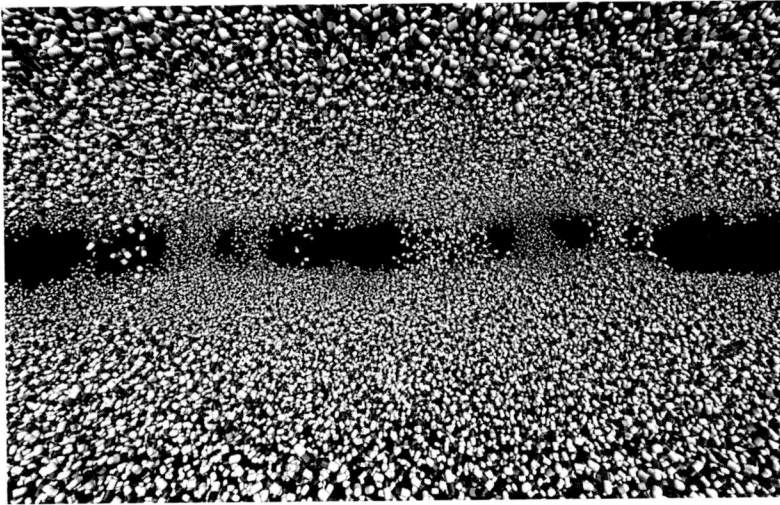

FIGURE 8.6 Water columns bridging (electroporating) a vacuum gap in a molecular dynamics simulation. 14 × 14 × 7 nm SPC water box with a 4 nm gap (19,484 water molecules). $E = 600$ mV/nm.

provide tantalizing glimpses at the molecular structures and mechanisms involved in electropermeabilization, and we note that the pattern of electro-restructuring is similar for a wide variety of phospholipid bilayers, and that octane and even vacuum can be "electroporated" (Figure 8.6)! But there are also many differences among these systems, and with the increasing availability of high-speed computing systems, it has become possible to define investigational programs combining experiments and simulations that will reveal to us the details of all these varieties of permeabilization, so that we can then put them to better use, in the laboratory and in the clinic.

A key source of value arising from models and simulations is the simplifications they offer, both conceptually and in material resources, an important practical matter. But their limitations, the areas where they fall short of accurately representing the behavior of real systems, quickly become apparent. The challenge is to extend the scope of the models and at the same time to refine their features, to retain the perspective of abstraction while retaining a firm grip on the details. All of this must be driven by and constrained by studies at the experimental pole of the simulation–observation synergy, certainly in the obvious categories of accurate parameters and physical realism but equally importantly in the sense that the simulations must not only conform to observations, they must also be designed to point to new experiments that can be quickly done, to validate the behavior of the models and to raise new questions.

8.4.2 Nanoelectropermeabilization of Real Membranes

To achieve truly realistic simulations that can serve as efficient and accurate tools for the study of nanosecond electropermeabilization, the simplifying assumptions that underlie much of the early and still ongoing work must be removed, one by one, and important properties of real electrical stimuli and real cells, which are not now included in models of membranes in electric fields, must be added. One category of these essential refinements involves the characteristics of the electric pulse as it is presented to living cells in experimental systems. Real pulses have a finite rise and fall time, and often a complex shape. And pulses are often delivered in trains, adding the important variable repetition rate. Another set of simplifications and omissions is the way in which the membrane is represented. A living cell membrane is not a homogeneous phospholipid bilayer. It contains many phospholipids in addition to those in our simple systems, and, perhaps more importantly, many components that are not phospholipids at all—sphingolipids, cholesterol, proteins, and carbohydrates. Finally, the membrane is not a planar structure with a clean, sharp interface to the aqueous media inside and outside the cell. Real membranes are attached to internal cytoskeletal structures and to the glycocalyx on the external face of the cell, and are connected to other membranes through a variety of junctions.

8.4.3 Differential Nanoelectropermeabilization to YO-PRO-1, Propidium, and Trypan Blue—Answers Within Reach

One example of a permeabilization puzzle ready to be tackled is the variation in influx of different small molecules that is observed after exposure of cells to electropermeabilizing pulses. Why do cells become permeable to YP much more readily than they do to propidium and TB (Idziorek et al., 1995; Sugiyama et al., 2004)? Figure 8.2 provides some clues, suggesting that at least part of the explanation lies in simple steric factors and the size of the pores generated by the electropermeabilizing dose. But the tools are available now to answer the questions raised by this puzzle in great detail, and in the process to construct a reconciling synthesis of molecular and continuum models, membrane biophysics, fluorescence microscopy, and patch clamp electrophysiology.

References

Abidor IG, Arakelyan VB, Chernomordik LV, Chizmadzhev YA, Pastushenko VF, Tarasevich MR. 1979. Electric breakdown of bilayer lipid membranes I. Main experimental facts and their qualitative discussion. *Bioelectrochem Bioenerg* 6(1):37–52.

Benz R, Zimmermann U. 1980. Pulse-length dependence of the electrical breakdown in lipid bilayer membranes. *Biochim Biophys Acta* 597(3):637–642.

Buchner R, Barthel J, Stauber J. 1999. The dielectric relaxation of water between 0 degrees C and 35 degrees C. *Chem Phys Lett* 306(1–2):57–63.

Chang DC, Reese TS. 1990. Changes in membrane structure induced by electroporation as revealed by rapid-freezing electron microscopy. *Biophys J* 58(1):1–12.

Craviso GL, Chatterjee P, Maalouf G, Cerjanic A, Yoon J, Chatterjee I, Vernier PT. 2009. Nanosecond electric pulse-induced increase in intracellular calcium in adrenal chromaffin cells triggers calcium-dependent catecholamine release. *IEEE Trans Dielect Elect Inst* 16:1294–1301.

Chizmadzhev YA, Abidor IG. 1980. Bilayer lipid membranes in strong electric fields. *Bioelectrochem Bioenerg* 7(1):83–100.

DeBruin KA, Krassowska W. 1998. Electroporation and shock-induced transmembrane potential in a cardiac fiber during defibrillation strength shocks. *Ann Biomed Eng* 26(4):584–596.

Drago GP, Marchesi M, Ridella S. 1984. The frequency dependence of an analytical model of an electrically stimulated biological structure. *Bioelectromagnetics* 5(1):47–62.

Frey W, White JA, Price RO, Blackmore PF, Joshi RP, Nuccitelli R, Beebe SJ, Schoenbach KH, Kolb JF. 2006. Plasma membrane voltage changes during nanosecond pulsed electric field exposure. *Biophys J* 90(10):3608–3615.

Gabriel S, Lau RW, Gabriel C. 1996. The dielectric properties of biological tissues: II. Measurements in the frequency range 10 Hz to 20 GHz. *Phys Med Biol* 41(11):2251–2269.

Gowrishankar TR, Weaver JC. 2003. An approach to electrical modeling of single and multiple cells. *Proc Natl Acad Sci U S A* 100(6):3203–3208.

Gowrishankar TR, Esser AT, Vasilkoski Z, Smith KC, Weaver JC. 2006. Microdosimetry for conventional and supra-electroporation in cells with organelles. *Biochem Biophys Res Commun* 341(4):1266–1276.

Grosse C, Schwan HP. 1992. Cellular membrane potentials induced by alternating fields. *Biophys J* 63(6):1632–1642.

Hofmann F, Ohnimus H, Scheller C, Strupp W, Zimmermann U, Jassoy C. 1999. Electric field pulses can induce apoptosis. *J Membr Biol* 169(2):103–109.

Hu Q, Joshi RP, Schoenbach KH. 2005a. Simulations of nanopore formation and phosphatidylserine externalization in lipid membranes subjected to a high-intensity, ultrashort electric pulse. *Phys Rev E Stat Nonlin Soft Matter Phys* 72(3 Pt 1):031902.

Hu Q, Viswanadham S, Joshi RP, Schoenbach KH, Beebe SJ, Blackmore PF. 2005b. Simulations of transient membrane behavior in cells subjected to a high-intensity ultrashort electric pulse. *Phys Rev E Stat Nonlin Soft Matter Phys* 71(3 Pt 1):031914.

Hubbard J, Onsager L. 1977. Dielectric dispersion and dielectric friction in electrolyte solutions. I. *J Chem Phys* 67(11):4850–4857.

Idziorek T, Estaquier J, De Bels F, Ameisen JC. 1995. YOPRO-1 permits cytofluorometric analysis of programmed cell death (apoptosis) without interfering with cell viability. *J Immunol Methods* 185(2):249–258.

Kinosita K, Jr., Tsong TY. 1977. Voltage-induced pore formation and hemolysis of human erythrocytes. *Biochim Biophys Acta* 471(2):227–242.

Kotnik T, Miklavcic D. 2000. Second-order model of membrane electric field induced by alternating external electric fields. *IEEE Trans Biomed Eng* 47(8):1074–1081.

Lopez-Garcia JJ, Horno J, Gonzalez-Caballero F, Grosse C, Delgado AV. 2000. Dynamics of the electric double layer: Analysis in the frequency and time domains. *J Colloid Interface Sci* 228(1):95–104.

Martinsen OG, Grimnes S, Schwan HP. 2002. Interface phenomena and dielectric properties of biological tissue. In *Encyclopedia of Surface and Colloid Science*. Marcel Dekker, New York, pp. 2643–2652.

Neu JC, Krassowska W. 1999. Asymptotic model of electroporation. *Phys Rev E* 59(3):3471–3482.

Pakhomov AG, Kolb JF, White JA, Joshi RP, Xiao S, Schoenbach KH. 2007a. Long-lasting plasma membrane permeabilization in mammalian cells by nanosecond pulsed electric field (nsPEF). *Bioelectromagnetics* 28(8):655–663.

Pakhomov AG, Shevin R, White JA, Kolb JF, Pakhomova ON, Joshi RP, Schoenbach KH. 2007b. Membrane permeabilization and cell damage by ultrashort electric field shocks. *Arch Biochem Biophys* 465(1):109–118.

Pakhomov AG, Bowman AM, Ibey BL, Andre FM, Pakhomova ON, Schoenbach KH. 2009. Lipid nanopores can form a stable, ion channel-like conduction pathway in cell membrane. *Biochem Biophys Res Commun* 385(2):181–186.

Pethig R, Kell DB. 1987. The passive electrical properties of biological systems: Their significance in physiology, biophysics and biotechnology. *Phys Med Biol* 32(8):933–970.

Plonsey R, Altman KW. 1988. Electrical stimulation of excitable cells—A model approach. *Proc IEEE* 76(9):1122–1129.

Popescu D, Rucareanu C, Victor G. 1991. A model for the appearance of statistical pores in membranes due to self-oscillations. *Bioelectrochem Bioenerg* 25(1):91–103.

Pucihar G, Kotnik T, Miklavcic D, Teissie J. 2008. Kinetics of transmembrane transport of small molecules into electropermeabilized cells. *Biophys J* 95(6):2837–2848.

Rogers WR, Merritt JH, Comeaux JA, Kuhnel CT, Moreland DF, Teltschik DG, Lucas JH, Murphy MR. 2004. Strength-duration curve for an electrically excitable tissue extended down to near 1 nanosecond. *IEEE Trans Plasma Sci* 32(4):1587–1599.

Schoenbach KH, Beebe SJ, Buescher ES. 2001. Intracellular effect of ultrashort electrical pulses. *Bioelectromagnetics* 22(6):440–448.

Schoenbach KH, Joshi RP, Kolb JF, Chen NY, Stacey M, Blackmore PF, Buescher ES, Beebe SJ. 2004. Ultrashort electrical pulses open a new gateway into biological cells. *Proc IEEE* 92(7):1122–1137.

Schoenbach KH, Xiao S, Joshi RP, Camp JT, Heeren T, Kolb JF, Beebe SJ. 2008. The effect of intense subnanosecond electrical pulses on biological cells. *IEEE Trans Plasma Sci* 36(2):414–422.

Schwan HP, Foster KR. 1980. RF-field interactions with biological systems: Electrical properties and biophysical mechanisms. *Proc IEEE* 68(1):104–113.

Shepherd JC, Buldt G. 1978. Zwitterionic dipoles as a dielectric probe for investigating head group mobility in phospholipid membranes. *Biochim Biophys Acta* 514(1):83–94.

Sher LD, Kresch E, Schwan HP. 1970. On the possibility of nonthermal biological effects of pulsed electro-magnetic radiation. *Biophys J* 10(10):970–979.

Shilov VN, Delgado AV, Gonzalez-Caballero F, Horno J, Lopez-Garcia JJ, Grosse C. 2000. Polarization of the electrical double layer. Time evolution after application of an electric field. *J Colloid Interface Sci* 232(1):141–148.

Stuchly MA, Athey TW, Stuchly SS, Samaras GM, Taylor G. 1981. Dielectric properties of animal tissues in vivo at frequencies 10 MHz–1 GHz. *Bioelectromagnetics* 2(2):93–103.

Sugar IP, Neumann E. 1984. Stochastic model for electric field-induced membrane pores. Electroporation. *Biophys Chem* 19(3):211–225.

Sugiyama T, Kobayashi M, Kawamura H, Li Q, Puro DG. 2004. Enhancement of P2X$_7$-induced pore formation and apoptosis: An early effect of diabetes on the retinal microvasculature. *Invest Ophthalmol Vis Sci* 45(3):1026–1032.

Svanberg C, Berntsen P, Johansson A, Hedlund T, Axen E, Swenson J. 2009. Structural relaxations of phos-pholipids and water in planar membranes. *J Chem Phys* 130(3):035101.

Tarek M. 2005. Membrane electroporation: A molecular dynamics simulation. *Biophys J* 88(6):4045–4053.

Teissie J, Rols MP. 1993. An experimental evaluation of the critical potential difference inducing cell mem-brane electropermeabilization. *Biophys J* 65(1):409–413.

Tieleman DP. 2004. The molecular basis of electroporation. *BMC Biochem* 5(1):10.

Tieleman DP, Leontiadou H, Mark AE, Marrink SJ. 2003. Simulation of pore formation in lipid bilayers by mechanical stress and electric fields. *J Am Chem Soc* 125(21):6382–6383.

Timoshkin IV, MacGregor SJ, Fouracre RA, Crichton BH, Anderson JG. 2006. Transient electrical field across cellular membranes: Pulsed electric field treatment of microbial cells. *J Phys D-Appl Phys* 39(3):596–603.

Vasilkoski Z, Esser AT, Gowrishankar TR, Weaver JC. 2006. Membrane electroporation: The absolute rate equation and nanosecond time scale pore creation. *Phys Rev E* 74(2):021904.

Vernier PT, Sun Y, Marcu L, Craft CM, Gundersen MA. 2004a. Nanoelectropulse-induced phosphatidyl-serine translocation. *Biophys J* 86(6):4040–4048.

Vernier PT, Sun Y, Marcu L, Craft CM, Gundersen MA. 2004b. Nanosecond pulsed electric fields perturb membrane phospholipids in T lymphoblasts. *FEBS Lett* 572(1–3):103–108.

Vernier PT, Sun Y, Gundersen MA. 2006a. Nanoelectropulse-driven membrane perturbation and small molecule permeabilization. *BMC Cell Biol* 7(1):37.

Vernier PT, Ziegler MJ, Sun Y, Chang WV, Gundersen MA, Tieleman DP. 2006b. Nanopore formation and phosphatidylserine externalization in a phospholipid bilayer at high transmembrane potential. *J Am Chem Soc* 128(19):6288–6289.

Vernier PT, Ziegler MJ, Sun Y, Gundersen MA, Tieleman DP. 2006c. Nanopore-facilitated, voltage-driven phosphatidylserine translocation in lipid bilayers—in cells and in silico. *Phys Biol* 3(4):233–247.

Vernier PT, Sun Y, Chen MT, Gundersen MA, Craviso GL. 2008. Nanosecond electric pulse-induced cal-cium entry into chromaffin cells. *Bioelectrochemistry* 73(1):1–4.

Wang S, Chen J, Chen MT, Vernier PT, Gundersen MA, Valderrabano M. 2009. Cardiac myocyte excita-tion by ultrashort high-field pulses. *Biophys J* 96(4):1640–1648.

Weaver JC. 2003. Electroporation of biological membranes from multicellular to nano scales. *IEEE Trans Dielectr Electr Insul* 10(5):754–768.

Weaver JC, Chizmadzhev YA. 1996. Theory of electroporation: A review. *Bioelectrochem Bioenerg* 41(2):135–160.

Zimmermann U. 1982. Electric field-mediated fusion and related electrical phenomena. *Biochim Biophys Acta* 694(3):227–277.

Zimmermann U, Pilwat G, Riemann F. 1974. Dielectric breakdown of cell membranes. *Biophys J* 14(11):881–899.

III

Mechanisms of Electroporation of Cells

9

Nanopores: A Distinct Transmembrane Passageway in Electroporated Cells

Andrei G. Pakhomov

Olga N. Pakhomova

The phenomenon of electropermeabilization of cell plasma membrane has been known for decades and has been reviewed in multiple publications, including this book. Nonetheless, the exact mechanism(s) that enable transmembrane flow of normally impermeable ions and molecules is not fully understood. The most widely accepted, albeit continually debated, concept explains electropermeabilization by *de novo* formation of water-filled lipid pores that serve as transmembrane pathways for water-soluble chemical species (Neumann et al. 1989). Assuming this mechanism is correct, one can reasonably expect that the size of such electropores, their conductive properties, resealing time, etc., would depend greatly on the amplitude and duration of the electroporation pulses, as well as the cell type and composition of the extracellular medium. Unfortunately, this topic has only scarcely been explored, and no systematic data are available. Multiple applications of electroporation in biotechnology and medicine rely more on empirical (trial and error) optimization of treatment conditions rather than knowledge of lipid pores, how they function, and how they contribute to cell physiology and biochemistry.

This chapter is specifically focused on so-called nanopores, which are perhaps the most intriguing category of membrane pores. Nanopores are distinguished by the smallest diameter (about 1 nm or less) and are remarkably stable, with resealing time up to 10–15 min at room temperature (Pakhomov et al. 2007a, 2009). Nanopores show complex conductive properties previously thought unique to specialized, protein-made ion channels; other properties distinguish nanopores from both known ion channels and larger (conventional) electropores. In this study, nanopores were created in mammalian cells by applying ultrashort electric pulses (USEP, 60–600 ns duration, 1–12 kV/cm); however, we infer that longer electric pulses and nonelectrical (e.g., chemical or mechanical) stimuli under certain conditions also open nanopores.

9.1 Detection of Nanopores

Opening of nanopores can be established by one or more of the following experimental techniques: (1) differential uptake of fluorescent markers (dyes, reporter ions, or molecules) by electroporated cells, (2) immediate externalization of phosphatidylserine (PS) residues, (3) isoosmotic changes of the cell volume, and (4) analysis of membrane conductance using electrophysiological methods such as patch clamp.

9.1.1 Fluorescent Dye Uptake Techniques

Historically, USEP were introduced as a modality that can penetrate the cell plasma membrane, causing various intracellular effects directly (Schoenbach et al. 2001, 2002, 2007, Beebe et al. 2003a,b). Apart from theoretical calculations, this mechanism of action was supported by the apparent lack of uptake of membrane-impermeable dyes (propidium and Trypan blue) by USEP-treated cells (Beebe et al. 2003a, Vernier et al. 2003, 2004, Pakhomov et al. 2004, 2007b). To test for propidium uptake, cells were bathed in a solution containing 3–30 μM of propidium iodide and treated by USEP. Propidium is only weakly fluorescent in the extracellular solution, but will enhance emission 20–30-fold upon binding to the intracellular DNA and RNA. Propidium cations do not pass through undisturbed cell membranes, and therefore the lack of their uptake is commonly used as a convenient and sensitive marker of the membrane integrity. (*Note*: the terminology used in most publications refers to "propidium iodide (PI) uptake," which is detected by "changes in PI fluorescence." However, in water, PI dissociates into propidium cation and iodide anion; it is the propidium cation that penetrates the membrane and causes fluorescence changes, not the entire PI molecule. Therefore, it is more accurate to use the term "propidium uptake," especially when talking about the pore size.)

In fact, the lack of propidium uptake by USEP-treated cells did not necessarily mean that the cell membrane integrity was preserved. The USEP-opened membrane pores are too small to allow the passage of the propidium cation but allow passage of smaller ion species.

Membrane permeabilization by USEP was readily demonstrated by fluorescent microscopy and using entry of a small ion (Tl^+) as a reporter of the membrane integrity (Pakhomov et al. 2009). To do so, cells are preloaded with a FluxOR™ fluorophore (Invitrogen, Eugene, Oregon) and placed in a Tl^+-containing bath buffer. Intact lipid membrane is impermeable to Tl^+, but this ion can potentially enter the cell via "classic" voltage–gated K^+-selective ion channels upon their activation by USEP. To prevent the ion channel–mediated entry of Tl^+, the bath buffer contained high concentrations of K^+ channel blockers (tetraethylammonium and 4-aminopyridine [DeCoursey et al. 1985]). In addition, on many occasions, we used CHO cells, which do not express any endogenous voltage-gated K^+ channels.

The detection of Tl^+ uptake is based on the property of the FluxOR dye to increase its fluorescence greatly in the presence of Tl^+. Since Tl^+ cation is not found in living cells in any appreciable amount, the fluorescent method of Tl^+ detection is (a) highly sensitive and (b) always indicates Tl^+ entry through the plasma membrane (in contrast, for example, to Ca^{2+} detection, since Ca^{2+} can come into the cytosol from various intracellular depots). Figures 9.1 and 9.2 show that a single 600 ns pulse at 3.3 kV/cm caused an immediate surge of Tl^+-dependent cell fluorescence, which reflects the USEP-triggered Tl^+ entry into the cell. However, this electric pulse caused no change in propidium fluorescence (even at the highest sensitivity of the detector), thereby indicating that the USEP-opened pores were too small for propidium passage into the cell. Boundbox dimensions (between the centers of the outmost atoms) and van der Waals volume of the propidium molecule (which is the volume impenetrable for other molecules) are presented in

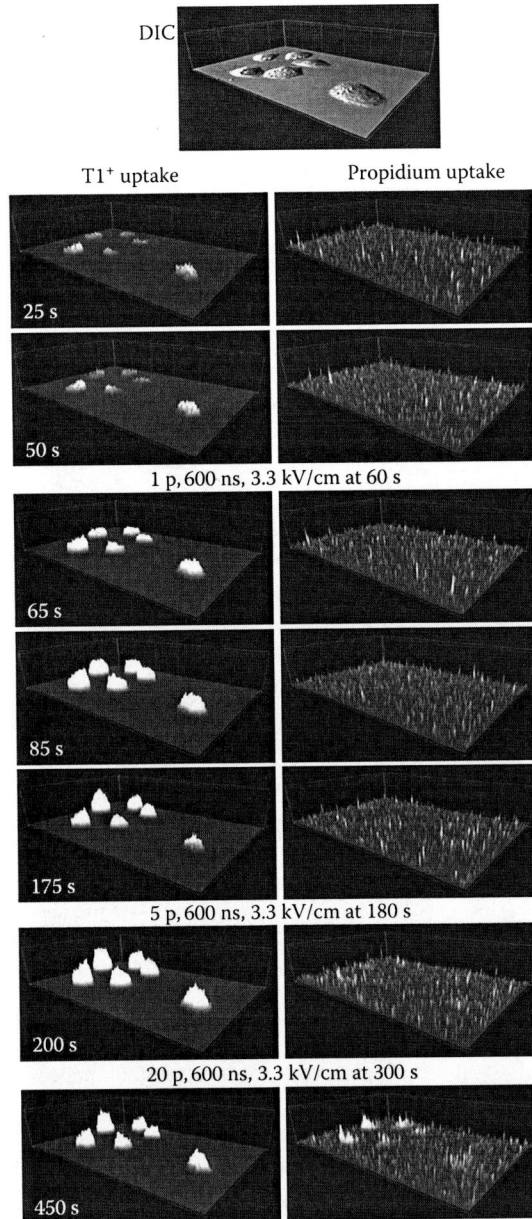

FIGURE 9.1 Fluorescent detection of plasma membrane permeabilization by ultrashort electric pulses. CHO cells were loaded with a Tl^+-sensitive dye (FluxOR) and incubated in a buffer containing 16 mM Tl^+ and 30 μg/mL of PI. Membrane permeabilization was detected as a change in the respective fluorescence signal (Tl^+: excitation 488 nm, emission 530 nm; propidium: excitation 532 nm, emission 630 nm). Images were taken every 5 s throughout the experiment; shown are representative images before the treatment (25 and 50 s); after a single 600 ns, 3.3 kV/cm pulse (65, 85, and 175 s); after additional 5 pulses (200 s), and 20 more pulses (450 s). Localization of the fluorescence signal and its intensity (arbitrary units) are coded by both the height and the brightness of "peaks" along Z-axis. A single electric pulse caused intense uptake of Tl^+, and five additional pulses initiated the second wave of Tl^+ uptake. However, the uptake of propidium became evident only after 26 pulses (450 s, right column). Note that the sensitivity of the propidium channel was set to maximum, resulting in significant background noise.

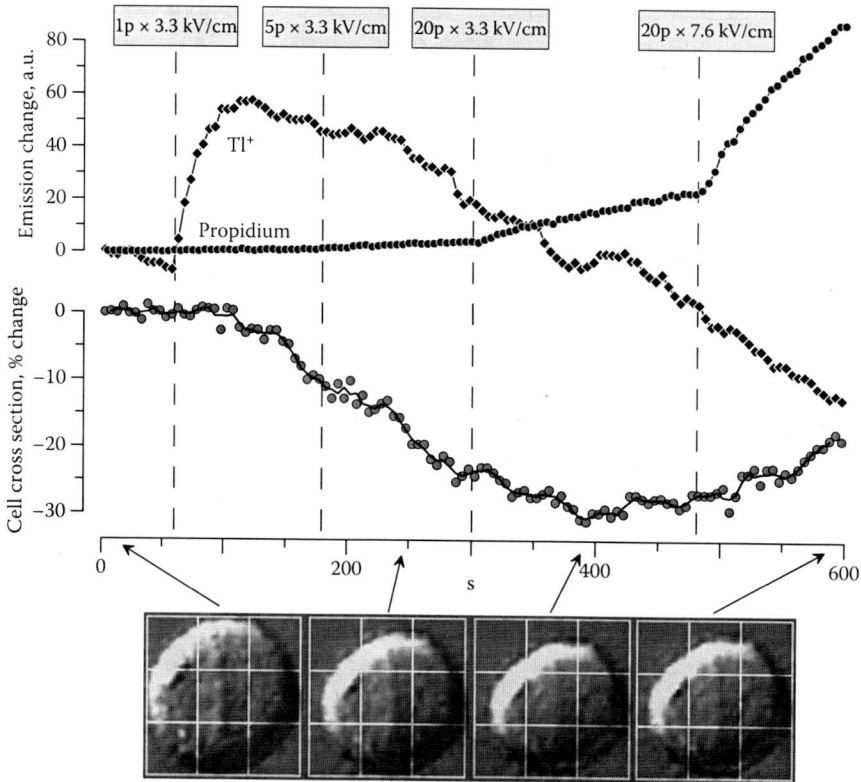

FIGURE 9.2 Effect of 600 ns pulses at 3.3 kV/cm on plasma membrane permeability and cell size. Concurrent measurements of Tl^+ and propidium uptake were performed in an individual CHO cell preloaded with a Tl^+-sensitive dye FluxOR. The bath buffer was composed of (in mM): 122 Na gluconate, 5 Na-EGTA, 4 $MgSO_4$, 8 Tl_2SO_4, 10 HEPES, 10 glucose (pH 7.4); it also contained 30 μg/mL of PI. Insets at the bottom show images of the cell taken at the selected timepoints; the grid size (small squares) are 5 × 5 μm. Note the uptake of Tl^+ after a single pulse and uptake of propidium only after multiple pulses. Note also that "mild" membrane permeabilization caused cell shrinking, which was reverted by a more severe damage. See text for more detail.

Figure 9.3. It is probably the van der Waals dimensions that will define the minimum size of a membrane opening permeable to propidium; in the simplest case, the minimum diameter of a round-shaped opening should be larger than two orthogonal dimensions of the molecule. For propidium, such opening should be larger than 1.4 nm in diameter (this number was obtained by adding the van der Waals radii of two singly bonded hydrogens (0.117 nm on each side) to the second-smallest boundbox dimension, 1.15 nm).

Therefore, we postulate that USEP-opened pores are typically less than 1.4 nm in diameter, but at least a considerable fraction of them is larger than 0.4 nm (i.e., they are permeable to Tl^+ with van der Waals diameter of 0.392 nm). Here, we do not take into account any ion hydration layers, which could potentially increase the minimum pore diameter requirement (the effective size of hydrated ions is a "fuzzy concept" [Hille 2001], which is beyond the scope of this chapter). As the best estimate of the diameter of USEP-opened pores is on the order of a nanometer or less, they were named "nanopores" (Vernier et al. 2006a, Pakhomov et al. 2009).

Although the lack of propidium uptake by USEP-treated cells was noted in multiple studies, this should not be interpreted as a doctrine. The truth is that propidium fluorescence of cells exposed to even tens or hundreds of USEP is still far weaker (by orders of magnitude) than in a typical "dead" cell (e.g., in a cell permeabilized with digitonin). Therefore, when a fluorescence detector is tuned to distinguish dead and live cells, it will not detect any propidium signal in USEP-treated cells (Beebe

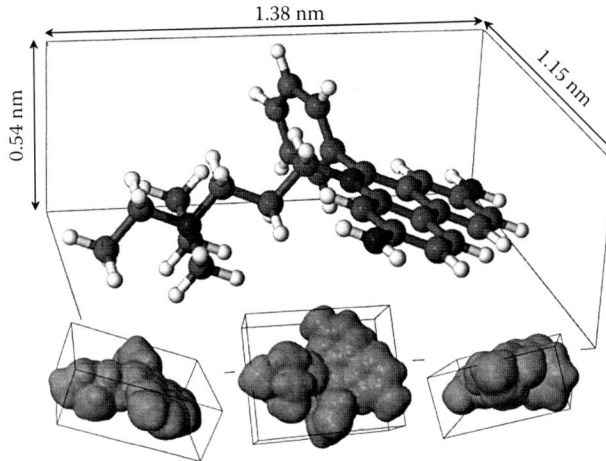

FIGURE 9.3 Boundbox dimensions of the propidium molecule (top) and several views of its van der Waals surface (bottom). All molecular views were generated from the molecular structure found in PDB library (http://www. rcsb.org/) using Jmol open-source Java viewer for chemical structures in 3D (http://www.jmol.org/). Boundbox dimensions (between the centers of the outmost atoms) were calculated using MOLEMAN2 (Uppsala Software Factory, Uppsala, Sweden, http://xray.bmc.uu.se/usf/). Van der Waals surface represents the impenetrable molecular volume (Creighton 1993), which extends out of the boundbox by about 0.12 nm in each direction.

et al. 2003a, Pakhomov et al. 2007b, Nuccitelli et al. 2009). Boosting the detector sensitivity to its highest limits may reveal propidium uptake, albeit rather modest and usually only after multiple USEP (e.g., Figures 9.1, 9.2, and 9.4; see also Vernier et al. 2006a, Pakhomov et al. 2009). This modest propidium entry can reflect thermal fluctuations of nanopore diameter, or it can be a sign of nanopores' evolution into larger pores as the cell degrades after the USEP insult.

9.1.2 Phosphatidylserine Externalization

PS is a membrane phospholipid typically found only on the internal face of the plasma membrane; its appearance on the external surface is believed to mark early stages of apoptosis, and can also be caused by cell activation (platelets) and maturation (sperm) (Bevers et al. 1999, Vance and Steenbergen 2005). Externalized PS can be detected by its specific binding to FITC-labeled protein Annexin V, although some indirect fluorescent detection techniques have also been reported (Vernier et al. 2006b).

In the case of USEP exposure, PS externalization can be detected within seconds, which is too fast for an apoptotic process and points to a direct effect of the externally applied electric field. The most likely explanation is the formation of lipid pores, which provide a continual lipid–water interface from the inside to the outside of the cell, and allow for the lateral diffusion (drift) of PS residues to the external membrane surface (Vernier et al. 2004, 2006b). Immediate PS externalization after USEP can occur in the absence of any detectable propidium uptake (Vernier et al. 2006b, Pakhomov et al. 2009), or when the propidium uptake is just marginally noticeable at the highest sensitivity of the fluorescence detector (Figure 9.4). In these cases, PS externalization can be interpreted as formation of lipid nanopores that are completely or mostly impermeable to propidium.

Compared with the Tl+ uptake assay, detection of Annexin V-FITC binding to externalized PS typically requires a more intense USEP treatment. It is not known yet whether this difference is explained simply by relatively low affinity and/or emission of Annexin V-FITC, or is related to the size or other properties of USEP-opened membrane pores. One way or another, the importance of the PS externalization phenomenon is that it points to the lipid nature of the pores. With Tl+ uptake, one can contemplate its entry by some unknown/unidentified ion channels that are activated by a USEP and are not inhibited

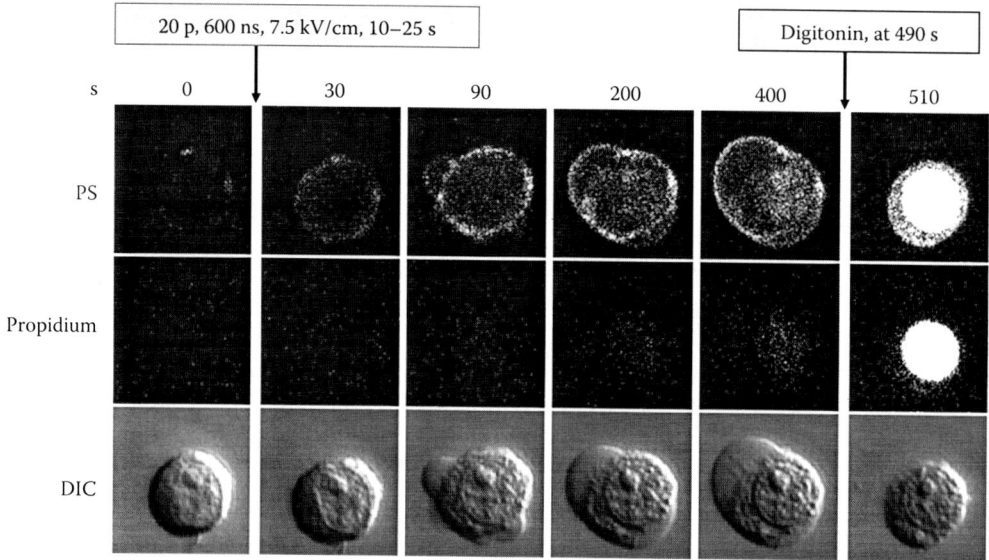

FIGURE 9.4 Ultrashort electric pulses cause PS externalization, cell swelling, and blebbing, but only modest uptake of propidium. Shown are sequential images of a GH3 cell subjected to a train of 20 (600 ns, 7.5 kV/cm) pulses from 10 to 25 s into the experiment. The bath buffer contained (in mM): 135 NaCl, 5 KCl, 4 MgCl$_2$, 2 CaCl$_2$, 10 HEPES, and 10 Glucose (pH 7.4), and was supplemented with 30 µg/mL of PI and with Annexin V-FITC (BD Biosciences, Franklin Lakes, NJ) at 10× dilution for PS detection. Both dyes were excited with a 488 nm laser, and emission was recorded at 530 nm for PS-Annexin V-FITC (top row) and at 605 nm for propidium (center). The bottom row shows differential interference contrast (DIC) images of the cell. As a positive control, at 490 s, the cell was permeabilized by adding digitonin to the bath (to the final concentration of 0.03%). Note that, after the addition of digitonin, the propidium emission became so intense that it caused detector saturation at both 605 and 530 nm. Also note that opening of large membrane pores by the addition of digitonin caused implosion of blebs and abolished cell swelling.

by the employed K$^+$-channel blockers. However, protein-made ion channels do not form a continual water–lipid interface pathway from the inside to the outside of the cell, hence activation of such ion channels can unlikely result in PS drift to the outside membrane face.

9.1.3 Cell Volume Change

It has been reported by many authors that an electric pulse exposure of mammalian cells can cause profound cell swelling and blebbing (Pakhomov et al. 2007b, Schoenbach et al. 2007, Tekle et al. 2008). In fact, depending on both the electric pulse parameters and composition of the bath medium, cell membrane permeabilization can cause either cell swelling (due to water uptake) or shrinking (due to water expulsion). The mechanism of cell volume changes is reasonably well understood and results from Donnan-type colloid osmotic pressure (Saulis 1999, Okada 2004).

Mammalian cells have osmolality of about 300 mOsm, and, out of this number, 30 mOsm or more are due to the presence of "colloid" (large molecules like DNA, RNA, proteins, organic anions, etc.). These large molecules are unlikely to leave the cell via nanopores or even through much larger pores. Most of the commonly used bath buffers (e.g., Hank's balanced salt solution or HBSS) are isoosmotic to cells (~300 mOsm), but only contain small ions and molecules.

So, what happens if an electric pulse opens pores that are fully permeable to all solutes in the bath buffer, but not to the large molecules inside the cell? In a simplified scenario, that also disregards any active volume control mechanisms (Okada 2004, Okada et al. 2006); small ions inside and outside the cell will come to a concentration equilibrium, e.g., K^+ will leave the cell and dilute to its concentration in the bath buffer, whereas Na^+ and Cl^- will enter the cell, eventually making their intra- and extracellular concentrations equal. The volume of the bath buffer usually is much larger than the volume of permeabilized cells, therefore the ion concentrations in the bath buffer will remain largely unaffected by cell permeabilization, and its osmolality will stay at 300 mOsm. However, the intracellular compartment will now be composed of both the extracellular ion species (providing for the osmolality of 300 mOsm) and the large molecules that were unable to leave the cell (providing for additional 30 mOsm). Hence, the combined intracellular osmolality will be at 330 mOsm, which is higher than in the extracellular buffer; this will force water entry into the cell and cause cell swelling. Noteworthy, the inflow of water will only decrease but never fully eliminate the osmolality gradient; hence the cell swelling will continue until pores are closed or the cell is lysed. This phenomenon was the basis of elegant experiments that quantified electroporation in erythrocytes by measuring elevation of the hemoglobin content in the bath buffer due to cell swelling that eventually culminated in lysis (Saulis 1999).

The increase of the intracellular osmolality, followed by water uptake and cell swelling, can be prevented by replacing the small ions in the bath buffer with an isoosmotic amount of larger (pore-impermeable) molecules or ions. In the above example, replacing 15 mM of NaCl (which is equivalent to ~30 mOsm) in the bath buffer with an isoosmotic amount of, say, sucrose will prevent the formation of an osmolality gradient if membrane pores are too small for sucrose passage. Moreover, making the bath content of a pore-impermeable compound higher than the intracellular "colloid" concentration will replace swelling of electroporated cells with their shrinking.

This volume change phenomenon can be readily employed for estimation of the pore size in the membrane. For example, if sucrose in the bath buffer protects cells from swelling, but an isoosmotic amount of mannitol does not do it, one can conclude that membrane pores are permeable to mannitol, but not to the larger molecule of sucrose. In real-life experiments, partial permeability and water mobility of tested compounds should also be taken into account. A large but still pore-permeable molecule added to the bath buffer at a high concentration will exert lower water mobility than smaller ions leaving the cell, so the osmotic imbalance will develop with a slight delay after the electric pulse (typically no more than several seconds), and the cell may display a brief, transient shrinking followed by swelling.

An example of the pore size–dependent change in cell size is given in Figure 9.2 (lower curve and the inset). For faster data acquisition, we only measured the area of the cell cross section, assuming that, in general, the cell volume is changing proportionally. Instead of NaCl, the bath buffer contained 122 mM of sodium gluconate, which is a relatively large anion (boundbox dimensions are 0.94 × 0.43 × 0.34 nm) with a "borderline" permeability through nanopores. (Note also that nanopores are far less permeable to anions compared to cations, see Section 9.2.5.) In the illustrated experiment, exposure to a single USEP caused significant cell shrinking, which is explained by the fact that USEP-opened nanopores were either too small for gluconate passage, or this passage was too slow. However, more intense exposures later in the experiment opened larger membrane pores (and/or more of them), which were permeable both to propidium and gluconate. As a result, cell shrinking was abolished and replaced with gradual volume increase.

9.1.4 Direct Measurements of Membrane Currents Using Patch Clamp

Patch clamp is arguably the most informative method to analyze membrane currents and establish functional properties of nanopores. However, application of patch clamp in cells exposed to USEP is not straightforward: (1) the patch-clamp recording pipette can cause electric field distortion; (2) high-voltage electroporating pulse can be picked up by the patch-clamp amplifier, causing saturation of the current-recording circuitry, recording artifacts, and potential damage to the amplifier; (3) lowering of the USEP-delivering electrodes into the bath buffer can cause a shift and drift of the zero potential; (4) USEP-generating equipment and additional ground loop(s) may increase electrical interference at the line frequency (50 or 60 Hz); and (5) the applied electric pulse may be damaging to the gigaohmic seal and reduce the quality of the patch-clamp recording configuration. The possibility of artifacts should be carefully considered and, whenever possible, the data should be verified by non-electrophysi-ological methods. It is also a good practice to compare patch-clamp findings in cells that were "patched" prior to USEP application with those in cells that were first exposed to USEP and patched afterward. In the latter case, membrane currents are compared in populations of exposed and sham-exposed cells. The population-based protocol is immune to the above artifacts, but it is more laborious, does not allow collection of pre-exposure data for each cell, and introduces a delay of at least 60–90 s between the USEP application and measurement of membrane currents (Pakhomov et al. 2007a,b, Ibey et al. 2009).

In fact, numerous experimental observations suggested that possible damage to the gigaohmic seal by USEP is less of a concern than it was originally thought. First, various conductive properties of nano-pores, including inward rectification (see below), did not depend on whether the whole-cell record-ing configuration was established prior to USEP exposure, or only after exposure (since there was no gigaohmic seal at the time of exposure, it could not be damaged by the pulse). Second, patch-clamp recordings concurrent with fluorescent dye uptake detection showed that the uptake typically starts at a cell location facing one of the USEP delivering electrodes, but not in the vicinity of the recording pipette; hence, the gigaohmic seal area remains intact (Pakhomov et al. 2009). Finally, USEP did not reduce the efficiency of the command voltage steps applied through the recording pipette in a whole-cell configura-tion: The actual membrane potential, as measured with a potential-sensitive dye d-8-ANEPPS, followed the pipette voltage equally well before and after nanopore opening (Figure 9.5). In case of gigaohmic seal damage, this response would be reduced or eliminated, but it was not the case after the USEP treatment.

As seen from the current–voltage (I–V) curves in Figure 9.5, the USEP-induced increase of the mem-brane conductance was not uniform across the range of studied transmembrane potentials. Instead, the maximum conductance increase occurred at the most negative potentials and corresponded to the maximum inward current. (*Note*: by a convention, the current direction in patch-clamp recording is determined by the flow of cations, so the "inward current" means that cations are entering the cell, and/or that anions are leaving it.) Such differential enhancement of the inward current, or inward rectification, proved to be a typical characteristic of nanopores and distinguishes them from larger, propidium-permeable pores, which show a linear I–V dependence. We found that cells which display inward rectification after USEP exposure remain completely or largely impermeable to propidium, whereas the onset of the propidium uptake also marks the reduction and subsequent loss of the rectification (Pakhomov et al. 2009).

Although the extent of rectification varies, it can be seen in most of USEP-treated cells, in different cell lineages, and with different pipette/bath solution combinations. Inward rectification is also present under the symmetrical solution conditions, when the ionic composition of the pipette and bath buffers is identical (Figure 9.6). Furthermore, the inward current increases nonlinearly with increasing negative voltage across the membrane, i.e., the conductance of nanopores is voltage dependent (Figures 9.5 and 9.6). Numerous experiments performed to date suggest that the inward rectification is an intrinsic and universal property of nanopores, which also points to their functional and structural asymmetry. The induction of inward rectification by USEP can be used as a nanopore "signature," i.e., as an indirect, but reliable and sensitive manifestation of nanopore formation.

Of note, a number of classic ion channels also display inward-rectifying properties (e.g., Ghamari-Langroudi and Bourque 2001, Varghese et al. 2006). It is not clear why and how would USEP cause activation of these channels, and pharmacological profiles of these channels are also different from that of nanopores. Nonetheless, possible confusion of nanopore response after USEP with currents of certain types of ion channels cannot always be ruled out.

To date, electrophysiological measurements proved more sensitive for the detection of nanopores than any fluorescent dye detection approach tested. With a single pulse of either 60 or 600 ns duration, and using different pulse voltages, the extent of membrane permeabilization was shown to be propor-tional to the absorbed dose, with the threshold at about 10 mJ/g (Ibey et al. 2009). Sample heating at this absorbed dose is less than 0.01°C, which clearly shows that membrane electropermeabilization is a nonthermal effect.

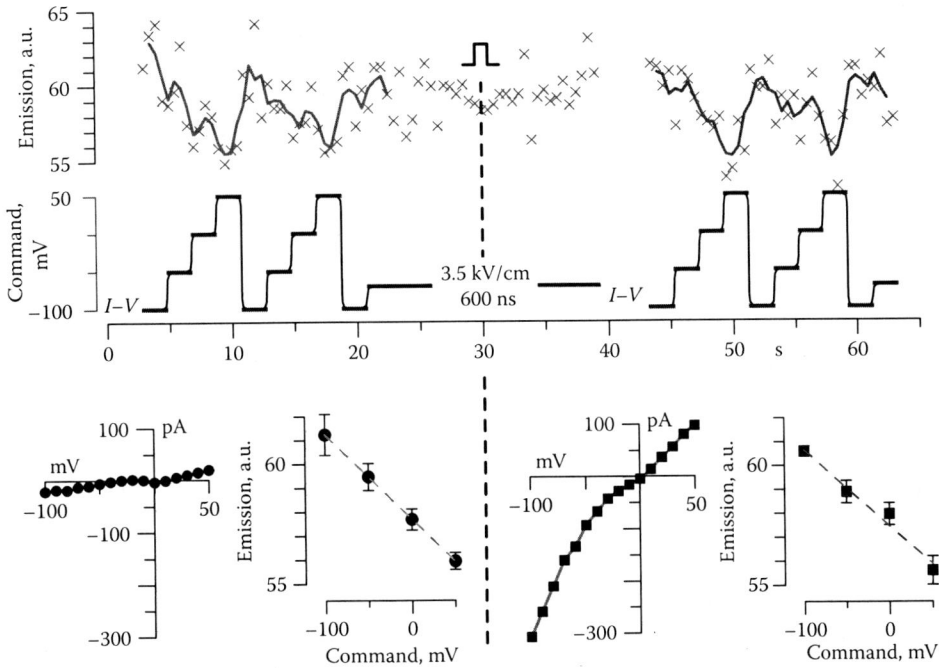

FIGURE 9.5 Opening of nanopores with a 600 ns, 3.6 kV/cm pulse does not destroy the whole-cell patch-clamp recording configuration. A voltage-sensitive dye d-8-ANEPPS was incorporated into the plasma membrane of a CHO cell, in order to test the efficiency of controlling the membrane potential by the command voltage applied through the recording pipette. The top graph shows fluctuations of the dye emission (actual datapoints and their "running average") in response to the command voltage steps, before and after electroporation (at 30 s). The bottom graphs show the current–voltage (*I–V*) characteristic of the membrane and the average (±s.e.) response of the dye to the command voltages before (left) and after electroporation (right). The *I–V* data were collected at 0 and at 40 s by applying brief ascending voltage steps; these steps were too short to record fluorescence, so the respective intervals on the emission curve (top) are blank. The electric pulse caused profound changes in the whole-cell current response (increased the membrane conductance and caused inward rectification), but the response of the dye to the command voltage remained unchanged. Hence, the command voltage controlled the actual membrane potential equally efficiently before and after electroporation. See text for more detail.

9.2 Experimentally Established Functional Properties of Nanopores

9.2.1 Extended Pore Lifetime

Studies that employed conventional electroporation (using milli- and microsecond-duration pulses) showed that the permeabilized state of the plasma membrane persists from fractions of a second to minutes and even hours after the treatment (see for review: Saulis et al. 1991, Saulis 1999). Such data dispersion could be a result of different methods of pore creation and detection. The resealing time could also vary with the cell type, and be affected by the composition and temperature of the extracellular medium. Counterintuitively, the smallest pores (~1 nm in diameter) were reported to have relatively long lifetime, probably because of the existence of a significant energy barrier which prevents their complete closing (Saulis et al. 1991).

However, first studies of plasma membrane poration with USEP introduced nanopores as "nanosecond-duration, nanometer-diameter openings in the membrane" (Vernier et al. 2004). In the same study, observations of the effect of high-rate pulse trains indicated longer nanopore lifetime, possibly "on the order of microseconds" (however, no direct measurements were performed).

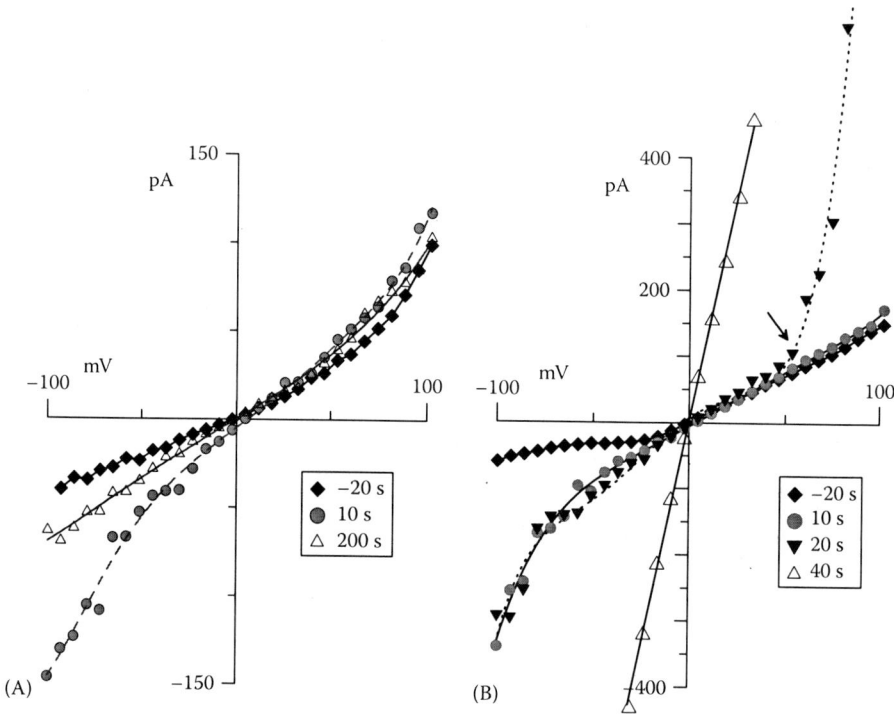

FIGURE 9.6 Ultrashort electric pulses open inward-rectifying membrane pores that gradually reseal (A) or break down into more conductive, non-rectifying pores (B). A and B are two different GH3 cells. The cells were exposed to a single 600 ns pulse at 1.6 kV/cm (A) or 2.4 kV/cm (B) at timepoint 0 s. Whole-cell currents were recorded under symmetrical solution conditions, starting at 20 s prior to exposure (−20 s) and at indicated time intervals after the pulse. The bath and pipette buffers contained (in mM): 150 K-acetate, 1 $MgCl_2$, and 5 HEPES (A), or 140 Cs-acetate, 5 Cs-EGTA, 4 $MgCl_2$, and 10 HEPES (B); pH 7.4 for both buffers. The exposure profoundly enhanced the inward current (at negative membrane potentials), but had little if any effect on the outward current. The arrow in (B) indicates the instance of apparent breakdown of nanopores into larger, non-rectifying pores (at 20 s after the pulse, when applying a +50 mV command voltage step).

On the contrary, later studies using patch clamp suggested the nanopore resealing time as long as up to 10–15 min after exposure to a 60 ns pulse (Pakhomov et al. 2007a), which was more in line with the conventional electroporation findings. This result was corroborated by fluorescent detection of Tl^+ entry into cells when this ion was added into the bath buffer at various intervals after exposure (Pakhomov et al. 2009). We now conclude that nanopores are remarkably stable, regardless of whether they were originally created as nanopores (when using USEP), or whether they formed by the shrinking of larger pores (when using longer electric pulses). One can also expect that nanopores created by nonelectrical insults (e.g., mechanical or chemical) will have the same properties, but this presumption needs experimental confirmation.

It is primarily the long lifetime of nanopores that determines their physiological significance. Open nanopores can profoundly affect transmembrane ion and water balance, causing cell volume changes and interfering with a broad range of cell functions. Timely resealing of nanopores is of paramount importance and may make the difference between cell recovery to the normal function and progression to death. Recent studies of chemically and mechanically induced membrane defects demonstrate that cells can actively eliminate such defects by either endo- or exocytosis (Idone et al. 2008a,b). Similar active mechanisms may assist resealing and elimination of nanoelectropores, but no experimental data are currently available to support this conjecture.

9.2.2 Fluctuation of Nanopore Currents and Single-Pore Opening Events

Original traces of the whole-cell current in USEP-exposed cells typically show large asynchronous fluctuations at the most negative membrane potentials (Pakhomov et al. 2009). This "noise" could reasonably be attributed to the chaotic fluctuations of the conductance of individual nanopores, which can be resolved better in USEP-treated excised membrane patches (Figure 9.7A). USEP stimulation did not significantly affect the holding current, but induced multiple spikes that varied in amplitude and duration. Contrary to classic ion channels, nanopores showed no discrete open states: Instead, they displayed smooth transitions between the electrically silent state and multiple conductive states, a reasonable expectation for lipid pores that do not have a fixed conductance. If isolated peaks in Figure 9.7A reflect the activity of individual nanopores, their peak conductance and mean open time can be estimated at about 100 pS and 1–2 ms, respectively. This single-pore conductance value falls in the same range as reported for most classic ion channels (tens to a few hundreds of pS), which also have the pore diameter on the order of 1 nm or less. On the other hand, the open times of a few milliseconds appear inconsistent with the minute-long increase in the whole-cell conductance after USEP stimulation; the most likely way to explain it is that nanopores do not close completely, but become too narrow for ion passage and turn electrically silent ("quasi-open" state). A *de novo* opening and a complete closure of a nanopore would both require crossing over a significant energy barrier, but the transition from quasi-open to open state could occur simply by thermal energy fluctuations. A similar quasi-open state of electropores has been demonstrated in planar lipid bilayers (Melikov et al. 2001).

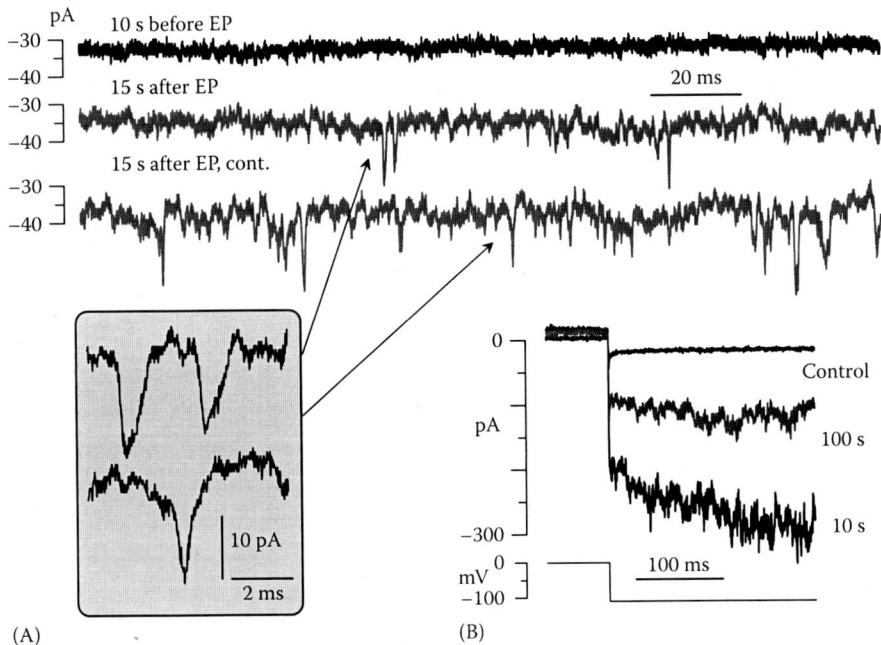

FIGURE 9.7 Spontaneous fluctuations of nanopore currents (A) and gradual enhancement of the inward current at a constant membrane potential (B). (A) Current traces in an outside-out patch of a GH3 cell before and 15 s after a 1.6 kV/cm, 600 ns electric pulse (EP). The patch was held at −100 mV in symmetrical solutions (in mM: 140 K-acetate, 5 K-EGTA, 4 $MgCl_2$, 10 HEPES, pH 7.2). Stimulation did not shift the holding current, but triggered spike-like activity, supposedly due to conductance fluctuations of individual pores. The inset shows selected pore opening events at a higher resolution. (B) Whole-cell currents in a CHO cell induced by a step from 0 to −110 mV, recorded before (control), and at 10 and 100 s after a 600 ns, 2.4 kV/cm EP. Post-EP records, especially at 10 s after the treatment, show continued increase of current at a constant voltage during the step.

9.2.3 Electric Current Sensitivity of Nanopores

In addition to the voltage sensitivity and inward rectification (discussed in Section 9.1.4), nanopores may change their electrical conductance even at a constant voltage, in response to the inward electric current (Figure 9.7B). Prior to USEP stimulation, the current induced by a hyperpolarization step was constant throughout the duration of the step. USEP greatly increased the current in response to the same voltage, but, in addition, the current continued to increase even though the voltage was held constant; this gradual increase continued during tens or even hundreds of milliseconds. This phenomenon indicates that nanopores are sensitive to the current they conduct, i.e., the current flowing through nanopores can further increase their electrical conductance.

Such gradual increase of nanopore current resembles activation of the delayed rectifier K^+ channels, but occurs at negative voltages and is not dependent on the presence of K^+. The time course of nanopore current also resembles opening of hyperpolarization-activated ion channels; however, these channels are efficiently blocked by Cs^+ ions (Ghamari-Langroudi and Bourque 2001), whereas nanopores are highly permeable to Cs^+ (Figure 9.8). The electric current sensitivity of nanopores gradually weakens and disappears as cells recover after USEP exposure; it also disappears if nanopores "break" into propidium-permeable larger pores.

9.2.4 Nanopore Breakdown

As described in Section 9.2.1, plasma membrane conductance in USEP-treated cells may gradually (over minutes) recover to its pre-exposure level. Alternatively, with a more severe USEP damage, or with already weakened cells, nanopores may suddenly open wider to allow propidium uptake. Electrophysiologically,

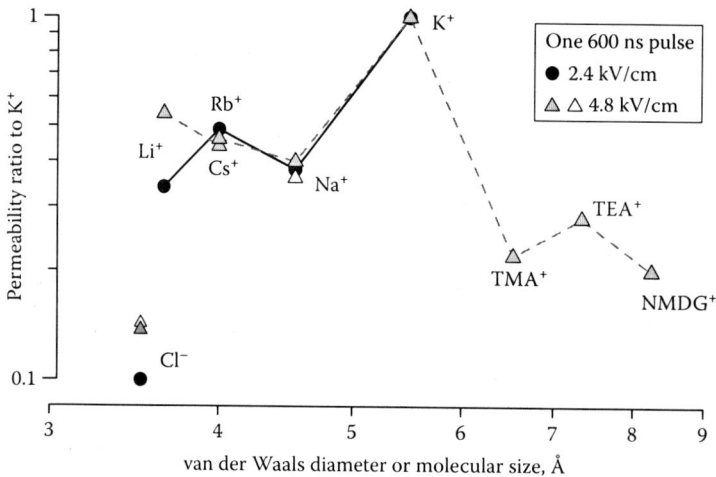

FIGURE 9.8 Relative permeability of nanopores to selected monovalent cations and Cl^-. Ion permeabilities were measured in GH3 cells exposed to a single 600 ns pulse at either 2.4 or 4.8 kV/cm. The cells were exposed intact; the whole-cell patch-clamp configuration was established within 1–2 min after the exposure, and whole-cell currents in different test solutions were measured within 2–4 min. All ion permeabilities were expressed as a ratio to K^+ permeability. To determine each permeability value, currents were measured and averaged in pulse-treated cells and control cells (8–18 cells per group), and the latter value was subtracted from the former one; see text for detail. Na^+ and Cl^- permeabilities were additionally measured in an independent set of experiments (open triangles), producing essentially the same results as before. The large cations tested were tetramethylammonium (TMA^+), tetraethylammonium (TEA^+), and N-methyl-D-gluconate ($NMDG^+$). For these cations, the van der Waals size was determined as the second smallest boundbox dimension, with addition of 2.4 Å to account for the radii of singly bound hydrogens on each side.

this transition is characterized by a sudden increase of the outward current at positive transmembrane voltages, followed by disappearance of the inward rectification and of the other nanopore-specific properties (Pakhomov et al. 2009, see also Figure 9.6B). In most cases, the nanopore breakdown was irreversible, and it typically (if not always) happened at a positive membrane potential. When applying a simple voltage-step protocol to explore membrane currents in USEP-treated cells (as in Figures 9.5 and 9.6), limiting the positive membrane potentials to just 20–40 mV greatly decreases the probability of nanopore breakdown and increases the chance of cell recovery.

9.2.5 Ion Selectivity of Nanopores

Aqueous pores having diameter much larger than ions passing through them are unlikely to have any ion selectivity, and relative permeabilities of different ions will be proportional to their water mobility. In contrast, USEP-opened nanopores display preferential permeability to certain ion species. This effect was found both in cells that were exposed intact and "patched" afterward, and in those that were "patched" prior to the exposure. Figure 9.8 illustrates relative permeabilities of several cations and Cl^- with respect to K^+. Intact (not "patched") GH3 cells were subjected to a 600 ns pulse at 0 (control), 2.5, or 4.8 kV/cm. Current–voltage (*I–V*) data in different bath buffers were collected within 2–4 min after the pulse; each condition was tested in 8–18 cells. To identify nanopore currents, mean *I–V* values for control experiments were subtracted from the respective values in cells subjected to USEP, and the best fit linear function was calculated for the data in the range from −80 to −50 mV. (The reason for choosing this range was higher nanopores' conductance at negative membrane potentials, and the lack of concurrent activation of voltage-gated channels.) The reversal potentials were determined from the intercept of the best fit line with the abscissa, and permeability ratios for different ions were calculated using Goldmann–Hodgkin–Katz equation (Hille 2001).

These experiments established that nanopores are preferentially permeable to cations. Although K^+ and Cl^- ions have the same mobility in water, Cl^- anion was about 10-fold less permeable, and this result was reproduced in several independent series of experiments. Furthermore, Cl^- had lower permeability compared to cations such as tetramethylammonium (TMA^+), tetraethylammonium (TEA^+), and *N*-methyl-D-gluconate ($NMDG^+$). Overall, the observation of the cationic selectivity of nanopores is consistent with expectations that they are lined up with negatively charged hydrophilic phosphate groups of membrane-forming phospholipids (Vernier et al. 2004).

The data also pointed to K^+ as the most permeable cation tested, but this finding needs to be interpreted with caution. GH3 cells express a variety of endogenous K^+ channels (Stojilkovic et al. 2005), and one cannot exclude that USEP, either directly or indirectly, could cause activation of some of these channels. In this case, higher K^+ permeability could have resulted from measuring mixed currents of nanopores and USEP-activated K^+ channels.

9.2.6 Blockage of Nanopores with Gd^{3+}

Within studied limits, plasma membrane conductance in nanoporated cells was not blocked by tested ion channel inhibitors such as Cs^+, Cd^{2+}, ruthenium red, amiloride, tetraethylammonium, 4-aminopyridine, or tetrodotoxin. The only potent inhibitor of nanopore conductance found to date is Gd^{3+}, either alone or in cocktail with La^{3+}. When Gd^{3+} was present in the bath buffer before USEP exposure, it attenuated the USEP effect two- to threefold; when Gd^{3+} was added after the exposure, it promptly restored the plasma membrane conductance (Pakhomov et al. 2007b). Recently, we also showed that even sub-micromolar concentrations of Gd^{3+} in the bath during an intense USEP treatment can significantly improve cell survival (data not shown).

Lanthanides are well-known blockers of diverse membrane transport mechanisms, including voltage-gated calcium channels, stretch-activated channels, several types of TRP channels, and several ionotropic and metabotropic ligand-gated channels; they affect Ca^{2+}-ATPase and Mg^{2+}-ATPase, block amine transporters and Na^+/Ca^{2+} exchangers (Hamill and McBride 1996, Iwamoto and Shigekawa 1998, Palasz and Czekaj 2000, Lipski et al. 2006). The mechanism of the blocking action of lanthanides can be related to the similarity of their size to Ca^{2+} ions (Palasz and Czekaj 2000). However, more recent studies suggested a direct effect of Gd^{3+} on the structure of the lipid bilayer (Ermakov et al. 2001, Tanaka et al. 2001, 2002), which is consistent with its exceptionally broad potency as a blocker and may also potentially explain the inhibitory effects on nanopores.

9.3 Hypothetical Structure of Nanopores

The most intriguing question about nanopores is how can they exert complex conductive behaviors that are thought to be a prerogative of protein ion channels with sophisticated, specialized organization?

A possible answer to this question comes from fundamental experiments with synthetic nanopores that were engineered in a polymer foil by heavy particle bombardment and subsequent etching of the pore from one side (Siwy et al. 2002). The resulting conically shaped synthetic nanopores behaved very similar to nanoelectropores in live cells: they passed ion currents preferentially in one direction, changed their conductance in response to applied voltage, and showed ion selectivity and single-pore current fluctuations. The properties of synthetic nanopores in foil have been detailed in later publications (Cervera et al. 2006, Kosinska 2006, Ramirez et al. 2007), and these findings appear to be at least partially applicable to USEP-opened nanopores in cells. Surprisingly, a simple assumption of asymmetrical, conical shape of USEP-opened nanopores can explain their unique behaviors remarkably well (Figure 9.9). We must emphasize, however, that the model presented in Figure 9.9 is entirely hypothetical and needs to be verified by direct experiments. It also remains to be explored if nanopores are made of membrane phospholipids only, or other membrane components, like structural proteins and

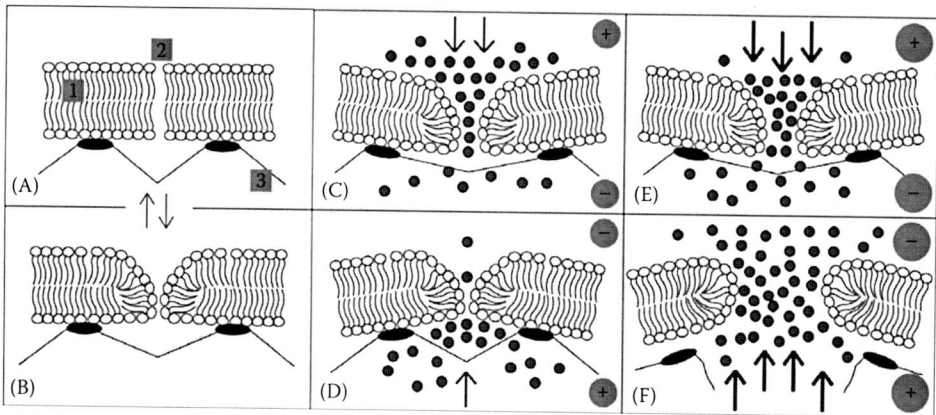

FIGURE 9.9 Hypothetical structure and states of a nanopore. The drawings show lipid bilayer of the plasma membrane (1), a pore (2), and cytoskeleton (3). For clarity, water molecules, membrane proteins, and anions are not shown; also, some hydrophobic tails of the membrane-forming phospholipid molecules are omitted. (A, B) A defect in the membrane eventually gets filled with water, and phospholipids line up the pore with polar heads facing the water column. The negative charge of the polar heads of phospholipids explains pore selectivity to cations. The water column prevents the pore from collapse, but allows thermal fluctuations of the pore shape and diameter. On the intracellular side, such fluctuations are limited by membrane bonds to the cytoskeleton, eventually resulting in a conical, funnel-like pore shape. (C, D) This shape facilitates the inward cation flow but restricts the outward flow, resulting in inward rectification. (E) Sustained inward current pushes the pore walls wider to further increase the inward current, which could explain both the current and voltage sensitivities of nanopore. (F) At high outward driving force, pressure of cations may detach the membrane from the cytoskeleton, causing nanopore "breakdown" into a larger, non-rectifying, propidium-permeable pore. (Drawings are courtesy of Dr. B. L. Ibey, Air Force Research Laboratory, Brooks City-Base, TX.)

cholesterol, contribute to nanopore formation and long-term stability. The model presented here is only intended to demonstrate that, at least in principle, even a simple pore structure can be capable of complex functions such as voltage and current sensitivity.

9.4 Lipid Nanopores or "Classic" Ion Channels?

De novo formation of lipid nanopores by USEP is a logical extension of the conventional electroporation phenomenon, and a smaller size of USEP-opened pores was predicted by modeling (Hu et al. 2005, Gowrishankar and Weaver 2006, Smith et al. 2006, Smith and Weaver 2008). In Section 9.3, we showed that complex changes in the electrical conductance may take place in structures far simpler than "classic" (protein-made) ion channels. On the other hand, there is still no consensus even about the conventional electroporation, whether it just opens large lipid pores or disrupts the membrane in some other way that has not been well understood (Teissie et al. 2005). Furthermore, the similarity of the functional properties of nanopores with some types of ion channels makes it reasonable to question it again whether (a) USEP causes activation of endogenous ion channel(s), rather than opening of *de novo* nanopores, and (b) opening of nanopores by stimuli other than USEP could be mistakenly ascribed by investigators as activation of ion channels (and of so-called nonspecific cation channels (NSCC) in particular)?

In general, the possibility that USEP-opened nanopores are confused with endogenous ion channels appears unlikely. First, a suspected ion channel should be voltage gated (because it is activated by an electric stimulus); however, we found that CHO cells, which do not express any endogenous voltage-gated channels, respond to USEP in the same manner as other cells. Furthermore, the suspected channel should exert broad permeability to various ions, including tetraethylammonium and N-methyl-D-glucamine, whereas most voltage-gated channels are highly selective. Lack of closure or inactivation for many minutes after the stimulus would also be highly unusual for classic ion channels. It would be hard to explain how ion channels break into propidium-permeable pores, and why it happens at the pole of the cell that faces one of the USEP-delivering electrodes (Pakhomov et al. 2009). One may note that the inward-rectifying I–V characteristic of nanopores is similar to that of hyperpolarization-activated channels (Robinson and Siegelbaum 2003, Varghese et al. 2006, Pena and Ordaz 2008); however, these channels are not voltage-gated, exclude Li^+, and are blocked by Cs^+, whereas nanopores are opened by voltage pulses and efficiently conduct both Li^+ and Cs^+. Finally, fast externalization of PS after USEP (see Section 9.1.2) suggests the formation of a continual lipid–water interface from the inside to the outside of the membrane, thereby providing perhaps the strongest indication of the lipid nature of USEP-opened nanopores.

Having said that, we should note that opening of nanopores is not necessarily the only effect of USEP, and activation of ion channels can also take place. Opening of voltage-gated channels can occur as an immediate result of membrane depolarization by USEP, as well as a secondary effect due to the loss of membrane potential caused by leak currents through nanopores. As an example of such response, nanosecond-duration electric pulses were reported to activate L-type voltage-gated Ca^{2+} channels in chromaffin cells (Vernier et al. 2008).

The use of USEP is a convenient way to create nanopores "on demand" at a selected location on the cell membrane, but there is no reason to believe that it is the only or unique way to do it. For instance, it is not unusual to see the emergence of "spontaneous" nanopore-like electric noise when performing routine patch-clamp measurements; this noise is commonly interpreted as a sign of gigaohmic seal deterioration, and the preparations displaying this noise are usually discarded. It makes sense, however, that deterioration of the gigaohmic seal starts with local formation of nanopores; then, it comes as no surprise that their currents are indeed similar to the currents of USEP-opened nanopores.

Likewise, certain cell responses that are believed to reflect opening of ion channels can equally well be explained by nanopore formation. Multitudes of identified and unidentified NSCC have been implicated in a widest range of physiological processes (Clapham 2003, Ramsey et al. 2006, Numata et al. 2007, Pena and Ordaz 2008). For example, unidentified NSCC are thought to be responsible for cell volume changes and necrotic transformation after exposure to ROS donors (Barros et al. 2001a,b) or following ATP depletion (Gabai et al. 1997). However, formation of nanopores could be an additional (if not an alternative) mechanism responsible for these effects, and similar examples can be continued. Overall, nanopores appear adequately equipped for many functions that are traditionally attributed to ion channels, but do not require complicated machinery to synthesize, assemble, and transport them to the membrane. One can reasonably expect that nanopores may form under various pathological and physiological conditions and play significant role in transmembrane ion traffic, especially in phylogenetically old, less-specialized functions like cell volume control and programmed cell death.

Acknowledgments

We thank Dr. V. Gabai (Boston University Medical School, Boston, Massachusetts) for discussions and help with data interpretation, Dr. S. Xiao (Old Dominion University, Norfolk, Virginia) for designing the nanosecond pulse source and E-field calculations, Dr. B. L. Ibey (Air Force Research Laboratory, Brooks City-Base, Texas) for valuable comments on the manuscript and figures, and Angela Bowman (ODU) for help with the experiments. The work was supported in part by R01CA125482 from the National Cancer Institute.

References

Barros LF, Hermosilla T, Castro J. 2001a. Necrotic volume increase and the early physiology of necrosis. *Comp Biochem Physiol A Mol Integr Physiol* 130(3):401–409.

Barros LF, Stutzin A, Calixto A, Catalan M, Castro J, Hetz C, Hermosilla T. 2001b. Nonselective cation channels as effectors of free radical-induced rat liver cell necrosis. *Hepatology* 33(1):114–122.

Beebe SJ, Fox PM, Rec LJ, Willis EL, Schoenbach KH. 2003a. Nanosecond, high-intensity pulsed electric fields induce apoptosis in human cells. *Faseb J* 17(11):1493–1495.

Beebe SJ, White J, Blackmore PF, Deng Y, Somers K, Schoenbach KH. 2003b. Diverse effects of nanosecond pulsed electric fields on cells and tissues. *DNA Cell Biol* 22(12):785–796.

Bevers EM, Comfurius P, Dekkers DW, Zwaal RF. 1999. Lipid translocation across the plasma membrane of mammalian cells. *Biochim Biophys Acta* 1439(3):317–330.

Cervera J, Schiedt B, Neumann R, Mafe S, Ramirez P. 2006. Ionic conduction, rectification, and selectivity in single conical nanopores. *J Chem Phys* 124(10):104706.

Clapham DE. 2003. TRP channels as cellular sensors. *Nature* 426(6966):517–524.

Creighton TE. 1993. *Proteins: Structures and Molecular Properties.* New York: W.H. Freeman and Company.

DeCoursey TE, Chandy KG, Gupta S, Cahalan MD. 1985. Voltage-dependent ion channels in T-lymphocytes. *J Neuroimmunol* 10(1):71–95.

Ermakov YA, Averbakh AZ, Yusipovich AI, Sukharev S. 2001. Dipole potentials indicate restructuring of the membrane interface induced by gadolinium and beryllium ions. *Biophys J* 80(4):1851–1862.

Gabai VL, Meriin AB, Mosser DD, Caron AW, Rits S, Shifrin VI, Sherman MY. 1997. Hsp70 prevents activation of stress kinases. A novel pathway of cellular thermotolerance. *J Biol Chem* 272(29):18033–18037.

Ghamari-Langroudi M, Bourque CW. 2001. Ionic basis of the caesium-induced depolarisation in rat supraoptic nucleus neurones. *J Physiol* 536(Pt 3):797–808.

Gowrishankar TR, Weaver JC. 2006. Electrical behavior and pore accumulation in a multicellular model for conventional and supra-electroporation. *Biochem Biophys Res Commun* 349(2):643–653.

Hamill OP, McBride DW Jr. 1996. The pharmacology of mechanogated membrane ion channels. *Pharmacol Rev* 48(2):231–252.

Hille B. 2001. *Ionic Channels of Excitable Membranes.* Sunderland, MA: Sinauer Associates.

Hu Q, Joshi RP, Schoenbach KH. 2005. Simulations of nanopore formation and phosphatidylserine externalization in lipid membranes subjected to a high-intensity, ultrashort electric pulse. *Phys Rev E Stat Nonlin Soft Matter Phys* 72(3 Pt 1):031902.

Ibey BL, Xiao S, Schoenbach KH, Murphy MR, Pakhomov AG. 2009. Plasma membrane permeabilization by 60- and 600-ns electric pulses is determined by the absorbed dose. *Bioelectromagnetics* 30:92–99.

Idone V, Tam C, Andrews NW. 2008a. Two-way traffic on the road to plasma membrane repair. *Trends Cell Biol* 18(11):552–559.

Idone V, Tam C, Goss JW, Toomre D, Pypaert M, Andrews NW. 2008b. Repair of injured plasma membrane by rapid Ca2+-dependent endocytosis. *J Cell Biol* 180(5):905–914.

Iwamoto T, Shigekawa M. 1998. Differential inhibition of Na+/Ca2+ exchanger isoforms by divalent cations and isothiourea derivative. *Am J Physiol* 275(2 Pt 1):C423–C430.

Kosinska ID. 2006. How the asymmetry of internal potential influences the shape of *I–V* characteristic of nanochannels. *J Chem Phys* 124(24):244707.

Lipski J, Park TIH, Li D, Lee SCW, Trevarton AJ, Chung KKH, Freestone PS, Bai J-Z. 2006. Involvement of TRP-like channels in the acute ischemic response of hippocampal CA1 neurons in brain slices. *Brain Res* 1077(1):187–199.

Melikov KC, Frolov VA, Shcherbakov A, Samsonov AV, Chizmadzhev YA, Chernomordik LV. 2001. Voltage-induced nonconductive pre-pores and metastable single pores in unmodified planar lipid bilayer. *Biophys J* 80(4):1829–1836.

Neumann E, Sowers AE, Jordan CA (eds.) 1989. *Electroporation and Electrofusion in Cell Biology*. New York: Plenum Press.

Nuccitelli R, Chen X, Pakhomov AG, Baldwin WH, Sheikh S, Pomicter JL, Ren W et al. 2009. A new pulsed electric field therapy for melanoma disrupts the tumor's blood supply and causes complete remission without recurrence. *Int J Cancer* 125(2):438–445.

Numata T, Shimizu T, Okada Y. 2007. TRPM7 is a stretch- and swelling-activated cation channel involved in volume regulation in human epithelial cells. *Am J Physiol Cell Physiol* 292(1):C460–C467.

Okada Y. 2004. Ion channels and transporters involved in cell volume regulation and sensor mechanisms. *Cell Biochem Biophys* 41(2):233–258.

Okada Y, Shimizu T, Maeno E, Tanabe S, Wang X, Takahashi N. 2006. Volume-sensitive chloride channels involved in apoptotic volume decrease and cell death. *J Membr Biol* V209(1):21–29.

Pakhomov AG, Kolb JF, White JA, Joshi RP, Xiao S, Schoenbach KH. 2007a. Long-lasting plasma membrane permeabilization in mammalian cells by nanosecond pulsed electric field (nsPEF). *Bioelectromagnetics* 28:655–663.

Pakhomov AG, Phinney A, Ashmore J, Walker K, Kolb JF, Kono S, Schoenbach KS, Murphy MR. 2004. Characterization of the cytotoxic effect of high-intensity, 10-ns duration electrical pulses. *IEEE Trans Plasma Sci* 32(4):1579–1585.

Pakhomov AG, Shevin R, White JA, Kolb JF, Pakhomova ON, Joshi RP, Schoenbach KH. 2007b. Membrane permeabilization and cell damage by ultrashort electric field shocks. *Arch Biochem Biophys* 465(1):109–118.

Pakhomov AG, Bowman AM, Ibey BL, Andre FM, Pakhomova ON, Schoenbach KH. 2009. Lipid nanopores can form a stable, ion channel-like conduction pathway in cell membrane. *Biochem Biophys Res Commun* 385(2):181–186.

Palasz A, Czekaj P. 2000. Toxicological and cytophysiological aspects of lanthanides action. *Acta Biochim Pol* 47(4):1107–1114.

Pena F, Ordaz B. 2008. Non-selective cation channel blockers: Potential use in nervous system basic research and therapeutics. *Mini Rev Med Chem* 8(8):812–819.

Ramirez P, Gomez V, Cervera J, Schiedt B, Mafe S. 2007. Ion transport and selectivity in nanopores with spatially inhomogeneous fixed charge distributions. *J Chem Phys* 126(19):194703.

Ramsey IS, Delling M, Clapham DE. 2006. An introduction to TRP channels. *Annu Rev Physiol* 68(1):619–647.

Robinson RB, Siegelbaum SA. 2003. Hyperpolarization-activated cation currents: From molecules to physiological function. *Annu Rev Physiol* 65:453–480.

Saulis G. 1999. Kinetics of pore disappearance in a cell after electroporation. *Biomed Sci Instrum* 35:409–414.

Saulis G, Venslauskas MS, Naktinis J. 1991. Kinetics of pore resealing in cell membranes after electroporation. *Bioelectrochem Bioenerg* 26:1–13.

Schoenbach KH, Beebe SJ, Buescher ES. 2001. Intracellular effect of ultrashort electrical pulses. *Bioelectromagnetics* 22(6):440–448.

Schoenbach KH, Katsuki S, Stark RH, Buesher ES, Beebe SJ. 2002. Bioelectrics—New applications for pulsed power technology. *IEEE Trans Plasma Sci* 30(1):293–300.

Schoenbach KS, Hargrave B, Joshi RP, Kolb J, Osgood C, Nuccitelli R, Pakhomov AG et al. 2007. Bioelectric effects of nanosecond pulses. *IEEE Trans Dielectr Electr Insul* 14(5):1088–1109.

Siwy Z, Gu Y, Spohr HA, Baur D, Wolf-Reber A, Spohr R, Apel P, Korchev YE. 2002. Rectification and voltage gating of ion currents in a nanofabricated pore. *Europhys Lett* 60(3):349–355.

Smith KC, Gowrishankar TR, Esser AT, Stewart DA, Weaver JC. 2006. Spatially distributed, dynamic transmembrane voltages of organelle and cell membranes due to 10 ns pulses: Predictions of meshed and unmeshed transport network models. *IEEE Trans Plasma Sci* 34(4):1394–1404.

Smith KC, Weaver JC. 2008. Active mechanisms are needed to describe cell responses to submicrosecond, megavolt-per-meter pulses: Cell models for ultrashort pulses. *Biophys J* 95(4):1547–1563.

Stojilkovic SS, Zemkova H, Van Goor F. 2005. Biophysical basis of pituitary cell type-specific Ca2+ signaling-secretion coupling. *Trends Endocrinol Metab* 16(4):152–159.

Tanaka T, Li SJ, Kinoshita K, Yamazaki M. 2001. La(3+) stabilizes the hexagonal II (H(II)) phase in phosphatidylethanolamine membranes. *Biochim Biophys Acta* 1515(2):189–201.

Tanaka T, Tamba Y, Masum SM, Yamashita Y, Yamazaki M. 2002. La(3+) and Gd(3+) induce shape change of giant unilamellar vesicles of phosphatidylcholine. *Biochim Biophys Acta* 1564(1):173–182.

Teissie J, Golzio M, Rols MP. 2005. Mechanisms of cell membrane electropermeabilization: A minireview of our present (lack of ?) knowledge. *Biochim Biophys Acta* 1724(3):270–280.

Tekle E, Wolfe MD, Oubrahim H, Chock PB. 2008. Phagocytic clearance of electric field induced 'apoptosis-mimetic' cells. *Biochem Biophys Res Commun* 376:256–260.

Vance JE, Steenbergen R. 2005. Metabolism and functions of phosphatidylserine. *Prog Lipid Res* 44(4):207–234.

Varghese A, Tenbroek EM, Coles J Jr., Sigg DC. 2006. Endogenous channels in HEK cells and potential roles in HCN ionic current measurements. *Prog Biophys Mol Biol* 90(1–3):26–37.

Vernier PT, Sun Y, Chen MT, Gundersen MA, Craviso GL. 2008. Nanosecond electric pulse-induced calcium entry into chromaffin cells. *Bioelectrochemistry* 73(1):1–4.

Vernier PT, Sun Y, Gundersen MA. 2006a. Nanoelectropulse-driven membrane perturbation and small molecule permeabilization. *BMC Cell Biol* 7:37.

Vernier PT, Sun Y, Marcu L, Craft CM, Gundersen MA. 2004. Nanoelectropulse-induced phosphatidylserine translocation. *Biophys J* 86(6):4040–4048.

Vernier PT, Sun Y, Marcu L, Salem S, Craft CM, Gundersen MA. 2003. Calcium bursts induced by nanosecond electric pulses. *Biochem Biophys Res Commun* 310 286–295.

Vernier PT, Ziegler MJ, Sun Y, Gundersen MA, Tieleman DP. 2006b. Nanopore-facilitated, voltage-driven phosphatidylserine translocation in lipid bilayers—In cells and in silico. *Phys Biol* 3(4):233–247.

10

Model of Cell Membrane Electroporation and Transmembrane Molecular Transport

Damijan Miklavčič

Leila Towhidi*

10.1 Introduction

Electroporation is a technique in which, under the influence of electric field, the permeability of the cell membrane increases due to the formation of pores in the cell membrane, providing pathways for molecular transport (Abidor et al. 1979, Neumann et al. 1998, Tsong 1991, Weaver and Chizmadzhev 1989, Zimmermann 1982).

As electroporation is usually evaluated by indirect indications (Gehl and Mir 1999, He et al. 2008, Hibino et al. 1991, Kotnik et al. 2000, Neumann and Rosenheck 1972, Pavlin et al. 2007, Pliquett and Weaver 2007, Pucihar et al. 2007, Rols and Teissié 1998, Saulis 1997, Saulis et al. 2007, Schwister and Deuticke 1985), some experimental observations have not yet been fully explained. For instance, in a number of experiments using single-pulse electroporation protocol, increased conductivity and reduced transmembrane voltage were observed (Hibino et al. 1991, 1993), depending on the amplitude and duration of electric field (Kotnik et al. 2003, Rols and Teissié 1998). In multiple-pulse electroporation protocols, however, significant increase in cell molecular uptake is observed due to pulse fractionation, which, however, also depends on pulse repetition frequency. To explain these observations, knowing the exact mechanism of electroporation is crucial.

---

* The work was done during her stay at the University of Ljubljana.

In spite of a wide use of electroporation with high reproducibility and effectiveness, the exact mechanisms of pore formation and closure, and, more importantly, resealing to membrane transport mechanism are not completely understood. Different theoretical models have been developed in order to interpret experimental results in a quantitative manner. Modeling, in addition to making the optimization of the process easier and reducing the experimental work and related costs, can serve as a test for our understanding of the mechanisms and processes behind the observed phenomena.

In this chapter, we present a *dynamic* model based on the reaction scheme used for pore formation and closure during and after the electric field exposure, while self-consistency is taken into account. In addition, by analyzing previously published experimental data (Glogauer et al. 1993, Puc et al. 2003, Pucihar et al. 2007, Rols et al. 1995, Zimmermann et al. 1990), we suggest two specific phenomena for resealing of the cell membrane—relaxation and memory effect (Teissié et al. 2005)—corresponding to two distinct transport processes: (1) interactive diffusion and (2) endocytotic-like transport. We investigated the number of loaded molecules to cells using our model. Simultaneously, the results of induced transmembrane voltage (*ITV*), the distribution of pores on the membrane surface, membrane conductivity changes, resealing behavior, and molecular uptake are obtained temporally and spatially. Using our model, we can describe and explain the results of single- and multiple-pulse electroporation protocols also taking into account pulse-repetition frequency and the number of pulses that has not been possible previously.

10.2 Model Descriptions

10.2.1 Membrane Structure and Conductivity Changes

When a cell is exposed to an external electric field, the *ITV* starts to increase based on the Laplace equation, which leads to the structural changes of the cell membrane. Based on a previously suggested (Neumann et al. 1999, Schmeer et al. 2004, Tsong 1991), and recently confirmed (Böckmann et al. 2008), kinetic model, in the first step the intact closed lipids (*C*) transform to tilted lipid head groups (*C1*). In the second step, the prepores (*P1*) are formed and, finally, in the last step, the final pores (*P2*) are formed. The sequential reaction can be described by Equation 10.1:

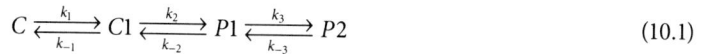

$$C \underset{k_{-1}}{\overset{k_1}{\rightleftarrows}} C1 \underset{k_{-2}}{\overset{k_2}{\rightleftarrows}} P1 \underset{k_{-3}}{\overset{k_3}{\rightleftarrows}} P2 \tag{10.1}$$

where *C1* is one intermediate closed or prepore state. The state *P1* denotes the pore structures of negligibly small permeability (transient or rapidly closing pore), while only *P2* is responsible for molecular uptake (stable or slowly closing pore). k_i and k_{-i} ($i = 1, 2, 3$) denote rate coefficients for pore formation and closure, respectively. For simplicity, the rate coefficients k_1, k_2, and k_3 are considered equal ($k_1 = k_2 = k_3 = k_p$) (Neumann et al. 1998). The rate laws for constituting steps in the scheme (Equation 10.1) are, respectively:

$$\frac{d\left[C(\vec{r},t)\right]}{dt} = -k_p\left[C(\vec{r},t)\right] + k_{-1}\left[C1(\vec{r},t)\right] \tag{10.2}$$

$$\frac{d\left[C1(\vec{r},t)\right]}{dt} = -k_p\left(\left[C1(\vec{r},t)\right] - \left[C(\vec{r},t)\right]\right) - k_{-1}\left[C1(\vec{r},t)\right] + k_{-2}\left[P1(\vec{r},t)\right] \tag{10.3}$$

$$\frac{d\left[P1(\vec{r},t)\right]}{dt} = -k_p\left(\left[P1(\vec{r},t)\right] - \left[C1(\vec{r},t)\right]\right) - k_{-2}\left[P1(\vec{r},t)\right] + k_{-3}\left[P2(\vec{r},t)\right] \tag{10.4}$$

$$\frac{d\left[P2(\vec{r},t)\right]}{dt} = k_p\left[P1(\vec{r},t)\right] - k_{-3}\left[P2(\vec{r},t)\right] \tag{10.5}$$

where

t denotes time

$\vec{r}$ is a vector representing the point on the membrane

$[C]$, $[C1]$, $[P1]$, and $[P2]$ show the normalized distribution of each membrane lipid state to the initial value of closed state $[C(\vec{r},0)]$

Actually, $[C(\vec{r},0)]=[C_0]$ is a specific value for cell system under observation and is independent of field duration and strength (Neumann et al. 1998, 1999). In addition, at pulse switch-on time, the initial condition is $[C1(\vec{r},0)]=[P1(\vec{r},0)]=[P2(\vec{r},0)]=0$. As demonstrated and discussed in the following sections, pore-formation-rate coefficient k_p depends on time and position on the membrane (i.e., it is field dependent), while the closure-rate coefficients (k_{-1}, k_{-2}, and k_{-3}) are constant (Gowrishankar et al. 1999, Neumann et al. 1998).

The rate coefficients of the chemical-kinetics model can be described through equilibrium constant (K) by $K = k_i/k_{-i}$ (Kakorin et al. 1996). In the previous studies, the dependence of equilibrium constant in the membrane field (E_m) is given by the van't Hoff relationship (Kakorin et al. 1996). Using $E_m = ITV/d_m$, where d_m is the thickness of the membrane, the equilibrium constant can be obtained, using

$$K = K_0 \exp\left(\frac{\Delta V_p \varepsilon_0 (\varepsilon_W - \varepsilon_L)}{2k_B T d_m^2} ITV^2 \right) \qquad (10.6)$$

where

K_0 is the value of K at $E = 0$

ΔV_p is the mean volume change due to pore formation

ε_0 is the permittivity of the vacuum

ε_W and ε_L are the dielectric constants of water and lipids, respectively

k_B is Boltzmann constant

T is temperature

Whenever electroporation occurs, an increase in conductivity during the pulse is observed (Chernomordik et al. 1987, Glaser et al. 1988), which can be explained by the formation of pores in the cell membrane. Based on the trapezium barrier model for the image forces (Glaser et al. 1988, Kakorin and Neumann 2002), the intrinsic pore conductivities $\sigma_{p,i}$ (i = 1 and 2 represents $P1$ and $P2$ pores, respectively) are expressed as follows:

$$\sigma_{p,i} = \sigma_{p,i}^0 \exp\left(\alpha_{p,i} n |ITV| \frac{F}{RT} \right) \qquad (10.7)$$

where

$$\sigma_{p,i}^0 = \frac{\sigma_{ex} + \sigma_{in}}{2} \exp\left(\frac{-\varphi_{im,i}^0 F}{RT} \right) \quad \text{and} \quad \alpha_{p,i} = 1 - \frac{RT}{F\varphi_{im,i}^0} \qquad (10.8)$$

where

σ_{ex} and σ_{in} are the extracellular and intracellular conductivities, respectively

n is the geometrical parameter of the trapezium model for energy barrier (Kakorin and Neumann 2002)

F is Faraday constant

$\varphi_{im,i}^0$ is the intrinsic pore barrier potential

TABLE 10.1   Values of Parameters Used in the Simulations

Parameter	Symbol	Value	References
Membrane thickness	d_m	5e−9 m	Neumann et al. (1998)
Extracellular conductivity	σ_{ex}	0.14 S/m[a]	Neumann et al. (1998) and Pucihar et al. (2006)
Intracellular conductivity	σ_{in}	0.3 S/m[b]	Kotnik et al. (1998) and Neumann et al. (1998)
Initial conductivity of membrane	σ_{m0}	5e−7 S/m	Plonsey and Barr (1988) and Pucihar et al. (2008)
Extracellular permittivity	ε_o	7.1e−10 As/Vm	Neumann et al. (1998)
Intracellular permittivity	ε_i	7.1e−10 As/Vm	Neumann et al. (1998)
Membrane permittivity	ε_m	4.4e−11 As/Vm[c]	Neumann et al. (1998)
Water relative dielectric constant	ε_w	80 As/Vm	Neumann et al. (1998)
Lipid relative dielectric constant	ε_l	2 As/Vm	Neumann et al. (1998)
Free diffusion coefficient	D_0	5e−10 m²/s	Neumann et al. (1998)
Zero-field equilibrium constant	K_0	2e−2	Neumann et al. (1998)
Mean average aqueous pore volume	ΔV_p	9e−27 m³	Neumann et al. (1998)
Intrinsic barrier potential of P1 state	φ_{im1}^0	0.13 V	Kakorin and Neumann (2002)
Intrinsic barrier potential of P2 state	φ_{im2}^0	0.084 V	Kakorin and Neumann (2002)
A geometrical parameter	N	0.12	Schmeer et al. (2004)
Decay-rate coefficient for C1	k_{-1}	10^5 s⁻¹	Gowrishankar et al. (1999), Hibino et al. (1993), and Pucihar et al. (2002)
Decay-rate coefficient for P1 pores	k_{-2}	2000 s⁻¹	Bier (2002), Chernomordik et al. (1987), Gowrishankar et al. (1999), and Hibino et al. (1993)
Decay-rate coefficient for P2 pores	k_{-3}	2 s⁻¹	Chernomordik et al. (1987), Ghosh et al. (1993), and Gowrishankar et al. (1999), Neu et al. (1999)
Decay-rate coefficients for endocytotic-like process	k_f and k_s	0.044, 0.003 s⁻¹	Neumann et al. (1998)

[a] This is for spinner minimum essential medium (SMEM). The range of extracellular medium is quite large.
[b] Reported between 0.2 and 0.55 S/m.
[c] Reported between 4.4 and 5×10^{-11} As/Vm.

The values of constants are taken from related papers and given in Table 10.1.

Therefore, considering the normalized distribution of *P*1 and *P*2 pores, the initial (i.e., of nonelectroporated membrane) conductivity of membrane (σ_{m0}) and conductivity related to each kind of pores, the conductivity of membrane σ_m can be obtained as

$$\sigma_m(\vec{r},t) = \sigma_{m0} + \left[P1(\vec{r},t)\right] \times \sigma_{p,1} + \left[P2(\vec{r},t)\right] \times \sigma_{p,2} \tag{10.9}$$

10.2.2 Molecular Uptake

The experimentally obtained characteristic recovery times of the cell membrane fall into a wide range (Saulis et al. 1991). One reason for this wide range can be ascribed to different experimental methods that were used to obtain these data. The recovery time reported, based on measuring voltage and current (Bier 2002, Chernomordik et al. 1993, Ghosh et al. 1993, He et al. 2008), patch clamp (Ryttsén et al. 2000), ultrarapid video microscopy (Gabriel and Teissié 1999, Sowers 1986), and pulsed-laser fluorescence microscopy (Hibino et al. 1993, Kinosita and Ikegami 1988), is in the range of milliseconds. In studies based on finding the percentage of cells still incorporating various marker molecules at different times after electroporation (Glaser et al. 1986, Khine et al. 2007, Saulis 1997, Shirakashi et al. 2004, Teissié and Ramos 1998), the resealing time is reported to be about 20 min. Moreover, other experimental results were reported in which authors obtained the uptake of particular marker molecules under conditions when marker molecules were either present or absent at the time of cell exposure to electric pulses. It was demonstrated that in the case of the probe presence at the pulse time, probe entrance to cells was uniform, whereas it was localized in large vesicles when the probe was added after the exposure (Glogauer et al. 1993, Rols et al. 1995, Zimmermann et al. 1990). In addition, the internalized components of the membrane after electroporation were observed (Glogauer et al. 1993). The authors suggested that in addition to passive diffusion through pores, an endocytosis-like (or macropinocytosis-like) transport also partially contributes to the loading of the molecule after the pulse at the electroporated area of membrane. This mechanism was suggested to occur due to the initiation of the membrane ruffling after the electroporation (Escoffre et al. 2007, Jones 2007, Lambert et al. 1990). The very short characteristic time of resealing for artificial lipid membranes (in the range of microseconds and millisecond), compared to biological cells (in the range of hours) (Glaser et al. 1986, Navarrete and Sacchi 2006), supports the involvement of the suggested endocytosis-like transport.

Based on the above considerations, we define two transport mechanisms: the first one is transport through the created pores on the membrane with relatively fast relaxation due to pore closure, and the second one is transport due to enhanced membrane perturbation and ruffling with quite slow resealing (Glogauer et al. 1993). The attributed mechanisms for uptake in each mentioned case are suggested to be interactive diffusion (Neumann et al. 1998) and endocytotic-like transport, respectively, which will be explained further in following sections.

10.2.2.1 Pores and Interactive Diffusion

It is believed that there are three mechanisms involved in molecular transport through the pores: diffusion, electrophoresis, and electroosmosis. Because of a short duration of electric pulses, diffusion is considered to be the predominant transport mechanism for small molecule cell uptake (Prausnitz et al. 1995, Puc et al. 2003).

From Equation 10.9 it can be seen that membrane conductivity changes depend on both P1 and P2 pores; therefore, the closure of pores is responsible for decreasing the conductivity of the membrane after the pulse. Due to the transient contacts of the molecule with the lipids of the pore edges, the transport of molecules through the membrane in this case is not free diffusion but interactive diffusion, which is reflected in diffusion coefficient with slightly lower value than free diffusion coefficient (Neumann et al. 1998). We should also consider that membrane conductivity is decreased in part also by this interactive diffusion as pores are occluded by molecules being transported through the pores.

10.2.2.2 Perturbed Area and Endocytotic-Like Transport

Although a strong decrease in the flow of molecules was seen shortly after the pulse (Gabriel and Teissié 1999), the observed increased membrane permeability is a long-lasting phenomenon (Glaser et al. 1986, Neumann et al. 1998, Saulis 1997, 2005, Teissié and Ramos 1998). In our study, this long-lasting phenomenon corresponds to the perturbed area of the membrane and is assigned to endocytosis-like transport, which has already been suggested previously (Glogauer et al. 1993, Rols et al. 1995, Zimmermann et al. 1990). Memory effect resealing in some studies was observed to fit simple exponentially decaying function (Rols and Teissié 1990) or a more complex exponential behavior (Glaser et al. 1986, Khine et al. 2007, Neumann et al. 1998, Pucihar et al. 2008). In our model, we considered a dual exponential decay function for returning the cell membrane to its normal state:

$$[M] = [P2]_e \left(B \exp(-k_f t) + (1 - B) \exp(-k_s t) \right) \tag{10.10}$$

where
   $[M]$ shows the normalized distribution of perturbed area due to electroporation
   $[P2]_e$ is the normalized distribution of pores at the end of pulse
   k_f and k_s are decay-rate coefficients for endocytotic-like transport
   B is a constant

k_f, k_s, and B are obtainable from the experimental results and depend on the type and size of transported molecules (Neumann et al. 1998).

10.2.2.3 Transmembrane Molecular Transport

In our model, two different transport coefficients are assigned to two transport mechanisms (interactive diffusion and endocytotic-like transport). With respect to these, the permeability of the membrane can be written as the sum of two distinct contributions:

$$P_m(\vec{r},t) = \frac{[P2(\vec{r},t)]D_p}{d_m} + \frac{[M(\vec{r},t)]D_r}{d_m}$$

(10.11)

where D_p and D_r are the diffusion coefficients for interactive transport and transport coefficient related to endocytosis-like transport, respectively.

It should be noted that D_p is about 0.1–0.3 of the free diffusion coefficient D_0 (depending on the type of molecule passing through the membrane) (Neumann et al. 1999, Pavlin et al. 2005), whereas D_r has been assigned a very low value (about $D_0/10{,}000$) (Neumann et al. 1999).

While the membrane is being permeabilized due to electric field, the molecules pass through the membrane based on concentration gradient. A quantitative description of diffusion is contained in Fick's first law (Plonsey and Barr 1988). The flux through the cell membrane (j) can be approximated by $j = P_m(c^{out} - c^{in})$, where c^{out} and c^{in} are the outside and inside concentrations adjacent to the membrane, and P_m is membrane permeability coefficient as described in Equation 10.11. In order to obtain the total moles transported to the cell (N_{mol}), an integration of j over the surface and time is performed as

$$N_{mol} = \int_{t=0}^{\tau}\int_S j\, dS\, dt$$

(10.12)

where
  S is the surface of the cell membrane
  τ is the time at which the quantity of transported molecules is to be determined.

The number of molecules transported through the membrane (N) can be then obtained by $N = N_A N_{mol}$.

10.3 Construction of the Model

The simulations in this study were performed using COMSOL 3.3 package (COMSOL Inc., Burlington, Massachusetts) based on finite element method. In order to construct a favorable geometrical model, a spherical cell with a radius of 5.6 μm was located inside a cylinder. The two circular electrodes were positioned at the bases of the cylinder, which are shown shaded in Figure 10.1A. The cylinder was chosen to take advantage of symmetry for simplifying the 3D geometrical model to a 2D-axial symmetry, and therefore saving time and memory during simulations (Figure 10.1B). With the purpose of complying with reality, the applied voltage to the electrodes was considered as a smoothed step function with the rising and falling times of 2 μs.

Incorporating extremely thin membrane compared to the cell size is problematic in meshing and solving the problem. Therefore, we assigned boundary condition to the membrane as previously described (Pucihar et al. 2006). In our calculations, the resting voltage was considered to be negligible with respect to the *ITV*.

The interactive diffusion coefficient was taken to be $D_p = D_0/5$, and the attributed transport coefficient to induced endocytosis-like transport process was set to $D_r = D_0/10{,}000$ (Neumann et al.

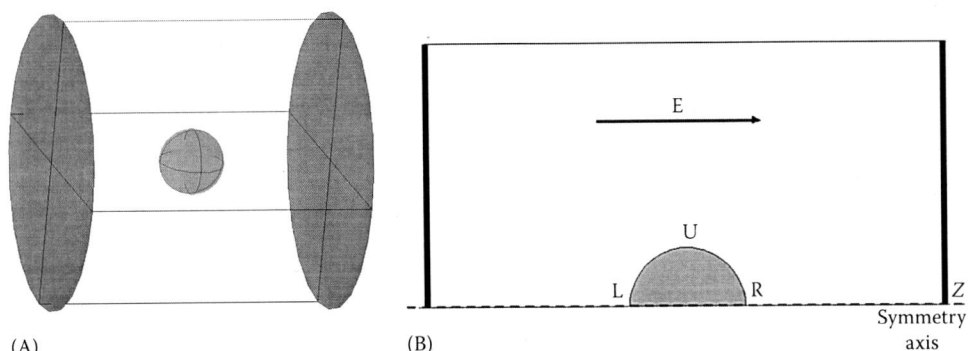

FIGURE 10.1 Simulation geometrical model. Schematic of a spherical cell placed between electrodes. The cell and electrodes are shaded. (A) 3D geometry. (B) Simplified 2D axial symmetry of the model with dashed line as the symmetry axis. The right pole, left pole, and the upper equator are indicated by R, L, and U, respectively. The arrow denotes the direction of the electric field.

1999) for a typical molecule (e.g., Serva Blue G). It is obvious that this value depends on the size and type of transported molecules through the transporting pathways of the membrane. The initial concentrations of intracellular and extracellular probe were set to be 0 and 10 mM, respectively, and the temperature was considered to be 20°C. As the changes near the membrane are more pronounced and important, we increased the number of meshing elements for finite element solver at these regions. Our simulation was designed to solve the Laplace equation in the region between two electrodes containing the cell to obtain the *ITV*, considering all the adjoined equations in this model. This set of equations should be solved for *ITV* and membrane conductivity, considering self-consistency of the parameters. In other words, as the unknowns are tightly involved, these equations should be solved simultaneously. The total number of internalized molecules to the cell in our model was obtained by integrating all the transported molecules from the entire cell membrane area for about 1000 s (i.e., approximately 20 min) after the pulse, after which the transport of molecules across the cell membrane becomes negligible.

The necessary parameters and the reference sources used for the simulation are listed in Table 10.1. All the simulations were performed on a PC equipped with a 2.8 GHz Pentium IV processor and 3 GB RAM. Each simulation run lasts between about 1 and 10 min, depending on the number of pulses and the elapsed time after the exposure of the cell to the electric field. Following the simulations, all the spatial and temporal quantities related to membrane conductivity, *ITV*, concentration, and molecular uptake are available.

10.4 Results

10.4.1 Structural Changes in the Membrane

After the start of the pulse, structural alterations in the form of pore formation start to appear without any threshold being imposed explicitly (Figure 10.2). The results for distribution of structural changes related to P1 and P2 pores on the cell membrane surface at the end of the pulse with three different field strengths and 100 μs duration are shown in Figure 10.2A and B. It is evident that in the cap regions (L and R in Figure 10.1), in which *ITV* is higher, pore formation is most pronounced. It can also be observed that the pore formation at 600 V/cm is negligible compared to the higher values of 800 and 1000 V/cm.

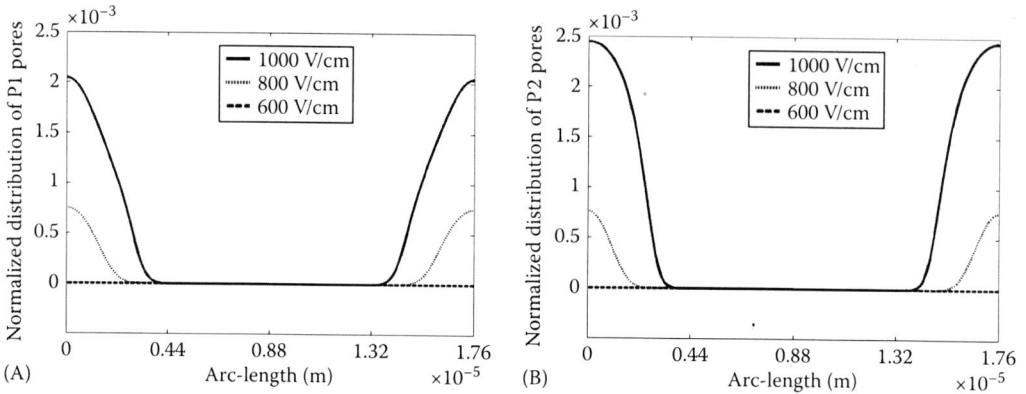

FIGURE 10.2 Normalized distribution of pores on the membrane. Normalized distribution of (A) P1 pores and (B) P2 pores on the cell membrane at the end of pulses of 600, 800, and 1000 V/cm with duration of 100 μs. The right pole, left pole, and the upper equator are indicated by R, L, and U, respectively.

10.4.2 Membrane Conductivity

Figure 10.3A displays the temporal membrane conductivity averaged over the cell membrane during 100 μs pulses considered in Figure 10.2 (considered voltages were 600, 800, and 1000 V/cm). It can be observed that the conductivity increase due to applied 600 V/cm is very small. Membrane conductivity related to 1000 V/cm first increases sharply, then a much slower increase is observed. At the end of pulse, the membrane conductivity is increased by a factor of 1000. The overall conductivity starts to disappear after the pulse termination (Figure 10.3B).

The effect of pulse duration on membrane conductivity increase for a 1000 V/cm pulse can be observed in Figure 10.3C. This figure reveals that the longer the pulse, the more efficient electroporation.

10.4.3 Induced Transmembrane Voltage

After the smoothed-step pulse is switched on, the membrane as a capacitor starts charging and *ITV* starts to increase that in turn results in membrane conductivity increase (Figure 10.3A). In some cases, even a decrease of the *ITV* value may be observed as a dip at poles (Figure 10.4C). Figure 10.4A shows the *ITV* evolution at one of the poles of the cell. It can be observed that for the field amplitudes more than 800 V/cm, the membrane capacitor charging shows a nonlinear behavior. Figure 10.4B demonstrates spatial *ITV* over the cell membrane at the end of 100 μs pulses of different pulse strength. The stronger the pulse, the steeper is the conductivity changes (Figure 10.3A), so that *ITV* does not exceed a certain value (Figure 10.4B) except at initial peak overshoot. Figure 10.4C displays the *ITV* distribution on the cell membrane for a 1000 V/cm and 100 μs pulse at times 1, 2, and 100 μs. It can be observed that at the beginning, a dip for *ITV* is created at the poles, which however flattens out with time.

10.4.4 Membrane Permeability

The permeability of the membrane during the pulse in our model is attributed only to P2 pores that are formed during and close after the pulse (Figure 10.5A). After the pulse, the permeability of membrane starts to decrease (Equation 10.11) due to the P2 pores closure and disappearance of endocytosis-like transport (Neumann et al. 1998). Figure 10.5B shows membrane permeability after the pulses that depends strongly on the field strength and pulse duration.

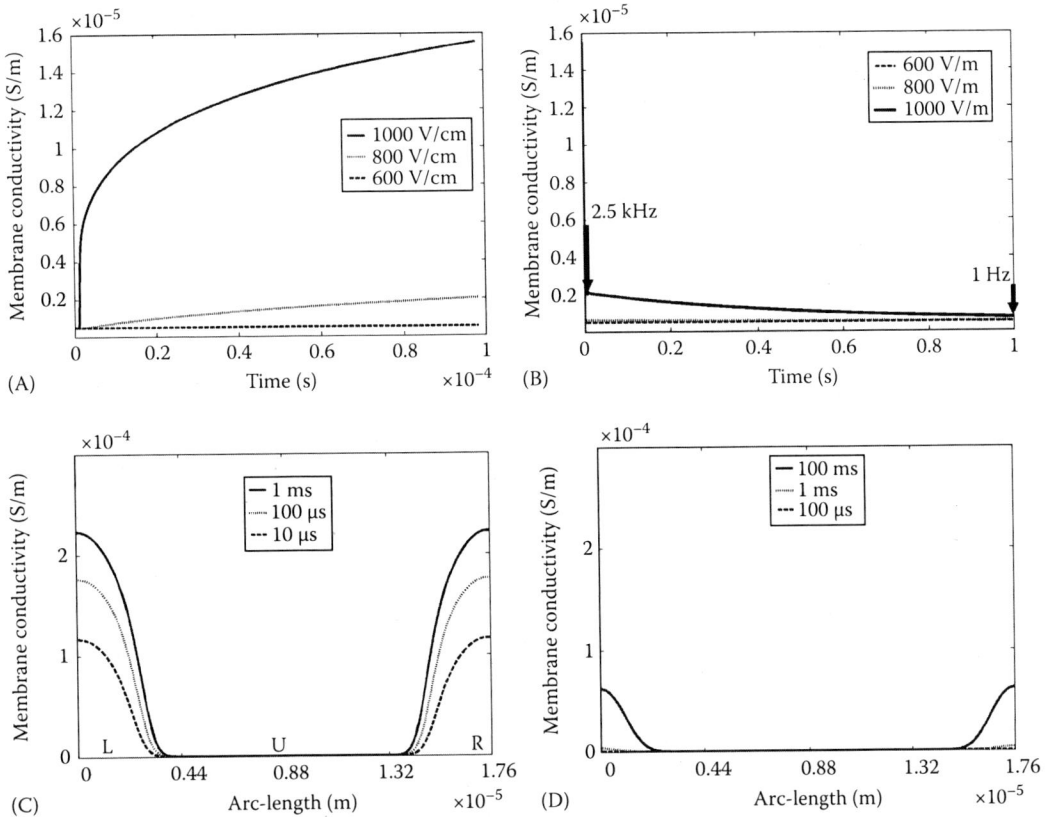

FIGURE 10.3 Temporal and spatial behavior of membrane conductivity. Temporal evolution of the overall membrane conductivity (A) during and (B) after a single 100 μs pulse of different amplitudes 600, 800, and 1000 V/cm. The arrows show the times at which the second pulse of the train pulses with frequencies 2.5 kHz and 1 Hz is to be applied. Spatial distribution of membrane conductivity at the end of (C) a 1000 V/cm pulse of durations 10 μs, 100 μs, and 1 ms and (D) a 700 V/cm pulse of 100 μs, 1 ms, and 100 ms. The right pole, left pole, and the upper equator are indicated by R, L, and U, respectively.

10.4.5 Number of Internalized Molecules to the Cell

The cellular uptake of marker molecule or chemotherapeutic agents is a determinant factor in the efficiency of electroporation-based applications. The results on the number of molecules internalized by the cells with respect to the pulse amplitude and duration, number of pulses, and pulse-repetition frequency are presented.

10.4.5.1 Effect of Pulse Strength and Duration

As Figure 10.6A displays, the amplitude of pulses significantly affects the internalization of molecules, which almost stops at about 16 min after the pulse. It can be seen from the figure that 600 V/cm pulse has no observable effect on the molecular uptake. The effect of pulse duration on the uptake however is not as strong as the influence of pulse amplitude. Figure 10.6B shows the differences in uptake for 1 ms and 100 μs pulses as a function of pulse strength.

10.4.5.2 Effect of Pulse Fractionation (Multiple-Pulse Protocols)

In Figure 10.7A, the uptake of molecules after the exposure of cells to a single 1 ms pulse of 800 V/cm is displayed by a dashed-line curve and the uptake after the fractionation of the pulses to 10 pulses with a duration of 100 μs and frequency of 1 Hz are shown in solid line for each subsequent pulse.

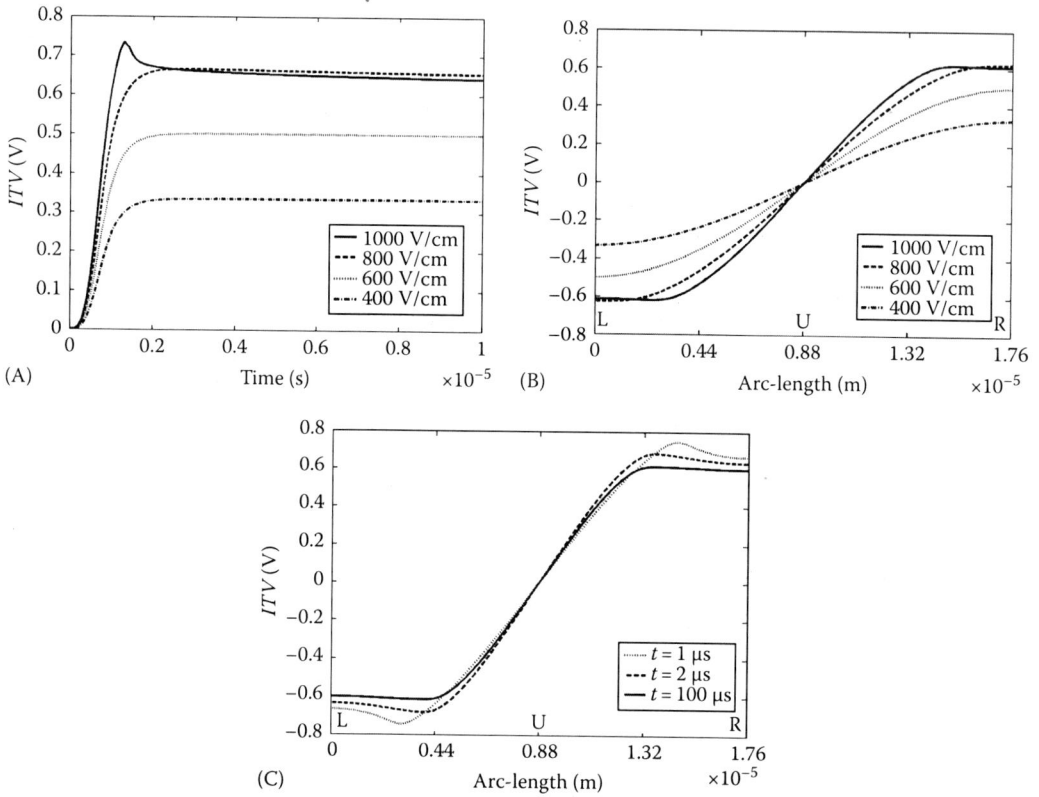

FIGURE 10.4 Temporal and spatial behavior of *ITV*. (A) Time evolution of *ITV* at the left pole (L) for a 100 μs pulse of amplitudes 400, 600, 800, and 1000 V/cm. (B) Spatial distribution of *ITV* on the cell membrane at the end of 100 μs pulses with the same pulse amplitudes of part (A). (C) Spatial distribution of *ITV* for a 1000 V/cm pulse demonstrated at 1, 2, and 100 μs after the pulse starts. The right pole, left pole, and the upper equator are indicated by R, L, and U, respectively.

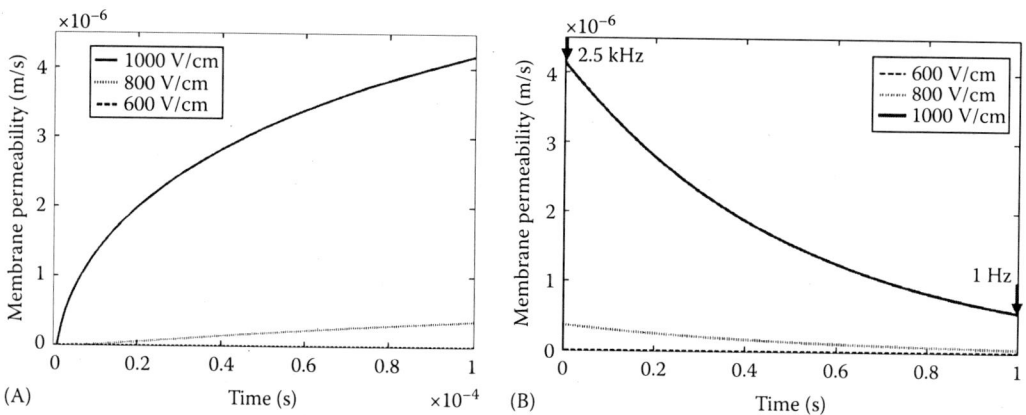

FIGURE 10.5 Temporal evolution of membrane permeability. Temporal evolution of the overall membrane permeability (A) during and (B) after a single 100 μs pulse of different amplitudes 600, 800, and 1000 V/cm. The arrows show the times at which the second pulse of the train pulses with frequencies 2.5 kHz and 1 Hz is to be applied.

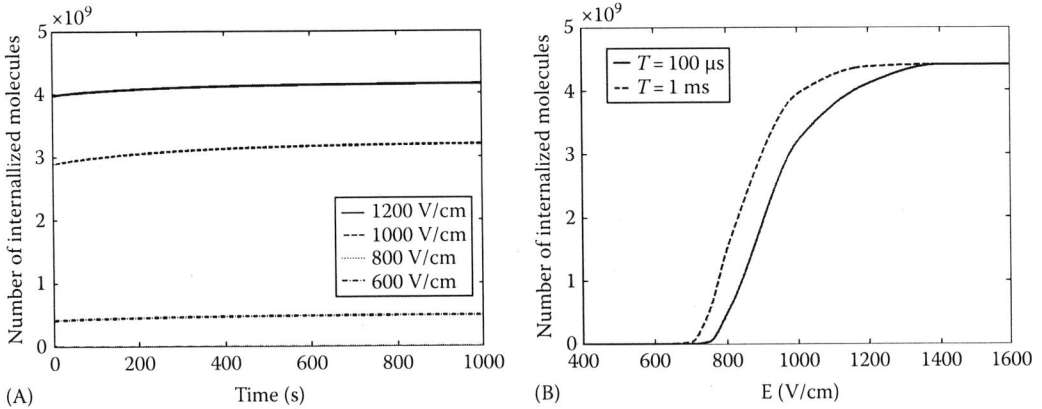

FIGURE 10.6 Dependence of uptake to pulse amplitude and duration. (A) Time evolution of molecule uptake in 16 min after pulse turn-off. Four different pulse amplitudes of 600, 800, 1000, and 1200 V/cm with the same duration of 100 μs are considered. (B) Electric-field amplitude dependence of molecule uptake, 16 min after pulse cessation for two different pulse durations of 100 μs and 1 ms.

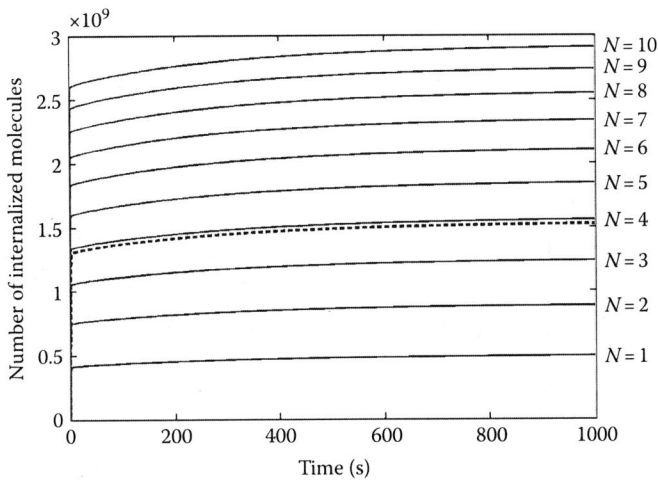

FIGURE 10.7 Dependence of uptake to pulse fractionation. Effect of pulse fractionation. The dashed line demonstrates the uptake due to a single pulse with 800 V/cm amplitude and 1 ms duration. The solid lines indicate the uptake after fractionation of pulsing to 10 pulses of the same amplitude with duration of 100 μs and frequency of 1 Hz. N displays the number of pulses.

This difference in the internalization of molecules after the exposure of cells to single and multiple pulses is attributed to the sharp increase of permeability at the beginning of each pulse followed by a smooth increase (Figure 10.5A).

10.4.5.3 Effect of Pulse Frequency and Number of Pulses

There are two distinct factors that need to be taken into account when considering the effect of pulse repetition frequency on molecular uptake: (1) membrane conductivity and (2) the level of membrane permeability, at the start of each consecutive pulse in the pulse train.

In our simulations, we considered 1 Hz and 2.5 kHz pulse-repetition frequency for which the time interval between subsequent pulses is 1 s and 400 μs, respectively. The arrows in Figures 10.3B and 10.5B show the time at which the second pulse of considered pulse repetition frequency is applied. It is evident

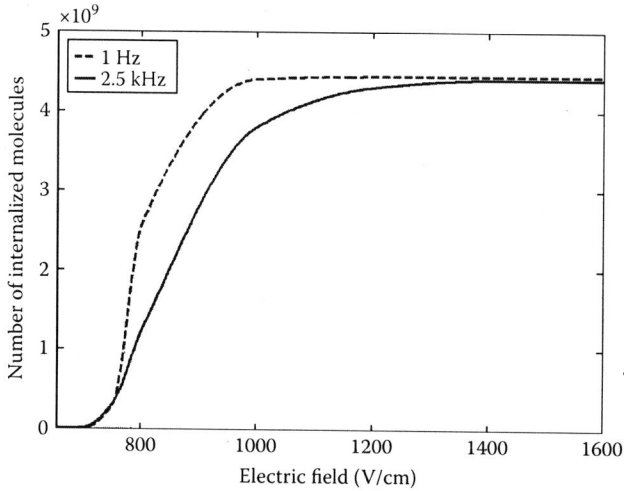

FIGURE 10.8 Dependence of uptake to pulse repetition frequency. Comparison of effectiveness for two trains of pulses consisting of eight pulses with $100\,\mu s$ duration and 1 Hz and 2.5 kHz frequencies for different pulse strengths.

from Figure 10.3B that when a sequence of pulses is applied, residual membrane conductivity from the preceding pulses results in the lower *ITV* of the cell membrane and subsequently less uptake increase. This leads us to conclude that pulses with higher pulse-repetition frequency are less efficient. That is, for certain electric field strength, the number of internalized molecules for lower pulse-repetition frequency is higher than the related number of internalized molecules for higher frequency. Besides that, the cell membrane appears to be at different states of permeabilization at the beginning of each consecutive pulse based on Figure 10.5B. It can be stipulated that the higher the pulse-repetition frequency, the larger the permeability of the membrane when the next pulse is to be applied. This factor is not as effective as the first one. Figure 10.8 shows the molecular uptake for two pulse-repetition frequencies considered 1 Hz and 2.5 kHz for different pulse amplitudes, which demonstrates that with increasing pulse-repetition frequency, the uptake of subsequent pulses decreases.

10.5 Discussion

The focus of this modeling was to study the effect of electric pulse, duration of exposure, number of pulses, and pulse-repetition frequency on molecular transport using a self-consistent model based on previously proposed chemical-kinetics electroporation scheme. At the same time, the described model enables the determination and prediction of *ITV*, membrane conductivity, and permeability temporarily and spatially for single and multiple electroporation pulse protocols.

Among different theoretical models for electroporation, some are well established such as resistive-capacitive model, electromagnetic bulk model, and energy models. In the resistive–capacitive (RC) transport lattices (Gowrishankar and Weaver 2003, Stewart et al. 2004), determining the spatial distribution of *ITV* over the cell membrane and the dynamical behavior of the membrane after the breakdown, however, is not possible. In electromagnetic bulk model, the Laplace equation is used to determine the *ITV* (Kotnik et al. 1997). Although with this model spatial distribution is available, it only shows which regions undergo electroporation, and does not consider the variations of *ITV* after electroporation occurs. One of the frequently used models for electroporation is based on Smoluchowski equation together with pore creation energy definition (DeBruin and Krassowska 1999, Neu and Krassowska 1999) in which pores growth and shrinkage (Joshi et al. 2002, 2004, Smith et al. 2004) and self-consistency (Krassowska and Filev 2007) can be taken into account, but it is not applicable to different cell shapes and pulse waveforms. Finally, there is also a model in which conductivity increase due to electroporation is defined by relations based on experimental findings (Hibino et al. 1993, Pavlin et al. 2005). Other models that allow the prediction of molecular uptake by the cells based on electroporation were also published. A study has been performed on a model for molecules transported after the pulse ends (Pliquett and Weaver 2007). Moreover, a model of diffusion-driven transmembrane transport has been constructed using a pharmacokinetic model (Puc et al. 2003). None of the above-mentioned models however deals with the dynamic behavior and conductivity changes of the cell membrane during and after the pulses and molecular transport through the membrane at the same time.

In a chemical-kinetics model (Böckmann et al. 2008, Neumann et al. 1998), a number of closed and porous states were introduced in order to match the experimental results, and the equations were obtained based on electrochemical considerations so that the pore formation and resealing of the membrane and also molecule uptake can be predicted considering the angular and time average of the parameters (Neumann et al. 1996, 1998, 1999). These models are only suitable for spherical cells and the distribution of pores in the electroporated regions from which the whole transport took place was assumed uniform. Besides, the dynamic behavior of other parameters such as *ITV* and membrane conductivity and permeability has not been considered.

In electroporation experiments, a wide range of resealing times has been reported. It has been observed practically that the electroporated cells continue to take up molecules for minutes and even hours after the pulse, while the membrane conductivity measurements have demonstrated very short time constants (about seconds). Regarding this observation, together with the experimental observations of vesicles during uptake at different times after the pulse (Glogauer et al. 1993, Rols et al. 1995, Zimmermann et al. 1990), we made our model based on the suggestion that all the molecular uptake does not occur through the open pores P2. Rather, we suggest two distinct mechanisms to be involved in transmembrane molecular transport: interactive diffusion through the pores and endocytosis-like transport trough the perturbed area characterized by increased membrane permeability (Equation 10.11).

Our results show that the structural changes of the membrane start at the beginning of the pulse and form continuously. The pore formations are considerable only if the electric field is high enough, which is in agreement with experimental observations (Kotnik et al. 2000, Teissié and Ramos 1998). Therefore, membrane conductivity and permeability change from the very beginning of the pulse, but only become detectable at large-enough electric-field amplitudes (Figure 10.3A). This is in accordance with apparent threshold reported in different experimental studies and should thus be considered as a detectability threshold (Gabriel and Teissié 1999, Kotnik et al. 2000, Rols and Teissié 1990). Even for smaller electric-field amplitudes, increasing the pulse duration may change the electroporation state from undetectable to detectable (Figure 10.3D). This has been observed in experiments but quite commonly explained as a decrease of the threshold transmembrane voltage by increasing the duration of pulses (Canatella et al. 2001, Gowrishankar et al. 1999, Pucihar et al. 2006, 2008). In addition, the obtained temporal and spatial variation of *ITV* (Figure 10.4) is in accordance with experimental observations (Hibino et al. 1993).

According to our model, after the pulse, membrane conductivity decreases based on pore's closure, while increased membrane permeability still allows marker molecules or drugs to be transported through the perturbed area of the membrane. Based on our results (and in agreement with experimental observations (Kotnik et al. 2003, Miklavčič et al. 2000, Puc et al. 2003, Pucihar et al. 2007, He et al. 2007)), the number of internalized molecules to the cell increases with pulse amplitude and duration (Figure 10.6). Another factor that appeared to be important in molecule uptake is the concentration gradient of the molecule at the time of pulse application (data not shown). One important result of our study, which is in line with the experiments and was confirmed by experiments (Hibino et al. 1991), is the description of pulse fractionation effectiveness in molecular uptake for multi-pulse protocols (Figure 10.7). At the same time, our results show that molecular uptake depends on pulse-repetition frequency. Besides, it was shown (Figure 10.6B) that the number of pulses can significantly affect the pulse uptake but only before reaching the saturation, which depends on field strength (Figure 10.6A) and pulse-repetition frequency (Figure 10.8). This has also been observed in previously published experimental studies (Rols and Teissié 1990). However, in experimental results, at large electric field strengths, the molecule uptake decreases due to irreversible electroporation (Kotnik et al. 2003, 2007). As in our model, irreversible permeabilization was not taken into account, and this decrease cannot be observed in the graphs.

Therefore, using the presented self-consistent model based on chemical-kinetics scheme, we can explain and predict the effect of all pulse parameters, i.e., amplitude, duration, pulse-repetition frequency, and pulse shape. Namely, in our model we can apply realistic pulse shapes and waveforms and also realistic cell geometry.

To have a precise description of molecular transport using this model for different molecules, we need, however, to determine molecular-specific parameters from experiments. Namely, for each different molecule having different size and charge and characteristics, we would need to determine corresponding diffusion coefficients. In our model, we considered uncharged molecules for transport through the membrane. For the macromolecules and highly charged molecules, the electrophoresis however also applies and becomes important in the observed molecular uptake. Thus, the mechanism for the uptake during the pulse is electro-diffusion. Also, in this study, resting transmembrane voltage was considered negligible with respect to *ITV*. These can both be easily introduced in our model and can thus be considered in our future development of this model.

Acknowledgments

This work was supported by the Slovenian Research Agency. The authors wish to thank Professor Eberhard Neumann for numerous stimulating and enlightening discussions.

References

Abidor IG, Arakelyan VB, Chernomordik LV, Chizmadzhev YA, Pastushenko VF, Tarasevich MR. 1979. Electrical breakdown of BLM: Main experimental facts and their qualitative discussion. *Bioelectrochem Bioenerg* 6:37–52.

Bier M. 2002. Resealing dynamics of a cell membrane after electroporation. *Phys Rev E* 66:062905.

Böckmann RA, Groot BL, Kakorin S, Neumann E, Grubmüller H. 2008. Kinetics, statistics, and energetics of lipid membrane electroporation studied by molecular dynamics simulations. *Biophys J* 95(4):1837–1850.

Canatella PJ, Karr JF, Petros JA, Prausnitz MR. 2001. Quantitative study of electroporation-mediated molecular uptake and cell viability. *Biophys J* 80(2):755–764.

Chernomordik LV, Sukharev SI, Popov SV, Pastushenko VF, Sokirko AV, Abidor IG, Chizmadzhev YA. 1987. The electrical breakdown of cell and lipid membranes: The similarity of phenomenologies. *Biochim Biophys Acta* 902(3):360–373.

Chernomordik LV, Vogel SS, Sokoloff A, Onaran HO, Leikina EA, Zimmerberg J. 1993. Lysolipids reversibly inhibit Ca²⁺-, GTP- and pH-dependent fusion of biological membranes. *FEBS Lett* 318:71–76.

DeBruin KA, Krassowska W. 1999. Modeling electroporation in a single cell. II. Effects of ionic concentrations. *Biophys J* 77(3):1225–1233.

Escoffre JM, Dean DS, Hubert M, Rols MP, Favard C. 2007. Membrane perturbation by an external electric field: A mechanism to permit molecular uptake. *Eur Biophys J* 36(8):973–983.

Gabriel B, Teissié J. 1999. Time courses of mammalian cell electropermeabilization observed by millisecond imaging of membrane property changes during the pulse. *Biophys J* 76:2158–2165.

Gehl J, Mir LM. 1999. Determination of optimal parameters for in vivo gene transfer by electroporation, using a rapid in vivo test for cell permeabilization. *Biochem Biophys Res Commun* 261(2):377–380.

Ghosh PM, Keese CR, Giaever I. 1993. Monitoring electropermeabilization in the plasma membrane of adherent mammalian cells. *Biophys J* 64:1602–1609.

Glaser RW, Wagner A, Donath E. 1986. Volume and ionic composition changes in erythrocytes after electric breakdown—Simulation and experiment. *Bioelectrochem Bioenerg* 16:455–467.

Glaser RW, Leikin SL, Chernomordik LV, Pastushenko VF, Sokirko AI. 1988. Reversible electrical breakdown of lipid bilayers: Formation and evolution of pores. *Biochim Biophys Acta* 940:275–287.

Glogauer M, Lee W, McCulloch CA. 1993. Induced endocytosis in human fibroblasts by electrical fields. *Exp Cell Res* 208(1):232–240.

Gowrishankar TR, Weaver JC. 2003. An approach to electrical modeling of single and multiple cells. *Proc Natl Acad Sci USA* 100(6):3203–3208.

Gowrishankar TR, Pliquett U, Lee RC. 1999. Dynamics of membrane sealing in transient electropermeabilization of skeletal muscle membranes. *Ann NY Acad Sci* 888:195–210.

He H, Chang DC, Lee YK. 2007. Using a micro electroporation chip to determine the optimal physical parameters in the uptake of biomolecules in HeLa cells. *Bioelectrochemistry* 70:363–368.

He H, Chang DC, Lee YK. 2008. Nonlinear current response of micro electroporation and resealing dynamics for human cancer cells. *Bioelectrochemistry* 72(2):161–168.

Hibino M, Shigemori M, Itoh H, Nagayama K, Kinosita K. 1991. Membrane conductance of an electroporated cell analyzed by submicrosecond imaging of transmembrane potential. *Biophys J* 59:209–220.

Hibino M, Itoh H, Kinosita K. 1993. Time courses of cell electroporation as revealed by submicrosecond imaging of transmembrane potential. *Biophys J* 64:1789–1800.

Jones AT. 2007. Macropinocytosis: Searching for an endocytic identity and role in the uptake of cell penetrating peptides. *J Cell Mol Med* 11(4):670–684.

Joshi RP, Hu Q, Schoenbach KH, Bebe SJ. 2002. Simulations of electroporation dynamics and shape deformations in biological cells subjected to high voltage pulses. *IEEE Trans Plasma Sci* 30:1536–1546.

Joshi RP, Hu Q, Schoenbach KH. 2004. Modeling studies of cell response to ultrashort, high-intensity electric fields—Implications for intracellular manipulation. *IEEE Trans Plasma Sci* 32:1677–1686.

Kakorin S, Neumann E. 2002. Ionic conductivity of electroporated lipid bilayer membranes. *Bioelectrochemistry* 56:163–166.

Kakorin S, Stoylov SP, Neumann E. 1996. Electro-optics of membrane electroporation in diphenylhexatriene-doped lipid bilayer vesicles. *Biophys Chem* 16;58(1–2):109–116.

Khine M, Zanetti CI, Blatz A, Wang LP, Lee LP. 2007. Single-cell electroporation arrays with real-time monitoring and feedback control. *Lab Chip* 7:457–462.

Kinosita K Jr, Ikegami A. 1988. A dynamic structure of membranes and subcellular components revealed by optical anisotropy decay method. *Subcell Biochem* 13:55–88.

Kotnik T, Bobanovic F, Miklavcic D. 1997. Sensitivity of transmembrane voltage induced by applied electric fields—A theoretical analysis. *Bioelectrochem Bioenerg* 43:285–291.

Kotnik T, Miklavcic D, Slivnik T. 1998. Time course of transmembrane voltage induced by time-varying electric fields—A method for theoretical analysis and its application. *Bioelectrochem Bioenerg* 45:3–16.

Kotnik T, Macek-Lebar A, Miklavcic D, Mir LM. 2000. Evaluation of cell membrane electropermeabilization by means of nonpermanent cytotoxic agent. *Biotechniques* 28:921–926.

Kotnik T, Pucihar G, Rebersek M, Mir LM, Miklavcic D. 2003. Role of pulse shape in cell membrane electropermeabilization. *Biochim Biophys Acta* 1614:193–200.

Krassowska W, Filev PD. 2007. Modeling electroporation in a single cell. *Biophys J* 92:404–417.

Lambert H, Pankov R, Gauthier J, Hancock R. 1990. Electroporation-mediated uptake of proteins into mammalian cells. *Biochem Cell Biol* 68:729–734.

Miklavcic D, Semrov D, Mekid H, Mir LM. 2000. A validated model of in vivo electric field distribution in tissues for electrochemotherapy and for DNA electrotransfer for gene therapy. *Biochim Biophys Acta* 1523:73–83.

Navarrete EG, Sacchi JS. 2006. On the effect of prestin on the electrical breakdown of cell membranes. *Biophys J* 90(3):967–974.

Neu JC, Krassowska W. 1999. Asymptotic model of electroporation. *Phys Rev E* 59:3471–3482.

Neumann E, Rosenheck K. 1972. Permeability changes induced by electric impulses in vesicular membranes. *J Membr Biol* 10(3):279–290.

Neumann E, Kakorin S, Tsoneva I, Nikolova B, Tomov T. 1996. Calcium-mediated DNA adsorption to yeast cells and kinetics of cell transformation by electroporation. *Biophys J* 71:868–877.

Neumann E, Toensing K, Kakorin S, Budde P, Frey J. 1998. Mechanism of electroporative dye uptake by mouse B cells. *Biophys J* 74(1):98–108.

Neumann E, Kakorin S, Toensing K. 1999. Fundamentals of electroporative delivery of drugs and genes. *Bioelectrochem Bioenerg* 48:3–16.

Pavlin M, Kanduser M, Rebersek M, Pucihar G, Hart FX, Magjarevic R, Miklavcic D. 2005. Effect of cell electroporation on the conductivity of a cell suspension. *Biophys J* 88:4378–4390.

Pavlin M, Leben V, Miklavcic D. 2007. Electroporation in dense cell suspension—Theoretical and experimental analysis of ion diffusion and cell permeabilization. *Biochim Biophys Acta* 1770:12–23.

Pliquett U, Weaver JC. 2007. Feasibility of an electrode-reservoir device for transdermal drug delivery by noninvasive skin electroporation. *IEEE Trans Biomed Eng* 54:536–538.

Plonsey R, Barr RC. 1988. *Bioelectricity. A Quantitative Approach*. New York, Plenum Press.

Prausnitz MR, Corbett JD, Gimm JA, Golan DE, Langer R, Weaver JC. 1995. Millisecond measurement of transport during and after an electroporation pulse. *Biophys J* 68(5):1864–1870.

Puc M, Kotnik T, Mir LM, Miklavcic D. 2003. Quantitative model of small molecules uptake after in vitro cell electropermeabilization. *Bioelectrochemistry* 60:1–10.

Pucihar G, Mir LM, Miklavcic D. 2002. The effect of pulse repetition frequency on the uptake into electropermeabilized cells in vitro with possible applications in electrochemotherapy. *Bioelectrochemistry* 57:167–172.

Pucihar G, Kotnik T, Valic B, Miklavcic D. 2006. Numerical determination of transmembrane voltage induced on irregularly shaped cells. *Ann Biomed Eng* 34:642–652.

Pucihar G, Kotnik T, Teissié J, Miklavcic D. 2007. Electroporation of dense cell suspensions. *Eur Biophys J* 36:173–185.

Pucihar G, Kotnik T, Miklavcic D, Teissié J. 2008. Kinetics of transmembrane transport of small molecules into electropermeabilized cells. *Biophys J* 95:2837–2848.

Rols MP, Teissié J. 1990. Electropermeabilization of mammalian cells. *Biophys J* 58:1089–1098.

Rols MP, Teissié J. 1998. Electropermeabilization of mammalian cells to macromolecules: Control by pulse duration. *Biophys J* 75:1415–1423.

Rols MP, Femenia P, Teissié J. 1995. Long-lived macropinocytosis takes place in electropermeabilized mammalian cells. *Biochem Biophys Res Commun* 208:26–38.

Ryttsén F, Farre C, Brennan C, Weber SG, Nolkrantz K, Jardemark K, Chiu DT, Orwar O. 2000. Characterization of single-cell electroporation by using patch-clamp and fluorescence microscopy. *Biophys J* 79(4):1993–2001.

Saulis G. 1997. Pore disappearance in a cell after electroporation: Theoretical simulation and comparison with experiments. *Biophys J* 73(3):1299–1309.

Saulis G. 2005 The loading of human erythrocytes with small molecules by electroporation. *Cell Mol Biol Lett* 10:23–35.

Saulis G, Venslauskas MS, Naktinis J. 1991. Kinetics of pore resealing in cell membranes after electroporation. *Bioelectrochem Bioenerg* 26:1–13.

Saulis G, Satkauskas S, Praneviciute R. 2007. Determination of cell electroporation from the release of intracellular potassium ions. *Anal Biochem* 360(2):273–281.

Schmeer M, Seipp T, Pliquett U, Kakorin S, Neumann E. 2004. Mechanism for the conductivity changes caused by membrane electroporation of CHO cell—pellets. *Phys Chem Chem Phys* 6:5564–5574.

Schwister K, Deuticke B. 1985. Formation and properties of aqueous leaks induced in human erythrocytes by electrical breakdown. *Biochim Biophys Acta* 816:332–348.

Shirakashi R, Sukhorukov VL, Tanasawa I, Zimmermann U. 2004. Measurement of the permeability and resealing time constant of the electroporated mammalian cell membranes. *Int J Heat Mass Transfer* 47(21):4517–4524.

Smith KC, Neu JC, Krassowska W. 2004. Model of creation and evolution of stable electropores for DNA delivery. *Biophys J* 86:2813–2826.

Sowers AE. 1986. A long-lived fusogenic state is induced in erythrocytes ghosts by electric pulses. *J Cell Biol* 102:1358–1362.

Stewart DA Jr, Gowrishankar TR, Weaver JC. 2004. Transport lattice approach to describing cell electroporation: Use of a local asymptotic model. *IEEE Trans Plasma Sci* 32(4):1696–1708.

Teissié J, Ramos C. 1998. Correlation between electric field pulse induced long-lived permeabilization and fusogenicity in cell membranes. *Biophys J* 74(4):1889–1898.

Teissié J, Golzio M, Rols MP. 2005. Mechanisms of cell membrane electropermeabilization: A mini review of our present (lack of?) knowledge. *Biochim Biophys Acta* 1724(3):270–280.

Tsong TY. 1991. Electroporation of cell membranes. *Biophys J* 60:297–306.

Weaver JC, Chizmadzhev YA. 1989. Theory of electroporation: A review. *Bioelectrochem Bioenerg* 41:135–160.

Zimmermann U. 1982. Electric field-mediated fusion and related electrical phenomena. *Biochim Biophys Acta* 694:227–277.

Zimmermann U, Schnettler R, Klöck G, Watzka H, Donath E, Glaser RW. 1990. Mechanisms of electro-stimulated uptake of macromolecules into living cells. *Naturwissenschaften* 77:543–545.

11

Kinetics of Pore Formation and Disappearance in the Cell during Electroporation

Gintautas Saulis

11.1 Introduction

The electroporation of biological membranes has numerous applications in molecular biology, biotechnology, and medicine (Gehl, 2003; Orlowski and Mir, 1993). These include cell fusion (Senda et al., 1979; Zimmermann and Vienken, 1982), the loading of cells with various substances (Haritou et al., 1998; Mohr et al., 2006; Saulis, 2005), cell transfection (gene electrotransfer) (Neumann et al., 1982; Potter, 1988), transdermal drug delivery (Jaroszeski et al., 2000), electrochemotherapy and gene therapy (Jaroszeski et al., 2000; Li, 2008), and numerous other ones. The electroporation of the cell membrane is the primary cause of microorganism inactivation by pulsed electric fields (Kekez et al., 1996; Sale and Hamilton, 1967; Tsong, 1991; Wouters et al., 2001); it is also used in the lab-on-a-chip systems for cell disruption purposes (Brown and Audet, 2008; Lu et al., 2006; Wang et al., 2006).

However, experimental procedures have not yet been optimized, as the mechanisms of pore formation under the influence of an electric field and their subsequent resealing are not fully understood. Also, the models that would allow predicting the consequences of the exposure of cells to a particular electric treatment have to be developed. For example, to implement food processing by pulsed electric fields as a nonthermal food-processing technique, the parameters of electric treatment assuring the best yield of microbial inactivation have to be determined (Saulis and Wouters, 2007).

There have been many theoretical papers on pore formation and disappearance processes (Gowrishankar et al., 2006; Joshi and Schoenbach, 2000; Krassowska and Filev, 2007; Pastushenko and Chizmadzhev, 1983; Vasilkoski et al., 2006; Weaver and Chizmadzhev, 1996), but in the majority of them, the formation of pores in planar lipid bilayer membranes (BLMs) is investigated. Besides, in those studies where the electroporation of a cell or vesicle was analyzed (Gowrishankar et al., 2006;

Krassowska and Filev, 2007; Pastushenko and Chizmadzhev, 1983, 1985; Sugar and Neumann, 1984), few characteristics of this process that can be determined experimentally are derived theoretically.

Up to now, the numerous attempts of deriving the equations correlating the parameters of electric treatment, e.g., electric field strength E_0 and/or the pulse duration τ_i (or total treatment time), with the fraction of electropermeabilized cells (Joersbo et al., 1990; Lebar et al., 1998) and/or cells killed by an electric pulses (Hülsheger and Niemann, 1980; Hülsheger et al., 1981) have been made. In most cases, the models used for this purpose are purely empirical and based on simple approximations of the experimentally obtained dependences by some of the well-known mathematical equations, such as a sigmoid (Alvarez et al., 2003; Peleg, 1995) or logarithmic (so-called *W. D. Bigelow* function (Bigelow, 1921)); (Rodrigo et al., 2003a) curves, Weibull frequency distribution function (Abram et al., 2003; Alvarez et al., 2003; Gomez et al., 2005; Zhang et al., 1994), polynomial (Giner et al., 2001), or other functions (Abram et al., 2003; Alvarez et al., 2003; Elez-Martinez et al., 2005; Gomez et al., 2005; Rodrigo et al., 2003a,b).

In this study, the processes of pore formation and disappearance in a cell are investigated theoretically. Several characteristics of cell electroporation process have been derived. Most of the dependences described theoretically can be obtained experimentally. Thus, we hope that theoretical analysis of the process of pore formation in a cell plasma membrane, the purpose of which is to obtain measurable characteristics, will be useful. In addition to theoretical simulations, the developed theory is compared with experimental data available in the literature, and the useful information that can be obtained from this comparison is discussed.

11.2 Theory

11.2.1 A Porated Cell

The electroporation of a cell can be regarded as an all-or-nothing event, i.e., a cell is either porated or not. The term "porated" will often be used here, since the possibility of the spontaneous appearance of metastable hydrophilic pores in the absence of a transmembrane potential cannot be excluded (Markin and Kozlov, 1985; Powell and Weaver, 1986; Taupin et al., 1975).

We can state that the transmembrane potential $\Delta\Phi_m$ has made an influence on the cell plasma membrane only if we can detect that some of the membrane properties, e.g., permeability or electrical conductivity, has changed. The exposure of cells to an electric pulse leads to an increase in membrane permeability for various species. However, a membrane which is impermeable for certain molecules (e.g., trypan blue, propidium iodide, sucrose) can at the same time be permeable for smaller species (e.g., Na^+, Tl^+, Rb^+, K^+) (Pakhomov et al., 2007; Saulis, 1999; Saulis and Venslauskas, 1993a; Serpersu et al., 1985).

Therefore, it is natural to consider poration with respect to the membrane barrier to the smallest species, i.e., ions such as Rb^+, Na^+, Tl^+, and K^+. So, the cell is porated when the increase in membrane permeability (under the influence of an electric pulse or spontaneously) to the smallest ions (Na^+, K^+, Rb^+, Tl^+) is such that it can be distinguished from the background level of permeability (Saulis and Venslauskas, 1993a).

Cell electroporation is consistent with the formation of transient aqueous pores and the membrane permeability is an essentially continuous function through both the number and sizes of pores changing (Barnett and Weaver, 1991). Hence, the membrane can achieve a "threshold" permeability level at various pore populations, and it is not evident which pore population satisfies the above condition. The simplest case is when the presence of only one hydrophilic pore of a minimum size in the cell is sufficient.

Similarly, cell death induced by pulsed electric fields can also be considered as an "all-or-nothing" effect in that the cell either dies or survives normally following an electric treatment (Simpson et al., 1999). Because the initial pores created by an electric pulse are very small and the leakage of intracellular

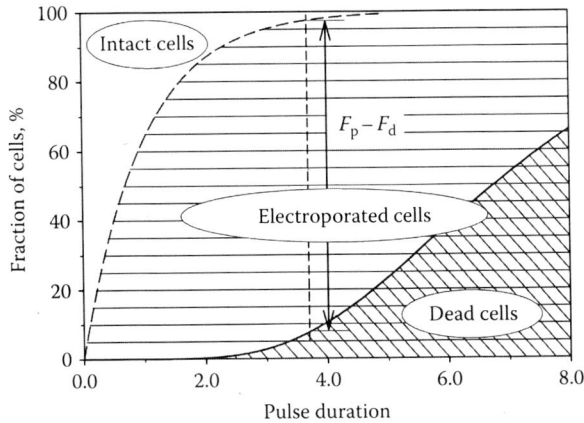

FIGURE 11.1 The schematic representation of the dependences of the fraction of electroporated and dead cells on the duration of an electric pulse. In the majority cases, there is some difference between the fraction of electroporated and dead cells $F_p - F_d$. The area between these two curves represents the cells that were electroporated but retained their viability due to pore resealing—the damage of the cells, induced by pulsed electric fields, is sublethal.

substances important for cell viability (ATP, proteins, enzymes) is limited, it cannot be expected that the appearance of such small pores would necessarily lead to the cell death (Saulis and Wouters, 2007).

Quantitatively, this can be depicted by two curves describing the dependence of the fraction of electroporated and dead cells on the parameters of an electric treatment as shown in Figure 11.1. The "electroporation" curve can be obtained by measuring the loss of intracellular K^+ (Kinosita and Tsong, 1977b; Saulis and Praneviciute, 2005) and the "death or inactivation" curve—by the viability tests, e.g., by cells clonogenic or the MTT (3-(4,5-dimethylthiazol-2-yl)-2,5-diphenyltetrazolium bromide) assays (Kotnik et al., 2000). In this chapter, the attention will be focused only on the processes of creation and resealing of pores in the cell, i.e., only the change in the number of pores in the cell will be investigated ("electroporation" curve in Figure 11.1), without taking into account how the radii and consequently the conductances and permeabilities of the pores changes and whether these changes lead to the cell death or not ("death" curve in Figure 11.1).

11.2.2 Kinetic Model

Here it will be considered that pores appear in the cell membrane according to the following mechanism. Hydrophobic pores are spontaneously and randomly initiated owing to thermal agitations of lipid molecules (Abidor et al., 1979; Glaser et al., 1988). Once the hydrophobic pore radius exceeds some critical value r^*, the transformation from hydrophobic to hydrophilic pores becomes energetically favorable and consequently the hydrophilization of hydrophobic pores may occur (Glaser et al., 1988). This moment corresponds to the overcoming of the energy barrier to pore formation, $\Delta W_f(\Delta\Phi_m)$ (Figure 11.2). Even at zero transmembrane potential hydrophilic pores are metastable owing to the existence of an energy barrier to pore resealing, $W_r(\Delta\Phi_m)$, which prevents them from closing (Glaser et al., 1988; Saulis et al., 1991).

On the basis of this mechanism, let us consider the following (Saulis and Venslauskas, 1991, 1993a):

1. The creation of metastable hydrophilic pores is a random one-step process, i.e., the pores need only overcome one energy barrier (Figure 11.2) (Glaser et al., 1988).
2. The disappearance of metastable hydrophilic pores is a random one-step process, i.e., the pores need only overcome one energy barrier (Figure 11.2) (Freeman et al., 1994; Glaser et al., 1988; Saulis, 1997).

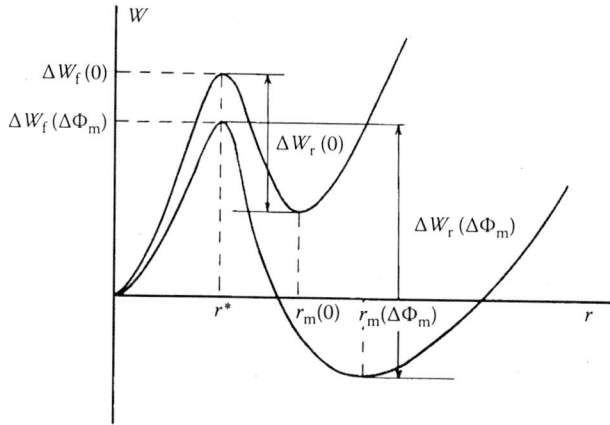

FIGURE 11.2 Pore energy W as a function of pore radius in the absence (*upper curve*) and presence (*lower curve*) of a transmembrane potential $\Delta\Phi_m$. Values of pore radius $r < r*$ correspond to hydrophobic and $r > r*$—to hydrophilic pores. Hydrophilic pores formed as a result of hydrophilization of hydrophobic pores are metastable, owing to the existence of an energy barrier $\Delta W_r(\Delta\Phi_m)$ that prevents their closing. At zero transmembrane potential, the energy of a membrane with one pore is greater than the energy of an intact membrane. Therefore, it can be expected that the energy barrier to pore disappearance should be smaller than the barrier to pore formation as it is illustrated here. The transmembrane potential diminishes the energy barrier to pore formation, $\Delta W_f(\Delta\Phi_m)$, and raises the energy barrier to pore disappearance, $\Delta W_r(\Delta\Phi_m)$. (Reproduced from Saulis, G. and Venslauskas, M.S., *Bioelectrochem. Bioenerg.*, 32, 221, 1993a. With permission.)

3. With increasing transmembrane potential $\Delta\Phi_m$, the rate of pore formation, k_f, increases (Glaser et al., 1988), whereas the rate of pore resealing, k_r, decreases (Saulis and Venslauskas, 1988).
4. The pores of only one type appear in the cell membrane. We consider pores of one type as pores with the same structure. This means that all pores have the same energy barrier $W_f(0)$, radius $r*$, and dependence of pore energy on transmembrane potential $\Delta\Phi_m$.
5. The pores do not interact with each other.
6. The time necessary for the transformation of hydrophobic to hydrophilic pores (hydrophilization time τ_h) is significantly less than the time constant of pore formation.
7. The time constant of the membrane-charging process is small compared with the time constant of the pore formation, that is, the analysis will be restricted by such a case of cell electroporation when metastable pores begin to appear only when the membrane-charging process is over.
8. After an electric pulse, the remaining transmembrane potential is small and the same at any point on the cell surface.

On the basis of these assumptions, the cell electroporation process can be described by the following general kinetic scheme (Saulis and Venslauskas, 1993a):

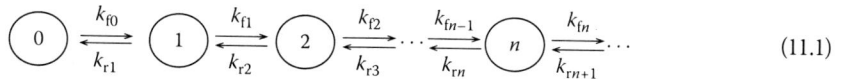

$$\textcircled{0} \underset{k_{r1}}{\overset{k_{f0}}{\rightleftarrows}} \textcircled{1} \underset{k_{r2}}{\overset{k_{f1}}{\rightleftarrows}} \textcircled{2} \underset{k_{r3}}{\overset{k_{f2}}{\rightleftarrows}} \cdots \underset{k_{rn}}{\overset{k_{fn-1}}{\rightleftarrows}} \textcircled{n} \underset{k_{rn+1}}{\overset{k_{fn}}{\rightleftarrows}} \cdots \tag{11.1}$$

where

the numerals refer to the number of pores in a cell

k_{fn} and k_{rn} are the rates of pore formation and resealing, respectively, in a cell with n pores

It should be mentioned that this kinetic scheme is a modification of the scheme used by Chizmadzhev's group to describe the electrical breakdown of bilayer lipid membranes (Pastushenko et al., 1979a,b).

Here this approach is used to analyze the kinetics not only of the pore formation in a cell membrane, but also the opposite process—the disappearance of pores after an electric treatment (see Section 11.2.3).

In our analysis, we have neglected the possibility of irreversible enlargement of the pore size, since, according to theoretical analysis, it seems that for closed membranes (cells and vesicles) the pore energy rises unlimitedly with increasing pore size (Pastushenko and Chizmadzhev, 1983; Sugar and Neumann, 1984) and only the electroporation when the transmembrane potential is not very high ($\Delta\Phi_m < 1$ V) will be analyzed.

On the basis of the kinetic scheme (11.1), the distribution function of cell poration times, $F_p(\tau_i)$, and the average number of pores per cell, $\bar{n}_i(t)$, as a function of time can be obtained. The distribution function of cell poration times, $F_p(\tau_i)$, shows the time dependence of the probability that a cell will become porated. On the basis of the definition of a porated cell given above, the distribution function $F_p(\tau_i)$ can be defined as (Saulis and Venslauskas, 1993a)

$$F_p(t) = \sum_{n=n_{cr}}^{\infty} P_n(t),$$ (11.2)

where

$P_n(t)$ is the probability that there are n pores in a cell at instant t

n_{cr} is the critical number of pores, i.e., the number of small metastable pores that is sufficient for a cell to be regarded as porated

Since

$$\sum_{n=0}^{\infty} P_n(t) = 1$$ (11.3)

we obtain

$$F_p(t) = 1 - \sum_{n=0}^{n_{cr}-1} P_n(t).$$ (11.4)

Here we will analyze only the case of sufficiently high transmembrane potential $\Delta\Phi_m$, when the resealing of pores can be ignored, i.e., $k_{fn} \gg nk_r$. The case of low transmembrane potential $\Delta\Phi_m$, when the process of pore disappearance must be taken into account, i.e., $k_{fn} \approx nk_r$, will be analyzed in Section 11.2.3.

The energy of both hydrophobic and hydrophilic pores decreases in the presence of a transmembrane potential $\Delta\Phi_m$ according to Abidor et al. (1979)

$$W(\Delta\Phi_m, r) = W(0, r) - 0.5\pi C_m \left(\frac{\varepsilon_w}{\varepsilon_m} - 1 \right) r^2 \Delta\Phi_m^2$$ (11.5)

where

r is the pore radius

C_m is the specific membrane capacity

ε_m and ε_w are the relative permittivities of the membrane and water inside the pore, respectively

Therefore, owing to the transmembrane potential, the energy barrier to pore formation, $\Delta W_f(\Delta\Phi_m)$, decreases and the rate of formation of metastable hydrophilic pores, k_f, increases (Glaser et al., 1988). It is likely that $r* < r_m$ (Figure 11.2), so the decrease in pore energy is greater at $r = r_m$ than at $r = r*$. Hence, an external electric field increases the energy barrier to pore resealing, $\Delta W_r(\Delta\Phi_m)$ (Figure 11.2), which in turn leads to a reduction in k_r. Thus, at high-enough transmembrane potential the condition $k_{fn} \gg nk_r$ is satisfied and the kinetic scheme (11.1) simplifies to (Saulis and Venslauskas, 1993a)

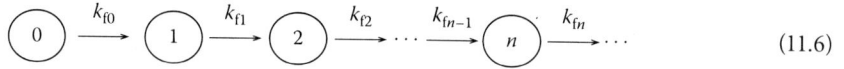

$$\underset{0}{\bigcirc} \xrightarrow{k_{f0}} \underset{1}{\bigcirc} \xrightarrow{k_{f1}} \underset{2}{\bigcirc} \xrightarrow{k_{f2}} \cdots \xrightarrow{k_{fn-1}} \underset{n}{\bigcirc} \xrightarrow{k_{fn}} \cdots \quad (11.6)$$

The probability of an additional pore forming does not depend upon how many pores already exist as long as the number of pores and their sizes are small, i.e., as long as the fraction of the membrane area occupied by pores is small (Barnett and Weaver, 1991). If there is no interaction between pores and the factors that can noticeably change the rate of pore formation can be ignored, in the first approximation, the rates satisfy the following conditions (Saulis and Venslauskas, 1993a):

$$k_{fn} = k_{fi} = k_f (i \neq n), \quad \text{and} \quad k_{rn} = nk_r. \quad (11.7)$$

In such a case, the distribution function of cell poration times, $F_p(t)$ is

$$F_p(t) = 1 - \sum_{n=0}^{n_{cr}-1} \frac{(k_f t)^n \exp(-k_f t)}{n!}. \quad (11.8)$$

Distribution functions $F_p(t)$ calculated using this expression for $n_{cr} = 1, 2, 5,$ and 10 are shown in Figure 11.3. From this figure it can be seen that the shape of the curve varies appreciably with n_{cr}. Equation 11.8 shows that at a fixed electric field strength increasing the electric field pulse length increases the fraction of electroporated cells. The rate of this increase is determined by the initial rate of pore formation in the cell, $k_{f0}(E_0)$.

For a spherical cell $k_f(E_0)$ can be calculated from the following expression (Saulis and Venslauskas, 1991, 1993a):

$$k_f(E_0) = \frac{2\pi\nu a^2}{a_1} \exp\left[-\frac{\Delta W_f(0)}{k_B T}\right] \int_{-1}^{1} \exp\left[\left(\pi C_m \frac{(\varepsilon_w/\varepsilon_m - 1)}{2k_B T} r^{*2}\right)(1.5 E_0 a y - \Delta\Phi_0)^2\right] dy \quad (11.9)$$

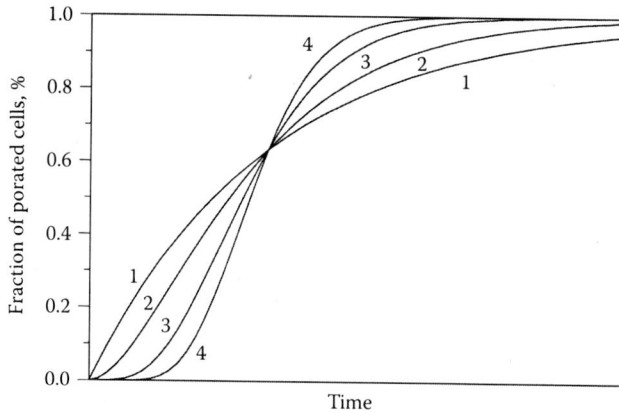

FIGURE 11.3 Distribution function of cell poration times, $F_p(\tau_i)$, in the case when the appearance of the first pores does not alter significantly the rate of their formation, for various values of critical number of pores, $n_{cr} = 1$ (curve 1), 2 (curve 2), 5 (curve 3), and 10 (curve 4). When $n_{cr} = 1$, the distribution function $F_p(\tau_i)$ is exponential. (Reproduced from Saulis, G. and Venslauskas, M.S., *Bioelectrochem. Bioenerg.*, 32, 221, 1993a. With permission.)

where

     v is the frequency of lateral fluctuations of lipid molecules

     a is the cell radius

     a_l is the area per lipid molecule

     k_B is Boltzmann's constant

     T is the absolute temperature

     $\Delta W_f(0)$ is the energy barrier to pore formation at $\Delta\Phi_m = 0$

     $r*$ is the radius of the pore corresponding to the top of this barrier

     $\Delta\Phi_0$ is the resting potential

     ε_m and ε_w are the relative permittivities of the membrane and the water inside the pore respectively

     C_m is the specific capacity of the membrane

Taking into account the dependence of the rate k_{f0} on the field strength E_0, Equation 11.8 can be rewritten as

$$F_p(E_0,t) = 1 - \sum_{n=0}^{n_{cr}-1} \frac{\left[k_{f0}(E_0)\tau_i\right]^n \exp\left[-k_{f0}(E_0\tau_i)\right]}{n!}, \tag{11.10}$$

where

     $k_{f0}(E_0)$ is given by Equation 11.9

     τ_i is the duration of the pulse

When $n_{cr} = 1$, we have

$$F_p(E_0,\tau_i) = 1 - \exp\left[-k_f(E_0)\tau_i\right]. \tag{11.11}$$

Consequently, we have obtained a general equation that describes the following important characteristics of the cell electroporation process: (1) the dependence of the probability that a cell is porated, F_p, on the pulse duration τ_i at a fixed electric field strength E_0 (distribution function of cell poration times, $F_p(\tau_i)$) and (2) the dependence of F_p on the field strength E_0 at a fixed pulse length τ_i (distribution function of cell poration field strengths, $F_p(E_0)$).

The latter dependence for pulse durations of 10 and 100 μs and $n_{cr} = 1$ and 3, plotted according to Equations 11.9 and 11.11, is shown in Figure 11.4. For the calculations the following set of parameters (the "standard cell") was used: $v = 10^{11}$ s^{-1}, $a_l = 0.6$ nm^2, $\Delta W_f(0) = 45\, k_B T$, $r* = 0.3$ nm (Glaser et al., 1988), $a = 3.5$ μm, $C_m = 1$ μF/cm^2, $\varepsilon_w = 81$, $\varepsilon_m = 2$, $T = 295$ K, $\Delta\Phi_0 = 25$ mV. It is seen that the slope of the dependence $F_p(\tau_i)$ is larger for a higher value of n_{cr}.

From Equations 11.9 and 11.11, it follows that increasing the amplitude of electric pulses is much more effective than lengthening the treatment time. This is consistent with the experimental data (Hülsheger et al., 1981; Qin et al., 1995).

The time dependence of the average number of pores per cell can be expressed as

$$\bar{n}(t) = \sum_{n=1}^{\infty} nP_n(t). \tag{11.12}$$

It is a function of both the pulse duration and electric field strength (Saulis and Venslauskas, 1993a):

$$\bar{n}(E_0,\tau_i) = k_{f0}(E_0)\tau_i \tag{11.13}$$

Using the expression for the rate of pore formation $k_{f0}(\Delta\Phi_m)$ in the spherical cell exposed to the electric field pulse (Equation 11.9), one can obtain theoretical relationships between the parameters of the electric

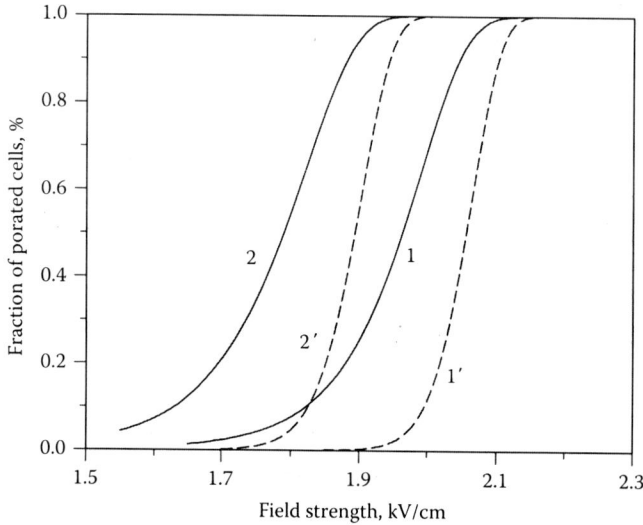

FIGURE 11.4 Theoretical dependence of the fraction of electroporated cells on the electric field strength E_0 (distribution function of cell poration electric field strengths, $F_p(E_0)$) calculated from Equations 11.9 and 11.11 at pulse lengths $\tau_i = 10$ (curves 1 and 1') and 100 μs (curves 2 and 2') for the "standard cell." Curves 1 and 2 are obtained assuming $n_{cr} = 1$, while curves 1' and 2' are calculated for $n_{cr} = 3$. (Reproduced from Saulis, G. and Venslauskas, M.S., *Bioelectrochem. Bioenerg.*, 32, 221, 1993a. With permission.)

pulse as a result of the exposure to which a definite number of pores, n, has appeared in the cell for any type of an electric treatment (exponential, square-wave, or other pulses). That is, one can find when

$$\int_0^\infty k_{f0}(E_0, t)dt = n. \qquad (11.14)$$

It can be expected that such a dependence represents to some extent the relationships between the parameters of the electric treatment resulting in cell electroporation ("electroporation" curves in Figure 11.1), e.g., the relationships between electric field strength, $E_{0.5}$, necessary for the poration of 50% of the cells and the duration or frequency of electric field pulses, which are usually determined experimentally.

The relationship between the external electric field strength required to create a single pore in the cell plasma membrane, which could be enough to consider the cell as a porated one, E_p, and the duration of a square-wave electric pulse, τ_i, calculated from Equations 11.9 and 11.14 by using the set of chosen parameters is shown in Figure 11.5. For the calculations, the following set of parameters (the "standard cell") was used: $\nu = 10^{11}$ s^{-1}, $a_1 = 0.6$ nm^2, $\Delta W_f(0) = 45$ $k_B T$, $r* = 0.3$ nm (Glaser et al., 1988; Saulis and Venslauskas, 1993b), $a = 3.5$ μm, $C_m = 1$ μF/cm^2, $\varepsilon_w = 81$, $\varepsilon_m = 2$, $T = 295$ K, $\Delta\Phi_0 = 25$ mV, and $\tau_m = 0.3$ μs. It is seen from Figure 11.5 that the shorter the pulse length, the higher the field strength should be to electroporate the cell. This dependence is much more pronounced for short pulses.

11.2.3 Kinetics of Pore Disappearance

Theoretical analysis and experimental data show that the resealing of the pore consists of several stages (Glaser et al., 1988; Hibino et al., 1993; Saulis et al., 1991), i.e., the stages of the fast reduction of pore size until the small value $r \approx 0.5$ nm and the stage of the slower, complete pore closure (overcoming the energy barrier to pore resealing, ΔW_r (0) (Figure 11.2) (Saulis et al., 1991). However, the time constant of the last stage is significantly greater than that of the first ones (Saulis et al., 1991), and therefore in the first approximation the first stages can be neglected. Then the disappearance of a metastable pore can be

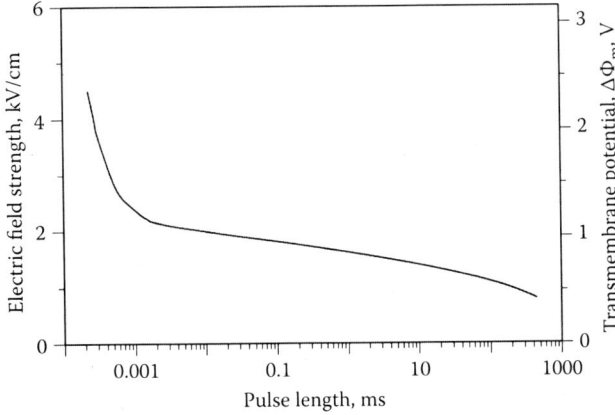

FIGURE 11.5 Theoretical dependence of the electric field strength necessary for electroporation, $E_{0.5}$, and corresponding transmembrane potential $\Delta\Phi_m$ on the duration of a square-wave electric pulse calculated according to Equations 11.9 and 11.14.

considered a one-step process. The rate of this process can be expressed as (Barnett and Weaver, 1991; Saulis et al., 1991)

$$k_r = \Lambda \exp\left[-\frac{\Delta W_r(\Delta\Phi_m)}{k_B T}\right],$$ (11.15)

where
    Λ is the pre-exponential factor with the dimension of velocity
    $\Delta W_r(\Delta\Phi_m)$ is the energy barrier to pore resealing at the transmembrane potential $\Delta\Phi_m$ (Figure 11.2)
    k_B is Boltzmann's constant
    T is the absolute temperature

The process of pore disappearance can be characterized by the distribution function of cell resealing times, $F_r(t)$, which shows the dependence on time of the probability that a cell is resealed. Obviously (Saulis, 1997),

$$F_r(t) = 1 - F_p(t),$$ (11.16)

where $F_p(t)$ is the time dependence of the probability that a cell is still porated after the end of the pulse, i.e., in the absence of an external electric field. By inserting $F_p(t)$ defined by Equation 11.4, we get (Saulis, 1997)

$$F_r(t) = \sum_{n=0}^{n_{cr}-1} P_n(t)$$ (11.17)

where
    $P_n(t)$ is the probability that there are n pores in the cell at an instant t
    n_{cr} is the critical number of pores

It has been shown, that even one small pore (radius 0.34–0.55 nm) is sufficient for a human erythrocyte to be regarded as porated (Saulis, 1999; Schwister and Deuticke, 1985). If $n_{cr} = 1$, the distribution function of cell resealing times, $F_r(t)$, shows the dependence of the probability that there are no pores in a cell on the post-pulse incubation time (Saulis, 1997):

$$F_r(t) = P_0(t). \tag{11.18}$$

Under certain conditions, electric-field-induced pores reseal after the electric pulse terminates. Usually, in considering the process of pore resealing after an electric pulse, it is assumed that additional pores do not appear in the membrane. If, in the absence of the transmembrane potential, the rate of pore resealing is significantly greater than the rate of pore formation ($k_r \gg k_f$), the latter can be neglected.

When pores do not interact with each other, the disappearances of pores are independent. In such a case, $k_{rn} = nk_r$ (Pastushenko et al., 1979a). Then, taking this into account, the kinetic scheme 11 simplifies to

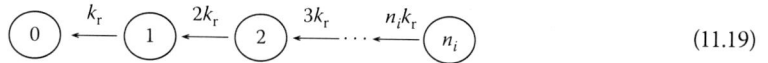

$$
\boxed{0} \xleftarrow{k_r} \boxed{1} \xleftarrow{2k_r} \boxed{2} \xleftarrow{3k_r} \cdots \xleftarrow{n_i k_r} \boxed{n_i} \tag{11.19}
$$

where n_i is the number of pores in a cell just after the electric pulse (the initial number).

For such a kinetic scheme (Saulis, 1997),

$$F_r(t) = \sum_{n_i=1}^{\infty} P_{n_i}(0) \left[1 - \exp(-k_r t) \right]^{n_i}. \tag{11.20}$$

where $P_{n_i}(0)$ is the probability that there are n_i pores in a cell just after the electric pulse. We consider the process of the appearance of pores to be stochastic. In Equation 11.20, this has been taken into account by assuming that the number of pores that have appeared in the cell membrane during the pulse (the initial number of pores, n_i, for the resealing process) is a random value.

Theoretical distribution functions $F_r(t)$ for $n_i = 1$, 3, and 10, plotted according to Equation 11.20, are shown in Figure 11.6 (*solid lines*). Dashed lines in this figure are distribution functions $F_r(t)$, calculated

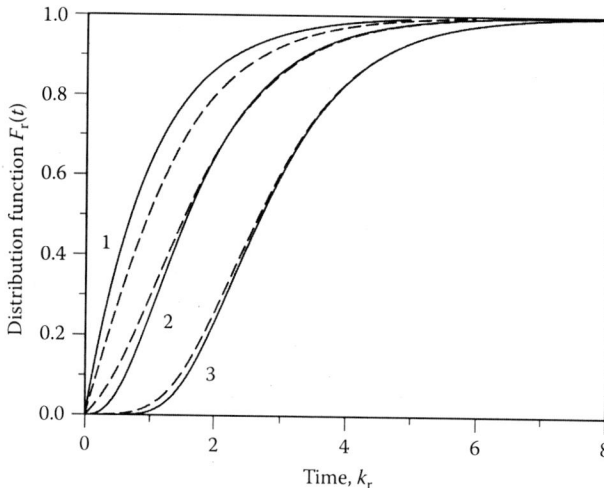

FIGURE 11.6 Distribution functions of cell resealing times, $F_r(t)$, in the case where the pore formation rate is significantly less than the pore resealing rate, calculated from Equation 11.20 for various values of the initial number of pores, $n_i = 1$ (curve 1), 3 (curve 2), and 10 (curve 3). When $n_i = 1$, the distribution function $F_r(t)$ is exponential. Dashed lines are distribution functions obtained assuming that n_i is a random value and is distributed according Poisson distribution (Equation 11.21), calculated from Equation 11.20 for the average number of pores $\bar{n}_i = 1$, 3, and 10. (Reproduced from Saulis, G., *Biophys. J.*, 73, 1299, 1997. With permission.)

according Equation 11.20, in which the probabilities $P_{n_i}(0)$ that n_i pores have appeared in a cell during the pulse are given by the Poisson distribution:

$$P_{n_i}(0) = \frac{(\bar{n}_i)^{n_i} \exp(-\bar{n}_i)}{n_i!},$$ (11.21)

where

$\bar{n}_i$ is the mean value of n_i
$n_i!$ is the factorial function $n_i! = 1.2 \ldots n_i$

From Figure 11.6 it can be seen that the shape of the distribution function $F_r(t)$ depends strongly on the initial number of pores in a cell, n_i. The more pores are created during the pulse, the longer the time required for the complete disappearance of pores in a cell. The deviations due to inhomogeneous distribution of the initial number of pores in the cells of a population are noticeable only for small n_i (Saulis, 1997).

Till now, in the analysis of the kinetics of pore disappearance, the pore formation process was neglected. This is because, in the absence of the external electric field, the rate of pore formation is usually much smaller than the rate of pore disappearance. However, in a more general case, the possibility of the spontaneous appearance of metastable hydrophilic pores cannot be excluded, even in the case of zero transmembrane potential (Popescu et al., 1991; Powell and Weaver, 1986; Saulis, 1997; Taupin et al., 1975). Some cells are permeable for membrane-impermeable substances without any exposure to an electric field (Joersbo et al., 1990; Marszalek et al., 1990) and the fraction of such cells can be as high as 18%–22% (Joersbo et al., 1990). In such a case, when analyzing pore disappearance, the process of pore formation has to be taken into account (Saulis, 1997). Note that after an electric pulse the transmembrane potential is small and the same at all points on the cell surface.

As long as the fraction of membrane area occupied by pores is small, the probability of an additional pore forming does not depend on how many pores already exist (Barnett and Weaver, 1991). Furthermore, it has already been stated that there is no interaction between pores. In such a case, we have the following kinetic scheme (Saulis, 1997):

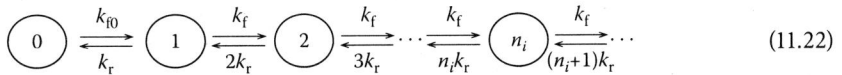

$$\boxed{0} \underset{k_r}{\overset{k_{f0}}{\rightleftharpoons}} \boxed{1} \underset{2k_r}{\overset{k_f}{\rightleftharpoons}} \boxed{2} \underset{3k_r}{\overset{k_f}{\rightleftharpoons}} \cdots \underset{n_i k_r}{\overset{k_f}{\rightleftharpoons}} \boxed{n_i} \underset{(n_i+1)k_r}{\overset{k_f}{\rightleftharpoons}} \cdots$$ (11.22)

From such a kinetic scheme, taking into account that the process of the appearance of pores is a stochastic one and therefore the initial number of pores, n_i is a random value, it has been obtained that the distribution function $F_r(t)$ is (Saulis, 1997):

$$F_r(t) = \exp\left\{-\frac{k_f}{k_r}\left[1 - \exp(-k_r t)\right]\right\} \times \sum_{n_i=1}^{\infty} P_{n_i}(0)\left[1 - \exp(-k_r t)\right]^{n_i}.$$ (11.23)

As can be seen from Equations 11.20 and 11.23, when, in the absence of an external electric field the rate of pore formation k_f is significantly less than the rate of pore resealing k_r ($k_f \ll k_r$), pores disappear completely, whereas, when the rate of pore formation is noticeable and thus cannot be neglected, the cell achieves the steady state with a nonzero number of pores. Substituting $t = \infty$ into Equation 11.23 and taking into account Equation 11.16, the fraction of cells that have pores in this state is

$$F_p(\infty) = 1 - \exp\left(-\frac{k_f}{k_r}\right).$$ (11.24)

For the kinetic scheme (11.22), one can find that the average number of pores per cell, $\bar{n}_i$, depends on time as

$$\bar{n}(t) = \left(\frac{k_f}{k_r}\right)\left[1 - \exp(-k_r t)\right] + n_i \exp(-k_r t) \tag{11.25}$$

From this equation it follows that at the stationary state (at $t \to \infty$),

$$\bar{n}(\infty) = \frac{k_f}{k_r}. \tag{11.26}$$

The theoretical time dependences of the average number of pores calculated according Equation 11.25 for $k_f/k_r = 0.05$ and $n_i = 0, 1, 3$, and 10 pores are presented in Figure 11.7. The stationary state that a cell enters after a long enough time does not depend on initial conditions (see Figure 11.7 and Equation 11.25). Only the time required to achieve this state increases with an increasing initial number of pores in a cell. So, if cell electroporation is fully reversible, the steady state should be the same for both a resealed cell and the cell unexposed to an electric pulse (Saulis, 1997).

Experimental data show that some cells undergo irreversible electroporation—the cells exposed to an electric pulse can die). For such a case, Equation 11.23 should be modified by taking into account possible irreversible electroporation (Saulis, 1997):

$$F_r(t) = (1 - F_{irr})\exp\left\{-\frac{k_f}{k_r}\left[1 - \exp(-k_r t)\right]\right\} \times \sum_{n_i=1}^{\infty} P_{n_i}(0)\left[1 - \exp(-k_r t)\right]^{n_i} - F_{irr} \tag{11.27}$$

where F_{irr} is the fraction of the cells that have been damaged irreversibly during electroporation.

In the simplest case, when $k_f \ll k_r$ and $n_i = 1$, Equation 11.27 yields (Saulis, 1997):

$$F_r(t) = (1 - F_{irr})\left[1 - \exp(-k_r t)\right] + F_{irr}. \tag{11.28}$$

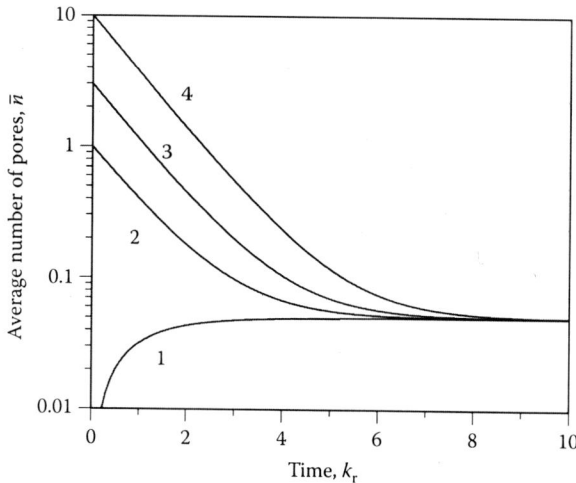

FIGURE 11.7 Theoretical dependences of the average number of pores in a cell, $\bar{n}_i$, on the time passed after the end of the electric pulse, calculated from Equation 11.25 for the initial number of pores, $n_i = 0$ (curve 1), 1 (curve 2), 3 (curve 3), and 10 (curve 4) assuming that $k_f/k_r = 0.05$. Irrespective of the value of the initial number of pores, a cell achieves the same steady state. In this steady state the number of pores is equal to k_f/k_r. (Reproduced from Saulis, G., *Biophys. J.*, 73, 1299, 1997. With permission.)

From Equations 11.16 and 11.28 we have (Saulis, 1997):

$$F_p(t) = [1 - F_{irr}]\exp(-k_r t) + F_{irr}. \tag{11.29}$$

Exactly the same equation, except for notations, was used by several groups to describe their experimental data (Muraji et al., 1993; Neumann and Boldt, 1990).

11.3 Comparison with Experiments and Discussion

Any theory should be compared with experiments. However, a comparison can be made only if the characteristics of the process, derived theoretically, can be obtained experimentally. In our case such characteristics are the distribution functions of cell poration and resealing times, $F_p(t)$ and $F_r(t)$, respectively. Luckily, both these functions can be determined experimentally (Saulis, 1997; Saulis and Venslauskas, 1991, 1993b).

The kinetics of pore formation in cellular and artificial membranes during an electric pulse and their disappearance after the pulse are mainly measured by studying the time course of the changes of (1) the membrane conductivity (Chen and Lee, 1994; Chernomordik et al., 1987; Hibino et al., 1993; Pakhomov et al., 2009; Pavlin et al., 2008; Pliquett and Wunderlich, 1983), (2) the total permeability of the membrane to small inorganic ions (Glaser et al., 1986; Schwister and Deuticke, 1985; Serpersu et al., 1985) or other substances (Kennedy et al., 2008; Saulis, 1999; Sowers and Lieber, 1986; Yumura et al., 1995), and (3) the fraction of cells permeable to small inorganic ions (Kinosita and Tsong, 1977b; Riemann et al., 1975; Saulis, 1997; Saulis et al., 1991) or certain membrane-impermeant compounds (Escande-Geraud et al., 1988; Gabriel and Teissie, 1995; Muraji et al., 1993; Neumann and Boldt, 1990).

In the first two methods, the parameters, the time dependences of which are recorded during experiments, depend on both the size and the number of pores. Thus, by measuring only the changes in the total conductivity or permeability during and/or after an electric pulse, it is impossible to determine whether this decrease is caused by the changes in the number of pores or by the changes in their sizes. In addition, these parameters are the sums of the values for individual pores. However, one large pore has the same conductivity or permeability as many small pores, and thus the enlargement or shrinkage of one pore is equivalent to the appearance or disappearance of many small pores. Moreover, the pore conductance depends nonlinearly not only on the pore radius, but also on the transmembrane potential $\Delta\Phi_m$ as well (Glaser et al., 1988; Vasilkoski et al., 2006). This makes the estimation of the rate of pore formation from the kinetics of the membrane conductivity during an electric pulse a rather complicated task.

Here, the kinetics of pore formation and resealing are estimated by determining the time courses of the fraction of porated cells (the cells permeable to small inorganic ions) during and after an electric pulse (Saulis, 1997; Saulis et al., 1991; Saulis and Venslauskas, 1993b). This approach can be used if the presence of a few small pores in the cell membrane can be detected experimentally. To ascertain whether this is the case, it is necessary to determine what pore population (the number of pores and their sizes) is sufficient for the cell to be regarded as porated.

11.3.1 A Porated Cell

From those studies in which the number and sizes of pores were estimated, it follows that the pores induced by exposure to an electric pulse leading to the poration of almost 100% of the cells are smaller than 0.5–1.0 nm (Kinosita and Tsong, 1977a; Saulis, 1999; Schwister and Deuticke, 1985; Serpersu et al., 1985). Nevertheless, the appearance of such small pores is sufficient for a cell to be regarded as porated.

For example, in Figure 11.8 the dependences of the fraction of human erythrocytes the membranes of which have become permeable to various substances, as a result of the exposure to an exponential electric pulse with the duration of 20 μs, are presented. It can be seen from this figure that the exposure

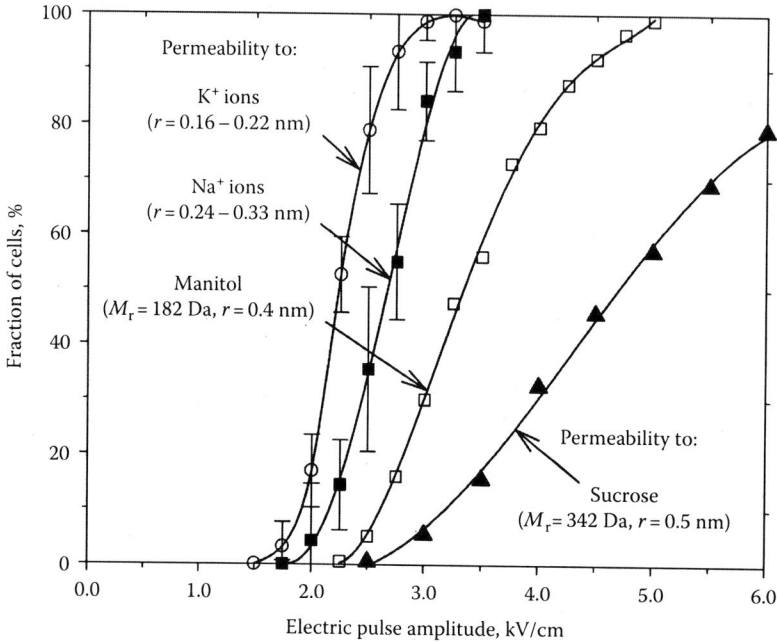

FIGURE 11.8 Dependence of the fraction of electroporated cells (permeable to K[+] ions) and the cells with pores through which ions of sodium (Na[+]), molecules of mannitol (M_r = 182.17 Da, radius about 0.35–0.43 nm), or sucrose (M_r = 342.3 Da, radius about 0.44–0.52 nm), can enter the cell on the electric pulse amplitude. Cells were subjected to an exponential electric pulse (τ = 20 µs) at varying field strengths at 20°C.

of the cells to a 3 kV/cm electric pulse leads to the poration of more than 80% of the cells, but in only 30% of the cells are there pores through which mannitol (M_r = 182.17 Da, radius about 0.35–0.43 nm (Schultz and Solomon, 1961)), can enter the cell. The pores permeable to sucrose (M_r = 342.3 Da, radius about 0.44–0.52 nm (Renkin, 1954; Schultz and Solomon, 1961)), only appeared in 5.5% of the cells (Saulis, 1999). This clearly indicates that the electroporation threshold is consistent with the appearance of small pores, the average radius of which is in the range of 0.2–0.5 nm (Saulis, 1999; Schwister and Deuticke, 1985).

The existence of a single pore with the radius in the range of 0.34–0.7 nm in the plasma membrane of erythrocytes can be detected. This has been confirmed by using erythrocytes treated by an electric pulse (Deuticke and Schwister, 1989; Saulis, 1999; Schwister and Deuticke, 1985), γ-irradiated (Soszynski and Bartosz, 1997) or exposed to hypochloric acid (Zavodnik et al., 2002). It has been shown that even one small pore (radius 0.34–0.55 nm) is sufficient for a human erythrocyte to be regarded as porated (Saulis, 1999; Schwister and Deuticke, 1985). The shapes of the experimental distribution functions of cell poration times, $F_p(t)$, and cell resealing times, $F_r(t)$, obtained for human erythrocytes also show that n_{cr} is close to unity for these cells (Saulis and Venslauskas, 1993b) (see below).

11.3.2 Kinetics of Pore Formation

The distribution functions $F_p(\tau_i)$ of erythrocyte poration times measured at electric field strengths varying from 1.1 to 1.5 kV/cm, are shown in Figure 11.9A. Each point corresponds to the mean value of four or five observations. The distribution function $F_p(\tau_i)$ was measured for pulse durations in the range 50 µs to 2 ms (Saulis and Venslauskas, 1993b).

It can be seen from Figure 11.9A that increasing the pulse length at a given electric field strength increases the fraction of porated cells. This means that the time necessary for the appearance of the

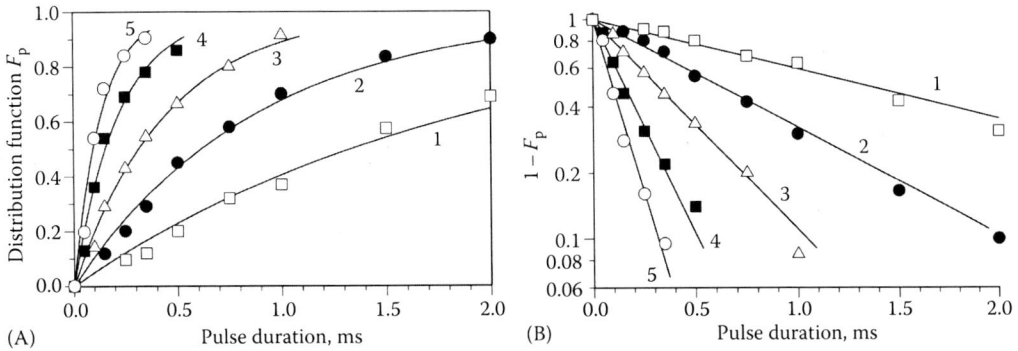

FIGURE 11.9 (A) Distribution functions $F_p(\tau_i)$ of human erythrocyte poration times at different electric field strength: (1) 1.1 kV/cm, (2) 1.2 kV/cm, (3) 1.3 kV/cm, (4) 1.4 kV/cm, and (5) 1.5 kV/cm. $F_p(\tau_i)$ was obtained by measuring the dependence of the fraction of the porated cells on the pulse duration τ_i. The fraction of the porated cells, i.e., the value of $F_p(\tau_i)$, was determined from the extent of hemolysis after prolonged incubation (20–24 h) in 0.9% NaCl solution at 4°C. Experimental data are approximated by the expression $F_p(\tau_i) = 1 - \exp(-k_{f0}(E_0)\tau_i)$. The values of the initial pore formation rate $k_{f0}(E_0)$ were calculated from the slope of the $\ln[1-F_p(\tau_i)]$ curves (B). (Reproduced from Saulis, G. and Venslauskas, M.S., *Bioelectrochem. Bioenerg.*, 32, 237, 1993b. With permission.)

critical number of pores (poration time) varies from cell to cell, and thus the formation of pores in the erythrocyte membrane is a random process (Saulis and Venslauskas, 1993a; Saulis and Venslauskas, 1993b). For example, in the presence of an electric field of 1.3 kV/cm, the first 20% of cells become electroporated even after less than 100 μs, but to electroporate the last 20% of the cells, it is necessary to wait for longer than 800 μs (Figure 11.9A).

Relationships $1 - F_p(\tau_i)$ plotted in a semilogarithmic plot (Figure 11.9B) can be approximated quite well by straight lines (correlation coefficient 0.984–0.998) in good agreement with the prediction of Equation 11.10. This indicates that the critical number of pores n_{cr} is close to unity for human erythrocytes (Saulis and Venslauskas, 1993b).

From the dependence of the fraction of electroporated cells on either the pulse duration τ_i at a fixed electric field strength E_0 (distribution function of cell poration times, $F_p(\tau_i)$) (Figure 11.3) or the field strength E_0 at a fixed pulse length τ_i (distribution function of cell poration field strengths, $F_p(E_0)$) (Figure 11.4), the relationship between the electric field strength $E_{0.5}$ (the amplitude of a square-wave electric pulse) at which 50% of the cells are porated and the pulse duration τ_i can be obtained. Such relationships obtained for human erythrocytes (Kinosita and Tsong, 1977b; Saulis and Praneviciute, 2007) and Chinese hamster ovary cells (Saulis and Saule, 2009) are plotted in Figure 11.10.

It is seen from Figure 11.10 that $E_{0.5}$ is dependent on the pulse duration not only for pulses shorter than 1–10 μs (due to the need to charge the cell plasma membrane), but also for pulses longer than 100–300 μs (due to the stochastic nature of pore formation) in agreement with theoretical analysis presented in Section 11.3.1 (see Figure 11.5) and in contrast with what was assumed by some authors (Riemann et al., 1975). For square-wave electric field pulse with the duration of 20 μs to 2 ms, at the values of the field intensity used in our investigations (0.6–2 kV/cm for human erythrocytes and 0.4–1.2 kV/cm for Chinese hamster ovary cells) the maximum transmembrane potential generated at the cell poles is of the order of 0.45–0.95 V (for $a = 3$ and 7 μm, respectively) (Saulis and Saule, 2009). The transmembrane potential was calculated from Schwan (1957):

$$\Delta\Phi_m = 1.5E_0 a, \tag{11.30}$$

where
E_0 is the electric field strength
a is the cell radius

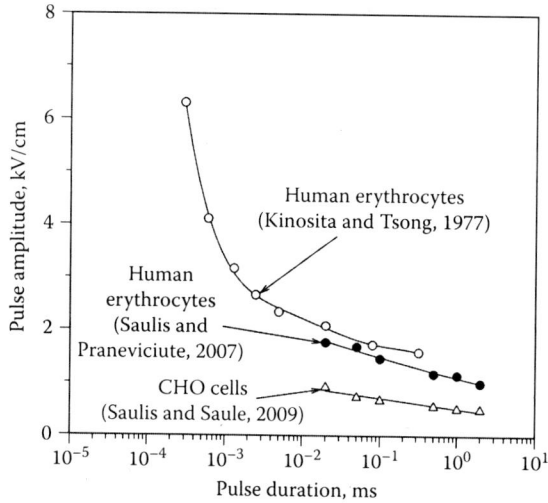

FIGURE 11.10 The dependences of the amplitude of a square-wave electric pulse required to electroporate 50% of cells on the pulse duration, obtained for human erythrocytes and Chinese hamster ovary cells.

The experimental dependence of the rate k_{f0} on the external electric field strength E_0 is plotted in Figure 11.11 (open circles). The solid curve in this figure shows the best fit for the dependence $k_{f0}(E_0)$ calculated from Equation 11.9 assuming $a = 3.5\,\mu m$, $\Delta\Phi_0 = 10\,mV$ (Glynn, 1956), $C_m = 0.7\,\mu F/cm^2$ (Takashima et al., 1988), $v = 10^{11}\,s^{-1}$, $a_l = 0.6\,nm^2$ (Glaser et al., 1988; Leikin et al., 1986), $\varepsilon_w = 79.4$ (Malmberg and Maryot, 1956), $\varepsilon_m = 2.1$ (Orme et al., 1988) and $T = 295\,K$. The best fit was obtained for $\Delta W_f(0) = 40\,k_B T$ (100 kJ/mol) and $r* = 0.34\,nm$ (Saulis and Venslauskas, 1993b).

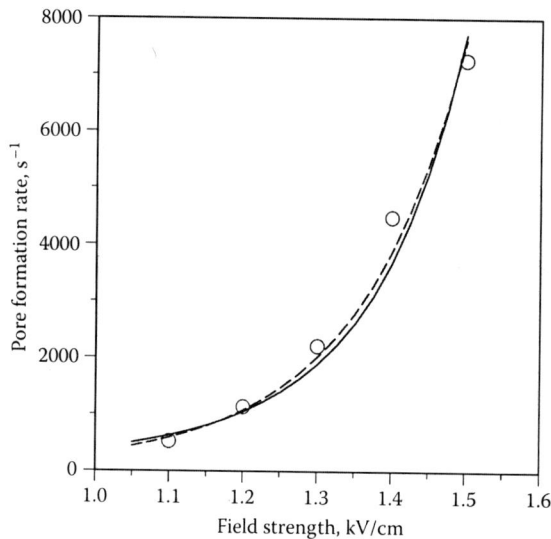

FIGURE 11.11 The experimental dependence of the initial pore formation rate k_{f0} in human erythrocytes on external electric field strength (open circles). The values of k_{f0} were calculated from the slopes of the relationship $\ln[1 - F_p(\tau_i)]$. The solid curve is the theoretical curve plotted in accordance with Equation 11.9. The parameters $\Delta W_f(0)$ and $r*$ were varied in order to obtain the best-fit curve. The best agreement was found for $\Delta W_f(0) = 40\,k_B T$ and $r* = 0.34\,nm$. The broken curve is the theoretical curve obtained when the dependence of the linear tension of the pore on the transmembrane potential is taken into account. (Reproduced from Saulis, G. and Venslauskas, M.S., *Bioelectrochem. Bioenerg.*, 32, 237, 1993b. With permission.)

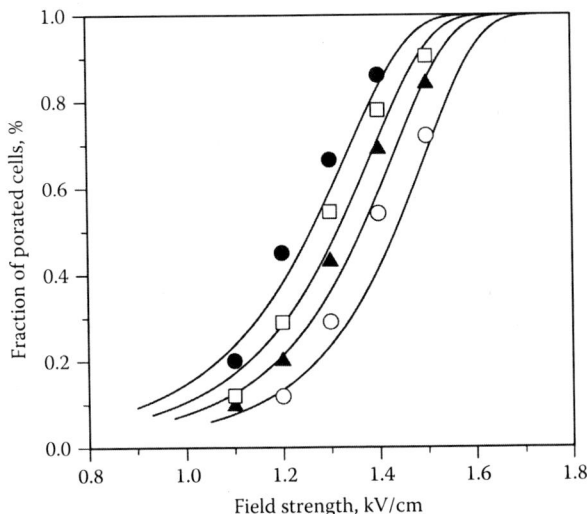

FIGURE 11.12 Dependences of the fraction F_p of electroporated human erythrocytes on electric field strength E_0 at various pulse lengths: ○ 150 μs, ▲ 250 μs, □ 350 μs, ● 500 μs. Experimental data are taken from Figure 11.9. Theoretical curves (solid lines) are calculated from Equation 11.10 for $\Delta W_f(0) = 40\,k_BT$ and $r* = 0.34$ nm, which were obtained from a comparison of the theoretical (Equation 11.9) and experimental (Figure 11.11) dependences $k_{f0}(E_0)$. (Reproduced from Saulis, G. and Venslauskas, M.S., *Bioelectrochem. Bioenerg.*, 32, 237, 1993b. With permission.)

Knowing the values of $\Delta W_f(0)$ and $r*$, we can use Equation 11.10 to obtain the relationship $F_p(E_0)$ between the fraction of electroporated cells and the electric field strength at a given pulse duration. These dependences calculated for various pulse lengths with $\Delta W_f(0) = 40\,k_BT$, $r* = 0.34$ nm, and other numerical values as cited above for Equation 11.9, and assuming that $F_p(E_0 = 0) = 0$, are shown in Figure 11.12. Experimental dependences are also presented in this figure and it can be seen that the theoretical curves describe them well. Thus, Equation 11.10, in which initial pore formation rate k_{f0} is given by Equation 11.9, provides a good description of the experimental dependences of the fraction of electroporated erythrocytes on pulse length at a fixed electric field strength and on electric field strength at a fixed pulse length.

It should be stressed that the statistical distribution of cells sizes has not been taken into account here. Despite this, the stochastic nature of the pore formation process governs a sigmoid dependence of the fraction of electroporated cells on the electric field strength. Meanwhile, there are attempts to derive a normalized dimensional distribution function for a cell population from electropermeabilization data without accounting for a stochasticity of the pore formation process (Puc et al., 2003; Turcu and Neamtu, 1995).

We attempted to estimate $\Delta W_f(0)$ and $r*$ assuming that $n_{cr} > 1$. Only a slightly lower value of the energy barrier $\Delta W_f(0)$ was obtained (38.5 k_BT at $n_{cr} = 4$), and the value of the critical radius $r*$ was unchanged. We also evaluated $\Delta W_f(0)$ and $r*$ taking into account the dependence of the linear tension of the pore on the transmembrane potential (Petrov, 1988). Slightly different values were obtained: $\Delta W_f(0) \approx 43\,k_BT$ and $r* \approx 0.2$ nm (Saulis and Venslauskas, 1993b). The agreement between the theoretical curve (dashed line in Figure 11.11) and experimental points was improved.

Despite that different membrane systems as well as the methods for the determination of the rate of pore formation were used, the values of the energy barrier $\Delta W_f(0)$ and the critical radius $r*$ obtained here for human erythrocytes are in good agreement with the values obtained for BLM—$\Delta W_f(0) = 40\text{--}45\,k_BT$ and $r* = 0.3\text{--}0.5$ nm (Glaser et al., 1988; Leikin et al., 1986).

The obtained values of $r*$ are close to 1 nm proposed from purely theoretical assumptions (Abidor et al., 1979; Barnett and Weaver, 1991; Glaser et al., 1988) and the experimental estimations of the pore size, which show that the smallest radius of created pores is in the range of 0.2–0.5 nm (Bier et al., 2004; Correa and Schelly, 1998; El-Mashak and Tsong, 1985; Pakhomov et al., 2007; Saulis, 1999; Saulis and Praneviciute, 2005; Schwister and Deuticke, 1985). In addition, recently, the spontaneous formation of hydrophilic pores on a nanosecond timescale have also been observed in molecular dynamics simulations of dipalmitoylphosphatidylcholine bilayers (Leontiadou et al., 2004; Tieleman et al., 2003). Under stress-free conditions, the radius of the pore that should correspond to the local minimum at $r \approx r_m$ (Figure 11.2) was ~0.7 nm (Leontiadou et al., 2004), that is, slightly larger than the radius of the pore corresponding to the top of the energy barrier to pore formation, $r* \approx 0.3–0.5$ nm, estimated for bilayer lipid membranes and human erythrocytes.

The results indicate that the theoretical analysis of the process of pore formation carried out in this chapter provides quite good quantitative descriptions of the main experimental observations.

11.3.3 Kinetics of Pore Disappearance

The kinetic model of cell electroporation presented in this chapter has been developed to describe the kinetics of the complete disappearance of pores in a cell. Fortunately, it is also applicable to the description of the disappearance of pores that are larger than a certain chosen size, i.e., those that are large enough to allow test molecules (e.g., propidium iodide, trypan blue) to pass through them (Saulis, 1997).

So, can the theoretical functions $F_r(t)$ obtained here describe the experimental dependences of the fraction of resealed cells on the time passed after the end of the pulse? In Figure 11.13 such dependences, taken from the papers published by Rols et al. (1990) (Figure 11.13A) and Muraji et al. (1993) (Figure 11.13B), are presented. Experimental points in Figure 11.13A show the dependences of the fraction of

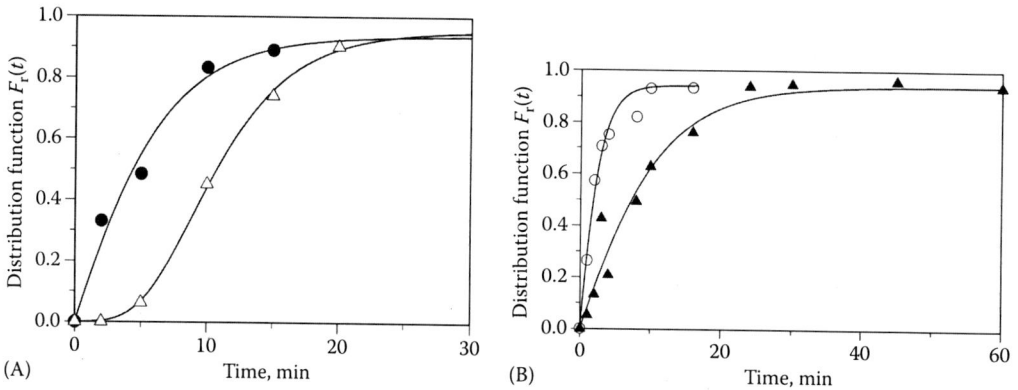

FIGURE 11.13 (A) Experimental points show the dependence of the fraction of Chinese hamster ovary cells, the membrane of which has restored its impermeability to trypan blue, on the time passed after the pulse. The data are taken from the paper published by Rols et al. (1990). Electroporation conditions were 10 square-wave pulses, $\tau_i = 100\,\mu s$, $\nu = 1$ Hz, $E_0 = 1.5$ (*filled circles*) and 1.8 (*open triangles*) kV/cm, and resealing was monitored at $T = 21°C$. Solid curves are the theoretical distribution functions plotted according Equation 11.27, assuming that the probabilities $P_{n_i}(0)$ that n_i pores have appeared in a cell during an electric treatment are given by the Poisson distribution (Equation 11.21) and $F_{irr} = 0$. The parameters $\bar{n}_i$, k_r, and the ratio k_f/k_r were varied to obtain the best fit of the theoretical dependences $F_r(t)$ to experimental points. (B) The experimental dependence of the fraction of yeast cells, the membrane of which has restored its impermeability to phloxine B, on the time of incubation at 25°C temperature. The data are taken from the work of Muraji et al. (1993). Cell suspension was subjected to a single exponential electric pulse with a time constant of 70 μs. The amplitude of the pulse was 3.7 (○) and 4.6 (▲) kV/cm. Solid curves are the theoretical distribution functions calculated from Equation 11.20, assuming that the probabilities $P_{n_i}(0)$ that n_i pores have appeared in a cell during an electric treatment are given by the Poisson distribution (Equation 11.21) and $F_{irr} = 0$. The value of the ratio $k_f/k_r = 0.06$ was obtained from the tails of the experimental dependences, and then the parameters $\bar{n}_i$ and k_r were varied to obtain the best fit. (Reproduced from Saulis, G., *Biophys. J.*, 73, 1299, 1997. With permission.)

Chinese hamster ovary cells, the membrane of which has restored its impermeability to trypan blue, on the time elapsed after the pulse. Electroporation conditions were 10 square-wave pulses, $\tau_i = 100\,\mu s$, $v = 1\,Hz$, $E_0 = 1.5$ (*filled circles*), and 1.8 (*open triangles*) kV/cm, and resealing was monitored at $T = 21°C$.

Solid lines in Figure 11.13A are the theoretical distribution functions calculated from Equation 11.27, assuming that the probabilities $P_{n_i}(0)$ that n_i pores have appeared in a cell during an electric treatment are given by the Poisson distribution (Equation 11.21) and $F_{irr} = 0$. The rest of the unknown parameters in Equations 11.21 and 11.27, were varied to obtain the best fit of the theoretical dependences $F_r(t)$ to experimental points.

The best agreement was obtained for $\bar{n}_i = 1$, $k_r = 3.9 \times 10^{-3}\,s^{-1}$, and $k_f/k_r = 0.06$ in the case where the amplitude of electric pulses was 1.5 kV/cm and for $\bar{n}_i = 9$, $k_r = 4.1 \times 10^{-3}\,s^{-1}$, and $k_f/k_r = 0.05$ when $E_0 = 1.8\,kV/cm$ (Saulis, 1997).

In Figure 11.13B, the experimental dependences of the fraction of yeast cells permeable to phloxine B on the time of incubation at 25°C, taken from Muraji et al. (1993), are shown. In this work, cell suspension was subjected to a single exponential electric pulse with a time constant of 70 μs. The amplitude of the pulse was 3.7 (*open circles*) and 4.6 (*filled triangles*) kV/cm. Solid lines in this figure are the theoretical distribution functions calculated from Equation 11.27, assuming that the probabilities $P_{n_i}(0)$ are given by the Poisson distribution (Equation 11.21) and $F_{irr} = 0$. The best agreement was obtained for $k_f/k_r = 0.06$, $\bar{n}_i = 1$, $k_r = 9 \times 10^{-3}\,s^{-1}$ for $E_0 = 3.7\,kV/cm$ and $k_f/k_r = 0.06$, $\bar{n}_i = 1$, $k_r = 2.3 \times 10^{-3}\,s^{-1}$ for $E_0 = 4.6\,kV/cm$ (Saulis, 1997).

It can be seen from Figure 11.13 that increasing the time of incubation at elevated temperature increases the fraction of resealed cells. This shows that the time necessary for the resealing varies from cell to cell, and thus the disappearance of pores from the cell membrane after electroporation is a random process. Although the shape of experimental relationships depends on the electroporation conditions, they can be described quite well by theoretical curves. Estimated numerical values of the parameters show that increasing the amplitude of an electric pulse increases either the apparent number of pores created during the pulse with an unchanged rate of pore resealing (experimental data from Rols et al., 1990) or the rate of pore resealing with a constant average number of pores (experimental data of Muraji et al., 1993).

Equation 11.15 shows that the process of pore resealing should depend strongly on temperature. This has been observed experimentally (Kinosita and Tsong, 1977b; Saulis et al., 1991). From Equation 11.15 it follows that

$$\ln[k_r(T)] = \ln\Lambda - \frac{\Delta W_r(0)}{k_B T} \tag{11.31}$$

where $\Delta W_r(0)$ is the energy barrier to pore resealing at zero transmembrane potential. Thus, by plotting the temperature dependence of the rate, $k_r(T)$, of the complete disappearance of pores in Arrhenius coordinates, the energy barrier to pore resealing, $\Delta W_r(0)$ and the preexponential factor Λ can be estimated.

The energy barrier to pore resealing, $\Delta W_r(0)$, was estimated by using the time course of the complete resealing of human erythrocytes exposed to a single exponential electric pulse ($E_0 = 3.25\,kV/cm$, $\tau_i = 22\,\mu s$), measured at two different temperatures, 32°C and 37°C, (the experimental data taken from Saulis et al. (1991)) (Figure 11.14). It was assumed that the initial number of pores is distributed according the Poisson distribution and $F_{irr} = 0$. The best coincidence between theoretical and experimental curves was obtained for $\bar{n}_i = 1$. The estimated values of k_r were $9.6 \times 10^{-4}\,s^{-1}$ at 32°C and $2.1 \times 10^{-3}\,s^{-1}$ at 37°C (Saulis, 1997).

The value $\Delta W_r(0) = 49\,k_B T$ was obtained. Such a value is on the same order of magnitude as the value of the energy barrier to pore formation $\Delta W_f(0) = 40–45\,k_B T$, estimated for bilayer lipid membranes (Glaser et al., 1988; Leikin et al., 1986) and human erythrocytes (Saulis and Venslauskas, 1993b) from the dependence of the rate of pore formation on the transmembrane potential (see above). Almost exactly the same numerical value of the energy barrier to pore disappearance ($\Delta W_r(0) = 50\,k_B T$) was used in modeling pore formation (Freeman et al., 1994).

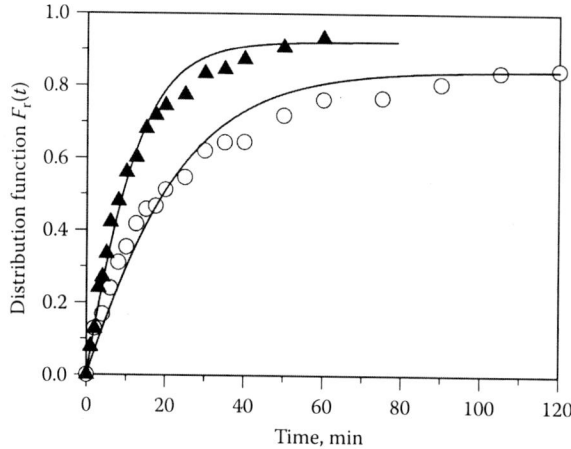

FIGURE 11.14   Distribution functions $F_r(t)$ of human erythrocyte resealing times at 32°C (O) and 37°C (▲) temperatures. Cells were subjected to a n exponential 3.25 kV/cm pulse (time constant $\tau_i = 22\,\mu s$) at 20°C, and immediately mixed with five volumes of incubation medium at 37°C or 32°C. An aliquot was taken at regular intervals and the fraction of cells, which, after incubation at high temperature, have no pores was determined from the extent of hemolysis after prolonged incubation (20–24 h) in 0.9% NaCl solution at 4°C (Saulis et al., 1991). The solid curves are the theoretical curves plotted in accordance with Equation 11.27, assuming that the initial number of pores is distributed according the Poisson distribution and $F_{irr} = 0$. The values of the ratio k_f/k_r were estimated from the last few experimental points, and then the average number of pores, $\bar{n}_i$, and the rate of pore disappearance, k_r, were estimated by obtaining a best fit of Equation 11.27 to the experimental points. The best coincidence was obtained for $\bar{n}_i = 1$. The estimated values of k_r were $9.6 \times 10^{-4}\,s^{-1}$ at 32°C and $2.1 \times 10^{-3}\,s^{-1}$ at 37°C. (Reproduced from Saulis, G., *Biophys. J.*, 73, 1299, 1997. With permission.)

Because any pore should shrink at first before it can disappear completely, it is likely that the rapid decrease in conductivity or permeability that is observed just after an electric pulse is due mainly to the decrease in the size of pores (Barnett and Weaver, 1991; Saulis et al., 1991). This may also explain why the time course for the decrease in conductivity and permeability to inorganic ions after electroporation does not obey first-order kinetics, as was observed more than once (Deuticke and Schwister, 1989; Glaser et al., 1988; Hibino et al., 1993).

11.4 Conclusions

In this chapter, the analysis of the processes of pore formation in a cell during an electric pulse and pore disappearance after the pulse has been terminated is carried out on the basis of a kinetic model in which the creation and disappearance of metastable pores are considered random one-step processes. All the results of the comparison of theoretical analysis with experimental data presented in this chapter are in favor of such a mechanism of cell electroporation. The theoretical expressions for the distribution function of cell poration times, $F_p(t)$, and cell resealing times, $F_r(t)$, derived on the basis of this model, describe quite well the experimental data, namely, the dependences of the fractions of porated cell on both the amplitude and the duration of the electric pulse, as well as the dependence of resealed cells on the post-pulse incubation time. This indicates that the processes of pore formation during an electric treatment and their annihilation after the pulse are fundamentally stochastic (Saulis, 1997; Saulis and Venslauskas, 1993b).

Acknowledgment

This work was in part supported by grant T-39/09 from the Lithuanian State Science and Studies Foundation.

References

Abidor IG, Arakelyan LV, Chernomordik LV, Chizmadzhev Y, Pastushenko VF, Tarasevich MR. 1979. Electric breakdown of bilayer lipid membranes. I. The main experimental facts and their qualitative discussion. *Bioelectrochem Bioenerg* 6: 37–52.

Abram F, Smelt JPPM, Bos R, Wouters PC. 2003. Modelling and optimization of inactivation of *Lactobacillus plantarum* by pulsed electric field treatment. *J Appl Microbiol* 94: 571–579.

Alvarez I, Virto R, Raso J, Condon S. 2003. Comparing predicting models for the *Escherichia coli* inactivation by pulsed electric fields. *Innov Food Sci Emerg Technol* 4: 195–202.

Barnett A, Weaver JC. 1991. A unified, quantitative theory of reversible electrical breakdown and rupture. *Bioelectrochem Bioenerg* 25: 163–182.

Bier M, Gowrishankar TR, Chen W, Lee RC. 2004. Electroporation of a lipid bilayer as a chemical reaction. *Bioelectromagnetics* 25: 634–637.

Bigelow WD. 1921. The logarithmic nature of thermal death time curves. *J Infect Dis* 29: 528–536.

Brown RB, Audet J. 2008. Current techniques for single-cell lysis. *J R Soc Interface* 5 (Suppl. 2): S131–S138.

Chen W, Lee RC. 1994. An improved double vaseline gap voltage clamp to study electroporated skeletal muscle fibers. *Biophys J* 66: 700–709.

Chernomordik LV, Sukharev SI, Popov SV, Pastushenko VF, Sokirko AV, Abidor IG, Chizmadzhev YA. 1987. The electrical breakdown of cell and lipid membranes: The similarity of phenomenologies. *Biochim Biophys Acta* 902: 360–373.

Correa NM, Schelly ZA. 1998. Dynamics of electroporation of synthetic liposomes studied using a pore-mediated reaction, $Ag^+ + Br^- \rightarrow AgBr$. *J Phys Chem B* 103: 9319–9322.

Deuticke B, Schwister K. 1989. Leaks induced by electric breakdown in the erythrocyte membrane. In: Neumann E, Sowers AE, Jordan CA (eds.), *Electroporation and Electrofusion in Cell Biology*. New York: Plenum Press, pp. 127–148.

El-Mashak EM, Tsong TY. 1985. Ion selectivity of temperature-induced and electric field induced pores in dipalmitoylphosphatidylcholine vesicles. *Biochemistry* 24: 2884–2888.

Elez-Martinez P, Escola-Hernandez J, Soliva-Fortuny RC, Martin-Belloso O. 2005. Inactivation of *Lactobacillus brevis* in orange juice by high-intensity pulsed electric fields. *Food Microbiol* 22: 311–319.

Escande-Geraud ML, Rols MP, Dupont MA, Gas N, Teissie J. 1988. Reversible plasma membrane ultra-structural changes correlated with electropermeabilization in Chinese hamster ovary cells. *Biochim Biophys Acta* 939: 247–259.

Freeman SA, Wang MA, Weaver JC. 1994. Theory of electroporation of planar bilayer membranes: Predictions of the aqueous area, change in capacitance, and pore-pore separation. *Biophys J* 67: 42–56.

Gabriel B, Teissie J. 1995. Control by electrical parameters of short- and long-term cell death resulting from electropermeabilization of Chinese hamster ovary cells. *Biochim Biophys Acta* 1266: 171–178.

Gehl J. 2003. Electroporation: Theory and methods, perspectives for drug delivery, gene therapy and research. *Acta Physiol Scand* 177: 437–447.

Giner J, Gimeno V, Barbosa-Canovas GV, Martin O. 2001. Effects of pulsed electric field processing on apple and pear polyphenoloxidases. *Food Sci Technol Int* 7: 339–345.

Glaser RW, Wagner A, Donath E. 1986. Volume and ionic composition changes in erythrocytes after electric breakdown. Simulation and experiment. *Bioelectrochem Bioenerg* 16: 455–467.

Glaser RW, Leikin SL, Chernomordik LV, Pastushenko VF, Sokirko AI. 1988. Reversible electrical breakdown of lipid bilayers: Formation and evolution of pores. *Biochim Biophys Acta* 940: 275–287.

Glynn IM. 1956. Sodium and potassium movements in human red cells. *J Physiol* 134: 278–310.

Gomez N, Garsia D, Alvarez I, Raso J, Condon S. 2005. A model describing the kinetics of inactivation of *Lactobacillus plantarum* in a buffer system of different pH and in orange and apple juice. *J Food Eng* 70: 7–14.

Gowrishankar TR, Esser AT, Vasilkoski Z, Smith KC, Weaver JC. 2006. Microdosimetry for conventional and supra-electroporation in cells with organelles. *Biochem Biophys Res Commun* 341: 1266–1276.

Haritou M, Yova D, Koutsouris D, Loukas S. 1998. Loading of intact rabbit erythrocytes with fluorophores and the enzyme pronase by means of electroporation. *Clin Hemorheol Microcirc* 19: 205–217.

Hibino M, Itoh H, Kinosita K Jr. 1993. Time courses of cell electroporation as revealed by submicrosecond imaging of transmembrane potential. *Biophys J* 64: 1789–1800.

Hülsheger H, Niemann EG. 1980. Lethal effect of high-voltage pulses on *E. coli* K12. *Radiat Environ Biophys* 18: 281–288.

Hülsheger H, Potel J, Niemann EG. 1981. Killing of bacteria with electric pulses of high field strength. *Radiat Environ Biophys* 20: 53–65.

Jaroszeski MJ, Heller R, Gilbert R (eds.) 2000. *Electrochemotherapy, Electrogenetherapy, and Transdermal Drug Delivery*. Totowa, NJ: Humana Press Inc.

Joersbo M, Brunstedt J, Floto F. 1990. Quantitative relationship between parameters of electroporation. *J Plant Physiol* 137: 169–174.

Joshi RP, Schoenbach KH. 2000. Electroporation dynamics in biological cells subjected to ultrafast electrical pulses: A numerical simulation study. *Phys Rev E* 62: 1025–1033.

Kekez MM, Savic P, Johnson BF. 1996. Contribution to the biophysics of the lethal effects of electric field on microorganisms. *Biochim Biophys Acta* 1278: 79–88.

Kennedy SM, Ji Z, Hedstrom JC, Booske JH, Hagness SC. 2008. Quantification of electroporative uptake kinetics and electric field heterogeneity effects in cells. *Biophys J* 94: 5018–5027.

Kinosita K, Tsong TY. 1977a. Formation and resealing of pores of controlled sizes in human erythrocyte membrane. *Nature* 268: 438–441.

Kinosita K, Tsong TY. 1977b. Voltage-induced pore formation and hemolysis of human erythrocytes. *Biochim Biophys Acta* 471: 227–242.

Kotnik T, Macek L, Miklavcic D, Mir LM. 2000. Evaluation of cell membrane electropermeabilization by means of a nonpermeant cytotoxic agent. *Biotechniques* 28: 921–926.

Krassowska W, Filev PD. 2007. Modeling electroporation in a single cell. *Biophys J* 92: 404–417.

Lebar AM, Kopitar NA, Ihan A, Sersa G, Miklavcic D. 1998. Significance of treatment energy in cell electropermeabilization. *Electromagn Biol Med* 17: 255–262.

Leikin SL, Glaser RW, Chernomordik LV. 1986. Mechanism of pore formation under electrical breakdown of membranes. *Biol Membr* 3: 944–951.

Leontiadou H, Mark AE, Marrink SJ. 2004. Molecular dynamics simulations of hydrophilic pores in lipid bilayers. *Biophys J* 86: 2156–2164.

Li S (ed.) 2008. *Electroporation Protocols: Preclinical and Clinical Gene Medicine*. Totowa, NJ: Humana Press.

Lu KY, Wo AM, Lo YJ, Chen KC, Lin CM, Yang CR. 2006. Three dimensional electrode array for cell lysis via electroporation. *Biosens Bioelectron* 22: 568–574.

Malmberg CG, Maryot AA. 1956. *J Res Natl Bur Stan* 56: 1.

Markin VS, Kozlov MM. 1985. Pores statistics in bilayer lipid membranes. *Biol Membr* 2: 205–223 (in Russian).

Marszalek P, Liu D-S, Tsong TY. 1990. Schwan equation and transmembrane potential induced by alternating electric field. *Biophys J* 58: 1053–1058.

Mohr JC, de Pablo JJ, Palecek SP. 2006. Electroporation of human embryonic stem cells: Small and macromolecule loading and DNA transfection. *Biotechnol Prog* 22: 825–834.

Muraji M, Tatebe W, Konishi T, Fujii T. 1993. The effect of electrical energy on the electropermeabilization of yeast cells. *Bioelectrochem Bioenerg* 31: 77–84.

Neumann E, Boldt E. 1990. Membrane electroporation: The dye method to determine the cell membrane conductivity. *Prog Clin Biol Res* 343: 69–83.

Neumann E, Schaefer-Ridder M, Wang Y, Hofschneider PH. 1982. Gene transfer into mouse lyoma cells by electroporation in high electric fields. *EMBO J* 1: 841–845.

Orme FW, Moronne MM, Macey RI. 1988. Modification of the erythrocyte membrane dielectric constant by alcohols. *J Membr Biol* 104: 57–68.

Orlowski S, Mir LM. 1993. Cell electropermeabilization: A new tool for biochemical and pharmacological studies. *Biochim Biophys Acta* 1154: 51–63.

Pakhomov AG, Shevin R, White JA, Kolb JF, Pakhomova ON, Joshi RP, Schoenbach KH. 2007. Membrane permeabilization and cell damage by ultrashort electric field shocks. *Arch Biochem Biophys* 465: 109–118.

Pakhomov AG, Bowman AM, Ibey BL, Andre FM, Pakhomova ON, Schoenbach KH. 2009. Lipid nano-pores can form a stable, ion channel-like conduction pathway in cell membrane. *Biochem Biophys Res Commun* 385: 181–186.

Pastushenko VF, Chizmadzhev YA. 1983. Electrical breakdown of lipid vesicles. *Biofizika* 28: 1036–1039.

Pastushenko VF, Chizmadzhev YA. 1985. Breakdown of lipid vesicles by an external electric field: A theory. *Biol Membr* 2: 1116–1129.

Pastushenko VF, Arakelyan VB, Chizmadzhev YA. 1979a. Electric breakdown of bilayer lipid membranes. VI. A stochastic theory taking into account the processes of defect formation and death: Membrane life time distribution function. *Bioelectrochem Bioenerg* 6: 89–95.

Pastushenko VF, Arakelyan VB, Chizmadzhev YA. 1979b. Electric breakdown of bilayer membranes: VII. A stochastic theory taking into account the process of defect formation and death: Statistical properties. *Bioelectrochem Bioenerg* 6: 97–104.

Pavlin M et al. 2008. Electroporation of planar lipid bilayers and membranes. In: Leitmanova Liu A (ed.), *Advances in Planar Lipid Bilayers and Liposomes*, Vol. 6. Amsterdam, the Netherlands: Elsevier, pp. 165–226.

Peleg M. 1995. A model of microbial survival after exposure to pulse electric fields. *J Sci Food Agric* 67: 93–99.

Petrov AG. 1988. Generalized lipid asymmetry and instability phenomena in membranes. In: Kuczera J, Przestalski S (eds.), *Biophysics of Membrane Transport*. Wroclaw, Poland: Agricultural University of Wroclaw, pp. 67–86.

Pliquett F, Wunderlich S. 1983. Relationship between cell parameters and pulse deformation due to these cells as well as its change after electrically induced membrane breakdown. *Bioelectrochem Bioenerg* 10: 467–475.

Popescu D, Rucareanu C, Victor G. 1991. A model for the appearance of statistical pores in membranes due to self oscillations. *Bioelectrochem Bioenerg* 25: 91–103.

Potter H. 1988. Electroporation in biology: Methods, applications, and instrumentation. *Anal Biochem* 174: 361–373.

Powell KT, Weaver JC. 1986. Transient aqueous pores in bilayer membranes: A statistical theory. *Bioelectrochem Bioenerg* 15: 211–227.

Puc M, Kotnik T, Mir LM, Miklavcic D. 2003. Quantitative model of small molecules uptake after *in vitro* cell electropermeabilization. *Bioelectrochemistry* 60: 1–10.

Qin B, Zhang Q, Barbosa-Canovas GV, Swanson BG, Pedrow PD. 1995. Pulsed electric field treatment chamber design for liquid food pasteurization using a finite element method. *Trans ASAE* 38: 557–565.

Renkin EM. 1954. Filtration, diffusion, and molecular sieving through porous cellulose membranes. *J Gen Physiol* 38: 225–243.

Riemann F, Zimmermann U, Pilwat G. 1975. Release and uptake of haemoglobin and ions in red blood cells induced by dielectric breakdown. *Biochim Biophys Acta* 394: 449–462.

Rodrigo D, Barbosa-Canovas GV, Martinez A, Rodrigo M. 2003a. Weibull distribution function based on an empirical mathematical model for inactivation of *Escherichia coli* by pulsed electric fields. *J Food Prot* 66: 1007–1012.

Rodrigo D, Ruiz P, Barbosa-Canovas GV, Martinez A, Rodrigo M. 2003b. Kinetic model for the inactivation of *Lactobacillus plantarum* by pulsed electric fields. *Int J Food Microbiol* 81: 223–229.

Rols M-P, Dahhou F, Mishra KP, Teissie J. 1990. Control of electric field induced cell membrane permeabilization by membrane order. *Biochemistry* 29: 2960–2966.

Sale AJH, Hamilton WA. 1967. Effects of high electric fields on microorganisms. I. Killing of bacteria and yeasts. *Biochim Biophys Acta* 148: 781–788.

Saulis G. 1997. Pore disappearance in a cell after electroporation: Theoretical simulation and comparison with experiments. *Biophys J* 73: 1299–1309.

Saulis G. 1999. Cell electroporation: Estimation of the number of pores and their sizes. *Biomed Sci Instrum* 35: 291–296.

Saulis G. 2005. The loading of human erythrocytes with small molecules by electroporation. *Cell Mol Biol Lett* 10: 23–35.

Saulis G, Praneviciute R. 2005. Determination of cell electroporation in small volume samples by using a mini potassium-selective electrode. *Anal Biochem* 345: 340–342.

Saulis G, Praneviciute R. 2007. Determination of cell electroporation in small-volume samples. *Biomed Sci Instrum* 43: 306–311.

Saulis G, Saule R. 2009. Comparison of electroporation of different cell lines *in vitro*. *Acta Phys Pol A* 115: 1056–1058.

Saulis G, Venslauskas MS. 1988. Asymmetrical electrical breakdown of the cells. Hypothesis of origin. *Biol Membr* 5: 1199–1204.

Saulis G, Venslauskas MS. 1991. Electrical breakdown of erythrocytes. Estimation of the energy barrier of pore formation. *Biol Membr* 5: 468–485.

Saulis G, Venslauskas MS. 1993a. Cell electroporation. Part 1. Theoretical simulation of the process of pore formation in the cell. *Bioelectrochem Bioenerg* 32: 221–235.

Saulis G, Venslauskas MS. 1993b. Cell electroporation. Part 2. Experimental measurements of the kinetics of pore formation in human erythrocytes. *Bioelectrochem Bioenerg* 32: 237–248.

Saulis G, Wouters PC. 2007. Probable mechanism of microorganism inactivation by pulsed electric fields. In: Lelieveld HLM, Notermans S, De Haan SWH (eds.), *Food Preservation by Pulsed Electric Fields: From Research to Application*. Cambridge, U.K.: Woodhead Publishing Limited, pp. 138–155.

Saulis G, Venslauskas MS, Naktinis J. 1991. Kinetics of pore resealing in cell membrane after electroporation. *Bioelectrochem Bioenerg* 26: 1–13.

Schultz SG, Solomon AK. 1961. Determination of the effective hydrodynamic radii of small molecules by viscometry. *J Gen Physiol* 44: 1189–1199.

Schwan HP. 1957. Electrical properties of tissue and cell suspensions. *Adv Biol Med Phys* 5: 147–209.

Schwister K, Deuticke B. 1985. Formation and properties of aqueous leaks induced in human erythrocytes by electrical breakdown. *Biochim Biophys Acta* 816: 332–348.

Senda M, Takeda I, Abe S, Nakamura T. 1979. Induction of cell fusion of plant protoplasts by electric stimulation. *Plant Cell Physiol* 20: 1491–1493.

Serpersu EH, Kinosita K, Tsong TY. 1985. Reversible and irreversible modification of erythrocyte membrane permeability by electric field. *Biochim Biophys Acta* 812: 770–785.

Simpson RK, Whittington R, Earnshaw RG, Russell NJ. 1999. Pulsed high electric field causes 'all or nothing' membrane damage in *Listeria monocytogenes* and *Salmonella typhimurium*, but membrane H^+-ATPase is not the primary target. *Int J Food Microbiol* 48: 1–10.

Soszynski M, Bartosz G. 1997. Effect of postirradiation treatment on the radiation-induced haemolysis of human erythrocytes. *Int J Radiat Biol* 71: 337–343.

Sowers AE, Lieber MR. 1986. Electropore diameters, lifetimes, numbers, and locations in individual erythrocyte ghosts. *FEBS Lett* 205: 179–184.

Sugar IP, Neumann E. 1984. Stochastic model for electric field-induced membrane pores. Electroporation. *Biophys Chem* 19: 211–225.

Takashima S, Asami K, Takahashi Y. 1988. Frequency domain studies of impedance characteristics of biological cells using micropipet technique. I. Erythrocyte. *Biophys J* 54: 995–1000.

Taupin C, Dvolaitzky M, Sauterey C. 1975. Osmotic pressure induced pores in lipid vesicles. *Biochemistry* 14: 4771–4775.

Tieleman DP, Leontiadou H, Mark AE, Marrink SJ. 2003. Molecular dynamics simulation of pore forma-tion in phospholipid bilayers by mechanical force and electric fields. *J Am Chem Soc* 124: 6382–6383.

Tsong TY. 1991. Electroporation of cell membranes. *Biophys J* 60: 297–306.

Turcu I, Neamtu S. 1995. Dimensional distribution of human erythrocytes obtained from electroperme-abilization experiments. *Biochim Biophys Acta* 1238: 81–85.

Vasilkoski Z, Esser AT, Gowrishankar TR, Weaver JC. 2006. Membrane electroporation: The absolute rate equation and nanosecond time scale pore creation. *Phys Rev E* 74: 021904-1–021904-12.

Wang HY, Bhunia AK, Lu C. 2006. A microfluidic flow-through device for high throughput electrical lysis of bacterial cells based on continuous DC voltage. *Biosens Bioelectron* 22: 582–588.

Weaver JC, Chizmadzhev YA. 1996. Theory of electroporation: A review. *Bioelectrochem Bioenerg* 41: 135–160.

Wouters PC, Bos AP, Ueckert J. 2001. Membrane permeabilization in relation to inactivation kinetics of *Lactobacillus* species due to pulsed electric fields. *Appl Environ Microbiol* 67: 3092–3101.

Yumura S, Matsuzaki R, Kitanishi-Yumura T. 1995. Introduction of macromolecules into living *Dictyostelium* cells by electroporation. *Cell Struct Funct* 20: 185–190.

Zavodnik LB, Zavodnik IB, Lapshyna EA, Buko VU, Bryszewska MJ. 2002. Hypochlorous acid-induced membrane pore formation in red blood cells. *Bioelectrochemistry* 58: 157–161.

Zhang Q, Monsalve-Gonzalez A, Qin BL, Barbosa-Canovas GV, Swanson BG. 1994. Inactivation of *Saccharomyces cerevisiae* in apple juice by square-wave and exponential-decay pulsed electric fields. *J Food Process Eng* 17: 469–478.

Zimmermann U, Vienken J. 1982. Electric field-induced cell-to-cell fusion. *J Membr Biol* 67: 165–182.

12

The Pulse Intensity-Duration Dependency for Cell Membrane Electroporation

Damijan Miklavčič

Gorazd Pucihar

Alenka Maček Lebar

Jasna Krmelj

Leila Towhidi

12.1 Introduction

External electric field, which is applied to cells, can, under suitable field parameters, induce local distortions and structural rearrangements of lipid molecules in the cell membrane. Depending on the field parameters, the membrane hence becomes either transiently or permanently permeable even after the field has ceased, allowing molecules that are otherwise deprived of transport mechanisms to cross the membrane and reach the cytosol. This phenomenon is often referred to as electroporation or electropermeabilization.

The method is today successfully used in different applications, such as the introduction of molecules into cells (Rols and Teissié 1998, Neumann et al. 1999, Canatella et al. 2001, Maček Lebar and Miklavčič 2001), transdermal drug delivery (Prausnitz 1996, Denet et al. 2004, Pavšelj and Préat 2005), fusion of cells (Zimmerman 1982, Ogura et al. 1994, Ušaj et al. 2009), electroinsertion of proteins into membranes (Elouagari et al. 1995, Teissié 1998), sterilization (Knorr 1999, Rowan et al. 2000, Teissié et al. 2002, El Zakhem et al. 2007), and tissue ablation (Davalos et al. 2005, Lavee et al. 2007). The main clinical success of electroporation was achieved in the treatment of cutaneous and subcutaneous tumors, where chemotherapeutic drugs in combination with electric pulses were delivered to tumor cells (electrochemotherapy) (Heller et al. 1999, Mir and Orlowski 1999, Marty et al. 2006, Serša et al. 2008), while another application, a nonviral delivery of nucleic acids to cells (gene electrotransfection), is also gaining increasing interest (Jaroszeski et al. 1999, Šatkauskas et al. 2002, Golzio et al. 2007).

For each specific application of electroporation, the electric pulses applied to target cells have to be of appropriate amplitude, duration, number, and repetition frequency, and these parameters also need to be adjusted for the particular cell type, size, orientation, and density of cells. When molecules are to be transported into electroporated cells, the characteristics of these molecules must also be considered when adjusting the pulse parameters. Using numerous different parameters of electric pulses, the advantageous effects were reported for different applications of electroporation. Lack of systematic studies addressing the role of electric pulse parameters in the efficiency of electroporation is the reason that the relation between application effectiveness and certain combination of electric pulse parameters is still unclear. Such a relation can be extremely useful in electroporation-based treatment planning. Thus, we focus in this chapter on possible mathematical relation between the parameters of major importance—the amplitude of the electric field and the field duration. Therefore, the experimental data from successful electroporation-based applications are reviewed and possible mathematical relations between the amplitude of the electric field and the field duration are proposed.

12.2 Overview of the Pulse Parameters Used in Biomedical and Biotechnological Applications of Electroporation

To date, a vast number of different pulse parameters have been reported for various biomedical and biotechnological applications of electroporation. It is practically impossible to present all these parameters in one place; however, we tried to summarize some of the most typical parameters in Table 12.1, at least to illustrate their diversity. In the table, the amplitude of the pulses delivered to the electrodes is given as a voltage-to-distance ratio, which roughly equals to the value of the electric field if the field distribution between the electrodes is homogeneous. Depending on the recovery of cells after electroporation, the table was divided into two parts: (i) applications resulting in reversible electroporation, such as the uptake of small and large molecules, and nanoelectroporation, and (ii) applications resulting in irreversible electroporation, such as tissue ablation and sterilization.

From Table 12.1, it becomes clear that in different applications of electroporation, diverse parameters are used, but also, that these parameters can vary even for the same application. This can partly be attributed to the experimental setup (e.g., different cell lines, media, characteristics of molecules) and partly to the fact that different pulse parameters yield similar outcomes. The latter is the focus of our chapter and the description of these parameters is given in more detail below.

For efficient uptake of small molecules, such as lucifer yellow, propidium iodide, and bleomycin into cells, the electric fields for electroporation, which are usually given as voltage-to-distance ratio, are in the range of 1 kV/cm with durations extending from hundred microseconds to milliseconds (Mir et al. 1988, Wolf et al. 1994, Heller et al. 1996a, 1999, Kotnik et al. 2000, Maček Lebar and Miklavčič 2001, Marty et al. 2006, Serša 2006, Snoj et al. 2006). Such pulses were most often used in experiments involving electrochemotherapy, where small chemotherapeutic agents (bleomycin or cisplatin) were delivered into tumor cells by means of electric pulses.

Larger molecules are, especially due to their size, more difficult to introduce into electroporated cells. In general, four different pulse protocols were applied: (i) high electric field amplitudes from one to few kV/cm, lasting from few microseconds to hundred microseconds (Neumann et al. 1982, Potter et al. 1984, Heller et al. 1996b); (ii) low electric field amplitudes of few hundred volts per centimeter but longer durations ranging into tens of milliseconds (Suzuki et al. 1998, Mir et al 1999); (iii) a combination of high and low field amplitudes, with the former in the range of 1 kV/cm and duration of hundreds of milliseconds, and the latter with approximately 100 V/cm with duration from 10 ms to few hundred milliseconds (Sukharev et al. 1992, Bureau et al. 2000, Pavšelj and Préat 2005, Šatkauskas et al. 2005a, André et al. 2008, Kandušer et al. 2009, Villemejane and Mir 2009); and (iv) the most recent approach, where the pulse protocol described in (iii) was followed by a long train of 40 kV/cm pulses with nanosecond

TABLE 12.1 A Summary of Typical Pulse Parameters Used in Different Applications of Electroporation

	Applications	Pulse Parameters	References
Reversible electroporation	Uptake of small molecules	8×1300 V/cm, 100 μs $1–64 \times (0.25–1$ kV/cm$)$, 20 μs to 1 ms $1–8 \times (0.5–1.3$ kV/cm$)$, 100 μs to 1 ms $1–10 \times 0.9$ kV/cm, 1 ms 8×1.5 kV/cm, 100 μs	Mir et al. (1988), Heller et al. (1996a), Maček Lebar and Miklavčič (2001), Serša (2006), Marty et al. (2006), Snoj et al. (2006), Spugnini et al. (2009), Kotnik et al. (2000), and Wolf et al. (1994)

	Uptake of large molecules	(a) *HV and short*
		    $(1–3) \times 4–8$ kV/cm, 5 μs
		    6×1 kV/cm, 99 μs
			(a) Neumann et al. (1982), Potter et al. (1984), Heller et al. (1996b), and Wolf et al. (1994)

		(b) *LV and long*
		    8×200 V/cm, 20 ms
		    8×250 V/cm, 50 ms
			(b) Mir et al. (1999) and Suzuki et al. (1998)

		(c) *Combination of HV + LV*
		    1×1.5 kV/cm, 100 μs + 1×75 V/cm, 40 ms
		    4×1 kV/cm, 200 μs + 1×75 V/cm, 100 ms
		    1×0.8 kV/cm, 100 μs + 4×80 V/cm, 100–400 ms
		    1×1 kV/cm, 100 μs + 1×140 V/cm, 400 ms
		    1×0.7 kV/cm 100 μs + 200 V/cm, 400 ms
		    1×6 kV/cm, 10 μs + 1×200 V/cm, 10 ms
			(c) Kandušer et al. (2009), Bureau et al. (2000), Šatkauskas et al. (2005a), André et al. (2008), Villemejane and Mir (2009), Pavšelj and Préat (2005), and Sukharev et al. (1992)

		(d) *Combination of HV + LV + nEP* HV + LV + $30,000 \times 40$ kV/cm, 10 ns	(d) Villemejane et al. (2009)
	Nanoporation	1×12 kV/cm, 60 ns	Pakhomov et al. (2007)
		1×26 kV/cm, 60 ns, 10×180 kV/cm, 10 ns; 1×65 kV/cm, 10 ns	Schoenbach et al. (2004)
		1×80 kV/cm, 4 ns	Vernier et al. (2008)
		3–5 pulses of 50 kV/cm, 60 ns	Schoenbach et al. (2001)
		10×25 kV/cm, 30 ns; 50×25 kV/cm, 7 ns	Vernier et al. (2003)
Irreversible electroporation	Tissue ablation	1×1000 V/cm, 20 ms	Edd et al. (2006)
		10×3800 V/cm, 100 μs	Maor et al. (2008)
		90×250 V/cm, 100 μs	Rubinsky et al. (2008)
		8, 16, 32×3800 V/cm, 100 μs	Lavee et al. (2007)
	Sterilization	>15 kV/cm, μs to ms	Knorr (1999)
		20×550 V/cm, 10 ms	Teissié et al. (2002)
		1000×5 kV/cm, 1 ms	El Zakhem et al. (2007)

Notes: The pulse parameters are given as no. of pulses × electric field, pulse duration. The electric field is given as voltage-to-distance ratio. HV, high voltage pulse; LV, low voltage pulse; nEP, nanosecond pulse.

duration, minutes later (Villemejane et al. 2009). These four protocols were typically used for transporting fragments of DNA into cells *in vitro* or *in vivo* for efficient gene electrotransfer.

While pulse protocols described above were mostly targeted to the plasma cell membrane, very strong fields of short, nanosecond duration were reported to affect also membranes of cell organelles. For this purpose, pulses exceeding 10 kV/cm (sometimes even 100 kV/cm) were used with durations from few nanoseconds to tens of nanoseconds (Schoenbach et al. 2001, 2004, Vernier et al. 2003, 2008, Pakhomov et al. 2007).

In tissue ablation, cells were irreversibly electroporated with either (i) high number of pulses with a low amplitude (a few hundred volts per centimeter) and hundred microsecond durations (Rubinsky et al. 2008); (ii) low number of pulses with few kV/cm lasting hundred microseconds (Lavee et al. 2007, Maor et al. 2008); or (iii) with pulses of 1 kV/cm and long duration of tens of milliseconds (Edd et al. 2006).

For sterilization, electric fields larger than 15 kV/cm and lasting from microseconds to milliseconds were delivered to irreversibly electroporate and destruct the membranes of microorganisms (Knorr 1999, Teissié et al. 2002). Alternatively, lower fields from few hundred volts per centimeter to few kV/cm and durations of milliseconds to tens of milliseconds were used for water sterilization (Teissié et al. 2002, El Zakhem et al. 2007).

In general, from the pulse parameters summarized in Table 12.1, it follows that the values of the field amplitudes and durations needed for efficient electroporation of cells (either for the successful uptake of molecules or destruction of cells) are inversely proportional. Namely, with higher pulse amplitudes, the pulse duration causing a similar effect can be shorter and vice versa. This observation is not new and was reported by many authors before. From Table 12.1, it can also be seen that the change in the value of a certain pulse parameter can be compensated by carefully adjusting the values of other parameters, e.g., reduction in the pulse amplitude can be compensated by increasing the pulse duration or number of pulses. This observation can be extremely useful in applications, where the limitations of the electrical devices, such as pulse generator (or electroporator), have to be taken into account.

Furthermore, there have been many contradicting reports on the threshold value of the electric field required for electroporation of cells. Electroporation of the cell membrane is usually associated with an increase of the electric field–induced transmembrane voltage above a certain threshold, which is estimated to be between 200 mV and 1 V (Tsong 1991, Teissié and Rols 1993, Towhidi et al. 2008). Therefore, a threshold value of the electric field that must be exceeded in order to load specific molecules into the cell or to affect the cell viability is expected. However, many theoretical studies, especially those, which were based on the theory of electroporation (Neumann et al. 1982), predicted that electroporation is not a threshold phenomenon. According to theoretical studies, the presence of the electric field only increased the probability of the occurrence of water pores in the membrane (hence the term electroporation). The discrepancy between experiments and theory can be explained by the fact that in experiments, electroporation of the cell membrane was determined by the detection of the internalized molecules, and this can be limited by the sensitivity of the experimental setup and the characteristics of these molecules. For example, several authors reported that structural changes in the membrane and the related increased permeability of the membrane became detectable within a few microseconds after the onset of the electric field (Benz et al. 1979, Kinosita and Tsong 1979, Hibino et al. 1993, Griese et al. 2002, Kakorin and Neumann 2002, Pavlin et al. 2005), but the occurrence of these changes did not coincide with the detection of the transmembrane transport of molecules, which was detected milliseconds or seconds after the pulse (Dimitrov and Sowers 1990, Tekle et al. 1990, 1994, Gabriel and Teissié 1999). The transport of single-atom ions and molecules continues for seconds or even minutes after electroporation, until the cell membrane completely recovers (reseals) or until the equilibrium in concentration of ions and molecules is obtained.

TABLE 12.2 Most Frequently Used Molecules for Studying Electroporation

Molecule	Molecular Weight (Da)	The Number of Molecules Inside the Cell Relative to the Outside (%)
Lucifer yellow	457	100
Propidium iodide	660	90
Bleomycin	1,500	30
Fragments of oligonucleotides	3,000	20
Ribosome inactivating protein PAP	30,000	10
FITC dextran	70,000	<1
Antibodies	150,000	0.1

Source: Maček Lebar et al., *Med. Razgl.*, 37, 339, 1998.

Note: The relative number of molecules was determined for the optimal electroporation parameters.

In contrast, membrane conductivity, which is elevated during the pulse, returns close to the initial value much faster, milliseconds after the exposure (Kinosita and Tsong 1979, Hibino et al. 1993, Schmeer et al. 2004). To a certain extent, these differences can be attributed to the size and the charge of molecules, which hinders the permeation of molecules through an electroporated cell membrane. Table 12.2 demonstrates the influence of the size of the molecules (molecular weight) on the transport into electroporated cells.

12.3 Mathematical Description of the Relation between the Field Strength and the Field Duration

In most studies, cell viability or the uptake of molecules into cells was measured only for a very limited range of pulse parameters. Furthermore, the mathematical relation between the parameters of major importance, field strength and pulse duration, was addressed only by few authors.

Krassowska et al. (2003) tested a train of six pulses with 10 different pulse durations (between 50 μs and 16 ms) and 9–14 different field strengths to determine the magnitude of the electric field required to kill 50% of the cells (E_{50}). When plotted in logarithmic scale, the relationship between E_{50} and pulse duration (T) was

$$E_{50} = kT^A. \tag{12.1}$$

On the basis of their experimental results, they reported a value of A as being very close to -1, indicating that E_{50} is inversely proportional to T. The relation proposed by Equation 12.1 does not have a threshold value for extremely long pulse durations; i.e., the limit of E_{50} as T approaches infinity is 0.

A similar relation was found by Butterwick et al. (2007) who studied the threshold for tissue damage as a function of pulse duration. They used balanced biphasic pulses of seven different pulse durations ranging from 6 μs to 6 ms. The authors reported that for a single pulse, the damage threshold current density j, scaled with pulse duration (T), can be approximated by $1/T^{0.5}$.

He et al. (2007) performed more than 2000 single-cell measurements on five molecules of different sizes to describe a functional dependence between the threshold electric field for loading molecules into cells, E_{EP}, and pulse duration T. Thirteen pulse durations ranging from 400 μs to 15 ms were used and corresponding threshold electric field E_{EP} for each molecular size was determined. They demonstrated that if the electric field is smaller than the threshold value E_0, cell electroporation cannot be detected, not even by extending the pulse duration. They concluded that the values of the threshold electric field, E_{EP}, for loading different molecules formed a three-parameter exponential function of pulse duration T:

$$E_{EP} = E_0 + E_1 e^{-T/\tau}, \tag{12.2}$$

where
    E_0 is the threshold value of the electric field at the longest duration
    E_1 is the difference of threshold electric field at longest duration and shortest duration
    τ is the time constant of exponential decay curve

The limit of E_{EP} as T approaches infinity is E_0; meaning that the proposed E_0 is a threshold value of the electric field that must be exceeded in order to load the specific molecule into cells. The larger the size of the molecule (molecular weight), the higher the value of E_0 is.

Šatkauskas et al. (2005b) electroporated the tumors *in vivo* with eight square wave electric pulses of variable durations (0.1, 0.25, 0.5, and 1 ms) delivered at 1 Hz. Using tumor doubling time as a criteria, they observed three-parameter exponential relation between pulse strength (defining the same tumor doubling time and consequently the same efficiency of electrochemotherapy) and pulse duration, similar to what was reported by He et al. (Equation 12.2).

According to the studies mentioned above, we can summarize that a threshold value of electric field needed for irreversible electroporation with extremely long pulse durations does not exist. On the other hand, a threshold value of electric field needed for reversible electroporation with extremely long pulse durations depends on the size of molecules; the smaller the size of the molecule, the lower the threshold value of the electric field is. For the realistic limiting case of molecular size, namely, single-atom ions, a threshold value of electric field at extremely long pulse durations should be very low or, hypothetically, even disappear. To investigate this, a threshold value of the electric field needed to load calcium ions into cells with a single pulse of different pulse durations was studied.

12.4 Materials and Methods

12.4.1 Cells

Chinese hamster ovary cells (CHO-K1) were plated in Lab-Tek II chambers (Nalge Nunc International, United States) at 2×10^5 cells/chamber in the culture medium HAM-F12 supplemented with 8% fetal calf serum, 0.15 mg/mL L-glutamine (all three from Sigma-Aldrich, Steinheim, Germany), 200 units/mL benzylpenicillin (penicillin G), and 16 mg/mL gentamicin and incubated in 5% CO_2 at 37°C. The experiments were performed 18–24 h after plating, when most cells were firmly attached to the surface of the chamber (see Figure 12.1) and most of them did not yet divide.

12.4.2 Monitoring Electroporation

To monitor electroporation of cells, a fluorescent calcium indicator Fura-2AM was used (Grynkiewicz et al. 1985). Fura2-AM enters the cell through an intact membrane, and is transformed in the cytosol into Fura2, the active and membrane-nonpermeant ratiometric dye (excitation 340/380 nm, emission 540 nm). Electroporation results in the entry of Ca^{2+} ions into the cells, where their binding to Fura2 causes the change in the fluorescence of the dye. Under moderate pulse parameters, the cell recovers after electroporation, stores the excess Ca^{2+} into the intracellular reservoirs or excludes the Ca^{2+} from the cytoplasm. The fluorescence thus decreases with time back to the initial value, allowing for another repetition of the experiment on the same cells.

The procedure for staining the cells with Fura was already described elsewhere (Towhidi et al. 2008). Briefly, prior to experiments, the cells were washed with Spinner's Minimum Essential Medium (SMEM, Gibco, United States; a calcium-depleted modification of EMEM), left at room temperature for 25 min in SMEM containing 2 µM Fura2-AM, and subsequently washed again with SMEM to remove the dye from the extracellular medium. Finally, SMEM was replaced with culture medium HAM-F12, which

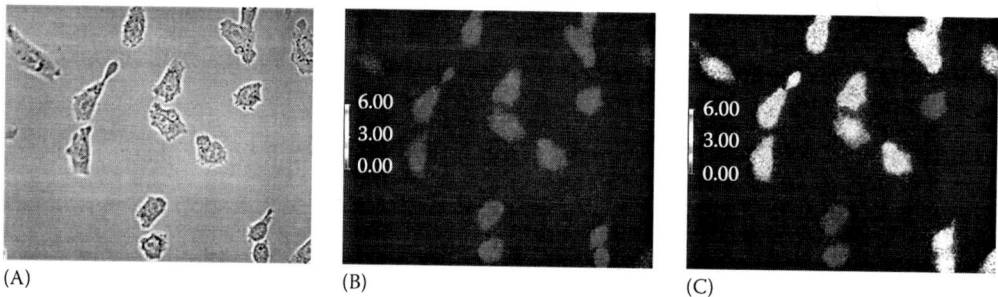

(A)   (B)   (C)

FIGURE 12.1   (A) CHO cells: bright field. (B) Fluorescence of cells in control (nonporated cells). (C) Cells 1 min after electroporation with a 100 V (250 V/cm), 10 ms pulse. Brighter cells were electroporated. Field direction was from top to bottom.

contains Ca^{2+} ions (app 1 mM). Thus, the calcium ions were present in the extracellular medium but were nearly absent from the cytosol, as they do not readily cross the nonporated cell membrane.

12.4.3 Pulse Delivery

Single rectangular electric pulse with duration of 0.03, 0.1, 1, 10, or 50 ms was generated with a Cliniporator⁻ device (IGEA s.r.l., Carpi, Modena, Italy). For a given pulse duration, the pulse amplitude was increased stepwise with each consecutive pulse (a 10 V step) until approximately 70% of the cells were electroporated (Figure 12.1). The second pulse with higher amplitude was delivered with at least 5 min delay to enable cell recovery (verified in a separate experiment, see below), except if cells already became fluorescent (electroporated). In this case, the next pulse was delivered 5 min after the fluorescence of cells returned to the initial value. To speed up the resealing, the cells were kept at 37°C.

The pulses were delivered to a pair of parallel Pt/Ir wire electrodes with 0.8 mm diameter and 4 mm distance between them, which were positioned at the bottom of the chamber. Cells were monitored under a fluorescence microscope (40× objective, AxioVert 200, Zeiss, Germany), equipped with a CCD camera and a monochromator (both Visitron, Germany). The presence of calcium was determined ratiometrically using MetaFluor 5.1 software (Molecular Devices, GB), with the excitation wavelengths set at 340 and 380 nm, and the emission measured at 540 nm for both excitation wavelengths.

12.4.4 Verification of Cell Recovery after Electroporation

To test if cells completely recovered 5 minutes after electroporation, we performed an additional experiment. Cells were prepared and incubated with the dye as described above. However, they were not electroporated in the culture medium but in SMEM, which does not contain Ca^{2+} ions. The pulse with the amplitude leading to 70% of electroporated cells was delivered (see above) and 5 min later the Ca^{2+} ions were added to SMEM to obtain a 1 mM Ca^{2+} concentration in the chamber.

Since the fluorescence of the cells remained low and unchanged, Ca^{2+} ions apparently did not enter the cells, meaning that the cells fully recovered after electroporation. The same pulse was applied again, and since the Ca^{2+} was now present in the extracellular solution, the fluorescence of the cells increased, confirming that the pulse was indeed electroporating the cells. Similar results were obtained for all investigated pulse durations.

12.4.5 Numerical Modeling of the Transport of Small Molecules into Electroporated Cells

Optimization of the pulse parameters for efficient electroporation is usually related to time-consuming experiments. An alternative is to use an appropriate model for this phenomenon and calculate the influence of different parameters on the transport. The models are based on the assumption that the electric field increases the probability of the occurrence of structural changes in the lipid bilayer of the membrane (water pores). Currently, the models are able to predict the influence of only a few pulse parameters (e.g., amplitude, duration) on pore formation and closure (DeBruin and Krassowska 1999, Neu and Krassowska 1999, Neumann et al. 1999, Joshi et al. 2002, Gowrishankar and Weaver 2003), while a descriptive model of the electroporation-mediated molecular transport is missing. We recently developed a *dynamic* model of the transport of molecules into electroporated cells of arbitrary shapes, which is able to predict the efficiency of electroporation for different pulse parameters. The model (Miklavčič and Towhidi 2008) is described in detail in one of the chapters of this book; below, we only give its brief description.

The model is based on previously suggested (Neumann et al. 1998) and recently confirmed (Böckmann et al. 2008) model, where the occurrence of structural changes in the membrane due to the external electric field is described with a transition from the closed/initial lipid state (C) to the open or porous state (P_2), as depicted in Equation 12.3. The transition occurs with two intermediate states, C_1, where lipid head groups are tilted, and P_1, which is a prepore or a transient pore state.

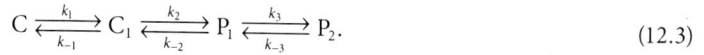

$$C \underset{k_{-1}}{\overset{k_1}{\rightleftharpoons}} C_1 \underset{k_{-2}}{\overset{k_2}{\rightleftharpoons}} P_1 \underset{k_{-3}}{\overset{k_3}{\rightleftharpoons}} P_2. \tag{12.3}$$

The permeability of the membrane in P_1 state is negligibly small so P_2 is predominantly responsible for the molecular transport. Pore formation and closure are denoted by the rate coefficients k_1, k_2, and k_3, which were considered as being equal ($k_1 = k_2 = k_3 = k_p$) (Neumann et al. 1998) but field dependent. This is in contrast to previous studies, where the average k was considered (Neumann et al. 1996, 1998, 1999), but allowed the calculation of the spatial and temporal distribution of pores on the membrane by this model.

The model was supplemented with equations describing transport to determine the molecular uptake into a single cell. The total uptake was computed with integration of transported molecules through the cell membrane over the time and the cell surface. Considering the geometry of the cell of interest, the solutions of self-consistent equations of suggested model were calculated using COMSOL 3.3 package based on finite-element method.

12.5 Results and Discussion

Figure 12.2 shows the experimentally determined electric field (E) needed to electroporate 70% of cells (filled circles) at different pulse durations (T). An exponential function with three constants (Equation 12.2; dotted line), proposed by He et al., and function proposed by Krassowska et al. (Equation 12.1; dash-dotted line) were fitted to the data using SigmaPlot 8.0. It can be seen that the mathematical relation described by Equation 12.1 fits the data much better ($R^2 = 0.9983$) than the relation described by Equation 12.2 ($R^2 = 0.9908$). The value of parameter A in mathematical relation described by Equation 12.1, which fits best the experimental data, is −0.1661. This is roughly six times smaller than the value of A given by Krassowska. A smaller value of A was expected, because we were reversibly electroporating cells in contrast to Krassowska, which performed irreversible electroporation, and also, we were loading smaller ions into cells.

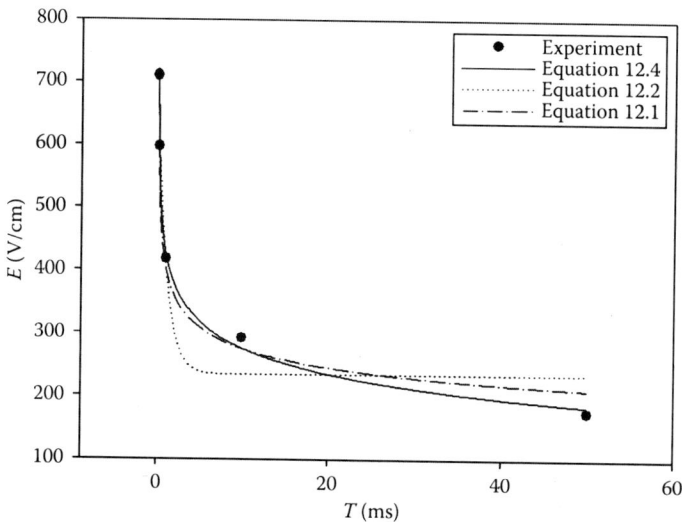

FIGURE 12.2 Experimentally determined electric field (E) needed to load calcium ions into 70% of cells at different pulse durations (T). Mathematical relations described by Equations 12.1, 12.2, and 12.4 were fitted to the data (filled circles) and are presented by dash-dotted line, dotted line, and solid line, respectively.

An even better fit ($R^2 = 0.9996$; solid line) is obtained with modified version of Equation 12.1 written as

$$E = \left[c_1 + c_2 \ln(T) \right] T^m. \tag{12.4}$$

Equation 12.4 is the solution of the second-order Euler–Cauchy equation:

$$T^2 \frac{d^2 E}{dT^2} + aT \frac{dE}{dT} + bE = 0, \tag{12.5}$$

for the case, when characteristic equation has just one real repeated root m; i.e., $a = 1 - 2m$ and $b = m^2$. The fit to experimental data gives $m = 0.0345$.

Similar results are obtained if the predictions of the numerical model are taken into consideration (Figure 12.3). Mathematical relations described by Equations 12.1, 12.2, and 12.4 were fitted to the numerical data (filled circles) and are presented by dash-dotted line, dotted line, and solid line, respectively. Again, Equation 12.4 gives the mathematical relation that fits the data best ($R^2 = 0.9991$), while the mathematical relation given by Equation 12.2 is the least adequate ($R^2 = 0.9911$).

Electroporation–based technologies and medical applications have already shown their laboratory and clinical relevance. *In vitro* electroporation is becoming a standard tool in biotechnology, and medical applications are in progress. Although the number of successful applications is increasing, several questions concerning the optimization of pulse parameters for specific application are still open. Among them is determination of appropriate amplitude, duration, number, and repetition frequency of electric pulses that assure successful application or treatment with minimal possible side effects. A review of the studies related to pulse parameters used in biomedical and biotechnological applications shows a palette of efficient pulse parameters combinations. It is apparent, that the change in the value of a certain pulse parameter can be compensated by carefully selected values of the other parameters. This observation can be extremely useful in the process of electroporation-based treatment planning, where limitations of the electrical devices have to be taken into account.

Systematic studies aimed at addressing the role of electric pulse parameters in efficiency of electroporation applications are scarce. Even scarcer are mathematical relations describing the pulse parameters.

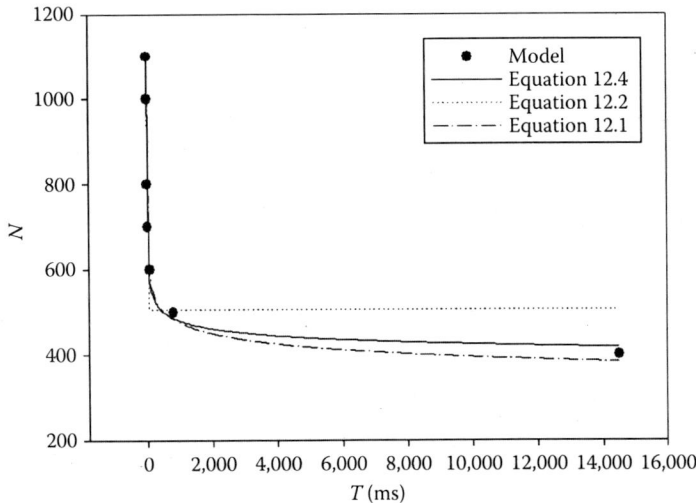

FIGURE 12.3 The calculated uptake of small molecules (N) at different pulse durations (T). Mathematical relations described by Equations 12.1, 12.2, and 12.4 were fitted to the data (filled circles) and are presented by dash-dotted line, dotted line, and solid line, respectively.

In this chapter, we focused on the mathematical relation between the electric field strength and pulse duration that assure loading of small ions into cells. The experimental results demonstrated that a threshold value of electric field needed to load small ions into the cells with extremely long pulse durations does not exist. The mathematical relationship proposed (Equation 12.4) agreed well with the experimental data and predictions of the model for a wide range of pulse durations (from µs to s). According to this relationship, we can select different pairs of electric field amplitude/duration that assure similar effectiveness of electroporation, which can represent an improvement in treatment planning.

Acknowledgment

This work was supported through various grants of the Slovenian Research Agency.

References

André FM, Gehl J, Serša G, Preat V, Hojman P, Eriksen J, Golzio M et al. 2008. Efficiency of high- and low-voltage pulse combinations for gene electrotransfer in muscle, liver, tumor, and skin. *Hum Gene Ther* 19: 1261–1271.

Benz R, Beckers F, Zimmermann U. 1979. Reversible electrical breakdown of lipid bilayer membranes—Charge-pulse relaxation study. *J Membr Biol* 48: 181–204.

Böckmann RA, Groot BL, Kakorin S, Neumann E, Grubmüller H. 2008. Kinetics, statistics, and energetics of lipid membrane electroporation studied by molecular dynamics simulations. *Biophys J* 95: 1837–1850.

Bureau MF, Gehl J, Deleuze V, Mir LM, Scherman D. 2000. Importance of association between permeabilization and electrophoretic forces for intramuscular DNA electrotransfer. *Biochim Biophys Acta* 1474: 353–359.

Butterwick A, Vankov A, Huie P, Freyvert Y, Palanker D. 2007. Tissue damage by pulsed electrical stimulation. *IEEE Trans Biomed Eng* 54: 2261–2267.

Canatella PJ, Karr JF, Petros JA, Prausnitz MR. 2001. Quantitative study of electroporation-mediated molecular uptake and cell viability. *Biophys J* 80: 755–764.

Davalos RV, Mir LM, Rubinski B. 2005. Tissue ablation with irreversible electroporation. *Ann Biomed Eng* 33: 223–231.

DeBruin KA, Krassowska W. 1999. Modeling electroporation in a single cell. II. Effects of ionic concentrations. *Biophys J* 77: 1225–1233.

Denet AR, Vanbever R, Preat V. 2004. Skin electroporation for transdermal and topical delivery. *Adv Drug Deliv Rev* 56: 659–674.

Dimitrov DS, Sowers AE. 1990. Membrane electroporation—Fast molecular-exchange by electroosmosis. *Biochim Biophys Acta* 1022: 381–392.

Edd JF, Horowitz L, Davalos RV, Mir LM, Rubinsky B. 2006. In vivo results of a new focal tissue ablation technique: Irreversible electroporation. *IEEE Trans Biomed Eng* 53: 1409–1415.

Elouagari K, Teissié J, Benoist H. 1995. Glycophorin A protects K562 cells from natural-killer-cell attack—Role of oligosaccharides. *J Biol Chem* 270: 26970–26975.

El Zakhem H, Lanoiselle JL, Lebovka NI, Nonus M, Vorobiev E. 2007. Influence of temperature and surfactant on *Escherichia coli* inactivation in aqueous suspensions treated by moderate pulsed electric fields. *Int J Food Microbiol* 120: 259–265.

Gabriel B, Teissié J. 1999. Time courses of mammalian cell electropermeabilization observed by millisecond imaging of membrane property changes during the pulse. *Biophys J* 76: 2158–2165.

Golzio M, Mazzolini L, Ledoux A, Paganin A, Izard M, Hellaudais L, Bieth A et al. 2007. In vivo gene silencing in solid tumors by targeted electrically mediated siRNA delivery. *Gene Ther* 14: 752–759.

Gowrishankar TR, Weaver JC. 2003. An approach to electrical modeling of single and multiple cells. *Proc Natl Acad Sci USA* 100: 3203–3208.

Griese T, Kakorin S, Neumann E. 2002. Conductometric and electrooptic relaxation spectrometry of lipid vesicle electroporation at high fields. *Phys Chem Chem Phys* 4: 1217–1227.

Grynkiewicz G, Poenie M, Tsien RY. 1985. A new generation of Ca^{2+} indicators with greatly improved fluorescence properties. *J Biol Chem* 260: 3440–3450.

He H, Chang DC, Lee YK. 2007. Using a micro electroporation chip to determine the optimal physical parameters in the uptake of biomolecules in HeLa cells. *Bioelectrochemistry* 70: 363–368.

Heller R, Jaroszeski MJ, Glass LF, Messina JL, Rapaport DP, DeConti RC, Fenske NA, Gilbert RA, Mir LM, Reintgen DS. 1996a. Phase I/II trial for the treatment of cutaneous and subcutaneous tumors using electrochemotherapy. *Cancer* 77: 964–971.

Heller R, Jaroszeski M, Atkin A, Moradpour D, Gilbert R, Wands J, Nicolau C. 1996b. In vivo gene electroinjection and expression in rat liver. *FEBS Lett* 389: 225–228.

Heller R, Gilbert R, Jaroszeski MJ. 1999. Clinical applications of electrochemotherapy. *Adv Drug Deliv Rev* 35: 119–129.

Hibino M, Itoh H, Kinosita K. 1993. Time courses of cell electroporation as revealed by submicrosecond imaging of transmembrane potential. *Biophys J* 64: 789–800.

Jaroszeski MJ, Gilbert R, Nicolau C, Heller R. 1999. In vivo gene delivery by electroporation. *Adv Drug Deliv Rev* 35: 131–137.

Joshi RP, Hu Q, Schoenbach KH, Bebe SJ. 2002. Simulations of electroporation dynamics and shape deformations in biological cells subjected to high voltage pulses. *IEEE Trans Plasma Sci* 30: 1536–1546.

Kakorin S, Neumann E. 2002. Ionic conductivity of electroporated lipid bilayer membranes. *Bioelectrochemistry* 56: 163–166.

Kandušer M, Miklavčič D, Pavlin M. 2009. Mechanisms involved in gene electrotransfer using high- and low-voltage pulses—An in vitro study. *Bioelectrochemistry* 74: 265–271.

Kinosita K, Tsong TY. 1979. Voltage-induced conductance in human erythrocyte membranes. *Biochim Biophys Acta* 554: 479–497.

Knorr D. 1999. Novel approaches in food-processing technology: New technologies for preserving foods and modifying function. *Curr Opin Biotechnol* 10: 485–491.

Kotnik T, Maček Lebar A, Miklavčič D, Mir LM. 2000. Evaluation of cell membrane electropermeabilization by means of nonpermeant cytotoxic agent. *Biotechniques* 28: 921–926.

Krassowska W, Nanda GS, Austin MB, Dev SB, Rabussay DP. 2003. Viability of cancer cells exposed to pulsed electric fields: The role of pulse charge. *Ann Biomed Eng* 31: 80–90.

Lavee J, Onik G, Mikus P, Rubinski B. 2007. A novel nonthermal energy source for surgical epicardial atrial ablation: Irreversible electroporation. *Heart Surg Forum* 10: 92–101.

Maček Lebar A, Miklavčič D. 2001. Cell electropermeabilization to small molecules in vitro: Control by pulse parameters. *Radiol Oncol* 35: 193–202.

Maček Lebar A, Serša G, Čemažar M, Miklavčič D. 1998. Elektroporacija. *Med Razgl* 37: 339–354.

Maor E, Ivorra A, Leor J, Rubinsky B. 2008. Irreversible electroporation attenuates neointimal formation after angioplasty. *IEEE Trans Biomed Eng* 55: 2268–2274.

Marty M, Serša G, Garbay JR, Gehl J, Collins CG, Snoj M, Billard V et al. 2006. Electrochemotherapy—An easy, highly effective and safe treatment of cutaneous and subcutaneous metastases: Results of ESOPE (European Standard Operating Procedures of Electrochemotherapy) study. *Eur J Cancer Suppl* 4: 3–13.

Miklavčič D, Towhidi L. 2008. Time course of electrical and diffusional parameters during and after electroporation. In: *IFMBE Proceedings*, vol. 22. Springer, Berlin, Germany, pp. 2659–2663.

Mir LM, Orlowski S. 1999. Mechanisms of electrochemotherapy. *Adv Drug Deliv Rev* 35: 107–118.

Mir LM, Banoun H, Paoletti C. 1988. Introduction of definite amounts of nonpermeant molcules into living cells after electropermeabilization—Direct access to the cytosol. *Exp Cell Res* 175: 15–25.

Mir LM, Bureau MF, Gehl J, Rangara R, Rouy D, Caillaud JM, Delaere P, Branellec D, Schwartz B, Scherman D. 1999. High-efficiency gene transfer into skeletal muscle mediated by electric pulses. *Proc Natl Acad Sci USA* 96: 4262–4267.

Neu JC, Krassowska W. 1999. Asymptotic model of electroporation. *Phys Rev E* 59: 3471–3482.

Neumann E, Schaeferridder M, Wang Y, Hofschneider PH. 1982. Gene-transfer into mouse lyoma cells by electroporation in high electric-fields. *EMBO J* 1: 841–845.

Neumann E, Kakorin S, Tsoneva I, Nikolova B, Tomov T. 1996. Calcium-mediated DNA adsorption to yeast cells and kinetics of cell transformation by electroporation. *Biophys J* 71: 868–877.

Neumann E, Toensing K, Kakorin S, Budde P, Frey J. 1998. Mechanism of electroporative dye uptake by mouse B cells. *Biophys J* 74: 98–108.

Neumann E, Kakorin S, Tönsing K. 1999. Fundamentals of electroporative delivery of drugs and genes. *Bioelectrochem Bioenerg* 48: 3–16.

Ogura A, Matsuda J, Yanagimachi R. 1994. Birth of normal young after electrofusion of mouse oocytes with round spermatids. *Proc Natl Acad Sci USA* 91: 7460–7462.

Pakhomov AG, Kolb JF, White JA, Joshi RP, Xiao S, Schoenbach KH. 2007. Long-lasting plasma membrane permeabilization in mammalian cells by nanosecond pulsed electric field (nsPEF). *Bioelectromagnetics* 28: 655–663.

Pavlin M, Kanduser M, Rebersek M, Pucihar G, Hart FX, Magjarević R, Miklavčič D. 2005. Effect of cell electroporation on the conductivity of a cell suspension. *Biophys J* 88: 4378–4390.

Pavšelj N, Préat V. 2005. DNA electrotransfer into the skin using a combination of one high- and one low-voltage pulse. *J Control Release* 106: 407–415.

Potter H, Weir L, Leder P. 1984. Enhancer-dependent expression of human kappa-immunoglobulin genes introduced into mouse pre-b lymphocytes by electroporation. *Proc Natl Acad Sci USA* 81: 7161–7165.

Prausnitz MR. 1996. The effects of electric current applied to skin: A review for transdermal drug delivery. *Adv Drug Deliv Rev* 18: 395–425.

Rols MP, Teissié J. 1998. Electropermeabilization of mammalian cells to macromolecules: Control by pulse duration. *Biophys J* 75: 1415–1423.

Rowan NJ, MacGregor SJ, Anderson JG, Fouracre RA, Farish O. 2000. Pulsed electric field inactivation of diarrhoeagenic *Bacillus cereus* through irreversible electroporation. *Lett Appl Microbiol* 31: 110–114.

Rubinsky J, Onik G, Mikus P, Rubinsky B. 2008. Optimal parameters for the destruction of prostate cancer using irreversible electroporation. *J Urol* 180: 2668–2674.

Šatkauskas S, Bureau MF, Puc M, Mahfoudi A, Scherman D, Miklavčič D, Mir LM. 2002. Mechanisms of in vivo DNA electrotransfer: Respective contributions of cell electropermeabilization and DNA electrophoresis. *Mol Ther* 5: 133–140.

Šatkauskas S, André F, Bureau MF, Scherman D, Miklavčič D, Mir LM. 2005a. Electrophoretic component of electric pulses determines the efficacy of in vivo DNA electrotransfer. *Hum Gene Ther* 16: 1194–1201.

Šatkauskas S, Batiuškaite D, Šalomskaite-Davalgiene S, Venslauskas MS. 2005b. Effectiveness of tumor electrochemotherapy as a function of electric pulse strength and duration. *Bioelectrochemistry* 65: 105–111.

Schmeer M, Seipp T, Pliquett U, Kakorin S, Neumann E. 2004. Mechanism for the conductivity changes caused by membrane electroporation of CHO cell-pellets. *Phys Chem Chem Phys* 6: 5564–5574.

Schoenbach KH, Beebe SJ, Buescher ES. 2001. Intracellular effect of ultrashort electrical pulses. *Bioelectromagnetics* 22: 440–448.

Schoenbach KH, Joshi RP, Kolb JF, Chen NY, Stacey M, Blackmore PF, Buescher ES, Beebe SJ. 2004. Ultrashort electrical pulses open a new gateway into biological cells. *Proc IEEE* 92: 1122–1137.

Serša G. 2006. The state-of-the-art of electrochemotherapy before the ESOPE study; advantages and clinical uses. *Eur J Cancer Suppl* 4: 52–59.

Serša G, Miklavčič D, Čemažar M, Rudolf Z, Pucihar G, Snoj M. 2008. Electrochemotherapy in treatment of tumours. *Eur J Surg Oncol* 34: 232–240.

Snoj M, Rudolf Z, Paulin-Košir SM, Čemažar M, Snoj R, Serša G. 2006. Long lasting complete response in melanoma treated by electrochemotherapy. *Eur J Cancer Suppl* 4: 26–28.

Spugnini EP, Vincenzi B, Citro G, Tonini G, Dotsinsky I, Mudrov N, Baldi A. 2009. Electrochemotherapy for the treatment of squamous cell carcinoma in cats: A preliminary report. *Vet J* 179: 117–120.

Sukharev SI, Klenchin VA, Serov SM, Chernomordik LV, Chizmadzhev YA. 1992. Electroporation and electrophoretic DNA transfer into cells - the effect of DNA interaction with electropores. *Biophys J* 63: 1320–1327.

Suzuki T, Shin BC, Fujikura K, Matsuzaki T, Takata K. 1998. Direct gene transfer into rat liver cells by in vivo electroporation. *FEBS Lett* 425: 436–440.

Teissié J. 1998. Transfer of foreign receptors to living cell surfaces: The bioelectrochemical approach. *Bioelectroch Bioenerg* 46: 115–120.

Teissié J, Rols MP. 1993. An experimental evaluation of the critical potential difference inducing cell membrane electropermeabilization. *Biophys J* 65: 409–413.

Teissié J, Eynard N, Vernhes MC, Benichou A, Ganeva V, Galutzov B, Cabanes PA. 2002. Recent biotechnological developments of electropulsation. A prospective review. *Bioelectrochemistry* 55: 107–112.

Tekle E, Astumian RD, Chock PB. 1990. Electropermeabilization of cell-membranes—Effect of the resting membrane-potential. *Biochem Biophys Res Commun* 172: 282–287.

Tekle E, Astumian RD, Chock PB. 1994. Selective and asymmetric molecular-transport across electroporated cell-membranes. *Proc Natl Acad Sci USA* 91: 11512–11516.

Towhidi L, Kotnik T, Pucihar G, Firoozabadi SMP, Mozdarani H, Miklavčič D. 2008. Variability of the minimal transmembrane voltage resulting in detectable membrane electroporation. *Electromagn Biol Med* 27: 372–385.

Tsong TY. 1991. Electroporation of cell membranes. *Biophys J* 60: 297–306.

Ušaj M, Trontelj K, Hudej R, Kandušer M, Miklavčič D. 2009. Cell size dynamics and viability of cells exposed to hypotonic treatment and electroporation for electrofusion optimization. *Radiol Oncol* 43: 108–119.

Vernier PT, Sun YH, Marcu L, Salemi S, Craft CM, Gundersen MA. 2003. Calcium bursts induced by nanosecond electric pulses. *Biochem Biophys Res Commun* 310: 286–295.

Vernier PT, Sun YH, Chen MT, Gundersen MA, Craviso GL. 2008. Nanosecond electric pulse-induced calcium entry into chromaffin cells. *Bioelectrochemistry* 73: 1–4.

Villemejane J, Mir LM. 2009. Physical methods of nucleic acid transfer: General concepts and applications. *Br J Pharmacol* 157: 207–219.

Villemejane J, Joubert V, Mir LM. 2009. Nanosecond electric pulses-induced enhancement of reporter gene expression after electrotransfer in skeletal muscle in vivo. In: *XX International Symposium on Bioelectrochemistry and Bioenergetics*, Romanian Society of Pure and Applied Biophysics, Sibiu, Romania, 136.

Wolf H, Rols MP, Boldt E, Neumann E, Teissié J. 1994. Control by pulse parameters of electric field-mediated gene transfer in mammalian cells. *Biophys J* 66: 524–531.

Zimmerman U. 1982. Electric field-mediated fusion and related electrical phenomena. *Biochim Biophys Acta* 694: 227–277.

IV

Mechanisms of Electroporation in Tissues

13

Drug-Free, Solid Tumor Ablation by Electroporating Pulses: Mechanisms That Couple to Necrotic and Apoptotic Cell Death Pathways

Axel T. Esser

Kyle C. Smith

Thiruvallur R. Gowrishankar

James C. Weaver

13.1 Introduction

Cancer treatments often fail at the individual cellular and tumor levels. At the cellular level, cancer cells frequently do not respond to pharmaceutical treatments because of acquired multiple drug resistance by active pump mechanisms (Szakcas et al. 2006), inhibitors of apoptosis (Dean et al. 2005, Putt et al. 2006), and inhibition of signaling molecules (Sebolt-Lepold and English 2006). Molecular mechanisms inhibiting apoptosis pathways may also hinder physical therapies, such as localized ionizing radiation as well as systemic pharmaceutical interventions (Galluzzi et al. 2006, Reed 2006a,b). At the multicellular tissue level, there are additional barriers within solid tumors to drug therapies, which may arise from the inability of drugs to fully penetrate abnormal, heterogeneous, and irregularly vascularized tumor tissue and thereby fail to reach all of the cancer cells at their therapeutic levels (Jain 1996, 2001, Padera et al. 2004, Minchinton and Tannock 2006).

Accordingly, local physical therapies that universally ablate, or alternatively keep in check, all cells within unwanted tissue are potentially valuable as minimally invasive clinical tools for surgical, and increasingly also for cosmetic applications (Livraghi et al. 2005). Tissue ablation may be achieved by thermal impact (heat or cold) due to focused ultrasound, RF electric fields, or cryo-ablation. These conventional ablation techniques all lead to cell death by necrosis. Strong electric field pulses causing electroporation (EP) of cellular membranes—a universal mechanism by which a membrane may become permeable to drugs, molecules, and genetic material (Weaver and Chizmadzhev 1996, Weaver 2003)—may alternatively be used for tissue ablation. Drug-assisted EP methods, such as electro-chemotherapy (ECT) (Mir and Orlowski 1999) and electro-gene therapy (Jaroszeski et al. 2000), are examples of EP-based tissue ablation methods with drugs. There is now evidence, however, that drugs actually may not be necessary, since the impact of particular strong electric fields, such as irreversible EP (IRE) pulses (Davalos et al. 2005, Miller et al. 2005, Edd et al. 2006, Al-Sakere et al. 2007, Rubinsky et al. 2007, Maor et al. 2009) and nanosecond pulsed electric fields (nsPEFs) (Beebe et al. 2002, Nuccitelli et al. 2006, Garon et al. 2007, Nuccitelli et al. 2009) alone may be sufficient to trigger molecular pathways that lead to cellular death.

In short, tissue ablation by IRE involves the application of a few pulses with durations of typically hundreds of microseconds and electric field strengths on the order of a few kilovolts per centimeter (kV/cm). Al-Sakere et al., for example, reported IRE tumor ablation in mice by using 80 pulses of 2.5 kV/cm and 100 μs duration, at a repetition frequency of 0.3 Hz (Al-Sakere et al. 2007). In a further report of IRE-ablation in vascular smooth muscle cells by Maor et al. (2009), e.g., 10 pulses with 100 μs duration and field strengths of 0.5–3.5 kV/cm at 0.3 Hz were used. These IRE pulses lead to conventional EP, that is, pores in the cell plasma membrane (PM), which are sufficiently large to facilitate molecular release and uptake. While those pores are reversible and reseal in most EP-based biotechnological and medical applications, IRE pulses are designed in such a way that at least some PM pores do not reseal. Thus, the PM could be said to be irreversibly electroporated. Therefore, the barrier function of the PM is immediately and irreversibly lost, a situation defined as cellular death (Kroemer et al. 2009). Specifically, solid tumor treatment by IRE is observed to cause cell swelling (Granot et al. 2009), a typical morphological signature of necrosis in the targeted cells (Al-Sakere et al. 2007, Kroemer et al. 2009). This mechanism has been used to kill microorganisms (Sale and Hamilton 1967) and is also known to contribute to non-thermal electrical injury (Lee and Kolodney 1987, Lee et al. 2000).

Tissue ablation by nsPEFs, on the other hand, is based on the use of Blumlein line or MOSFET-based pulsed power technologies and involves hundreds of ultrashort pulses with durations of ten to several hundreds of nanoseconds and field strengths of tens or even hundreds of kV/cm (Beebe et al. 2002, Nuccitelli et al. 2006, Garon et al. 2007, Nuccitelli et al. 2009). The use of those sub-microsecond pulses was motivated by the observation of pronounced intracellular effects (Schoenbach et al. 2001, Beebe et al. 2002, 2003a,b, Vernier et al. 2003, 2004, White et al. 2004) and led to the development of the supra-EP hypothesis (Stewart et al. 2004, Gowrishankar and Weaver 2006, Gowrishankar et al. 2006), which is a different degree of EP. The concept of using nsPEFs as a therapeutic tool for solid tumor treatments was first demonstrated by Beebe et al. (2002), reporting the induction of apoptosis in solid tumors (mouse fibrosarcoma) ex vivo and the reduction of fibrosarcoma tumor size *in vivo* by nsPEF pulses up to 300 kV/cm and durations from 10 to 300 ns. Morphologic signs of apoptotic cell death comprise cell shrinkage and persistent externalization of phosphatidylserine (PS) at the PM, supplemented by biochemical changes, such as the activation of caspases and the fragmentation of DNA (Beebe et al. 2002). Apoptosis induction by nsPEF pulses has also been observed for mammalian cancer cells *in vitro* (Vernier et al. 2005).

In addition, Nuccitelli et al. (2006, 2009) used, e.g., 400 electric field pulses with durations of about 300 ns and electric field strengths of 20 and 40 kV/cm, respectively at 0.5 Hz to demonstrate the self-destruction of melanomas. Specifically, the reduction of cellular and nuclear cell volume (pyknosis) was reported (Nuccitelli et al. 2006, 2009), which is a classical morphological signature of apoptosis. Also biochemical features of the apoptotic pathway were found, such as DNA fragmentation. The nonthermal mechanism by which the melanoma cells die was therefore interpreted as apoptosis.

However, those apoptotic features may also be mixed with necrotic biochemical and morphological features such that a clear and absolute distinction between apoptosis and necrosis may not always be possible. For example, Garon et al. (2007) used up to 1000 electric field pulses that were even one order of magnitude shorter, namely 20 ns or less, and field strengths up to 60 kV/cm at 20 Hz. Their results showed decreased cell viability of a variety of human cancer cells *in vitro*, induction of tumor regression and hemorrhagic necrosis *in vivo*, and the successful treatment of a human subject with a basal cell carcinoma for which they found a "complete pathologic response." Cell death does not appear to be due to immediate PM destruction, as above for IRE pulses. Instead, it appears to be the result of delayed effects, which may be caused by the efflux of Ca^{2+} from intracellular stores (e.g., endosplasmic reticulum) and may eventually cause apoptotic cell death (Garon et al. 2007).

Thus, the above two drug-free, EP-based methods with nsPEF and IRE pulses lead to different cellular death mechanisms, necrosis and apoptosis. This may be related to the different electrical and permeability degrees of conventional versus supra-EP. In order to understand this difference, it is important to recognize that EP is not simply punching nanometer-sized holes through cellular membranes. Instead, a widely accepted hypothesis of EP involves dynamic transient pores (Weaver and Chizmadzhev 1996, Weaver 2003). The dynamics of these pores are related to their electrical formation, their agile expansion and contraction in response to actual values of the time-dependent local PM transmembrane voltage, $\Delta\psi_{PM}$, and eventually their stochastic destruction. In addition, the resulting membrane permeability must be based upon the resulting pore sizes and is fundamental to understanding necrotic cell death by IRE pulses versus apoptotic cell death by nsPEFs.

However, an initial basic distinction needs to be made: irreversibility can be manifested at the membrane level or, if membrane EP is itself reversible, at the cellular level. In this sense, there are two distinct mechanisms that may lead to necrotic cell death: (1) the evolution of a pore population, such that a significant number or size of pores become irreversibly trapped or held open permanently (for instance, by the interaction between a pore and channel), and therefore lead to an irreversible lethal biochemical imbalance, and (2) evolution of a pore population, such that a lethal biochemical change occurs within the cellular compartment, for example, by Ca^{2+}-induced activation of calpains (Kroemer et al. 2009) even if all pores vanish in a relatively short time (Weaver 2003), and thus, EP at the PM itself is reversible. It is presently unknown which of the above two mechanism is most relevant for necrotic cell death by IRE. We believe, however, that electrical impedance tomography (EIT), as suggested by Granot et al. (2009) may be used to rule out one scenario over the other. Exposures by nsPEFs, on the other hand, almost certainly lead to reversible resealing of membrane pores, such that irreversible behavior seems to be the result of intracellular biochemical and morphological changes.

The consideration of both IRE pulses and nsPEFs thus provides the tantalizing prospect of designing specific electric field pulses and treatment protocols, which may even lead to a choice of preferential cell death: necrosis with IRE and apoptosis with nsPEFs. The ability to target different cellular death mechanisms by an appropriate choice of electric field parameters may, at the present time, seem like an unnecessary choice for a patient in urgent need of tumor treatment. However, the apoptotic cell death pathway may provide certain advantages. Specifically, if secondary necrosis can be avoided, then it should be possible to bypass nonspecific damage to nearby tissue due to e.g., inflammation and/or scarring. It might then also be possible to avoid the tumor lysis syndrome resulting from massive tumor necrosis.

The object of the present contribution is therefore to seek an underlying mechanistic understanding of the tissue response to both EP conditions. This is achieved, here, by using a multicellular model and a tissue model (Esser et al. 2007, 2009a), for which we determine the redistribution and local magnitudes of electric fields and currents in the treated tissue, both intracellularly and extracellularly, and then by studying the biophysical response in terms of pore densities and pore sizes leading to different cell permeability changes for IRE and nsPEF conditions. Throughout, we attempt to make objective comparisons between responses to the many, ultrashort and very large nsPEFs and the fewer, longer and weaker IRE pulses.

13.2 Methods

We use a transport lattice (TL) method, which allows for a convenient description of the electrical, chemical, and thermal behavior in complex biological geometries that may contain inhomogeneities and anisotropies. Basic features of the TL method have been presented elsewhere (Gowrishankar and Weaver 2003, 2006, Gowrishankar et al. 2004, 2006, Stewart et al. 2004, Smith et al. 2006). Here we describe two system models, a multicellular model of irregular cells, and a tissue model and consider their electrical responses to some characteristic IRE and nsPEF conditions. Each system model represents rabbit liver tissue but on a different spatial scale (Esser et al. 2009, 2010).

13.2.1 Multicellular Model

The system geometry, shown in Figure 13.1A, is based on a drawing motivated by a tissue section image. It features a layer of 20 liver cells with 14% interstitial fluid volume. The hepatocytes have an average cell

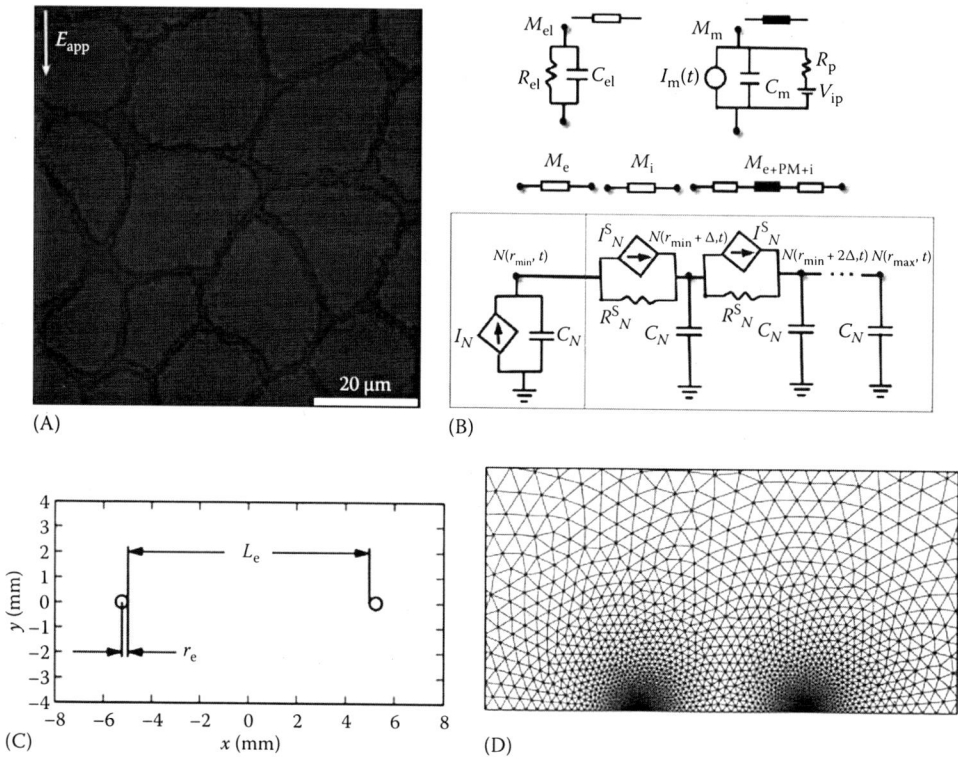

(A)　(B)

(C)　(D)

FIGURE 13.1 **(See color insert following page 268.)** (A) Geometry of the multicellular model of a region of liver tissue with 14% interstitial space and an average cell dimension of 21.7 μm. The extracellular electrolyte is shown in dark gray, the cells in lighter gray (see online color version). Also given is the direction of the applied electric field E_{app}. (B) TL circuits: Functional local models that represent electrolyte (M_{el}), membrane (M_m), and the PM-electrolyte interface (M_{e+PM+i}) in the 101 × 101 TL; the lattice spacing l, as well as the depth of the system model, is 1 μm, leading to a spatial scale of 100 μm × 100 μm; more details described elsewhere (Gowrishankar and Weaver 2003, Gowrishankar et al. 2006). The equivalent EP circuit representing the asymptotic EP model (left gray box) (Stewart et al. 2004) and its extension to the SE (Esser et al. 2010) is solved at every local membrane site. The local pore distributions are discretized, and pore drift and diffusion determine the associated non-Ohmic pore conductance G_m that is input for the time-dependent membrane current I_m (t) in module M_m. (C) Two circular electrodes in a 200 mm × 100 mm tissue region (only a small subregion of the entire simulation region is shown). The electrodes have radii r_e = 0.25 mm and separation L_e = 10 mm (edge-to-edge). (D) The tissue system mesh (only a small subregion of the mesh close to the needle electrodes is shown here).

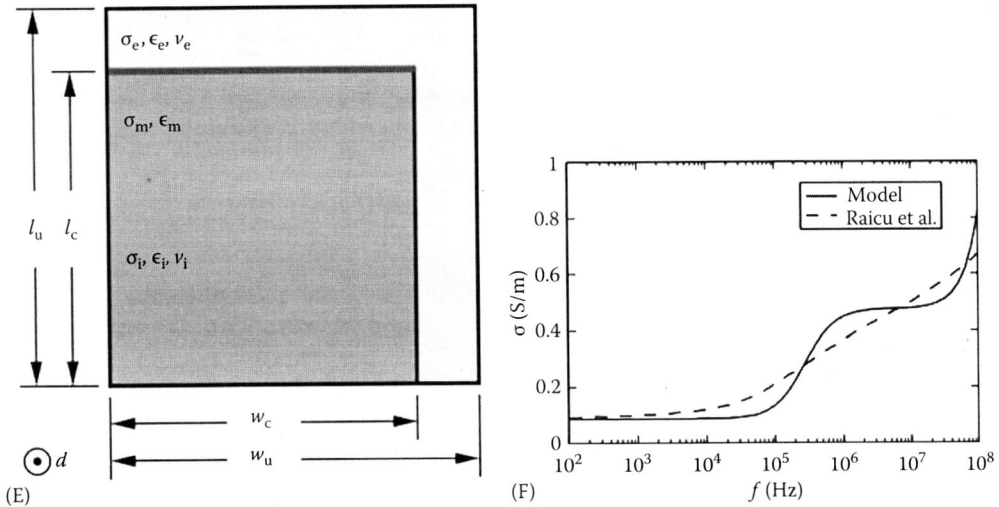

FIGURE 13.1 (continued) (E) The tissue model cell unit comprises extracellular (e), membrane (m), and intracellular (i) regions in series and a parallel shunt region. The spatial and electrical parameters of the regions are labeled. (F) The passive frequency-dependent tissue conductivity of the tissue model, obtained from an AC-frequency sweep in SPICE, is compared with experimental measurements on rat liver (Raicu et al. 1998). (From Esser, A.T. et al., *Technol. Cancer Res. Treat.*, 6, 263, 2007.)

size diameter of 21.7 µm (Esser et al. 2009) and a corresponding TL (101 nodes × 101 nodes) was constructed as a large electric circuit comprising passive (resistors, capacitors) and active (pumps, channels, EP) elements at the membrane. The linked local membrane and electrolyte models are distributed spatially and are connected to their nearest neighbors on a Cartesian lattice (Gowrishankar and Weaver 2003, Stewart et al. 2004). The local membrane models are interconnected at the regularly spaced nodes, with submodels that represent the PM and two contacting regions of electrolytes (Figure 13.1B) (Gowrishankar and Weaver 2003, Stewart et al. 2004). There is no transport in the *z*-direction in this two-dimensional model. The lattice spacing *l* and the depth of the system model is 1 µm, and the system has a spatial scale of 100 µm × 100 µm. Voltages applied along the top and bottom boundary of the system model provide the applied uniform field. The multicellular cell system model has ~10^4 interconnected local models for charge transport and storage within electrolytes and nonlinear charge transport at the membranes (Figure 13.1B).

13.2.2 Multiscale Tissue Model

The tissue system comprises a large tissue region and two ideal cylindrical electrodes, each with a radius of $r_e = 0.25$ mm, separated by $L_e = 10$ mm (Figure 13.1C). The nominal applied electric field, E_{app}, is defined here as the voltage difference between the electrodes, V_{app}, divided by the electrode spacing, L_e. Although it is well known that needle electrodes actually have spatially varying fields, the term "nominal electric field" is used for convenience by some and is also consistent with the original definition in the limit of infinite needle radii (i.e., planar electrodes). The tissue system with a scale of 200 mm × 200 mm is symmetric about $y = 0$. Thus, a no-flux boundary was placed at $y = 0$ and only the region $y \geq 0$ was actually meshed and simulated. Nodes were optimally distributed throughout the tissue system using a meshing algorithm developed by Persson and Strang (2004). Figure 13.1D shows the mesh near the electrodes. A multiscale tissue model is developed that accounts for the electrical response at both the microscopic (e.g., PM EP) and macroscopic (e.g., needle geometry) scales and the interplay between the two (Smith 2006). The spatial scale of the tissue system is orders of magnitude larger than the spatial scale of the cells in the tissue and therefore a discretization of the system could not realistically resolve individual cells and membranes. The multiscale model uses representative simple cell models distributed throughout the

TABLE 13.1   Tissue Model Parameters

Parameter	Description	Value
r_e	Electrode radius [mm]	0.25
L_e	Electrode distance [mm] (edge-to-edge)	10
$\sigma_e = \sigma_i$	Electrolyte conductivity [S/m]	1.2
σ_m	Initial membrane conductivity [S/m]	9.5×10^{-9}
$\varepsilon_e = \varepsilon_i$	Electrolyte permittivity	$72 \times \varepsilon_0$
d_m	Membrane thickness [nm]	5
ν_e	Tortuosity (external)	1
ν_i	Tortuosity (internal)	3
f_e	Extracellular volume fraction	0.14
$w_c = l_c$	Cell dimension [μm]	21.7
$w_u = l_u$	Unit cell dimensions [μm]	23.4

system model to calculate the local cell and membrane response, and the macroscopic electrical transport properties are determined by the distributed models.

13.2.2.1 Tissue Level

The impedance between two nodes that may comprise a number of cells in a small local tissue region is equal to the impedance of a cell scaled to have the same relative dimensions, assuming that total tissue volume comprises a uniform grid of such cells. Thus, we use the simple cell model shown in Figure 13.1E, which has a membrane enclosed region of intracellular electrolytes surrounded by extracellular electrolytes (Smith 2006). An advantage of this simple model is that it can be straightforwardly translated into an equivalent circuit. The membrane and each region of electrolytes have an associated conductivity σ and permittivity ε (Table 13.1). Additionally, each electrolyte region has a tortuosity to account for the structural complexity of tissue otherwise not represented by the model. The relative sizes were chosen such that 14% of the total volume was extracellular. The tortuosities ν were used as free parameters in fitting the frequency-dependent liver tissue conductivity to that measured experimentally by Raicu et al. (1998). The passive electrical properties of tissue, which are relevant for weak electric fields, are frequency-dependent (Foster and Schwan 1996) as higher frequencies lead to stronger displacement currents at the cellular membranes and therefore allow for access to subcellular compartments. This behavior is reproduced in our tissue model; see Figure 13.1F. The equivalent circuit for the simple cell model (Figure 13.1E) is placed between each pair of adjacent nodes in the mesh with electrical components scaled in accordance with the local mesh geometry. The effective conductance of the membrane changes in accordance with the local degree of membrane EP as determined by the distributed cell models.

13.2.2.2 Cell Level

An equivalent circuit for a single cell is created for each node in the mesh to determine the cellular response to the local electric field. Each of these circuits is distinct from the primary, macroscopic tissue-level circuit network, but all of the circuits are solved simultaneously. The asymptotic model of EP is used as described below. The voltage across the simple cell unit is equal to the local electric field magnitude, as determined from the electric potential of the nodes in the mesh, multiplied by the cell unit length l_u. The distributed cell models determine the transmembrane voltage and pore density throughout the tissue domain as functions of time. The pore density determines the membrane conductance used in the macroscopic transport network. Thus, there is continual feedback between the macroscopic (tissue level) and microscopic (cellular level) models: the macroscopic behavior determines the local electric field in the microscopic model and the subsequent behavior at the microscopic scale (e.g., EP), which then determines the local electrical properties at the macroscopic scale.

13.2.3 Electrolyte and Membrane Models

The passive electric components for the electrolyte are resistors and capacitors (Gowrishankar and Weaver 2003) (Figure 13.1B). The membrane circuits (Figure 13.1B) include components for charge storage and conduction, resting potential, and an EP model (Stewart et al. 2004) (see below). This provides a convenient means for including the $d_m = 5$ nm thick membrane in a TL of a much larger scale. Close to a pore, the membrane dielectric is treated as pure lipid and assigned a dielectric constant, ε_l, of 2.1. This choice recognizes that the local membrane properties are relevant to pore formation. In contrast, the PM capacitance involves a spatial average over membrane lipid and protein regions resulting in a relative permittivity, ε_m, of 5. The extracellular electrolyte has a conductivity, σ_e, of 1.2 S/m while the medium inside the cell has a conductivity, σ_i, of 0.4 S/m. These conductivity values are identical to the validated macroscopic tissue model, wherein the effective conductivities are the ratio of the electrolyte conductivity and the tissue tortuosity (see Figure 13.1F and Table 13.1). Following Läuger (1991), we use a simplified, single resting potential source model (Figure 13.1B) comprised of an active voltage source, V_{ip}, and source series resistance, R_{ip} (Stewart et al. 2004). Here, the fixed quantities V_{ip} and R_{ip}, together with a negligible conductance of the equilibrium pores (Vasilkoski et al. 2006), determine the membrane resting potential, $\Delta\psi_{PM,rest}$, in the absence of applied electric fields.

13.2.4 EP Models

The transient aqueous pore hypothesis of EP is based on the continuum models of membrane pores, electrostatic energy differences, and thermal fluctuations, usually in the form of the Smoluchowski equation (SE) (Powell and Weaver 1986, Barnett and Weaver 1991, Freeman et al. 1994, Vasilkoski et al. 2006):

$$\frac{\partial n}{\partial t} = \frac{\partial}{\partial r_p}\left[D_p\left(\frac{\partial n}{\partial r_p} + \frac{n}{k_B T}\frac{\partial W}{\partial r_p}\right)\right] \qquad (13.1)$$

The SE describes the evolution of the local PM pore distribution n in terms of the number of hydrophilic pores and pore radius r_p given by the diffusion constant D_p, and is used in the multicellular model. The asymptotic EP model (Neu and Krassowska 1999) used in the tissue model is an approximation to the SE-based EP models that disregard pore size change and can be used to describe cell and tissue responses to nanosecond timescale pulses (Gowrishankar and Weaver 2006, Gowrishankar et al. 2006). For longer ECT and IRE pulses, the asymptotic EP model approximates the system electrical response, whereby more 0.8 nm pores are created to increase the membrane conductance in response to elevated transmembrane voltages (Vasilkoski et al. 2006). These pores readily transport Na^+, Cl^-, and K^+ ions that dominate extracellular and cytosolic conductivity, but not significantly larger molecules.

The implementation of the asymptotic EP model in terms of an equivalent circuit is described in detail elsewhere (Stewart et al. 2004). This equivalent circuit can readily be generalized to include dynamic pore expansion and contraction based on the SE (Esser et al. 2007, 2009, 2010). The extended equivalent circuit is given in Figure 13.1B and represents drift and diffusion in the pore radius space from the minimum pore radius ($r_{min} = 0.8$ nm) to a maximal pore radius (considered here as $r_{max} = 3$ nm). From the pore distribution, we find the local non-Ohmic conductivity σ_m (Vasilkoski et al. 2006), which, together with the local $\Delta\psi_{PM}(t)$, determines the time-dependent membrane current $I_m(t)$ as input into the membrane circuit M_m. This approach may describe the PM response for all types of EP pulses, starting from nanosecond pulses that lead to supra-EP to microsecond or longer pulses that give rise to conventional EP. Pore lifetimes reported in the literature vary over many orders of time (from milliseconds for lipid bilayers (Melikov et al. 2001) to minutes and hours for cells in suspension (Weaver 2003)). Because there is presently no mechanistic understanding of this large range, we use an illustrative experimental value of $\tau_p = 3$ ms for the mean pore lifetime (Melikov et al. 2001). Note, however, that we only consider electric and permeability conditions during the pulses for which the value of the pore lifetime is not directly relevant. All parameters of this standard model of EP are given elsewhere (Vasilkoski et al. 2006).

13.2.5 TL Solution

The system model circuits are solved for the electric potential ϕ by means of Kirchhoff's laws using Berkeley SPICE 3f5. Spice-generated solutions are processed and displayed by MATLAB® as distributed equipotentials, distributions of electroporated regions, and pore histograms.

13.3 Results

13.3.1 Multicellular Model

We first consider IRE conditions and show in Figure 13.2 the temporal and spatially distributed electric response of the multicellular model to a trapezoidal 700 V/cm pulse of 100 μs duration with 1 μs rise and fall times. Specifically, on the left of Figure 13.2, equipotential field lines (black lines) are given by the local values of the electric potential, ϕ, inside and outside the cells. The local electric fields are

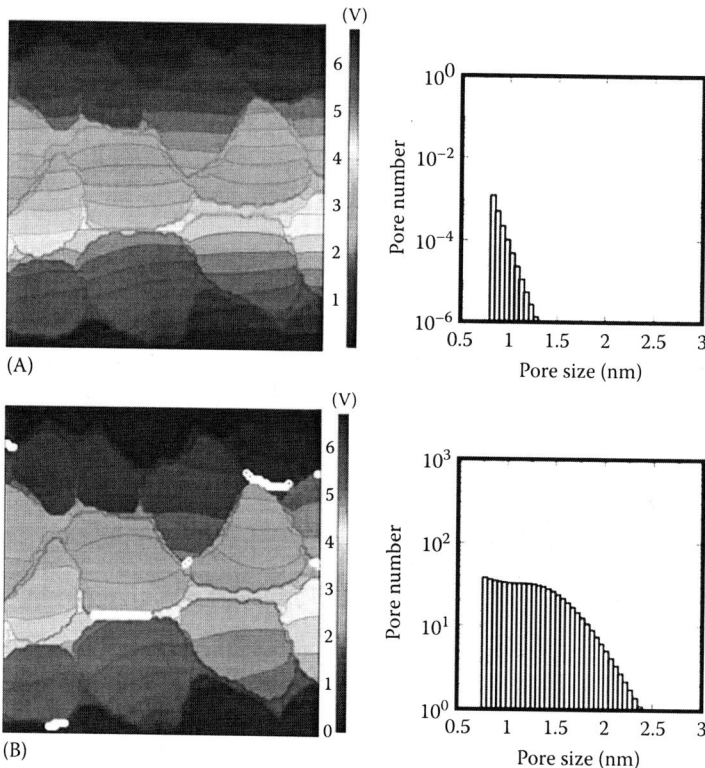

FIGURE 13.2 (See color insert following page 268.) Electrical response of multicellular model to 700 V/cm pulse (1 μs rise and fall times, 98 μs pulse plateau) at different timepoints, and pore histograms of the entire model (pore size interval of 0.05 nm). (A) Charging phase at $t = 1$ μs: Displacement currents at the PM lead to intracellular electric fields. Pore histogram has different scale here, as no significant EP has occurred yet. (B) EP has occurred in distinct cells at $t = 1.7$ μs, a significant number of pores are created at the minimum size radius of 0.8 nm, and some pores have already grown towards larger radii. White dots in the left panels correspond to pore sites with more than 1 pore (equivalent to a pore density of 10^{12} pores/m²). Equipotentials within cells indicate rising intracellular electric fields as a consequence of EP and the resulting nonlinear increase of membrane conductivity.

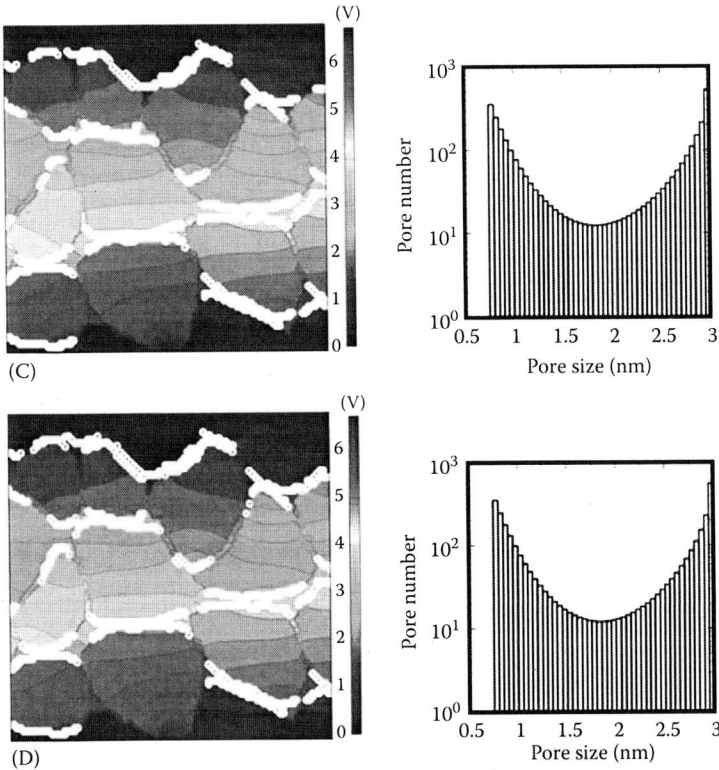

FIGURE 13.2 (continued) (C) All cells are electroporated at $t = 21\,\mu s$. The pore histogram shows both many small and many large pores in the multicellular model, resulting in distinct permeability properties for molecular uptake and release. The pore accumulation at the maximum radius of 3 nm considered here is not a physical effect but a technical limitation of the present model. It means that those pores eventually would reach an even larger size. (D) Distributed response does not change qualitatively, thus at the end of the pulse plateau at $t = 99\,\mu s$ we find a similar pattern to (C). (From Esser, A.T. et al., *Technol. Cancer Res. Treat.*, 6, 267, 2007.)

perpendicular to those equipotential field lines. White dots in the left panels correspond to local PM sites with at least one local pore (corresponding to a pore density of $10^{12}\,\mathrm{m}^{-2}$). The corresponding pore histograms, on the right panels of Figure 13.2, show the total number of pores in all cells of the multicellular model in pore size intervals of 0.05 nm. There is essentially no EP at the end of the IRE pulse rise time of $1\,\mu s$ (Figure 13.2A), and the pore histogram (note the logarithmic histogram scale) reflects the response of the equilibrium pore distribution to elevated transmembrane voltages, $\Delta\psi_{PM}$, of up to 0.75 V. With no pores present at $t = 1.0\,\mu s$, the displacement currents at the PMs give rise to the significant intracellular electric fields, E_{int}.

At $t = 1.7\,\mu s$, shown in Figure 13.2B, EP has occurred at distinct cell sites starting at a minimum pore size of 0.8 nm (Vasilkoski et al. 2006), and the pore histogram indicates that some pores already had time to expand. In addition, part of the electric field and current is driven through the individual cells, as indicated by the intracellular equipotentials (Figure 13.2B) although E_{int} is smaller at this point in time than in Figure 13.2A. This happens because the applied pulse does not change in time during the pulse plateau and thus the displacement currents decay. However, the intracellular electric fields start to rise again in magnitude due to the EP-induced conductive pore currents at the PM.

Figure 13.2C shows that during the IRE pulse more pores are created, which expand to larger radii and thus increase the membrane conductivity. Thus, an increasing intracellular electric field E_{int} is the consequence. At the end of the pulse at $t = 99\,\mu s$, shown in Figure 13.2D, all cells are electroporated, and the pore histogram essentially exhibits two subpopulations of pores: many small pores and many large pores. These two subpopulations within the pore distribution have also been found in an isolated spherical cell model (Krassowska and Filev 2007, Esser et al. 2010). This behavior results from the behavior of the pore energy W at elevated transmembrane voltages (~0.5 V) favoring small and large pores, but intermediate pore sizes around 2 nm are less likely (Esser et al. 2010). The number of large pores is enhanced in this multicellular model because of the presence of nearby cells or, equivalently, by the lack of sufficient extracellular space. This results in local plateau values of $\Delta\psi_{PM}$, which are noticeably higher than for isolated cells.

Next, we consider nsPEF conditions for comparison. Figure 13.3 shows, on the left, a strikingly different distributed electrical response to a trapezoidal 40 kV/cm pulse with 55 ns rise and fall times and 215 ns plateau duration (Nuccitelli et al. 2006, 2009). Figure 13.3A shows a timepoint during the pulse rise at $t = 34$ ns, which corresponds to a field strength of 23 kV/cm. Here, displacement and conduction currents at the PM are strong and lead to essentially equal intracellular and extracellular electric fields. This can be judged by the almost parallel equipotentials throughout the entire multicellular model rendering the electric field uniform. In this sense, the cellular structure is essentially electrically transparent. This is in strong contrast to the IRE conditions in Figure 13.2, for which the cellular shape is still visible. The electroporative creation of pores has already started at this timepoint at almost all PM sites, regardless of the cell size, but also regardless of whether or not the local membrane site faces

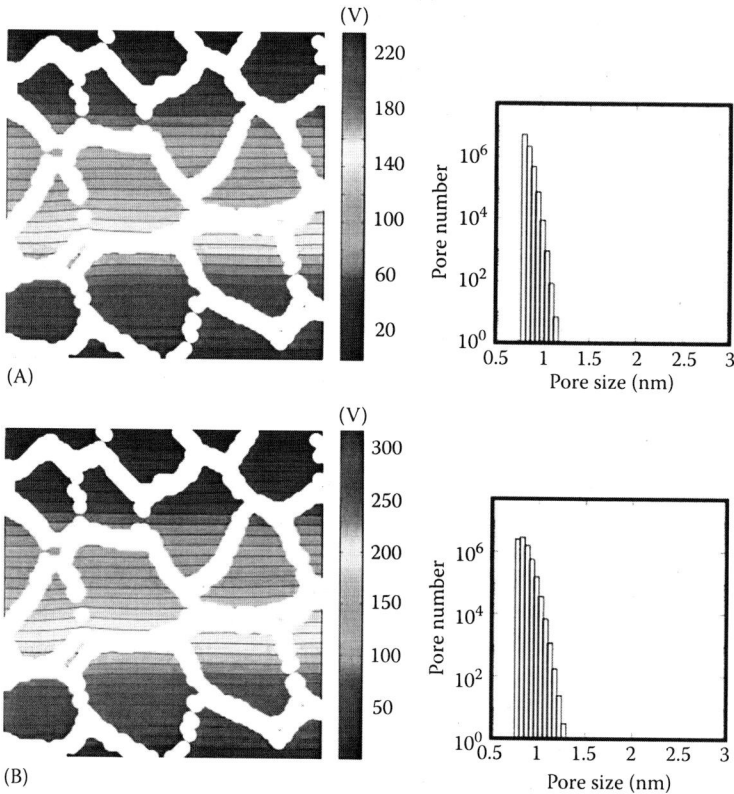

FIGURE 13.3 (**See color insert following page 268.**) Electrical response of multicellular model to a trapezoidal 40 kV/cm pulse (55 ns rise and fall times, 215 ns pulse plateau) and corresponding pore histograms showing the total number of pores (pore size interval of 0.05 nm). (A) Membrane charging phase during pulse rise time at $t = 34$ ns (when the applied field strength is about 23 kV/cm): Due to the larger E_{app} compared to Figure 13.2 pore creation has occurred at essentially all PM sites. White dots in the left panels correspond to pore sites with more than one pore (equivalent to a pore density of 10^{12} pores/m²). (B) Still during the pulse rise time at $t = 45$ ns (when the applied field strength is about 30 kV/cm): Elevated PM transmembrane voltages lead to pulse-duration limited pore expansion.

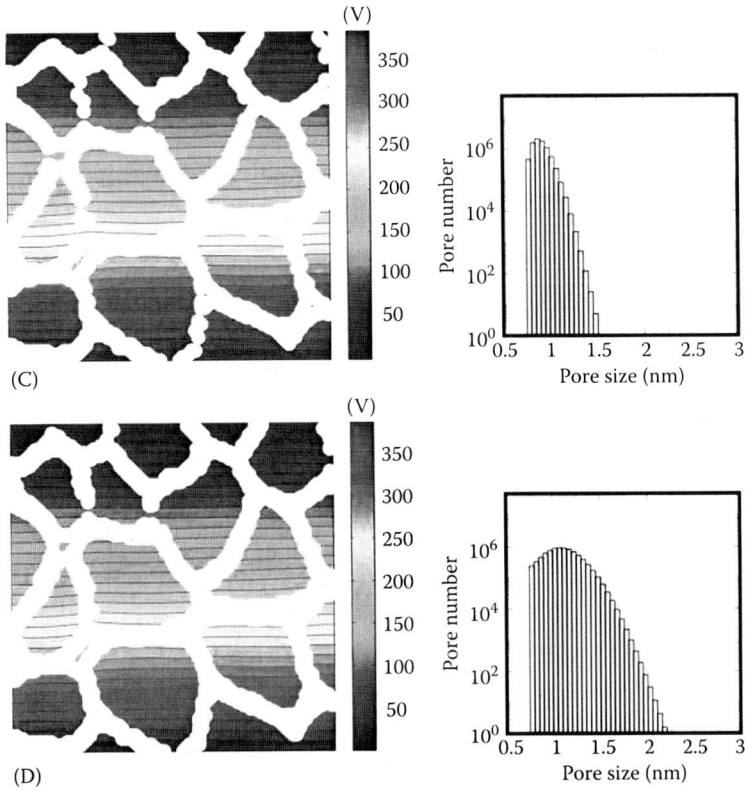

FIGURE 13.3 (continued) (C) At $t = 67$ ns during the pulse and afterwards, basically no new pores are created at the PM, but the elevated PM transmembrane voltage of 0.6–1 V cause the present pores to further grow in size. (D) Distributed electric response at $t = 270$ ns at the end of the pulse plateau does not change qualitatively from (C). The pore histogram shows still further pore expansion, but the short overall duration of the pulse restricts the pores to an averaged size of 1.15 nm. (From Esser, A.T. et al., *Technol. Cancer Res. Treat.*, 8, 295, 2009a.)

the electrodes (polar regions). This is a hallmark of supra-EP, which does not depend on cell size and cell orientation with regard to the field direction (Stewart et al. 2004, Gowrishankar and Weaver 2006).

Figure 13.3B shows the distributed electric response at a timepoint during the pulse rise time at $t = 45$ ns, which corresponds to a present field strength of 30 kV/cm. Further pores have been created and the overall number of pores is two to three orders of magnitude higher than for the IRE conditions in Figure 13.2. At $t = 67$ ns during the pulse plateau, the pore histogram in Figure 13.3C shows that basically no new pores are created at the PM, but that the elevated local PM transmembrane voltages in the range of 0.6–1.1 V (not shown) cause the small pores that are present to grow in size. The distributed electric response in Figure 13.3D at the end of the pulse plateau at $t = 270$ ns again does not change qualitatively from Figure 13.3C. The pores have further grown somewhat in size during the pulse, but the short overall duration of the pulse restricts the pores to small sizes, such that an averaged size of only $\langle r_p \rangle = 1.15$ nm is achieved. In particular, about 100 pores have grown to a size of 2 nm for all cells in the model (corresponding to a density of about 6×10^{10} m^{-2}). These findings about the pore size distribution provide further evidence that the asymptotic EP model, which neglects pore expansion, is a good approximation to describe the membrane responses to nsPEFs (Vasilkoski et al. 2006).

As seen from Figure 13.3, the electric current is always being driven through both the interstitial and the intracellular space because of the strong contribution of both displacement and conduction currents. As such, the applied electric field also perturbs the organelles not explicitly considered here. This also holds true for longer pulses (Esser et al. 2007), such as those under IRE conditions, but the resulting intracellular electric field is typically smaller than for nsPEF conditions. Thus, the stronger and direct exposure of the cell interior to electric fields is a unique property of nsPEFs, and arguably is responsible for the observed intracellular effects (Schoenbach et al. 2001, Beebe et al. 2002, 2003a,b, Vernier et al. 2003, 2004, White et al. 2004). Turning this argument around, it is therefore important to determine the resulting intracellular electric field, E_{int}, for any electric exposure to evaluate the possibility of cellular effects.

Figure 13.4 shows, for an additional nsPEF condition, the distributed electrical response and correspondent pore histograms of the multicellular model to a trapezoidal 60 kV/cm pulse with 2.5 ns rise and fall times and a 15 ns plateau duration (Garon et al. 2007). Figure 13.4A shows the distributed electric response at $t = 2.8$ ns, which is already during the pulse plateau. Displacement currents at the PM are even stronger here because of the higher applied field strength and lead to intracellular electrical fields that are essentially equal to the extracellular electric fields. Pore creation has started at this time at distinct membrane sites. Notably, EP may not necessarily start at the polar side of a cell facing the electrodes, as is well known for isolated cells (Weaver 2003), but rather next to the interstitial space where current flow is strongest. Figure 13.4B shows a timepoint at $t = 3.6$ ns, at which EP has occurred on almost all PM sites with essentially the same spatial pore distribution as in Figure 13.3. At $t = 11.2$ ns, shown in Figure 13.4C, only a few more pores have been created at the PM, as can be best seen from the maximum value of the pore histogram. Elevated PM transmembrane voltage $\Delta\psi_{PM}$ values in the range of 0.6–1.1 V (not shown)

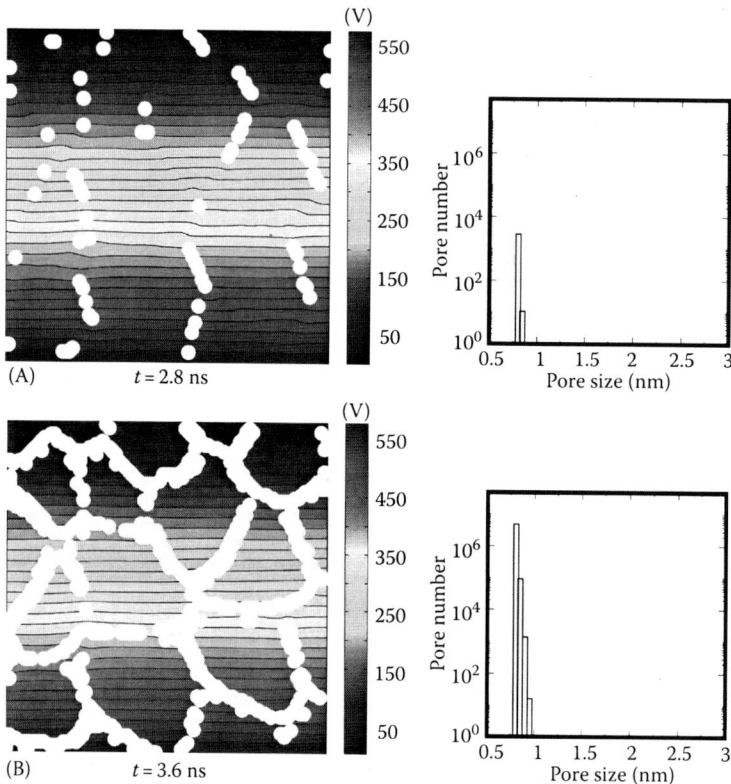

FIGURE 13.4 (See color insert following page 268.) Electrical response of multicellular model to a trapezoidal 60 kV/cm pulse (2.5 ns rise and fall times, 17 ns pulse plateau) and correspondent pore histograms showing the total number of pores (pore size interval of 0.05 nm). (A) At $t = 2.8$ ns already during the pulse plateau: Displacement currents at the PM dominate and lead to intracellular electrical fields essentially equal to extracellular electric fields. Pore creation due to EP has started at this time at some membrane sites, predominantly in the nonpolar regions of the cells. White dots in the left panels correspond to pore sites with more than one pore (equivalent to a pore density of 10^{12} pores/m^2). (B) At $t = 3.6$ ns: EP has occurred on essentially all PM sites, including the membrane sites not facing the polar side of the cell.

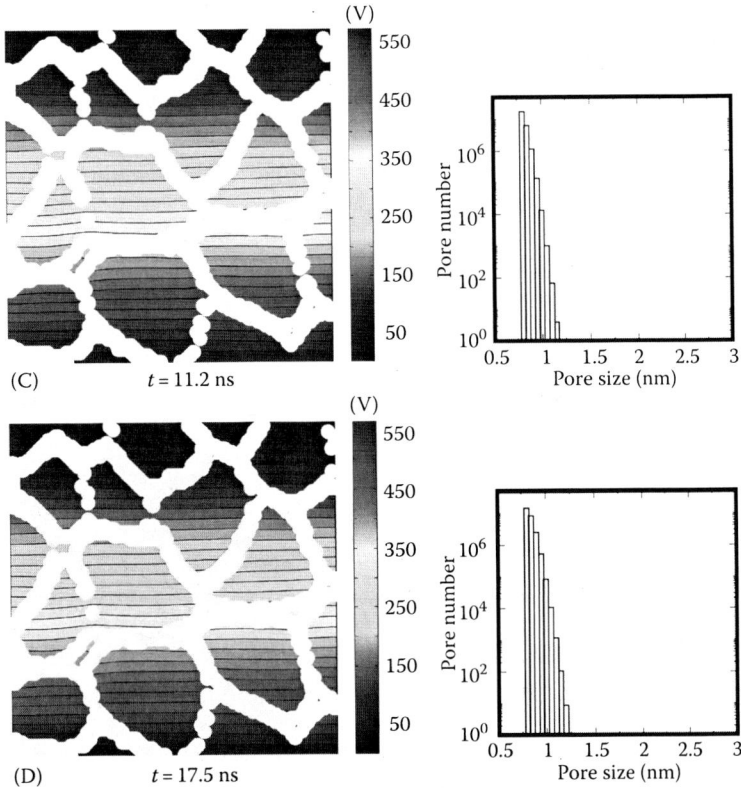

FIGURE 13.4 (continued) (C) At $t = 11.2$ ns during the pulse plateau, basically no new pores are created at the PM, but the elevated PM transmembrane voltage of 0.6–1 V cause the pores to grow. (D) Distributed electric response at $t = 17.5$ ns at the end of the pulse plateau does not change qualitatively from (C). The pore histogram shows some further pore expansion, but the even shorter duration of the pulse compared to Figure 13.3 restricts the pores to an averaged size of 0.85 nm. (From Esser, A.T. et al., *Technol. Cancer Res. Treat.*, 8, 296, 2009a.)

cause the small pores present to start growing in size. The distributed electric response in Figure 13.4D at the end of the pulse plateau at $t = 17.5$ ns does not change qualitatively from Figure 13.4C. The pore histogram indicates that some more pores have grown in size, but the shorter overall duration of this pulse, as compared with Figure 13.3, restricts the pores to even smaller sizes, such that an averaged size of only $<r_p> = 0.95$ nm is reached. No pores with a size of 2 nm are reached in this case.

The results in Figures 13.3 and 13.4 are in contrast to literature statements claiming that the PM is not affected by nsPEF pulses (Beebe et al. 2002, 2003a,b, Vernier et al. 2003). According to our supra-EP hypothesis (Stewart et al. 2004), the PM must respond to avoid unsustainable transmembrane voltages of several volts. All molecular dynamics (MD) simulations of EP (Tieleman 2004) and the experimental results of Frey et al. (2006) are consistent with our models that show supra-EP. The PM is thus very strongly affected by nsPEFs. Resulting pore densities are much higher than those expected for longer IRE-type conventional EP pulses. Supra-EP, however, is a different response as conventionally known, since all pores remain of nanometer size only causing different electropermeability values than for conventional EP.

The application of electric field pulses to tissue leads to the dissipation of energy into heat and thus a temperature increase, ΔT. While this effect is wanted for, e.g., RF tumor ablation and causes thermal damage leading to necrotic cell death, EP-based ablation methods typically cause only small ΔT values. The pulse-related increase in ΔT may be found in the multicellular model by calculating the distributed specific absorption rate (SAR) as input into, e.g., the Pennes bioheat equation that includes thermal relaxation (Esser et al. 2007, 2009b). Even though EP increases the resulting ΔT values, those values remain small if a sufficiently long time is chosen between the multiple pulses. For this reason, the EP-mediated effects are nonthermal.

13.3.2 Multiscale Tissue Model

Local solid tumor treatment by electric field pulses is controlled by an appropriate placement of electrodes. In order to control the spatial extent of the tissue region being treated by nsPEF pulses, it is essential to mathematically model the electric field redistribution in tissue, which results from the dynamic behavior of EP. Figures 13.5 through 13.7 show the distributed electric response in the tissue

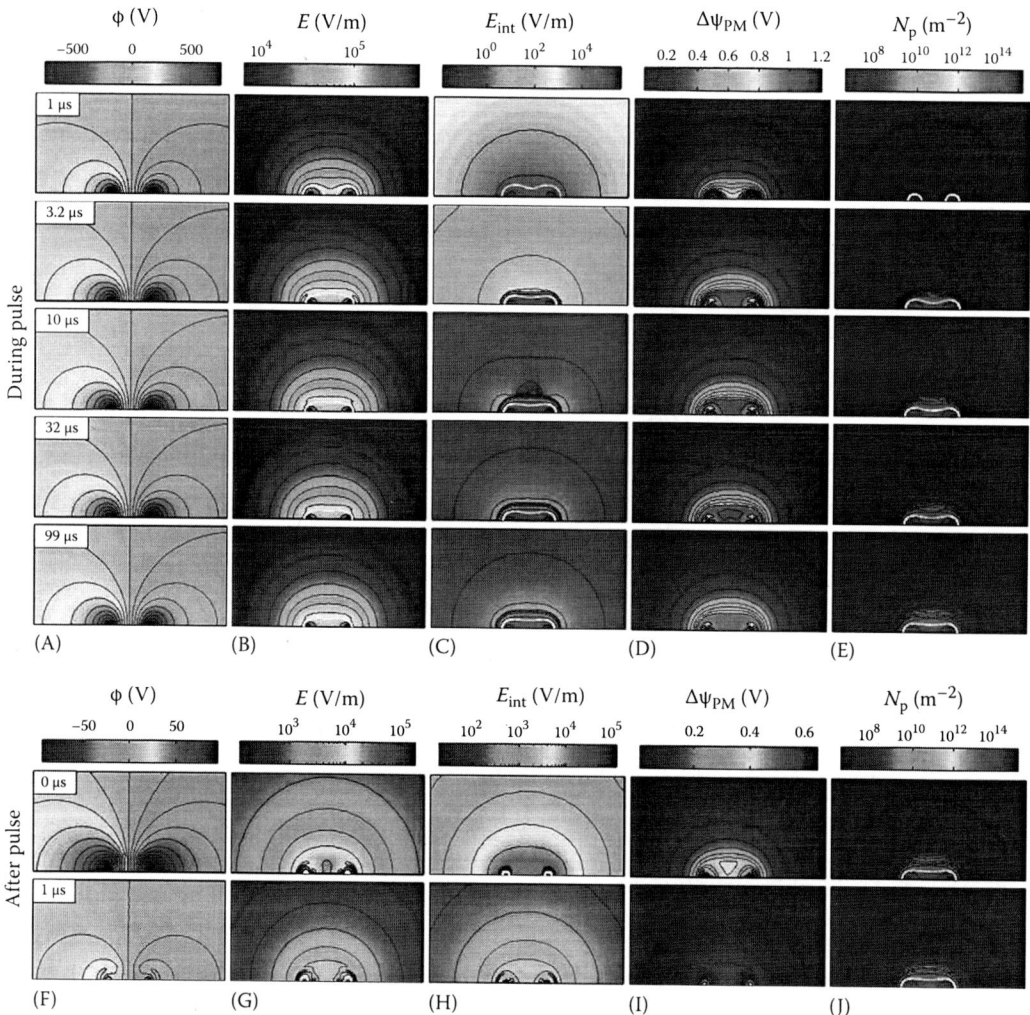

FIGURE 13.5 **(See color insert following page 268.)** Spatial tissue response to the 1.5 kV/cm trapezoidal pulse with 100 μs duration and 1 μs rise and fall times showing the (A, F) electric potential ϕ, (B, G) averaged electric field magnitude E, (C, H) intracellular electric field E_{int}, (D, I) transmembrane voltage $\Delta\psi_{PM}$, and (E, J) pore density N_p, as calculated from the asymptotic EP model, near the electrodes during (A)–(E) and after (F)–(J) the pulse, along the centerline ($y = 0$ mm). On each plot, 21 (for ϕ) or 11 (for E, E_{int}, $\Delta\psi_{PM}$, and N_p) contour lines are spaced evenly between the extreme values of the associated colorscale bar. Times shown are 1, 3, 10, 32, and 99 μs during the pulse, as well as 0 and 1 μs after the pulse. The white contour in (B) denotes an electric field strength E of 0.7 kV/cm. (From Esser, A.T. et al., *Technol. Cancer Res. Treat.*, 6, 268, 2007.)

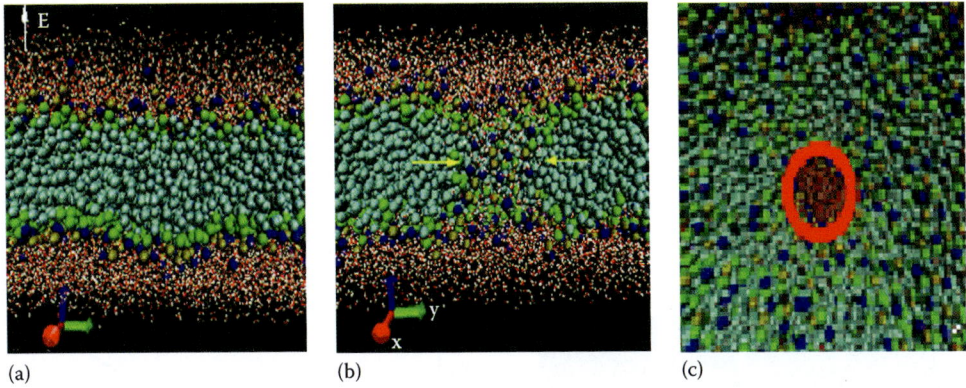

FIGURE 2.9 Molecular dynamics results showing nano-pore formation at a membrane by an external 0.5 V/nm pulsed electric field. (a) Before application of electric field, (b) 3.3 ns after application of pulsed field (cross section), and (c) same, but end-on view. (From Schoenbach, K.H. et al., *IEEE Trans. Dielect. Electr. Insul.*, 14, 1088, 2007. With permission.)

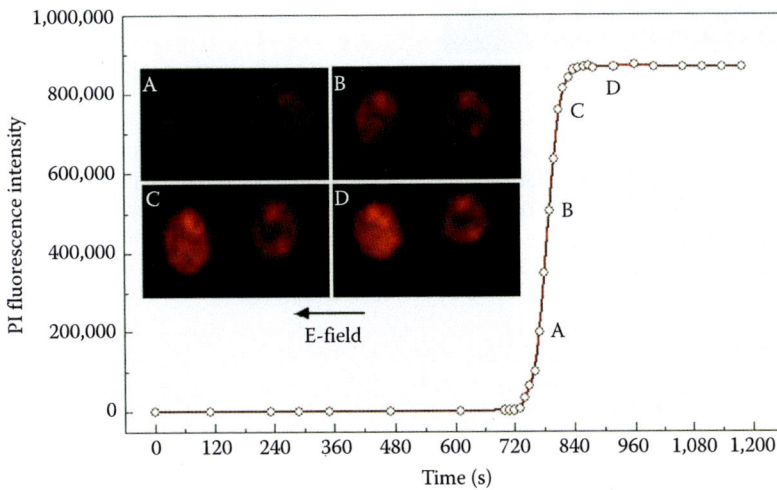

FIGURE 2.11 The temporal development of the PI uptake and microscopic real-time images of typical HL-60 cells undergoing PI uptake (A) 770 s, (B) 790 s, (C) 810 s, and (D) 920 s, following a 60 ns, 26 kV/cm electric pulse. The electric field orientation is marked. (From Chen, N. et al., *Biochem. Biophys. Res. Commun.*, 317, 421, 2004. With permission.)

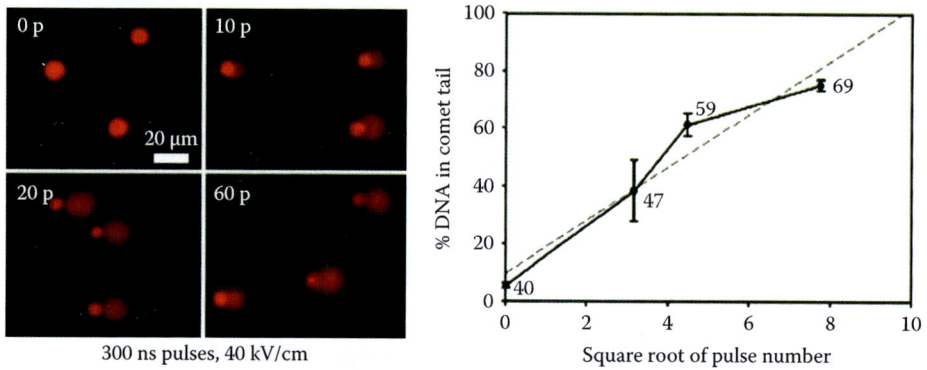

FIGURE 2.14 Comet assay using B16 cells *in vitro*. 40 kV/cm pulses 300 ns long were used and the pulse number is indicated on each figure. Quantification of propidium iodide fluorescence allows us to estimate the percentage of total DNA in the comet tail. When plotted against the square root of the pulse number, a linear dependence is revealed that predicts 100% DNA fragmentation when cells are exposed to 100 pulses. The straight line is a least squares fit to the four data points and the error bars represent the SEM with the number of cells averaged in each point written next to it. (From Nuccitelli, R. et al., *Int. J. Cancer*, 125, 438, 2009. With permission.)

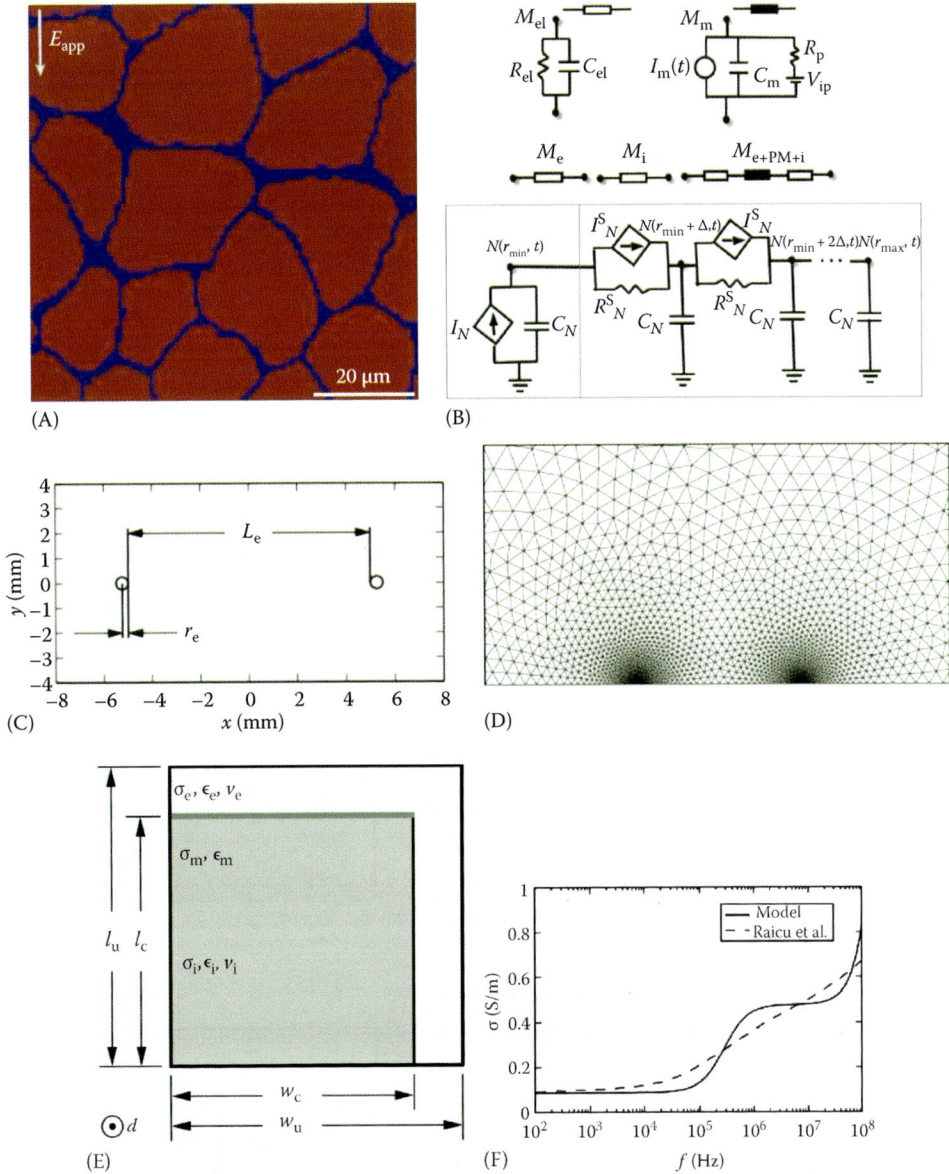

(A)

(B)

(C)

(D)

(E)

(F)

FIGURE 13.1 (A) Geometry of the multicellular model of a region of liver tissue with 14% interstitial space and an average cell dimension of 21.7 μm. The extracellular electrolyte is shown in dark gray, the cells in lighter gray (see online color version). Also given is the direction of the applied electric field E_{app}. (B) TL circuits: Functional local models that represent electrolyte (M_{el}), membrane (M_m), and the PM-electrolyte interface (M_{e+PM+i}) in the 101 × 101 TL; the lattice spacing l, as well as the depth of the system model, is 1 μm, leading to a spatial scale of 100 μm × 100 μm; more details described elsewhere (Gowrishankar and Weaver 2003, Gowrishankar et al. 2006). The equivalent EP circuit representing the asymptotic EP model (left gray box) (Stewart et al. 2004) and its extension to the SE (Esser et al. 2010) is solved at every local membrane site. The local pore distributions are discretized, and pore drift and diffusion determine the associated non-Ohmic pore conductance G_m that is input for the time-dependent membrane current I_m (t) in module M_m. (C) Two circular electrodes in a 200 mm × 100 mm tissue region (only a small subregion of the entire simulation region is shown). The electrodes have radii $r_e = 0.25$ mm and separation $L_e = 10$ mm (edge-to-edge). (D) The tissue system mesh (only a small subregion of the mesh close to the needle electrodes is shown here). (E) The tissue model cell unit comprises extracellular (e), membrane (m), and intracellular (i) regions in series and a parallel shunt region. The spatial and electrical parameters of the regions are labeled. (F) The passive frequency-dependent tissue conductivity of the tissue model, obtained from an AC-frequency sweep in SPICE, is compared with experimental measurements on rat liver (Raicu et al. 1998). (From Esser, A.T. et al., *Technol. Cancer Res. Treat.*, 6, 263, 2007.)

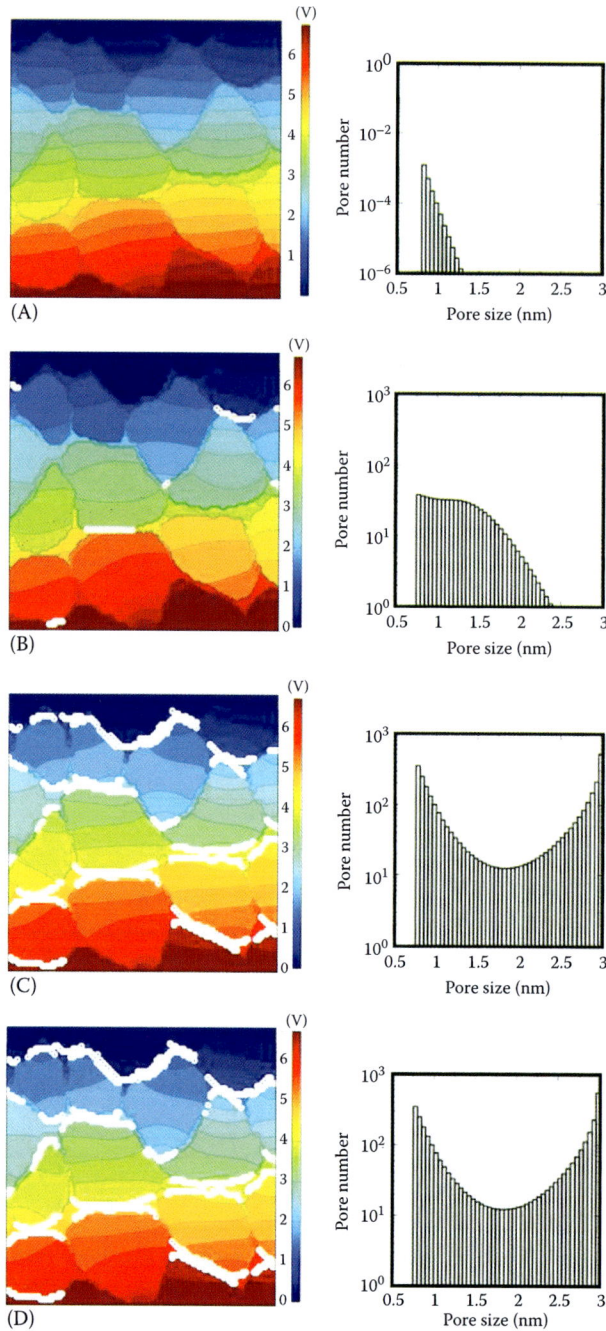

FIGURE 13.2 Electrical response of multicellular model to 700 V/cm pulse (1 μs rise and fall times, 98 μs pulse plateau) at different timepoints, and pore histograms of the entire model (pore size interval of 0.05 nm). (A) Charging phase at $t = 1$ μs: Displacement currents at the PM lead to intracellular electric fields. Pore histogram has different scale here, as no significant EP has occurred yet. (B) EP has occurred in distinct cells at $t = 1.7$ μs, a significant number of pores are created at the minimum size radius of 0.8 nm, and some pores have already grown towards larger radii. White dots in the left panels correspond to pore sites with more than 1 pore (equivalent to a pore density of 10^{12} pores/m^2). Equipotentials within cells indicate rising intracellular electric fields as a consequence of EP and the resulting nonlinear increase of membrane conductivity. (C) All cells are electroporated at $t = 21$ μs. The pore histogram shows both many small and many large pores in the multicellular model, resulting in distinct permeability properties for molecular uptake and release. The pore accumulation at the maximum radius of 3 nm considered here is not a physical effect but a technical limitation of the present model. It means that those pores eventually would reach an even larger size. (D) Distributed response does not change qualitatively, thus at the end of the pulse plateau at $t = 99$ μs we find a similar pattern to (C). (From Esser, A.T. et al., *Technol. Cancer Res. Treat.*, 6, 267, 2007.)

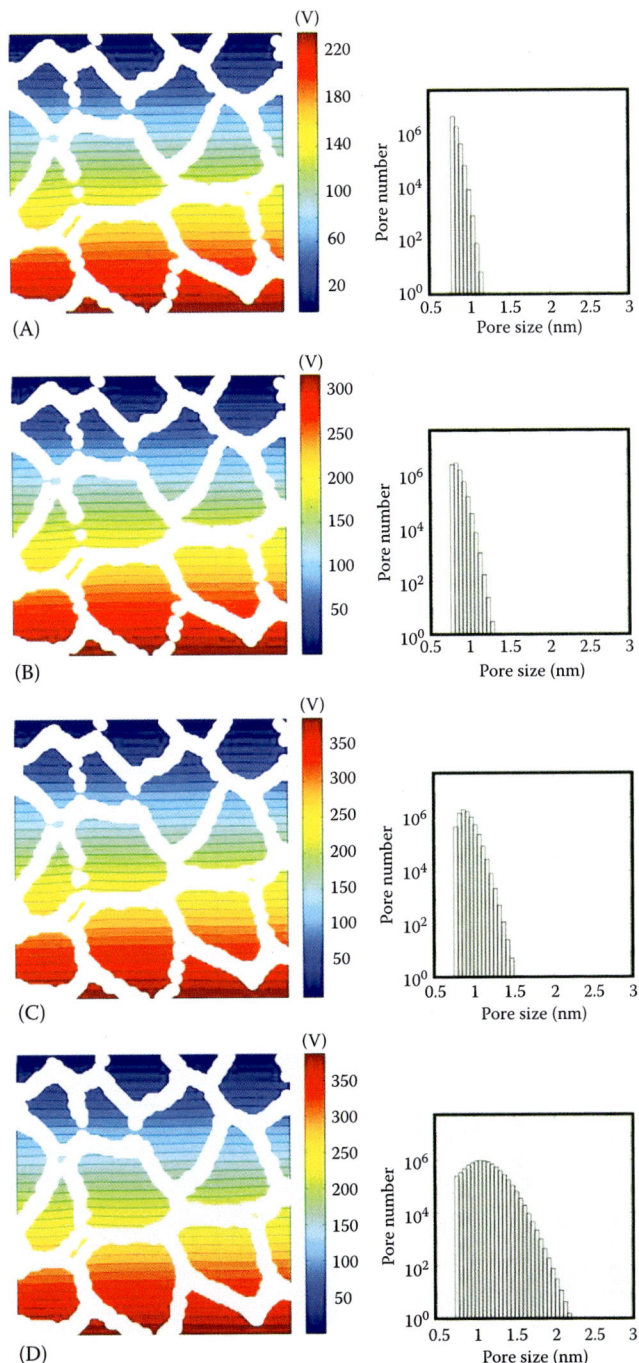

FIGURE 13.3 Electrical response of multicellular model to a trapezoidal 40 kV/cm pulse (55 ns rise and fall times, 215 ns pulse plateau) and corresponding pore histograms showing the total number of pores (pore size interval of 0.05 nm). (A) Membrane charging phase during pulse rise time at $t = 34$ ns (when the applied field strength is about 23 kV/cm): Due to the larger E_{app} compared to Figure 13.2 pore creation has occurred at essentially all PM sites. White dots in the left panels correspond to pore sites with more than one pore (equivalent to a pore density of 10^{12} pores/m²). (B) Still during the pulse rise time at $t = 45$ ns (when the applied field strength is about 30 kV/cm): Elevated PM transmembrane voltages lead to pulse-duration limited pore expansion. (C) At $t = 67$ ns during the pulse and afterwards, basically no new pores are created at the PM, but the elevated PM transmembrane voltage of 0.6–1 V cause the present pores to further grow in size. (D) Distributed electric response at $t = 270$ ns at the end of the pulse plateau does not change qualitatively from (C). The pore histogram shows still further pore expansion, but the short overall duration of the pulse restricts the pores to an averaged size of 1.15 nm. (From Esser, A.T. et al., *Technol. Cancer Res. Treat.*, 8, 295, 2009a.)

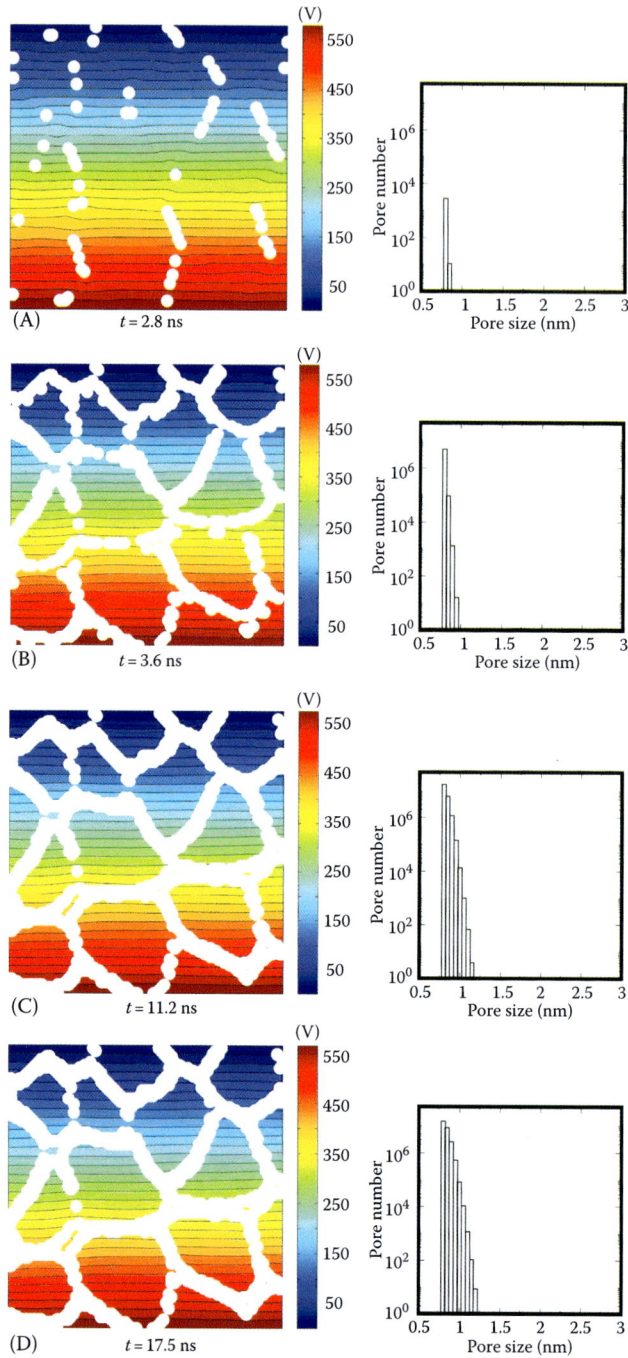

FIGURE 13.4 Electrical response of multicellular model to a trapezoidal 60 kV/cm pulse (2.5 ns rise and fall times, 17 ns pulse plateau) and correspondent pore histograms showing the total number of pores (pore size interval of 0.05 nm). (A) At $t = 2.8$ ns already during the pulse plateau: Displacement currents at the PM dominate and lead to intracellular electrical fields essentially equal to extracellular electric fields. Pore creation due to EP has started at this time at some membrane sites, predominantly in the nonpolar regions of the cells. White dots in the left panels correspond to pore sites with more than one pore (equivalent to a pore density of 10^{12} pores/m²). (B) At $t = 3.6$ ns: EP has occurred on essentially all PM sites, including the membrane sites not facing the polar side of the cell. (C) At $t = 11.2$ ns during the pulse plateau, basically no new pores are created at the PM, but the elevated PM transmembrane voltage of 0.6–1 V cause the pores to grow. (D) Distributed electric response at $t = 17.5$ ns at the end of the pulse plateau does not change qualitatively from (C). The pore histogram shows some further pore expansion, but the even shorter duration of the pulse compared to Figure 13.3 restricts the pores to an averaged size of 0.85 nm. (From Esser, A.T. et al., *Technol. Cancer Res. Treat.*, 8, 296, 2009a.)

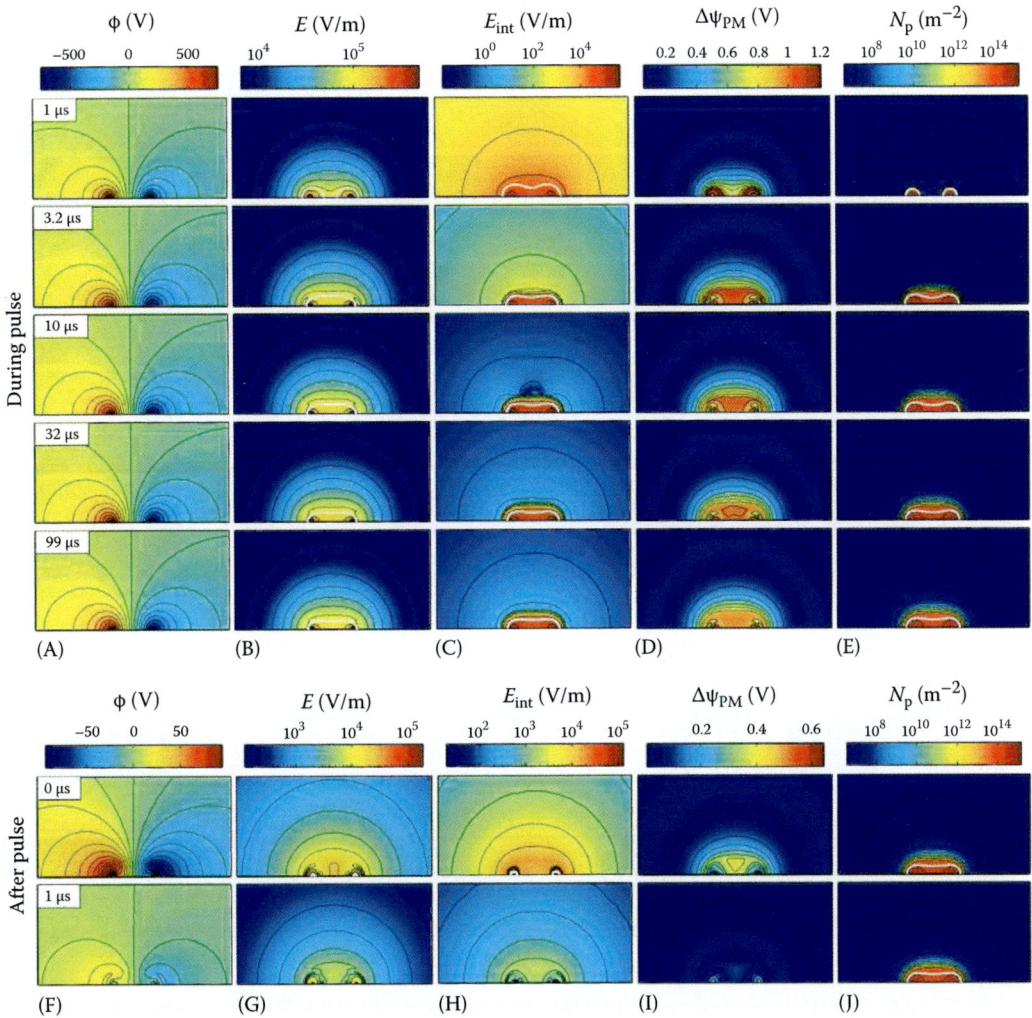

FIGURE 13.5 Spatial tissue response to the 1.5 kV/cm trapezoidal pulse with 100 μs duration and 1 μs rise and fall times showing the (A, F) electric potential ϕ, (B, G) averaged electric field magnitude E, (C, H) intracellular electric field E_{int}, (D, I) transmembrane voltage $\Delta\psi_{PM}$, and (E, J) pore density N_p, as calculated from the asymptotic EP model, near the electrodes during (A)–(E) and after (F)–(J) the pulse, along the centerline ($y = 0$ mm). On each plot, 21 (for ϕ) or 11 (for E, E_{int}, $\Delta\psi_{PM}$, and N_p) contour lines are spaced evenly between the extreme values of the associated colorscale bar. Times shown are 1, 3, 10, 32, and 99 μs during the pulse, as well as 0 and 1 μs after the pulse. The white contour in (B) denotes an electric field strength E of 0.7 kV/cm. (From Esser, A.T. et al., *Technol. Cancer Res. Treat.*, 6, 268, 2007.)

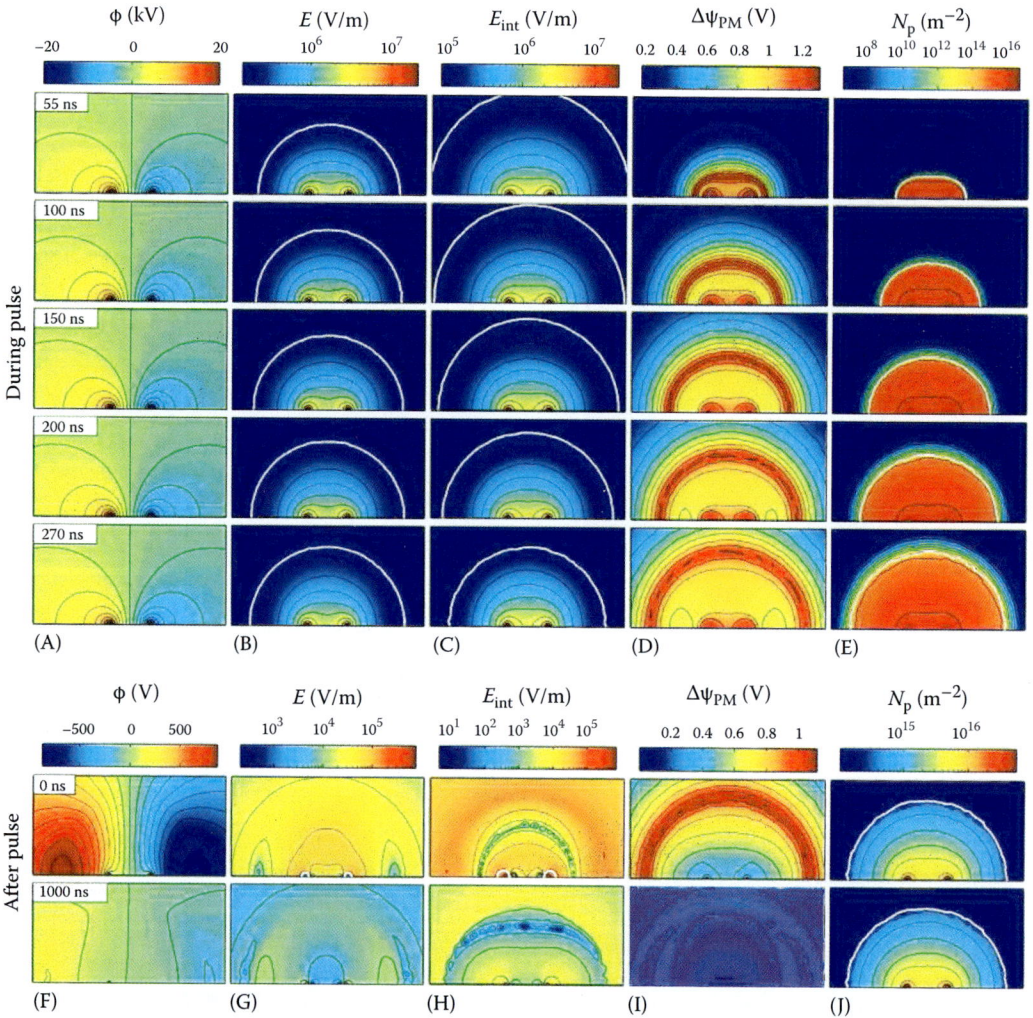

FIGURE 13.6 Spatial tissue response to the 40 kV/cm trapezoidal pulse with 325 ns duration and 55 ns rise and fall times showing the (A, F) electric potential ϕ, (B, G) averaged electric field magnitude E, (C, H) intracellular electric field E_{int}, (D, I) transmembrane voltage $\Delta\psi_{PM}$, and (E, J) pore density N_p, as calculated from the asymptotic EP model, near the electrodes during (A)–(E) and after (F)–(J) the pulse, along the centerline ($y = 0$ mm). On each plot, 21 (for ϕ) or 11 (for E, E_{int}, $\Delta\psi_{PM}$, and N_p) contour lines are spaced evenly between the extreme values of the associated colorscale bar. Times shown are 55, 100, 150, 200, and 270 ns during the pulse, as well as 0 and 1000 ns after the pulse. The white contour in (B) denotes an electric field strength E of 1.81 kV/cm, while the white contour in (C) denotes an intracellular electric field E_{int} of 0.81 kV/cm, and the white contour in (E) denotes a pore density N_p of 10^{14} m^{-2}. Note that the three white contours change their spatial position during the pulse but by construction all fall onto the same position at the end of the pulse. (From Esser, A.T. et al., *Technol. Cancer Res. Treat.*, 8, 301, 2009a.)

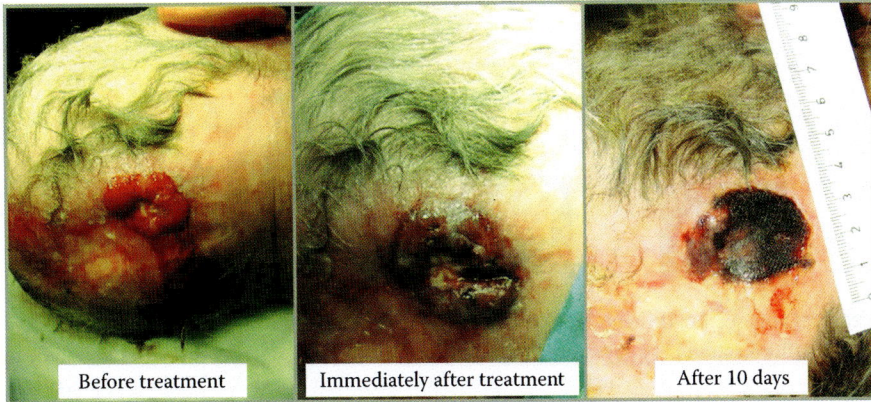

FIGURE 21.7 Use of ECT for palliative treatment of bleeding tumors. Bleeding squamous cell carcinoma metastasis was treated by ECT with bleomycin. Immediately after the treatment the bleeding was stopped. Within 10 days a scab formed and good antitumor effectiveness was evident.

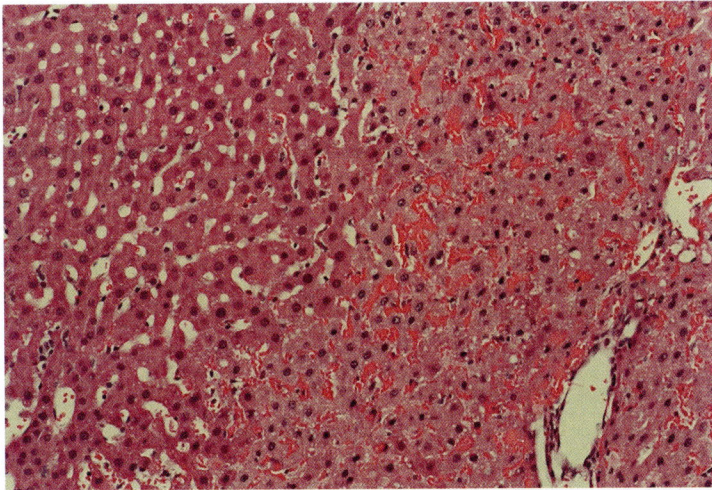

FIGURE 25.1 H&E stained liver that has undergone NTIRE. The left-hand side is the normal liver and the right-hand side is the electroporated liver. Note the sharp line of distinction between the treated and untreated areas.

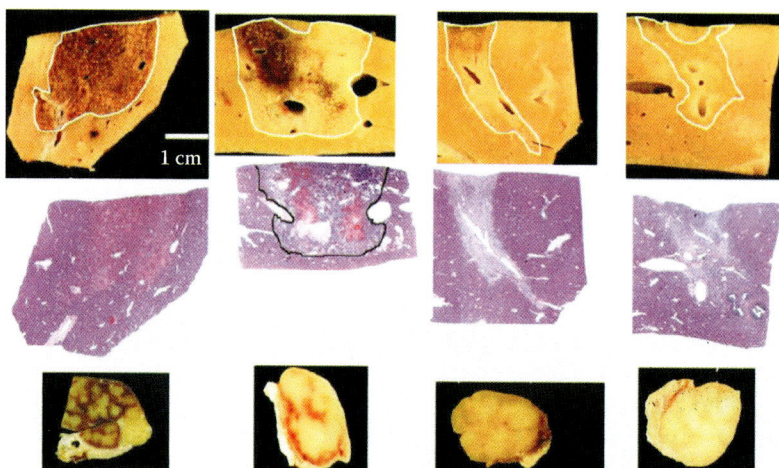

FIGURE 25.2 Cross section through NTIRE treated pig liver after: 24h (first column from left), 3 days (second column), 7 days (third column) 14 days (fourth column). Top row—macroscopic cross section, middle row—H&E stained section, and bottom row—lymph nodes. (From Rubinsky, B. et al., *Technol. Cancer Res. Treat.*, 6, 37, 2007. With permission.)

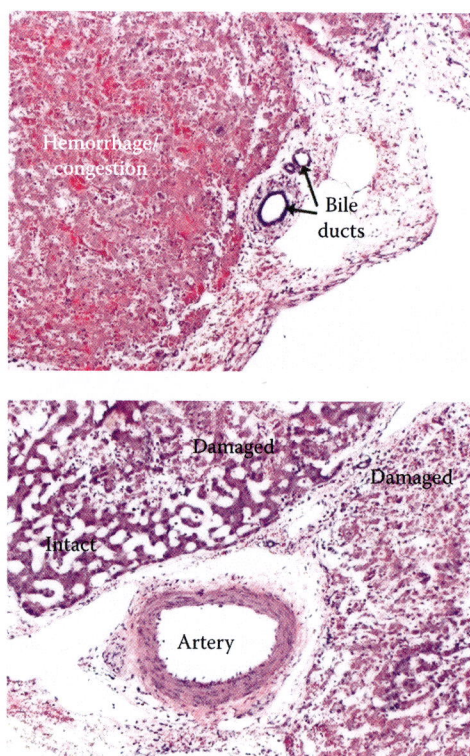

FIGURE 25.3 Intact bile ducts and arteries within NTIRE treated tissue. (From Rubinsky, B. et al., *Technol. Cancer Res. Treat.*, 6, 37, 2007. With permission.)

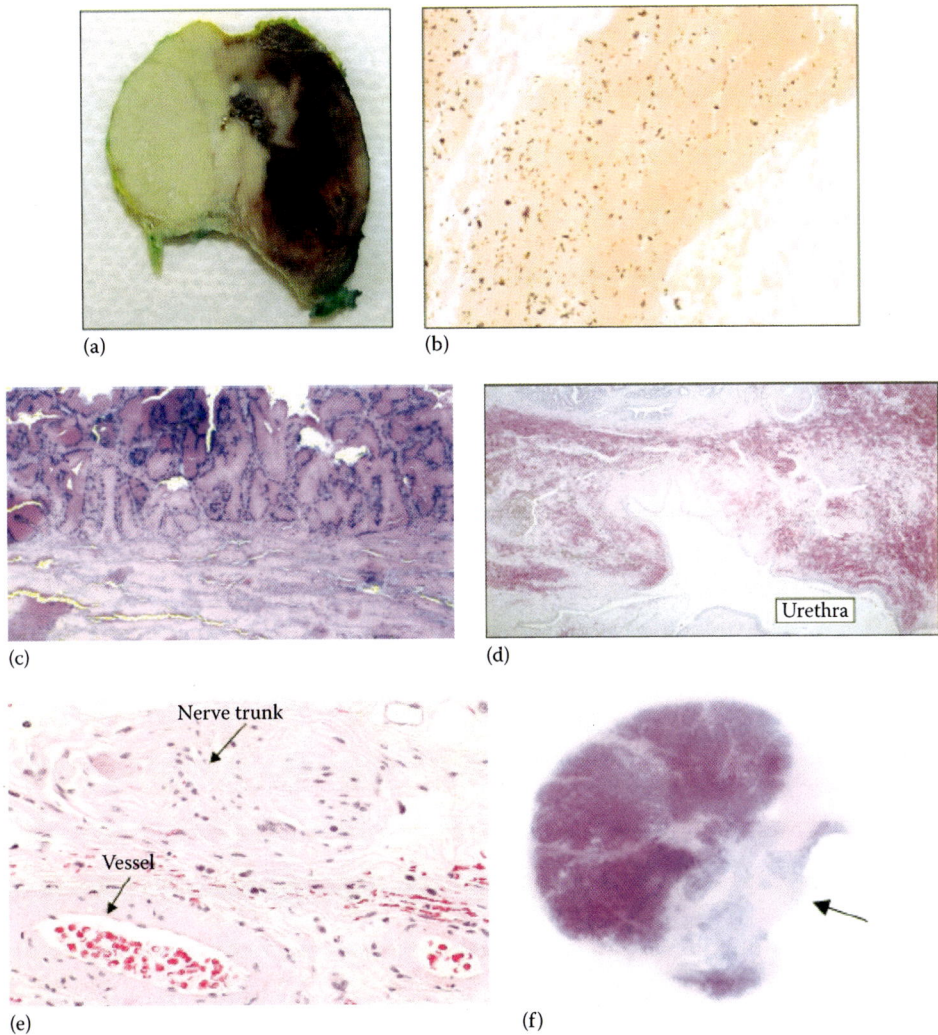

FIGURE 25.4 Dog prostate treated with NTIRE. (From Onik, G. et al., *Technol. Cancer Res. Treat.* 6, 295, 2007. With permission.) (a) Gross pathology of the IRE lesion at 24 h. The right side of the gland is hemorrhagic (pulses = 8, Kv = 1). (b) Photomicrograph of prostate tissue that has been electroporated at 24 h. No glandular elements are visible (pulses = 8, Kv = 1). (c) Photomicrograph at the margin of the IRE lesion. A very narrow zone of transition between normal and necrotic tissue is noted at the margin (pulses = 8, Kv = 1). (d) The urethra is noted at the center of the micrograph as the open space at 24 h. Sub-mucosal hemorrhage is noted but the integrity of the urethra is still intact (pulses = 8, Kv = 1). (e) Photomicrograph of the neurovascular bundle after electroporation at two weeks (pulses = 80, Kv = 1.5). It can be seen that both the vessel and the nerve trunk show no evidence for necrosis. (f) Whole mount slide of a prostate where the right side of the gland was electroporated 2 weeks prior. There is marked shrinkage of the lobe with replacement by fibrous tissue (pulses = 80, Kv = 1.5).

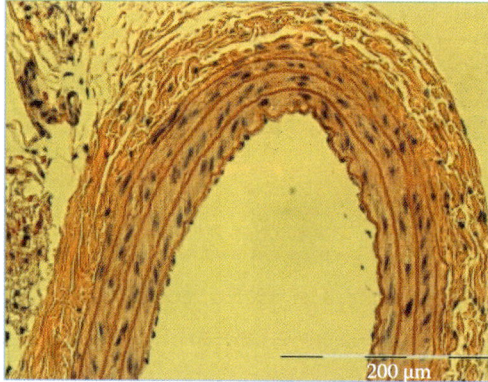

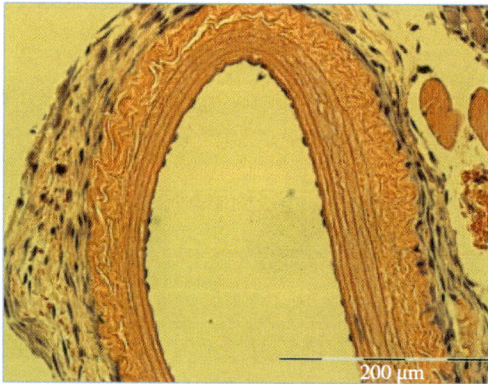

FIGURE 25.5 Cross section through a rat carotid artery. Top panel shows a normal artery. The vascular smooth muscles as well as the endothelial cells around the lumen are evident. An NTIRE treated artery a week after the procedure. Note the complete absence of the smooth muscles as well as the beginning of the formation of the endothelial cells around the lumen.

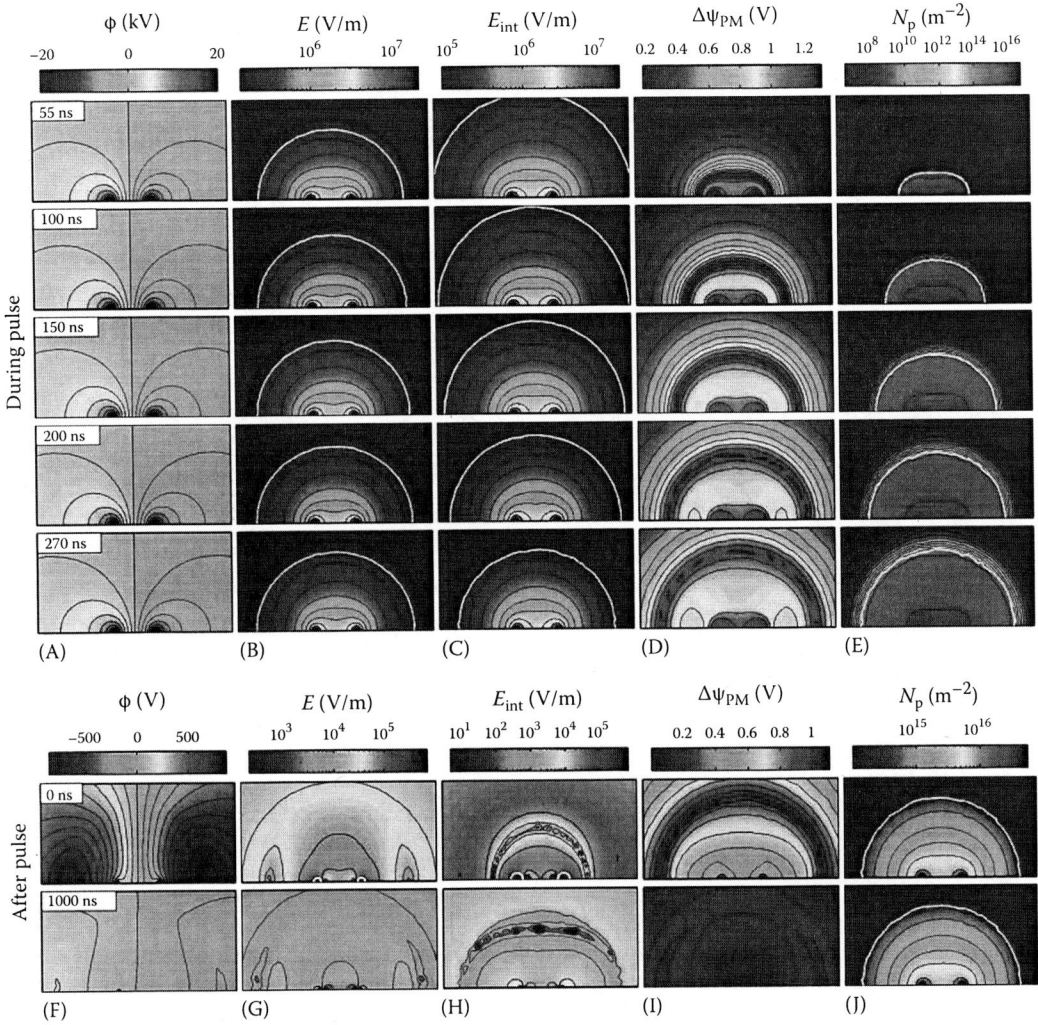

FIGURE 13.6 **(See color insert following page 268.)** Spatial tissue response to the 40 kV/cm trapezoidal pulse with 325 ns duration and 55 ns rise and fall times showing the (A, F) electric potential ϕ, (B, G) averaged electric field magnitude E, (C, H) intracellular electric field E_{int}, (D, I) transmembrane voltage $\Delta\psi_{PM}$, and (E, J) pore density N_p, as calculated from the asymptotic EP model, near the electrodes during (A)–(E) and after (F)–(J) the pulse, along the centerline ($y = 0$ mm). On each plot, 21 (for ϕ) or 11 (for E, E_{int}, $\Delta\psi_{PM}$, and N_p) contour lines are spaced evenly between the extreme values of the associated colorscale bar. Times shown are 55, 100, 150, 200, and 270 ns during the pulse, as well as 0 and 1000 ns after the pulse. The white contour in (B) denotes an electric field strength E of 1.81 kV/cm, while the white contour in (C) denotes an intracellular electric field E_{int} of 0.81 kV/cm, and the white contour in (E) denotes a pore density N_p of 10^{14} m^{-2}. Note that the three white contours change their spatial position during the pulse but by construction all fall onto the same position at the end of the pulse. (From Esser, A.T. et al., *Technol. Cancer Res. Treat.*, 8, 301, 2009a.)

model to IRE and nsPEFs conditions in time and space, respectively. Needle electrodes are placed as shown in Figure 13.1C. The distributed electric response is represented in Figures 13.5 through 13.7 during and after the application of the pulse by the electric potential ϕ (A, F); the local averaged electric field E (B, G); the intracellular electric field E_{int} (C, H), supplemented by the local transmembrane voltages $\Delta\psi_{PM}$ (D, I); and the local pore densities N_p (E, J). The latter is calculated using the asymptotic EP model wherein pore expansion is neglected.

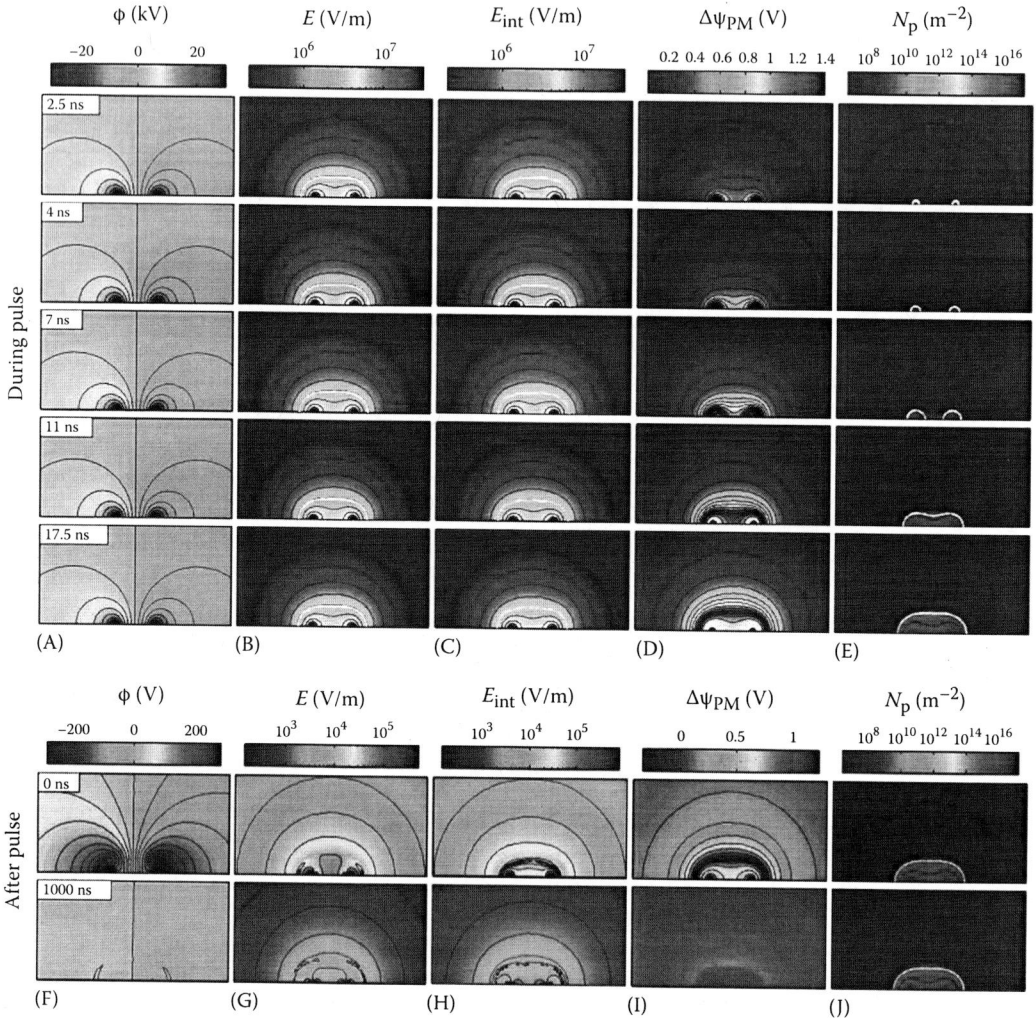

FIGURE 13.7 Spatial tissue response to the 60 kV/cm trapezoidal pulse with 15 ns duration and 2.5 ns rise and fall times showing the (A, F) electric potential ϕ, (B, G) averaged electric field magnitude E, (C, H) intracellular electric field E_{int}, (D, I) transmembrane voltage $\Delta\psi_{PM}$, and (E, J) pore density N_p, as calculated from the asymptotic EP model, near the electrodes during (A)–(E) and after (F)–(J) the pulse, along the centerline ($y = 0$ mm). On each plot, 21 (for ϕ) or 11 (for E, E_{int}, $\Delta\psi_{PM}$, and N_p) contour lines are spaced evenly between the extreme values of the associated colorscale bar. Times shown are 2.5, 4, 7, 11, and 17.5 ns during the pulse, as well as 0 and 1000 ns after the pulse. The white contour in (B) denotes an electric field strength E of 1.6 kV/cm, while the white contour in (C) denotes an intracellular electric field E_{int} of 1.57 kV/cm, and the white contour in (E) denotes a pore density N_p of 10^{14} m^{-2}. Note that the three white contours change their spatial position during the pulse but by construction all fall onto the same position at the end of the pulse. (From Esser, A.T. et al., *Technol. Cancer Res. Treat.*, 8, 302, 2009a. With permission.)

Figure 13.5 shows the spatial tissue response to a 100 μs IRE pulse with a nominal electric field of 1.5 kV/cm that has 1 μs rise and fall times. The field strength for this IRE condition is now higher than in Figure 13.2 since the field pulse magnitude and timescale determines the spatial extent of the tissue region in which the IRE is triggered. The membrane charging rate, and by extension the time of onset of significant EP and reversible electrical breakdown (REB), is determined by the local electric field magnitude, which is largest near the electrodes and drops off quickly with distance. As such, the membranes in the regions of tissue nearest the electrodes charge fastest and electroporate first. Subsequently, a wave of elevated $\Delta\psi_{PM}$ and pore creation moves outward from the electrodes into the central region of the tissue between the electrodes, leaving $\Delta\psi_{PM} \leq 1V$ in its wake. The field distribution is thus controlled by electrode position and electrode geometry. IRE treatments attempt to map or image iso-electric field lines onto the peripheral extensions of a tumor, or any other unwanted tissue region, to treat tissue within that predefined area (Rubinsky et al. 2007). The equi-electric field line with 700 V/cm, indicating the borderline between a tissue region that is reportedly subject to reversible EP and IRE, is highlighted by a white contour in Figure 13.5B.

The spatial extent of EP changes little after ~5 μs (Figure 13.5) because the membranes in the unelectroporated regions of tissue have essentially reached their maximal $\Delta\psi_{PM}$. In those regions in which $\Delta\psi_{PM}$ does not exceed 1 V, little EP will occur on the timescale of the 100 μs pulse (Vasilkoski et al. 2006). We believe that this is a general feature of EP: local fields tend towards uniformity during a pulse. At the electrode interface, $\Delta\psi_{PM}$ peaks at 1.26 V at 0.32 μs and N_p reaches a pore density of 1.3×10^{15} m^{-2}. At the center of the tissue region, $\Delta\psi_{PM}$ peaks at the smaller value of 1.18 V at 1.41 μs and N_p reaches a smaller value of 1.5×10^{14} m^{-2}. Transmembrane voltages of 1 V and above can be maintained only for a short time, as massive pore creation leads to REB, which in turn causes the PM transmembrane potential $\Delta\psi_{PM}$ to decrease even during the pulse (Freeman et al. 1994, Vasilkoski et al. 2006). Consequently, while the tissue conductivity increases throughout the tissue region between the electrodes, it increases most near the electrodes and least in the central region of the tissue. Because of this gradient in tissue conductance, the electric field becomes more uniform between the electrodes by the end of the pulse. Following the pulse, ϕ, E, E_{int}, and $\Delta\psi_{PM}$ rapidly decrease (Figure 13.5E–G) with a complex discharge pattern. N_p remains elevated (Figure 13.5H) and decays with the assumed 3 μs time constant. As such, the perturbation of the tissue is long-lived, lasting much longer than the duration of the applied pulse, and molecular uptake or release may persist after the pulse.

Figures 13.6 and 13.7 show tissue response to nsPEF conditions. Again, the membranes in the tissue region nearest the electrodes electroporate first, reaching pore densities of about 10^{16} m^{-2}, which is about two orders of magnitude higher than for the IRE condition in Figure 13.5. In contrast to IRE pulses, however, $\Delta\psi_{PM}$ does not relax to values significantly below 1 V during the pulse because of the overall short timescale of the pulse. After the pulse, all electroporated cell membranes in the tissue are depolarized for the lifetime of the pores. As such, the overall electric potential is essentially zero throughout the tissue, resulting in a lower number of equipotential lines on the scale presented in Figures 13.6 and 13.7F 1000 ns after the pulse.

The intracellular electric field, E_{int}, shown in Figures 13.5 through 13.7 may be an additional key to understanding the different effects observed for IRE pulses and nsPEFs. It may, in particular, provide insights about the extent of electric perturbation of organelle membranes, the influence on organelle channels, and the likelihood of organelle EP (Esser et al. 2009b). The averaged E_{int} has been estimated in our tissue model from the macroscopic electric field, E, and the cell parameters according to

$$E_{int} = E - \frac{2\Delta\psi_{PM}}{l_{cell}} \tag{13.2}$$

where $\Delta\psi_{PM}$ again is the PM transmembrane voltage and l_{cell} is the cell size (see Figure 13.1).

For values of $E = 10$ kV/cm, $\Delta\psi_{PM} = 1$ V and $l_{cell} = 22$ μm, for example, we find $E_{int} = 0.9$ kV/cm. This timescale is sufficient to electroporate the large organelles in cells, such as the endoplasmic reticulum (Gowrishankar et al. 2006). While E_{int} is about an order of magnitude smaller for the IRE conditions of Figure 13.5, in the case of the applied field strengths of 40–60 kV/cm in Figures 13.6 and 13.7, E_{int} becomes more and more equal to E.

Figures 13.6 and 13.7 show that the spatial extent of EP changes during the entire nsPEF. This is in strong contrast to IRE pulse conditions, shown in Figure 13.5, which reach a stationary profile of the pore density relatively early during the pulse plateau (Esser et al. 2007). It is also noteworthy that the electroporated region is larger for the nsPEFs than for the IRE pulse (Esser et al. 2009). As demonstrated here, the electroporated tissue is not only positioned between the first and second electrode, but spreads out radially from both needle electrodes. This is significant and has implications for the targeted tissue volume, which should be treated, and to avoid tissue, which should not be treated.

IRE treatments attempt to arrange electrodes and apply IRE pulses such that a particular tissue volume, which corresponds with the solid tumor, experiences a specific field strength threshold, e.g., 700 V/cm. In particular, an iso-electric field line with 700 V/cm was suggested (Rubinsky et al. 2007) to differentiate between a tissue region subject to reversible EP ($E < 700$ V/cm) and IRE ($E > 700$ V/cm). But the tissue region defined by this particular IRE threshold is only an empirical region and is based solely on the particular IRE pulses applied to liver tissue (Rubinsky et al. 2007). However, if a different pulse with a different field strength and pulse duration were to be applied, such as nsPEFs, or if the tissue had cells not 22 μm on average in size, this approach would not be robust. Then, if a clinician were to repeat the treatment of some tissue with an nsPEF (Nuccitelli et al. 2006), it can be seen from Figures 13.6 and 13.7 that there is clearly a region in the tissue that is exposed to the 700 V/cm threshold. However, this region is not necessarily electroporated because the timescale of the pulse duration is much too short. This shows that the concept of "critical potentials" in EP is incorrect. Instead, EP involves a "critical event" (pore creation), which depends at least on time as well as on $\Delta\psi_{PM}$.

Figure 13.8 presents the total current I_t (per system depth) through the tissue model as a function of time for the trapezoidal IRE pulse with 1.5 kV/cm and 100 μs duration (top) of Figure 13.5, as well as the trapezoidal 40 kV/cm nsPEF with 325 ns duration (middle) of Figure 13.6, and the trapezoidal 60 kV/cm nsPEF with 20 ns duration (bottom) of Figure 13.7. We compare the total current I_t in Figure 13.8 for an active tissue model that includes EP with a passive tissue model, which does not consider EP (Esser et al. 2007).

The total current in Figure 13.8A is approximately twice as large for the active model (solid line) than for the passive model (dashed line) because EP in the active model increases the tissue conductivity by that factor. This is an average factor, as there are gradients in tissue conductivity as discussed in Figure 13.5. This means that the Joule heating is about two times bigger than in the passive model. After the initial current spike due to displacement currents that also coincide with the maximum in the pore creation rate, the passive model has an essentially flat current plateau, whereas the asymptotic EP model shows a characteristic slope in the current that indicates the slower creation of additional pores. We argue here that the EP model based on the SE would tend to increase this slope even more as pore expansion during the pulse leads to further membrane conductivity increases.

Figure 13.8B (middle) shows for the active tissue model an increase of tissue current during the pulse rise, followed by a peak, and then a gradual reduction of I_t during the trapezoidal pulse plateau followed by the decline of I_t during the pulse fall. The stepwise changes in the slope of the applied trapezoidal electric field pulse at, e.g., $t = 0, 55, 270,$ and 325 ns cause abrupt changes in the total system current as a result of stepwise changes in displacement currents. The passive model (no EP), in contrast, shows a strong decay of the tissue current even during the pulse plateau. Although this pulse is only about 300 ns long, this duration is sufficiently long for some of the high frequency current components to decay. As this happens, the membrane becomes increasingly significant in determining the total current flowing through the system. In the active model, the membrane impedance remains small because the many pores allow the flow of ionic conduction currents, and the total system current remains relatively steady. In the passive model, however, the impedance of the membrane (and therefore that of the tissue) grows as the high frequency components of the displacement currents decay, the large resistance of the membrane prevents a significant conduction current, and the total system current decreases.

In other words, while the large, EP-based membrane conductance in the active tissue model permits a large transcellular current, the small membrane conductance of the passive model increasingly excludes

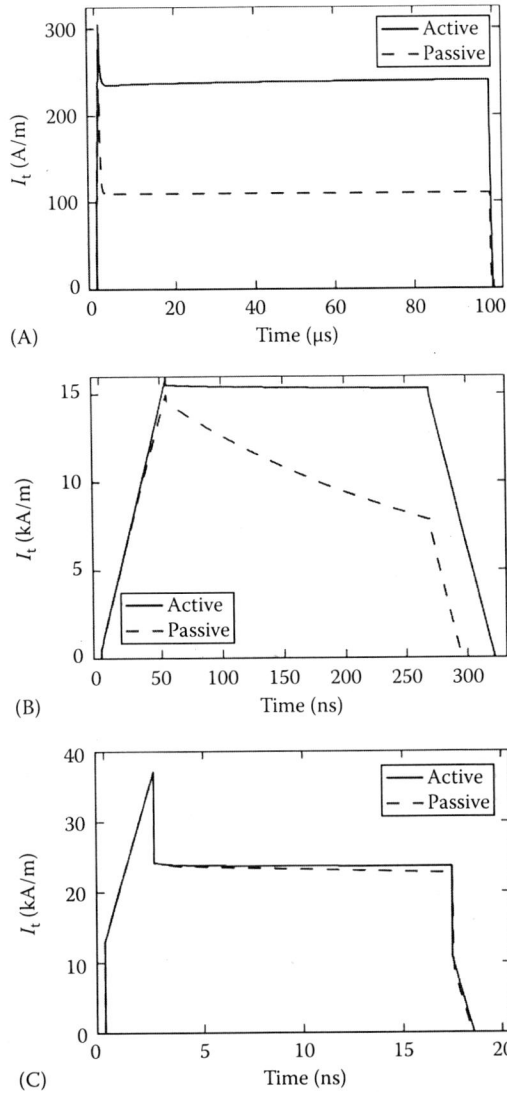

FIGURE 13.8 The total current (per system depth) flowing through the tissue system is shown as a function of time for the 1.5 kV/cm pulse (top), 40 kV/cm pulse (middle), and the 60 kV/cm pulse for a passive model, which is the tissue model without explicit EP, and for an active model, which is the tissue model with EP being taken account. Displacement currents dominate the total tissue current for the short timescale (high frequencies) of the pulse rise in (A) and (B) and the total duration of the pulse (C). The longer the duration of the pulse and the more the displacement currents decay in time, the less agreement is obtained between the active and the passive model. (From Esser, A.T. et al., *Technol. Cancer Res. Treat.*, 8, 303, 2009a. With permission.)

transcellular current and a larger fraction of the total current I_t must flow through the shunt outside each cell (see Figure 13.1E). This can also be understood from the frequency-dependent tissue conductivity shown in Figure 13.1F. At the beginning of the pulse, the system basically starts off with tissue conductivity on the right side at $f = 10^8$ Hz at the beginning of the pulse and then, during the pulse, moves to the left to lower frequencies. The longer the duration of the pulse, therefore, the less agreement is observed between a passive and active tissue model prediction for the tissue current. Overall, the tissue current is always higher for the active tissue model, since EP increases the tissue conductivity by a large increase of the PM conductance. EP thus always leads to an increased Joule heating in the tissue in comparison with a passive model (Esser et al. 2007, 2009).

On the even shorter timescale of the pulse in Figure 13.8C (bottom), the impedance of the membrane is much smaller than the electrolyte impedance. As such, the shift from displacement to conduction-dominated transmembrane currents (as pores form in the active model) has relatively little effect on the total system, and there is relatively little difference between the total system current flowing through the active and passive model systems.

The above predictions for the tissue current in the passive and active tissue models appear to be in somewhat better agreement for nsPEFs than for longer IRE pulses (Esser et al. 2007, 2009). This, however, is entirely due to the strong contribution of displacement currents during this temporal regime. Compared with the IRE pulse condition, nsPEFs lead to a more complex interplay between conduction and displacement currents to the total current (Smith and Weaver 2008). In particular, the PM impedance is initially largely determined by the PM dielectric properties because of the extremely low conductance of the PM, when no EP has occurred yet, and the high-frequency content of the pulse rising edge. The large increase in membrane conductance, which results from the creation of minimum-sized pores, causes a shift from a dielectric-dominated PM impedance to a conductance-dominated PM impedance. Subsequently, the high PM conductance then leads to a continued penetration of the electric field and electric currents into the intracellular space, even when the high-frequency components decay (Smith and Weaver 2008), as demonstrated in Figure 13.8.

Therefore, the alleged similarity between the passive and active tissue currents shown in Figure 13.8B and C should not be misunderstood. Specifically, it is only superficially suggested that passive tissue models are more appropriate to study nanosecond EP pulses than it is for IRE pulses. On the contrary, passive models are inadequate to describe the response at the PM where EP occurs to self-limit the transmembrane voltage to values of about 1 V and instead predict tens of volts for the transmembrane voltages, far in excess of what a biological membrane can sustain (Smith and Weaver 2008).

13.4 Discussion and Conclusions

We have presented multicellular and tissue models and their electric responses to some characteristic nsPEFs and IRE conditions. Such increasingly realistic in silico models involve complex and dynamic electrical interactions throughout the biological system on a sub-nanosecond to seconds timescale and nanometer to micrometer length scale and allow for a rapid screening of many electrical conditions in cells. Our approach is in contrast to other mathematical tissue models, which assume only a time- or frequency-independent tissue conductivity (Davalos et al. 2005) or spatial variations in tissue conductivity, which are based on a series of static models (Miklavcic et al. 2000, Davalos et al. 2004, Sel et al. 2005) and require a mapping between electric field and tissue conductivity, based on the very experimental results these models seek to predict. Those models, therefore, lack a mechanistic hypothesis of what is causing the dynamic conductivity changes and may lead to misleading results if displacement currents are not accounted for. They do not provide predictive power, for example, with respect to molecular transport in tissue, or, when different pulse conditions are considered.

In particular, we have estimated the response of tissue to pulsed electric fields by taking into account dynamic displacement and dynamic conduction currents. Our membrane models assume neither where EP will occur, nor the magnitude of the transmembrane voltage, $\Delta\psi_{PM}$, or time at which significant numbers of pores will be created. Instead, EP emerges on the basis of the underlying physics-based EP model as the result of electric interactions throughout the entire system model. As demonstrated here, dynamic EP of irregular cell membranes in a multicellular environment can now also be described, providing new insights into the conditions on the microscopic scale and leading towards a mechanistic understanding of tissue under EP conditions, which is critical for pursuing tissue ablation by EP. Both of our models demonstrate strong EP-driven redistribution of fields and currents and give rise to different degrees of EP, conventional EP with IRE, and supra-EP with nsPEF with consequences for electrical, molecular transport, and thermal behavior.

A full description of EP for exposures to IRE and nsPEF conditions thus leads to three major consequences: first, dramatic conductance changes at the membrane occur for both exposures. This leads to elevated values of the intracellular electric fields, E_{int}, which in itself may cause organelle membrane perturbations, organelle channel gating, and EP of organelles (Esser et al. 2010) and thus a biochemical effect inside the cell leading to cell death by either necrosis or apoptosis. While pore expansion during IRE conditions allows for higher conductance changes of the membrane and therefore larger intracellular electric fields, it needs time. The nsPEF-induced conductance changes are, in the absence of sufficient time, not the result of pore expansion, but due to the creation of many more nanometer-sized pores. The conductance change then is so large that the intracellular electric field is essentially the same as the extracellular electric field. Therefore, nsPEFs cause pronounced intracellular effects, which presumably are responsible for triggering a different cell death mechanism (apoptosis) than longer IRE pulses (necrosis). Supra-EP at the PM leads to a conductance change of the membrane that exceeds that of longer IRE pulses. Because of the inevitability of intracellular perturbations, the magnitude of the resulting intracellular electric field needs to be quantified and may vary from unimportant to tremendously important.

Second, effective permeability changes may be found from the time-dependent pore population with dynamic pore size changes during the entire pulse and after, thereby altering the transport of molecules of different sizes and charges into and out of each cell. Electropermeabilization is thus a major consequence of EP. Thus, the local permeability $P_m(r_s)$ with respect to a certain set of molecules "s" should be quantified (Esser et al. 2007). For example, nsPEFs seem to cause larger changes in tissue conductivity and strong local electric fields, but the membrane permeability to molecules like adenosine triphosphate (ATP), DNA, or traditional EP markers, such as propidium iodide, may, as judged from the pore histograms, remain negligibly small. In contrast, conventional electric field pulses, as discussed here for IRE, will lead to much less dramatic conductivity changes at the PM, and, typically, smaller intracellular electric fields, but the $P_m(r_s)$ to those larger molecules is significant. ATP loss, for example, leads to necrosis. Thus, the conductance change does not translate into a permeability change to molecules that are larger than the nanometer-sized pores. Hence, the permeability change due to supra-EP for larger molecules falls short to that of longer, conventionally used pulses and only the electropermeability of small ions and molecules is significantly enhanced. The significant difference in the resulting electropermeability for small and large molecules that is caused by IRE and nsPEFs is a key to understanding the different cell fates. In other words, the number of pores does not necessarily translate into a membrane permeability. This level of detail is critical for understanding the pathways by which IRE causes necrotic cell death and nsPEFs causes apoptotic cell death.

And third, a greater electric dissipation by Joule heating occurs due to the EP-based change of tissue conductance. Thus, a thermal threshold for the amount of tissue heating caused by IRE and nsPEF conditions cannot be obtained by passive tissue models alone. Instead, an increase in electrical conductivity by EP of the PM must be included and is significant. Post-pulse conductivity measurements (Edd et al. 2006) do not show the full conductivity increase that is obtained during the pulses as some pores may have smaller sizes then and some may be destroyed by the time of measurement.

But what is the mechanism to trigger either the necrotic or apoptotic cell death pathway? One answer is likely related to the resulting abundance of intracellular energy stores. While apoptosis seems to require some intracellular ATP, necrosis is, in general, accompanied by the total depletion of ATP (Nicotera and Orrenius 1998). Thus, the resulting ATP-permeability at the PM for IRE pulses may indeed provide a pathway for sufficient ATP-efflux. The resulting ATP-permeability at the PM for nsPEFs is, on the other hand, anticipated to be vanishingly small. In addition, intracellular calcium overload may lead to both necrosis and apoptosis. Specifically, intracellular Ca^{2+} concentrations of a few μM promote cell death through apoptosis, whereas larger intracellular Ca^{2+} concentrations may lead to cell death through necrosis (Nicotera and Orrenius 1998, Rizzuto et al. 2003).

In general, several pulses are employed for drug-free tumor ablation by electroporating pulses. IRE protocols employ a few identical pulses, in contrast to nsPEF protocols, which seem to require hundreds of pulses. Due to the memory effect of EP, a second pulse may not give rise to more PM pores, irrespective of pulse repetition frequency and pore lifetime. But this is true only if the cell extension or even cell orientation does not change after a pulse—an unlikely scenario, since cell swelling and necrotic death (Granot et al. 2009) or cell shrinkage and apoptotic death (Nuccitelli et al. 2006) is a more likely scenario. A high-conductance state due to EP, i.e., the specific number of pores and their size, is determined by the pulse amplitude and duration. A second pulse will therefore lead to electrical interaction with a higher conductance membrane. Only as many pores are added to the pre-existing number of pores, which last from the first pulse and have not yet decayed, such that the same total number of pores are created as with the first pulse, or even fewer if pore expansion is accounted for. Therefore, a few pulses in IRE may lead to more irreversible pores, which in turn may allow for more release of, e.g., ATP or more influx of, e.g., Ca^{2+}. On the other hand, the requirement of hundreds of nsPEFs may have another basis. In this case, a cumulative intracellular transport is likely to occur, for example, a repeated release of Ca^{2+} from internal stores, such as the endoplasmic reticulum (Nicotera and Orrenius 1998, Rizzuto et al. 2003). Thus, one likely consequence of hundreds of nsPEFs is Ca^{2+} overload, leading to apoptotic cell death.

Conventional EP, as resulting from IRE protocols, is cell-size dependent. In general, larger or more extended cells require smaller electric fields than smaller cells to cause a similar effect. In a more homogeneous tissue, such as liver, all cells are expected to experience similar pore densities for a specific pulse. Solid tumors, however, may have heterogeneous cell size distributions, such that the size of the small cells will dictate the IRE pulse conditions. This cell size limitation is minimized for nsPEFs since the temporal onset of EP is shifted into the linear charging regime of the PM, which does not depend on cell size (Stewart et al. 2004, Gowrishankar et al. 2006). But the resulting pore size distribution is limited to small nanometer pores only. Therefore, nsPEF treatments have the capability of working on a variety of tumors with diverse cell morphologies. Whether or not this variety of tumors all experience nsPEF-induced apoptosis is an open question, and will depend on the specific target of the short electric pulses and whether the resulting mechanism is sufficient to bypass other competing signaling pathways. A more practical question of nsPEFs is related to the cost, reliability, and safety of the necessary very high voltage and fast switching pulser technology.

Solid tumor treatment by IRE pulses and nsPEFs are examples of where the application of electric field pulses leads to drug-free, EP-mediated cell death. In thinking towards future applications of solid tumor ablation by electric field pulses, further research may indeed find more critical insights about the cell-death mechanisms and may provide both surgeons and patients with the choice of a preferred cell death, an opportunity no other ablation technique can thus far offer.

Acknowledgments

This work has been supported by NIH grant RO1-GM63857, Aegis Industries, and a Graduate Fellowship from the National Science Foundation to K.C.S. We thank T. Vernier for helpful discussions, K.G. Weaver for computer support, and R.S. Son for technical support.

References

Al-Sakere, B., Andre, F., Bernat, C. et al. 2007. Tumor ablation with irreversible electroporation. *PLOS One* 11: e1135.

Barnett, A. and Weaver, J. C. 1991. Electroporation: A unified, quantitative theory of reversible electrical breakdown and rupture. *Bioelectrochem. Bioenerg.* 25: 163–182.

Beebe, S. J., Fox, P. M., Rec, L. J., Somers, K., Stark, R. H., and Schoenbach, K. H. 2002. Nanosecond pulsed electric field (nsPEF) effects on cells and tissues: Apoptosis induction and tumor growth inhibition. *IEEE Trans. Plasma Sci.* 30: 286–292.

Beebe, S. J., White, J., Blackmore, P. F., Deng, Y., Somers, K., and Schoenbach, K. H. 2003a. Diverse effects of nanosecond pulsed electric fields on cells and tissues. *DNA Cell Biol.* 22: 785–796.

Beebe, S. J., Fox, P. M., Rec, L. J., Willis, L. K., and Schoenbach, K. H. 2003b. Nanosecond, high intensity pulsed electric fields induce apoptosis in human cells. *FASEB J.* 17: 1493–1495.

Davalos, R. V., Rubinsky, B., Mir, L. M., and Otten, D. M. 2004. Electrical impedance tomography for imaging tissue electroporation. *IEEE Trans. BME* 51: 761–767.

Davalos, R. V., Mir, L. M., and Rubinsky, B. 2005. Tissue ablation and irreversible electroporation. *Ann. Biomed. Eng.* 33: 223–231.

Dean, M., Fojo, T., and Bates, S. 2005. Tumor stem cells and drug resistance. *Nat. Rev. Cancer* 5: 275–284.

Edd, J. F., Horowitz, L., Davalos, R. V., Mir, L. M., and Rubinsky, B. 2006. In vivo results of a new focal tissue ablation technique: Irreversible electroporation. *IEEE Trans. Biomed. Eng.* 53: 1409–1415.

Esser, A. T., Smith, K. C., Gowrishankar, T. R., and Weaver, J. C. 2007. Towards solid tumor treatment by irreversible electroporation: Intrinsic redistribution of fields and currents in tissue. *Technol. Cancer Res. Treat.* 6(4): 261–273.

Esser, A. T., Gowrishankar, T. R., Smith, K. C., and Weaver, J. C. 2009. Towards solid tumor treatment by nanosecond pulsed electric fields. *Technol. Cancer Res. Treat.* 8(4): 289–306.

Esser, A. T., Smith, K. C., Gowrishankar, T. R., and Weaver, J. C. 2010. Mechanism for intracellular effects by conventional electroporation pulses *Biophysical J.* (accepted).

Foster, K. R. and Schwan, H. P. 1996. Dielectric properties of tissues, in: Polk, C. and Postow, E. (eds.), *Handbook of Biological Effects of Electromagnetic Fields*, 2nd edn., pp. 25–102. CRC Press, Boca Raton, FL.

Freeman, S. A., Wang, M. A., and Weaver, J. C. 1994. Theory of electroporation for a planar bilayer membrane: Predictions of the fractional aqueous area, change in capacitance and pore-pore separation. *Biophys. J.* 67: 42–56.

Frey, W., White, J. A., Price, R. O. et al. 2006. Plasma membrane voltage changes during nanosecond pulsed electric field exposures. *Biophys. J.* 90: 3608–3615.

Galluzzi, L., Larochette, N., Zamzami, N., and Kroemer, G. 2006. Mitochondria as therapeutic targets for cancer chemotherapy. *Oncogene* 25: 4812–4830.

Garon, E. B., Sawcer, D., Vernier, P. T. et al. 2007. In vitro and in vivo evaluation and a case report of intense nanosecond pulsed electric field as a local therapy for human malignancies. *Int. J. Cancer* 121: 675–682.

Gowrishankar, T. R. and Weaver, J. C. 2003. An approach to electrical modeling of single and multiple cells. *Proc. Natl. Acad. Sci.* 100: 3203–3208.

Gowrishankar, T. R. and Weaver, J. C. 2006. Electrical behavior and pore accumulation in a multicellular model for conventional and supra-electroporation. *Biochem. Biophys. Res. Commun.* 349: 643–653.

Gowrishankar, T. R., Stewart, D. A., Martin, G. T., and Weaver, J. C. 2004. Transport lattice models of heat transport in skin with spatially heterogeneous, temperature-dependent perfusion. *Biomedical. Eng. Online* 3: 42.

Gowrishankar, T. R., Esser, A. T., Vasilkoski, Z., Smith, K. C., and Weaver, J. C. 2006. Microdosimetry for conventional and supra-electroporation in cells with organelles. *Biochem. Biophys. Res. Commun.* 341: 1266–1276.

Granot, Y., Ivorra, A., Maor, E., and Rubinsky, B. 2009. In vivo imaging of irreversible electroporation by means of electrical impedance tomography. *Phys. Med. Biol.* 54: 4927–4943.

Jain, R. K. 1996. Delivery of molecular medicine to solid tumors. *Science* 271: 1079–1080.

Jain, R. K. 2001. Delivery of molecular and cellular medicine to solid tumors. *Adv. Drug Deliv. Rev.* 46: 149–168.

Jaroszeski, M. J., Gilbert, R., and Heller, R. 2000. *Electrically Mediated Delivery of Molecules to Cells: Electrochemotherapy, Electrogenetherapy and Transdermal Delivery by Electroporation.* Humana Press, Totowa, NJ.

Krassowska, W. and Filev, P. D. 2007. Modeling electroporation in a single cell. *Biophys. J.* 92: 404–417.

Kroemer, G., Galluzzi, L., and Vandenabeele, P. et al. 2009. Classification of cell death: Recommendations of the nomenclature committee on cell death. *Cell Death Differ.* 16: 3–11.

Läuger, P. 1991. *Electrogenic Ion Pumps.* Sinauer Associates, Sunderland, U.K.

Lee, R. C. and Kolodney, M. S. 1987. Electrical injury mechanisms: Electrical breakdown of cell membranes. *Plast. Reconstr. Surg.* 80: 672–679.

Lee, R. C., Zhang, D., and Hannig, J. 2000. Biophysical injury mechanisms in electrical shock trauma. *Ann. Rev. Biomed. Eng.* 2: 477–509.

Livraghi, T., Mueller, P., Silverman, S., van Sonnenberg, E., McMullen, W., and Solbiati, L. 2005. *Tumor Ablation: Principles and Practice.* Springer, New York.

Maor, E., Ivorra, A., and Rubinsky, B. 2009. Nonthermal irreversible electroporation: Novel technology for vascular smooth muscle cells ablation. *PLOS One* 4: e4757.

Melikov, K. C., Frolov, V. A., Shcherbakov, A., Samsonov, A. V., Chizmadzhev, Y. A., and Chernomordik, L. V. 2001. Voltage-induced nonconductive pre-pores and metastable pores in unmodified planar bilayer. *Biophys. J.* 80: 1829–1836.

Miklavcic, D., Semrov, D., Mekid, H., and Mir, L. M. 2000. A validated model of in vivo electric field cc distribution in tissues for electrochemotherapy and for DNA electrotransfer for gene therapy. *Biochim. Biophys. Acta* 1523: 73–83.

Miller, L., Leor, J., and Rubinsky, B. 2005. Cancer cells ablation with irreversible electroporation. *Technol. Cancer Res. Treat.* 4: 699–705.

Minchinton, A. I. and Tannock, I. F. 2006. Drug penetration in solid tumors. *Nat. Rev. Cancer* 6: 583–592.

Mir, L. M. and Orlowski, S. 1999. Mechanisms of electrochemotherapy. *Adv. Drug Deliv. Rev.* 35: 107–118.

Neu, J. C. and Krassowska, W. 1999. Asymptotic model of electroporation. *Phys. Rev. E* 59: 3471–3482.

Nicotera, P. and Orrenius, S. 1998. The role of calcium in apoptosis. *Cell Calcium* 23: 173–180.

Nuccitelli, R., Pliquett, U., Chen, X. et al. 2006. Nanosecond pulsed electric fields cause melanomas to self-destruct. *Biochem. Biophys. Res. Commun.* 343: 351–360.

Nuccitelli, R., Chen, X., Pakhomov, A. G. et al. 2009. A new pulsed electric field therapy for melanomas disrupts the tumor's supply and caused complete remission without recurrence. *Int. J. Cancer* 125(2): 438–445.

Padera, T. P., Stoll, B. R., Tooredman, J. B. et al. 2004. Cancer cells compress intratumor vessels. *Nature* 427: 695.

Persson, P. O. and Strang, G. 2004. A simple mesh generator in MATLAB. *SIAM Rev.* 46(2): 329–345.

Powell, K. T. and Weaver, J. C. 1986. Transient aqueous pores in bilayer membranes: A statistical theory. *Bioelectrochem. Bioelectroenerg.* 15: 211–227.

Putt, K. S., Chen, G. W., Pearson, J. M. et al. 2006. Small-molecule activation of procaspase-3 to caspase-3 as a personalized anticancer strategy. *Nat. Rev. Biol.* 2: 543–550.

Raicu, V., Saibara, T., Enzan, H., and Irimajiri, A. 1998. Dielectric properties of rat liver in vivo: A non-invasive approach using an open-ended coaxial probe at audio/radio frequencies. *Bioelectrochem. Bioenerg.* 47(2): 325–332.

Reed, J. C. 2006a. Drug insight: Cancer therapy strategies based on restoration of endogenous cell death mechanism. *Nat. Clin. Pract. Oncol.* 3: 388–398.

Reed, J. C. 2006b. Proapoptotic multidomain Bcl-2/Bax-family proteins: Mechanisms, physiological roles, and therapeutic opportunities. *Cell Death Differ.* 13: 1378–1386.

Rizzuto, R., Pinton, P., Ferrari, D. et al. 2003. Calcium and apoptosis: Facts and hypothesis. *Oncogene* 22: 8619–8627.

Rubinsky, B., Onik, G., and Mikus, P. 2007. Irreversible electroporation: A new ablation modality—Clinical implications. *Technol. Cancer Res. Treat.* 6: 37–48.

Sale, A. J. H. and Hamilton, A. 1967. Effects of high electric fields on microorganisms: I. Killing of bacteria and yeasts. *Biochem. Biophys. Acta* 148: 781–788.

Schoenbach, K. H., Beebe, S. J., and Buescher, E. S. 2001. Intracellular effect of ultrashort pulses. *Bioelectromagnetics* 22: 440–448.

Sebolt-Lepold, J. S. and English, J. M. 2006. Mechanisms of drug inhibition of signaling molecules. *Nature* 441: 457–462.

Sel, D., Cukjati, D., Batiuskaite, D., Slivnik, T., Mir, L. M., and Miklavcic, D. 2005. Sequential finite element model of tissue electropermeabilization. *IEEE Trans. Biomed. Eng.* 52: 816–827.

Smith, K. C. 2006. Modeling cell and tissue electroporation. SM thesis. Massachusetts Institute of Technology, Cambridge, MA (online: http://hdl.handle.net/1721.1/35301).

Smith, K. C. and Weaver, J. C. 2008. Active mechanisms are needed to describe cell responses to submicrosecond, megavolt-per-meter pulses: Cell models for ultrashort pulses. *Biophys. J.* 95: 1547–1563.

Smith, K. C., Gowrishankar, T. R., Esser, A. T., Stewart, D. A., and Weaver, J. C. 2006. Spatially distributed, dynamic transmembrane voltages of organelle and cell membranes due to 10 ns pulses: Predictions of meshed and unmeshed transport network models. *IEEE Trans. Plasma Sci.* 34: 1394–1404.

Stewart, D. A., Gowrishankar, T. R., and Weaver, J. C. 2004. Transport lattice approach to describing cell electroporation: Use of a local asymptotic model. *IEEE Trans. Plasma Sci.* 32: 1696–1708.

Szakcas, G., Paterson, J. K., Ludwig, J. A., Booth-Genthe, C., and Gottesman, M. M. 2006. Targeting a multidrug resistance in cancer. *Nat. Rev. Drug Discov.* 5: 219–234.

Tieleman, D. P. 2004. The molecular basis of electroporation. *BMC Biochem.* 5: 10.

Vasilkoski, Z., Esser, A. T., Gowrishankar, T. R., and Weaver, J. C. 2006. Membrane electroporation: The absolute rate equation and nanosecond timescale pore creation. *Phys. Rev. E* 74: 021904.

Vernier, P. T., Sun, Y., Marcu, L., Salemi, S., Craft, C. M., and Gundersen, M. A. 2003. Calcium bursts induced by nanosecond electric pulses. *Biochem. Biophys. Res. Commun.* 310: 286–295.

Vernier, P. T., Sun, Y., Marcu, L., Craft, C. M., and Gundersen, M. A. 2004. Nanoelectropulse-induced phosphatidylserine translocation. *Biophys. J.* 86: 4040–4048.

Vernier, P. T., Sun, Y., Wang, J. et al. 2005. Nanoelectropulse intracellular perturbation and electropermeabilization technology: Phospholipid translocation, calcium bursts, chromatin rearrangement, cardiomyocyte activation, and tumor cell sensitivity. *Eng. Med. Biol.* 6: 5850–5853.

Weaver, J. C. 2003. Electroporation of biological membranes from multicellular to nanoscales. *IEEE Trans. Dielectric. Elect. Insulator* 10: 754–768.

Weaver, J. C. and Chizmadzhev, Y. A. 1996. Theory of electroporation: A review. *Bioelectrochem. Bioenerg.* 41: 135–160.

White, J. A., Blackmore, P. F., Schoenbach, K. H., and Beebe, S. J. 2004. Stimulation of capacitive calcium entry in HL-60 cells by nanosecond pulsed electric fields (nsPEFs). *J. Biol. Chem.* 279: 22964–22972.

14

Gene Electrotransfer: From Basic Processes to Preclinical Applications

Jean-Michel Escoffre

Aurélie
Paganin-Gioanni

Elisabeth Bellard

Muriel Golzio

Marie-Pierre Rols

Justin Teissié

14.1 Introduction

Plasmid-DNA-based gene transfer is an elegant strategy for gene therapy because it suppresses the need for a biological (viral) vector, although its use is limited by the lack of efficient and safe delivery methods (Gill et al., 2009). When compared with viral vector, the advantages include the absence of immunogenicity and integration into the host genome (Gill et al., 2009). One physical method that has emerged as a way to improve the *in vivo* delivery of plasmid DNA is electropermeabilization (or electroporation). Since its first demonstration 25 years ago (Neumann et al., 1982), this method is now routinely used for *in vitro* transfection (Escoffre et al., 2009; Mir, 2009). The application of controlled electric pulses causes a transient permeabilization of the plasma membrane and thus allows exogenous molecules to enter the cells (Teissié et al., 2005; Escoffre et al., 2009). *In vivo*, electropermeabilization is efficient for enhanced plasmid DNA delivery and expression (Heller et al., 1996; Rols et al., 1998). *In vivo*, electropermeabilization is effective on many tissues: skeletal muscle (Aihara and Miyazaki, 1998; Mir et al., 1999), liver (Heller et al., 1996), skin (Vandermeulen et al., 2007), brain (Kondoh et al., 2000), testis (Huang et al., 2000), and tumor (Rols et al., 1998). The use of *in vivo* gene electrotransfer has seen tremendous growth, including the initiation of clinical trials (Daud et al., 2008; Horton et al., 2008; Gehl, 2008).

Gene expression levels, patterns, and kinetics after *in vivo* gene electrotransfer can be controlled by modifying electrode configurations, electrical parameters, tissue features, and plasmid constructs (Heller and Heller, 2006). The resulting flexibility in *in vivo* expression is a distinct advantage of electropulsation. Most of these parameters should be carefully selected to obtain specific gene expression associated for desired therapeutic benefit. The increased use of *in vivo* gene electrotransfer has established its potential for many therapeutic applications such as cancer therapy and DNA vaccination. Useful information on *in vivo* gene electrotransfer can be found in several recent reviews (Heller and Heller, 2006; Mir, 2009).

This chapter focuses on the basic processes involved in preclinical application of gene electrotransfer. The biophysical mechanisms of electrically mediated gene transfer will first be described. The relative contributions of physical and biochemical parameters on the efficiency and safety of *in vivo* gene electrotransfer will also be delineated.

14.2 Biophysical Considerations

14.2.1 Present Knowledge on the *In Vitro* Electrotransfer

Gene electrotransfer to mammalian cells is obtained by mixing cells and plasmids in a biocompatible buffer, then by applying a well-controlled electric field pulse train (shape of pulses; choice of field strength, E; pulse duration, T; number of pulses, N; pulse repetition frequency, f), and finally bringing the mixture in a culture medium (Neumann et al., 1982; Klenchin et al., 1991; Wolf et al., 1994). Gene electrotransfer is applicable whatever the cell types (bacteria, yeast, and mammalian cells) (Golzio et al., 1998; Eynard et al., 1992; Ganeva et al., 1995).

14.2.1.1 Electric Field Strength Effects

Gene electrotransfer is only detected for electric field values leading to membrane permeabilization ($E_c > E_p$). Transfection threshold values (E_t) are the same as the ones for cell permeabilization (E_p) when milliseconds pulses are applied (Rols and Teissié, 1998). Field strength is observed to have a critical role. Plasmid molecules, negatively charged, migrate when submitted to an electric field (Neumann et al., 1992; Wolf et al., 1994). In the "low" electric field regime (i.e. $E_c < E_p$), no membrane permeabilization occurs and plasmid simply electrophoretically flows along the cell membrane toward the anode. However, above the critical permeabilizing field value ($E_c > E_p$), two main processes occur: (1) plasma membrane is permeabilized (μs); and (2) plasmid undergoes the electrophoretic migration (ms) and interacts with permeabilized membrane (Golzio et al., 2002; Teissié et al., 2008).

14.2.1.2 Timescale

No free plasmid diffusion into the cytoplasm is detected though it was proposed in older reports (Klenchin et al., 1991). The biophysical structure of the plasmid/membrane complex has to be elucidated. In the minutes following pulse applications, plasmids leave the complex and diffuse in the cytoplasm. Then, plasmids can be observed at the nucleus surface a few hours after electropulsation, but only a small fraction crosses the nuclear envelope to be expressed (Golzio et al., 2002). These intracellular steps remain rather poorly understood, as already mentioned (Teissié et al., 2008).

Under permeabilizing field conditions, the pulse duration plays a critical role in the formation of the plasmid/membrane complexes. These complexes are easily detected when the pulse duration is at least about 1 ms. Furthermore, this interpretation is supported by the observation that the plasmid content in the complex is under the control of the field strength (E), the number of successive pulses (N), and the pulse duration (T) (Golzio et al., 2002). The time for reaction with the membrane of the plasmid dragged against permeabilized membrane under the electrophoretic migration is increased by long pulse durations. This again is involved in the positive role of the pulse duration in gene electrotransfer. This contribution of the pulse duration to the plasmid/membrane interaction has already been illustrated by a complex dependence of the gene expression (Wolf et al., 1994). The practical conclusion is that *in vitro* an effective transfer is obtained by using long pulses in order to drive the plasmid toward the permeabilized membrane but with low field strength to preserve the cell viability (Kubiniec et al., 1990; Rols and Teissié, 1998). But one should keep in mind that plasmids must be mixed with cells before the application of electric pulses.

14.2.1.3 Biological Parameters

The dependence on the plasmid concentration is rather complex. Expression levels increase with the plasmid concentrations. But high levels of plasmids appear to be toxic (Rols et al., 1992).

Because there are different physical barriers and heterogeneous geometries within tissue, *in vitro* pseudo-tissue models such as dense cell suspensions (Pucihar et al., 2007) and multicellular tumor spheroid (Canatella et al., 2004; Wasungu et al., 2009) have been developed to understand the biophysical processes of electropermeabilization and gene transfer in tissues. These studies showed a perturbation of local electrical field on dense cell suspensions (Susil et al., 1998; Pavlin et al., 2002), affecting both permeabilization and gene expression (Pucihar et al., 2007) and a hindrance to gene delivery (Wasungu et al., 2009) related to the self-organization of cells in pseudo-tissues. Indeed, close contacts between cells and extracellular matrix (1) modify the electric field distribution and (2) act as physical barriers that limit the diffusion of DNA plasmid (steric hindrance) and therefore its access to cells present in the core of the tissue. The systematic comparison of biophysical studies from isolated cells to 3D spheroid model allows the development and the optimization of *in vivo* gene electrotransfer procedures.

14.2.2 *In Vivo* Gene Electrotransfer

14.2.2.1 Electropulsation Protocols

The increased use of *in vivo* gene electrotransfer is related to both the efficiency and the flexibility (i.e. easy to perform, fast, reproducible, and safe) of this physical method. Plasmid DNA has been successfully delivered to both internal and surface tissues and organs (Heller and Heller, 2006). Because the gene expression characteristics (i.e., level and duration) can be varied by meticulous selection of electrical parameters, the selection of those parameters is important to achieve a compatible profile of gene expression with a therapeutic benefit. In agreement with *in vitro* mechanism of gene electrotransfer, these electrical parameters have to be adapted to induce cell permeabilization within the tissue, electrophoresis of plasmid molecules, and no damage to the tissue. Seven different protocols of electropulsation have been reported to obtain suitable gene expression into tissues (Table 14.1). Appropriate electrical parameter selection is dependent on the tissue being targeted. One should notice that in most protocols, the voltage term (V) while inaccurate is used for the field. This "V" terminology will be retained.

14.2.2.1.1 Long Pulse and Medium Voltage (LPMV)

As for *in vitro* protocols, the use of square-wave long and medium pulses is the most popular pulse design for *in vivo* gene electrotransfer. The first successful efficient gene transfer in melanoma tumors was obtained with electric pulses of a few ms duration (Rols et al., 1998). Low and long field strength pulses are shown to induce high and long gene expression in skeletal muscle (Aihara and Miyazaki, 1998; Mir et al., 1999; Lucas and Heller, 2001), skin (Pedron-Mazoyer et al., 2007), and liver (Suzuki et al., 1998). As with *in vitro* results, DNA plasmid had to be present during the electric pulses. The application of long and low pulses induced the tissue permeabilization and localized electrophoresis of DNA plasmid within the target tissues. Muscle damage may be present. Interestingly, a pulse duration control of the delivered voltage was demonstrated to reduce muscle damage while maintaining the same level of gene transfection (Cukjati et al., 2007).

TABLE 14.1   *In Vivo* Gene Electrotransfer Protocols

References	No. of Pulses	Shape, Polarity	Electric Field	Pulse Length (ms)	Frequency (Hz)	Tissue
Aihara and Miyazaki (1998), Draghia-Akli et al. (1999)	4–6	Square, monopolar (alternate polarity)	60–400 V/cm	50	1	Skeletal muscle
Rols et al. (1998), Mir et al. (1999), MacMahon et al. (2001)	8–10	Square, monopolar	100–200 V/cm	20	1–8	Skeletal muscle, melanoma
Mathiesen (1999), Rizzuto et al. (1999)	8,000–10,000	Square, bipolar	100 V/cm	0.2	1000	Skeletal muscle
Vicat et al. (2000), Heller et al. (1996)	5	Square, monopolar	1600 V/cm	0.1	1	Skeletal muscle, liver
Bureau et al. (2000), Satkauskas et al. (2002, 2005) Pavselj and Préat (2005)	HV: 1–8 LV: 1–16	Square, monopolar	HV: 800 V/cm LV: 60–100 V/cm	HV: 0.01–0.5 LV: 50–800	HV: 1 LV: 5–100	Skeletal muscle, liver, skin, tumor
Liu et al. (2007), Simon et al. (2008)	1–9	Alternating current sine wave	10–30 V/cm	100–900	60	Skeletal muscle, liver, skin, tumor
Khan et al. (2005), Hirao et al. (2008)	2–3	Square, monopolar	0.1–0.5 A[a]	20–52	1	Skeletal muscle

[a] Current.

14.2.2.1.2 *Short Pulses and High Voltage (SPHV)*

In hepatocellular carcinoma, high expression is obtained with short and high field strength pulses (Heller et al., 1996). Vicat and collaborators showed that short and high field strength pulses also induced high gene expression in the skeletal muscle. Gene electrotransfer with short and high field strength pulses was reported to provide sustained and long-lasting gene expression (Vicat et al., 2000), a conclusion in conflict with a later observation (Lucas and Heller, 2001) where short expression was associated to the use of short pulses.

14.2.2.1.3 *High Voltage and Low Voltage (HVLV)*

The studies on mechanisms of gene electrotransfer *in vitro* highlighted that electrically mediated delivery included a permeabilizing component that permeabilized the plasma membrane and an electrophoretic component that facilitated transport of molecules to the permeabilized membrane. These studies suggested that this delivery method should include *in vivo* two basic components that could be obtained by combining two pulse types (Mir, 2009). This combination includes a short and high field strength pulse (i.e., permeabilizing component), followed by long and low field strength pulses (i.e., electrophoretic component). This hypothesis was evaluated for gene transfer to skeletal muscle (Bureau et al., 2000; Durieux et al., 2002; Satkauskas et al., 2002, 2005; André et al., 2008), tumor, skin (Pavselj and Préat, 2005), and liver (André et al., 2008; Cemazar et al., 2009). The combined pulses allow for the use of pulsing parameters, which could reduce potential discomfort from the electropulsation procedure. A large range of LV strengths can be used to obtain a significant luciferase gene expression in muscles Additional variables that should be considered include pulse number (the LV can be a train of pulses) and pulse frequency (which was always 1 Hz).

14.2.2.1.4 Short Pulse and High Frequency (SPHF)

An alternative approach, which consisted of applying short bipolar pulses (0.2 ms) at high repetition frequency (100–1000 Hz), was found to be also effective (Mathiesen, 1999; Rizzuto et al., 1999, 2000). Rizzuto and collaborator demonstrated that high-frequency trains of electric pulses cause less damage than single long pulses for the same cumulative pulse duration (Rizzuto et al., 1999; Zampaglione et al., 2005). On the basis of previous studies, bipolar pulses should be chosen for future applications, as they do not in fact cause a net movement of charges and thus side effects such as electrochemical effects should be reduced. SPHF electropulsations are a safe and efficient method to erythropoietin gene transfer into skeletal muscles of mice (Rizzuto et al., 1999, 2000; Mennuni et al., 2002; Cappelletti et al., 2003; Fattori et al., 2005), rabbit (Fattori et al., 2005; Zampaglione et al., 2005), and nonhuman primates (Fattori et al., 2005; Zampaglione et al., 2005) in preclinical trials of anemia related to kidney failure. Moreover, this electropulsation procedure is successfully used in preclinical development of genetic vaccination against tumors (Cipriani et al., 2008; Dharmapuri et al., 2009; Peruzzi et al., 2009). This protocol opens a new field for basic research, as the electrophoretic long range drift of DNA is not present. The bipolar field gives only a local movement on the submillisecond timescale.

14.2.2.1.5 Alternating Current Sine Waves (ACSW)

Currently, all commercial electropulse generators are designed to provide direct current (DC) square-wave pulses, which cause membrane permeabilization and are believed to be critical also for providing the electrophoretic force needed for DNA movement to the cell membrane (Bureau et al., 2000; Miklavcic et al., 2000; Satkauskas et al., 2002, 2005). Typically, achieving high levels of gene expression using the traditional DC square-wave pulses requires electric field strengths, prone to result in irreversible tissue damage (Muramatsu et al., 1998; Durieux et al., 2004; MacMahon and Wells, 2004). Studies in human volunteers on electroporation of skin (Wallace et al., 2001) and muscle (Tjelle et al., 2006) have shown that the high field strength of DC square-wave pulses at 1 Hz leads to an augmented sensation of pain. Liu and collaborators showed that efficient *in vivo* gene transfer is achieved using the low and safe pulses of ACSW with a frequency of 60 Hz (Liu et al., 2007). Compared with traditional DC square-wave pulses, the ACSW pulses used in this study established no net electrophoretic force but resulted in higher gene transfer efficiency, supporting their previous finding that electrophoretic forces are not involved in gene electrotransfer *in vivo* (Liu et al., 2006). Importantly, the field strength required to obtain high level of gene transfer was with less toxicity observed than with conventional DC square-wave pulses.

14.2.2.1.6 Constant-Current Electroporation (CCV)

Conventional electropulsation technologies are based upon constant-voltage concepts. Due to decreases in the tissue resistance during the electropermeabilization process, a clamped voltage pulse brings an increase in the current flowing through the tissue during the duration of the pulse and may result in loss of the perfect square-wave function, tissue damage, and reduced plasmid transfer and expression (Gehl and Mir, 1999; Gehl et al., 1999). Previous studies focused that the tissue resistance varies from subject to subject, from tissue to tissue within the same animal and during the electropermeabilization due to cellular uptake of the formulation volume (Khan et al., 2005). The constant-current electropulsation setup is able to measure the tissue resistance before, between, and during the electric field pulses and adjust the voltage to account for these individual changes. This feature allows the device to maintain a true constant-current delivery and a square wave through the tissue during electropermeabilization, preventing heating of tissue (Bloquel et al., 2004) and consequently reducing tissue damage and pain, as well as contributing to the overall increase in plasmid transfer and expression (Fattori et al., 2005; Khan et al., 2005). Constant-current electropulsation is a very efficient technology to develop DNA vaccination on small and high animals (Curcio et al., 2008; Draghia-Akli et al., 2008; Hirao et al., 2008).

14.2.2.2 Applicator Types

The choice of applicator types used for electropermeabilization is a crucial step in an efficient and safe gene transfer. A key parameter of gene electrotransfer is the local electric field strength. As the field results from a voltage applied between two electrodes, the electrode configuration is obviously controlling the field distribution and resulting transfection efficiency (Gehl and Mir, 1999). Electrode configurations for therapeutic purposes are parallel plates, contact wires, contact plates, needle pairs, and needle arrays (Gilbert et al., 1997; Jaroszeski et al., 1997; Ramirez et al., 1998; Mazères et al., 2009). Electrode configuration controls electric field distribution in tissue. However, due to its anatomy and its electrical properties, the tissue reacts to the applied external electric field. If the applied external electric field is high enough, local permeabilization of the tissue occurs, i.e., electric field distribution strongly controls permeabilization (Miklavcic et al., 1998). But if the local electric field is too high, an irreversible local alteration of cell membrane occurs. This may result in local burns. In gene therapy, it is very important to obtain a large volume of permeabilized tissue, within the tissue being subjected to electropulsation while preserving cell viability. A safe approach in the design of optimal electrode configuration is to compute the electric field distribution in tissues by means of numerical modeling. Modeling of electric field distribution in tissue is difficult due to heterogeneous material properties of tissue and its shape (Mossop et al., 2006; Pavselj and Miklavcic, 2008). Therefore, tissue electrical heterogeneity was never taken into account in the simulation, (Gowrishankar and Weaver, 2003). Due to the swelling, the volume fraction was affected (Deng et al., 2003). Such a geometrical change should affect the field distribution in tissue and the value of the field at the cell level in the tissue (Pavlin et al., 2002). Numerical modeling on homogeneous phantom tissues has been successfully used and also validated by comparison between computed and measured consequences of electric field distribution (Miklavcic et al., 1998, 2000). Electropermeabilization induces a membrane conductance change as previously described (Kinosita and Tsong, 1979; Abidor et al., 1994) and observed in tissue (Miklavcic et al., 2000). A precise simulation of the time-dependent field distribution in the tissue is clearly needed for different electrode geometries to optimize this electro-technical aspect of the biological treatment (Pucihar et al., 2009).

14.2.2.2.1 Plate Parallel Electrodes

Plate parallel electrodes are the most frequently used in gene electrotransfer of different kinds of tissues such as skeletal muscle and tumor. Their limit is that the tissue must be pinched between the electrodes. These electrodes offer the advantages that electric pulses can be applied transcutaneously and that electric field between the electrodes is quite homogeneous (Gehl et al., 1999). Nevertheless, these electrodes have two main limitations: (1) the small gap between the electrodes, which is limited by the electrical power of electropulsators and (2) high field at the contact of the electrode with the skin, which can induce electrical burns.

14.2.2.2.2 Needle Electrodes

The needle electrodes enable deeper penetration of the electric field into the tissue. The electric field distribution is not homogenous resulting in higher field intensity around the needles bringing local tissue necrosis. The heterogeneous field distribution is under the control of the diameter of each electrode (Miklavcic et al., 2000). Furthermore, a syringe electrode device for simultaneous injection and electropermeabilization showed that lower electric field strength compared to plate electrodes is required for the same transfection efficiency, also reducing the muscle wound (Liu and Huang, 2002).

14.2.2.2.3 New Designs

Therefore, new electrodes are designed and tested to minimize tissue damages (Babiuk et al., 2003; Dona et al., 2003). Spatula electrodes used for electropermeabilization of mouse skeletal muscle were shown to induce less tissue damage compared to plate and needle electrodes (Dona et al., 2003). Needle-free patch and Meander contact electrodes were proved to be effective and safe for gene delivery in the skin (Babiuk et al., 2003). Wire contact electrodes are highly user friendly for the treatment of large skin surfaces (Mazères et al., 2009).

14.3 Biological Aspects

14.3.1 Design of Plasmid Vectors

Plasmid DNA molecules are covalent closed circles of double-stranded DNA with no associated proteins. Plasmid DNA molecules are simpler, easier to mass-produce, and potentially safer than viral vectors (Gill et al., 2009). Low immunogenicity and lack of integration of plasmid DNA make them a highly attractive molecule for gene therapy provided that an efficient, safe, and targeted delivery can be achieved (Gill et al., 2009).

14.3.1.1 Plasmid Sequences

Huge constructs (3.5–20 kbp) can be electrotransferred but the size of the plasmid also has a role in effectiveness of gene electrotransfer (Wang et al., 2005). Injection of small plasmids alone and in combination with electropulsation induced better transfection efficiency compared to larger plasmid (Molnar et al., 2004; Wang et al., 2005). The relationship between the plasmid size and transfection efficiency was linear (Bloquel et al., 2004). Moreover, a new form of supercoiled plasmid DNA, called minicircle (1.5–3 kbp), has been developed, without any bacterial sequences and antibiotic resistance markers (Darquet et al., 1997; Gill et al., 2009). The bacterial sequences contain unmethylated bacterial CG dinucleotides, which are immunostimulator motives decreasing the duration of gene expression. A reduction or elimination of CpGs from plasmid DNA (CpG-free plasmid) leads to improvements in the level and persistence of gene expression (Hyde et al., 2008).

14.3.1.2 Nuclear Targeting

The mechanism, by which DNA migrates toward and into the nucleus in *in vivo* conditions by electropermeabilization, still has to be elucidated. To overcome the limitations regarding the migration of the plasmid inside the nucleus, several approaches have been used. One example is the inclusion of sequence that binds to transcription factor (TF) to facilitate intracellular trafficking and nuclear import. A region of smooth muscle γ-actin (SMGA) promoter that contains a number of smooth muscle-specific TF-binding sites have been shown to improve the nuclear uptake of plasmid DNA and also dramatically increases gene expression in vivo (Miller and Dean, 2008). When plasmid DNA containing the SMGA sequence was electrotransferred, gene expression was specifically detected into rat smooth muscle vasculature compared to plasmid DNA with SV40 sequence (Young et al., 2008).

14.3.1.3 Promoters

Promoter selection is crucial to the level and persistence of gene expression in different tissues. The preference for viral promoters (Vandermeulen et al., 2009), capable of high-level but often short-lived gene expression, has recently shifted toward selecting constitutively expressing (Matsuda and Cepko, 2007) or tissue-specific endogenous promoters (Durieux et al., 2005). Use of the human polyubiquitin C promoter has resulted in sustained expression in the mouse lung, following the delivery of naked plasmid DNA through electroporation (Gazdhar et al., 2006). Tissue-specific promoters may offer improved specificity and safety for gene electrotransfer (Durieux et al., 2005). For example, skin-specific promoters are used to restrict the expression of DNA vaccine to specific cell types (Vandermeulen et al., 2009). The use of plasmids with tissue-specific promoter resulted in significant, but very low protein expression, as compared to that obtained with ubiquitous and strong promoter, e.g., CMV and CAG promoters, plasmids. Nevertheless, for the success of gene therapy in clinics, it is essential to develop gene regulation systems (Rubenstrunk et al., 2005; Matsuda and Cepko, 2007).

Rubenstrunk and collaborators reported a regulation strategy based on the murine metallothionein promoter in a plasmid context using electric pulses delivery as an inducer (Rubenstrunk et al., 2005).

14.3.2 Injection and Biodistribution of Plasmid Vectors

Plasmid solutions should reach the target tissue localized between the electrodes, where electric pulses are applied. Gene expression depends on the amount of injected plasmid (range of plasmid amount: 20–100 μg) (Mathiesen, 1999; Mir et al., 1999) and on the volume of injection (Dupuis et al., 2000). For preclinical applications, different routes of injection are defined on the target: intramuscular (i.m.), intradermal (i.d.), intratumoral (i.t.), and intravenous (i.v.) (Lucas et al., 2002). The volume of injection is limited by the size of the target to avoid a dramatic and damaging tissue swelling. The injection speed is seldom taken into account. A recent work suggested that the injection speed is a key parameter to gene delivery into skeletal muscle and liver (André et al., 2006). In case of mice skeletal muscle, injection speed of 1.5 μL/s is associated with a high-level expression (Golzio et al., 2004). The delay between injection and electropulsation depends on the tissue. In the case of murine B16 melanoma tumors, a short delay (<1 min) leads to a high level of transfection (Rols et al., 1998). Whereas, in murine skeletal muscle, a delay between a few seconds and 4 h does not change the transfection efficiency (Satkauskas et al., 2001).

The space, which surrounds the cells, called extracellular space, contains several macromolecules, polysaccharides, or glycosaminoglycanes, fibrous proteins, salts, and water, called extracellular matrix (ECM). This ECM forms a structured gel (Berrier and Yamada, 2007). The presence of ECM into the tissue does not perturb the distribution of electric field but limits the diffusion, the electrophoretic movement, and the distribution of plasmids into the target tissue (Cappelletti et al., 2003; Henshaw and Yuan, 2008). Previous works show that delivery of plasmid DNA to cells in the target tissue such as skeletal muscles (Bureau et al., 2004), tumors (Netti et al., 2000; Pluen et al., 2001; Alexandrakis et al., 2004), skin (Vandermeulen et al., 2007), and respiratory tissues (Walther et al., 2005), is affected by the amount of nucleases, which degrade plasmid DNA, and by the amount of ECM components such as collagen (Netti et al., 2000; Pluen et al., 2001) and hyaluronic acid (Alexandrakis et al., 2004). Consequently, in order to increase the efficiency of gene transfer, some strategies have been developed to increase diffusion of plasmid DNA and to limit its degradation into the target tissues. The main strategy to increase the diffusion and distribution of plasmid DNA into the tissue consisted in controlled degradation of ECM by using enzymes such as hyaluronidase and collagenase. This approach is more efficient in the gene transfer into skeletal muscle (MacMahon et al., 2001; Mennuni et al., 2002; Molnar et al., 2004; Schertzer et al., 2006; Evans et al., 2008) and tumors (Mesojednik et al., 2007) but inefficient in skin (Vandermeulen et al., 2007). Indeed, some works showed that the pretreatment with hyaluronidase induced a 10- to 25-fold increase of gene expression in mice and rabbit skeletal muscle after i.m. injection and electropulsation compared to without pretreatment (MacMahon et al., 2001; Mennuni et al., 2002; Molnar et al., 2004; Schertzer et al., 2006). To limit the degradation of plasmid DNA, the main approach was to protect the plasmid DNA by using chemical formulation or to inhibit the exogenous DNAses. Nicol and collaborators showed that plasmid DNA formulated with poly-L-glutamate induced a 4- to 12-fold increase of gene expression in skeletal muscle after i.m. injection and electropulsation in comparison to saline injection alone (Nicol et al., 2002). Other works showed that the plasmid DNA formulated with poloxamers such as SP1017 increased gene expression by about 10-fold and maintained higher gene expression into skeletal muscle after gene electrotransfer compared with naked DNA alone (Riera et al., 2004). These data suggested that chemical formulation might enhance plasmid DNA expression by protecting the plasmid DNA from degradation by nucleases (Lemieux et al., 2000). The second approach consisted in the use of nuclease inhibitors such as aurintricarboxylic acid (ATA) (Glasspool-Malone and Malone, 2002).

14.4 Preclinical Applications

14.4.1 Production of Ectopic Proteins

The skeletal muscle is the preferred target for gene electrotransfer. Indeed, the skeletal muscle has interesting physiological properties being multinucleated (i.e., high number of expression machineries) and presenting long-lived fibers (i.e., long-lasting gene expression), which open several applications in gene therapy (Aihara and Miyazaki, 1998; Mir et al., 1999). Moreover, its rich vascularization makes muscle a secreting organ of therapeutic proteins (Trollet et al., 2008). Indeed, i.m. electrotransfer of plasmid encoding erythropoietin (EPO) induces the production of EPO therapeutic doses to treat the anemia related to kidney diseases (Rizzuto et al., 1999; Hojman et al., 2007). In the same way, TGFR-β2 receptor and VEGF-164 factor expressions treat respectively, the lung injury and fibrosis (Yamada et al., 2007) and diabetic neuropathy (Murakami et al., 2006).

Gene electrotransfer allowed the development of antitumoral immunotherapies. Indeed, electrotransfer of plasmid encoding suicide gene (e.g., ePNP/fludarabine) (Deharvengt et al., 2007), cytokines (e.g., IL-12) (Daud et al., 2008), antiangiogenesis factors (e.g., vasostatin) (Jazowiecka-Rakus et al., 2006), and tumor suppressors (e.g., p53) (Kusumanto et al., 2007) into skeletal muscle or tumors stimulated or activated the immune responses against the tumors. This immunotherapy induced a decrease of tumor growth and limited the tumor recovery and metastasis progression.

14.4.2 DNA Vaccination

Gene electrotransfer allowed the development of genetic electrovaccination. Genetic vaccination results from a direct injection of plasmid encoding vaccinal protein into the muscle or the skin (Wolff et al., 1990; Rice et al., 2008). This protein induces the host response and activates the immune system. Compared to gene therapy, genetic vaccination requires only low and transient gene expression in a few cells (Rice et al., 2008). Several works showed that gene electrotransfer increases the immune response against the antigen compared to the injection alone (Dupuis et al., 2000; Widera et al., 2000; Babiuk et al., 2002; Dayball et al., 2003). The electrovaccination allowed the development of genetic vaccines against bacterial infections such as *Mycobacterium tuberculosis* (Zhang et al., 2007) and viral infections such as HIV (Martinon et al., 2009). Recently, electrovaccination showed its efficiency to induce immune response against tumors (Kalat et al., 2002; Buchan et al., 2005; Curcio et al., 2008; Seo et al., 2009). Improvement and modulation of the immune response induced by DNA vaccines are observed by boosting DNA vaccine expression with plasmid encoding cytokines such as interleukin-12 (Hirao et al., 2009). An open question is the putative positive effect of muscle fiber electrodamages that may play a role of adjuvants (Widera et al., 2000; Chiarella et al., 2008).

14.5 Conclusions

This chapter attempts to evaluate the contributions of different parameters on the efficiency of nucleic acid delivery by electropulsation. A rather fair control of gene electrotransfer can be obtained by selecting appropriate physical as well as biochemical parameters, and optimization can be performed in order to obtain highly efficient expression both *in vitro* and *in vivo* for preclinical studies. At the present state of our knowledge, no negative consequence of gene electrotransfer has been reported *in vivo*. Even if clinical protocols now are under trial (Daud et al., 2008), other experimental studies (such as preclinical research on pets: dog, cat, and horse) are needed to evaluate other potential side effects.

Acknowledgments

This contribution was supported by grants from Region Midi-Pyrénées, ANR Cemirbio, DGA (REI2), and AFM.

References

Abidor IG, Li LH, Hui SW (1994) Studies of cell pellets: II. Osmotic properties, electroporation, and related phenomena: Membrane interactions. *Biophys J* 67:427–435.

Aihara H, Miyazaki JI (1998) Gene transfer into muscle by electroporation in vivo. *Nat Biotechnol* 16:867–870.

André FM, Cournil-Henrionnet C, Vernerey D, Opolon P, Mir LM (2006) Variability of naked DNA expression after direct local injection: The influence of the injection speed. *Gene Ther* 13:1619–1627.

André FM, Gehl J, Sersa G, Préat V, Hojman P, Eriksen J et al. (2008) Efficiency of high- and low-voltage pulse combinations for gene electrotransfer in muscle, liver, tumor and skin. *Hum Gene Ther* 19:1261–1271.

Alexandrakis G, Brown EB, Tong RT, McKee TD, Campbell RB, Boucher T et al. (2004) Two-photon fluorescence correlation microscopy reveals the two-phase nature of transport in tumors. *Nat Med* 10:203–207.

Babiuk S, Baca-Estrada ME, Foldvari M, Storms M, Rabussay D, Widera G, Babiuk LA (2002) Electroporation improves the efficacy of DNA vaccines in large animals. *Vaccine* 20:3399–3408.

Babiuk S, Baca-Estrada ME, Foldvari M, Baizer L, Stout R, Storms M et al. (2003) Needle-free topical electroporation improves gene expression from plasmids administrated in porcine skin. *Mol Ther* 8:992–998.

Berrier AL, Yamada KM (2007) Cell-matrix adhesion. *J Cell Physiol* 213:565–573.

Buchan S, Gronevik E, Mathiesen I, King CA, Stevenson FK, Rice J (2005) Electroporation as a "prime/boost" strategy for naked DNA vaccination against a tumor antigen. *J Immunol* 174:6292–6298.

Bloquel C, Fabre E, Bureau FM, Scherman D (2004) DNA electrotransfer for intracellular and secreted proteins expression: New methodological developments and applications. *J Gene Med* 6(suppl 1):s11–s23.

Bureau FM, Gehl J, Deleuze V, Mir LM, Scherman D (2000) Importance of association between permeabilization and electrophoretic forces for intramuscular DNA electrotransfer. *Biochim Biophys Acta* 1474:353–359.

Bureau MF, Naimi S, Torero Ibad R, Seguin J, Georger C, Arnould E, Maton L, Blanche F, Delaere P, Scherman D (2004) Intramuscular plasmid DNA electrotransfer: Biodistribution and degradation. *Biochim Biophys Acta* 1676:138–148.

Canatella PJ, Black MM, Bonnichsen DM, Mckenna C, Prausnitz MR (2004) Tissue electroporation: Quantification and analysis of heterogeneous transport in multicellular environments. *Biophys J* 86:3260–3268.

Cappelletti M, Zampaglione I, Rizzuto G, Ciliberto G, La Monica N, Fattori E (2003) Gene electro-transfer improves transduction by modifying the fate of intramuscular DNA. *J Gene Med* 5:324–332.

Cemazar M, Golzio M, Sersa G, Hojman P, Kranjc S, Mesojednik S, Rols MP, Teissié J (2009) Control of pulse parameters of DNA electrotransfer into solid tumors in mice. *Gene Ther* 16:635–644.

Chiarella P, Massi E, De Robertis M, Sibilio A, Parrella P, Fazio VM, Signori E (2008) Electroporation of skeletal muscle induces danger signal release and antigen-presenting cell recruitment independently of DNA vaccine administration. *Expert Opin Biol Ther* 8:1645–1657.

Cipriani B, Fridman A, Bendtsen C, Dharmapuri S, Mennuni C, Pak I et al. (2008) Therapeutic vaccination halts disease progression in BALB-neuT mice: The amplitude of elicited immune response is predictive of vaccine efficacy *Hum Gene Ther* 19:670–680.

Cukjati D, Batiuskaite D, André F, Miklavcic D, Mir LM (2007) Real time electroporation control for accurate and safe *in vivo* non-viral gene therapy. *Bioelectrochemistry* 70:501–507.

Curcio C, Khan AS, Spadaro M, Quaglino E, Cavallo F, Forni G, Draghia-Akli R (2008) DNA immunization using constant-current electroporation affords long-term protection from autochthonous mammary carcinomas in cancer-prone transgenic mice. *Cancer Gene Ther* 15:108–114.

Darquet AM, Cameron B, Wils P, Scherman D, Crouzet J (1997) A new DNA vehicle for nonviral gene delivery: Supercoiled minicircle. *Gene Ther* 4:1341–1349.

Daud AI, DeConti RC, Andrews S, Urbas P, Riker AI, Sondak VK et al. (2008) Phase I trial of interleukin-12 plasmid electroporation in patients with metastatic melanoma. *J Clin Oncol* 26:5896–5903.

Dayball K, Millar J, Miller M, Wan YH, Bramson J (2003) Electroporation enables plasmid vaccines to elicit CD8+ T cell responses in the absence of CD4+ T cells. *J Immunol* 171:3379–3384.

Deharvengt S, Rejuba S, Wack S, Aprahamian M, Hajri A (2007) Efficient electrogene therapy for pancreatic adenocarcinoma treatment using the bacterial purine nucleoside phosphorylase suicide gene with fludarabine. *Int J Oncol* 30:1397–1406.

Deng J, Schoenbach KH, Buescher ES, Hair PS, Fox PM, Beebe SJ (2003) The effects of intense submicrosecond electrical pulses on cells. *Biophys J* 84:2709–2714.

Dharmapuri S, Aurisicchio L, Neumer P, Verdirame M, Ciliberto G, La Monica N (2009) An oral TLR7 agonist is a potent adjuvant of DNA vaccination in transgenic mouse tumor models. *Cancer Gene Ther* 16:462–472.

Dona M, Sandri M, Rossini K, Dell'Aica I, Podhorska-Okolow M, Carraro U (2003) Functional in vivo gene transfer into the myofibers of adult skeletal muscle. *Biochem Biophys Res Commun* 312:1132–1138.

Draghia-Akli R, Fiorotto ML, Hill LA, Malone PB, Deaver DR, Schwartz RJ (1999) Myogenic expression of an injectable protease-resistant growth hormone-releasing hormone augments long-term growth in pigs. *Nat Biotechnol* 17:1179–1183.

Draghia-Akli R, Khan AS, Brown PA, Pope MA, Wu L, Hirao L, Weiner DB (2008) Parameters for DNA vaccination using adaptive constant-current electroporation in mouse and pig models. *Vaccine* 26:5230–5237.

Dupuis M, Denis-Mize K, Woo C, Goldbeck C, Selby MJ, Chen M et al. (2000) Distribution of DNA vaccines determines their immunogenicity after intramuscular injection in mice. *J Immunol* 165:2850–2858.

Durieux AC, Bonnefoy R, Manissolle C, Freyssenet D (2002) High-efficiency gene electrotransfer into skeletal muscle: Description and physiological applicability of a new pulse generator. *Biochem Biophys Res Commun* 296:443–450.

Durieux AC, Bonnefoy R, Russo T, Freyssenet D (2004) In vivo gene electrotransfer into skeletal muscle: Effects of plasmid DNA on the occurrence and extent of muscle damage. *J Gene Med* 6:809–816.

Durieux AC, Bonnefoy R, Freyssenet D (2005) Kinetic of transgene expression after electrotransfer into skeletal muscle: Importance of promoter origin/strength. *Biochim Biophys Acta* 1725:403–409.

Escoffre JM, Portet T, Wasungu L, Teissié J, Dean D, Rols MP (2009) What is (still not) known of the mechanism by which electroporation mediates gene transfer and expression in cells and tissues. *Mol Biotechnol* 41:286–295.

Evans V, Foster H, Graham IR, Foster K, Athanasopoulos T, Simons JP, Dickson G, Owen JS (2008) Human apolipoprotein E expression from mouse skeletal muscle by electrotransfer of nonviral DNA (plasmid) and pseudotyped recombinant adeno-associated virus (AAV2/7). *Hum Gene Ther* 19:569–578.

Eynard N, Sixou S, Duran N, Teissié J (1992) Fast kinetics studies of Escherichia coli electrotransformation. *Eur J Biochem* 209:431–436.

Fattori E, Cappelletti M, Zampaglione I, Mennuni C, Arcuri M, Rizzuto G, Costa P, Perretta G, Ciliberto G, La Monica N (2005) Gene electro-transfer of an improved erythropoietin plasmid in mice and non-human primates. *J Gene Med* 7:228–236.

Ganeva V, Galutzov B, Teissié J (1995) Fast kinetics studies of plasmid DNA transfer in intact yeast cells mediated by electropulsation. *Biochem Biophys Res Commun* 214:825–832.

Gazdhar A, Bilici M, Pierog J, Ayuni EL, Gugger M, Wetterwald A, Cecchini M, Schmid RA (2006) In vivo electroporation and ubiquitin promoter-a protocol for sustained gene expression in the lung. *J Gene Med* 8:910–918.

Gehl J (2008) Electroporation for drug and gene delivery in the clinic: Doctors go electric. *Methods Mol Biol* 423:351–359.

Gehl J, Mir LM (1999) Determination of optimal parameters for in vivo gene transfer by electroporation, using a rapid in vivo test for cell permeabilization. *Biochem Biophys Res Commun* 261:377–380.

Gehl J, Sorensen TH, Nielsen K, Raskmark P, Nielsen SL, Skovsgaard T, Mir LM (1999) In vivo electroporation of skeletal muscle: Threshold, efficacy and relation to electric field distribution. *Biophys Biochim Acta* 1428:233–240.

Gilbert RA, Jaroszeski MJ, Heller R (1997) Novel electrode designs for electrochemotherapy. *Biochim Biophys Acta* 1334:9–14.

Gill DR, Pringle IA, Hyde SC (2009) Progress and prospects: The design and production of plasmid vectors. *Gene Ther* 16:165–171.

Glasspool-Malone J, Malone RW (2002) Enhancing direct in vivo transfection with nuclease inhibitors and pulsed electrical fields. *Methods Enzymol* 346:72–91.

Golzio M, Mora MP, Raynaud C, Delteil C, Teissié J, Rols MP (1998) Control by osmotic pressure of voltage-induced permeabilization and gene transfer in mammalian cells. *Biophys J* 74:3015–3022.

Golzio M, Teissié J, Rols MP (2002) Direct visualization at the single-cell level of electrically mediated gene delivery. *Proc Natl Acad Sci USA* 99:1292–1297.

Golzio M, Rols MP, Teissié J (2004) In vitro and in vivo electric-field-mediated permeabilization, gene transfer, and expression. *Methods* 33:126–135.

Gowrishankar TR, Weaver JC (2003) An approach to electrical modelling of single and multiple cells. *Proc Natl Acad Sci USA* 100:3203–3208.

Heller LC, Heller R (2006) In vivo electroporation for gene therapy. *Hum Gene Ther* 17:890–897.

Heller R, Jaroszeski M, Atkin A, Moradpour D, Gilbert R, Wands J, Nicolau C (1996) In vivo gene electro-injection and expression in rat liver. *FEBS Lett* 389:225–228.

Henshaw JW, Yuan F (2008) Field distribution and DNA transport in solid tumors during electric field-mediated gene delivery. *J Pharm Sci* 97:691–711.

Hirao LA, Wu L, Khan AS, Hokey DA, Yan J, Dai A, Betts MR, Draghia-Akli R, Weiner DB (2008) Combined effects of IL-12 and electroporation enhances the potency of DNA vaccination in macaques. *Vaccine* 26:3112–3120.

Hirao LA, Wu L, Khan AS, Satishchandran A, Draghia-Akli R, Weiner DB (2008) Intradermal/subcutaneaous immunization by electroporation improves plasmid vaccine delivery and potency in pigs and rhesus macaques. *Vaccine* 26:440–448.

Hojman J, Gissel H, Gehl J (2007) Sensitive and precise regulation of haemoglobin after gene transfer of erythropoietin to muscle tissue using electroporation. *Gene Ther* 14:950–959.

Horton HM, Lalor PA, Rolland AP (2008) IL-2 plasmid electroporation: From preclinical studies to phase clinical trial. *Methods Mol Biol* 423:361–372.

Huang Z, Tamura M, Sakurai T, Chuma S, Saito T, Nakatsuji N (2000) In vivo transfection of testicular germ cells and transgenesis by using the mitochondrially localized jellyfish fluorescent protein gene. *FEBS Lett* 487:248–251.

Hyde SC, Pringle IA, Abdullah S, Lawton AE, Davies LA, Varathalingam A et al. (2008) CpG-free plasmids confer reduced inflammation and sustained pulmonary gene expression. *Nat Biotechnol* 26:549–551.

Jaroszeski MJ, Gilbert RA, Heller R (1997) In vivo antitumor effects of electrochemotherapy in a hepatoma model. *Biochim Biophys Acta* 1334:15–18.

Jazowiecka-Rakus J, Jarosz M, Szala S (2006) Combination of vasostatin gene therapy with cyclophosphamide inhibits growth of B16(F10) melanoma tumours. *Acta Biochim Pol* 53:199–202.

Kalat M, Kupcu Z, Schuller S, Zalusky D, Zehetner M, Paster W, Schweighoffer T (2002) In vivo plasmid electroporation induces tumor antigen-specific CD8+ T-cell responses and delays tumor growth in a syngeneic mouse melanoma model. *Cancer Res* 62:5489–5494.

Khan AS, Pope MA, Draghia-Akli R (2005) Highly efficient constant-current electroporation increases in vivo plasmid expression. *DNA Cell Biol* 24:810–818.

Kinosita K Jr, Tsong TY (1979) Voltage-induced conductance in human erythrocyte membranes. *Biochim Biophys Acta* 554:479–497.

Klenchin VA, Sukharev SI, Serov SM, Chernomordik LV, Chizmadzev YuA (1991) Electrically induced DNA uptake by cells is a fast process involving DNA electrophoresis. *Biophys J* 60:804–811.

Kondoh T, Motooka Y, Bhattacharjee AK, Kokunai T, Saito N, Tamaki N (2000) In vivo gene transfer into the periventricular region by electroporation. *Neurol Med Chir* 40:618–622.

Kubiniec RT, Liang H, Hui SW (1990) Effects of pulse length and pulse strength on transfection by electroporation. *Biotechniques* 8:16–20.

Kusumanto YH, Mulder NH, Dam WA, Losen M, De Baets MH, Meijer C, Hospers GA (2007) Improvement of in vivo transfer of plasmid DNA in muscle: Comparison of electroporation versus ultrasound. *Drug Deliv* 14:273–277.

Lemieux P, Guérin N, Paradis G, Proulx R, Chistyakova L, Kabanov A, Alakhov V (2000) A combination of poloxamers increases gene expression of plasmid DNA in skeletal muscle. *Gene Ther* 7:987–991.

Liu F, Huang L (2002) Electric gene transfer to the liver following systemic administration of plasmid DNA. *Gene Ther* 9:1116–1119.

Liu F, Heston S, Shollenberger LM, Sun B, Mickle M, Lovell M, Huang L (2006) Mechanism of in vivo DNA transport into cells by electroporation: Electrophoresis across the plasma membrane may not be involved. *J Gene Med* 8:353–361.

Liu F, Sag D, Wang J, Shollenberger LM, Niu F, Yuan X, Li SD, Thompson M, Monahan P (2007) Sine-wave current for efficient and safe in vivo gene transfer. *Mol Ther* 15:1842–1847.

Lucas ML, Heller R (2001) Immunomodulation by electrically enhanced delivery of plasmid DNA encoding IL-12 to murine skeletal muscle. *Mol Ther* 3:47–53.

Lucas ML, Heller L, Coppola D, Heller R (2002) IL-12 plasmid delivery by in vivo electroporation for the successful treatment of established subcutaneaous B16.F10 melanoma. *Mol Ther* 5:668–675.

MacMahon JM, Wells DJ (2004) Electroporation for gene transfer to skeletal muscles: Current status. *BioDrugs* 18:155–165.

MacMahon JM, Signori E, Wells KE, Fazio VM, Wells DJ (2001) Optimisation of electrotransfer of plasmid into skeletal muscle by pretreatment with hyaluronidase-increased expression with reduced muscle damage. *Gene Ther* 8:1264–1270.

Martinon F, Kaldma K, Sikut R, Culina S, Romain G, Tuomela M et al. (2009) Persistent immune responses induced by a HIV DNA vaccine delivered in association with electroporation in the skin of nonhuman primates. *Hum Gene Ther* 20:1291–1307.

Mathiesen I (1999) Electropermeabilization of skeletal muscle enhances gene transfer in vivo. *Gene Ther* 6:508–514.

Matsuda T, Cepko CL (2007) Controlled expression of transgenes introduced by in vivo electroporation. *Proc Natl Acad Sci USA* 104:1027–1032.

Mazères S, Sel D, Golzio M, Pucihar G, Tamzali Y, Miklavcic D, Teissié J (2009) Non invasive contact electrodes for in vivo localized cutaneous electropulsation and associated drug and nucleic acid delivery. *J Control Release* 134:125–131.

Mennuni C, Calvaruso F, Zampaglione I, Rizzuto G, Rinaudo D, Dammassa E, Ciliberto G, Fattori E, La Monica N (2002) Hyaluronidase increases electrogene transfer efficiency in skeletal muscle. *Hum Gene Ther* 13:355–365.

Mesojednik S, Pavlin D, Sersa G, Coer A, Kranjc S, Grosel A, Tevz G, Cemazar M (2007) The effect of the histological properties of tumors on translocation efficiency of electrically assisted gene delivery to solid tumors in mice. *Gene Ther* 14:1261–1269.

Miklavcic D, Beravs K, Semrov D, Cemazar M, Demsar F, Sersa G (1998) The importance of electric field distribution for effective in vivo electroporation of tissues. *Biophys J* 74:2152–2158.

Miklavcic D, Semrov D, Mekid H, Mir LM (2000) A validated model of in vivo electric field distribution in tissues for electrochemotherapy and for DNA electrotransfer for gene therapy. *Biochim Biophys Acta* 1523:73–83.

Miller AM, Dean DA (2008) Cell-specific nuclear import of plasmid DNA in smooth muscle requires tissue-specific transcription factors and DNA sequences. *Gene Ther* 15:1107–1115.

Mir LM (2009) Nucleic acids electrotransfer-based gene therapy (electrogenetherapy): Past, current, and future. *Mol Biotechnol* 43:167–176.

Mir LM, Bureau MF, Gehl J, Rangara R, Rouy D, Caillaud JM, Delaere P, Branellec D, Schwartz B, Scherman D (1999) High-efficiency gene transfer into skeletal muscle mediated by electric pulses. *Proc Natl Acad Sci USA* 96:4262–4267.

Molnar MJ, Gilbert R, Lu Y, Liu AB, Guo A, Larochelle N et al. (2004) Factors influencing the efficacy, longevity, and safety of electroporation-assisted plasmid-based gene transfer into mouse muscles. *Mol Ther* 10:447–455.

Mossop BJ, Barr RC, Henshaw JW, Zaharoff DA, Yuan F (2006) Electric fields in tumors exposed to external voltage sources: Implication for electric field-mediated drug and gene delivery. *Ann Biomed Eng* 34:1564–1572.

Murakami T, Arai M, Sunada Y, Nakamura A (2006) VEGF 164 gene transfer by electroporation improves diabetic sensory neuropathy in mice. *J Gene Med* 8:773–781.

Muramatsu T, Nakamura A, Park HM (1998) In vivo electroporation: A powerful and convenient means of nonviral gene transfer to tissues of living animals. *Int J Mol Med* 1:55–62.

Neumann E, Schaefer-Ridder M, Wang Y, Hofschneider PH (1982) Gene transfer into mouse lyoma cells by electroporation in high electric fields. *EMBO J* 1:841–845.

Neumann E, Werner E, Sprafke A, Kruger K (1992) Electroporation phenomena. Electrooptics of plasmid DNA and of lipid bilayer vesicles. In: Jennings BR, Stoylov SP, eds., *Colloid and Molecular Electro-Optics*. Bristol, U.K.: IOP publishing Ltd., pp. 197–206.

Netti PA, Berk DA, Swartz MA, Grodzinsky AJ, Jain RK (2000) Role of extracellular matrix assembly in interstitial transport in solid tumors. *Cancer Res* 60:2497–2503.

Nicol F, Wong M, MacLaughlin FC, Perrard J, Wilson E, Nordstrom JL, Smith LC (2002) Poly-L-glutamate, an anionic polymer, enhances transgene expression for plasmids delivered by intramuscular injection with in vivo electroporation. *Gene Ther* 9:1351–1358.

Pavlin M, Pavselj N, Miklavcic D (2002) Dependence of induced transmembrane potential on cell density, arrangement, and cell position inside a cell system. *IEEE Trans Biomed Eng* 49:605–612.

Pavselj N, Miklavcic D (2008) Numerical modeling in electroporation-based biomedical applications. *Radiol Oncol* 42:159–168.

Pavselj N, Préat V (2005) DNA electrotransfer into the skin using a combination of one high- and one low-voltage pulse. *J Control Release* 106:407–415.

Pedron-Mazoyer S, Plouet J, Hellaudais J, Teissié J, Golzio M (2007) New anti angiogenesis developments through electro-immunization: Optimization by in vivo optical imaging of intradermal electro gene transfer. *Biochim Biophys Acta* 1770:137–142.

Peruzzi D, Mori F, Conforti A, Lazzaro D, De Rinaldis E, Ciliberto G, La Monica N, Aurisicchio L (2009) MMP11: A novel target antigen for cancer immunotherapy. *Clin Cancer Res* 15:4104–4113.

Pluen A, Boucher Y, Ramanujan S, McKee TD, Gohongi T, di Tomaso E et al. (2001) Role of tumor-host interactions in interstitial diffusion of macromolecules: Cranial vs subcutaneous tumors. *Proc Natl Acad Sci USA* 98:4628–4633.

Pucihar G, Kotnik T, Teissié J, Miklavcic D (2007) Electropermeabilization of dense cell suspensions. *Eur Biophys J* 36:173–185.

Pucihar G, Miklavcic D, Kotnik T (2009) A time-dependent numerical model of transmembrane voltage inducement and electroporation of irregularly shaped cells. *IEEE Trans Biomed Eng* 56:1491–1501.

Ramirez LH, Orlowski S, An D, Bindoula H, Dzodic R, Ardouin P et al. (1998) Electrochemotherapy on liver tumours in rabbits. *Br J Cancer* 77:2104–2111.

Rice J, Ottensmeier CH, Stevenson FK (2008) DNA vaccines: Precision tools for activating effective immunity against cancer. *Nat Rev Cancer* 8:108–120.

Riera M, Chillon M, Aran JM, Cruzado JM, Torras J, Grinyo JM, Fillat C (2004) Intramuscular SP1017-formulated DNA electrotransfer enhances transgene expression and distributes hHGF to different rat tissues. *J Gene Med* 6:111–118.

Rizzuto G, Cappelletti M, Malone D, Savino R, Lazzaro D, Costa P et al. (1999) Efficient and regulated erythropoietin production by naked DNA injection and muscle electroporation. *Proc Natl Acad Sci USA* 96:6417–6422.

Rizzuto G, Cappelletti M, Malone D, Mennuni C, Wiznerowicz M, De Martis A, Maione D, Ciliberto G, La Monica N, Fattori E (2000) Gene electrotransfer results in a high-level transduction of rat skeletal muscle and corrects anemia of renal failure. *Hum Gene Ther* 11:1891–1900.

Rols MP, Teissié J (1998) Electropermeabilization of mammalian cells to macromolecules: Control by pulse duration. *Biophys J* 75:1415–1423.

Rols MP, Coulet D, Teissié J (1992) Highly efficient transfection of mammalian cells by electric field pulses. Application to large volumes of cell culture by using a flow system. *Eur J Biochem* 206:115–121.

Rols MP, Delteil C, Golzio M, Dumond P, Cros S, Teissié J (1998) In vivo electrically mediated protein and gene transfer in murine melanoma. *Nat Biotechnol* 16:168–171.

Rubenstrunk A, Trollet C, Orsini C, Scherman D (2005) Positive in vivo heterologous gene regulation by electric pulses delivery with metallothionein I gene promoter. *J Gene Med* 7:1565–1572.

Satkauskas S, Bureau MF, Mahfoudi A, Mir LM (2001) Slow accumulation of plasmid in muscle cells: Supporting evidence for a mechanism of DNA uptake by receptor-mediated endocytosis. *Mol Ther* 4:317–323.

Satkauskas S, Bureau MF, Puc M, Mahfoudi A, Scherman D, Miklavcic D, Mir LM (2002) Mechanisms of in vivo DNA electrotransfer: Respective contributions of cell electropermeabilization and DNA electrophoresis. *Mol Ther* 5:133–140.

Satkauskas S, André F, Bureau MF, Scherman D, Mikkavcic D, Mir LM (2005) Electrophoretic component of electric pulses determines the efficacy of in vivo DNA electrotransfer. *Hum Gene Ther* 16:1194–1201.

Schertzer JD, Plant DR, Lynch GS (2006) Optimizing plasmid-based gene transfer for investigating skeletal muscle structure and function. *Mol Ther* 13:795–803.

Seo SH, Jin HT, Park SH, Youn JI, Sung YC (2009) Optimal induction of HPV DNA vaccine-induced CD8(+) T cell responses and therapeutic antitumor effect by antigen engineering and electroporation. *Vaccine* 27:5906–5912.

Simon AJ, Casimirio DR, Finnefrock AC, Davies ME, Tang A, Chen M, Chastain M, Kath GS, Chen L, Shiver JW (2008) Enhanced in vivo transgene expression and immunogenicity from plasmid vectors following electrostimulation in rodents and primates. *Vaccine* 26:5202–5209.

Susil R, Semrov D, Miklavcic D (1998) Electric field induced transmembrane potential depends on cell density and organization. *Electro. Magnetobiol* 17:391–399.

Suzuki T, Shin BC, Fujikura K, Matsuzaki T, Takata K (1998) Direct gene transfer into rat liver cells by in vivo electroporation. *FEBS Lett* 425:436–440.

Teissié J, Golzio M, Rols MP (2005) Mechanisms of cell membrane electropermeabilization: A minireview of our present (lack of ?) knowledge. *Biochim Biophys Acta* 1724:270–280.

Teissié J, Escoffre JM, Golzio M, Rols MP (2008) Time dependence of electric field effects on cell membranes. A review for a critical selection of pulse duration for therapeutical applications. *Radiol Oncol* 42:196–206.

Tjelle TE, Salte R, Mathiesen I, Kjeken R, Scheerlinck JP, Karlis J (2006) A novel electroporation device for gene delivery in large animals and humans. *Vaccine* 24:4667–4670.

Trollet C, Scherman D, Bigey P (2008) Delivery of DNA into muscle for treating systemic diseases: Advantages and challenges. *Methods Mol Biol* 423:199–214.

Vandermeulen G, Staes E, Vanderhaeghen ML, Bureau, MF, Scherman D, Préat V (2007) Optimisation of intradermal DNA electrotransfer for immunisation. *J Control Res* 124:81–87.

Vandermeulen G, Richiardi H, Escriou V, Ni J, Fournier P, Schirrmacher V, Scherman D, Préat V (2009) Skin-specific promoters for genetic immunization by DNA electroporation. *Vaccine* 27:4272–4277.

Vicat JM, Boisseau S, Jourdes P, Lainé M, Wion D, Bouali-Benazzouz R, Benabid AL, Berger F (2000) Muscle transfection by electroporation with high-voltage and short-pulse currents provides high-level and long-lasting gene expression. *Hum Gene Ther* 11:909–916.

Wallace MS, Ridgeway B, Jun Z, Schulteis G, Rabussay D, Zhang L (2001) Topical delivery of lidocaine in healthy volunteers by electroporation, electroincorporation, or iontophoresis: An evaluation of skin anesthesia. *Reg Anesth Pain Med* 26:229–238.

Walther W, Stein U, Siegel R, Fichtner I, Schlag PM (2005) Use of the nuclease inhibitor aurintricarboxylic acid (ATA) for improved non-viral intratumoral in vivo gene transfer by jet-injection. *J Gene Med* 7:477–485.

Wang XD, Tang JG, Xie XL, Yang JC, Shuai L, Ji JG, Gu J (2005) A comprehensive study of optimal conditions for naked plasmid DNA transfer into skeletal muscle by electroporation. *J Gene Med* 7:1235–1245.

Wasungu L, Escoffre JM, Valette A, Teissié J, Rols MP (2009) A 3D in vitro spheroid model as a way to study the mechanisms of electroporation. *Int J Pharm* 379:278–284.

Widera G, Austin M, Rabussay D, Goldbeck C, Barnett SW, Chen M et al. (2000) Increased DNA vaccine delivery and immunogenicity by electroporation in vivo. *J Immunol* 164:4635–4640.

Wolf H, Rols MP, Boldt E, Neumann E, Teissié J (1994) Control by pulse parameters of electric field-mediated gene transfer in mammalian cells. *Biophys J* 66:524–531.

Wolff JA, Malone RW, Williams P, Chong W, Acsadi G, Jani A, Felgner PL (1990) Direct gene transfer into mouse muscle in vivo. *Science* 247:1465–1468.

Yamada M, Kuwano K, Maeyama T, Yoschimi M, Hamada N, Fukumoto J, Egashira K, Hiasa K, Takayama K, Nakanishi Y (2007) Gene transfer of transforming growth factor type II receptor by in vivo electroporation attenuates lung injury and fibrosis. *J Clin Pathol* 60:916–920.

Young JL, Zimmer WE, Dean DA (2008) Smooth muscle-specific gene delivery in the vasculature based on restriction of DNA nuclear import. *Exp Biol Med* 233:840–848.

Zampaglione I, Arcuri M, Cappelletti M, Ciliberto G, Perretta G, Nicosia A, La Monica N, Fattori E (2005) In vivo DNA gene electro-transfer: A systematic analysis of different electrical parameters. *J Gene Med* 7:1475–1481.

Zhang X, Divangahi M, Ngai P, Santosuosso M, Millar J, Zganiacz A, Wang J, Bramson J, Xing Z (2007) Intramuscular immunization with a monogenic plasmid DNA tuberculosis vaccine: Enhanced immunogenicity by electroporation and co-expression of GM-CSF transgène. *Vaccine* 25:1342–1352.

V

Technical Considerations

15

Modeling Electric Field Distribution *In Vivo*

Nataša Pavšelj

Anže Županič

Damijan Miklavčič

15.1 Introduction

The application of electric pulses to cells, either in suspension or tissue, causes the electroporation of the cell membrane, increasing its permeability and making it possible for larger molecules that otherwise cannot cross the membrane, such as drug molecules or DNA, to enter the cell. If the pulse is of adequate amplitude, the electric field and consequently the induced transmembrane voltage are high enough to cause cell membrane permeabilization. For any given cell, the induced transmembrane voltage is proportional to the electric field; more precisely, it is proportional to the local electric field in which the cell is placed. More details on induced transmembrane voltage and electroporation on the cell level are given in Chapter 3, titled "Induced transmembrane voltage—Theory, modeling, and experiments" by Kotnik and Pucihar. In this chapter, the focus is on the electroporation on a tissue level, more specifically on how the electric field is distributed in different electrode-tissue setups in the applied use of electroporation.

Numerous experiments, both *in vitro* and *in vivo*, have to be performed before a biomedical application is put to practical use in the clinical environment. As a complementary work to *in vivo* experimenting, analytical and numerical models can be used to represent, as realistically as possible, real biological phenomena. In this way, we can better understand some of the processes involved and analyze and explain the experimental results. Different electrical parameters can be evaluated in advance, such as pulse amplitude, duration, and number of pulses. All of that can help us plan new protocols, design electroporation devices, facilitate the design of electrodes and their placement with respect to target tissue, and plan new experiments and treatments (Šemrov and Miklavčič 1998, Brandisky and Daskalov 1999,

Miklavčič et al. 2000, Dev et al. 2003, Miklavčič et al. 2006a, Šel et al. 2007, Čorović et al. 2008b, Županič et al. 2008). Of course, models have to be validated by experiments and, if necessary, improved. Experimenting with such models is easier and sometimes the only possible or ethically acceptable alternative to experimenting on real biological systems. Both experimental work and numerical modeling combined give us valuable information and help us to understand the underlying mechanisms of the process(es) we are aiming to describe.

As a simple definition, a mathematical model is a representation of the chosen essential aspects of a real system (may it be a living, engineering, or social system), described by a set of variables and a set of equations that establish relationships between the variables. Mathematical models represent an important tool in the study of the effects of the electromagnetic fields and accompanying coupled phenomena on cells, tissues, and organs (Fear and Stuchly 1998, Debruin and Krassowska 1999a,b, Miklavčič et al. 2000, Šel et al. 2005, Pavšelj and Miklavčič 2008a). These biological systems are often geometrically highly intricate, so analytical methods are, in most cases, entirely replaced by numerical methods. In continuation, we provide the basics of electromagnetic field theory, describe the characteristics of biological tissues, explain the basic steps in constructing numerical models of tissue electroporation, and give some reference to numerical modeling–based treatment planning.

15.2 Electromagnetic Field Theory

15.2.1 Maxwell's Equations

In 1865, Maxwell had put forward a set of equations that describe the properties of the macroscopic electric and magnetic fields and relate them to their sources: Ampere's law (Equation 15.1), Faraday's law of induction (Equation 15.2), Gauss's laws (Equation 15.3), Gauss's law for magnetism (Equation 15.4), and the continuity equation (Equation 15.5)

$$\nabla \times \vec{B} = \mu_0 \vec{J} + \mu_0 \varepsilon_0 \frac{\partial \vec{E}}{\partial t} \tag{15.1}$$

$$\nabla \times \vec{E} = -\frac{\partial \vec{B}}{\partial t} \tag{15.2}$$

$$\nabla \cdot \vec{E} = \frac{\rho}{\varepsilon_0} \tag{15.3}$$

$$\nabla \cdot \vec{B} = 0 \tag{15.4}$$

$$\nabla \cdot \vec{J} = -\frac{\partial \rho}{\partial t} \tag{15.5}$$

where
$\vec{B}$ is the magnetic flux density
$\vec{J}$ is the total current density
$\vec{E}$ is the electric field
ρ is the electric charge
μ_0 is the permeability of free space
ε_0 is the permittivity of free space

When complemented by the constitutive relations pertaining to the media under consideration and by their relevant boundary conditions, these equations are suitable for initiating the numerical or analytical solution of a given problem. Today, numerical calculations of the distribution of macroscopic electric and magnetic fields are usually performed using different sets of equations (Equations 15.6 and 15.7), usually derived from Maxwell's equations and the definitions of the electric potential V (Equation 15.8) and the magnetic vector potential A (Equation 15.9):

$$\nabla^2 V + \frac{\partial}{\partial t}\left(\nabla \cdot \vec{A}\right) = -\frac{\rho}{\varepsilon_0} \tag{15.6}$$

$$\left(\nabla^2 \vec{A} - \frac{1}{c^2}\frac{\partial^2 \vec{A}}{\partial t^2}\right) - \nabla\left(\nabla \cdot \vec{A} + \frac{1}{c^2}\frac{\partial V}{\partial t}\right) = -\mu_0 \vec{J} \tag{15.7}$$

$$\vec{E} = -\nabla V - \frac{\partial \vec{A}}{\partial t} \tag{15.8}$$

$$\vec{B} = \nabla \times \vec{A} \tag{15.9}$$

By working with potentials instead of fields, the number of degrees of freedom of the calculations is reduced, as V and A only have four components to be solved for instead of six for E and B.

15.2.2 Constitutive Relations

When electromagnetic fields are applied to matter, the polarization and magnetization of bound charges and currents take place. By considering the constitutive relations for dielectric and magnetic materials (Equations 15.10 and 15.11)

$$\vec{D} = \varepsilon \vec{E} \tag{15.10}$$

$$\vec{B} = \mu \vec{H} \tag{15.11}$$

where
  D is the electric flux density
  H is the magnetic field intensity
  ε is the permittivity
  μ is the permeability

A new set of Maxwell's equation is derived (Equations 15.12 through 15.15)

$$\nabla \times \left(\frac{\vec{B}}{\mu}\right) = \vec{J}_f + \varepsilon\frac{\partial \vec{E}}{\partial t} \tag{15.12}$$

$$\nabla \times \vec{E} = -\frac{\partial \vec{B}}{\partial t} \tag{15.13}$$

$$\nabla \cdot \left(\varepsilon \vec{E}\right) = \rho_f \tag{15.14}$$

$$\nabla \cdot \vec{B} = 0 \tag{15.15}$$

where
  J_f is the free current
  ρ_f is the free charge

It is worth noting that neither ε nor μ are necessarily constants, rather they are functions that can depend on position, field strength, field direction, or frequency. The same is true for the electrical conductivity that describes the relation between electric fields and electric currents in matter—Ohm's law (Equation 15.16). We focus on the physical properties of biological materials relevant to electroporation in Section 15.3 of this chapter.

$$\vec{J} = \sigma \vec{E} \tag{15.16}$$

15.2.3 Boundary Conditions

Since calculation of electromagnetic fields is usually limited to a finite region of space and time, it is necessary to use boundary and initial conditions. In modeling electric fields in biological tissues, the fields are introduced into the region of interest via Dirichlet (Equation 15.17) and Neumann (Equation 15.18) boundary conditions:

$$V = V_0 \tag{15.17}$$

$$\frac{\partial V}{\partial n} = q_0 \tag{15.18}$$

While the Dirichlet boundary condition specifies the value that the solution (in our case electric potential) takes on the boundary, the Neumann boundary condition specifies the value of the derivative of the solution on the boundary.

15.2.4 Electric Field Calculations for Electroporation

According to the theory of electroporation (see Chapter 3 by Kotnik and Pucihar in this book), when a cell is exposed to an external electric field, a transmembrane potential proportional to the field is induced on the cell plasma membrane. Since the magnitude of the induced transmembrane potential is related to the level of membrane permeabilization, bulk electroporation can be related to the local electric field distribution.

Most often, the electric fields used for electroporation are delivered in the form of unipolar rectangular electric pulses. These pulses are much longer than the membrane charging time; therefore, the induced transmembrane voltage reaches its final value long before the end of the pulse. This means that the electric field distribution can be modeled in its steady-state, disregarding the transients that occur during the pulse rise time. In practice, this means that equations used to calculate the local electric field distribution in electroporation modeling become much simpler. Note that the equations are still nonlinear, the nonlinearity being hidden in the material properties. The equations are reduced to the Laplace steady-state equation (Equation 15.19), which can be derived from Equations 15.5, 15.8, and 15.16 by taking into account that all time derivatives are equal to zero.

$$\nabla \cdot (\sigma \nabla V) = 0 \tag{15.19}$$

15.3 Biological Tissues

Biological tissues perform different physiological functions, which are reflected in a number of specific characteristics that have to be considered when representing them in a model at both the cellular and higher organizational level. These differences are also clearly reflected in highly different bulk properties of biological materials (Gabriel et al. 1996a,b, Miklavčič et al. 2006b). They define the current densities and pathways that result from an applied electric stimulus and are thus very important in the analysis of a wide range of biomedical applications used for diagnosis and treatment. Biological tissues are, in general, inhomogeneous and nonlinear.

15.3.1 Biological Tissues in Electric Field

The electrical properties of any material, including biological tissue, can be broadly separated into two categories: conducting and insulating. In a conductor, the electric charges move freely in response to the applied electric field whereas in an insulator (dielectric) the charges are fixed and are not free to move. A more detailed discussion of the fundamental processes underlying the electrical properties of tissue can be found in Foster and Schwan (1989).

If a conductor is placed in an electric field, charges will move within the conductor until the interior field is zero. In the case of an insulator, there are no free charges; therefore, the net migration of charge does not occur. In polar materials, however, the positive and negative charge centers in the molecules do not coincide, which causes an electric dipole moment, p. An applied field, E_0, tends to orient the dipoles and produces a field inside the dielectric, E_p, which opposes the applied field. This process is called polarization. Most materials contain a combination of orientable dipoles and relatively free charges so that the electric field is reduced in any material. The net field inside the material, E, is then

$$\vec{E} = \vec{E}_0 - \vec{E}_p \tag{15.20}$$

The net field is lowered by a significant amount relative to the applied field if the material is an insulator and is essentially zero for a good conductor. This reduction is characterized by a factor ε_r, which is called the relative permittivity or dielectric constant, according to

$$\vec{E} = \frac{\vec{E}_0}{\varepsilon_r} \tag{15.21}$$

In practice, most materials, including biological tissue, actually display some characteristics of both insulators and conductors because they contain dipoles as well as free charges that can move, but in a restricted manner. For materials that are heterogeneous in structure, charges may become trapped at interfaces. Because positive and negative ions move in opposite directions in the applied field, internal charge separations can then result within the material, producing an effective internal polarization that acts like a very large dipole.

On the macroscopic level, we describe the material as having a permittivity, ε, and a conductivity, σ. The permittivity characterizes the material's ability to trap or store charge or to rotate molecular dipoles whereas the conductivity describes its ability to transport charge (Grimnes and Martinsen 2000):

$$\varepsilon = \varepsilon_r\varepsilon_0 \tag{15.22}$$

Consider a sample of material that has a thickness, d, and cross-sectional area, A. If the material is an insulator, then we treat the sample as a capacitor with a capacitance of

$$C = \varepsilon \cdot \frac{A}{d} \tag{15.23}$$

If it is a conductor, then we treat it as a conductor with a conductance of

$$G = \sigma \cdot \frac{A}{d} \tag{15.24}$$

If a constant (direct current, DC) voltage V is applied across this parallel combination, then a conduction current $I_C = GV$ will flow and an amount of charge $Q = CV$ will be stored.

Suppose, instead, we apply an alternating (alternating current, AC) voltage:

$$V(t) = V_0 \cos(\omega t) \tag{15.25}$$

where
  V_0 is the amplitude of the voltage
  $\omega = 2\pi f$, where f is the frequency of the applied signal

The charge on the capacitor plates is now changing with frequency f. This change is associated with a flow of charge or current in the circuit. We characterize this flow as a displacement current:

$$I_d = \frac{dQ}{dt} = -\omega C V_0 \sin(\omega t) \tag{15.26}$$

The total current flowing through the material is the sum of the conduction and displacement currents that are separated in phase by 90°. This phase difference can be expressed as

$$V(t) = V_0 e^{i\omega t} \quad \text{where } i = \sqrt{(-1)} \tag{15.27}$$

taking its real part for physical significance. The total current is $I = I_c + I_d$ (I_c being the conductive and I_d being the displacement current), hence

$$I = GV + C \cdot \frac{dV}{dt} = (\sigma + i\omega\varepsilon)A \cdot \frac{V}{d} \tag{15.28}$$

The actual material, then, can be characterized as having an admittance, Y^*, given by

$$Y^* = G + i\omega C = \left(\frac{A}{d}\right)(\sigma + i\omega\varepsilon) \tag{15.29}$$

where * indicates a complex-valued quantity. In terms of material properties, we define a corresponding, complex-valued conductivity or admittivity as

$$\sigma^* = (\sigma + i\omega\varepsilon) \tag{15.30}$$

Describing a material in terms of its admittance emphasizes its ability to transport current. Alternatively, we could emphasize its ability to restrict the flow of current by considering its impedance $Z^* = 1/Y^*$, or for a pure conductance, its resistance, $R = 1/G$.

Factoring $i\omega\varepsilon_0$ in Equation 15.28 yields

$$I = \left(\varepsilon_r - \frac{i\sigma}{\omega\varepsilon_0}\right) i\omega\varepsilon_0 A \cdot \frac{V}{d} = C\frac{dV}{dt} \tag{15.31}$$

We can define a complex-valued relative permittivity as

$$\varepsilon^* = \varepsilon_r - \frac{i\sigma}{\omega\varepsilon_0} = \varepsilon_r' - i\varepsilon_r'' \tag{15.32}$$

with
  $\varepsilon_r' = \varepsilon_r$
  $\varepsilon'' = \sigma/(\omega\varepsilon_0)$

The complex conductivity and complex permittivity are related by

$$\sigma^* = i\omega\varepsilon^* = i\omega\varepsilon_0\varepsilon_r^* \tag{15.33}$$

In physical terms, we can consider the conductivity of a material as a measure of the ability of its charge to be transported throughout its volume due to the applied electric field. Similarly, its permittivity is a measure of the ability of its dipoles to rotate or its charge to be stored by an applied external field. Note that if the permittivity and conductivity of the material are constant, the displacement current will increase with frequency whereas the conduction current does not change. At low frequencies, the material will behave like a conductor, but capacitive effects will become more important at higher frequencies. For most materials, however, these material properties are not constant, but vary with the frequency of the applied signal. σ^* and ε^* are frequency-dependent.

15.3.2 Biological Tissue in DC Electric Field

The electrical response of biological tissues when stimulated with DC electroporative pulses can be seen as quasi-stationary. Namely, for any material whose electric properties are in the range of those of biological tissues or organs and whose dimensions do not exceed 1 m and the frequency of the electric field is low, the electrical behavior in any given moment as a response to electric current can be numerically described with a set of equations describing stationary fields. Although the impedance of biological tissue has a capacitive component, the electric field can be considered as time independent, thus, the capacitive effects and the finite propagation of the electric current in the biological tissue are disregarded.

The electric field in a tissue and electric current passing through the tissue are coexisting and are related by Ohm's law (Equation 15.16). The corresponding integral values are electric current I, conductance G (which is the reverse of resistance R), and voltage U. Ohm's law then takes the form of

$$U = R \cdot I \tag{15.34}$$

or

$$I = G \cdot U \tag{15.35}$$

Current passes through the tissue if a potential difference exists between two points in the tissue, and the current loop is closed. In practice, we generate the potential difference (voltage) on the electrodes with an electric pulse generator. When both electrodes (one needs at least two electrodes to close the loop) are placed on/in the tissue (which is a conductive material where charge carriers are ions as in electrolyte solutions), the current loop is closed and the current passes through the tissue.

As the electric current passes through a biological tissue, it is distributed through different parts of the tissue, depending on their electrical conductivity. In general, highly perfused tissues have higher conductivity; blood is highly conductive, as well as muscles, whereas bone and fatty tissue have low conductivity. The current will flow more easily and for the same voltage in higher proportion through more conductive tissues (e.g., muscles). On the contrary, the electric field in these tissues will be lower than in tissues with low conductivity for the same current.

Nevertheless, as the electric current takes the shortest and easiest path through the tissue, the current will be contained predominantly between the electrodes if they are close enough to each other. This property allows for relatively good control and containment of electric field distribution predominantly between the electrodes (Miklavčič et al. 1998).

Even though the pulses usually used in electroporation are DC, the capacitive properties of the biological material cannot always be disregarded. This holds true for the cases where the transient of cell membrane charging may also be interesting to study. Namely, cell membrane charging time is on the order of microseconds, and typical pulses used for electropermeabilization of the cell membrane are 100 μs long, with the amplitude of around 500 V/cm (Kotnik et al. 1997, 1998). It has been found that if pulses of much higher amplitude (e.g., 50 kV/cm) and much shorter duration are used—in the order of tens of ns—the charging effect also becomes pronounced on the membranes of intracellular organelles (Schoenbach et al. 2001, Tekle et al. 2005, Kotnik and Miklavčič 2006). For a qualitative analysis of these processes, the time courses of organelle and cell plasma membrane charging become important. Thus, the capacitive component describing the electrical properties of the cell, its organelle(s), and their membranes can no longer be neglected.

15.3.3 Some Biological Tissue Properties Important in the Applied Use of Electroporation

15.3.3.1 Tissue Anisotropy

When the properties of a material are the same in all directions, the material is said to be isotropic. However, some biological materials are distinctly anisotropic. Typical anisotropic tissues are, for example, skeletal muscle and tendon. Therefore, when referring to published electrical property data, the information about the orientation of the electrodes relative to the major axis of the tissue during impedance measurements is important (longitudinal, transversal, or a combination of both).

The conductivity of the material (in units: S/m) can, in the case of an anisotropic conductor, be represented by a tensor:

$$\sigma = \begin{bmatrix} \sigma_{xx} & \sigma_{xy} & \sigma_{xz} \\ \sigma_{yx} & \sigma_{yy} & \sigma_{yz} \\ \sigma_{zx} & \sigma_{zy} & \sigma_{zz} \end{bmatrix}$$

Whenever the material's conductivity can be described in the orthogonal Cartesian system and its spatial dependence can be aligned with the axes, the electric field and the current density can be described in the same way; the nondiagonal elements of the matrix equal zero, hence the matrix becomes diagonal:

$$\sigma = \begin{bmatrix} \sigma_{xx} & 0 & 0 \\ 0 & \sigma_{yy} & 0 \\ 0 & 0 & \sigma_{zz} \end{bmatrix}$$

If we take skeletal muscle for example, two different conductivity values can be measured in two different directions: one for the direction along the length of the muscle fibers and one that is perpendicular to it. If the muscle tissue is aligned with one of the axes of the coordinate system, two diagonal elements in the above matrix have the same value.

Tissue anisotropy is often related to the structure and physiological properties of the tissue. Skeletal muscles are composed of fibers that are very large, highly elongated individual cells and are aligned in the direction of muscle contraction. Electrical conduction along the length of the fiber is thus significantly easier than conduction between the fibers (the difference is about sevenfold) (Reilly 1998). The longitudinal conductivity is significantly higher than the transverse conductivity, especially in the low frequency range. Tissue anisotropy is also frequency-dependent (Hart et al. 1999). If the frequency of the current is high enough, the anisotropic properties disappear (specifically for skeletal muscle that happens in the MHz frequency range). At higher frequencies, charge movement takes place over shorter distances so large-scale structures become less important and capacitive coupling across membranes becomes more important.

15.3.3.2 Nonlinear Behavior

With respect to the intrinsic characteristics of a system or equations describing it, an important consideration is whether the system is linear. Mathematically speaking, a nonlinear system does not satisfy the superposition principle stating that the response of a system caused by two or more input stimuli is the sum of the responses, which would have been caused by each stimulus individually. In terms of equations, a nonlinear system is any system where the variable(s) to be solved for cannot be written as a linear sum of independent components. Unfortunately, most physical systems are inherently nonlinear in their nature and, unfortunately, biological tissues are not an exception. Often, the physical property of a material is changed during a process (material's temperature coefficients, conductivity changes during tissue electroporation). For some applications, a linear approximation of a nonlinear function can be found at (or around) a given point, for specific input values. However, if the model has to cover the whole (or larger) range of input values, the nonlinearities have to be considered in the model.

If we speak strictly about tissue properties exposed to electric current, at least two important nonlinearities need to be considered.

One is the increase in tissue conductivity (σ) due to an increased electric field (E) causing cell membrane electroporation (Pliquett and Weaver 1996, Pavšelj et al. 2005, Šel et al. 2005). This change of material properties has two more nonlinear characteristics. First, it is considered to be a threshold phenomenon, meaning that the electric field has to reach a certain value, termed reversible electropermeabilization threshold E_{rev}, in order to cause conductivity changes. Second, for the duration of the pulse, this conductivity change is an irreversible phase transition process. More specifically, once the conductivity is increased in a given tissue volume, it cannot be changed back to its lower value during pulse delivery, even if the electric field strength drops below the threshold due to changed conductivities. Here, we would like to point out that one has to be careful to distinguish between the reversibility of cell electroporation (provided the electric field was below the irreversible threshold) after the cessation of electric pulses; and the irreversible nature of the conductivity changes during pulse delivery—the change is only possible in one direction, tissue conductivity can only increase (Pavlin et al. 2005).

The second nonlinearity comes from the electrical–thermal coupling (Pliquett 2003). Once a part of a tissue is permeabilized, it becomes more conductive and the current density increases several times, causing resistive heating. In turn, tissue conductivity increases even more, as the temperature coefficient of electrical conductivity of most biological materials is positive—in the range of 1%–3% °C^{-1} (Duck 1990).

15.3.3.3 Electric Field Threshold of Biological Tissues

The cell membrane is permeabilized when the threshold transmembrane potential is reached, thus, when the external electric field is above the threshold value. This increased cell membrane permeability is reversible, provided the electric field is not too high. However, if cells are exposed to an electric field above the irreversible threshold, they suffer permanent damage. For electroporation-based applications such as gene delivery (Golzio et al. 2004, André et al. 2008) or transdermal drug delivery (Prausnitz 1999, Denet and Préat 2003, Denet et al. 2004), this is not a desirable effect as the cells have to be viable after the treatment. On the other hand, for applications based on irreversible electroporation, the target cells can be irreversibly destroyed within a narrow range while leaving neighboring cells unaffected. This technique represents a promising new treatment for cancer, heart disease, and other conditions that require tissue ablation (Davalos et al. 2005, Lavee et al. 2007, Onik et al. 2007, Rubinsky et al. 2007).

In any case, it is important to determine the needed amplitude of electric pulses at a given electrode-tissue setup to achieve an electric field distribution in the tissue that is adequate for a given application. Electric field reversible and irreversible thresholds are both inherent characteristics of the tissue (also different for different tissues), no matter what kind of electrodes we use or if inhomogeneous or composed tissues are involved. Of course, the electropermeabilization process as well as cell viability depend on electrical parameters, i.e., pulse amplitude, pulse duration, and the number of pulses (see Figure 15.1) (Maček-Lebar et al. 2002, Puc et al. 2003).

However, accomplishing an adequate electric field distribution in the tissue is much more complex than merely calculating the voltage we need at a given electrode separation (U/d). Mathematically, this ratio gives an electric field only when delivering pulses to a homogeneous tissue through parallel plate electrodes whose surface is large (infinite) compared with the electrode separation (see Figure 15.2a). It can still be used as an approximation of the electric field in the area between the parallel plate electrodes

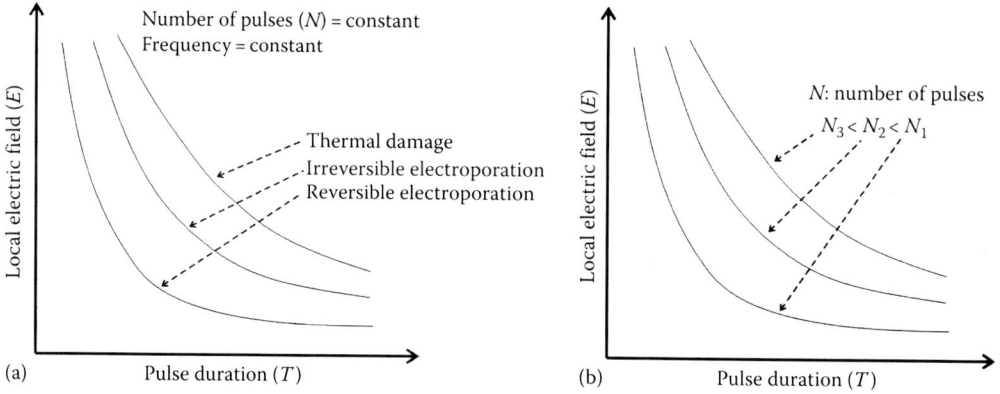

FIGURE 15.1 Electroporation process is (for a given tissue-electrode geometry) controlled by pulse parameters. (a) At constant number of pulses (N) and their frequency, lengthening pulse duration requires lower local electric field (pulse amplitude) for the same effect. If both are increased, the effects on the tissue become irreversible, or, at even higher values, tissue thermal damage can be observed, due to excessive resistive heating. (b) Similarly, for any of the curves, if number of applied pulses is larger, the same effect can be achieved with a lower pulse amplitude and/or duration.

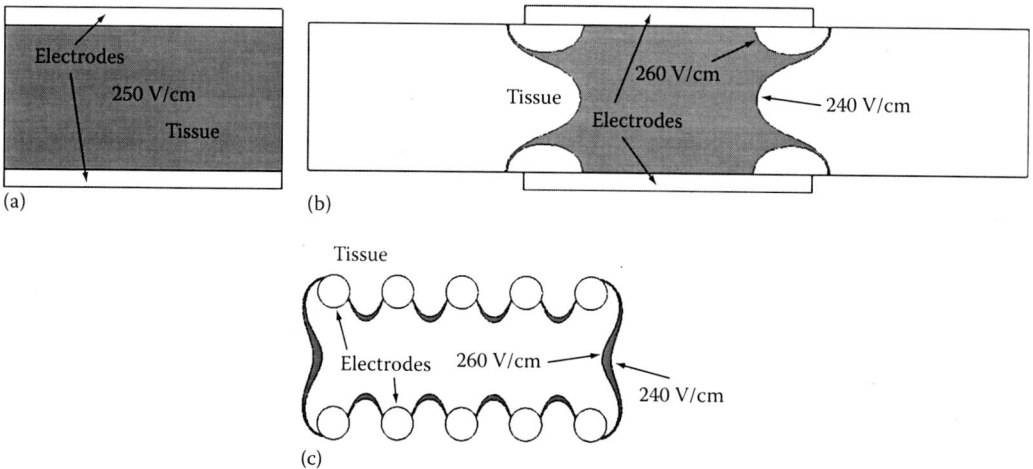

FIGURE 15.2 Curves of the same electric field in a homogeneous tissue in a section plane perpendicular to electrodes. (a) The electric field equals the ratio U/d (voltage/distance between electrodes) only in the theoretical case where electric pulses are delivered through plate electrodes of infinite surface. Here, only a portion of this infinite structure is modeled with boundary conditions set to represent an infinite volume. The distance between the electrodes (d) is 4 mm, the applied voltage (U) is 100 V, and thus the electric field equals U/d = 250 V/cm throughout the tissue between the electrodes. (b) A real situation where electrodes are of finite dimension. The electric field in the gray area between the two black isocontours is between 240 and 260 V/cm, so the voltage to distance ratio in this area is a good approximation, except near the edges of the electrodes. The U/d approximation is valid in a greater portion of the tissue between the electrodes if the electrode surface is increased or the distance between them is smaller. (c) Two rows of needle electrodes are used instead of plate electrodes. The length of the electrode array and the distance between the rows is the same as in (b). The gray area between the two black isocontours denoting electric field between 240 and 260 V/cm is very small and limited to a few narrow stripes. Throughout most of the area inside the electrode array the electric field is higher (around 300 V/cm or higher) and thus cannot be satisfactorily approximated by voltage to distance ratio.

of finite dimensions, away from their edges (Figure 15.2b). The *U/d* ratio is also often used to estimate the electric field between two parallel rows of needle electrodes. Some are referring to the *U/d* ratio as a "nominal" field; however, the approximation is extremely rough (Figure 15.2c). In the case of any other electrode geometry using plate, needle, microneedle, or surface electrodes or if more than one tissue is involved, a numerical analysis has to be performed beforehand, as a part of treatment planning, in order to choose the right electrode configuration and the pulse amplitude.

15.4 Modeling

Having acquired some basic knowledge about the electromagnetic field theory and specificities of biological tissues, we can set about constructing mathematical models representing different aspects of electroporation.

The starting point of the modeling process is deciding on the mathematical approach to adequately describe the modeled system by first acquiring enough observable and measurable information about it. Typically, more than one modeling approach is possible and choosing the most suitable one depends on the modeler's or end user's objective needs and personal preferences, as well as the physical and geometrical characteristics of the modeled system. In some cases, using more than one modeling approach can be beneficial in terms of model verification and validation. More than one phenomenon can determine a system, which is especially the case with biological systems. Generalizations and simplifications are possible and, in most cases, cannot be avoided. The model can refer only to some aspects of the real system, while disregarding the ones that either have a very limited influence on its accuracy or are out of the scope of this particular model. Prior to building a model, we need to define its scope, apply necessary simplifications while being aware of the circumstances or the range of input variables for which the model is valid.

Deciding on the right level of complexity for our model is not always an easy task as it involves a trade-off between simplicity and accuracy. As a general guideline, if we are choosing from different models giving comparable results, the simplest one is the most desirable. Namely, we need to be aware that adding complexity can make the model difficult to understand and experiment with and can pose computational problems.

Analytical methods are rather complicated and are only feasible for use on problems where the geometry, material properties, and boundary conditions can easily be described in a defined coordinate system (Cartesian, cylindrical, or spherical). Simple analytical models can have certain advantages over numerical models. First, the input data needed is typically less extensive than that of numerical models. Also, analytical solutions have no numerical and discretization errors. The obvious limitation of analytical models is that only the simple and uniform geometries, boundaries, and initial conditions can easily be modeled. In the last decades, analytical models have mostly been replaced by numerical models based on boundary element, finite difference, finite volume, or finite element methods, due to the miniaturization and accessibility of both computer hardware and software. Of these methods, the latter is preferred in the modeling of electroporation, due to its relatively easy implementation and its ability to handle more intricate geometries. The principle behind this method is the discretization of the geometry into smaller elements where the quantity to be determined is approximated with a function or is assumed to be constant throughout the element. Discrete elements can be of different shapes and sizes, which allows for the modeling of intricate geometries. In such models, the excitations can be changed easily, being that it only involves changing the boundary conditions on the same model. The model geometry, however, takes time and precision to be built and generalizations and simplifications need to be used when possible.

15.4.1 Building a Geometry

When designing a numerical model, we must decide for the appropriate details to be included. Geometrically more detailed models will inevitably consume more of both the modeler's and the computer's time, but do not necessarily produce better quality results, as the inclusion of geometrical details depends on the purpose of the model.

15.4.1.1 Geometrical Symmetries

Taking advantage of the geometrical symmetries of the system we are modeling allows us to analyze a structure or a system by modeling only a portion of it by applying appropriate boundary conditions. This approach can be used when the same symmetry can be observed in both the geometry as well as the sources (in our case of electric current). It reduces the size of the model and consequently the analysis run time as well as the demands on computer resources. Alternatively, modeling only a portion of the whole geometry allows us to include more details in the model, when needed, thereby obtaining better results that would not have been possible with the full geometry (Pavšelj et al. 2007). For example, when representing a tumor with a simplified elliptical shape, supposing also the symmetry of the electric stimulation), only a quarter of the tumor can be modeled (see Figure 15.3a). A similar approach can be used when modeling a geometrical structure with a repetitive infinite or quasi-infinite pattern (Figure 15.3b). In this case, only a small portion of the whole array, a unit cell, needs to be modeled by applying the appropriate periodic boundary conditions (Susil et al. 1998, Pavlin et al. 2002).

Similarly, sometimes we are able to represent a 3D structure by a single 2D plane. For example, in Cartesian coordinates, a structure stretched along a straight line (*Z*-direction) can be represented with a structure in the *X*–*Y* plane, while the model is assumed to be uniform in the perpendicular *Z*-direction (see Figure 15.4a). Similarly, if the structure is axisymmetrical, the plane of symmetry is the cross section anywhere around the axis of symmetry. In this case, we are using a single 2D slice (*r*–*z* cylindrical coordinates) to represent the whole 360° of the structure (see Figure 15.4b).

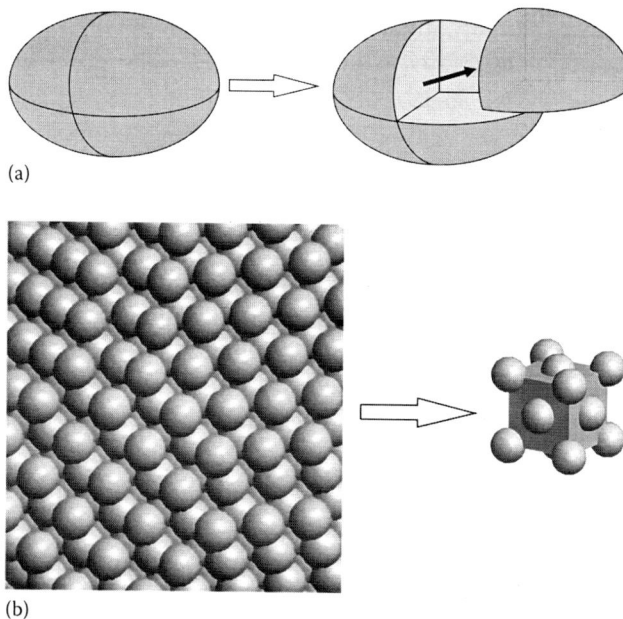

(a)

(b)

FIGURE 15.3   Representation of a large structure with only a portion of its geometry. (a) When modeling an elliptically shaped tumor during electrochemotherapy, supposing the electric stimulation exerts the same symmetry, only a quarter of the tumor need be modeled. (b) A finite representation of an infinite 3D lattice, such as cells in a cluster.

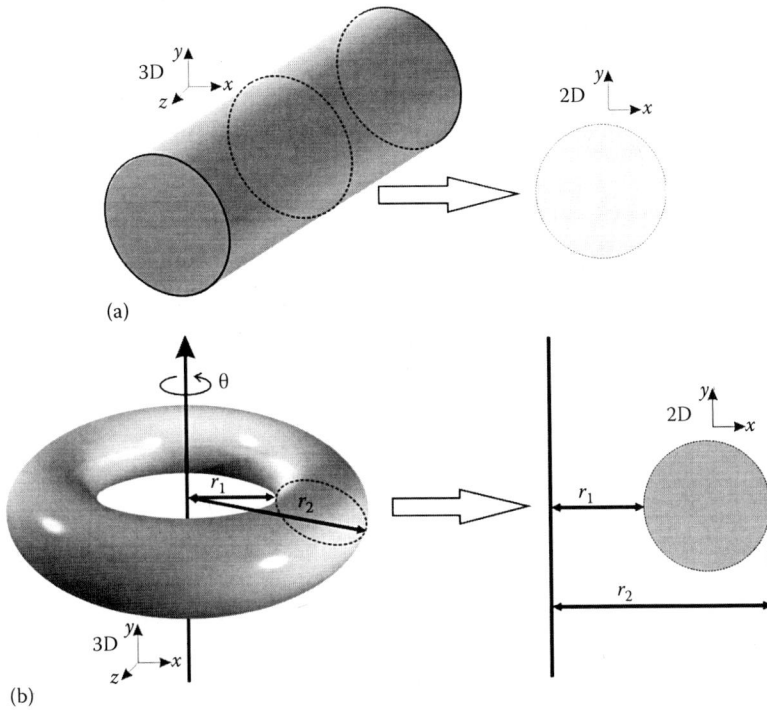

FIGURE 15.4 Representation of a 3D structure by a single 2D plane. (a) 2D representation of a 3D geometry. In this case, the cylinder was represented by its cross section. (b) 2D representation of a 3D axisymmetrical structure.

15.4.1.2 The Size of the Modeled Volume/Area

In some cases, the modeled system has no borders electrically insulating it from its surroundings; the electrical quantities are simply diminishing with increasing distance from the source. One such example is needle electrodes inserted in a tissue (Figure 15.5). In such cases, the outer boundaries of the model need to be far enough from the source(s), in order not to restrain the natural flow of the electric current (see Figure 15.5a). Namely, when modeling such a system, the borders of the model are artificially electrically insulated from the surroundings. This effectively means that the boundary condition is set in such a way that no electric current flows in or out of the enclosing box—only tangential components of the electric current exist on the outer tissue borders while the normal component equals zero. If these borders are too close to the source(s) of the electric current, such as in Figure 15.5b, the electric field and current distribution is deformed and does not reflect the true situation. The safest way to choose the right distance of the model borders from the source(s) is by changing the dimensions of the enclosing box and observing its effect on the results. The enclosing box is large enough when, if further increased, the effect on the calculated scalar and vector fields is negligible.

15.4.1.3 Modeling of Biological Entities

Numerous examples could be given to illustrate either the importance or futility of including physiological details in the geometry of the model. Already at the cell level, *in vitro* observations on cell suspensions can be represented numerically on different levels. For example, when studying the magnitude and the distribution of the electric field in a cell suspension, a material with homogeneous properties is an adequate model (Pavlin and Miklavčič 2003). However, if the aim of our research is to study the phenomena on the cell level or if we are looking into interactions between cells, the influence of their size, shape, density, and orientation, individual cells rather than bulk material have to be modeled (Susil et al. 1998, Pavlin et al. 2002, Valič et al. 2003, Pucihar et al. 2006, Pavlin and Miklavčič 2008, Towhidi et al.

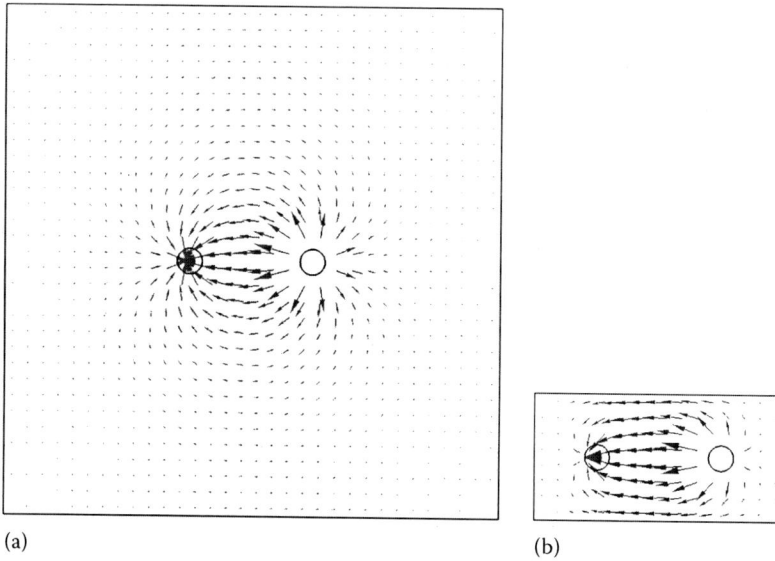

(a) (b)

FIGURE 15.5 The electric current density—*J* (gray arrows) in a homogeneous material, electric pulse is delivered through needle electrodes (the two black circles). The electric current density is shown in a section plane cut through the material, perpendicular to electrodes. (a) The insulated borders of the model are far enough from the electrodes, which allows for the natural flow of the electric current, the electric field and the electric current near model borders are very close to zero. (b) The borders of the model are too close to the electrodes, the electric field close to the border is not zero, electric current is artificially constrained.

2008). Even further, if we are using nanopulses, cell organelles must be added to the geometry (Kotnik and Miklavčič 2006). Still, by using smart approaches, such as replacing a thin, non-zero conductivity cell membrane by a boundary condition between the cytoplasm and the exterior, getting rid of complexities while maintaining accuracy of the model is possible (Pucihar et al. 2006).

Similar observations hold true on the tissue level. Different levels of complexity and inhomogeneity can be observed in different tissues; however, including their particularities depends strongly on the purpose of the model. Skin, for example, is a very intricate tissue due to its highly inhomogeneous structure, leading to inhomogeneous electric properties. It consists of different layers in terms of dimensions (thickness) and electrical properties: the outer thin layer of dead flat skin cells, the stratum corneum, the viable epidermis, dermis, and the subcutaneous tissue (Yamamoto and Yamamoto 1976a,b, Chizmadzhev et al. 1998). If the aim of the model is to study the electroporation of skin as a target tissue, this layered structure needs to be included in the geometrical representation (Pavšelj et al. 2007). Moreover, even this bulk layered structure might sometimes prove inadequate. Smaller structures, such as hair follicles, sweat glands, and blood vessels, or local transport regions as a result of skin electroporation (Pavšelj et al. 2008a) may have to be added in order to study the processes on the microscale, where they occur, and only then compare them to bulk observations. On the one hand, such details, unavoidably adding to the overall complexity of the model, can be omitted in cases where skin electroporation is not studied directly, such as any application where electric pulses are delivered with external electrodes to tissues beneath the skin (Pavšelj et al. 2005). On the other hand, other structures, such as major blood vessels may be important and need to be included in the model when studying the mechanisms of the electrochemotherapy of tumors (Serša et al. 2008).

15.4.2 Setting the Physics of the Model

After the geometry of the model has been constructed, the next step in the modeling process is setting the physics of the model, such as underlying equations, material properties, boundary, and initial conditions.

15.4.2.1 Frequency-Dependent Component

First, the material's response will be different when exposed to either direct (DC) or alternating current (AC). If our material is purely resistive, the system exerts no frequency dependency; the current is proportional to the voltage irrespective of the frequency. However, in general, materials have their capacitive or inductive component so the voltage to current ratio does depend on frequency and is termed impedance (Z). Impedance is a complex quantity consisting of a resistance R (the real part) and a reactance X (the imaginary, frequency-dependent part):

$$\vec{Z} = R + jX \tag{15.36}$$

The resistance can only be positive, while the reactance can be either positive (inductive character, current lagging behind voltage—X_L) or negative (capacitive character, voltage lagging behind current—X_C).

$$X_L = j\omega L \tag{15.37}$$

$$X_C = -j\frac{1}{\omega C} \tag{15.38}$$

Voltage and current can be considered as vectors in the complex plane (phasors) that are out of phase, so the voltage to current ratio—the impedance—can also be given by its magnitude and phase angle:

$$\vec{Z} = |Z| \cdot e^{j\Theta} \tag{15.39}$$

$$|Z| = \sqrt{R^2 + X^2} \tag{15.40}$$

$$\Theta = arctg\frac{X}{R} \tag{15.41}$$

Resistance is only a special case of impedance, when the material we are considering exerts no or negligible capacitive or inductive character ($jX = 0$). Further, if a system is exposed to DC, the frequency-dependent part—the reactance X—plays no role when the system is in steady-state, after all the transients have faded out. It does, however, dictate the course of the transient of the system, which poses the next question in the modeling process: Are we interested only in the steady-state of our system or are we studying transient phenomena—changes over time from $t = 0$ until the system has reached its steady-state?

15.4.2.2 Transient vs. Steady-State

Transient behavior occurs when the magnitude and direction of electrical quantities change with time. On the contrary, if they are constant with time throughout the entire volume, the system is already in its steady-state. To avoid any ambiguity, the steady-state does not mean the absence of movement or flow in the system! If we take electric currents in a material as an example, it simply means that the "amount" of electric current in the system does not change within an observed time; the magnitude of the current exiting the system equals the current magnitude flowing into the system at any time when in steady-state. In other words, time becomes an irrelevant variable for the analysis, since the recently observed behavior of the system will continue into the future.

In many systems, steady-state is not achieved until sufficient time has elapsed after the system is started or stimulated (externally or internally). The situation after the occurrence of the described changes of the system and before all internal quantities (states) of the system reach the steady-state is defined as the transient state. As an example, in an electrical system of purely resistive character, no transients occur at $t = 0$, when the electrical stimulation is turned on. However, if the imaginary part (the reactance) is present in the impedance of the electrical system, its behavior exerts inertia, meaning that the change of electrical quantities in the system is not instantaneous. The capacitive or the inductive component opposes the sudden change at $t = 0$ (applied voltage or current) and enforces the transient state onto the system that will, however, eventually fade out.

15.4.2.3 Multiphysics

The effects of various physical phenomena can be investigated by separately analyzing each individual phenomenon without any consideration of the interaction between them. However, often we are dealing with two or more interacting, simultaneous phenomena, such as the coupling between the electric and the magnetic fields. An important coupling of physical phenomena in applications using electric pulses on biological tissues is heat transfer in tissue due to resistive heating (Tungjitkusolmun et al. 2000). This coupling may give rise to tissue conductivity changes (due to temperature increase), which in turn change the magnitude of the electric current. When constructing a model, the influence of such interactions have to be estimated and, if needed to obtain accurate results, mutual dependencies have to be included. To do so, we need data on how the material properties significant for one field (such as the electric field) vary with the magnitude of another field (such as temperature) and vice versa.

15.4.3 Interpretation of Results

When modeling the electroporation of biological tissues, much consideration has to be given to the interpretation of the results in relation to possible simplifications in the model or inherent characteristics of different biological tissues. Namely, some simplifications might not have much effect in isotropic, homogeneous tissues, such as the liver, but may yield useless results in inhomogeneous, composed biological structures, such as layered skin or subcutaneous tumors, where electric field distribution is much more complex (Pavšelj et al. 2005, Ivorra et al. 2008, Pavšelj and Miklavčič 2008b). To illustrate, when modeling electroporation in a homogeneous tissue, such as the liver, the results are still useful and comparable to experimental data even if the conductivity increase due to tissue electropermeabilization is neglected. In fact, early models did not take this nonlinear tissue behavior into account (Miklavčič et al. 2000). However, when more complicated electrode-tissue setups were being studied with numerical models, experimentally observed phenomena could not be satisfactorily modeled in this way. Namely, upon applying electric pulses on a composed or layered tissue with an inhomogeneous distribution of electrical conductivities, the voltage is divided among them proportionally to their electrical resistances (Pavšelj and Miklavčič 2008b). This leads to a more complex electric field distribution, meaning that some parts of the tissue, due to their low electrical conductivity (disproportionally lower than the rest of the tissue), are exposed to a much stronger electric field. The electric field is the highest in the layer with the highest resistivity (lowest conductivity). In the case of the subcutaneous tumor, this is the skin, which has the lowest electrical conductivity, and in the case of the skin fold, the highest electric field is in the nonconductive outermost skin layer, the stratum corneum. But more importantly, the electric field in the target tissues (tumor and viable skin layers) stays too low for successful electroporation. This fact raised the question of how the experimentally confirmed successful permeabilization of the target tissues theoretically is possible when external plate electrodes are used, which led to the inclusion of tissue conductivity changes due to electroporation in the numerical models.

15.4.4 Model Verification and Validation

The last, but nevertheless very important part of the modeling process is the verification and the validation of the constructed model, involving different aspects of evaluation. Mostly, these aspects should be taken into consideration from the very beginning of the process and can roughly be divided into three categories:

1. *Verification*: The main question here is whether we reached the aim of the model. Already in the planning phase of the modeling process, we have to set the scope of our model, the range of input data it should be valid for, as well as geometrical details to be included. However, as we build the model, some simplifications and trade-offs may have to be made. Comparing the actual result

with the requirements set during the planning phase will demonstrate whether our resulting model is still within the planned scope of the model or not.

2. *Descriptive realism*: Have we identified and explained the underlying physics? In cases where almost nothing is known about the phenomena describing the modeled system, we are dealing with the so-called black box problem that can only be treated in terms of its input and output characteristics. Our only option may be finding a curve that has the best fit to a series of data points, while respecting possible constraints without actual physical reference to the described process(es). However, different techniques of system identification (Ljung 1999) can be applied in order to identify the physics defining our "black box," which can then be modeled. Namely, the purpose of modeling is to gain insight and explain underlying phenomena, as well as using them for predicting the output at certain input data sets. We should therefore direct our efforts to turn the black box into a set of equations, if possible. Once again, as some trade-offs will most likely be necessary, we should assess if the modeled physical phenomena successfully explain the most important experimental observations.

3. *Validation*: Does our model agree with the empirical data? One way to justify the physics used in the model (sometimes the only way) is by comparing the output data obtained from the model to experimental data. Usually, or ideally, the experimental data can be divided into two groups: the training data and the validation data. The former is used to identify the process and to construct a model with its relevant parameters and constraints, while the latter is used to assess if the model is valid for any range of input parameters within the defined constraints.

15.5 Treatment Planning

When electroporation is used in biomedical experiments and medical treatments, the (steady-state) electric field distribution inside the target tissues is one of the most important predictors of success (Miklavčič et al. 1998). Models have helped us to understand that the electric field in tissue changes its magnitude during pulses, as the conductivity increases due to electroporation (Pavšelj et al. 2005, Šel et al. 2005). Also, modeling has shown the importance of ensuring good surface contact between the electrodes and tissue, when plate electrodes are used to deliver electric pulses (Čorović et al. 2008a), and the importance of the depth of insertion, when needle electrodes are used (Čorović et al. 2008b). It has also been shown that for a known number and duration of applied electric pulses, the electric field has to be higher than a threshold value ($E > E_{th}$) for electroporation to occur (Šemrov and Miklavčič 1998). As such, the electric field distribution can serve as a predictor of treatment outcome. As has been mentioned in the previous sections, the local electric field distribution inside biological tissue is very hard to predict without numerical models. If a specific distribution is needed, as is the case in electroporation-based medical treatments, several attempts are needed before the right electrode positions relative to the target tissue and voltages between the electrodes are found. The more complex the case, the more time is needed to determine the treatment parameters by trial and error (forward planning), therefore, numerical optimization techniques (inverse planning) have to be used: a desired electric field distribution can be set and appropriate treatment parameters (electrode positions, voltages) can be determined by numerical optimization (Župančič et al. 2008).

In practice, these more complex cases include target tissues that are located deep in the body or close to the vital organs. In such cases, it is important to control the magnitude and distribution of the electric field so that a minimum volume of vital tissue is compromised by the treatment. Numerical modeling may also be necessary for treatment planning in tissues with highly anisotropic properties and highly nonhomogeneous tissues (Pavšelj and Miklavčič 2008b). In such cases, the treatment planning procedure has to be applied individually for each patient and the electric field distribution has to be sculpted carefully to guarantee that the entire target tissue is exposed to a high enough electric field, while vital tissues are as unaffected as possible.

Treatment planning does not consist solely of optimization and modeling, but is instead an integral part of the whole treatment process. Normally, medical imaging (CT, MRI) is first used to obtain information about the anatomical details of the treated volumes. The images are converted into mathematical representations and used in the numerical model to calculate the electric field distribution. The numerical model is used in the optimization algorithm to calculate the best treatment parameters (electrode positions and voltages). Additional measures can be taken to ensure that the treatment plan is successfully executed, e.g., the insertion of electrodes can be controlled by ultrasound imaging and the extent of electroporation can be monitored by current and voltage measurements (Cukjati et al. 2007) or electrical impedance imaging (Davalos and Rubinsky 2004, Ivorra and Rubinsky 2007). Therefore, it can be argued that the quality of numerical treatment planning for electroporation-based treatment depends on the quality of medical imaging, target and normal tissue identification, detailed knowledge of the biological effects of the electric field (changes in tissue properties, threshold values), and the quality of the underlying model. At present, not all of the requisites are met. For one, electroporation thresholds are tissue specific and are not yet readily available. Furthermore, several electric pulse parameters that affect the threshold values: pulse duration, number of pulses, and to some extent also pulse repetition frequency (Pucihar et al. 2002, Edd and Davalos 2007) have not yet been included in numerical models. Therefore, prior to any treatment planning, data on thresholds for all relevant tissues should be available for a range of electroporation parameters.

We present an example of treatment planning of the electrochemotherapy of a tumor nodule located near a vital organ using numerical modeling and a genetic optimization algorithm. The goal is to determine the best possible configuration and electric potentials of six electrodes surrounding a subcutaneous tumor—the target tissue. The treatment parameters must irreversibly electroporate ($E > E_{rev}$) the entire tumor volume while sparing the hypothetical spherical vital organ ($E < E_{irr}$) situated next to the tumor (Figure 15.6).

The steady-state numerical model of electroporation is used, taking into account the changes in tissue conductivities because of electroporation; i.e., electric field distribution in the tissue caused by an electric pulse is determined by solving the Laplace equation for static electric currents (2.19) with $\sigma(E)$. All tissues are considered isotropic and homogeneous. The assigned conductivity values and electroporation thresholds are given in Table 15.1. These values are mostly taken from existing literature (Gabriel et al. 1996a,b, Davalos et al. 2005, Pavšelj et al., 2005, Cukjati et al. 2007) or, in cases where data cannot be found in existing literature, are educated guesses, meant only for demonstration purposes.

A genetic algorithm is used for the optimization procedure. A population of possible solutions (treatment plans consisting of the positions and direction of each electrode [x, y, z, φ, θ] and all used voltages) is first randomly chosen. The solutions then evolve in iterations by mathematical operation cross-over and mutation according to their fitness function

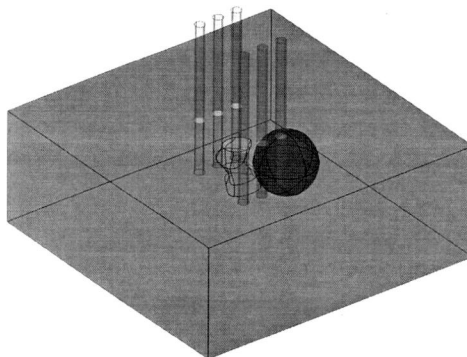

FIGURE 15.6 Model geometry: healthy tissue, tumor (between the electrodes)—geometry taken from Šel et al. (2007), vital organ (sphere). Needle electrodes are inserted into the tissue and appropriate electric potentials are assigned to each electrode so that the entire tumor volume is reversibly electroporated and the least possible volume of the vital organ is irreversibly electroporated.

TABLE 15.1 Tissue Properties Used in the Numerical Model

Tissue	σ_1 (S/m)	σ_2 (S/m)	E_{rev} (V/m)	E_{irr} (V/m)
Tumor	0.2	0.7	400	900
Vital organ	0.15	0.5	250	600
Healthy tissue	0.15	0.5	250	600

$$F = 10 \cdot V_{Trev} - 2 \cdot V_{VOir}, \tag{15.42}$$

where

F is the fitness

V_{Trev} is the fraction of tumor volume subjected to local electric fields above reversible thresholds $(E > E_{rev})$

V_{VOir} is the fraction of volume of the organ at risk subjected to $E > E_{irrev}$

The weights in the fitness function (importance factors) are set arbitrarily, but with respect to the importance of the individual parameters for efficient electroporation. Namely, the most important endpoint of the treatment is the reversible electroporation of the entire tumor (weight 10 in Equation 15.42), while sparing the vital organ is not as important (weight 2). In the clinical environment, these importance factors would have to be set by an experienced physician. In our case, the algorithm optimizes 36 different parameters (positions [x, y, z, φ, θ] and electric potentials of all six electrodes). It is presumed that it is possible to achieve a good enough electric field distribution using these parameters.

The final treatment plan is presented in Figure 15.7. The resulting electric field distribution is not at all homogeneous; the field around the electrodes and in some parts of the tumor is much higher than in other parts of the tissue. Nevertheless, the tumor is completely reversibly electroporated, while 0.3% of the vital organ is irreversibly electroporated. From that we can conclude that the treatment planning goals have been met.

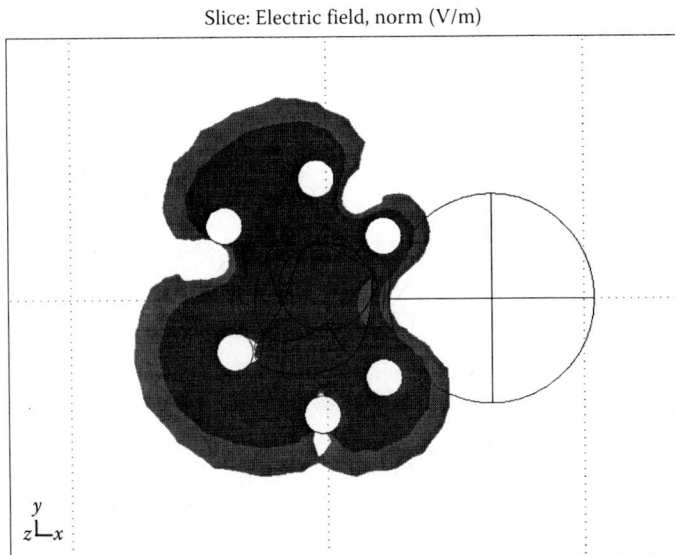

Slice: Electric field, norm (V/m)

FIGURE 15.7 Electric field distribution for the treatment plan 2 is shown in the XY plane through the center of the tumor. Light gray areas are reversible electroporated ($E > E_{rev}$), while dark gray areas are irreversibly electroporated ($E > E_{irr}$).

15.6 Summary

This chapter explains in detail the process of electric field distribution modeling in biological tissues. The equations used in modeling are derived from Maxwell's equation of the electromagnetic field, while taking into account the dynamics of the induced transmembrane potential compared with the duration of electric pulses and the behavior of biological tissue in the presence of external electric fields. Biological tissues, which can be heterogeneous, anisotropic, and nonlinear, are included in the equations in the form of electric fields and direction-dependent tissue properties. The model geometry has to be chosen carefully, according to the modeling aims—the simplest possible geometry and form of equations that give good results should be used to make the calculation as fast as possible and also easier to interpret. Finally, the numerical modeling of electric field distribution is not only useful for explaining the experimental results and hypothesis testing, but also in the clinical setting, where it can be used together with optimization techniques in the inverse treatment planning of electroporation-based treatments.

Acknowledgments

The authors would like to thank the Slovenian Research Agency and the European Commission for financial support.

References

André F, Gehl J, Serša G, Préat V, Hojman P, Eriksen J, Golzio M et al. 2008. Efficiency of high- and low-voltage pulse combinations for gene electrotransfer in muscle, liver, tumor, and skin. *Human Gene Ther* 19:1261–1271.

Brandisky K, Daskalov I. 1999. Electrical field and current distributions in electrochemotherapy. *Bioelectrochem Bioenerg* 48:201–208.

Chizmadzhev YA, Indenbom AV, Kuzmin PI, Galichenko SV, Weaver JC, Potts RO. 1998. Electrical properties of skin at moderate voltages: Contribution of appendageal macropores. *Biophys J* 74:843–856.

Ćorović S, Al Sakere B, Haddad V, Miklavčič D, Mir LM. 2008a. Importance of contact surface between electrodes and treated tissue in electrochemotherapy. *Technol Cancer Res Treat* 7:393–399.

Ćorović S, Županič A, Miklavčič D. 2008b. Numerical modeling and optimization of electric field distribution in subcutaneous tumor treated with electrochemotherapy using needle electrodes. *IEEE Trans Plasma Sci* 36:1665–1672.

Cukjati D, Batiuskaite D, André F, Miklavčič D, Mir LM. 2007. Real time electroporation control for accurate and safe in vivo non-viral gene therapy. *Bioelectrochemistry* 70:501–507.

Davalos RV, Rubinsky B. 2004. Electrical impedance tomography of cell viability in tissue with application to cryosurgery. *J Biomech Eng* 126(2):305–309.

Davalos RV, Mir LM, Rubinsky B. 2005. Tissue ablation with irreversible electroporation. *Ann Biomed Eng* 33(2):223–231.

Debruin KA, Krassowska W. 1999a. Modeling electroporation in a single cell. I. Effects of field strength and rest potential. *Biophys J* 77:1213–1224.

Debruin KA, Krassowska W. 1999b. Modeling electroporation in a single cell. II. Effects of ionic concentrations. *Biophys. J.* 77:1225–1233.

Denet A-R, Préat V. 2003. Transdermal delivery of timolol by electroporation through human skin. *J Control Release* 88:253–262.

Denet A-R, Vanbever R, Préat V. 2004. Skin electroporation for transdermal and topical delivery. *Adv Drug Deliv Rev* 56(5):659–674.

Dev SB, Dhar D, Krassowska W. 2003. Electric field of a six-needle array electrode used in drug and DNA delivery in vivo: Analytical versus numerical solution. *IEEE Trans Biomed Eng* 50(11):1296–1300.

Duck FA. 1990. *Physical Properties of Tissue: A Comprehensive Reference Book*. Academic Press, London, U.K.

Edd JF, Davalos RF. 2007. Mathematical modeling of irreversible electroporation for treatment planning. *Technol Cancer Res Treat* 6(4):275–286.

Fear EC, Stuchly MA. 1998. Modeling assemblies of biological cells exposed to electric fields. *IEEE Trans Biomed Eng* 45(10):1259–1271.

Foster KR, Schwan HP. 1989. Dielectric properties of tissues and biological materials: A critical review. *Crit Rev Biomed Eng* 17:25–104.

Gabriel C, Gabriel S, Corthout E. 1996a. The dielectric properties of biological tissues: I. Literature survey. *Phys Med Biol* 41:2231–2249.

Gabriel S, Lau RW, Gabriel C. 1996b. The dielectric properties of biological tissues: II. Measurements in the frequency range 10 Hz to 20 GHz. *Phys Med Biol* 41:2251–2269.

Golzio M, Rols MP, Teissié J. 2004. In vitro and in vivo electric field-mediated permeabilization, gene transfer, and expression. *Methods* 33(2):126–135.

Grimnes S, Martinsen OG. 2000. *Bioimpedance & Bioelectricity Basics*. Academic Press, London, U.K.

Hart FX, Berner NJ, McMillen RL. 1999. Modelling the anisotropic electrical properties of skeletal muscle. *Phys Med Biol* 44:413–421.

Ivorra A, Rubinsky B. 2007. In vivo electrical impedance measurements during and after electroporation of rat liver. *Bioelectrochemistry* 70(2):287–295.

Ivorra A, Al-Sakere B, Rubinsky B, Mir LM. 2008. Use of conductive gels for electric field homogenization increases the antitumor efficacy of electroporation therapies. *Phys Med Biol* 53:6605–6618.

Kotnik T, Miklavčič D. 2006. Theoretical evaluation of voltage inducement on internal membranes of biological cells exposed to electric fields. *Biophys J* 90:480–491.

Kotnik T, Bobanović F, Miklavčič D. 1997. Sensitivity of transmembrane voltage induced by applied electric fields—A theoretical analysis. *Bioelectrochem Bioenerg* 43:285–291.

Kotnik T, Miklavčič D, Slivnik T. 1998. Time course of transmembrane voltage induced by time-varying electric fields—A method for theoretical analysis and its application. *Bioelectrochem Bioenerg* 45:3–16.

Lavee J, Onik G, Mikus P, Rubinsky B. 2007. A novel nonthermal energy source for surgical epicardial atrial ablation: Irreversible electroporation. *Heart Surg Forum* 10(2):96–101.

Ljung L. 1999. *System Identification—Theory for the User*, 2nd edn., PTR Prentice Hall, Upper Saddle River, NJ.

Maček-Lebar A, Serša G, Kranjc S, Grošelj A, Miklavčič D. 2002. Optimisation of pulse parameters in vitro for in vivo electrochemotherapy. *Anticancer Res* 22:1731–1736.

Miklavčič D, Beravs K, Šemrov D, Čemažar M, Demšar F, Serša G. 1998. The importance of electric field distribution for effective in vivo electroporation of tissues. *Biophys J* 74:2152–2158.

Miklavčič D, Šemrov D, Mekid H, Mir LM. 2000. A validated model of in vivo electric field distribution in tissues for electrochemotherapy and for DNA electrotransfer for gene therapy. *Biochim Biophys Acta* 1523:73–83.

Miklavčič D, Čorović S, Pucihar G, Pavšelj N. 2006a. Importance of tumour coverage by sufficiently high local electric field for effective electrochemotherapy. *Eur J Cancer Suppl* 4:45–51.

Miklavčič D, Pavšelj N, Hart FX. 2006b. *Electric Properties of Tissues*. Wiley Encyclopedia of Biomedical Engineering, John Wiley & Sons, New York.

Onik G, Mikus P, Rubinsky B. 2007. Irreversible electroporation: Implications for prostate ablation. *Technol Cancer Res Treat* 6(4):295–300.

Pavlin M, Miklavčič D. 2003. Effective conductivity of a suspension of permeabilized cells: A theoretical analysis. *Biophys J* 85:719–729.

Pavlin M, Miklavčič D. 2008. Theoretical and experimental analysis of conductivity, ion diffusion and molecular transport during cell electroporation—Relation between short-lived and long-lived pores. *Bioelectrochemistry* 74:38–46.

Pavlin M, Pavšelj N, Miklavčič D. 2002. Dependence of induced transmembrane potential on cell density, arrangement, and cell position inside a cell system. *IEEE Trans Biomed Eng* 49:605–612.

Pavlin M, Kandušer M, Reberšek M, Pucihar G, Hart FX, Magjarević R, Miklavčič D. 2005. Effect of cell electroporation on the conductivity of a cell suspension. *Biophys J* 88:4378–4390.

Pavšelj N, Miklavčič D. 2008a. Numerical models of skin electropermeabilization taking into account conductivity changes and the presence of local transport regions. *IEEE Trans Plasma Sci* 36:1650–1658.

Pavšelj N, Miklavčič D. 2008b. Numerical modeling in electroporation-based biomedical applications. *Radiol Oncol* 42:159–168.

Pavšelj N, Bregar Z, Cukjati D, Batiuskaite D, Mir LM, Miklavčič D. 2005. The course of tissue permeabilization studied on a mathematical model of a subcutaneous tumor in small animals. *IEEE Trans Biomed Eng* 52:1373–1381.

Pavšelj N, Préat V, Miklavčič D. 2007. A numerical model of skin electropermeabilization based on in vivo experiments. *Ann Biomed Eng* 35:2138–2144.

Pliquett U. 2003. Joule heating during solid tissue electroporation. *Med Biol Eng Comput* 41(2):215–219.

Pliquett U, Weaver JC. 1996. Electroporation of human skin: Simultaneous measurement of changes in the transport of two fluorescent molecules and in the passive electrical properties. *Bioelectrochem Bioenerg* 39(1):1–12.

Prausnitz MR. 1999. A practical assessment of transdermal drug delivery by skin electroporation. *Adv Drug Deliv Rev* 35:61–76.

Puc M, Kotnik T, Mir LM, Miklavčič D. 2003. Quantitative model of small molecules uptake after in vitro cell electropermeabilization. *Bioelectrochemistry* 60:1–10.

Pucihar G, Mir LM, Miklavčič D. 2002. The effect of pulse repetition frequency on the uptake into electropermeabilized cells in vitro with possible applications in electrochemotherapy. *Bioelectrochemistry* 57:167–172.

Pucihar G, Kotnik T, Valič B, Miklavčič D. 2006. Numerical determination of transmembrane voltage induced on irregularly shaped cells. *Ann Biomed Eng* 34:642–652.

Reilly JP. 1998. *Applied Bioelectricity, from Electrical Stimulation to Electropathology*, Springer-Verlag, New York.

Rubinsky B, Onik G, Mikus P. 2007. Irreversible electroporation: A new ablation modality—Clinical implications. *Technol Cancer Res Treat* 6(1):37–48.

Schoenbach KH, Beebe SJ, Buescher ES. 2001. Intracellular effect of ultrashort electrical pulses. *Bioelectromagnetics* 22:440–448.

Šel D, Cukjati D, Batiuskaite D, Slivnik T, Mir LM, Miklavčič D. 2005. Sequential finite element model of tissue electropermeabilization. *IEEE Trans Biomed Eng* 52:816–827.

Šel D, Maček-Lebar A, Miklavčič D. 2007. Feasibility of employing model-based optimization of pulse amplitude and electrode distance for effective tumor electropermeabilization. *IEEE Trans Biomed Eng* 54:773–781.

Šemrov D, Miklavčič D. 1998. Calculation of the electrical parameters in electrochemotherapy of solid tumors in mice. *Comput Biol Med* 28:439–448.

Serša G, Jarm T, Kotnik T, Coer A, Podkrajšek M, Šentjurc M, Miklavčič D et al. 2008. Vascular disrupting action of electroporation and electrochemotherapy with bleomycin in murine sarcoma. *Br J Cancer* 98:388–398.

Susil R, Šemrov D, Miklavčič D. 1998. Electric field induced transmembrane potential depends on cell density and organization. *Electro Magnetobiol* 17:391–399.

Tekle E, Oubrahim H, Dzekunov SM, Kolb JF, Schoenbach KH, Chock PB. 2005. Selective field effects on intracellular vacuoles and vesicle membranes with nanosecond electric pulses. *Biophys J* 89:274–284.

Towhidi L, Kotnik T, Pucihar G, Firoozabadi SMP, Mozdarani H, Miklavčič D. 2008. Variability of the minimal transmembrane voltage resulting in detectable membrane electroporation. *Electromagn Biol Med* 27:372–385.

Tungjitkusolmun S, Woo EJ, Cao H, Tsai J-Z, Vorperian VR, Webster JG. 2000. Thermal-electrical finite element modelling for radio frequency cardiac ablation: Effects of changes in myocardial properties. *Med Biol Eng Comput* 38:562–568.

Valič B, Golzio M, Pavlin M, Schatz A, Faurie C, Gabriel B, Teissié J, Rols MP, Miklavčič D. 2003. Effect of electric field induced transmembrane potential on spheroidal cells: Theory and experiment. *Eur Biophys J* 32:519–528.

Yamamoto T, Yamamoto Y. 1976a. Electrical properties of the epidermal stratum corneum. *Med Biol Eng* 14(2):151–158.

Yamamoto T, Yamamoto Y. 1976b. Dielectric constant and resistivity of epidermal stratum corneum. *Med Biol Eng* 14(5):494–500.

Županič A, Čorović S, Miklavčič D. 2008. Optimization of electrode position and electric pulse amplitude in electrochemotherapy. *Radiol Oncol* 42:93–101.

16

Concepts of Electroporation Pulse Generation and Overview of Electric Pulse Generators for Cell and Tissue Electroporation

Matej Reberšek

Damijan Miklavčič

16.1 Introduction

It is well accepted now that the efficacy of electroporation, a phenomenon that occurs in lipid membranes exposed to strong electric fields, depends on several physical and biological parameters (Wong and Neumann, 1982; Mir, 2001; Rols, 2006). These are parameters of the electric field (i.e., pulse amplitude, pulse duration, number of pulses, pulse repetition frequency, pulse shape, and electric field direction) (Rols and Teissie, 1998; Vernhes et al., 1999; Kotnik et al., 2001a,b, 2003; Macek-Lebar and Miklavcic, 2001; Canatella et al., 2001; Pucihar et al., 2002; Faurie et al., 2004; Rebersek et al., 2007) and cell parameters that define the state of cells, their surroundings, and cell geometry (i.e., cell size and shape, temperature, osmotic pressure, etc.) (Rols and Teissie 1992; Rols et al. 1994; Kotnik et al., 1997; Bobrowska-Hagerstrand et al., 1998; Golzio et al., 1998; Pucihar et al., 2001). Cell parameters are diverse and usually cannot be controlled. Therefore, in electroporation applications, the parameters of the electroporation signal are optimized to specific cells, tissues, and most of all to achieving electroporation objectives. For example, in DNA transfection, the pulse amplitude is optimized to the specific cell size to achieve reversible cell membrane electroporation and to the pulse duration to allow plasmid DNA membrane complex formation, which then leads to gene expression. Electroporation can be reversible or irreversible (Figure 16.1), where reversibility/irreversibility is related to cell survival/death. Reversible

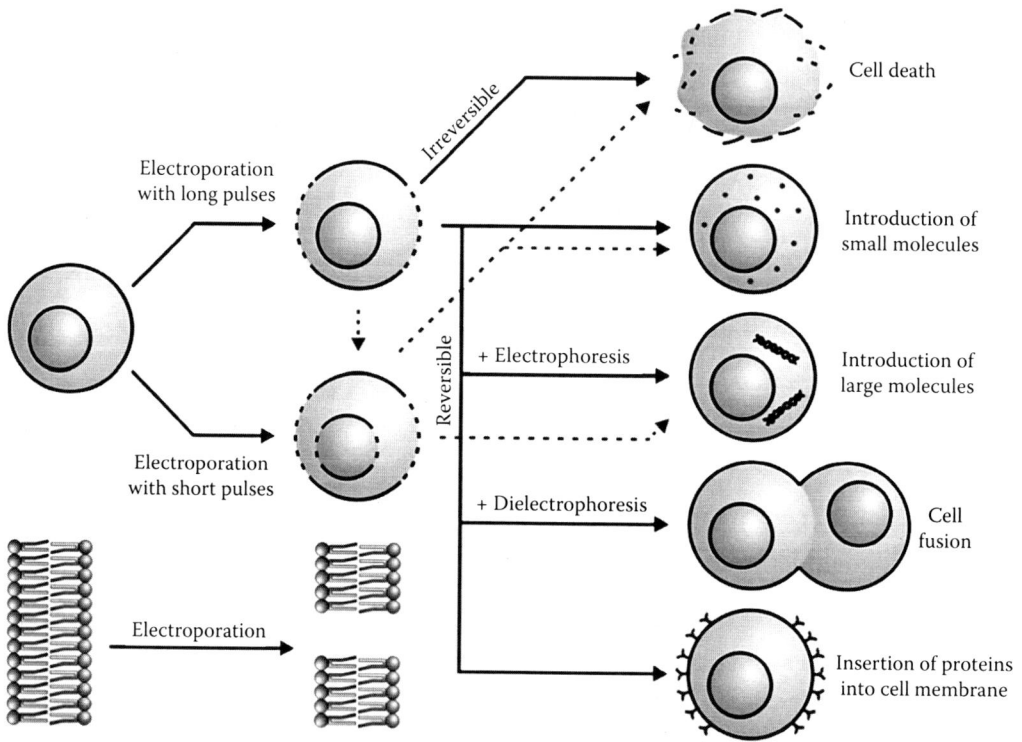

FIGURE 16.1 Electroporation of a cell may be reversible or irreversible. Reversible electroporation is character-ized by cell survival and can be even furthermore optimized for introduction of small and large molecules, fusion of cells, and insertion of proteins into cell membrane. For some of these specific applications an auxiliary pulses are sometimes used such as electrophoretic and dielectrophoretic pulses. Electroporation may also be used on mem-brane model systems such as planar lipid bilayer and vesicles.

electroporation can be even further optimized for specific objectives, for example, the introduction of small and large molecules, the fusion of cells, and the insertion of proteins into the cell membrane. At this optimization, auxiliary pulses are sometimes used, such as electrophoretic pulses for DNA and dielectrophoretic pulses for cell pearl chain formation in fusion.

Nowadays, electroporation is widely used in various biological, medical, and biotechnological appli-cations (Table 16.1) (Kanduser and Miklavcic, 2008). It is used for electrochemotherapy (Mir et al., 1995,

TABLE 16.1 Pulse Parameters for Different Electroporation Applications

Application	Amplitude	Duration	Auxiliary Pulses
Electrochemotherapy	~kV	μs, usually $8 \times 100\,\mu s$	—
Gene electrotherapy	~kV	μs–ms	Electrophoretic pulses <500 V, >ms
Electroinsertion	<kV	ms–s	—
Transdermal drug delivery	<kV	ms	—
Electrofusion	~kV	μs	Dielectrophoretic pulses <200 V, >s, ~MHz
Pasteurization	≫kV	μs	—
Tissue ablation	>kV	μs–ms	—
Single cell electroporation	>mV	μs	—
Electroporation research	mV–kV	ns–s	—

2006; Heller et al., 1999; Rols et al., 2002; Sersa, 2006), gene electrotransfer (Ferber, 2001; Mir, 2001; Satkauskas et al., 2002; Herweijer and Wolff, 2003; Prud'homme et al., 2006), cell fusion (Zimmermann, 1982; Ramos and Teissie, 2000; Trontelj et al., 2008), insertion of proteins into cell membranes (Mouneimne et al., 1989), transdermal drug delivery (Vanbever et al., 1994; Pavselj and Preat 2005), water treatment (Teissie et al., 2002), food preservation (Beveridge et al., 2002), tissue ablation (Rubinsky et al. 2007), and electroporation of cell organelles (Schoenbach et al., 2001; Rebersek et al., 2009). With microelectrodes, electroporation can be used on a single cell (Agarwal et al., 2009). Electroporation is sometimes also used or studied on simpler lipid membranes such as planar lipid bilayers or multilayers, vesicles, etc. (Montal and Mueller, 1972; Dimova et al., 2009; Kramar et al., 2009). Pulse amplitude is lower in the case of planar bilayer membranes and higher in the case of vesicles.

The diversity of the electrodes and target cells (either single cells, cell suspensions *in vitro*, or cells in tissues *in vivo*) is vast and only a rough introduction of the most important parameters of the electric field can be given. Pulse amplitudes that are used for reversible electroporation at electric pulse durations of 100 µs vary from a few V up to a few 100 V, for irreversible electroporation they vary from a few 100 V to several kV, and for electroporation of lipid bilayers they vary from 10 mV to V (Figure 16.2a).

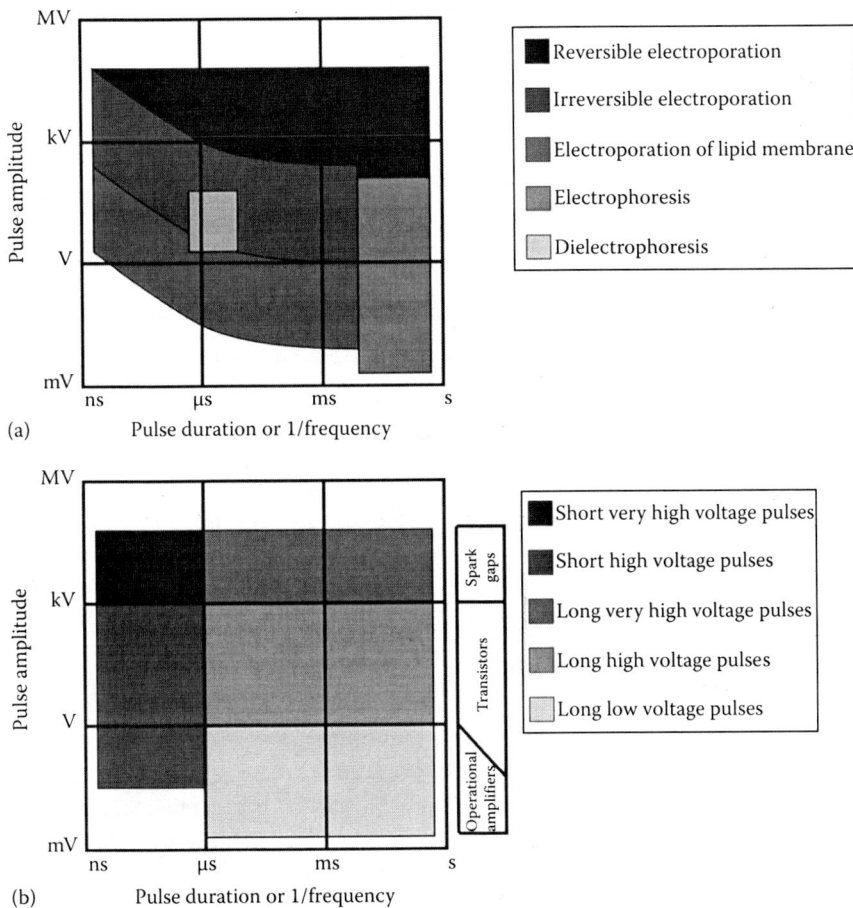

FIGURE 16.2 Areas of amplitude and duration of electrical pulses which are used in the research of electroporation and related effects (a). Five different areas of electroporation pulse generation (b). To amplify or to generate very-high-voltage electroporation pulses (over a few kV) spark gaps and similar elements are used, for high-voltage (a few V to a few kV) transistors and for low-voltage operational amplifiers are used. Nanosecond (short) pulses are generated with different techniques than pulses longer than 1 µs.

For shorter pulses, mainly in the nanosecond range, much higher amplitudes have to be used because of the lipid membrane time constant. On the other hand, for longer pulses, slightly lower amplitudes have to be used for a comparable electroporation effect. The auxiliary signals for gene electrotransfer and cell fusion, namely electrophoretic and dielectrophoretic pulses, have considerably lower amplitudes from the ones needed for electroporation. Electrophoretic pulses are at least a few milliseconds long and dielectrophoretic pulses are in the frequency range from a few 100 kHz to a few MHz. In summary, electroporation pulses (depending on the application) have amplitudes ranging from mV to kV and with frequency content from Hz to GHz. Such a variety of signals/pulses can evidently not be generated by a single generator.

16.2 Concepts of Electroporation Pulse Generation

The effectiveness of electroporation depends on the distribution of electric fields inside the treated sample (Sersa et al., 1996; Miklavcic et al., 1998; Gehl et al., 1999). To achieve (effective) electroporation, we have to use an appropriate set of electrodes (e.g., needle, parallel plates, cuvettes, etc.) and an electroporation device—an electroporator that generates the required voltage or current signals. Although both parts of the mentioned equipment are important for the effectiveness of electroporation, the electroporator is a more complicated device as it has to be able to generate the required signal and deliver the necessary current/power, which is defined by the load (Puc et al., 2001; Pavselj and Miklavcic, 2008).

To design or purchase an electroporator, one has to know the application for which the electroporator will be used. For example, clinical devices have to meet medical safety standards [IEC-60601], gene electrotransfer requires high- and low-voltage pulses (Satkauskas et al., 2002; Kanduser et al., 2009), cell electrofusion requires dielectrophoretical pulses (Pilwat et al., 1981), multi-needle electrodes require electrode commutators (Rebersek et al., 2007), etc. However, it is also important to know the range of electrical parameters that will be used. The maximal voltage and current of electroporation pulses define the switching elements that will be used in the design, because a particular switching element works as required only in the specific voltage and current range. For very high-voltage electroporation pulses (over a few kV), spark gaps and similar elements are used; for high-voltage electroporation pulses (from several V to a few kV), transistors are used; and for low-voltage electroporation pulses (below several V), operational amplifiers are used (Figure 16.2b).

The next parameter that has a crucial role in electroporation is the duration of the electroporation pulses. At pulses shorter than 1 μs, the rise time of the pulse is usually shorter than the electrical length between the source and the load. Therefore, it is very important to match the impedance of the load to the impedance of the generator, so that there are no strong pulse reflections and consequently pulse prolongations. However, for pulses longer than 1 μs, it is much more important to consider the fact that the load impedance varies during the pulse or sample-to-sample and that the load conduction is not predefined. Generators of nanosecond pulses are thus considerably different than for classical electroporation and are described in this book (Kolb, 2010). Another important parameter to be considered is the pulse shape. For very-high-voltage pulses, only exponential and square wave pulses can be used, as very-high-voltage switches work discretely. However, for lower voltages, arbitrary pulses can be generated as switches at these voltages can also operate linearly (Flisar et al., 2003). There are a variety of different concepts of signal generation for electroporation available. Each concept allows for the generation of different pulse shapes and has specific advantages and disadvantages and is thus more or less appropriate for specific electroporation applications (Table 16.2). Signal generators for electroporation can be divided into four major groups: capacitor discharge, square wave generators, modular square wave generators, and analog generators. For special applications, electroporators can be upgraded with additional modules, for example, low-voltage modules, dielectrophoretic modules, and a commutator.

TABLE 16.2 Comparison of Different Concepts of Signal Generation for Electroporation

Concept	Application	Advantages	Disadvantages
Capacitor discharge	Bacterial and yeast gene electrotransfer	Simple and inexpensive construction Simple control system High voltages	Poor flexibility and control of parameters Low cell survival
Square wave generators	Tissue ablation, electrochemotherapy, electroinsertion	Simple control system High currents Good control and flexibility of time parameters	Amplitude drop during the pulse Low amplitude flexibility
Modular square wave generators	Electrochemotherapy, gene electrotherapy, electroporation control, electroporation research	Wide flexibility of pulse parameters Arbitrary signal shape Electroporation control High currents	Limited amplitude resolution Complex control system Price
Analog generators	Electrochemotherapy, gene electrotherapy, electroporation control, electroporation research	Wide flexibility of pulse parameters Arbitrary signal shape Electroporation control	Complex control system Limitation of power dissipation
Additional low-voltage module	Gene electrotherapy, transdermal drug delivery	Electrophoretic drag Wide flexibility of amplitude and pulse duration	Complex control system
Additional dielectrophoretic module	Cell fusion	Pearl chain formation	Complex control system
Additional commutator	Large tumor treatment, gene electrotransfer, cell fusion	Multi-needle electrodes Polarity control Electric field direction control Higher safety	Complex user interface and control system

16.2.1 Capacitor Discharge

This is the oldest concept of electroporation pulse generation primarily used *in vitro* but also *in vivo* (Neumann and Rosenhec, 1972; Okino and Mohri, 1987). The concept comprises a variable high-voltage power supply (V), a capacitor (C), a switch (S), and optionally a resistor (R) (Figure 16.3). The generator operates in two phases, charge and discharge, and generates exponentially decaying pulses. During the charge phase, the switch (S) is in position 1 and the variable high-voltage power supply (V) charges the capacitor (C) to the preset voltage. In the discharge phase, the switch is in position 2 and the

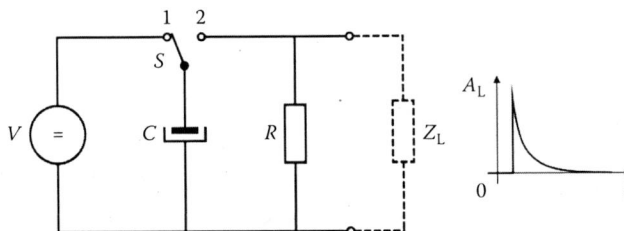

FIGURE 16.3 Capacitor discharge circuit for generation of exponentially decaying pulses. The concept comprises a variable high-voltage power supply (V), a capacitor (C), a switch (S), and optionally an internal resistor (R). The generator operates in two phases, charge (switch is in position 1 and capacitor charges to the preset voltage) and discharge (switch is in position 2 capacitor discharges through the load connected to the electrodes).

capacitor discharges through the load connected to the output. The time constant of discharge τ can be approximated by the product $Z_L C$, where C is the capacitance of the capacitor and Z_L is the absolute value of the load impedance. However, the impedance of biological load reduces during the pulse delivery (Pavlin et al., 2002, 2005; Davalos and Rubinsky, 2004; Granot et al., 2009). This also means that the time constant changes during the pulse. Therefore, most commercially available capacitor discharge-based electroporators have built-in resistances that are connected in parallel to the load. Their main purpose is to better define the time constant of the discharge. Namely, if additional resistors are connected in parallel to the load, the time constant of discharge is defined by $(R\|Z_L)C$, where R is the resistance of the internal resistors. If the absolute value of the impedance of load Z_L is at least 10 times larger than the resistance R ($Z_L \geq 10R$), the time constant can be approximated by the RC product (Equation 16.1):

$$(R \| Z_L)C = \left(\frac{RZ_L}{R+Z_L}\right)C = \left(\frac{10R}{11R}\right)C \cong RC; \quad (Z_L = 10R) \tag{16.1}$$

The presented capacitor discharge concept is very simple and inexpensive for construction. The exponentially decaying pulse generated can be used even for gene transfection as it includes the high-voltage part for permeabilization and low-voltage electrophoretic part (Satkauskas et al., 2002; Kanduser et al., 2009). However, the flexibility of such high- and low-voltage pulse composition is very poor, as the electrical parameters of the high-voltage part cannot be changed without affecting the low-voltage part and vice versa. Moreover, the low-voltage part is usually undesired in other electroporation applications, as it mainly affects the cell viability (Danfelter et al., 1998). Also, the repetition frequency of such pulse generation is low due to the relatively long charge phase.

16.2.2 Square Wave Pulse Generators

For better control of electric field parameters, square wave pulse generators have been introduced. The concept is very similar to the capacitor discharge concept, except that the voltage power supply (V) constantly charges the capacitor (C) and the power switch (S) is capable of fast switching (Figure 16.4) (Tokmakci, 2006; Bertacchini et al., 2007). Usually, fast power metal oxide semiconductor field effect transistors (MOSFETs) or insulated gate bipolar transistors (IGBTs) are used as the switch. The output amplitude of the pulse is defined by the amplitude of variable power supply, while pulse duration, pulse repetition frequency, and possibly the number of pulses are defined by the switching sequence of the fast power switch. As the switching sequence is faster and more complex, the control unit of the generator must also be faster and more complex than the capacitor discharge generator.

Despite improved control over the electric field parameters, this concept still has drawbacks that limit the flexibility and accuracy of pulse parameters available to the user. The main problem is that

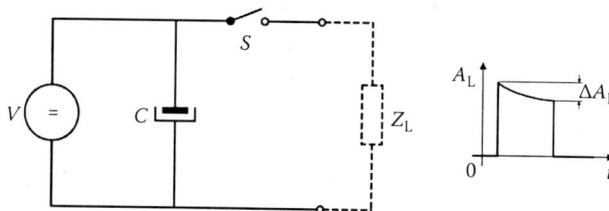

FIGURE 16.4 High-voltage power supply switching circuit for generation of square wave pulses. The concept comprises a variable high-voltage power supply (V), a capacitor (C), and a fast switch (S). The variable high-voltage power supply continuously charges the capacitor that stores energy required during the pulse. To deliver the pulse to load, the fast switch is turned on for the duration of the square wave pulse.

electroporation pulses have high power but are very short. Thus, a power supply cannot generate the energy for the pulse during the pulse generation, but it has to be generated and stored into the capacitor before the generation of the pulse. This usually results in a voltage drop (ΔA_1) during the pulse. In order to minimize this voltage drop, a very large capacitance is needed. However, as a consequence of a very large capacitance, it is now harder to change the amplitude between the pulses. Therefore, square wave pulse generators usually generate pulses with only one (preset) voltage. Nevertheless, at very high loads (very high current flow) voltage on the capacitor will inevitably decrease during the pulse generation. As it is usually required that each pulse has the same amplitude as the first one that was generated, the next pulse can only be delivered after the capacitor is recharged to the preset voltage. Therefore, limited power supply also defines the highest pulse repetition frequency.

16.2.2.1 Modular Square Wave Pulse Generator

This concept was designed to improve the output amplitude flexibility of the square wave pulse generator. The concept was described previously in detail (Petkovsek et al., 2002). Briefly, it consists of several (N) square wave pulse generators connected in a series (Figure 16.5). The generators have galvanically insulated high-voltage power supplies (V). Each square wave pulse generator is controlled individually based on the principle of a digital-to-analog (D/A) converter, thus the amplitude of the particular source V_N is twice as high as the predecessor ($V_N = 2V_{N-1}$). The voltage of the individual generator is constant and can contribute to the generation of a common output pulse at any time. With an appropriate control of switches S_1–S_N, a total of 2^N different output voltage levels with the resolution of V_1 are obtained (Petkovsek et al., 2002). Although the design of each individual source is similar to the design of the previously described square wave pulse generator, the individual power supply used in this concept has constant (not variable) output voltage. As constant voltage power supplies do not discharge during the pulse delivery, they are simpler to be designed to sustain the maximum possible current during the pulse generation. In this way, the output amplitude does not decrease during the pulse delivery.

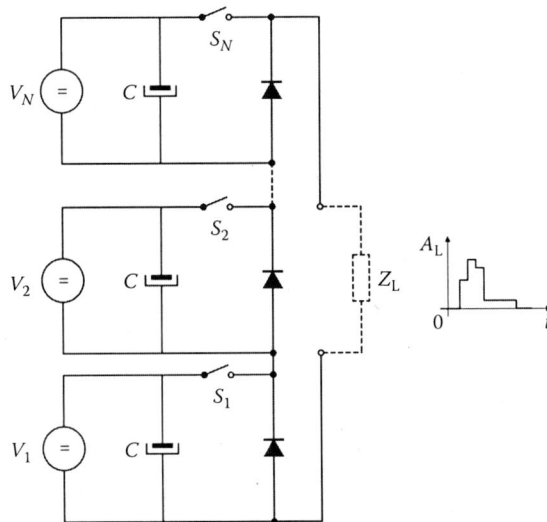

FIGURE 16.5 Modular square wave pulse generator. Operation of the device is based on a principle of D/A converter, thus the device comprises several (N) individually controlled electrically insulated DC voltage modules, where the amplitude of the particular voltage source V_N is twice as high as in the preceding module. With an appropriate control of switches S_1–S_N the modules are connected in series and a total of 2^N different output voltage levels with the resolution of V_1 are obtained.

The presented modular concept can generate well-defined pulses as its rise and fall times are fast and the amplitude is stable. However, to have enough output voltage levels, many square wave pulse generators are needed (depending on "voltage" steps/resolution), which consequently increases the cost of the device.

16.2.3 Analog Generators

Although square wave and exponentially decaying pulses were and probably still are the most frequently used signals for electroporation, analog generators definitely have some advantages over them. The concept of analog generators was designed to generate arbitrarily shaped electroporation pulses and to improve the output amplitude stability of the square wave pulse generator (Figure 16.6) (Petkovsek et al., 2002; Rebersek et al., 2007). The concept comprises variable high-voltage power supplies (V), capacitors (C), signal generators (F_G), linear switches (Q), and resistor decades (R_1 and R_2).

Energy for the pulse is stored in the capacitor by setting the power supply voltage higher than the maximal generated amplitude. Therefore, lower capacitance is needed to store the energy for the pulse and the amplitude of the pulse will not drop during the pulse generation, unless the capacitor voltage drops below the expected output amplitude. Usually, the preset voltage is at least 10% or 50 V higher than the maximal generated amplitude.

The pulse shape is first generated by the signal generator, which is usually a computer with a D/A converter. This signal is then amplified by a linear amplifier. Usually, an amplifier with a common source and galvanically separated input is used (Figure 16.6), as it is noninverting voltage and current amplifier. This amplifier needs a galvanic separation between the driving and the power supply circuit. This, however, is definitely not a drawback, as all electroporators should have galvanically insulated output for safety reasons. The linear amplifier consists of a linear switch (usually MOSFET or IGBT) and resistor decade. A resistor decade is used as a feedback for the regulation of the output amplitude. The signal, reduced for a voltage threshold of the linear amplifier (U_T), is therefore amplified by the factor $(R_1 + R_2)/R_1$. If the current amplification in this stage is not high enough, a current amplifier (common source) can be added to the output.

This design allows wide flexibility of all electrical parameters and electroporation control (Cukjati et al., 2007), yet some drawbacks still exist. The driving stage is much more complex than in previously described concepts and the rise and fall times of the pulses cannot be as fast as with a square wave pulse generator. Nevertheless, the major drawback of this concept is the safe operation area (SOA; voltage, current, power, and energy limitations) of linear transistors. Therefore, the duration, voltage, and current of the pulse are limited as there is high power dissipation when the transistor is working in its

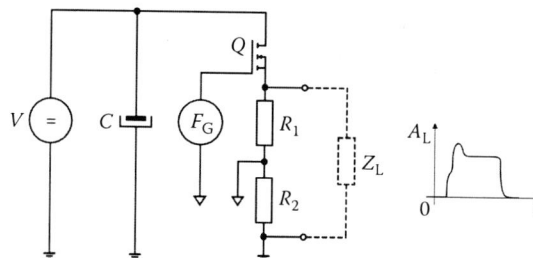

FIGURE 16.6 Analog generator for generation of arbitrary pulses comprises a variable high-voltage power supply (V), a capacitor (C), a signal generator (F_G), a linear switch (Q), and a resistor decade (R_1 and R_2). Energy for the pulse is stored in the capacitor by setting the power supply voltage higher than the maximal generated amplitude. The pulse shape is first generated by the signal generator, which is usually a computer with a D/A converter. This signal is then amplified by a linear amplifier (Q, R_1, and R_2). Resistor decade is used as a feedback for regulation of the output amplitude.

linear area. Analog generators can also be designed to generate bipolar pulses by means of a push-pull amplifier (Flisar et al., 2003). However, for these generators, the current between the pulses (zero driving voltage) has to be minimized.

16.2.4 High-Voltage, Low-Voltage, and Dielectrophoretic Signals

Electroporators that are designed for gene electrotransfer ideally would have to generate high-voltage pulses for reversible electroporation and low-voltage pulses for the electrophoretic drag of DNA toward the permeabilized plasma membrane (Satkauskas et al., 2002; Kanduser et al., 2009). Electroporators that are designed for cell fusion ideally should generate high-voltage pulses for reversible electroporation and sinusoidal signals for the dielectrophoretic force (Pilwat et al., 1981). For low-voltage electroporators, these pulses can be generated by means of one analog generator. However, for high-voltage electroporators, a modular concept has to be used. A modular concept means that the electroporator is made of two or more generators, each having a special capability. Thus, an electroporator for gene electrotransfer is usually made of two square wave pulse generators. Two generators are used so that the power supply voltage of a generator does not change between the pulses and square wave pulse generators are used as they have good control over electric field parameters with respect to the expenses. Similarly, an electroporator for cell fusion is usually made of one square wave pulse generator for reversible electroporation and one bipolar analog generator for sinusoidal signal. In an electroporator designed for gene electrotransfer and cell fusion, all three modules (high voltage, low voltage, and dielectrophoretic) can be integrated into one device. The outputs of the modules are connected together with the diodes if all the output signals are unipolar or with the relays if one of the modules is generating bipolar pulses. The control of different signals using this modular approach is good, but the price of such a device can be considerable.

16.2.5 Multi-Electrode and Polarity Control

Electroporators (E_p) can be supplemented with an output commutator that can deliver electroporation pulses to multiple electrodes (E_X) and can control the polarity of the pulse (Figure 16.7). A typical commutator has an integrated one-half bridge switch for each electrode. The half bridge switch is designed with two switches, one for the positive pole (S_p) and one for the negative (S_N) pole. These two switches, however, should never be active at the same time, as they can short-circuit the electroporator. The half bridge can be used to connect the electrode with the positive or negative pole or to disconnect the electrode. Namely, if the positive switch S_p is active, a positive pole is connected to the electrode; if the negative switch S_N is active, a negative pole is connected to the electrode; and if both switches are inactive, the electrode is in a high impedance state. Bipolar pulses are used to increase the

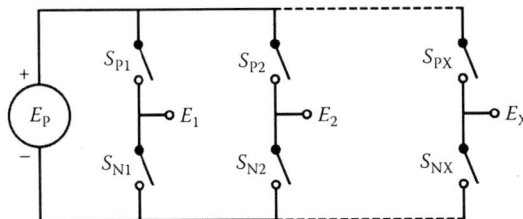

FIGURE 16.7 Multi-electrode and polarity control. Electroporators (E_p) can be supplemented with an output commutator that can deliver electroporation pulses to multiple electrodes (E_X). Usual commutator has one half bridge switch integrated for each electrode. The half bridge switch is designed of two switches, one for positive (S_p) and one for negative (S_N) pole. These two switches should however never be active at the same time, as they can short-circuit the electroporator. The half bridge can be used to connect the electrode with the positive or negative pole or to disconnect the electrode.

efficiency of electroporation (Kotnik et al., 2001a) and to reduce electrolytic contamination (Kotnik et al., 2001b). Multiple needle electrodes are used to treat larger tumors (Sel et al., 2007; Rebersek et al., 2008; Zupanic et al., 2008).

16.3 Safety

Irrespective of how simple or complex an electroporator prototype may be, safety must be the primary objective throughout their development. The most significant technical standard that should be considered during the development is the standard issued by the International Electrotechnical Commission IEC-60601, *Medical Electrical Equipment*. This standard is adopted by the European Union (EU) as EN-60601 and by the United States as UL-60601. Other countries have also adopted this standard; however, the EU and the United States are well known for their regulations that ensure medical device efficacy and safety. In the EU, a series of directives establish the requirements that manufacturers of medical devices must meet before they can obtain *CE marking* for their product to authorize their sale and use.

According to IEC-60601, a possible risk for electrical shock is present whenever a patient or operator can be exposed to a voltage exceeding $25\,V_{RMS}$ or $60\,V_{DC}$. Patient and operator safety must be ensured under both normal and single-fault conditions. Obviously, the enclosure of the device is the first barrier of protection that can protect the operator or patient from intentional or unintentional contact with these hazards. Beyond the electrical protection supplied by the enclosure, however, the circuit of the medical instrument must be designed with other safety barriers to maintain leakage currents within the limits allowed by the safety standards. The leakage current in the electroporator is minimized by the galvanical separation of electroporator output and the ground. The separation can be made in the power supply circuit or at the output of the electroporator. The galvanical separation is made by an insulation transformer for power signals or an optocoupler for data signals.

Medical devices must also meet requirements for electromagnetic interference (EMI) and electromagnetic compatibility (EMC). Therefore, electroporators should not emit electromagnetic signals that can interfere with other devices and should have sufficient immunity to operate as intended in the presence of interference (electrostatic discharge, power line transients, electromagnetic radiation, short circuit, etc.) (Prutchi and Norris, 2005).

16.4 Commercial Electroporators

Electroporation is mostly used with high-voltage electroporation pulses in the range of 100 V to 3 kV (Figure 16.2b). Reviews of commercially available electroporators were published previously (DeFrancesco, 1997; Puc et al., 2004), however, we now present an updated list of the available electroporators and their respective electrical parameters, biological applications, and signal generation techniques (Table 16.3).

Electroporators are grouped by the manufacturer and each device is presented with the following characteristics: output signal, voltage range, current (I)/load (R), time constant (τ)/pulse length (T), and charge time (t)/pulse repetition frequency (f). The value of the last two parameters (τ/T and t/f) depends on the output characteristics; if the device produces exponentially decaying pulses, the time constant and charge time are given as parameters. On the other hand, if the device generates square wave pulses, the pulse length and pulse repetition frequency are given as parameters. The data in Table 16.3 was checked and collected on the manufacturers' home pages (as written in the table) in September 2009.

Cliniporator, Cliniporator VITAE, and NanoKnife are for now the only devices approved for clinical use. CythorLab is the only device that can control the progress of electroporation, can generate arbitrarily shaped electroporation pulses, and is great in its flexibility and full control of pulse offered to the

TABLE 16.3 List of Commercially Available Electroporators with Their Parameters and Biological Applications

Company/Product	Output Characteristics	Voltage Range	Current (I)/ Load (R)	Time Constant (τ)/Pulse Length (T)	Charge Time (t)/ Pulse Repetition Frequency (f)	Biological Applications	Possible Signal Generation Technique
ADITUS MEDICAL, http://www.aditusmedical.com							
CythorLab	Arbitrary	LV: 0 V–600 V$_{PP}$; HV: 0 V–3000 V$_{PP}$	NA	LV: T = up to 400 ms; HV: T = up to 5 ms	NA	*In vitro, in vivo*	NA
ANGIODYNAMICS, http://www.angiodynamics.com							
NanoKnife	Square wave	100–3000 V	$I < 50$ A	$T = 20$–100 μs	$f = 1$ Hz–10 kHz	Irreversible electroporation, clinical device	Square wave
BIORAD, http://www3.bio-rad.com							
Micro Pulser	Exponential	200–3000 V	$R > 600\,\Omega$	$\tau = 1$–5 ms	$t = 5$ s	Bacterial, yeast	Capacitor discharge
Gene Pulser Xcell	Exponential Square wave	10–3000 V	$R > 600\,\Omega$	$\tau = 0.5$ ms–3.3 s; $T = 0.05$–10 ms	$t = 5$ s; $f = 0.1$–10 Hz	All cell types	Capacitor discharge
MXcell	Exponential Square wave	10–500 V	NA	$\tau = 1.3$ ms–3.7 s; $T = 0.05$ ms–1 s	$f = 0.1$–10 Hz	Eukaryotic cells	Capacitor discharge
BTX, http://www.btxonline.com							
ECM 399	Exponential	LV: 2–500 V; HV: 10–2500 V	$R > 150\,\Omega$	LV: $\tau = 157$ ms; HV: $\tau = 5.4$ ms	$t < 5$ s	All cell types	Capacitor discharge
ECM 630	Exponential	LV: 10–500 V; HV: 50–2500 V	$I < 6000$ A; $I < 3000$ A	LV: $\tau = 25$ μs–5 s; HV: $\tau = 625$ μs–78 ms	$t < 5$ s	All cell types, *in vivo*, plant	Capacitor discharge
ECM 830	Square wave	LV: 5–500 V; HV: 30–3000 V	$I < 500$ A	LV: $T = 10$ μs–10 s; HV: $T = 10$–600 μs	$f = 0.1$–10 Hz	All cell types, *in vivo*, plant	Square wave generator Pulse transformer
ECM 2001	Square wave Sinus (AC)	LV: 10–500 V; HV: 10–3000 V; 0 V–150 V$_{PP}$	NA	LV: $T = 10$ μs–99 ms; HV: $T = 1$–99 μs; $f_{AC} = 1$ MHz	NA	All cell types, cell fusion	Square wave generator Pulse transformer

(continued)

TABLE 16.3 (continued)   List of Commercially Available Electroporators with Their Parameters and Biological Applications

Company/Product	Output Characteristics	Voltage Range	Current (I)/Load (R)	Time Constant (τ)/Pulse Length (T)	Charge Time (t)/Pulse Repetition Frequency (f)	Biological Applications	Possible Signal Generation Technique
CLONAID, http://www.clonaid.com							
RMX2010	Square wave	5–200 V	NA	$T = 10$–$990\,\mu s$	$f = 1$–$1000\,Hz$	Embryonic cell fusion	Square wave generator
CYTO PULSE SCIENCES, http://www.cytopulse.com							
PA-3000	Square wave	20–1100 V	$R > 10\,\Omega$	$T = 1\,\mu s$–$10\,ms$	$f = 2.5$–$8\,Hz$	*In vitro, in vivo,* ex vivo	Square wave generator
PA-4000	Square wave	LV: 20–400 V; HV: 20–1100 V	$R > 10\,\Omega$	LV: $T = 1\,\mu s$–$40\,ms$; HV: $T = 1\,\mu s$–$1\,ms$	$f = 2$–$8\,Hz$	*In vitro, in vivo,* ex vivo	Square wave generator
PA-4000/PA-101	Square wave; Sinus (AC)	20–800 V; 15–75 V_{pp}	$R > 10\,\Omega$	$T = 1$–$100\,\mu s$; $f_{AC} = 0.2$–$2\,MHz$	$f = 8\,Hz$	Cell fusion	Square wave generator
PA-4000/PA-201	Square wave	LV: 20–400 V; HV: 20–1100 V	$R > 10\,\Omega$	LV: $T = 1\,\mu s$–$40\,ms$; HV: $T = 1\,\mu s$–$1\,ms$	$f = 2$–$8\,Hz$	Multi-needle electrodes	Square wave generator
PA-4000/PA-301	Square wave	20–3000 V	$R > 10\,\Omega$	$T = 1$–$100\,\mu s$	$f = 2.5$–$8\,Hz$	Bacterial, yeast	Square wave generator
Derma vax	Square wave	50–1000 V	$R > 10\,\Omega$	$T = 0.05$–$10\,ms$	$f = 1\,Hz$–$5\,kHz$	Intra-dermal	Square wave generator
Hybrimune	Square wave; Sinus (AC)	100–1000 V; 10–150 V_{pp}	$R > 10\,\Omega$	$T = 20$–$1000\,\mu s$; $f_{AC} = 0.2$–$2\,MHz$	NA	Cell fusion	Square wave generator
CytoLVT	Square wave	50–1000 V	$R > 10\,\Omega$	$T = 0.05$–$10\,ms$	$f = 1\,Hz$–$5\,kHz$	Large volume	Square wave generator
Onco vet	Square wave	50–1000 V	$R > 10\,\Omega$	$T = 0.05$–$10\,ms$	$f = 1\,Hz$–$5\,kHz$	Veterinary device	Square wave generator
EPPENDORF SCIENTIFIC, http://www.eppendorf.com							
Electroporator 2510 Multiporator	Exponential	200–2500 V	$R > 600\,\Omega$	$\tau = 5\,ms$	$t < 8\,s$	Bacterial, yeast	Capacitor discharge
Eukaryotic module	Exponential	20–1200 V	NA	$\tau = 15$–$500\,\mu s$	$t < 60\,s$	Eukaryotic cells, plant	Capacitor discharge
Bacteria and yeast module	Exponential	200–2500 V	$R > 600\,\Omega$	$\tau = 5\,ms$	NA	Bacterial, yeast	Capacitor discharge
Fusion module	Square wave; Sinus (AC)	5–300 V; 1–10 V_{pp}	NA	$T = 15$–$300\,\mu s$; $f_{AC} = 0.2$–$2\,MHz$	$f = 1\,Hz$	Cell fusion	Square wave generator

IGEA, http://www.igea.it						
Cliniporator	Square wave	LV: 20–200 V HV: 50–1000 V	$I < 16\,A$	LV: $T = 10\,\mu s$–20 ms HV: $T = 30$–200 μs	$f = 1\,Hz$–10 kHz	Electrochemotherapy; gene transfection, clinical device	Analog generator
Cliniporator VITAE	Square wave	500–3000 V	$I < 50\,A$	$T = 50$–1000 μs	$f = 1\,Hz$–10 kHz	Electrochemotherapy clinical device	Square wave
INOVIO, http://www.inovio.com							
MedPulser	Square wave	NA	NA	NA	NA	Gene transfection	NA
Elgen	Square wave	NA	NA	NA	NA	Gene transfection	NA
Cellectra	Square wave	NA	NA	NA	NA	Gene transfection	NA
LONZA, http://www.lonzabio.com							
Nucleofector	NA	NA	NA	NA	NA	*In vitro* transfection	NA
CLB-Transfection	NA	NA	NA	NA	NA	*In vitro* transfection	NA
PROTECH INTERNATIONAL, http://www.protechinternational.com							
CUY21EDIT	Square wave	1–500 V	$I < 5\,A$	$T = 0.1\,ms$–1 s	$f = 1\,Hz$–10 kHz	*In vitro, in vivo*, ex vivo	Square wave generator
CUY21SC	Square wave	0.1–100 V	$I < 1.6\,A$	$T = 0.05$–100 ms	$f = 1\,Hz$–10 kHz	Single cell	Square wave generator
LF101	Square wave Sinus (AC)	0–999 V 0–60 V_{PP}	$R > 50\,\Omega$	$T = 5$–99 μs $f_{AC} = 0.1$–3.9 MHz	$f = 0.1\,Hz$–10 kHz	Cell fusion	Square wave generator
LF201	Square wave Sinus (AC)	0–1200 V 0–75 V_{PP}	$R > 50\,\Omega$	$T = 1$–100 μs $f_{AC} = 1\,MHz$	$f = 0.1\,Hz$–10 kHz	Cell fusion	Square wave generator
TRITECH RESEARCH, http://www.tritechresearch.com							
Mammo Zapper	Exponential	NA	NA	NA	$t = 15\,s$[a]	Mammalian	Capacitor discharge
Bacto Zapper	Exponential	<2000 V[a]	NA	$\tau = 10\,ms$[a]	$t = 5\,s$[a]	Bacterial	Capacitor discharge

Source: Puc, M. et al. *Bioelectrochemistry*, 64, 113, 2004.

Note: NA, not available.

[a] Data was not available at the time this chapter was written so it has not been rechecked but has been taken from Puc et al. (2004).

experimenter. ECM 2001, PA - 4000/PA – 101, Hybrimune, Multiporator: Fusion module, LF101, and LF201 are designed to also generate a dielectrophoretic signal for cell fusion. Although the prices are not readily available and vary considerably between models and producers, they range from €10,000 to €150,000.

16.5 Conclusion

The choice of an electroporator is always driven by the objectives and application. These define the requirements for electric pulse parameters (i.e., pulse amplitude, pulse duration, number of pulses, pulse repetition frequency, pulse shape, and electric field direction). However, for electroporation research, it is very useful to have wide range, flexibility, and control over pulse parameters; therefore, such electroporators are expensive and not easy to obtain. Usually, the interesting results in electroporation research start where the electrical parameters of available electroporators are ending/out of range. For the specific application, the choice, in principle, is easier as the pulse parameters have been optimized before; the load is well characterized and so the range of parameters is narrowed. There are quite a number of laboratory electroporators available on the market but only a few are available for use in treating patients.

Acknowledgments

The authors want to thank the Slovenian Research Agency (ARRS) and European Commission for their support through various grants.

References

Agarwal A, Wang MY, Olofsson J, Orwar O, Weber SG. 2009. Control of the release of freely diffusing molecules in single-cell electroporation. *Anal Chem* 81(19):8001–8008.

Bertacchini C, Margotti PM, Bergamini E, Lodi A, Ronchetti M, Cadossi R. 2007. Design of an irreversible electroporation system for clinical use. *Technol Cancer Res Treat* 6(4):313–320.

Beveridge JR, MacGregor SJ, Marsili L, Anderson JG, Rowan NJ, Farish O. 2002. Comparison of the effectiveness of biphase and monophase rectangular pulses for the inactivation of micro-organisms using pulsed electric fields. *IEEE Trans Plasma Sci* 30(4):1525–1531.

Bobrowska-Hagerstrand M, Hagerstrand H, Iglic A. 1998. Membrane skeleton and red blood cell vesiculation at low pH. *Biochim Biophys Acta Biomem* 1371(1):123–128.

Canatella PJ, Karr JF, Petros JA, Prausnitz MR. 2001. Quantitative study of electroporation-mediated molecular uptake and cell viability. *Biophys J* 80(2):755–764.

Cukjati D, Batiuskaite D, Andre F, Miklavcic D, Mir LM. 2007. Real time electroporation control for accurate and safe in vivo non-viral gene therapy. *Bioelectrochemistry* 70(2):501–507.

Danfelter M, Engstrom P, Persson BRR, Salford LG. 1998. Effect of high voltage pulses on survival of Chinese hamster V79 lung fibroblast cells. *Bioelectrochem Bioenerg* 47(1):97–101.

Davalos R, Rubinsky B. 2004. Electrical impedance tomography of cell viability in tissue with application to cryosurgery. *J Biomech Eng* 126(2):305–309.

DeFrancesco L. 1997. Shock jocks. *Scientist* 11(15):19–21.

Dimova R, Bezlyepkina N, Jordo MD, Knorr RL, Riske KA, Staykova M, Vlahovska PM, Yamamoto T, Yang P, Lipowsky R. 2009. Vesicles in electric fields: Some novel aspects of membrane behavior. *Soft Matter* 5(17):3201–3212.

Faurie C, Phez E, Golzio M, Vossen C, Lesbordes JC, Delteil C, Teissie J, Rols MP. 2004. Effect of electric field vectoriality on electrically mediated gene delivery in mammalian cells. *Biochim Biophys Acta Biomembr* 1665(1–2):92–100.

Ferber D. 2001. Gene therapy: Safer and virus-free? *Science* 294(5547):1638–1642.

Flisar K, Puc M, Kotnik T, Miklavcic D. 2003. Cell membrane electropermeabilization with arbitrary pulse waveforms. *IEEE Eng Med Biol* 22(1):77–81.

Gehl J, Sorensen TH, Nielsen K, Raskmark P, Nielsen SL, Skovsgaard T, Mir LM. 1999. In vivo electroporation of skeletal muscle: Threshold, efficacy and relation to electric field distribution. *Biochim Biophys Acta Gen Subj* 1428(2–3):233–240.

Golzio M, Mora MP, Raynaud C, Delteil C, Teissie J, Rols MP. 1998. Control by osmotic pressure of voltage-induced permeabilization and gene transfer in mammalian cells. *Biophys J* 74(6):3015–3022.

Granot Y, Ivorra A, Maor E, Rubinsky B. 2009. In vivo imaging of irreversible electroporation by means of electrical impedance tomography. *Phys Med Biol* 54(16):4927–4943.

Heller R, Gilbert R, Jaroszeski MJ. 1999. Clinical applications of electrochemotherapy. *Adv Drug Deliv Rev* 35(1):119–129.

Herweijer H, Wolff JA. 2003. Progress and prospects: Naked DNA gene transfer and therapy. *Gene Ther* 10(6):453–458.

Kanduser M, Miklavcic D. 2008. Electroporation in biological cell and tissue: An overview. In: Vorobiev E, Lebovka N (eds.), *Electrotechnologies for Extraction from Food Plants and Biomaterials*. New York: Springer Science, pp. 1–37.

Kanduser M, Miklavcic D, Pavlin M. 2009. Mechanisms involved in gene electrotransfer using high- and low-voltage pulses—An in vitro study. *Bioelectrochemistry* 74(2):265–271.

Kolb J. 2010. Generation of ultrashort pulses. In: Pakhomov AG, Miklavcic D, Markov MS (eds.), *Advanced Electroporation Techniques in Biology and Medicine*. Boca Raton, FL: CRC Press.

Kotnik T, Bobanovic F, Miklavcic D. 1997. Sensitivity of transmembrane voltage induced by applied electric fields—A theoretical analysis. *Bioelectrochem Bioenerg* 43(2):285–291.

Kotnik T, Mir LM, Flisar K, Puc M, Miklavcic D. 2001a. Cell membrane electropermeabilization by symmetrical bipolar rectangular pulses—Part I. Increased efficiency of permeabilization. *Bioelectrochemistry* 54:83–90.

Kotnik T, Miklavcic D, Mir LM. 2001b. Cell membrane electropermeabilization by symmetrical bipolar rectangular pulses—Part II. Reduced electrolytic contamination. *Bioelectrochemistry* 54:91–95.

Kotnik T, Pucihar G, Rebersek M, Mir LM, Miklavcic D. 2003. Role of pulse shape in cell membrane electropermeabilization. *Biochim Biophys Acta* 1614:193–200.

Kramar P, Miklavcic D, Lebar AM. 2009. A system for the determination of planar lipid bilayer breakdown voltage and its applications. *IEEE Trans Nanobiosci* 8(2):132–138.

Macek-Lebar A, Miklavcic D. 2001. Cell electropermeabilization to small molecules in vitro: Control by pulse parameters. *Radiol Oncol* 35(3):193–202.

Miklavcic D, Beravs K, Semrov D, Cemazar M, Demsar F, Sersa G. 1998. The importance of electric field distribution for effective in vivo electroporation of tissues. *Biophys J* 74(5):2152–2158.

Mir LM. 2001. Therapeutic perspectives of in vivo cell electropermeabilization. *Bioelectrochemistry* 54:1–10.

Mir LM, Orlowski S, Belehradek J, Teissie J, Rols MP, Sersa G, Miklavcic D, Gilbert R, Heller R. 1995. Biomedical applications of electric pulses with special emphasis on antitumor electrochemotherapy. *Bioelectrochem Bioenerg* 38(1):203–207.

Mir LM, Gehl J, Sersa G, Collins CG, Garbay JR, Billard V, Geertsen PF, Rudolf Z, O'Sullivan GC, Marty M. 2006. Standard operating procedures of the electrochemotherapy: Instructions for the use of bleomycin or cisplatin administered either systemically or locally and electric pulses delivered by the Cliniporator (TM) by means of invasive or non-invasive electrodes. *Eur J Cancer Suppl* 4(11):14–25.

Montal M, Mueller P. 1972. Formation of bimolecular membranes from lipid monolayers and a study of their electrical properties. *Proc Natl Acad Sci USA* 69(12):3561–3566.

Mouneimne Y, Tosi PF, Gazitt Y, Nicolau C. 1989. Electro-insertion of xeno-glycophorin into the red blood cell membrane. *Biochem Biophys Res Commun* 159(1):34–40.

Neumann E, Rosenhec K. 1972. Permeability changes induced by electric impulses in vesicular membranes. *J Membr Biol* 10(3–4):279–290.

Okino M, Mohri H. 1987. Effects of a high-voltage electrical impulse and an anticancer drug on in vivo growing tumors. *Jpn J Cancer Res* 78(12):1319–1321.

Pavlin M, Slivnik T, Miklavcic D. 2002. Effective conductivity of cell suspensions. *IEEE Trans Biomed Eng* 49(1):77–80.

Pavlin M, Kanduser M, Rebersek M, Pucihar G, Hart FX, Magjarevic R, Miklavcic D. 2005. Effect of cell electroporation on the conductivity of a cell suspension. *Biophys J* 88(6):4378–4390.

Pavselj N, Preat V. 2005. DNA electrotransfer into the skin using a combination of one high- and one low-voltage pulse. *J Control Release* 106(3):407–415.

Pavselj N, Miklavcic D. 2008. Numerical modeling in electroporation-based biomedical applications. *Radiol Oncol* 42(3):159–168.

Petkovsek M, Nastran J, Voncina D, Zajec P, Miklavcic D, Sersa G. 2002. High voltage pulse generation. *Electron Lett* 38(14):680–682.

Pilwat G, Richter HP, Zimmermann U. 1981. Giant culture cells by electric field-induced fusion. *FEBS Lett* 133(1):169–174.

Prud'homme GJ, Glinka Y, Khan AS, Draghia-Akli R. 2006. Electroporation-enhanced nonviral gene transfer for the prevention or treatment of immunological, endocrine and neoplastic diseases. *Curr Gene Ther* 6(2):243–273.

Prutchi D, Norris M. 2005. Design of safe medical device prototypes. In: *Design and Development of Medical Electronic Instrumentation*. Hoboken, NJ: John Wiley & Sons, pp. 97–146.

Puc M, Flisar K, Rebersek S, Miklavcic D. 2001. Electroporator for in vitro cell permeabilization. *Radiol Oncol* 35:203–207.

Puc M, Corovic S, Flisar K, Petkovsek M, Nastran J, Miklavcic D. 2004. Techniques of signal generation required for electropermeabilization. Survey of electropermeabilization devices. *Bioelectrochemistry* 64(2):113–124.

Pucihar G, Kotnik T, Kanduser M, Miklavcic D. 2001. The influence of medium conductivity on electropermeabilization and survival of cells in vitro. *Bioelectrochemistry* 54(2):107–115.

Pucihar G, Mir LM, Miklavcic D. 2002. The effect of pulse repetition frequency on the uptake into electropermeabilized cells in vitro with possible applications in electrochemotherapy. *Bioelectrochemistry* 57(2):167–172.

Ramos C, Teissie J. 2000. Electrofusion: A biophysical modification of cell membrane and a mechanism in exocytosis. *Biochimie* 82(5):511–518.

Rebersek M, Faurie C, Kanduser M, Corovic S, Teissie J, Rols MP, Miklavcic D. 2007. Electroporator with automatic change of electric field direction improves gene electrotransfer in-vitro. *Biomed Eng Online* 6(25):1–11.

Rebersek M, Corovic S, Sersa G, Miklavcic D. 2008. Electrode commutation sequence for honeycomb arrangement of electrodes in electrochemotherapy and corresponding electric field distribution. *Bioelectrochemistry* 74:26–31.

Rebersek M, Kranjc M, Pavliha D, Batista-Napotnik T, Vrtacnik D, Amon S, Miklavcic D. 2009. Blumlein configuration for high-repetition-rate pulse generation of variable duration and polarity using synchronized switch control. *IEEE Trans Biomed Eng* 56(11):2642–2648.

Rols MP. 2006. Electropermeabilization, a physical method for the delivery of therapeutic molecules into cells. *Biochim Biophys Acta* 1758:423–428.

Rols MP, Teissie J. 1992. Experimental evidence for the involvement of the cytoskeleton in mammalian cell electropermeabilization. *Biochim Biophys Acta* 1111(1):45–50.

Rols MP, Teissie J. 1998. Electropermeabilization of mammalian cells to macromolecules: Control by pulse duration. *Biophys J* 75(3):1415–1423.

Rols MP, Delteil C, Serin G, Teissie J. 1994. Temperature effects on electrotransfection of mammalian cells. *Nucleic Acids Res* 22(3):540–545.

Rols MP, Tamzali Y, Teissie J. 2002. Electrochemotherapy of horses. A preliminary clinical report. *Bioelectrochemistry* 55:101–105.

Rubinsky B, Onik G, Mikus P. 2007. Irreversible electroporation: A new ablation modality—Clinical implications. *Technol Cancer Res Treat* 6(1):37–48.

Satkauskas S, Bureau MF, Puc M, Mahfoudi A, Scherman D, Miklavcic D, Mir LM. 2002. Mechanisms of in vivo DNA electrotransfer: Respective contributions of cell electropermeabilization and DNA electrophoresis. *Mol Ther* 5(2):133–140.

Schoenbach KH, Beebe SJ, Buescher ES. 2001. Intracellular effect of ultrashort electrical pulses. *Bioelectromagnetics* 22(6):440–448.

Sel D, Lebar AM, Miklavcic D. 2007. Feasibility of employing model-based optimization of pulse amplitude and electrode distance for effective tumor electropermeabilization. *IEEE Trans Biomed Eng* 54(5):773–781.

Sersa G. 2006. The state-of-the-art of electrochemotherapy before the ESOPE study; advantages and clinical uses. *Eur J Cancer Suppl* 4(11):52–59.

Sersa G, Cemazar M, Semrov D, Miklavcic D. 1996. Changing electrode orientation improves the efficacy of electrochemotherapy of solid tumors in mice. *Bioelectrochem Bioenerg* 39(1):61–66.

Teissie J, Eynard N, Vernhes MC, Benichou A, Ganeva V, Galutzov B, Cabanes PA. 2002. Recent biotechnological developments of electropulsation. A prospective review. *Bioelectrochemistry* 55:107–112.

Tokmakci M. 2006. A high-voltage pulse generation instrument for electrochemotherapy method. *J Med Syst* 30:145–151.

Trontelj K, Rebersek M, Kanduser M, Serbec VC, Sprohar M, Miklavcic D. 2008. Optimization of bulk cell electrofusion in vitro for production of human-mouse heterohybridoma cells. *Bioelectrochemistry* 74(1):124–129.

Vanbever R, Lecouturier N, Preat V. 1994. Transdermal delivery of metoprolol by electroporation. *Pharm Res* 11(11):1657–1662.

Vernhes MC, Cabanes PA, Teissie J. 1999. Chinese hamster ovary cells sensitity to localized electrical stresses. *Bioelectrochem Bioenerg* 48(1):17–25.

Wong TK, Neumann E. 1982. Electric field mediated gene transfer. *Biochem Biophys Res Commun* 107(2):584–587.

Zimmermann U. 1982. Electric field-mediated fusion and related electrical phenomena. *Biochim Biophys Acta* 694(3):227–277.

Zupanic A, Corovic S, Miklavcic D. 2008. Optimization of electrode position and electric pulse amplitude in electrochemotherapy. *Radiol Oncol* 42(2):93–101.

17

Generation of Ultrashort Pulses

Juergen F. Kolb

The response of cells and tissues to electrical stimuli is controlled by the parameters of the exposure with respect to the dielectric or electric characteristics of the target. Conversely, in pursuit of a given objective, exposure parameters need to be chosen accordingly. The possible desired outcomes can span a wide range and include tissue ablation (Davalos et al., 2004), stimulation of nerves and muscles (even including cardiac pacing) (Kroon et al., 2002; Hamid and Hayek, 2008), drug and gene delivery (Gothelf et al., 2003; Heller et al., 2005), and modification or instigation of specific cell functions (Scarlett et al., 2009). In general, the respective parameters of the electrical stimulus required to achieve different goals are the peak values or amplitude of the applied current or voltage; the way these are delivered, i.e., continuous, oscillating, or pulsed; and exposure times. For pulsed applications, exposure times need to be differentiated further, giving the duration of a single pulse and the number of pulses applied. (Often, equivalent parameters, such as duty cycle or repetition rate, are used instead.) In addition to these basic parameters, pulsed exposures are often crucially dependent on the specific shape of the stimulus pulse and then require additional parameters, such as pulse rise time, for their description.

Corresponding to the large variety of applications, a similarly large assortment of systems exists to generate and provide the required stimulus. Exposures with the aim of charging cellular membranes require high electric fields of some tens to hundreds of volts per meter for short durations that are, at most, on the order of milliseconds, but usually much shorter (Puc et al., 2004). Under these conditions, membranes can become temporarily permeable to large molecules (Weaver, 1993). With even shorter pulses—in the nanosecond range, and higher electric fields—on the order of megavolts per meter, subcellular membranes are affected and charged in a fashion similar to the plasma membrane (Schoenbach et al., 2004). As a result, transmembrane voltages across organelle membranes can likewise reach critical values (Kotnik and Miklavčič, 2006), which lead to conformational changes, such as the formation of pores (Gowrishankar et al., 2006). In addition, functional molecules embedded in membranes are exposed to high potential differences and can be expected to trigger certain responses. A prominent example of a subsequent subcellular effect is the induction of apoptosis, promising new possibilities for cancer therapies (Beebe et al., 2003).

Principally, the charging of subcellular membranes to critical values has to be accomplished before the accumulation of charges at the plasma membrane shields the cell interior from the applied electric field. Consequently, pulse durations shorter than the charging time of the outer cell membrane are sufficient. A simple electrical model, shown in Figure 17.1, allows for the comparison of the effect of different exposure parameters on cell membrane and subcellular structures. For simple subcellular geometries, analytical expressions can be derived that allow for calculating the development of the transmembrane voltage across the outer cell membrane (Equation 17.1) and organelle membranes (Equation 17.2), for example, along the

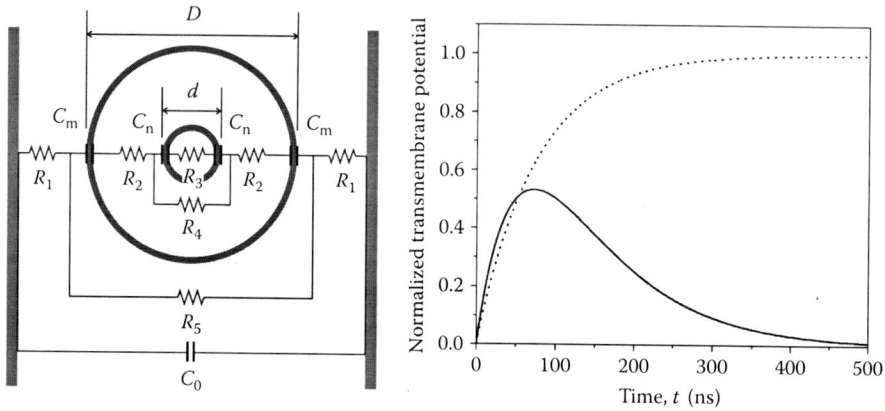

FIGURE 17.1 (Left) Circuit model for the charging of outer cell membrane and organelle membrane. Membranes themselves are assumed to be ideal insulators, while the cell suspension, cytoplasm, and organelle interior can be characterized by their resistance only. (Right) Changes in the transmembrane voltages of a spherical organelle (solid line) which is centered inside a likewise spherical cell with a diameter of 10 μm in comparison to the development of the transmembrane voltage across the cell membrane (dotted line). The organelle occupies up to 80% of the cell volume. The voltages are calculated according to Equations 17.1 and 17.2 and normalized against the steady-state value of the outer membrane. Only while the outer membrane is still charging, is the voltage across the organelle membrane different from zero.

direction of the applied electric field, E, cell diameter, D, or organelle diameter, d and the charging time constants, τ_c for the outer cell membrane and τ_o for the organelle membrane, respectively (Schoenbach et al., 2004). (The azimuthal dependency of the transmembrane voltage is described by the angle θ.)

$$\left|V_c(t)\right| = 1.5 \cdot E \cdot \frac{D}{2} \cdot \left(1 - \exp\left(\frac{-t}{\tau_c}\right)\right) \tag{17.1}$$

$$\left|V_o(t)\right| = (1.5)^2 \cdot E \cdot \frac{d}{2} \cdot \exp\left(\frac{-t}{\tau_c}\right) \cdot \frac{\tau_c}{\tau_c - \tau_o} \cdot \exp\left(\frac{t}{\tau_c} - \frac{t}{\tau_o}\right) \tag{17.2}$$

The charging time constants depend on the dielectric properties of the different membranes, and the resistivities of cell suspension, cytoplasm, and organelle interior. Assuming that these parameters are similar for all components, charging times remain only linearly dependent on the size of the cell or the organelle. For example, the charging time for a cell of 10 μm is 75 ns. It is only slightly lower (~70 ns) for a large organelle, such as the nucleus, which in many cells can occupy about 80% of the cell volume.

With typical charging time constants on the order of 100 ns (and practical charging times of less than 1 μs accordingly), intracellular manipulation is expected for applied electric field pulses of similar durations. Conversely, longer pulses will have a stronger effect on the outer cell membrane stronger.

The simple model that is summarized by Equations 17.1 and 17.2, as well as illustrated in Figure 17.1, cannot account for the intricate details of either exposure or target and hence is limited. Kotnik et al. have shown that membrane charging is correctly described by Equation 17.1 only for electric field stimulations with frequencies of less than 100 MHz, corresponding to pulse durations of 10 ns or longer (Kotnik and Miklavčič, 2000). In a subsequent analysis, Kotnik et al. have further asserted that the charging of subcellular membranes with respect to the simultaneous charge development across the outer cell membrane is only possible under very specific conditions (Kotnik and Miklavčič, 2006). In general, the membrane structure and molecular dynamics of the charging process is not addressable by analytical models. The problem can be alleviated by computer simulations (Hu et al., 2005, 2006; Tieleman, 2004; Böckmann et al., 2008), however, ultimately experimental efforts are required to verify models or provide the data to improve them (Frey et al., 2006).

From the charging characteristic, it further follows that for an effective intracellular manipulation, the rise time of the applied voltage pulse should be as short as possible. In the equivalent description in the frequency domain, this corresponds to a high frequency of the electrical stimulation, which can be capacitively coupled into the cell. For a gradual increase of the field strength, charging of outer and inner membranes will be slower, but in consequence, the subcellular space will be shielded more effectively from the external field. Moreover, charge-dependent processes will be able to follow the rise of the electric field more closely, and effects caused by a rapid change can become less pronounced. This characteristic can be seen in the Fourier spectrum of the applied pulse, which shows a less significant contribution of higher frequencies (Figure 17.2).

The discussion can be summarized to identify the requirements of the exposure conditions for the effective intracellular electromanipulation of cells. To charge internal membranes to voltages that are comparable to electroporation thresholds, applied electric fields have to be significantly higher than those usually applied to achieve electroporation. Since the charging of organelles has to be achieved before the cell interior is shielded from the external field, these electric fields have to be provided in times that are short compared with the charging time of the cell. In addition, for an optimum efficacy, the rise time of the applied field pulse should be as short as possible (but in any case, considerably shorter than the charging time constant of the cell).

A typical pulse generator for intracellular electromanipulation, therefore, needs to provide the following:

- Field strengths on the order of $10\,\text{MV/m}$
- Pulse durations of less than $1\,\mu\text{s}$
- Pulse rise times on the order of $10\,\text{ns}$

In general, these conditions cannot be provided by conventional electronic circuits, since most components are either too slow or cannot handle high enough voltages to generate the necessary electric fields in a relevant volume. However, the technological challenge can be addressed by methods known as pulsed power engineering. This way, simple and robust systems for the generation of ultrashort high voltage pulses can be provided.

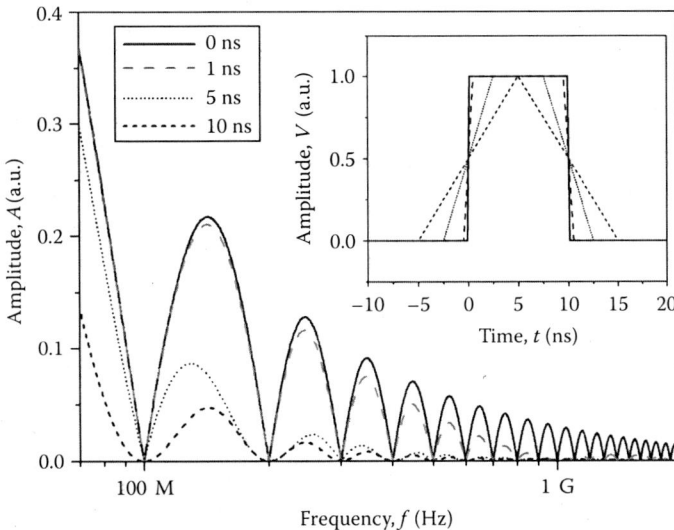

FIGURE 17.2 Fourier spectra for the frequency range above 100 MHz of trapezoidal pulses with different rise times (0, 1, 5, 10 ns) and a nominal 10 ns duration (determined as width at half the maximum amplitude, see inset). Higher frequency contributions become less significant with increasing pulse rise times.

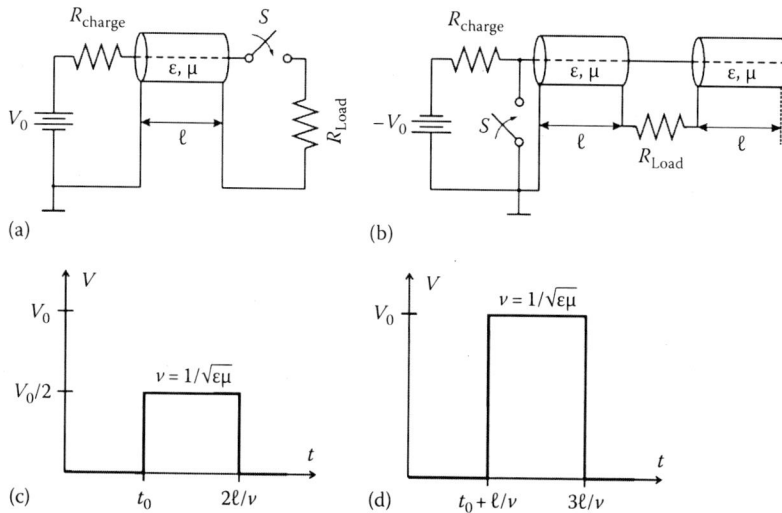

FIGURE 17.3 (a) Concept of a transmission line pulse generator. When the switch, S, is closed, an ideally rectangular pulse is applied across a load with a resistance that matches the impedance of the transmission line (shown in panel c). In this case, the pulse duration is determined by the length of the transmission line and the pulse amplitude equals half the charging voltage. (b) Concept of a Blumlein line pulse generator. When the switch, S, is closed, an ideally rectangular pulse is applied across a load with a resistance that matches the impedance of the transmission line (shown in panel d). In this case, the pulse duration is determined by the length of the transmission lines and the pulse amplitude equals the charging voltage, however, its polarity is reversed.

Trapezoidal (or ideally, rectangular) pulses, in particular, can be generated by a transmission line approach (Smith, 2002). The principle is illustrated in Figure 17.3a and c. With this system, the discharge characteristic is mainly determined by the relation of the cable impedance to the load resistance. If both values are the same, a single rectangular pulse with an amplitude equivalent to half the voltage to which the system is charged is provided to the load.

A low resistance of only a few ohms for cells in suspensions, and of several tens of ohms for tissue samples, is typical for biological loads. These values are either matched by the cable impedance directly (a common cable impedance is 50 Ω), or several identical transmission lines can be connected in parallel to reduce the overall value. The impedance of the pulse generator is then given by the value for an individual transmission line divided by the number of transmission lines. An example of a transmission line pulse generator of this kind is shown in Figure 17.4.

The pulse amplitude, in principle, is only limited by the dielectric strength and wall thickness of the dielectric that is used in the transmission line. Many commercial high voltage cables (with practical diameters of less than 20 mm) are specified with voltage-withstand values on the order of 10 kV (vrms), e.g., RG217 and RG213. In our experience, they can withstand direct current (DC) voltages that are 3–4 times higher, and the real challenge becomes the design of the connections between the load, the charging resistor, and the switch. Alternatively, transmission lines can also be custom-made. The geometry of striplines, for example, can easily be adjusted with respect to impedance and charging voltage. However, for practical reasons, this approach is only recommended for shorter pulse durations.

The charging resistor insulates the charging power supply from the pulse generator during the generation of a pulse. With a value much larger than the impedance of the generator, the charging time of the system is at least several orders of magnitude higher than the pulse duration. This also effectively limits the repetition rate of the system.

In the moment the switch, S, is closed (see Figure 17.3), electromagnetic waves are propagated along the transmission line, which are superimposed with their reflections from either end at the load. If the

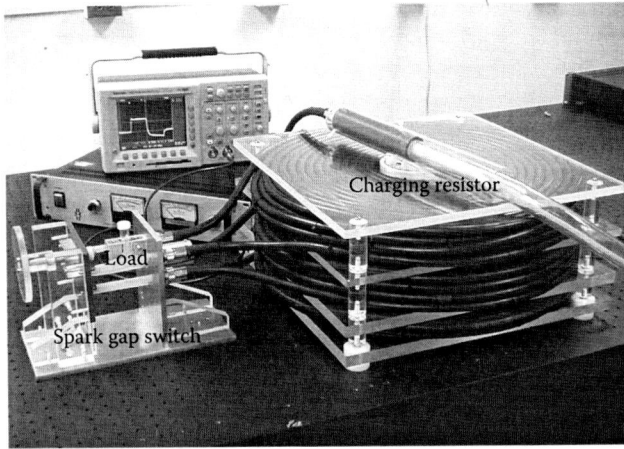

FIGURE 17.4 Transmission line pulse generator for the exposure of cell suspensions to rectangular pulses of 300 ns duration and voltages of 40 kV. Five coaxial cables (RG217) are connected in parallel, resulting in a pulse generator impedance of 10 Ω. (The oscilloscope shows the trace for a load resistance slightly lower than 10 Ω.)

waves experience the load as equal to the impedance of the line, they will be completely consumed by the load. Otherwise, the difference in impedances will lead to reflections, which will travel back and forth along the line and lead to a subsequent voltage drop across the load. The process repeats and leads to additional voltage pulses that are applied to the load with exponentially decreasing amplitudes, until the energy that was stored in the line is completely dissipated. The amplitudes of the first and subsequent pulses are thereby determined by the ratio of load resistance and pulse generator impedance. When the load impedance is higher than the pulse generator impedance, the amplitude of the first pulse in the train is larger than the value expected for the matched case. When the load impedance is lower than the pulse generator impedance, the amplitude of the first pulse is lower instead and the polarity of subsequent pulses is alternating. (For the limit of the load resistance approaching infinity, the discharge voltage across the load approaches the discharge characteristic of capacitive discharge.) Examples are shown in Figure 17.5.

The duration of an ideal pulse (or an individual pulse in a train) is determined by another characteristic of the transmission line dielectric—the permittivity. The square root of the product of permittivity and permeability of a medium determines the propagation velocity of electromagnetic waves in this medium.

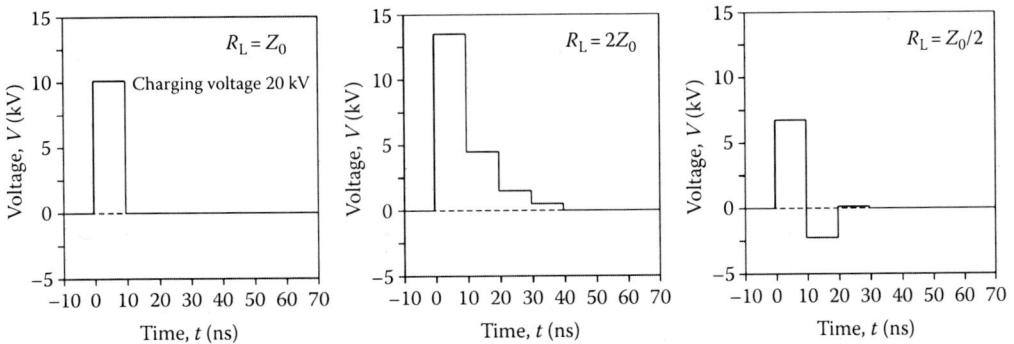

FIGURE 17.5 Pulse delivery characteristic for a transmission line pulse generator which is designed to provide a pulse with 10 ns duration to a load with a resistance equal to the pulse generator impedance Z_0. If the load resistance is larger (e.g., $R_L = 2Z_0$) or smaller (e.g., $R_L = Z_0/2$) than Z_0, a train of pulses with exponentially decreasing amplitudes is applied instead.

Common high voltage insulators, such as polyethylene and polytetrafluoroethylene, have a relative permeability of 1 and relative permittivities between 2 and 4. Consequently, the propagation velocity is approximately equal to two-thirds the speed of light in a vacuum. Hence, as a rule of thumb, the transmission line wave travels 1 m in 5 ns. A detailed analysis of the pulse generation with a transmission line pulse generator, according to the concept shown in Figure 17.3a, shows that the pulse is generated from the wave, hitting the load immediately after the switch closes, and a delayed second wave, which started in the opposite direction and reflected at the charging resistor, eventually arrives at the load (Smith, 2002). Consequently, the pulse duration for such a transmission line pulse generator can be calculated from its length, ℓ, as

$$\tau_{\text{Trans}} = 2\ell \cdot 5 \text{ ns/m} \qquad (17.3)$$

where the factor of 2 accounts for the sum of the voltage drops from direct and reflected waves.

The constraints imposed by the permittivity and the DC hold-off voltage of the dielectric also determine the physical setup of the pulse generator. If 'long' pulses are required for delivery to a low-impedance load, both the size and the weight of the system become problematic. (For example, the pulse generator shown in Figure 17.4 providing a pulse of 300 ns into a 10 Ω load weighs approximately 70 kg with an overall cable length of 150 m for all five cables in parallel.) More compact, and lighter, systems can be constructed by replacing the dielectric with materials of higher permittivity. One such alternative is the use of ceramics with high permittivities. However, these materials are neither readily available in large quantities, nor easy to machine. Another appealing solution is the use of polar liquids, such as water. With a relative permittivity close to 81, the length of the transmission line can be reduced by a factor of nine. Since polar liquids, in general, also have high DC conductivities, transmission line pulse generators require pulsed charging systems for their operation. However, this disadvantage is compensated for by the increased failure tolerance of the system. Whereas a high voltage breakdown in the solid dielectric destroys the system, the same phenomenon is not a problem with the self-restoring liquid dielectric. Accordingly, transmission lines using water as a dielectric, have been investigated as pulse generators for the exposure of biological samples (Sun et al., 2007). When the switch is integrated into the structure of the transmission line, the relatively high dielectric strength of water additionally offers the possibility of generating pulses of very short rise times for short charging pulses (Sun et al., 2007).

Next to the connection of the load to the pulse generator, the rise time of the pulse is primarily determined by the closing time of the switch that initiates the pulse. Ideally, the switch can hold off several kilovolts in the open state but drops to a conducting resistance close to zero instantaneously. For lower voltages of about 1 kV, this can be approximated by solid-state switches, such as MOSFETs or insulated gate bipolar transistors (IGBTs) that are commercially available. These can therefore successfully be employed in systems such as those used on microscopes that only require the generation of high electric fields across short electrode distances. In addition, the devices are easily synchronized with other diagnostic systems by conventional, low-voltage trigger signals (Frey et al., 2006).

Switches with a comparable performance are not readily available for higher voltages of several tens of kilovolts. Typical switches for this regime are thyratrons or stacked high-voltage IGBTs, which have commutation times of 30 ns or more. (The switching time is longer when the voltage that has to be switched is higher.)

A means of switching high voltages with closing times of only a few nanoseconds is the spark-gap switch. In its simplest configuration, the switch consists of two electrodes that are separated by a given distance, d, in a gas-filled volume. When the applied voltage exceeds the hold-off voltage of the gaseous dielectric, a plasma channel forms and closes the connection. The process is intrinsically fast and can provide particularly fast rise times when the switch can be integrated into the transmission line without introducing impedance discontinuities and when the electrode separation can be kept short. The concept is shown schematically in Figure 17.6.

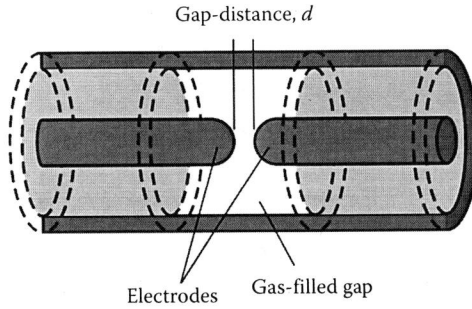

FIGURE 17.6 Configuration for a spark gap switch integrated into a coaxial line.

Hemispherical shapes (and other optimized curved surfaces) are often used for electrode profiles. The curvature radius and the size of the electrodes are usually larger than or similar to the gap distance. This way, the geometry gives rise to a slight field enhancement along the central axis of the spark gap, forcing the spark channel to develop along this path. The hold-off voltage of the gap is a function of gas pressure, p, and gap distance, d. The relation has been investigated for various gases, in particular for pressures in the range of up to fractions of 1 atm, and is known as Paschen's law (Raizer, 1997). Although the mechanistic model and the associated assumptions that are underlying this relation are only valid for lower pressures of no more than several hundred Torr, the relation is often extrapolated and used at higher pressures as well. Typically the actual breakdown voltage can be estimated within a margin of a few percent. For air, Pai and Zhang have presented the following formula (17.4) to calculate the breakdown voltage, V_{bd}, in kV when the pressure, p, and the gap distance, d, are entered in bar or cm, respectively (Pai and Zhang, 1995):

$$V_{bd} = 6.72\sqrt{pd} + 24.36\,pd \qquad (17.4)$$

The dependence on pressure and distance permits the adjustment of the breakdown voltage and therefore, the pulse amplitude, by changing either the gas pressure or the gap distance. The latter allows for the operation of "open" spark gaps, i.e., with ambient atmospheric pressure air. Rise times of less than 10 ns are easily achieved. To provide faster rise times, spark gaps can be operated at a fixed short gap on the order of 1 mm (thereby reducing the inductive circuit contribution of the spark channel) and can be adjusted by increasing the pressure. Rise times on the order of 1 ns have been demonstrated (Kolb et al., 2006). Even shorter gap distances and faster rise times are possible by replacing the gas with a liquid dielectric (Schoenbach et al., 2008).

Most technologies that require the application of high voltage pulses across extended volumes are only interested in the post-exposure biological response. In this case, synchronization with other equipment is not required, and the spark gap can be operated without a trigger signal in self-breakdown mode, i.e., a discharge will occur and the switch will close, whenever the transmission line capacitance has been charged to the breakdown voltage of the gap. How quickly the system can be recharged will further determine the repetition rate for multiple-pulse applications. The main constraints on this recharging are the power supply output current and the charging resistor, but the rate of the system can be increased by using a charging voltage higher than the breakdown voltage of the gap.

As versatile as the transmission line pulse generator concept is, it has two major disadvantages. One is the required placement of the switch close to the load. The other is that a voltage of twice the desired pulse amplitude is necessary to charge the system. Both issues are overcome by using a Blumlein line setup of the transmission line pulse generator, as shown in Figure 17.3b. In this configuration, the load is placed between two identical transmission lines. The duration of the pulse is again determined by the length of each individual line, according to Equation 17.3. As a result of this serial connection of lines, the impedance of the pulse generator is equal to twice the impedance of an individual line. The charging circuit and switch can be connected on either end of the line. When the load is matched to the

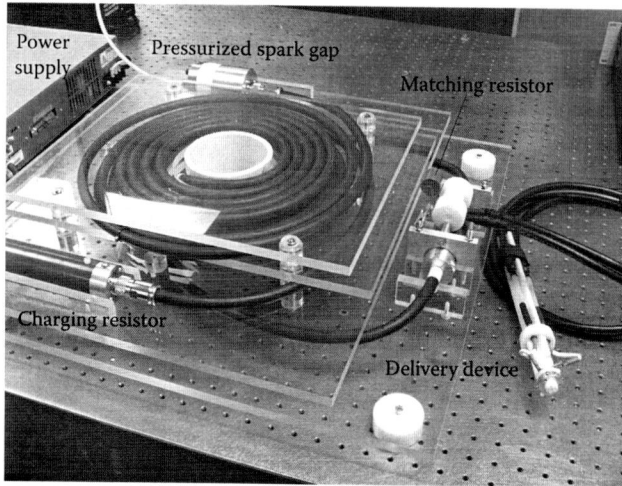

FIGURE 17.7 60 ns Blumlein line pulse generator for the exposure of tissue samples. To match the impedance of the load to the 100 Ω impedance of the system, a matching resistor is connected in parallel to the load between the output terminals of the pulse generator. A rise time that is considerably faster than the pulse duration is achieved by a pressurized spark gap switch with gap of 1 mm.

impedance of the pulse generator (twice the value of an individual line), a rectangular pulse is generated as the potential difference across the load from the superposition of the direct and reflected waves that are propagating along the lines (Smith, 2002). The amplitude of this pulse is equal to the charging voltage, but of the opposite polarity.

In the same fashion as with the simple transmission line pulse generators, the overall impedance of this system can be reduced by connecting additional lines in parallel on either side of the load. For very low impedance loads, this increases the complexity of the system considerably. However, the inherent advantage of the Blumlein line pulse generator is apparent for loads with an impedance equal to, or higher than, 100 Ω, i.e., twice the impedance of readily available coaxial high voltage cables. Similar to the transmission line pulse generator, the load can be matched to the system by using a resistor in parallel to the load. As a result, this method is particularly useful for the exposure of tissues, which generally have higher impedance values. An example of such a system, designed for the treatment of animals, is given in Figure 17.7.

A mismatch of the load with the system again leads to a train of pulses instead of a single pulse. However, due to the nature of the pulse generation, for the Blumlein line, the delay between individual pulses has a duration equal to the pulse duration (Figure 17.8). Also in this case, a load resistance approaching infinity will result in the amplitude of the first pulse being twice the charging voltage with exponentially decreasing amplitudes of the trailing pulses.

Both the simple transmission line configuration and Blumlein line configuration can only provide pulses with a fixed duration, which is determined by the length of the individual transmission lines. However, if the waves that are launched into the transmission lines on either side of the load can be controlled independently, the concept of the Blumlein line pulse generator can be modified to generate pulses of arbitrary duration. This is achieved by having a closing switch on either end of the Blumlein line setup (de Angelis et al., 2008; Reberšek et al., 2009). With individual trigger signals supplied to each switch, the pulse across the load is determined by the delay between the trigger signals. (With no delay, the pulse duration is maximal and corresponding to the length of the line.) Since the configuration requires the accurate timing of the switches, it has so far only been accomplished with fast MOSFET switches, which are currently available only for voltages not exceeding 1 kV.

The use of continuous transmission lines, such as commercially available coaxial cables, provides an efficient way to generate pulses that are almost rectangular, i.e., with negligible rise times. For longer

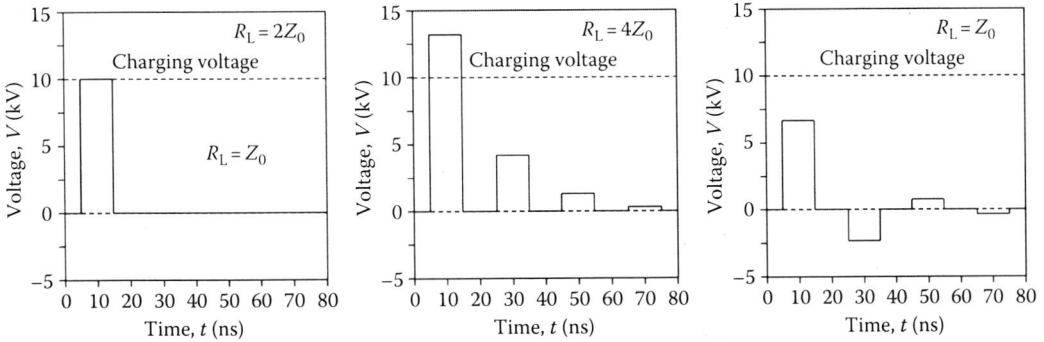

FIGURE 17.8 Pulse delivery characteristic for a Blumlein line pulse generator which is designed to provide a pulse with 10 ns duration to a load with a resistance equal to the pulse generator impedance $2Z_0$ (Z_0: characteristic impedance of an individual transmission line). If the load resistance is larger (e.g., $R_L = 4Z_0$) or smaller (e.g., $R_L = Z_0$) than $2Z_0$, a train of pulses with exponentially decreasing amplitudes is applied instead. (From Kolb, J. F., Kono, S. and Schoenbach, K. H., *Bioelectromagnetics*, 27, 172, 2006. With permission.)

pulse durations, however, the physical setup becomes laborious, mainly because of the cable length required. Since longer pulses are generally employed to achieve an enhanced effect on the cell membrane, they often do not require extremely fast rise times. Under these relaxed exposure conditions, pulse generators can be designed with lumped circuit elements (capacitors and inductors) that mimic the distributed characteristics of a transmission line. An example for such a pulse forming network in a Blumlein line configuration is shown in Figure 17.9. With this approach, pulse generators of moderate size can be designed that will provide pulses of several hundred nanoseconds, or even microseconds. The impedance of the pulse generator is determined by the values of the capacitors and inductors used, the flatness of the pulse, and the number of stages. Pulse generators of this type have recently been used very successfully in the treatment of solid cancers in animal models (Nuccitelli et al., 2006).

All the pulse generators presented so far are designed to provide almost rectangular pulses with rise times as short as possible. These conditions are especially favored for research systems with the goal of an unambiguous evaluation of exposure parameters, in particular, pulse duration and amplitude. Often, other parameters, such as cell population and environmental conditions (e.g., buffer solutions), are also carefully controlled. However, as the previous example of a pulse forming network already shows, these

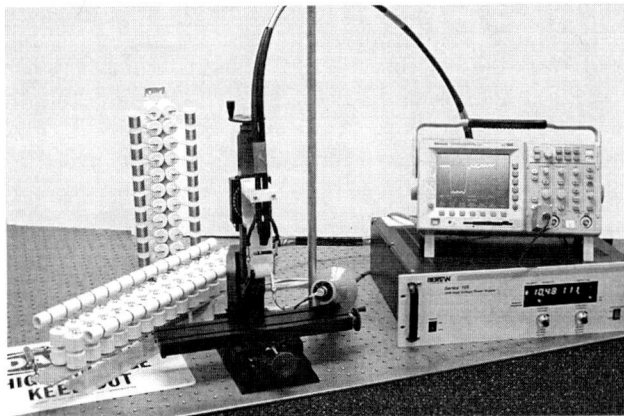

FIGURE 17.9 Blumlein pulse forming network used for the treatment of solid tumors in animals. The picture shows two models of 100 Ω impedance and of 300- or 600 ns duration, respectively.

stringent conditions are sometimes relaxed, mostly in *in vivo* studies. In this case, the practical considerations are more important and the exposure does not allow for the absolute control of many variables, anyway. Moreover, the observed results are mostly obtained after days and weeks and are consequently describing a synergistic and systemic response.

Accordingly, several concepts, other than the transmission line approach presented here, have been suggested for the generation of ultrashort pulsed electric field exposures, i.e., pulses in the nanosecond range or shorter. In general, they each give up some advantages inherent to the transmission line pulse generators (e.g., rectangular pulse shapes, low impedance), while they are superior in some other aspects (e.g., pulse amplitude, repetition rate). A general overview of short pulse generator technology was recently given by Mankowski and Kristiansen (2000). Often, several concepts, including transmission lines or pulse forming networks, are combined in subsequent stages in an actual pulse generator. Unfortunately, many of these pulse generator circuits have not yet been used in experiments with biological samples, although there are some notable exceptions. Kuthi et al. have described a nanosecond pulse generator for microscope-based experiments using fast recovery diodes (Kuthi et al., 2005). The generator provides triangular pulses with a width of 3.5 ns (full width half maximum) and maximum amplitudes of 1.2 kV into a 50 Ω load at repetition rates of up to 100 kHz. Another method that is often used for the generation of short high-voltage pulses is magnetic pulse compression. By combining LC modulating stages, magnetic pulse compression, and diode opening switch output stages, Tang et al. generated pulses of 20 and 5 ns and corresponding amplitudes of 4.5 and 7.5 kV, respectively, that were applied to a 10 Ω cell suspension with a repetition rate of 20 Hz (Tang et al., 2007). One more pulse generator concept, which allows for the application of extremely high voltages, is the Marx bank circuit. Pulses with a duration of 1.4 μs at total charging voltages of 350 kV have been delivered at 20 Hz to sugar beets in an industrial-scale facility by Sack et al. (2005). Combining the Marx bank with peaking switches and tailcut switches, the output pulse of a 4-stage Marx bank was shaped by Camp et al. into a 150 ps pulse of 60 kV amplitude, which was delivered to a 50 Ω cell suspension sample (Camp et al., 2008). With the possibility of applying even shorter pulses of subnanosecond duration and significant amplitude, new ways of manipulating cells become available. For this type of exposure, cell responses, in general, depend primarily on the dielectric mechanisms and dielectric characteristics of cell constituents. Resistive charging and shielding effects, by comparison, develop on much longer time scales. It is foreseeable that eventually this new technology will again lead to novel medical applications in diagnostic and therapies just as electroporation and nanosecond pulsed electric field exposures have done.

References

Beebe SJ, Fox PM, Rec LJ, Willis LK, Schoenbach KH. 2003. Nanosecond, high-intensity pulsed electric fields induce apoptosis in human cells, *FASEB J.* 17:1493–1495 (doi: 10.1096/fj.02-0859fje, www.fasebj.org/cgi/content/full/17/11/1493).

Böckmann RA, de Groot BL, Kakorin S, Neumann E, Grubmüller H. 2008. Kinetics, statistics, and energetics of lipid membrane electroporation studied by molecular dynamics simulations, *Biophys. J.* 95:1837–1850.

Camp JT, Xiao S, Schoenbach KH. 2008. Development of a high voltage, 150 ps pulse generator for biological applications, *Proceedings of 28th IEEE International Power Modulator and High Voltage Conference*, Las Vegas, NV, May 27–31, 2008, pp. 338–341.

Davalos RV, Mir LM, Rubinsky B. 2004. Ablation with irreversible electroporation, *Ann. Biomed. Eng.* 32:223–231.

De Angelis A, Kolb JF, Zeni L, Schoenbach KH. 2008. Kilovolt Blumlein pulse generator with variable pulse duration and polarity, *Rev. Sci. Instrum.* 79:044301.

Frey W, White JA, Price RO, Blackmore PF, Joshi RP, Nuccitelli R, Beebe SJ, Schoenbach KH, Kolb JF. 2006. Plasma membrane voltage changes during nanosecond pulsed electric field exposure, *Biophys. J.* 90:3608–3615.

Gothelf A, Mir LM, Gehl J. 2003. Electrochemotherapy: Results of cancer treatment using enhanced delivery of bleomycin by electroporation, *Cancer Treat. Rev.* 29:371–387.

Gowrishankar TR, Esser AT, Vasilkoski Z, Smith KC, Weaver JC. 2006. Microdosimetry for conventional and supra-electroporation in cells with organelles, *Biochem. Biophys. Res. Commun.* 341:1266–1276.

Hamid S, Hayek R. 2008. Role of electrical stimulation for rehabilitation and regeneration after spinal cord injury: An overview, *Eur. Spine J.* 17:1256–1269.

Heller LC, Ugen K, Heller R. 2005. Electroporation for targeted gene transfer, *Expert Opin. Drug Deliv.* 2:255–268.

Hu Q, Sridhara V, Joshi RP, Kolb JF, Schoenbach KH. 2006. Molecular dynamics analysis of high electric pulse effects on bilayer membranes containing DPPC and DPPS, *IEEE Trans. Plasma Sci.* 34:1405–1411.

Hu Q, Joshi RP, Schoenbach KH. 2005. Simulations of nanopore formation and phophatidylserine externalization in lipid membranes subjected to a high-intensity, ultrashort electric pulse, *Phys. Rev. E* 72:031902.

Kolb JF, Kono S, Schoenbach KH. 2006. Nanosecond pulsed electric field generators for the study of subcellular effects, *Bioelectromagnetics* 27:172–187.

Kotnik T, Miklavčič D. 2006. Theoretical evaluation of voltage inducement on internal membranes of biological cells exposed to electric fields, *Biophys. J.* 90:480–491.

Kotnik T, Miklavčič D. 2000. Second-order model of membrane electric field induced by alternating external electric fields, *IEEE Trans. Biomed. Eng.* 47:1074–1081.

Kroon JR, van der Lee JH, IJzerman MJ, Lankhorst GJ. 2002. Therapeutic electrical stimulation to improve motor control and functional abilities of the upper extremity after stroke: A systematic review, *Clin. Rehabil.* 16:350–360.

Kuthi A, Gabrielsson P, Behrend MR, Vernier PT, Gundersen MA. 2005. Nanosecond pulse generator using fast recovery diodes for cell electromanipulation, *IEEE Trans. Plasma Sci.* 33:1192–1197.

Mankowski J, Kristiansen M. 2000. A review of short pulse generator technology, *IEEE Trans. Plasma Sci.* 28:102–108.

Nuccitelli R, Pliquett U, Chen X, Ford W, Swanson RJ, Beebe SJ, Kolb JF, Schoenbach KH. 2006. Nanosecond pulsed electric fields cause melanomas to self-destruct, *Biochem. Biophys. Res. Commun.* 343:351–360.

Pai ST, Zhang Q. 1995. *Introduction to High Power Pulse Technology, Advanced Series in Electrical and Computer Engineering*, Vol. 10. Singapore: World Scientific Publishing Co. Pte. Ltd.

Puc M, Corovic S, Flisar K, Pekovšek M, Nastram J, Miklavčič D. 2004. Techniques of signal generation required for electropermeabilization. Survey of electropermeabilization devices. *Bioelectrochemistry* 64:113–124.

Raizer YP. 1997. *Gas Discharge Physics*. Berlin, Germany: Springer Verlag.

Reberšek M, Kranjc M, Pavliha D, Batista-Napotnik T, Vrtacnik D, Amon S, Miklavčič D. 2009. Blumlein configuration for high-repetition-rate pulse generation of variable duration and polarity using synchronized switch control, *IEEE Trans. Biomed. Eng.* 56:2642–2648.

Sack M, Schultheiss C, Bluhm H. 2005. Triggered Marx generators for the industrial-scale electroporation of sugar beets, *IEEE Trans. Ind. Appl.* 41:707–714.

Scarlett SS, White JA, Blackmore PF, Schoenbach KH, Kolb JF. 2009. Regulation of intracellular calcium concentration by nanosecond pulsed electric fields, *Biochim. Biophys. Acta* 1788:1168–1175.

Schoenbach K, Kolb J, Xiao S, Katsuki S, Minamitani Y, Joshi R. 2008. Electrical breakdown of water in microgaps, *Plasma Sources Sci. Technol.* 17:024010.

Schoenbach KH, Joshi RP, Kolb JF, Chen N, Stacey M, Blackmore PF, Buescher ES, Beebe SJ. 2004. Ultrashort electrical pulses open a new gateway into biological cells, *Proc. IEEE* 92:1122–1137.

Smith PW. 2002. *Transient Electronics, Pulsed Circuit Technology*. Chichester, England: John Wiley & Sons, Ltd.

Sun Y, Xiao S, White JA, Kolb JF, Stacey M, Schoenbach KH. 2007. Compact, nanosecond, high repetition rate pulse generator for bioelectric studies, *IEEE Trans. Dielectr. Electr. Insul.* 14:863–870.

Tang T, Wang F, Kuthi A, Gundersen MA. 2007. Diode opening switch based nanosecond high voltage pulse generators for biological and medical applications, *IEEE Trans. Dielectr. Electr. Insul.* 14:878–883.

Tieleman PD. 2004. The molecular basis of electroporation, *BMC Biochem.* 5:10 (doi: 10.1186/1471-2091-5-10, www.biomedcentral.com/1471-2091/5/10).

Weaver JC. 1993. Electroporation: A general phenomenon for manipulating cells and tissues, *J. Cell. Biochem.* 51:426–435.

18

Nanosecond Pulsed Electric Field Delivery to Biological Samples: Difficulties and Potential Solutions

Aude Silve

Julien Villemejane

Vanessa Joubert

Antoni Ivorra

Lluis M. Mir

Tip to the Reader

Non-bold italic symbols are employed here to denote scalars or time variables (e.g., a voltage signal in the time domain will be noted as $V(t)$ or V). Bold italic symbols are employed for complex variables (e.g., the Fourier transform of a voltage signal $V(t)$ will be noted as $V(\omega)$ or $\boldsymbol{V}$). On the other hand, bold non-italic symbols are employed for vectors such as the electric field, $\mathbf{E}$, and the current density, $\mathbf{J}$.

18.1 Introduction

Electroporation using microsecond- or millisecond-long electric pulses is a well-known technology frequently used on the bench (e.g., bacteria and eukaryotic cell transformation in research laboratories) as well as at the bed (e.g., tumor treatment by electrochemotherapy, nowadays routinely used in Europe). NsPEF is an emerging technology that encompasses a lot of promises because it opens new opportunities for cell "electromanipulation." Indeed, with "long" (μs and ms) pulses, changes can only be generated at the level of the cell membrane, while "short" (ns) pulses can also provoke modifications at the level of the internal membranes. So, coming from the experience of using "long" pulses, several groups have started to use "short" pulses on living cells. It is also expected that, in the future, when equipment will become easier to use and more accessible, many other groups will also apply this technology.

In this chapter, we discuss the particular features of the delivery systems of nsPEFs that must be taken into account when cells or tissues are exposed to the pulses. While the transition from the use of millisecond pulses to the delivery of microsecond pulses (and vice versa) is almost straightforward, the same cannot be said for the transition from the micro-/milli-second pulses to nsPEFs. The purpose of this chapter is to provide some advice to future users of this technology on how to apply it correctly. More precisely, the ultimate goal of the chapter is to give a warning message regarding the use of nsPEF technology: usage of this technology is not straightforward; the experimental setup has to be designed carefully and some nontrivial considerations must be taken into account to ensure experimental repeatability and reproducibility.

18.2 Taking Propagation into Account When Applying nsPEFs

Generally, in benchtop-sized electrical systems involving signals with frequencies below 1 MHz, or with durations longer than 1 μs, propagation-related phenomena through the electrical connections can be considered negligible and it can be assumed that all points in a good electrical conductor reach the same voltage simultaneously. This is, for example, the case of conventional electroporation in which square pulses with a duration longer than 1 μs (typically 100 μs or longer) are applied. However, for nsPEFs the duration of the signals may be shorter than the time of propagation through the connections and, as a result, propagation-related phenomena will be very relevant. Actually, as it will be shown here, even in the case of the long pulses employed in conventional electroporation, it is possible to notice the consequences of propagation-related phenomena.

In this section, we give a brief introduction to propagation-related phenomena and we point out how these phenomena can be relevant when performing experiments involving nsPEFs. More comprehensive explanations for propagation-related phenomena and concepts can be found in the appropriate textbooks.

18.2.1 When Propagation Is Relevant

A pair of long conductors separated by a dielectric forms a *transmission line*, which is a structure that allows the transmission of voltage signals between two points. In a lot of cases, such transmission appears to be instantaneous (e.g., switching on lights), but in others the voltage *wave* requires a significant amount of time to propagate from one point to the other one (e.g., wired electrical telegraphy). When transmission is considered to be instantaneous, each conductor of the transmission line is simply modeled as an ideal wire, which implies that all points along that wire have the same voltage value at the same time and that simple circuit theory is applicable (i.e., Kirchhoff's circuit laws). On the other hand, when propagation is relevant, different phenomena associated with propagation are manifested (e.g., delay, reflections, and stationary waves) and the ideal wire model is no longer valid.

Determining when propagation will be relevant or not is not always obvious. A continuous sinusoidal signal applied to one of the extremes of the transmission line will cause a sinusoidal pattern of voltages along the transmission line length as the signal propagates. The distance between the crests of this spatial sinusoidal pattern, measured in meters, is known as *wavelength*, λ, and it depends on the frequency of the excitation signal, f, expressed in hertz (Hz) and the propagation speed of the transmission line for that specific frequency, v, expressed in m/s:

$$\lambda = \frac{v}{f}$$

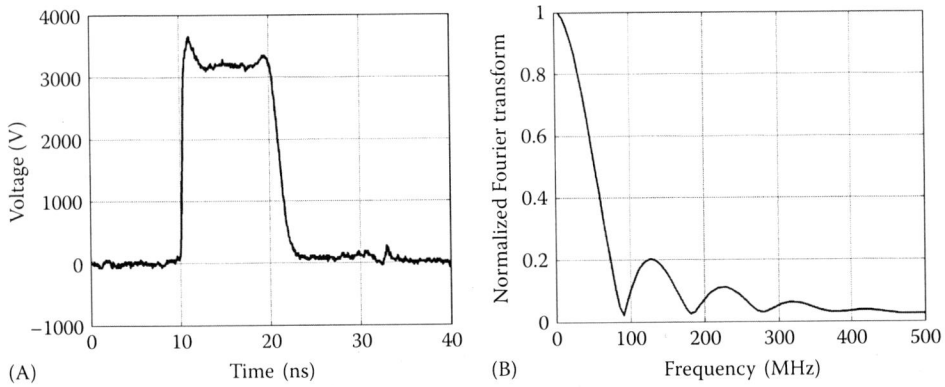

FIGURE 18.1 Example of a 10 ns pulse delivered by a high-voltage pulse generator. (A) Voltage signal measured at a 50 Ω load, and (B): normalized spectrum of the voltage signal in (A) computed by performing the Fourier transform.

Following this definition, in most textbooks it is stated that propagation phenomena have to be considered when the transmission line length, L, is larger than the wavelength of the signal in that particular line. This rule is not applicable for the case of pulse signals but, with some limitations, it can be transposed for those cases by saying: propagation-related phenomena will have to be considered when the transmission line is longer than the wavelength corresponding to the maximum frequency at which the power spectral density of the pulse signal is still significant. For instance, in Figure 18.1, it can be observed that the power spectral density of a 10 ns quasi-square pulse is still significant up to frequencies above 200 MHz. For this 200 MHz frequency, since the wavelength will be in the order of 1 m ($v \sim 2 \times 10^8$ m/s), according to the transposed rule it can be said that propagation phenomena have to be considered in benchtop-sized systems. The frequency content of a pulse signal will depend both on its duration (observe that in Figure 18.1 the first notch corresponds to the inverse of pulse duration) and on its rising time: the steeper the voltage rise is, the broader the frequency content will be.

Intuitively, it could be thought that an alternate, and simpler, rule for deciding when propagation will be relevant in the case of pulse signals could be the following: propagation has to be taken into account when the propagation time of the pulse through the length of the transmission line is longer, or in the same order, than the duration of the pulse (i.e., propagation is relevant when $(L/v) > T$, where T is the duration of the pulse). However, as it will be shown now, this rule is not valid.

As it has been mentioned, in conventional electroporation, propagation-related phenomena are generally neglected. This is not surprising when it is taken into consideration that cables between the electrodes and the generator are shorter than 1 or 2 m and that propagation speeds are in the order of 2×10^8 m/s; it would take 10 ns for an electrical signal to travel along 2 m of cable, which is negligible compared with the total duration of the pulse (i.e., $(L/v) \ll T$).

Nevertheless, it must be noted that the impact of propagation-related phenomena on pulse signals with a duration larger than 1 µs can actually be noticed if careful observations are performed. For instance, Figure 18.2 shows the result of a simulation in which a 100 µs pulse is delivered to a 10 Ω resistance through a cable with a propagation delay of 7.5 ns (equivalent to 1.5 m). At first glance, it can be observed that the voltage signals at the generator (V_{in}) and at the load (V_{out}) are apparently identical. However, if the attention is focused on the pulse edges, then it can be noticed that both signals are not strictly identical; V_{out} appears to be delayed and "softened." This distortion is not only due to mere propagation but also to a propagation-related phenomena that is later described: signal reflections occur at the connections between the three different elements of the system (generator, cable, and resistance).

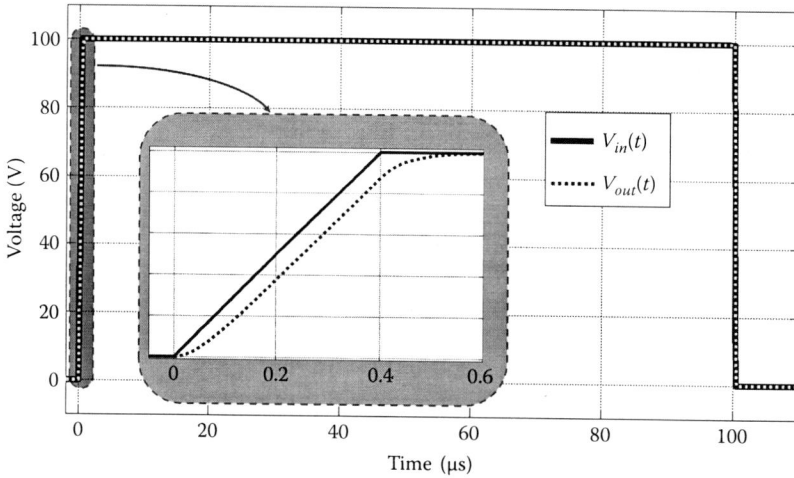

FIGURE 18.2 Simulated example of the influence of propagation related phenomena on the shape of a 100 μs pulse. This is a SPICE simulation (LTspice IV by Linear Technology Corp., Milpitas, California) in which the voltage at a 10 Ω load (V_{out}) was monitored when pulses (V_{in}) from a 0 Ω output impedance generator were delivered through a transmission line with propagation delay of 7.5 ns (equivalent to 1.5 m) and a characteristic impedance of 50 Ω (these concepts are later explained in Section 18.2.2).

Therefore, propagation-related phenomena actually do have an impact on the signals typically used in conventional electroporation. However, this impact is only significant on the pulse edges and, in terms of biological consequences, it has been demonstrated that the shape of those edges is not relevant in conventional electroporation (Kotnik et al. 2001a).

18.2.2 Behavior of a Transmission Line

18.2.2.1 Propagation Phenomenon

This section presents a classical electrical model for transmission lines (e.g., coaxial cables) that reveals the propagation phenomenon.

For the whole length of the transmission line, the model consists of a succession of small electric sub-circuits each of which has a infinitesimal length, dx, very small compared with the wavelength. Because of their short length, in each sub-circuit it is possible to ignore the propagation phenomenon and the simple circuit theory analysis can be applied. Each sub-circuit (Figure 18.3) is modeled by a series inductance ($L = l \cdot dx$ [H], l is the linear inductance [H/m]) and a parallel capacitance ($C = c \cdot dx$ [F], c is the linear capacitance [F/m]). In addition, more complete models include resistive elements (a series resistance with L and a parallel conductance with C) that are responsible for an attenuation of propagated signals. Nevertheless, signal attenuation, and hence those dissipative elements, are usually neglected if the transmission line length does not exceed the wavelength by several orders of magnitude, which is the case in nsPEF experimental setups.

Then, despite the fact that each sub-circuit does not model the propagation phenomenon, it is possible to demonstrate by means of differential equations that voltage signals do propagate without distortion along the transmission line length (x) with a speed equal to

$$v = \frac{1}{\sqrt{lc}} \quad [\text{m/s}]$$

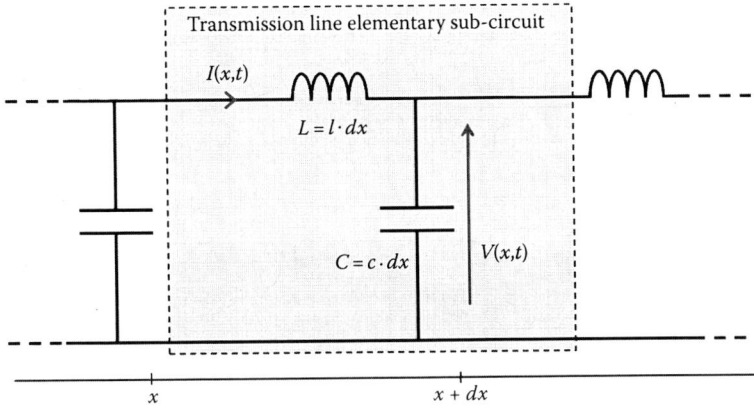

FIGURE 18.3 A transmission line can be modeled as a succession of elementary sub-circuits in which the propagation phenomenon can be neglected.

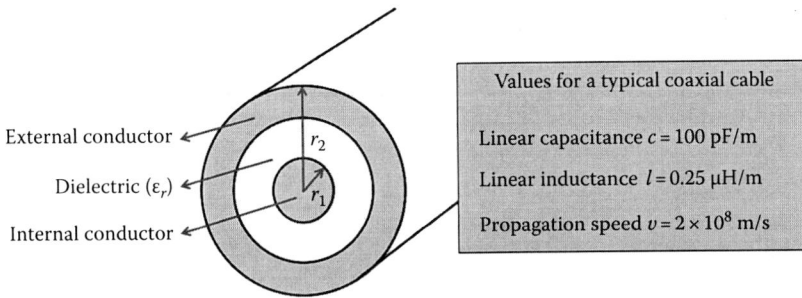

FIGURE 18.4 Electrical characteristics of a common coaxial cable.

The propagation speed is fixed by the linear capacitance and inductance of the transmission line, which are themselves imposed by the materials (especially the dielectric) that compose the line and by the geometry of the transmission line. In the case of a coaxial cable (presented in Figure 18.4), approximate values for linear capacitance and inductance can be evaluated by the following expressions:

$$c = \frac{2\pi\varepsilon_0\varepsilon_r}{\mathrm{Log}(r_1/r_2)} \; [\mathrm{F/m}] \quad \text{and} \quad l = \frac{\mu_0}{2\pi}\mathrm{Log}\!\left(\frac{r_1}{r_2}\right) [\mathrm{H/m}]$$

18.2.2.2 Characteristic Impedance and Reflections

It can be demonstrated that at any point in the transmission line, when the signal propagates from the pulse generator to the biological load (+ direction), the voltage $(V(x)_+)$ and current $(I(x)_+)$ are related according to

$$\frac{V(x)_+}{I(x)_+} = Z_c \quad \text{with } Z_c = \sqrt{\frac{l}{c}}.$$

The impedance introduced here, Z_c, is called the *characteristic impedance* of the transmission line and is expressed in ohms. A standard value for radio-communications coaxial cables is $50\,\Omega$.

The sample that is exposed to nsPEFs is commonly placed at the end of the cable and, from an electrical point of view, it can be modeled as a load with complex impedance Z (Figure 18.5). If this load matches the characteristic impedance of the cable, i.e., when $Z = Z_c$, then the voltages and the currents at the two sides

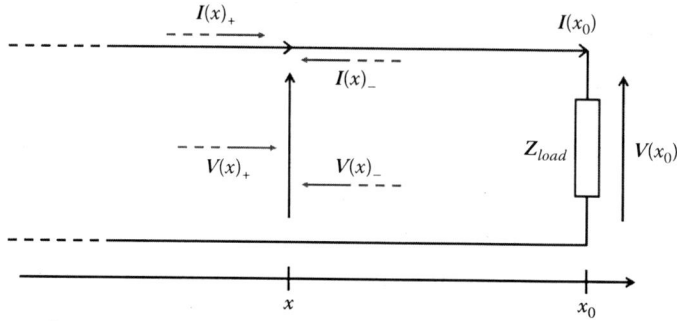

FIGURE 18.5   Reflection on the termination of the transmission line. When the voltage and current signals propagating in the positive x-direction ($V(x)_+$; $I(x)_+$) reach a mismatched load, they are partially reflected. The reflected signals ($V(x)_-$; $I(x)_-$) propagate in the opposite direction.

of the transmission line termination (i.e., at the load and at the transmission line) are the same. In this case, all the energy that was propagating in the line is absorbed by the termination load. On the contrary, if $Z \neq Z_c$, a reflection will occur and a voltage wave will start to propagate in the opposite direction through the transmission line (i.e., from the load towards the pulse generator). This reflected wave is produced by the following mechanism: the Ohm's law is valid at both sides of the transmission line termination (i.e., at the load and within the transmission line); therefore, since the voltage at both sides of the termination is also the same, when $Z \neq Z_c$, it is required that a current is generated in the opposite direction in order to obey Kirchhoff's current law. This current in turn generates the voltage wave ($V(x)_- = Z_c \cdot I(x)_-$). Once the reflected wave reaches the generator, another reflected wave (in the + direction) will be generated if the impedance of the generator (i.e., its output impedance) does not match the characteristic impedance.

When reflections occur, part of the energy contained in the pulse is absorbed by the load and part of it is reflected. The fraction of energy transmitted to the load can be evaluated. The results of the computation for a 50 Ω cable and a resistive termination R_{load}, are plotted in Figure 18.6. As displayed, transmitted power is maximum for the impedance matching condition and it rapidly decreases for lower or larger load resistance values.

It is important to mention that not only the amount of energy transmitted to the load will be affected by reflection. The levels and shapes of voltage and current signals can be dramatically different from the ones initially coming out of the generator. Indeed, the voltage at the load and the current that runs through it are the sum of the initial signals and the reflected ones:

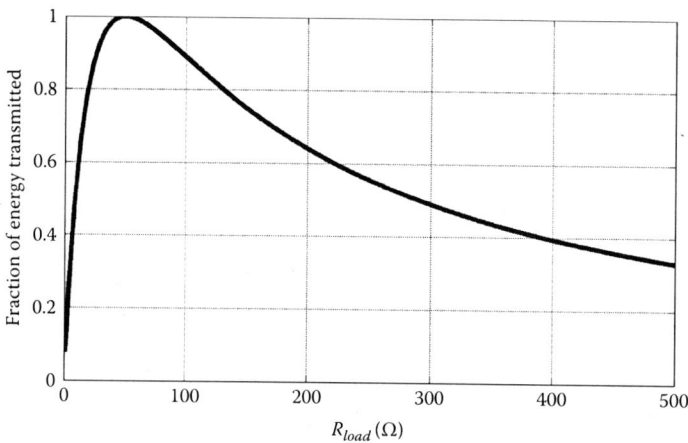

FIGURE 18.6   Fraction of the energy transmitted to the load in the case of a resistive termination and a 50 Ω cable.

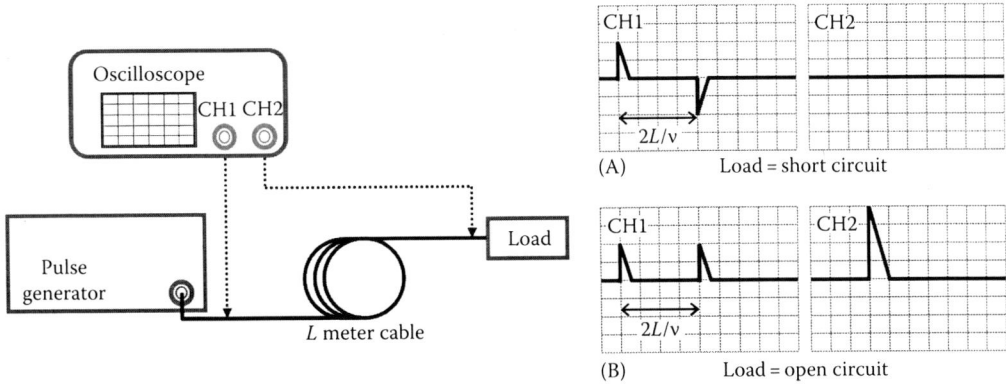

FIGURE 18.7 Illustration of reflection for two different loads. A pulse generator is connected to a load through a transmission line (length *L*; propagation speed v). If the load is a short circuit (A), the pulse is totally reflected in phase opposition and the resulting voltage on the load is null. If the load is an open circuit (B), the pulse is totally reflected in phase and the resulting voltage is twice the generated one.

$$I(x_0) = I(x_0)_+ + I(x_0)_- \quad \text{and} \quad V(x_0) = V(x_0)_+ + V(x_0)_-$$

These relations are also true in the frequency domain:

$$I(x_0) = I(x_0)_+ + I(x_0)_- \quad \text{and} \quad V(x_0) = V(x_0)_+ + V(x_0)_-$$

The magnitudes and phases of the reflected signals can be expressed as functions of the initial signals defining the complex reflection coefficients Γ_V and Γ_I:

$$\Gamma_V = \frac{V(x_0)_-}{V(x_0)_+} = \frac{Z_{load} - Z_c}{Z_{load} + Z_c} \quad \text{and} \quad \Gamma_I = \frac{I(x_0)_-}{I(x_0)_+} = \frac{Z_c - Z_{load}}{Z_{load} + Z_c} = -\Gamma_V$$

The most striking examples of impedance mismatch are observable in extreme cases: a perfect short-circuit and a perfect open circuit. These examples are illustrated in Figure 18.7. When the load is an open circuit, the voltage amplitude of the pulse at the load terminals is twice the amplitude of the pulse that was traveling from the generator. On the other hand, if the load is a short-circuit, the voltage amplitude at the load will be null.

As a consequence, it appears that for a given pulse generator and cable, the applied signals, and in particular the applied voltage, will be highly dependent on the impedance of the exposure device, which in turn will depend on the properties of the biological sample.

18.2.3 Impedance Matching for Exposure Devices

One of the common ways of doing *in vitro* experiments is to use commercial electroporation cuvettes filled with cell suspensions (Garon et al. 2007, Kolb et al. 2006). The typical dimensions for such cuvettes are: an electrode area *S* of 1 cm² and an electrode separation *l* of 1 mm (volume = 0.1 mL). With these dimensions, a cuvette filled with a solution with a conductivity of $\sigma = 1$ S/m would have a resistance *R* (Ω) of approximately

$$R = \frac{1}{\sigma} \cdot \frac{l}{S} = 10 \, \Omega$$

If a 10 kV pulse was applied through a 50 Ω cable to such a cuvette, the resulting voltage according to the reflection coefficients would be 3.3 kV. In order to minimize such voltage drop, it is possible to adjust the conductivity of the solution or to modify the geometrical characteristics of the cuvette so that the load impedance is better matched to the characteristic impedance of the cable. Unfortunately, the experimental conditions narrow the degrees of freedom for such kind of solutions. For instance, although it is possible to reduce the conductivity of a cell suspension medium without altering its osmolarity by adding non-ionic agents (e.g., sucrose), the biological consequences of an extreme depletion of extracellular ions will add uncertainty to the interpretation of the observations.

Alternative solutions can be based on modifying the other elements of the system. For instance, it is possible to interconnect multiple generators in order to change the total source impedance or even to build an adapted generator (Hall et al. 2007). Another feasible solution is to insert a matching impedance in parallel with the load in those cases in which the sample has a high impedance. Some examples of this approach can be found for single cell exposure (Pakhomov et al. 2007). In this configuration, most electrical power will be transmitted to the matching impedance and not directly to the biological sample.

In general, both the resistive component, R, and the reactive component, X, of the biological sample impedance ($\mathbf{Z} = R + jX$) will be significant at the frequencies of interest. Then, since the characteristic impedance of most transmission lines is purely resistive ($\mathbf{Z}_c = 50\,\Omega + j0\,\Omega$), some degree of pulse distortion will be almost unavoidable. In radio-communications, it is possible to match reactive impedances by inserting compensating circuits. However, such compensating circuits are only useful for a single frequency or for a narrow bandwidth, which is not the case in pulse transmission.

Finally, it must be mentioned that the electrical connection from the transmission line to the sample will also contribute by itself to the global impedance of the load. Indeed, a cuvette holder is necessary for connecting a cuvette to a coaxial cable for *in vitro* experiments. Similarly, for performing *in vivo* experiments, electrodes and connectors are necessary. These connections will basically add parasitic capacitances and inductances to the sample. An essential rule for limiting parasitic effects is to make the connection as short as possible both to avoid propagation phenomena and to minimize inductive and capacitive parasitic elements. In conclusion, even if all the issues described above are carefully addressed when implementing the experimental setup, distortion cannot be fully avoided. As a consequence, nsPEF delivery must be accompanied by monitoring of the signal waveforms.

18.3 Breakdown of Dielectric Materials

The so-called *dielectric strength* of an electrically insulating material indicates the maximum electric field that the material can withstand without experiencing an abrupt failure of its insulating properties, either reversible or irreversible. Such failure in isolation is known as an *electrical breakdown* and it has been observed to occur in dielectric solids, gases, and liquids. Dry air at atmospheric pressure has a dielectric strength of about 3 MV/m, polytetrafluoroethylene (Teflon™) has a dielectric strength higher than 60 MV/m, that value for alumina is about 13 MV/m, for polystyrene it is about 20 MV/m, and for fused silica it is possible to measure dielectric strength values above 400 MV/m. The breakdown process typically develops within a few nanoseconds (Beddow and Brignell 1966) and, for most materials, it is believed that the main phenomenon responsible for it is the *avalanche breakdown*: if the electric field is strong enough, free electrons in the dielectric material, either preexisting or freed by the electric field, are accelerated by the electric field and liberate additional electrons when colliding with the atoms of the material (i.e., ionization) so that the number of free charged particles is thus increased rapidly in an avalanche-like fashion. In the context of this book, it is interesting to note that cell membrane electroporation is a case in which avalanche breakdown does not seem to be involved in the insulation failure process (Crowley 1973).

Transmission lines consist of metallic conductors and dielectrics. It is obvious then that the maximum voltage that a transmission line will be able to withstand will depend on the geometrical features of the line (e.g., the separation distance between conductors) and the dielectric strength of the dielectric material. A typical RG-58/U coaxial cable employed in 10BASE2 Ethernet computer networks and for radio-communications is specified to withstand 1500 V_{DC} between the inner conductor and the shield, whereas a RG-194/U coaxial cable, intended for pulse transmission, can withstand DC voltages above

30 kV. Such an increase in voltage tolerance comes with the drawback of a larger diameter (5 mm for the RG-58/U against 50 mm for the RG-194/U) and an increase in mechanical stiffness.

The same kinds of considerations have to be applied to the connectors between the transmission lines and the exposure chamber or the pulse generator. Furthermore, an aspect that must be carefully taken into account is that the dielectric strength of air is particularly low (3 MV/m) and sparks (i.e., dielectric breakdown) may occur easily at open connectors or between the electrodes of the exposure chamber. These sparks will probably not be destructive but they will prevent the proper application of the pulses to the sample. On the other hand, it is convenient to mention that for solid dielectrics even a single breakdown event can severely permanently degrade the insulating capabilities of the material.

18.4 Monitoring of an Ultrashort Electric Signal

18.4.1 Choice of Suitable Probes

As mentioned earlier, it is advisable to capture and observe the signals that are actually applied to the exposure chamber. Besides the voltage signal, it can also be convenient to monitor the current that flows through the sample.

The first stage in the acquisition chain consists of the so-called *probes*. A *high-voltage probe* scales down the voltage to tolerable levels by the next stage in the acquisition chain (e.g., a 50 Ω oscilloscope input typically tolerates voltage amplitudes of about 1 V), whereas a current probe transforms the current signal into a voltage signal. High-voltage probes are usually based on resistive voltage dividers and high-current probes are generally based on the Hall effect. The specific features of the nsPEFs (high voltage, high current, and high bandwidth) impose some serious design constraints for the probes and, as consequence, only a handful of suppliers commercialize suitable probes. Probes useful for monitoring nsPEF signals are typically employed in applications that involve similar signal features (e.g., ultra large band radar and plasma generation).

A fundamental limitation of high-voltage probes and current probes is their bandwidth. Typically, these probes behave as first-order low-pass filters and their cutoff frequency will determine to a large extent the fidelity of the whole acquisition system. For instance, the simulations depicted in Figure 18.8

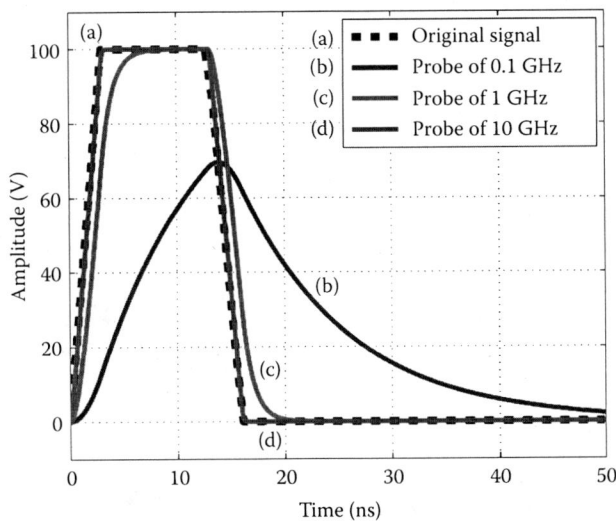

FIGURE 18.8 Time behavior of voltage probes with different cutoff frequencies. Responses (traces b, c, and d) have been obtained from SPICE simulations (MacSpice) under the assumption that probes behave as first-order low-pass filters. The simulated original signal (trace a) has an amplitude of 100 V and a duration of 16 ns (including 3 ns + 3 ns rise and fall times).

show the response to a 16 ns pulse (including 3 ns + 3 ns rise and fall times) when three voltage probes with different cutoff frequencies are employed: 0.1, 1, and 10 GHz. As it can be noticed, a voltage probe with a cutoff frequency of 0.1 GHz (trace b) would not only dramatically distort the shape of the original signal but it would also yield an underestimation of the pulse amplitude (70 V instead of 100 V).

As an alternative to high-voltage probes, it is possible to employ *tap-offs*, also known as *power dividers* or *splitters*, inserted in the transmission line. These components are particularly interesting as they allow a direct visualization of the reflections.

18.4.2 Monitoring Devices

Once the high-voltage and current signals have been transformed into low-voltage signals, capturing them and digitalizing them before proceeding to the subsequent monitoring stages is required (e.g., storage, parameterization, and visualization). Different solutions can be adopted in order to perform all the required steps but, in terms of convenience, modern digital oscilloscopes probably outperform all other approaches. Digital oscilloscopes integrate signal acquisition elements together with visualization elements (i.e., display), data storage, and parameterization mathematical tools (e.g., automatic quantification of pulse amplitude and width).

The particular features of the nsPEFs signals impose at least two specific characteristics of the monitoring system: (1) large acquisition bandwidth and (2) large memory.

Signal acquisition bandwidth is largely determined by the sampling rate of the analog to digital converter (ADC) embedded within the acquisition system. The Nyquist–Shannon theorem states that it is sufficient to sample at a rate twice higher than the bandwidth of the input signal in order to be able to accurately reconstruct the original waveform. However, oversampling, that is, digitalizing the signal at a higher frequency than twice the bandwidth of the signal being sampled, is commonly performed in signal acquisition systems as it provides some advantages in terms of noise cancellation and facilitates the design of the system. As a matter of fact, digital oscilloscopes generally have sampling rates well above three times the maximum input bandwidth.

A second important feature, especially in the case of acquisition of long trains of pulses, is the memory available for the storage of the acquired data. A single 10 ns pulse sampled at 50 GS/s will only require 500 points of memory. However, a single pause of 100 ms between two pulses in a train of pulses (i.e., pulse repetition frequency of 10 Hz) will fill 5×10^9 points of memory, which is well above the *memory depth* of current scopes. Fortunately, current high-performance oscilloscopes include a memory segmentation function that allows the recording of all the "interesting" points, whereas the "dull" ones are disregarded. When a pulse is detected (i.e., oscilloscope *trigger*), only a limited number of samples before and after the trigger are recorded so that only the time segments corresponding to the pulses are recorded sequentially in memory. Later on, all the useful information concerning the applied pulses (e.g., amplitude, duration, and rise and fall times) can be assessed.

18.5 Side Effects of the Application of Pulsed Electric Fields

The administration of pulsed electric fields to biological samples can be accompanied by phenomena not related to the direct effects of the electric fields on biological structures and their constituents. These phenomena, which could be labeled as *interfering phenomena*, must be taken into account and, if possible, avoided in order to simplify the interpretation of the experimental results. Phenomena such as cell electrodeformation (due to Maxwell stress) or ionization of macromolecules should not be included within this category as they are in fact a direct consequence of the pulses on the constituents of the biological sample.

Here we briefly introduce and discuss two groups of *interfering phenomena* caused by the application of high field pulses: (1) thermal effects and (2) electrochemical effects. Our description of both families of phenomena is focused on nsPEFs but most of what is said here can also be applied to pulses employed in conventional electroporation.

18.5.1 Thermal Effects

Electrical power is dissipated as heat at any conductor. This phenomenon is known as *Joule heating* but it is also referred to as *ohmic heating* or *resistive heating* because of its relationship with Ohm's law ($V = IR$)

$$P_{dissipated} = VI = I^2 R = \frac{V^2}{R} = V^2 G$$

where

$P_{dissipated}$ is the total amount of heat power generated in the whole conductor (expressed in watts [W])
V is the applied voltage (volts [V])
I is the current that flows through the conductor (amperes [A])
R is the resistance of the conductor (ohms [Ω])
G is the conductance of the conductor (inverse of the resistance, expressed in siemens [S])

The above equation is intended for two-terminal components (e.g., a piece of wire). For infinitesimal volumes, the Ohm's law can be expressed as

$$\rho = \frac{|\mathbf{E}|}{|\mathbf{J}|}$$

where

ρ is the resistivity of the material (ohms × meter, [Ω m])
$|\mathbf{E}|$ is the local magnitude of the electric field (V/m)
$|\mathbf{J}|$ is the magnitude of the current density (A/m^2)

And now it is possible to write down an expression for the dissipated power due to Joule heating in a unitary volume ($p_{dissipated}$, expressed in W/m^3)

$$p_{dissipated} = |\mathbf{E}||\mathbf{J}| = \frac{|\mathbf{E}|^2}{\rho} = \sigma |\mathbf{E}|^2$$

where σ is the conductivity of the material (inverse of ρ, [S/m]).

If it is assumed that no heat exchange occurs within the conductive material and between the material and its environment (i.e., the worst case scenario if sample heating is undesirable), then the increase in temperature at each sample point (ΔT) can be easily calculated as

$$\Delta T = \frac{U_{heat}}{cd} = \frac{\sigma |\mathbf{E}|^2 t}{cd}$$

where

t is the time (s)
U_{heat} is the applied thermal energy (= power × time, expressed in joules, [J] = [W s])
c is the specific heat capacity of the material (joules/(grams × Kelvin), [J/(g K)])
d is the mass density of the material (g/m^3)

If a single 10 ns pulse of 1 MV/m (10 kV/cm) is applied to a cell suspension in a highly conductive medium (σ = 1.5 S/m), the maximum temperature increase that can be expected is only 0.0036 K (c_{water} = 4.184 J/(g K), d_{water} = 0.997 × 10^6 g/m³), which is negligible under biological experimental conditions. On the other hand, if longer or larger pulses are considered, then the temperature jumps can start to be significant. For instance, a single pulse of 5 MV/m and 50 ns would produce a temperature rise of about 0.45°, which is probably still too low to cause observable biological effects; particularly taking into account that such an increase in temperature will only last for a few seconds or fractions of a second in the actual setups in which heat dissipates. However, it must be noticed that pulses are commonly applied at repetition frequencies of 1 Hz or higher and, in these cases, temperature can build up if heat does not dissipate rapidly enough. In such cases, it is highly advisable to perform thermal measurements (e.g., by means of fiber optic thermal sensors or infrared cameras, which are both fast and free of pulse interference) or to perform numerical thermal modeling of the system (Davalos et al. 2005).

The maximum tolerable temperature increase will depend on the experimental conditions and on the objectives of each biological study. As stated by Diller and Pearce (1999), damage to biological structures resulting from elevated temperatures is very sensitive to the value of the highest temperature that is reached and less so to the time of exposure, but, of course, both factors are important. Nevertheless, although some models have been implemented to describe such double dependency (Diller and Pearce 1999) and have been applied for electroporation (Maor et al. 2008), until now most studies in conventional electroporation and in the nsPEF field only consider a maximum value for the temperature increase as the criterion for tolerability. In particular, researchers in the nsPEF field consider that temperature rises of about some tenths of Kelvin (Schoenbach et al. 2001, Vernier et al. 2004) or of a few Kelvins (Deng et al. 2003, Nuccitelli et al. 2006) are perfectly tolerable in their studies.

It must be noted that Joule heating depends on the square of the electric field magnitude. This fact implies that field distribution heterogeneities will have a significant impact on heating. For instance, high electric fields around the electrodes due to the edge effect may cause local thermal burns after nsPEF administration in the same way that has been observed in conventional tissue electroporation (Edd et al. 2006). Moreover, Vernier et al. hypothesized that field distribution heterogeneities at the cell or sub-cellular level could cause thermal differences at the microscopic level, but those were not noticed in their study (Vernier et al. 2004).

Temperature plays an important role in ionic conductance: the viscosity of the solvent decreases as the temperature rises, increasing ion mobility and, consequently, increasing electrical conductivity. Although such dependence is not very significant (for most aqueous ionic solutions, the temperature dependence of conductivity is about +2%/K), this phenomenon can in turn have a slight effect on the temperature: as temperature increases due to Joule heating, conductivity increases and therefore Joule heating further increases, which accelerates the climb in temperature.

Finally, we want to point out another temperature-related phenomenon that may have some biological significance: water thermal expansion. When liquid water is heated, its volume increases and, if such volume increase is confined or cannot disperse sufficiently rapidly, as happens in the case of short thermal bursts, then pressure increases according to the equation

$$\Delta P = \frac{\alpha}{\beta} \Delta T$$

where
  ΔP is the resulting pressure increase
  α is the thermal expansion coefficient of water (207 × 10^{-6} K⁻¹)
  β is the compressibility coefficient of water (4.6 × 10^{-4} Pa⁻¹)
  ΔT is the applied temperature increase

This sudden increase in pressure may have some effect by itself at the location where it is created but it can also affect distant locations as it causes a shockwave that propagates. In fact, this phenomenon has been proposed as a mechanism to generate shockwaves for lithotripsy treatments as an alternative to underwater sparks (U.S. patent 6,383,152 B1). Nevertheless, it does not seem very plausible that this phenomenon could have significant biological consequences. A sudden increase in temperature of 0.5° would imply an increase in pressure of about 225 kPa (2.2 atm). This figure may seem important but in fact is almost negligible taking into account that lithotripsy shockwaves have pressures of about 100 atm and are only effective on hard materials (e.g., kidney stones) and not on soft tissues.

18.5.2 Electrochemical Effects

When DC currents are forced to flow through the interface between an electronic conductor (e.g., a metal) and an ionic conductor (e.g., biological media), oxidation and reduction chemical reactions ("redox" reactions) occur, which involve the molecules of the metal (electrode) and those of the ionic media (electrolyte) in a process called electrolysis. Some of the resulting chemical species liberated into the media can interfere with the biological processes up to the point of compromising cell viability. As a matter of fact, at present, electrochemical reactions are intentionally caused with low level currents applied through metallic needles for the ablation of solid tumors in a procedure called electrochemical treatment (EChT) (Nilsson et al. 2000).

In comparison with Joule heating, electrochemical effects are fundamentally accumulative: the chemical species created by electrolysis will accumulate pulse after pulse; although some molecules will finally escape as gas or will recombine with molecules in the media to form neutral species.

The main reactions that are believed to occur in biological samples when inert electrodes (e.g., platinum) are employed are (Nilsson et al. 2000, Saulis et al. 2005)

1. $2H_2O + 2e^- \Rightarrow H_2$ (gas) $+ 2OH^-$ at the cathode
2. $2H_2O \Rightarrow O_2$ (gas) $+ 4H^+ + 4e^-$ at the anode
3. $2Cl^- \Rightarrow Cl_2$ (gas) $+ 2e^-$ also at the anode

Furthermore, if instead of inert electrodes, electrochemically soluble electrodes (e.g., copper, aluminum, or stainless steel) are employed, then other oxidation reactions that release metallic ions can also occur at the anode (Kotnik et al. 2001b), such as (Saulis et al. 2005)

4. $2Al \Rightarrow 2Al^{+3} + 6e^-$

The biological significance of each one of the above reactions and their resulting species is still a matter of debate. Nevertheless, it is believed that H^+ production at the anode (i.e., pH decrease) is particularly important in EChT (Nilsson et al. 2000), whereas the release of Al^+ ions has been found to alter Ca^{2+} homeostasis in cell cultures (Loomis-Husselbee et al. 1991).

Faraday's laws of electrolysis indicate that the amount of substance (M), measured in moles (i.e., number of molecules), produced by the above electrochemical reactions will be proportional to the total electric charge injected by the external source (Q, measured in coulombs, [C])

$$M \propto Q = \int_{-\infty}^{t} I(t)dt$$

or, for a constant current pulse (I) with Δt duration,

$$M \propto Q = I\Delta t$$

Therefore, if it is assumed that electrochemically generated species freely diffuse through the whole sample, it seems reasonable to specify doses for electrochemical treatments as the amount of charge per unit of volume (C/m^3). Yen et al. (1999) reported that charge to volume ratios as low as $0.3\,C/mL$ (3×10^5 C/m^3) have an impact on cell growth. That figure can be put in the perspective of the nsPEFs with the following example: a single $1\,MV/m$ pulse of $10\,ns$ applied in a cuvette with an electrode area of $1\,cm^2$ and an electrode separation of $1\,mm$ (i.e., volume $= 1 \times 10^{-7}\,m^3 = 0.1\,mL$) would produce a charge of approximately $15\,\mu C/mL$ assuming that the medium had a conductivity of $1.5\,S/m$. Therefore, in order to reach the $0.3\,C/mL$ dose, 20,000 pulses would be required, which is much higher than the number of pulses that researchers usually apply in the nsPEF field. On the other hand, since the amount of charge is proportional both to the pulse duration and to the current, a single $5\,MV/m$ pulse of $50\,ns$ applied to the same setup would produce a charge of $375\,\mu C/mL$. In this case, for reaching the $0.3\,C/mL$ dose, 800 pulses would be required, which, although still higher than the number of pulses usually applied, comes closer to the actual experimental conditions. Furthermore, two facts must be noted here for precaution: (1) the $0.3\,C/mL$ dose was obtained under specific experimental conditions and quite probably the dose for other conditions and other biological effects is significantly lower and (2) at higher voltages, the kinetics of the electrochemical reactions may be significantly different and other chemical species could be produced. In other words, currently there is the need for experimental studies in which the dose to avoid electrochemical effects is obtained for conditions equivalent to those used with nsPEFs.

It is interesting to note that Vijh (2004) points at hydrogen cavitation at the cathode as a possible cause of erosive damage to tissues in electrochemical treatments. As far as we know, this hypothetical source of damage is not mentioned by other authors in the field, which is not surprising considering the low rate at which hydrogen is generated in standard electrochemical treatments. However, with nanosecond pulses of high intensity, this phenomenon appears to be more plausible because the amount of gas suddenly created is quite significant. Moreover, the gas bubbles created at the anode (O_2 and Cl_2) and the cathode (H_2) may not withstand the magnitude of the nsPEF and dielectric rupture (i.e., sparks) may occur accompanied by pressure shockwaves and flashes of light. In this case, the pressure of the shockwaves could be higher than that of those caused by the water thermal expansion phenomenon. Further research is required in this area.

18.6 Conclusion

A reliable nsPEF generator is a necessary but not sufficient prerequisite for reproducible experimentation in the field of nsPEFs delivery to biological samples. Here, our aim has been to point out the different aspects that can have a very significant impact on the experimental conditions. Our own experience tells us that those aspects can be easily overlooked, particularly by those research teams more focused on the biological aspects of nsPEFs delivery or those that come from the conventional electroporation field. Hence, this chapter should be taken as a caution message for those researchers interested in entering this field.

In this chapter, we have described first how the propagation phenomenon and particularly one of its associated phenomena, the reflection phenomenon, have a very relevant impact on the pulse field that is actually applied to the sample. Then we have explained that, up to a point, it is possible to control those reflections by matching the impedance of the exposure cell. Nevertheless, due to the uncertainties of this impedance matching process, we consider that voltage and current monitoring systems are highly advisable and we have given some indications about the use of these systems and their limitations. Furthermore, since high-voltage is a fundamental attribute of nsPEF signals, we have provided some details on the electrical breakdown phenomenon.

Finally, we have discussed two groups of side effects of the application of nsPEFs: electrochemical and thermal effects. Actually, both families of phenomena are also present in the case of conventional electroporation. The extremely short duration of the nsPEFs does not necessarily avoid the problems related to the electrochemical formation of chemical species at the electrode–sample interface. A rise in temperature because of the Joule effect will be particularly significant if the nsPEF application comprises a train of pulses at a high repetition frequency.

The nsPEF can be an extraordinary tool for intracellular electromanipulation. The use of nsPEFs may raise many hopes in terms of the improvement of our knowledge in cell biology and cell functioning, as well as in terms of the development of new biotechnological and biomedical applications. The future will confirm these hopes or prove them wrong. In any case, only well-controlled experimental conditions will result in robust and reproducible biological effects and in their translation into interesting applications.

Acknowledgments

The work of the authors is supported by the grants of the CNRS, Institut Gustave-Roussy, Univ. Paris-Sud, EU 6th FP (CLINIGENE - FP6, LSH-2004-018933), INCA (Institut National du Cancer, France—contract number 07/3D1616/Doc-54-3/NG-NC), DGA/D4S/MRIS (contract number 06 34 017), and the French National Agency (ANR) through the Nanoscience and Nanotechnology Program (Nanopulsebiochip ANR-08-NANO-024-01).

References

Beddow AJ, Brignell JE. 1966. Nanosecond breakdown time lags in a dielectric liquid. *Electron Lett* 2: 142–143.

Crowley JM. 1973. Electrical breakdown of bimolecular lipid membranes as an electromechanical instability. *Biophys J* 13: 711–724.

Davalos RV, Mir LM, Rubinsky B. 2005. Tissue ablation with irreversible electroporation. *Ann Biomed Eng* 33: 223–231.

Deng J, Schoenbach KH, Buescher ES, Hair PS, Fox PM, Beebe SJ. 2003. The effects of intense submicrosecond electrical pulses on cells. *Biophys J* 84: 2709–2714.

Diller KR, Pearce JA. 1999. Issues in modeling thermal alterations in tissues. *Ann NY Acad Sci* 888: 153–164.

Edd JF, Horowitz L, Davalos RV, Mir LM, Rubinsky B. 2006. In vivo results of a new focal tissue ablation technique: Irreversible electroporation. *IEEE Trans Bio-Med Eng* 53: 1409–1415.

Garon EB, Saweer D, Vernier PT, Tang T, Sun Y, Marcu L, Gundersen MA, Koeffler HP. 2007. *In vitro* and *in vivo* evaluation and a case report of intense nanosecond pulsed electric field as a local therapy for human malignancies. *Int J Cancer* 121: 675–682.

Hall EH, Schoenbach KH, Beebe SJ. 2007. Nanosecond pulsed electric fields have differential effects on cells in the S-phase. *DNA Cell Biol* 26: 160–171.

Kolb JF, Kono S, Schoenbach KH. 2006. Nanosecond pulsed electric field generators for the study of subcellular effects. *Bioelectromagnetics* 27: 172–187.

Kotnik T, Mir LM, Fisar K, Puc M, Miklavcic D. 2001a. Cell membrane electropermeabilization by symmetrical bipolar rectangular pulses. Part I. Increased efficiency of permeabilization. *Bioelectrochemistry* 54: 83–90.

Kotnik T, Miklavcic D, Mir LM. 2001b. Cell membrane electropermeabilization by symmetrical bipolar rectangular pulses. Part II. Reduced electrolytic contamination. *Bioelectrochemistry* 54: 91–95.

Loomis-Husselbee JW, Cullen PJ, Irvine RF, Dawson AP. 1991. Electroporation can cause artefacts due to solubilization of cations from the electrode plates. Aluminum ions enhance conversion of inositol 1,3,4,5-tetrakisphosphate into inositol 1,4,5-trisphosphate in electroporated L1210 cells. *Biochem J* 277: 883–885.

Maor E, Ivorra A, Rubinsky B. 2008. Intravascular irreversible electroporation: Theoretical and experimental feasibility study. *Conf Proc IEEE Eng Med Biol Soc* 2008: 2051–2054.

Nilsson E, Von Euler H, Berendson J, Thörne A, Wersäll P, Näslund I, Lagerstedt AS, Narfström K, Olsson JM. 2000. Electrochemical treatment of tumours. *Bioelectrochemistry* 51: 1–11.

Nuccitelli R, Pliquett U, Chen X, Ford W, James Swanson R, Beebe SJ, Kolb JF, Schoenbach KH. 2006. Nanosecond pulsed electric fields cause melanomas to self-destruct. *Biochem Biophys Res Commun* 343: 351–360.

Pakhomov AG, Kolb JF, White JA, Joshi RP, Xiao S, Schoenbach KH. 2007. Long-lasting plasma membrane permeabilization in mammalian cells by nanosecond pulsed electric field (nsPEF). *Bioelectromagnetics* 28: 655–663.

Saulis G, Lape R, Praneviciūte R, Mickevicius D. 2005. Changes of the solution pH due to exposure by high-voltage electric pulses. *Bioelectrochemistry* 67: 101–108.

Schoenbach KH, Beebe SJ, Buescher ES. 2001. Intracellular effect of ultrashort electrical pulses, *Bioelectromagnetics* 22: 440–448.

Vernier PT, Sun Y, Marcu L, Craft CM, Gundersen MA. 2004. Nanoelectropulse-induced phosphatidyl-serine translocation. *Biophys J* 86: 4040–4048.

Vijh AK. 2004. Electrochemical treatment (ECT) of cancerous tumours: Necrosis involving hydrogen cavitation, chlorine bleaching, pH changes, electroosmosis. *Int J Hydrogen Energy* 29: 663–665.

Yen Y, Li JR, Zhou BS, Rojas F, Yu J, Chou CK. 1999. Electrochemical treatment of human KB cells in vitro. *Bioelectromagnetics* 20: 34–41.

VI

Applications of Electroporation

19

Translation of Electroporation-Mediated DNA Delivery to the Clinic

Loree C. Heller

Richard Heller

Successful gene therapy depends on the capability of delivering the desired gene or cDNA to the tissue of interest, attaining the desired level of expression and, most importantly, the desired clinical response. *In vivo* gene therapy methods can be broadly divided into two categories, viral delivery and nonviral delivery. Both viral and nonviral delivery methods have uses in gene therapy, depending on the location, level, and expression time course necessary for the therapeutic transgene. However, each delivery type has associated problems. The primary issue with nonviral gene therapy is low transfection efficiency, while for viral delivery, the potential for insertional mutagenesis or the induction of an immune response are of concern.

In 1990, transgene expression after simple plasmid injection into mouse muscle was demonstrated (Wolff et al., 1990). This observation was one of the first indications that nonviral gene therapy was possible. In nonviral gene therapy, the gene or cDNA is encoded in a plasmid and delivered to specific tissues by simple injection or with chemical or physical assistance. To enhance clinical utility, optimization may be performed during plasmid design (Luckay et al., 2007; Kutzler and Weiner, 2008). Low levels of expression have been an issue for nonviral approaches. Several physical methods of enhancing delivered DNA expression have since been tested, including lipid or polymer conjugation, particle-mediated delivery, hydrodynamic delivery, ultrasound, and electroporation (Niidome and Huang, 2002; Wells, 2004; Lavigne and Gorecki, 2006).

The basis of electroporation is the induction of temporary permeabilized areas or pores by the combination of thermal and electric field energy (Weaver, 1995; Weaver and Chizmadzhev, 2007). *In vivo* electroporation was originally used to effectively deliver chemotherapeutic agents to tumors in animals and in humans. Clinical trials have been performed to test this therapy on melanoma, squamous cell carcinoma, and basal cell carcinoma (Mir et al., 1998; Gothelf et al., 2003). The first demonstration of *in vivo* electroporation for the delivery of plasmid DNA was to the skin of newborn mice reported in 1991 (Titomirov et al., 1991). However, transformant selection was performed *in vitro*, and later studies directly demonstrated tissue expression in rat liver (Heller et al., 1996) and orthotopic rat brain tumors

(Nishi et al., 1996). Since then, this technique has been applied to many other tissues in animal models including skin, kidney, liver, testis, brain, cartilage, arteries, prostate, cornea, and skeletal muscle as well as many tumor types (Heller and Heller, 2006; Favard et al., 2007).

19.1 Current *In Vivo* Electroporation Gene Therapy Trials

The first clinical trial delivering plasmid DNA using *in vivo* electroporation (www.clinicaltrials.gov identifier NCT00323206) was performed on patients with metastatic melanoma (Daud et al., 2008). This was a Phase I trial for safety and tolerability following the electroporation-mediated delivery of a plasmid encoding human interleukin 12 (hIL-12) directly into up to four surface tumors. After administration of lidocaine, six 100 μs direct current rectangular pulses at a field strength of 1300 V/cm were applied with a 6-needle array using a Medpulser DNA EPT System Generator (Inovio Biomedical Inc., San Diego, California). These deliveries were performed three times over 8 days. As expected, increased hIL-12 protein levels were observed in treated tumors. The major adverse affect was transient pain with the electroporation. Although this was designed as a safety trial, 10% of patients showed complete regression of all metastases, both treated and untreated, and 42% showed disease stabilization or partial responses.

A second trial tested the intramuscular delivery of a DNA vaccine rather than a direct intratumor cancer therapy (Low et al., 2009). This study utilized a plasmid encoding a tetanus toxin domain fused to prostate-specific membrane antigen, which was delivered to patients with recurrent prostate cancer. In this trial, injection and pulse delivery were performed using two needles. After injection, a train of five 20 ms pulses at 8.3 Hz were delivered with a maximum current of 250 mA using an Elgen Twinjector device (Inovio Biomedical Inc., San Diego, California). Patients received a total of five deliveries over 48 weeks. Brief pain was observed at the delivery site. Antibody responses to the tetanus toxin persisted up to 18 months. Data on anti-PSMA antibodies and possible clinical efficacy have not yet been published.

Currently, four additional active and/or recruiting cancer gene therapy clinical trials are registered in the www.clinicaltrials.gov database when "electroporation" or "electropermeabilization" is used as a search term. The clinicaltrials.gov Web site has an extensive database of active clinical trials and currently has 82,506 trials with locations in 170 countries. In addition to searching for trials with a keyword, specific trials can be found using the clinicaltrials.gov identifier (NCT followed by a number). These current trials have been extensively reviewed (Bodles-Brakhop et al., 2009); therefore, they will not be discussed here in detail. The cancer trials include direct tumor delivery of a plasmid encoding IL-2 (identifier NCT00223899) as well as delivery of DNA vaccines encoding rhesus prostate-specific antigen (NCT00859729), a xenogeneic tyrosinase (NCT00471133), or human telomerase reverse transcriptase (NCT00753415) for several solid tumors.

An additional electroporation-mediated therapeutic gene delivery field that has entered clinical trials is infectious disease vaccines. Three registered electroporation trials are evaluating the enhanced expression of plasmid-encoded viral antigens by intramuscular electroporation. A human immunodeficiency virus (HIV) trial (NCT00545987) expands on a previous trial in which plasmid DNA encoding HIV gag, env, pol, nef, and tat antigens was injected intramuscularly (NCT00249106). In a papillomavirus vaccine trial (NCT00685412), plasmids encoding the E6 and E7 genes of human papillomavirus (HPV) subtypes 16 and 18 are delivered. Finally, for a hepatitis C genotype 1 vaccine, a plasmid encoding the NS3/4A gene is delivered (NCT00563173). In each of these trials, the electrodes and electroporation parameters used for delivery are not detailed. These parameters are a critical part of the delivery criteria and should be clearly delineated when the results of each trial are published. Other trials may be in progress but not recorded in this registry.

As noted above, transient pain is a possible side effect of electroporation delivery and should be considered when designing trials. Several nontherapeutic clinical trials gauging the tolerability of delivery with electroporation have been performed. In a trial of 20 volunteers, a single exponentially decaying pulse was delivered with a surface electrode to the shaft and part of the glans of the penis using a Transdermal Delivery Device (Genetronics, San Diego, California) without drug or anesthetic (Zhang and Rabussay, 2002). This was a voltage escalation study utilizing a range from 50 to 80 V delivered for 3 ms. Hundred percent of subjects found 50 V to be tolerable (5 on a scale of 1–10). In a second study of five volunteers, 60 pulses of 1 ms duration at 150 V were delivered to the forearm skin with square contact electrode pads using a Gene Pulser Xcell (Bio-Rad, Hercules, California). Pain was reduced by using a decreased space between electrodes (range 0.5–6 mm) and by increasing the pulse interval (range 0.1–1.0 s) during delivery (Wong et al., 2006). In a trial of 40 healthy volunteers, while pain intensity did not vary with pulse frequency, subjects preferred a higher frequency, and therefore shorter delivery time (Zupanic et al., 2007). The discordance between these two pulse timing studies may be due to the subjectivity of the volunteers' responses or to the other parameters associated with each study such as the electrode used and the location of pulse application. A final study followed 24 subjects for 14 days after delivery of two 60 ms pulses at a field strength of approximately 200 V/cm using a MedPulser DDS (Inovio Biomedical Inc., San Diego, California) to the deltoid muscle (Wallace et al., 2009). The maximum discomfort was reported within 5 min of delivery and peaked at 2.3 on the McGill Pain Questionnaire scale of 1–5 or 4.75 on the Brief Pain Inventory scale of 1–10. In addition, erythema at the electroporation site was reported in 58.3% of subjects. While pain/discomfort is associated with the application of electric pulses, it is transient and can vary depending on the electrical parameters, electrode used, and site of electroporation. There are indications that the discomfort can be diminished or eliminated by either reducing the distance between electrodes (decreasing the applied voltage and area affected) or by varying the frequency.

19.2 Potential Future Electroporation Gene Therapy Clinical Trials

In vivo electroporation of genes other than reporters has been used for many nontherapeutic purposes, including promoter and enhancer characterization (Widlak et al., 2003; Fabre et al., 2006; Lagor et al., 2007; Langevin et al., 2007; Yao et al., 2007; Boone et al., 2009) or the elucidation of biochemical (Tanaka et al., 2000; Fox et al., 2006) and developmental pathways (Swartz et al., 2001). However, in this chapter, the focus is on preclinical studies using electroporation-mediated gene delivery that have demonstrated potential therapeutic value. For example, cancer studies in which complete tumor regression or effects on distant tumors was demonstrated, infectious disease vaccine studies in which protection was demonstrated, or protein replacement therapies in which a therapy for the genetic defect was demonstrated is discussed.

19.2.1 Cancer Therapy by Direct Tumor Delivery

Since the delivery of plasmid encoding IL-12 was demonstrated to be safe and surprisingly induced both local and distant responses in a Phase I clinical trial, it is possible that other immune modulators will also have clinical success. In preclinical cancer models, IFNα has induced long-term tumor regression as a single therapy for melanoma (Heller et al., 2002) or squamous cell carcinoma (Li et al., 2002), although side effects such as cachexia may be produced. Delivery of IL-15 in melanomas (Ugen et al., 2006) or IL-21 in mouse rectal carcinomas (Hanari et al., 2007) has induced long-term tumor regression. Combinations of immune modulators may also be effective. Intratumor delivery of plasmids encoding GM-CSF and B7.1 induced complete regression in a subcutaneous mouse fibrosarcoma model (Collins et al., 2006). Intratumor delivery of IL-12 and B7.1 eradicated 80% of tumors in mice bearing TRAMP or squamous cell carcinoma tumors (Liu et al., 2006).

Delivery of plasmids encoding immune modulators has been tested in combination with electrochemotherapy, which is the enhanced delivery of cell impermeable chemotherapeutic agents to tumors with electric pulses. Electrochemotherapy was originally combined with plasmids encoding IL-2 or GM-CSF to induce some long-term complete regressions in a melanoma model (Heller et al., 2000). Tumor eradication has been demonstrated by the combination with plasmids encoding IL-12 in mouse melanomas (Kishida et al., 2003) and squamous cell carcinomas and mammary tumors (Torrero et al., 2006). Electrochemotherapy was combined with plasmid p53 delivery to induce tumor regression in a mouse sarcoma model (Grosel et al., 2006). Electrochemotherapy has also been combined with another immune modulator, CpG oligodeoxynucleotides, inducing regression in both treated and distant untreated mouse fibrosarcomas and ovalbumin antigen–modified melanomas (Roux et al., 2008).

The combination of delivery of a plasmid encoding herpes simplex virus thymidine kinase and ganciclovir injection has also induced complete tumor regressions in a pancreatic model (Cascante et al., 2005) and in combination with a plasmid encoding IL-12 in a colon carcinoma model (Goto et al., 2004).

Tumor regression may also be induced by interference with tumor-signaling pathways. Intratumor delivery of plasmids encoding a dominant negative Stat 3 inhibitor (Niu et al., 1999) or the HIV protein VPR (McCray et al., 2004) induced tumor regression in a mouse melanoma model. In several tumor types implanted subcutaneously in nude mice, delivery of antisense oligonucleotides to polo-like kinase 1 induced tumor regression (Elez et al., 2003). In a rat mammary adenocarcinoma model, complete tumor regression was reported after intratumor delivery of an MHC gene–liposome complex with electroporation (Shimizu et al., 2009), potentially enhancing tumor antigenicity.

Clearly, a number of different therapeutic gene therapy approaches with clinical potential have been tested in animal cancer models and could be considered for human trials. These studies have included both single and combination therapies. The majority of these studies have utilized an immunotherapy approach, primarily utilizing tumor or muscle as the target tissue. These preclinical investigations are fair models for the protocols already being tested in the clinic. There are many more studies being conducted to test this delivery method for potential anticancer protocols; however, in this chapter only the potential therapies demonstrating a clear therapeutic response were included. As the development of electroporation delivery moves forward, it is highly possible that in the near future several new electroporation clinical trials for cancer therapy will be initiated.

19.2.2 Cancer Vaccines

Cancer antigen vaccines are active immunizations designed to treat growing tumors. These vaccines may be peptide or carbohydrate-based transfected cells, or viruses or plasmids encoding tumor antigens. This method is less successful against existing vascularized tumors and may be more effective as a therapy for residual disease, to reduce disease recurrence, or to prolong disease-free survival (Rosenberg et al., 2004). For example, the recently published clinical trial described above targets recurrent prostate cancer (Low et al., 2009).

A number of preclinical studies describing intradermal or intramuscular rather than intratumoral plasmid delivery by electroporation as a cancer therapy have been described. Plasmids encoding antigens, enzyme inhibitors, or antiangiogenesis factors have been described. As a rule, the most significant observations have been tumor growth inhibition, increased survival, or resistance to challenge with cancer cells. These studies show considerable promise, at this point most possibly as adjuvant therapies.

A few studies do demonstrate complete regression after intramuscular plasmid electroporation delivery: IL-12 plasmid for B-cell lymphomas (Lee et al., 2003), a tumor epitope-tetanus toxin fusion in a colon carcinoma model (Buchan et al., 2005) and a fusion protein of Flt3L and the E6 and E7 tumor oncoproteins in a mouse tumor model (Seo et al., 2009). A recent study compared intramuscular electroporation delivery of plasmid encoding E7 with intradermal delivery via the gene gun (Best et al., 2009). Both approaches resulted in significantly elevated antibody levels compared to intramuscular plasmid injection. Electroporation delivery induced higher levels of cellular immunity and resulted in better protection and survival. In a study evaluating intramuscular delivery of IL-24, increased survival was seen in a mouse ML-1 hepatoma model. Suppression of tumor growth and enhanced survival was observed when tumors were induced subcutaneously as well as when ML-1 cells were injected in the spleen leading to liver metastases (Chen et al., 2005). In each case, therapeutic delivery was performed either immediately or up to 3 days after cell injection. Although these studies technically demonstrate a therapeutic effect, regression of palpable tumors in response to muscle delivery has not been described.

Current clinical trials utilizing electroporation to deliver cancer vaccines are designed primarily to evaluate safety and immune stimulation against specific tumor antigens. As the development of this approach moves forward, it will be important to determine if there can be a demonstrated therapeutic effect against existing disease as has been demonstrated with immunotherapy approaches. The vaccine approach should find utility in patients with minimal or no disease as a means of preventing recurrence and increasing the duration of disease-free survival.

19.2.3 Infectious Disease Vaccines

As discussed above, electroporation-mediated delivery of DNA vaccines for HIV, HPV, and HCV have reached clinical trials. DNA vaccines for infectious diseases are safer than the conventional method, which may involve cultivation of the pathogen for the vaccine or infection in immunocompromised individuals. DNA vaccines can be produced quickly to deal with emerging outbreaks or with genetically changing organisms. Several mechanisms for plasmid delivery to muscle or skin have been described, including plasmid formulations and physical methods such as electroporation. DNA vaccines have been extensively reviewed (Kutzler and Weiner, 2008; Moss, 2009).

Vaccine development is a classic method of generating protection from infectious disease. High levels of genetic variation are observed in many viruses, where the ability to quickly modify a vaccine would be important. In addition, relatively few effective antiviral or antiparasitic agents have been developed. In the face of growing antimicrobial resistance, bacterial vaccines are also rapidly becoming more important.

Increased levels of specific cellular and humoral immunity have been demonstrated after electroporation-mediated delivery of virtually all the viral, bacterial, or parasitic antigens tested. Undoubtedly, a complete characterization of these induced immune responses is important, but beyond the scope of this chapter. In the few cases where a preclinical model is available, a simple protection against challenge is a straightforward indicator of potential clinical feasibility and when protection has been demonstrated, the study has been included in this discussion.

In mice, protection was demonstrated after intramuscular delivery of plasmids encoding influenza A (Chen et al., 2000; Kadowaki et al., 2000), influenza B (Fang et al., 2008), and H5N1 (avian) influenza A (Chen et al., 2009) antigens. This immunity tended to be subtype specific, although cross-protection was observed between H5N1 and H1N1 using a combination of antigens (Chen et al., 2009). Intramuscular delivery of synthetic H5N1 plasmid vaccines reduced morbidity and mortality in mouse and ferret models (Laddy et al., 2008) and reduced symptoms and viral load upon challenge in nonhuman primates (Laddy et al., 2009). Complete protection from vaccinia virus (Hooper et al., 2007) or 80% protection from Japanese encephalitis virus (Wu et al., 2004) challenge after antigen delivery was demonstrated in mice. Interestingly, after delivery of a vaccine encoding an HCV antigen, hepatitis C positive cells were eliminated in a mouse model (Ahlen et al., 2007). In a challenge experiment in rhesus macaques, decreased peak viremia and rapidly decreasing viral loads were observed after IM delivery of a plasmid encoding the majority of SIVmac239 proteins (Rosati et al., 2009). Finally, in a bacterial study, significant reduction in lung and spleen bacterial loads were observed in response to challenge after delivery of plasmids encoding a *Mycobacteria tuberculosis* antigen (Zhang et al., 2007).

Many infectious disease vaccines have considerable clinical potential due to the ease of production and the innate malleability of plasmid DNA. When discussing electroporation-mediated delivery, the primary issue is the discomfort associated with pulse delivery. Since vaccination may be optional for these subjects, it will be important to not only consider optimizing immune responses, but minimize discomfort as well.

19.2.4 Protein Replacement Therapy

Most studies of plasmid delivery with *in vivo* electroporation explore therapies that require systemic rather than localized expression. Replacement of a functional protein may be more difficult in the instance of electroporation due to the tight expression control and long-term expression necessary. In these cases, regulation by inducible promoters may be necessary. Some additional optimization of the use of these promoters is sometimes necessary. In an interesting rat study, sweetening the water containing the inducer doxycycline increased water intake and therefore transgene expression (Hojman et al., 2007a). An immune response to the transactivators may reduce the duration of transgene expression (Lena et al., 2005).

Long-term expression of erythropoietin as a correction for anemia and factors VIII and IX for hemophilia have been tested using *in vivo* electroporation. Erythropoietin levels extending as long as 6 months were initially described with significantly increased hematocrit levels (Kreiss et al., 1999; Rizzuto et al., 1999). Anemia (Maruyama et al., 2001) and β-thalassemia (Payen et al., 2001) after erythropoietin delivery in mouse models have been corrected. Recent papers have addressed the control of expression levels using careful management of an inducible promoter (Hojman et al., 2007b) or multiple small dose plasmid deliveries (Fabre et al., 2008). Factor IX therapy was originally tested in 2001 (Fewell et al., 2001), with a later demonstration of an observed tendency toward increased clotting times (Liu et al., 2007). Efficacy of Factor XIII has been demonstrated using the tail clip assay after intramuscular (Long et al., 2005) and liver (Jaichandran et al., 2006) plasmid delivery. Skin or muscle delivery may be optimal because it is noninvasive and offers an option for multiple treatments.

These recent studies utilizing electroporation have generated excitement as to the potential for this approach. Preclinical efficacy has been demonstrated for several indications with delivery to either muscle or skin. The control of expression utilizing inducible promoters will be an important consideration as these studies are translated for clinical testing.

19.2.5 Other Applications

Gene therapy for autoimmune diseases including arthritis generally focuses on either direct inhibition of specific inflammatory cytokines or broad downregulation of autoimmunity (Ghivizzani et al., 2008). As with other gene therapies, both viral and plasmid delivery have been tested. A number of successful preclinical studies exploring electroporation-mediated gene delivery to reduce inflammation in arthritis models have been published. Studies have focused on inhibition of TNFα by delivery of soluble receptor variants (Kim et al., 2003; Bloquel et al., 2004; Gould et al., 2004) or siRNA (Inoue et al., 2005; Schiffelers et al., 2005). Several other methods of immune modulation, singly or as combination therapy, have been tested (Celiker et al., 2002; Jeong et al., 2004; Ho et al., 2006; Kuroda et al., 2006; Onodera et al., 2007; Inoue et al., 2009; Nakagawa et al., 2009). One study focused on the inhibition of apoptosis (Grossin et al., 2006). Another interesting study focused on reducing the pain associated with arthritis (Chuang et al., 2004). A therapeutic effect was observed in each of these studies showing promise as potential clinical therapies.

Therapeutic approaches for ischemia have evaluated the delivery of factors that would induce angiogenesis. Electroporation approaches have evaluated delivery of plasmid DNA to areas of ischemia, including hindlimb and skin flap. Delivery of basic fibroblast growth factor (bFGF) was tested in a rat ischemic skin flap and rabbit hind limb models. In the rabbit ischemic limb model, increased blood pressure, angiographic score, blood flow, and capillary density of ischemic limbs were observed when compared to controls (Nishikage et al., 2004). In the ischemic flap, it was observed that treatment of the dorsal muscle in the recipient bed under the skin flap increased vascularization of the flap (Fujihara et al., 2005). Another approach utilized co-transfection of plasmids encoding angiopoietin and vascular endothelial factor (VEGF) delivered to ischemic hindlimbs of rats. Plasmids were delivered intramuscularly in the ischemic limbs with and without electroporation. An increased capillary density and decreased limb necrosis was seen when plasmids were delivered with electroporation. There was also an increase when both plasmids were delivered compared to either VEGF or angiopoietin-1 alone (Jiang et al., 2006). Intramuscular delivery of plasmid encoding adrenomedullin to mouse ischemic limb model resulted in enhanced blood flow when the plasmid was delivered with electroporation (Abe et al., 2006). These studies demonstrated that revascularization could be enhanced with this approach. Delivery needed to be near the ischemic area. Long-term follow-up to assess either potential therapeutic effect or possible adverse effects have not as yet been evaluated. These studies will need to be done prior to translation to the clinic.

Efficient healing of wounds is a critical process to prevent the loss of skin integrity. If healing is delayed, it can lead to scarring or impaired function (Branski et al., 2007). The healing process involves a complex sequence of cellular and molecular processes, including angiogenesis and reepithelialization. The process is regulated by many growth factors including transforming growth factor-b, keratinocyte growth factor, and vascular endothelial growth factor (Kunugiza et al., 2006). Several gene therapy approaches have evaluated delivery of these growth factors in an attempt to accelerate the healing process. Keratinocyte growth factor-1 was utilized to treat wounds in both a mouse diabetic model as well as a rat septic model. Intradermal plasmid delivery with electroporation resulted in an accelerated healing rate, increased density of new blood vessels, and increased quality of epithelialization when compared to injection of plasmid without electroporation (Marti et al., 2004; Lin et al., 2006). Delivery of bFGF to an ischemic flap in a rat model resulted in decreased necrosis of the flap when the plasmid was delivered with electroporation (Fujihara et al., 2005). In another study, a plasmid encoding transforming growth factor-beta 1 was delivered intradermally to a full thickness wound in a diabetic mouse model. When the plasmid was delivered with electroporation, the investigators observed an increased healing rate, increased angiogenesis, increased collagen synthesis, and enhanced quality of reepithelialization (Lee et al., 2004). Intradermal delivery of plasmid encoding VEGF was evaluated in a rat random flap model. Delivery of plasmid with electroporation resulted in increased healing, increased vascularization/ perfusion of the flap, and decreased necrosis when compared to injection only, electroporation alone, or no treatment (Ferraro et al., 2009). Significant progress in electroporation-mediated gene delivery for wound healing has occurred over the past few years. It is possible that a single agent therapy may show some success in clinical trials; however, it is most likely that clinical success will involve delivery of multiple factors over a timed sequence to coincide with the various phases of the healing process.

19.3 Summary

Over the past 10 years, the use of *in vivo* electroporation to deliver plasmid DNA has seen tremendous growth. During this time, numerous papers have been published evaluating delivery parameters and mechanisms as well as testing delivery to new tissue targets. As this approach has matured, there have been hundreds of published preclinical studies exploring potential clinically relevant therapies delivered by *in vivo* electroporation. The successes reported by these studies have fueled interest in electroporation and has demonstrated that it is a viable nonviral delivery method for plasmid DNA. The true measure of the success of this approach will not be based on the number of preclinical studies showing efficacy but the demonstration of safety and effectiveness in clinical studies. With the initiation of clinical trials utilizing electroporation, the true potential of this method will be determined. Recently, the first publications of these clinical trials have begun to shed light on that potential. While it is too early to draw any conclusions, these first reports are encouraging. Both showed that electroporation was safe and tolerable. The IL-12 trial showed signs of clinical efficacy not only locally but also at non-treated sites of disease. The vaccine study reported an immune stimulation against one of the delivered antigens. As more preclinical studies are translated to the clinic and current trials are concluded and reported, a true picture of the future of this approach will emerge. If the preclinical results are an indication of the potential, then it is possible that electroporation delivery of plasmid DNA will play an important role in the effective development of gene therapy for several indications.

References

Abe M, Sata M, Suzuki E, Takeda R, Takahashi M, Nishimatsu H, Nagata D et al. 2006. Effects of adreno-medullin on acute ischaemia-induced collateral development and mobilization of bone-marrow-derived cells. *Clin Sci (Lond)* 111:381–387.

Ahlen G, Soderholm J, Tjelle T, Kjeken R, Frelin L, Hoglund U, Blomberg P, Fons M, Mathiesen I, Sallberg M. 2007. In vivo electroporation enhances the immunogenicity of hepatitis C virus nonstructural 3/4A DNA by increased local DNA uptake, protein expression, inflammation, and infiltration of CD3 + T cells. *J Immunol* 179:4741–4753.

Best SR, Peng S, Juang CM, Hung CF, Hannaman D, Saunders JR, Wu TC, Pai SI. 2009. Administration of HPV DNA vaccine via electroporation elicits the strongest CD8 + T cell immune responses compared to intramuscular injection and intradermal gene gun delivery. *Vaccine* 27:5450–5459.

Bloquel C, Bessis N, Boissier MC, Scherman D, Bigey P. 2004. Gene therapy of collagen-induced arthritis by electrotransfer of human tumor necrosis factor-alpha soluble receptor I variants. *Hum Gene Ther* 15:189–201.

Bodles-Brakhop AM, Heller R, Draghia-Akli R. 2009. Electroporation for the delivery of DNA-based vaccines and immunotherapeutics: current clinical developments. *Mol Ther* 17:585–592.

Boone LR, Niesen MI, Jaroszeski M, Ness GC. 2009. In vivo identification of promoter elements and transcription factors mediating activation of hepatic HMG-CoA reductase by T3. *Biochem Biophys Res Commun* 385:466–471.

Branski LK, Pereira CT, Herndon DN, Jeschke MG. 2007. Gene therapy in wound healing: Present status and future directions. *Gene Ther* 14:1–10.

Buchan S, Gronevik E, Mathiesen I, King CA, Stevenson FK, Rice J. 2005. Electroporation as a "prime/boost" strategy for naked DNA vaccination against a tumor antigen. *J Immunol* 174:6292–6298.

Cascante A, Huch M, Rodriguez LG, Gonzalez JR, Costantini L, Fillat C. 2005. Tat8-TK/GCV suicide gene therapy induces pancreatic tumor regression in vivo. *Hum Gene Ther* 16:1377–1388.

Celiker MY, Ramamurthy N, Xu JW, Wang M, Jiang Y, Greenwald R, Shi YE. 2002. Inhibition of adjuvant-induced arthritis by systemic tissue inhibitor of metalloproteinases 4 gene delivery. *Arthritis Rheum* 46:3361–3368.

Chen Q, Kuang H, Wang H, Fang F, Yang Z, Zhang Z, Zhang X, Chen Z. 2009. Comparing the ability of a series of viral protein-expressing plasmid DNAs to protect against H5N1 influenza virus. *Virus Genes* 38:30–38.

Chen WY, Cheng YT, Lei HY, Chang CP, Wang CW, Chang MS. 2005. IL-24 inhibits the growth of hepatoma cells in vivo. *Genes Immun* 6:493–499.

Chen Z, Kadowaki S, Hagiwara Y, Yoshikawa T, Matsuo K, Kurata T, Tamura S. 2000. Cross-protection against a lethal influenza virus infection by DNA vaccine to neuraminidase. *Vaccine* 18:3214–3222.

Chuang IC, Jhao CM, Yang CH, Chang HC, Wang CW, Lu CY, Chang YJ, Lin SH, Huang PL, Yang LC. 2004. Intramuscular electroporation with the pro-opiomelanocortin gene in rat adjuvant arthritis. *Arthritis Res Ther* 6:R7–R14.

Collins CG, Tangney M, Larkin JO, Casey G, Whelan MC, Cashman J, Murphy J et al. 2006. Local gene therapy of solid tumors with GM-CSF and B7-1 eradicates both treated and distal tumors. *Cancer Gene Ther* 13:1061–1071.

Daud AI, DeConti RC, Andrews S, Urbas P, Riker AI, Sondak VK, Munster PN et al. 2008. Phase I trial of interleukin-12 plasmid electroporation in patients with metastatic melanoma. *J Clin Oncol* 26:5896–5903.

Elez R, Piiper A, Kronenberger B, Kock M, Brendel M, Hermann E, Pliquett U, Neumann E, Zeuzem S. 2003. Tumor regression by combination antisense therapy against Plk1 and Bcl-2. *Oncogene* 22:69–80.

Fabre EE, Bigey P, Beuzard Y, Scherman D, Payen E. 2008. Careful adjustment of Epo non-viral gene therapy for beta-thalassemic anaemia treatment. *Genet Vaccines Ther* 6:10.

Fabre EE, Bigey P, Orsini C, Scherman D. 2006. Comparison of promoter region constructs for in vivo intramuscular expression. *J Gene Med* 8:636–645.

Fang F, Cai XQ, Chang HY, Wang HD, Yang ZD, Chen Z. 2008. Protection abilities of influenza B virus DNA vaccines expressing hemagglutinin, neuraminidase, or both in mice. *Acta Virol* 52:107–112.

Favard C, Dean DS, Rols MP. 2007. Electrotransfer as a non viral method of gene delivery. *Curr Gene Ther* 7:67–77.

Ferraro B, Cruz YL, Coppola D, Heller R. 2009. Intradermal delivery of plasmid VEGF(165) by electroporation promotes wound healing. *Mol Ther* 17:651–657.

Fewell JG, MacLaughlin F, Mehta V, Gondo M, Nicol F, Wilson E, Smith LC. 2001. Gene therapy for the treatment of hemophilia B using PINC-formulated plasmid delivered to muscle with electroporation. *Mol Ther* 3:574–583.

Fox SA, Yang L, Hinton BT. 2006. Identifying putative contraceptive targets by dissecting signal transduction networks in the epididymis using an in vivo electroporation (electrotransfer) approach. *Mol Cell Endocrinol* 250:196–200.

Fujihara Y, Koyama H, Nishiyama N, Eguchi T, Takato T. 2005. Gene transfer of bFGF to recipient bed improves survival of ischemic skin flap. *Br J Plast Surg* 58:511–517.

Ghivizzani SC, Gouze E, Gouze JN, Kay JD, Bush ML, Watson RS, Levings PP et al. 2008. Perspectives on the use of gene therapy for chronic joint diseases. *Curr Gene Ther* 8:273–286.

Gothelf A, Mir LM, Gehl J. 2003. Electrochemotherapy: Results of cancer treatment using enhanced delivery of bleomycin by electroporation. *Cancer Treat Rev* 29:371–387.

Goto T, Nishi T, Kobayashi O, Tamura T, Dev SB, Takeshima H, Kochi M, Kuratsu J, Sakata T, Ushio Y. 2004. Combination electro-gene therapy using herpes virus thymidine kinase and interleukin-12 expression plasmids is highly efficient against murine carcinomas in vivo. *Mol Ther* 10:929–937.

Gould DJ, Bright C, Chernajovsky Y. 2004. Inhibition of established collagen-induced arthritis with a tumour necrosis factor-alpha inhibitor expressed from a self-contained doxycycline regulated plasmid. *Arthritis Res Ther* 6:R103–R113.

Grosel A, Sersa G, Kranjc S, Cemazar M. 2006. Electrogene therapy with p53 of murine sarcomas alone or combined with electrochemotherapy using cisplatin. DNA *Cell Biol* 25:674–683.

Grossin L, Cournil-Henrionnet C, Pinzano A, Gaborit N, Dumas D, Etienne S, Stoltz JF et al. 2006. Gene transfer with HSP 70 in rat chondrocytes confers cytoprotection in vitro and during experimental osteoarthritis. *FASEB J* 20:65–75.

Hanari N, Matsubara H, Hoshino I, Akutsu Y, Nishimori T, Murakami K, Sakata H, Miyazawa Y, Ochiai T. 2007. Combinatory gene therapy with electrotransfer of midkine promoter-HSV-TK and interleukin-21. *Anticancer Res* 27:2305–2310.

Heller L, Pottinger C, Jaroszeski MJ, Gilbert R, Heller R. 2000. In vivo electroporation of plasmids encoding GM-CSF or interleukin-2 into existing B16 melanomas combined with electrochemotherapy induces longterm antitumour immunity. *Melanoma Res* 10:577–583.

Heller LC, Heller R. 2006. In vivo electroporation for gene therapy. *Hum Gene Ther* 17:890–897.

Heller LC, Ingram SF, Lucas ML, Gilbert RA, Heller R. 2002. Effect of electrically mediated intratumor and intramuscular delivery of a plasmid encoding IFN alpha on visible B16 mouse melanomas. *Technol Cancer Res Treat* 1:205–209.

Heller R, Jaroszeski M, Atkin A, Moradpour D, Gilbert R, Wands J, Nicolau C. 1996. In vivo gene electroinjection and expression in rat liver. *FEBS Lett* 389:225–228.

Ho SH, Lee HJ, Kim DS, Jeong JG, Kim S, Yu SS, Jin Z, Kim S, Kim JM. 2006. Intrasplenic electro-transfer of IL-4 encoding plasmid DNA efficiently inhibits rat experimental allergic encephalomyelitis. *Biochem Biophys Res Commun* 343:816–824.

Hojman P, Eriksen J, Gehl J. 2007a. Tet-On induction with doxycycline after gene transfer in mice: Sweetening of drinking water is not a good idea. *Anim Biotechnol* 18:183–188.

Hojman P, Gissel H, Gehl J. 2007b. Sensitive and precise regulation of haemoglobin after gene transfer of erythropoietin to muscle tissue using electroporation. *Gene Ther* 14:950–959.

Hooper JW, Golden JW, Ferro AM, King AD. 2007. Smallpox DNA vaccine delivered by novel skin electroporation device protects mice against intranasal poxvirus challenge. *Vaccine* 25:1814–1823.

Inoue A, Takahashi KA, Mazda O, Arai Y, Saito M, Kishida T, Shin-Ya M et al. 2009. Comparison of anti-rheumatic effects of local RNAi-based therapy in collagen induced arthritis rats using various cytokine genes as molecular targets. *Mod Rheumatol* 19:125–133.

Inoue A, Takahashi KA, Mazda O, Terauchi R, Arai Y, Kishida T, Shin-Ya M et al. 2005. Electro-transfer of small interfering RNA ameliorated arthritis in rats. *Biochem Biophys Res Commun* 336:903–908.

Jaichandran S, Yap ST, Khoo AB, Ho LP, Tien SL, Kon OL. 2006. In vivo liver electroporation: optimization and demonstration of therapeutic efficacy. *Hum Gene Ther* 17:362–375.

Jeong JG, Kim JM, Ho SH, Hahn W, Yu SS, Kim S. 2004. Electrotransfer of human IL-1Ra into skeletal muscles reduces the incidence of murine collagen-induced arthritis. *J Gene Med* 6:1125–1133.

Jiang J, Jiangl N, Gao W, Zhu J, Guo Y, Shen D, Chen G, Tang J. 2006. Augmentation of revascularization and prevention of plasma leakage by angiopoietin-1 and vascular endothelial growth factor co-transfection in rats with experimental limb ischaemia. *Acta Cardiol* 61:145–153.

Kadowaki S, Chen Z, Asanuma H, Aizawa C, Kurata T, Tamura S. 2000. Protection against influenza virus infection in mice immunized by administration of hemagglutinin-expressing DNAs with electroporation. *Vaccine* 18:2779–2788.

Kim JM, Ho SH, Hahn W, Jeong JG, Park EJ, Lee HJ, Yu SS, Lee CS, Lee YW, Kim S. 2003. Electro-gene therapy of collagen-induced arthritis by using an expression plasmid for the soluble p75 tumor necrosis factor receptor-Fc fusion protein. *Gene Ther* 10:1216–1224.

Kishida T, Asada H, Itokawa Y, Yasutomi K, Shin-Ya M, Gojo S, Cui FD et al. 2003. Electrochemo-gene therapy of cancer: Intratumoral delivery of interleukin-12 gene and bleomycin synergistically induced therapeutic immunity and suppressed subcutaneous and metastatic melanomas in mice. *Mol Ther* 8:738–745.

Kreiss P, Bettan M, Crouzet J, Scherman D. 1999. Erythropoietin secretion and physiological effect in mouse after intramuscular plasmid DNA electrotransfer. *J Gene Med* 1:245–250.

Kunugiza Y, Tomita N, Taniyama Y, Tomita T, Osako MK, Tamai K, Tanabe T, Kaneda Y, Yoshikawa H, Morishita R. 2006. Acceleration of wound healing by combined gene transfer of hepatocyte growth factor and prostacyclin synthase with Shima Jet. *Gene Ther* 13:1143–1152.

Kuroda T, Maruyama H, Shimotori M, Higuchi N, Kameda S, Tahara H, Miyazaki J, Gejyo F. 2006. Effects of viral interleukin 10 introduced by in vivo electroporation on arthrogen-induced arthritis in mice. *J Rheumatol* 33:455–462.

Kutzler MA, Weiner DB. 2008. DNA vaccines: ready for prime time? *Nat Rev Genet* 9:776–788.

Laddy DJ, Yan J, Khan AS, Andersen H, Cohn A, Greenhouse J, Lewis M et al. 2009. Electroporation of synthetic DNA antigens offers protection in nonhuman primates challenged with highly pathogenic avian influenza virus. *J Virol* 83:4624–4630.

Laddy DJ, Yan J, Kutzler M, Kobasa D, Kobinger GP, Khan AS, Greenhouse J, Sardesai NY, Draghia-Akli R, Weiner DB. 2008. Heterosubtypic protection against pathogenic human and avian influenza viruses via in vivo electroporation of synthetic consensus DNA antigens. *PLoS One* 3:e2517.

Lagor WR, Heller R, de Groh ED, Ness GC. 2007. Functional analysis of the hepatic HMG-CoA reductase promoter by in vivo electroporation. *Exp Biol Med (Maywood)* 232:353–361.

Langevin LM, Mattar P, Scardigli R, Roussigne M, Logan C, Blader P, Schuurmans C. 2007. Validating in utero electroporation for the rapid analysis of gene regulatory elements in the murine telencephalon. *Dev Dyn* 236:1273–1286.

Lavigne MD, Gorecki DC. 2006. Emerging vectors and targeting methods for nonviral gene therapy. *Expert Opin Emerg Drugs* 11:541–557.

Lee PY, Chesnoy S, Huang L. 2004. Electroporatic delivery of TGF-beta1 gene works synergistically with electric therapy to enhance diabetic wound healing in db/db mice. *J Invest Dermatol* 123:791–798.

Lee SC, Wu CJ, Wu PY, Huang YL, Wu CW, Tao MH. 2003. Inhibition of established subcutaneous and metastatic murine tumors by intramuscular electroporation of the interleukin-12 gene. *J Biomed Sci* 10:73–86.

Lena AM, Giannetti P, Sporeno E, Ciliberto G, Savino R. 2005. Immune responses against tetracycline-dependent transactivators affect long-term expression of mouse erythropoietin delivered by a helper-dependent adenoviral vector. *J Gene Med* 7:1086–1096.

Li S, Xia X, Zhang X, Suen J. 2002. Regression of tumors by IFN-alpha electroporation gene therapy and analysis of the responsible genes by cDNA array. *Gene Ther* 9:390–397.

Lin MP, Marti GP, Dieb R, Wang J, Ferguson M, Qaiser R, Bonde P, Duncan MD, Harmon JW. 2006. Delivery of plasmid DNA expression vector for keratinocyte growth factor-1 using electroporation to improve cutaneous wound healing in a septic rat model. *Wound Repair Regen* 14:618–624.

Liu F, Sag D, Wang J, Shollenberger LM, Niu F, Yuan X, Li SD, Thompson M, Monahan P. 2007. Sine-wave current for efficient and safe in vivo gene transfer. *Mol Ther* 15:1842–1847.

Liu J, Xia X, Torrero M, Barrett R, Shillitoe EJ, Li S. 2006. The mechanism of exogenous B7.1-enhanced IL-12-mediated complete regression of tumors by a single electroporation delivery. *Int J Cancer* 119:2113–2118.

Long YC, Jaichandran S, Ho LP, Tien SL, Tan SY, Kon OL. 2005. FVIII gene delivery by muscle electroporation corrects murine hemophilia A. *J Gene Med* 7:494–505.

Low L, Mander A, McCann KJ, Dearnaley D, Tjelle TE, Mathiesen I, Stevenson FK, Ottensmeier CH. 2009. DNA vaccination with electroporation induces increased antibody responses in patients with prostate cancer. *Hum Gene Ther* 20:1269–1278.

Luckay A, Sidhu MK, Kjeken R, Megati S, Chong SY, Roopchand V, Garcia-Hand D et al. 2007. Effect of plasmid DNA vaccine design and in vivo electroporation on the resulting vaccine-specific immune responses in rhesus macaques. *J Virol* 81:5257–5269.

Marti G, Ferguson M, Wang J, Byrnes C, Dieb R, Qaiser R, Bonde P, Duncan MD, Harmon JW. 2004. Electroporative transfection with KGF-1 DNA improves wound healing in a diabetic mouse model. *Gene Ther* 11:1780–1785.

Maruyama H, Ataka K, Gejyo F, Higuchi N, Ito Y, Hirahara H, Imazeki I et al. 2001. Long-term production of erythropoietin after electroporation-mediated transfer of plasmid DNA into the muscles of normal and uremic rats. *Gene Ther* 8:461–468.

McCray AN, Cao CH, Muthumani K, Weiner DB, Ugen K, Heller R. 2004. Regression of established melanoma tumors through intratumoral delivery of HIV-1 Vpr using in vivo electroporation. *Mol Ther* 9:S363.

Mir LM, Glass LF, Sersa G, Teissie J, Domenge C, Miklavcic D, Jaroszeski MJ et al. 1998. Effective treatment of cutaneous and subcutaneous malignant tumours by electrochemotherapy. *Br J Cancer* 77:2336–2342.

Moss RB. 2009. Prospects for control of emerging infectious diseases with plasmid DNA vaccines. *J Immune Based Ther Vaccines* 7:3.

Nakagawa S, Arai Y, Mori H, Matsushita Y, Kubo T, Nakanishi T. 2009. Small interfering RNA targeting CD81 ameliorated arthritis in rats. *Biochem Biophys Res Commun* 388:467–472.

Niidome T, Huang L. 2002. Gene therapy progress and prospects: Nonviral vectors. *Gene Ther* 9:1647–1652.

Nishi T, Yoshizato K, Yamashiro S, Takeshima H, Sato K, Hamada K, Kitamura I et al. 1996. High-efficiency in vivo gene transfer using intraarterial plasmid DNA injection following in vivo electroporation. *Cancer Res* 56:1050–1055.

Nishikage S, Koyama H, Miyata T, Ishii S, Hamada H, Shigematsu H. 2004. In vivo electroporation enhances plasmid-based gene transfer of basic fibroblast growth factor for the treatment of ischemic limb. *J Surg Res* 120:37–46.

Niu G, Heller R, Catlett-Falcone R, Coppola D, Jaroszeski M, Dalton W, Jove R, Yu H. 1999. Gene therapy with dominant-negative Stat3 suppresses growth of the murine melanoma B16 tumor in vivo. *Cancer Res* 59:5059–5063.

Onodera S, Ohshima S, Tohyama H, Yasuda K, Nishihira J, Iwakura Y, Matsuda I, Minami A, Koyama Y. 2007. A novel DNA vaccine targeting macrophage migration inhibitory factor protects joints from inflammation and destruction in murine models of arthritis. *Arthritis Rheum* 56:521–530.

Payen E, Bettan M, Rouyer-Fessard P, Beuzard Y, Scherman D. 2001. Improvement of mouse beta-thalassemia by electrotransfer of erythropoietin cDNA. *Exp Hematol* 29:295–300.

Rizzuto G, Cappelletti M, Maione D, Savino R, Lazzaro D, Costa P, Mathiesen I et al. 1999. Efficient and regulated erythropoietin production by naked DNA injection and muscle electroporation. *Proc Natl Acad Sci USA* 96:6417–6422.

Rosati M, Bergamaschi C, Valentin A, Kulkarni V, Jalah R, Alicea C, Patel V et al. 2009. DNA vaccination in rhesus macaques induces potent immune responses and decreases acute and chronic viremia after SIVmac251 challenge. *Proc Natl Acad Sci USA* 106:15831–15836.

Rosenberg SA, Yang JC, Restifo NP. 2004. Cancer immunotherapy: Moving beyond current vaccines. *Nat Med* 10:909–915.

Roux S, Bernat C, Al Sakere B, Ghiringhelli F, Opolon P, Carpentier AF, Zitvogel L, Mir LM, Robert C. 2008. Tumor destruction using electrochemotherapy followed by CpG oligodeoxynucleotide injection induces distant tumor responses. *Cancer Immunol Immunother* 57:1291–1300.

Schiffelers RM, Xu J, Storm G, Woodle MC, Scaria PV. 2005. Effects of treatment with small interfering RNA on joint inflammation in mice with collagen-induced arthritis. *Arthritis Rheum* 52:1314–1318.

Seo SH, Jin HT, Park SH, Youn JI, Sung YC. 2009. Optimal induction of HPV DNA vaccine-induced CD8 + T cell responses and therapeutic antitumor effect by antigen engineering and electroporation. *Vaccine* 27:5906–5912.

Shimizu H, Nukui Y, Mitsuhashi N, Kimura F, Yoshidome H, Ohtsuka M, Kato A, Miyazaki M. 2009. Induction of antitumor response by in vivo allogeneic major histocompatibility complex gene transfer using electroporation. *J Surg Res* 154:60–67.

Swartz ME, Eberhart J, Pasquale EB, Krull CE. 2001. EphA4/ephrin-A5 interactions in muscle precursor cell migration in the avian forelimb. *Development* 128:4669–4680.

Tanaka S, Uehara T, Nomura Y. 2000. Up-regulation of protein-disulfide isomerase in response to hypoxia/brain ischemia and its protective effect against apoptotic cell death. *J Biol Chem* 275:10388–10393.

Titomirov AV, Sukharev S, Kistanova E. 1991. In vivo electroporation and stable transformation of skin cells of newborn mice by plasmid DNA. *Biochim Biophys Acta* 1088:131–134.

Torrero MN, Henk WG, Li S. 2006. Regression of high-grade malignancy in mice by bleomycin and interleukin-12 electrochemogenetherapy. *Clin Cancer Res* 12:257–263.

Ugen KE, Kutzler MA, Marrero B, Westover J, Coppola D, Weiner DB, Heller R. 2006. Regression of subcutaneous B16 melanoma tumors after intratumoral delivery of an IL-15-expressing plasmid followed by in vivo electroporation. *Cancer Gene Ther* 13:969–974.

Wallace M, Evans B, Woods S, Mogg R, Zhang L, Finnefrock AC, Rabussay D et al. 2009. Tolerability of two sequential electroporation treatments using MedPulser DNA delivery system (DDS) in healthy adults. *Mol Ther* 17:922–928.

Weaver JC. 1995. Electroporation theory. Concepts and mechanisms. *Methods Mol Biol* 55:3–28.

Weaver JC, Chizmadzhev YA. 2007. Electroporation. In: Barnes FS and Greenebaun B (eds.), *Handbook of Biological Effects of Electromagnetic Fields*. Boca Raton, FL: CRC Press, pp. 293–332.

Wells DJ. 2004. Gene therapy progress and prospects: Electroporation and other physical methods. *Gene Ther* 11:1363–1369.

Widlak W, Scieglinska D, Vydra N, Malusecka E, Krawczyk Z. 2003. In vivo electroporation of the testis versus transgenic mice model in functional studies of spermatocyte-specific hst70 gene promoter: A comparative study. *Mol Reprod Dev* 65:382–388.

Wolff JA, Malone RW, Williams P, Chong W, Acsadi G, Jani A, Felgner PL. 1990. Direct gene transfer into mouse muscle in vivo. *Science* 247:1465–1468.

Wong TW, Chen CH, Huang CC, Lin CD, Hui SW. 2006. Painless electroporation with a new needle-free microelectrode array to enhance transdermal drug delivery. *J Control Release* 110:557–565.

Wu CJ, Lee SC, Huang HW, Tao MH. 2004. In vivo electroporation of skeletal muscles increases the efficacy of Japanese encephalitis virus DNA vaccine. *Vaccine* 22:1457–1464.

Yao M, Stenzel-Poore M, Denver RJ. 2007. Structural and functional conservation of vertebrate corticotropin-releasing factor genes: evidence for a critical role for a conserved cyclic AMP response element. *Endocrinology* 148:2518–2531.

Zhang L, Rabussay DP. 2002. Clinical evaluation of safety and human tolerance of electrical sensation induced by electric fields with non-invasive electrodes. *Bioelectrochemistry* 56:233–236.

Zhang X, Divangahi M, Ngai P, Santosuosso M, Millar J, Zganiacz A, Wang J, Bramson J, Xing Z. 2007. Intramuscular immunization with a monogenic plasmid DNA tuberculosis vaccine: Enhanced immunogenicity by electroporation and co-expression of GM-CSF transgene. *Vaccine* 25:1342–1352.

Zupanic A, Ribaric S, Miklavcic D. 2007. Increasing the repetition frequency of electric pulse delivery reduces unpleasant sensations that occur in electrochemotherapy. *Neoplasma* 54:246–250.

20

Clinical Electrochemotherapy: The Italian Experience

Carlo Ricardo Rossi

Luca Giovanni
Campana

20.1 Introduction

Cutaneous metastases from solid tumors account for 0.7%–9.0% of all metastases (Rolz-Cruz and Kim 2008). Besides their dismal prognostic value, they can represent a distressing problem for patients. Oncologists can choose among several different options according to the patient's status and disease spread (exclusively superficial versus superficial plus visceral).

Electrochemotherapy (ECT) has been recently proposed as a novel and complementary therapeutic weapon for the control of the superficial disease. This therapeutic approach is a feasible alternative in both cases of inoperable tumors located in pre-irradiated areas or resistant to chemotherapy. There are a number of potential benefits in using ECT in the palliative setting range from tissue preservation to brief hospitalization, repeatability a favorable cost-benefit ratio. Since its development at the Institute Gustave Roussy (Mir et al. 1991), ECT has been quickly tested in the clinical setting but, despite its promising results, it has not entered clinical practice until recently. ECT has attracted recent attention thanks to the support of the EU Commission that has led to the development of a new device, the Cliniporator™ (IGEA, Modena, Italy), which simplifies the procedure. Furthermore, the ESOPE (European Standard Operating Procedures on Electrochemotherapy) project has favored the standardization of this treatment, thus providing the information necessary for the spread of the use of ECT in the clinical practice (Mir et al. 2006). The ESOPE study recruited 41 patients with cutaneous and subcutaneous nodules smaller than 3 cm. ECT proved to be effective both in melanoma and nonmelanoma tumor nodules with a complete response (CR) in 74% of tumors, a partial response (PR) in 11%, no change (NC) in 10% according to WHO criteria; local tumor control rate was 73%–88% 5 months after the treatment (Marty et al. 2006).

ECT was introduced in Italy in 2005. So far only a few centers have been able to collect an adequate caseload in order to evaluate the clinical impact of the procedure. Nowadays, many centers in Italy are equipped with the Cliniporator pulse generator and are ready to start treating patients with ECT. In this chapter, the clinical experience on the use of ECT gained so far in Italy and specifically at the university hospitals of Turin and Padua is discussed. So far this new treatment approach has been applied mainly in patients with superficial metastases generated from different tumor types, which were unsuitable for conventional treatments. The aim of this chapter is therefore to describe the results obtained with ECT in the clinical setting with a special emphasis on the patients' quality of life (QoL) and to comment on some open questions raised by its application. Finally, in the last section, the rationale and the design of the most relevant ongoing clinical studies on ECT in Italy are also reported.

20.2 Clinical Experiences in Italy: Turin and Padua

The Cliniporator device is the first CE-labeled pulse generator fully designed for clinical purpose. It provides the user with a complete immediate feedback that allows the clinician to verify the adequacy of the electric pulse applied to each and every single tumor lesion. ECT has quickly entered clinical practice in Italy but currently only the two cancer centers of Turin and Padua have collected sufficiently large series of patients. It is indeed necessary to collect large series within properly designed trials to make ECT fully accepted by the oncological community and in order to confirm the clinical results so far obtained, especially in terms of improved QoL.

20.2.1 The Turin Experience

Authors from the Dermatologic Clinic of the University of Turin have recently published a prospective nonrandomized study on 14 melanoma patients (Quaglino et al. 2008). The enrollment was carried out according to the following criteria: histologically confirmed cutaneous melanoma, cutaneous and/or subcutaneous recurrent disease not amenable to surgery, measurable cutaneous and/or subcutaneous tumor nodules suitable for application of electric pulses, age ≥18 years, Karnofsky performance status ≥70, life expectancy at least 3 months, and a washout period of at least 3 weeks after the previous treatments. Patients with symptomatic and/or rapidly progressive extracutaneous involvement, coagulation disorders, allergic reaction to bleomycin, epilepsy or peripheral neuropathy, chronic renal dysfunction, and arrhythmia or pacemaker were excluded from treatment. All patients had a cutaneous disease, which relapsed after one or more previous radical surgical treatments; moreover, four patients relapsed after isolated limb perfusion and two were treated with systemic chemotherapy. No patient had previously received bleomycin. All the treatments were carried out using the Cliniporator device. All patients were treated under mild general anesthesia. Three different types of electrodes were used. Type I electrodes are plate electrodes made up of two parallel stainless-steel plates, used for the treatment of small superficial lesions. Type II needle electrodes are suitable for treatment of thicker and deeper seated tumor nodules: the electrodes are made up of two parallel arrays of needles with a 4 mm gap between them for treatment of small nodules, whereas Type III are made up of a hexagonal array of electrodes (six needles forming a hexagon and one needle at its center with an 8 mm gap between them) for larger nodules (>1 cm in diameter) (Mir et al. 2006). Intravenous bleomycin (15 mg/m² of body surface area) was used in all cases, and administered within a time frame of 30 s to 1 min. Electric pulses were delivered from 8 until 28 min after bleomycin infusion to obtain the most response benefiting from the most appropriate drug concentrations in tissues.

The response to treatment was assessed 8 weeks after the ECT session. Treated lesions were measured on day 0 (pretreatment), at follow-up weeks 2, 4, 8, and 12, and then every 2 months. The clinical measurements were made using calipers and ultrasound imaging. For the evaluation of treatment response, tumor size was calculated as the product of the two largest perpendicular diameters. Four to seven measurable metastases larger than 5 mm in diameter were chosen as index lesions in each patient (index lesions are a limited number of lesions which are chosen among all tumor nodules in order to evaluate the response to the anticancer drug). The response was classified in accordance to the World Health

Organization (WHO) guidelines (Miller et al. 1981) as progressive disease (PD) for increase in tumor size >25%; no change (NC) for increase in tumor size <25% or decrease <50%; partial response (PR) for decrease in tumor size >50% for at least 4 weeks; and CR for total clinical disappearance of the tumor.

A total of 14 stage III patients were enrolled between October 2005 and June 2007. The response, scored 8 weeks after ECT, was obtained in 13/14 patients (93%) with a complete clinical regression of all skin metastases in 7 (50%) and a PR in 6 patients. Only one patient did not respond. The clinical CR was histologically confirmed in one index lesion from each patient. All clinically detectable skin metastases were treated. The average number of lesions per patient was 10 for a total of 160 treated metastases; four patients had more than 20 lesions.

20.2.1.1 Local Response

An objective response was obtained in 153/160 (95%) lesions, with a CR rate of 62%. No differences in response rate were observed between cutaneous and subcutaneous metastases. The lesion size was the most predictive parameter for response. Indeed, an objective response was obtained in 123/124 (99%) metastases sized $\leq 1\,cm^2$ and in 30/36 (83%) larger lesions (>1 cm^2).

Another ECT session was performed in seven patients on 73 new metastases developed outside the initial treatment area. The ECT activity was maintained though the response rate (86%) was significantly lower than that obtained after the initial treatment due to the presence of bigger (>1 cm^2 sized) metastases. Two hundred and thirty-three metastases were treated with a single ECT session. A reduction of 1 cm^2 in size was observed in 184 metastases (79%) with a median diameter of 7 mm (range 2–13 mm); 49 out of 233 metastases (21%) were >1 cm^2, with a median diameter of 15 mm (range 11–75 mm). The overall response rate obtained was 93% with a CR rate of 58%. The response was 98% for $\leq 1\,cm^2$ lesions and 73% for >1 cm^2 lesions; similarly, the CR rate was 68% and 22%, respectively. Twenty-nine lesions from three different patients (25 PR and 4 NC after the first ECT) underwent treatment a second session and five of them a third session in order to evaluate if the initial clinical response could be improved following retreatment. A further response was obtained in 21/29 metastases (72%). Nine CRs were obtained, five in lesions sized >1 cm^2. None of the metastases that achieved a CR relapsed after a median follow-up of 21 months (range 5–28 months); on the other hand, 54/80 PR lesions showed a >25% size increase. After a 2 years follow-up, the authors reported a local tumor control rate of 74.5%, with a median time to failure of 18.3 months.

20.2.1.2 Toxicity

ECT treatment was well tolerated and its toxicity was limited. Erythema and slight edema occurred at the site of treated lesions in three patients and disappeared within a few days. In addition, marks from the electrodes and superficial epidermal erosions were noted in all cases, followed by scars healing within a month of ECT. Patients did not report local pain or other subjective symptoms. No hematologic toxicity was observed.

20.2.2 The Padua Experience

The main aim of this study, which is the largest ever reported so far, was to evaluate not only the toxicity and the activity of bleomycin-based ECT performed according to the ESOPE guidelines but also the response duration and the impact of ECT on disease-related symptoms and the patients' functioning in everyday life (Campana et al. 2009). It is namely crucial to assess whether this treatment is useful to improve QoL, given that currently ECT application is mainly accepted by the oncologist in the palliative setting.

From July 2006 through May 2008, 52 consecutive patients with superficial metastases from different tumor types (34 affected by superficial melanoma metastases) and excluded from surgery or radiation therapy underwent ECT within the frame of a prospective nonrandomized phase II study. The inclusion criteria and the technical procedure were derived from the ESOPE: patients were to present measurable tumor nodules suitable for electrode application and an acceptable performance status.

Exclusion criteria included serious lung, heart or liver disease, epilepsy, short life expectancy (<3 months), active infection, previous treatment with bleomycin to maximal dosage, and different anti-cancer therapies administered within 2 weeks before ECT. The only deviance from the protocol was the inclusion of tumor nodules bigger than 3 cm. In this study, the tumor response was assessed through the RECIST criteria (Therasse et al. 2000). For each patient, one to eight lesions were selected as "target lesions" according to the number of tumor nodules: they were measured in their largest diameter with a millimetric ruler and registered. The largest lesions have been always recorded among the "target" ones. Ultrasound scan was used to measure subcutaneous lesions. The remaining nodules were registered, photographed but not measured.

Bleomycin was preferred to cisplatin due to the high number of tumor nodules and its favorable toxicity profile. It was administered either intravenously ($15\,mg/m^2$ in a bolus lasting $50''$–$60''$) or intratumorally injected at a dose depending on size of the nodule ($1000\,IU/cm^3$ for tumors smaller than $0.5\,cm^3$, $500\,IU/cm^3$ for tumors between 0.5 and $1\,cm^3$, $250\,UI/cm^3$ for tumors bigger than $1\,cm^3$). Electric pulses were delivered by means of two types of needle electrodes (parallel arrays (Type II) and hexagonal arrays (Type III) connected to the Cliniporator device.

Tumor response, local and systemic toxicity (according to the Common Toxicity Criteria, Version 2.0) (Trotti et al. 2000) were evaluated through physical/ultrasound examination and blood tests at 2, 4, 8, 12, and 16 weeks and then according to standard follow-up. Patients whose tumors were judged as "nonresponders" or "partial responders" after 4 weeks were scheduled for retreatment. As a local treatment, ECT is not responsible for the appearance of metastases in previously unaffected anatomical areas outside the treatment field; accordingly, these nodules were considered as "new lesions" (NL) and therefore an indication for additional ECT application. Response duration was considered from the date of response achievement to relapse/progression or the last follow-up in case of local disease-free status.

Patients' QoL was evaluated by means of a dedicated questionnaire before treatment (first ECT session) and 1 and 2 months thereafter (Table 20.1). The subjective evaluation was graded with a score (either ranging from 0 to 10 or yes versus no), according to eight items: (1) wound healing; (2) wound bleeding; (3) aesthetic impairment; (4) impairment on activity of daily living; (5) impairment on social relations; (6) pain control; (7) overall satisfaction of the treatment received; and (8) acceptance of retreatment, if needed.

TABLE 20.1   Quality of Life Questionnaire Used for Evaluating Superficial Disease-Related Complaints and Patients' Perception of ECT Treatment at the University of Padua

Item	Score	
Wound healing	0 1 2 3 4 5 6 7 8 9 10	
	0 = ulceration, requiring dressing > once/day	10 = perfectly healed
Wound bleeding	0 1 2 3 4 5 6 7 8 9 10	
	0 = abundant, requiring dressing > once/day	10 = no bleeding
Aesthetic impairment	0 1 2 3 4 5 6 7 8 9 10	
	0 = very negative	10 = no impairment
Impairment on activity of daily life	0 1 2 3 4 5 6 7 8 9 10	
	0 = impossibility of performing a specific activity in daily life	10 = no impairment
Impairment of social relations	0 1 2 3 4 5 6 7 8 9 10	
	0 = serious limitations	10 = no impairment
Pain control	0 1 2 3 4 5 6 7 8 9 10	
	0 = maximum pain	10 = no pain
Overall satisfaction	0 1 2 3 4 5 6 7 8 9 10	
	0 = unsatisfied	10 = completely satisfied
Acceptance of retreatment, if needed	No	Yes

The sums of the scores of the first six items were calculated before and after treatment (overall score). Patient population was mainly represented by melanoma patients ($n = 34$) and breast cancer—chest wall recurrence—patients ($n = 11$); patients with soft tissue sarcoma ($n = 5$), squamous carcinoma ($n = 2$) were treated. About 20% of tumor nodules were found to be bigger than 3 cm size. Thirty-four out of 52 patients enrolled, had in-transit melanoma metastases: of these, 20 were in AJCC stage III and 14 in stage IV (11 with parenchymal disease). Eleven patients had chest wall recurrence from breast carcinoma (5 of 8 had also visceral metastatic disease; 8/8 had previously received chemo- and radiotherapy). The remaining patients had locally recurrent soft tissue sarcomas ($n = 5$), locally advanced recurrence from head and neck carcinoma ($n = 1$), and locoregional relapsing skin epidermoid carcinoma located on the trunk ($n = 1$). None of these remaining seven patients had parenchymal tumor involvement. The tumor nodules mainly involved the head and neck region, trunk and limbs in 5, 28, 19 patients, respectively.

Thirty-two out of 52 patients (62%) had previously received at least two cycles of systemic chemo- or immunotherapy without response to the superficial disease and 12/39 (31%) patients with melanoma or sarcoma had received locoregional chemotherapy (isolated limb perfusion or stop-flow perfusion). At least 1 month after ECT, 25/52 (48%) patients received conventional treatments, mainly systemic chemotherapy, according to the oncologist's judgment.

Overall 97 courses were provided (mean 1.8/patient). Eleven sessions (11%) were performed under local anesthesia only; the addition of IV sedation was chosen by the clinician in the remaining 81 cases. Bleomycin was administrated intravenously, intratumorally injected or in a combined route in 24, 6, and 22 patients, respectively.

20.2.2.1 Local Response

Antitumor activity was observed in all tumor types and both in melanoma and nonmelanoma metastases (CR was obtained in 17/34 melanoma patients and in 9/18 nonmelanoma patients). One month after the first ECT application, 125 out of 267 (47%) target lesions showed CR, 126 (47%) PR, and 16 (6%) NC. With regard to tumor response-related factors, an inverse correlation between complete local response and the maximum diameter of the target lesion was observed: 66% for tumors <1.5 cm, 36% for nodules between 1.6 and 3 cm and 28% for those >3 cm. Tumor response does not appear to be affected by the drug administration route, as similar response rates were observed in patients who underwent ECT with intralesional, intravenous, and combined bleomycin injection.

Overall, 608 lesions were treated at the first therapeutic session of ECT (mean number for patients 12, range: 1 to >50), with dimensions ranging from 2 to 57 mm (mean 22 mm). Considering all treated nodules ($N = 608$), 262 (43%) had CR, 306 (50%) PR, and 40 (7%) NC as assessable at physical examination and from digital photos available.

The studies from Turin and Padua University, although accurate, are different in some features. First, the two equipes of oncologists have employed different tumor response criteria: the WHO (Miller et al. 1981) and the RECIST (Therasse et al. 2000), respectively. While the WHO criteria require a bidimensional tumor measure (the product of the longest diameter and its longest perpendicular diameter for each tumor), the RECIST criteria require only to record the largest diameter. Moreover, the categories of response are slightly different ("partial response" is defined as at least 50% or 30% decrease in tumor size in WHO and RECIST, respectively, while "progressive disease" is defined as an increase of 20% and 25%, respectively).

Second, the time for tumor response evaluation in Turin and Padua studies is only apparently different (8 and 4 weeks, respectively): in fact, according to the RECIST criteria, patients are evaluated after 4 weeks for tumor response: however, as per RECIST protocol, the response recorded at this time point must be confirmed after 4 more weeks (that is, 8 weeks after treatment). This makes the RECIST evaluation time point actually overlapping with that of the WHO criteria.

Finally, both the WHO and the RECIST criteria have proved reliable in the assessment of tumor response (Park et al. 2003). However, it is essential that in the future, clinical trials researchers adopt the same tumor response criteria in order to judge local tumor response in an homogeneous manner.

In fact, since they require only one diameter to be assessed, the RECIST criteria are easier to be met by investigators while guaranteeing accuracy: also for this reason, ongoing multicentric studies in Italy are adopting this evaluation scale to measure the efficacy of ECT.

20.2.2.2 Patients' Outcome

One month after the first ECT application, an objective response was obtained in 50 out of 52 patients (96%): in particular, according to the RECIST criteria, there were 26 complete (50%), 24 partial (46%) responses. In two patients (4%), there was no change in tumor size at 4 weeks.

Among 26 complete responders at the first ECT, 17 are locally disease-free after a median follow-up of 9 months (range: 2–21 months). Seven of these patients underwent retreatment because of the appearance of new lesions in untreated areas (after a median time of 10 weeks, range: 7–21). Two patients died because of systemic disease progression (without local recurrence after a local disease-free interval of 16 and 25 weeks, respectively), and one patient developed local recurrence after 7 months.

Among 27 patients who underwent a second course of ECT (14 because of PR, 7 because of NL, 4 for both PR and NL, 2 for a NC response), 18 obtained local complete response and 9 PR. Totally, 257 tumor nodules were treated during the second ECT application: 170 completely disappeared (66%), while 87 achieved PR (34%).

Of 14 patients retreated because of previous PR, 11 achieved CR, and 3 PR with some local benefit (less bleeding and better pain control). In this subgroup, a total number of 158 tumor nodules were retreated: 131 (83%) obtained CR, 27 (17%) PR.

Although the response rate in this series was remarkable (96%), retreatment was required in up to nine patients for the appearance of new metastases outside the treatment field, two patients undergoing up to five treatments for this reason. Overall, after a median follow-up of 9 months (range: 2–21), local tumor control (CR + PR) was achieved in 50 out of 52 patients (96%): 31 with local CR and 19 with PR.

The clinical status of the two patients without local control showed one local relapse (after two ECT and a 7 months lasting clinical CR) and one local relapse after 17 months lasting CR obtained by means of two ECT applications. In these two cases (both patients affected with melanoma), ECT failed only in one out of the several tumor nodules treated: a 40 mm metastasis of the chest wall and a large (54 mm) lesion of the leg in an 85-year-old woman. In the remaining tumor nodules of these two patients, ECT obtained a mixed response (i.e., some partial responses and some complete responses).

20.2.2.3 Toxicity

No serious adverse events were reported. The already known acceptable toxicity profile of bleomycin was confirmed (Belehradek et al. 1993, Gothelf et al. 2003). Four patients experienced postoperative complications including lipothymia ($n = 2$) and nausea/vomiting ($n = 2$). Treated lesions invariably developed an inflammatory reaction, which resolved at the end of the follow-up in 46 of 52 (88%) patients. The locoregional dermatologic toxicity was mild: injection site reactions graded I–II according to the Common Toxicity Criteria were reported in four patients; some forms of grade I–II rush, desquamation, or pigmentation were present at 4 weeks control in six patients. Notably, ulceration was present in some of the tumor nodules in 22 of 52 patients at the time of accrual, while it was present in five cases at the end of the study, either in patients receiving intratumoral and intravenous bleomycin. Three out of four patients with bleeding tumor nodules achieved good hemostasis within 4 weeks; the fourth one died of stroke due to CNS metastasis while her superficial tumor lesion was still ulcerated. Patients ($n = 5$) with tumor nodules in the head and neck region (four with melanoma, one with epithelial cancer) did not suffer significant local toxicity.

20.2.2.4 Treatment Compliance

Patient discomfort was minimal and no muscle relaxant drugs were administrated. Electric currents sometimes caused an unpleasant sensation (8/52 patients, 15%) and general sedation and some forms of intravenous intraoperative analgesia ensured a better pain control in 84/87 (96%) sessions. Posttreatment pain was mild and manageable with minor analgesics in 46 of 52 patients; the remaining were already treated with opioids at the time of accrual. Hospital stay lasted few hours for all but two patients (50/52) who were dismissed the day after the ECT procedure due to an episode of lipothymia.

20.2.2.5 Quality of Life

The questionnaire for the evaluation of the patients' perception was completed in 36 cases (69%). The local disease-related complaints, if present, ranged from difficulty in sitting (as a consequence of treatment of large tumors on the back, $n = 4$) to walking (due to a tumor nodule on the foot, $n = 2$); bathing (due to ulcerated disease, $n = 2$); dressing (because of oozing, $n = 7$); mastication (tumor nodule on the oral mucosa, $n = 1$); and pain ($n = 22$). Thirty-four out of 36 patients (94%) declared a positive impact of ECT on one or more of the items investigated before and after treatment (wound healing, bleeding, aesthetic impairment, activity of daily living, social relations, and pain control). Nearly all patients (93%) experienced a local benefit, as demonstrated by the improvement of the scores assessing ulceration and bleeding, with a better aesthetic appearance stated by 31 of 36 patients. Among these patients, five with tumor nodules on the head and neck region indicated an improvement in aesthetic appearance at 1 and 2 months survey. Nine patients indicated an improvement in some of the activities of daily living as a consequence of tumor treatment and six in social relations. Better pain control was registered in 9 out of 22 patients with painful tumor nodules. Overall, 34 of 36 patients were satisfied with the treatment received, 1 uncertain, 1 unsatisfied; 34 of 36 (94%) patients were keen to have sessions in ECT in case of recurrence. The sum of the scores obtained in the six items (bleeding, ulceration, aesthetics, pain, activity of daily living, social relations) registered after ECT (at 1 month and 2 months) were statistically different as compared to that registered before treatment (overall Score: 46, 52, and 55 at pretreatment, 1 month and 2 months, respectively; Kruskal–Wallis one-way analysis, $p < 0.005$).

20.2.2.6 Considerations on the Padua Experience

The findings from Padua University confirm that ECT performed with the Cliniporator device according to the ESOPE provide a high local tumor response although lower if compared to the European trial (CR in 74% of nodules) and other series (5, 9–13) (CR = 58%–89%). The lower response rate in this series may reflect, at least in part, the larger dimensions, and the higher numbers of tumor nodules that we treated with ECT: in particular, in our series the largest tumor nodule was greater than 30 mm in 14/52 (27%) patients; moreover, a single metastasis was present in only eight out of the 52 (15%) patients, with the mean diameter being 38 mm. In contrast, the ESOPE study included only tumors smaller than 30 mm, and one-third of patients had a single lesion. Overall, in the Padua's experience, the local response rate 1 month after the first ECT was remarkable, and increased further after a second ECT application. Our findings support the indication to treat with ECT even nodules larger than 3 cm given the high tumor burden (27% of patients with nodules bigger than 3 cm) even though, in these cases, multiple applications may be required in order to obtain tumor response. Local response was inversely correlated with the maximum diameter of the target lesion: the mean diameter of nodules achieving CR was of 15 mm, while that of partially regressed lesions was of 30 mm.

Unfortunately, the encouraging results obtained with ECT treatment do not modify the natural history of the disease, because a considerable number of patients experience visceral progression or the appearance of new superficial metastatic lesions. Nevertheless, they may still benefit from additional ECT applications and maintain good superficial tumor control: in fact, all patients who died of disease progression ($n = 9$) were still experiencing a CR ($n = 5$) or a PR ($n = 4$) at the last follow-up. Moreover, the mean local disease-free period was 7 months and only two patients (both affected by melanoma) relapsed in the field of treatment. The possible explanations for treatment failure in these two cases might be, in one case that the procedure was characterized by suboptimal electroporation due to difficult needle insertion into a previously radio-treated, fibrotic chest region; in the second patient, the correct insertion of the electrode into the lesion was difficult to assess, and some areas could have been inadequately electroporated. The hypothesis that the large tumor dimension alone does not entirely account for treatment failure may be supported by the observation that the only two patients with a NC response at first ECT (with adequate electroporation parameters) had both tumor nodules with limited dimensions (17 and 15 mm, respectively). Upon retreatment, these two patients obtained CR and PR, respectively.

At present, the main technical pitfalls for an optimal tumor electroporation are the following: (1) large tumor-involved anatomical areas, which require repeated and time-consuming electrode applications (considering the 20 min window available after drug injection according to the standard procedures); (2) large tumor size (>3 cm), which makes it impossible to reach the inner portion of the tumor at the first application with the currently available electrodes; and (3) previously irradiated fields, which cause a partial electrode needle penetration and consequently a suboptimal delivery of the electrical currents in the fibrotic tissue.

It would be advisable for surgeons, oncologists, biomedical engineers, and radiological physicists to work in a coordinate manner for the further implementation of ECT. For instance, one patient developed a symptomatic melanoma metastasis located in the oral mucosa: the collaboration with the engineers of the IGEA company allowed to develop a dedicated four-needle probe of small dimensions in order to apply electric pulses in this narrow anatomical region. The new device, a ring to be worn on the distal phalange of operator's index finger, proved to be safe; its small size did not hinder maneuvrability and allowed an effective treatment that resulted in a complete tumor regression (Figure 20.1).

Today, it is clear that deep metastatic localizations are not affected by this treatment strategy, but a combined approach could be a powerful way to boost the antitumor response (Heller et al. 2000, Serša et al. 2000, Roux et al. 2008). Experimental and clinical experiences where chemo-, immuno-, and radio-therapy play a concomitant role deserve further exploration. It will also be interesting to clarify whether an ECT-induced systemic immune activation could play a role in tumor response. In conclusion, the favorable outcome reported in this study indicates that ECT is a reliable treatment option that may improve the patients' functioning, thus enhancing the effect of the palliative setting. ECT can successfully destroy superficial metastases from different tumor types and its limited toxicity, even on areas otherwise requiring disfiguring resections, can permit a combination with other anticancer therapies. The benefit obtained in terms of QoL, although preliminary, should be regarded as a considerable result and need to be further assessed after a formal validation of the dedicated questionnaire.

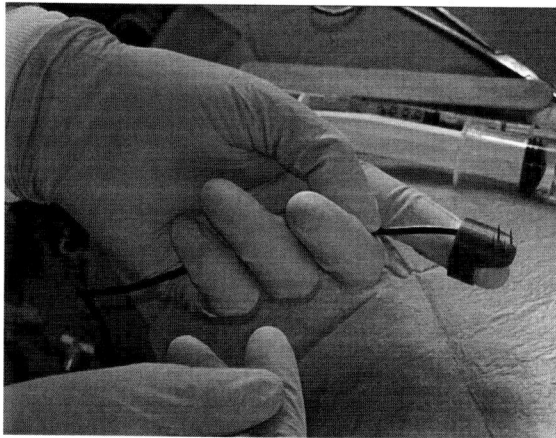

FIGURE 20.1 The finger-type electrode. The electrode consists of a ring to be worn on the distal phalange of one of the surgeon's fingers. This configuration allows maximum flexibility when reaching the desired location, especially in narrow anatomical regions (i.e., oral cavity, anal canal).

20.3 Open Questions

The clinical results reported above are mainly based on the ECT activity in melanoma patients; other histotypes seem to respond well (for example, breast cancer) while the clinical experience is still limited in less frequently treated tumors such as the Merkel cell carcinoma and the Kaposi sarcoma. Despite the promising results so far obtained in terms of local response, very little is known about the real impact of ECT on the long-term disease outcome, and in particular on the QoL after the treatment. It is still unclear whether this treatment could play a role in the earlier stages of the natural history of a tumor or in neo-adjuvant setting. Moreover, another interesting task, which has not been studied yet, is the association of ECT with different treatment approaches; this potential multimodal treatment deserves further investigation in the context of properly designed clinical studies. Nowadays, despite its documented high local activity, ECT is considered only in case of a palliative approach and is generally applied on advanced patients, when several other treatments have failed.

Finally, the local activity of ECT is related to the conformation of the technical devices (i.e., maximum needle electrodes length of 3 cm) available at present. Currently, ECT's role is largely palliative, and, despite the promising clinical results that have been so far achieved, the effectiveness of ECT in the treatment of earlier stages of disease has not been defined yet. However, even in advanced stages, the precise role of ECT in association with other treatment options is still at the discretion of the clinician and based on an individual basis. In this context, the presence of a disease-dedicated multidisciplinary team plays a vital role in coordinating the multiple and complex technologies now commonly employed in cancer diagnosis and treatment. This coordination involves direct patient care, documentation in the medical record, participation in therapy, symptom management, both patient and family education, as well as counseling throughout diagnosis, therapy, and follow-up.

20.4 Current Developments

The addressed questions prompted for ongoing new clinical research protocols in Italy. On one hand, these trials are trying to confirm the preliminary encouraging clinical results so far obtained with ECT and, on the other hand, they are designed to expand the indications concerning histotypes (i.e., non-melanoma skin cancers) and tumor location (i.e., deep-seated tumors).

20.4.1 The IMI-GIDO Italian Multicenter Study

Among the increasing number of centers that are implementing their activity with ECT in Italy, the Italian Melanoma Intergroup (IMI) launched a multicenter study that will prospectively collect data on 600 patients treated with ECT. This study, coordinated by the IMI and the Italian Group of Oncologic Dermatology (GIDO), has started in 2009 and the period of recruitment is 36 months. At present, patients' accrual has been activated at eight hospitals, but many other centers are willing to participate in this study. The rationale of this study is based on the observation that 80% of nodules treated with ECT respond completely, irrespective to the histotype and the patients' staging, but data on long-term clinical outcome are scarce.

In 2006, the results of the ESOPE study were published (Mir et al. 2006); they established the guidelines to safely apply the ECT in the clinical setting. Such data deserve further confirmation on a large number of patients: ECT activity should be assessed also in nonmelanoma histotypes, and its clinical benefit should be investigated in a larger study population. The main aims of this multicenter study are to collect a large series of patients treated with ECT and to assess local response and their long-term clinical outcome. The data collected will also allow to confirm the operating procedure previously elaborated by means of the ESOPE study in a limited number of patients; the QoL of patients treated with ECT will also be investigated, with an emphasis on superficial disease-related complaints. Finally, the association with other treatment options (radio-, chemo-, immuno-therapy) and its activity will be evaluated. Patients will be enrolled on the basis of inclusion criteria below and then treated according to the standardized procedures of the ESOPE study. Before treatment, a selection of the larger tumor nodules (≤3 cm depth from the floor skin) will be performed; they will be registered as "target lesions" up to a maximum of 7; the sum of the diameters of target lesions will also be calculated and it will represent the benchmark for measuring the response to the treatment, according to the method of assessment of the RECIST criteria. The same lesions will be measured individually for each subsequent checkup and a new sum of the diameters calculated. All other tumor nodules present at enrollment will represent the "nontarget" lesions, and they will be recorded but not measured and only their persistence or eventual regression will be recorded during the follow-up. The emergence of new tumor nodules in areas not treated with ECT will be carefully recorded, but it will not be considered as local progression of disease; this is justified considering the fact that the range of action of the ECT treatment is entirely local, i.e., limited to the insertion sites of the electrodes. However, the new lesions can be treated in a subsequent ECT session. Patients will be followed up with monitoring visits at 15, 30, 60, and 90 days after therapy administration; and then at 6 and 12 months. The local skin toxicity will also be recorded according to the Common Toxicity Criteria (CTC 3.0). The survey on issues related to QoL of the patient involves the administration of a validated questionnaire (QLQ-C30) and an assessment of pain in the treatment areas (in the hours immediately following ECT administration and in the following days) by a visual analogue scale. In the follow-up visit of the study, any benefits (physical, psychological, functional) connected with ECT administration will be registered along with any toxicity or limitations arising from the treatment itself through a dedicate questionnaire. Patients may undergo multiple sessions of ECT with an interval of at least 1 month, according to the opinion of the treating physician.

The criteria for inclusion of this multicenter study will be those established by the ESOPE trial (the only deviance is the inclusion of tumors bigger than 3 cm): inoperable skin tumors, in-transit metastases from melanoma which are unresponsive to or not candidate for conventional treatments, cutaneous metastases of any origin, maximum depth of the single lesion of 3 cm, patient's life expectancy of more than 3 months, normal hematology, hepatic and renal function, ECOG performance status of 0–2. The patient with any of the following characteristics will be excluded from the present study: history of allergic reactions to bleomycin or cisplatin exceeded maximum cumulative dose of bleomycin (250,000 IU/m²), peripheral neuropathy (only in case of use of cisplatin administration), hepatic or renal function impairment, history of epilepsy, patients with pacemakers (in case of disease located on the chest wall), severe cardiac arrhythmias, condition of pregnancy or lactation, impaired lung function (acute or chronic), and unavailability to follow-up visit.

20.4.2 Quality of Life Study

The validation of the QoL questionnaire has been set up as a collateral issue of the Italian Multicentre Study on ECT. This study aims at a formal construction of the interview applied in the preliminary experience at the Istituto Oncologico Veneto (IOV). Superficial metastases may have a detrimental impact on the patient's QoL, especially if requiring intensive dressing, so that their treatment could ensure the preservation of the patient's functioning and well-being. In the recent years, there has been a growing interest in taking into consideration the patients' perspectives when measuring the clinical benefit of cancer therapies (Wagner et al. 2007). Among the validated tools for the assessment of QoL, no questionnaire is specific for the evaluation of a local palliative treatment of superficial metastases. We designed an eight-item interview to evaluate the patients' perceptions of the impact of superficial disease and of its treatment in our study population. Almost all patients reported an improvement in local skin conditions, physical appearance, or some of daily activities and social relations. This positive perception of the treatment, along with the limited ECT-related toxicity, likely explains the high rate of retreatment acceptance. The underway study on the formal assessment of QoL during the follow-up is warranted to validate the interview and to confirm the favorable findings reported in the early clinical experience. Encouraging clinical results have been obtained with ECT as assessed by means of this non-validated tool. The following disease-related complaints seem to be effectively managed by the application of ECT: aesthetic appearance of superficial tumor lesions, tumor bleeding, wound healing, everyday activities, and social relations.

These clinical results prompted us to enterprise a formal validation process of the questionnaire. The phases of this process are the following: first, the set up of a patient focus group of volunteers recruited among those treated with ECT with the collaboration and supervision of a dedicated psycho-oncologist. This group has identified some superficial disease-related complaints that are felt important by the patients and some troubles connected with having superficial metastases that they generally expect to resolve (efficient palliation) with ECT even if they know that it is not a treatment option.

The second phase will be a pre-validation study on 20–25 patients by means of a few specific questions on their perception of the items explored by questionnaire; the answers collected will be used to set up a re-modulation of the dedicated questionnaire (i.e., the introduction, if necessary, of new items) according to the patients' criticism. Finally, the interview will be tested for consistency in a larger series of patients. This phase would assess coherence between the patients' and the clinician's perception on the items explored by the implemented questionnaire. Once validated, this tool could be used to better assess patients' perception and could be useful for a more appropriate patients' selection by all centers involved in offering ECT to patients.

20.4.3 ECT for Large and Deep-Seated Soft Tissue Tumors

The possible explanations for some treatment failures of ECT could be attributed to intrinsic tumor resistance or to technical pitfalls of the procedure. For instance, in our experience, the main reasons for an unsuccessful ECT have been the following: (1) large tumor-involved anatomical areas, which require repeated and time-consuming electrode applications (considering the 20 min window available after drug injection according to the standard procedures); (2) large tumor size (>3 cm), which makes it impossible to reach the inner portion of the tumor at the first application with the currently available probes; (3) tumor deepness (>3 cm); and (4) previously irradiated fields, which cause a partial electrode needle penetration and consequently a suboptimal delivery of the electrical currents in the fibrotic tissue.

The availability of new technical instrumentation (electrodes and hardware) might expand the indications for ECT allowing the treatment of bigger and deeper soft tissue tumors. In fact, in a recent review of 118 patients with soft tissue metastases (involving skeletal muscle or skeletal muscle and subcutaneous tissue) from a single academic medical center, those affected by tumors bigger than 3 cm were 38/118 (32%). Distant metastases to soft tissue are rarely reported in literature. The published case reports and small series provide limited perspective on base evaluation and management (Damron and Heiner 2000, Plaza et al. 2008). Magnetic resonance imaging (MRI) has become the preferred technique for distinguishing soft tissue metastases from sarcomas and other processes. MRI has revealed soft tissue metastases to have poorly defined margins, low signal on T1-weighted sequences, high signal on T2-weighted sequences, and enhancement with gadolinium. However, the prognosis of these patients must often be considered when evaluating the potential merits of the varying treatment options; the vast majority of patients where a metastatic soft tissue mass is diagnosed, particularly those with carcinoma, will die of their disease. The treatment options, depending upon the clinical setting, include radiotherapy, chemotherapy, and surgical excision. For the painful mass in the context of widespread metastatic disease, either chemotherapy or radiotherapy or a combination may be elected, depending upon the primary tumor, other organs involved, extent of involvement, symptoms attributable to the various sites, and patient's age, overall health, and goals. Excision may be indicated for carefully selected, isolated, soft tissue metastases, particularly after a long disease-free interval. Success in local control and elimination of pain after excision of soft tissue metastases has also been reported in the setting of widely disseminated disease. However, local recurrence has been reported, even after adjunctive radiotherapy. Preoperative planning before any attempt at excision of solitary metastatic disease should take into consideration the pathophysiology of the intramuscular spread. Because of the infiltrative borders without capsule or pseudocapsule formation, extremely wide margins are necessary to accomplish a complete excision.

The prognosis of the underlying disease in a single patient should always be considered carefully before embarking upon an operation for soft tissue metastases. In two large series of patients with soft tissue sarcomas of a variety of histological types, deep soft tissue metastases occurred in 1.2%–6.6% of cases (Vezeridis et al. 1983, Huth and Eilber 1998). No patient with distant subcutaneous metastasis was alive at 5 years. The median survival was 11.5 months, in contrast to 33% 5 year survival in patients with local recurrence only.

In another clinical study that has just started in Italy, we plan to test a new device for clinical electroporation on a different set of patients with large and deep soft tissue nodules. This is a monocentric pilot study conducted at the IOV of Padua. The study will be made possible by the use of new needle electrodes (Figure 20.2) linked to a machine that is able to assure adequate electric field in a conformational manner. The target of this study is represented by nodules larger and deeper (>3 cm) that cannot be currently managed with the available electrodes. Usually these kinds of lesions, especially if superficial, require multiple ECT applications through which a progressive tumor reduction is obtained. The improvement of ECT as a treatment option for large, primary or metastatic, soft tissue tumors, with a highly effective, minimally invasive and simple procedure, could benefit patients that would otherwise face burdensome surgery, extreme loss of QoL and reduction in performance status. Soft tissue metastases, especially those in the skeletal muscles, are frequently painful and ECT may represent a favorable alternative to surgical resection. Moreover, since its low impact on patient's status, ECT could represent a therapeutic option for those patients unable or unwilling to undergo surgery.

The working hypothesis on which this study is based is the assumption that the new independent electrodes and the flexible configuration geometry achieved by a flexible positioning system may generate adequate electric field able to encompass the whole tumor mass and its margins. The procedures to perform an adequate treatment planning are not established and will be the objective of the coordinate investigation of clinicians and radiologists This will allow to chose the most suitable imaging techniques, the proper manner of percutaneous insertion of needle electrodes; these technical improvements could finally lead to obtain a better tumor response and control even on large and deep soft tissue tumors with a "one-shoot" treatment that is still based on a mini-invasive approach. The main aim of this study is to assess the activity (tumor response) of ECT performed by means of new electrodes (longer than those currently applied in clinical practice) and positioning system. Secondarily, local toxicity as well as tumor response duration will be considered.

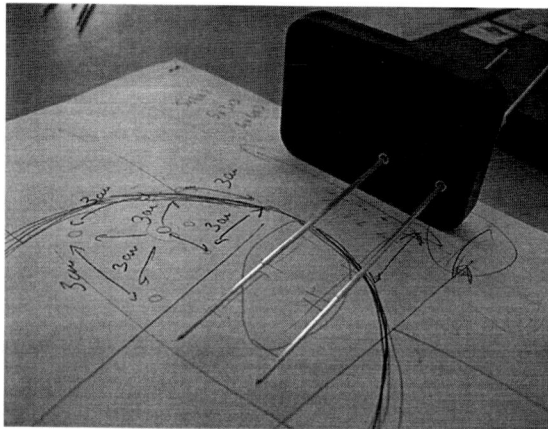

FIGURE 20.2 The new electrodes for deep-seated tumors. These electrodes are configured as long metal needles partially insulated: the central body is sheathed with an insulating layer to allow the application of the electric field limited to the terminal "active" part and preserve the healthy tissues surrounding the lesion.

The main endpoint will be the effectiveness of ECT in terms of response rate. An objective response rate of 40% or less will be considered to be insufficient to warrant further investigation. On the other hand, a probability of complete response rate (CRR) >60% would be clinically sufficient indicating that further investigation of procedure is appropriate. This level of activity has been chosen according to our previous experience with ECT in patients with smaller lesions and considering the literature data of activity in this clinical context. With a sample size of 38 patients, the risk of erroneously recommending a procedure whose CRR is inadequate amounts to 5%, while the chance of erroneously rejecting the treatment is less than 20% in case of truly promising activity set to 60% CRR. The procedure will be recommended for further studies if at least 27 complete responses will be recorded. Technical improvements in electrode positioning and optimization of the electric field are expected along the study, by affording case-by-case problems related to different tumor anatomical sites.

Patients will be consecutively enrolled among those affected by histologically proven deep soft tissue tumors (primary or metastatic) from any type (at least one lesion between 3 and 6 cm or deeper than 3 cm) not suitable for conventional treatments. One tumor mass with the above-mentioned features will be selected as "target lesion" and treated with the new instrumentation (Cliniporator-VITAE device and new needle electrodes). The target lesion will be measured with a flexible millimetric ruler, when superficial, or by means of US scan or MR imaging; its pathology, anatomical site, clinical presentation, and previous treatments (local or systemic) will also be registered. The presence and the amount of other superficial or parenchymal tumor deposits will be evaluated through physical examination and radiological imaging.

Patients suitable for this study will have to satisfy the following inclusion criteria: measurable soft tissue tumor nodules (cytology/histology proven) manageable with new electrodes, at least one tumor size ranging from 3 to 6 cm or tumor deepness >3 cm, ECOG performance status scale ≤2.

Patients with serious lung, heart or liver disease, epilepsy, short life expectancy (<3 months), active infection, previous treatment with bleomycin to maximal dosage, different systemic anticancer therapies administered within 4 weeks before ECT, and abnormalities in coagulation tests will be excluded from the study.

20.5 Conclusions

The Italian clinical trials so far performed have confirmed the ESOPE's results in terms of feasibility and activity, mainly in melanoma patients. Both trials (Turin and Padua) have reported a high local response (complete response up to 58%, further increasing after retreatment) an acceptable local disease control (74% at 2 years) with a very low toxicity. Often, these results have been achieved by administering multiple ECT cycles (up to five in the University of Padua's study) either in order to obtain a tumor response in large nodules or to treat newly appeared tumor nodules. In fact, the tumor size was the most important predictive factor for local response.

The efficacy of ECT on less frequent histotypes (breast cancer, Merkel, etc.) is still under evaluation. The IMI-GIDO Multicentric Italian Study on ECT should allow collecting a large series of patients and confirming results on melanoma and other histotypes in order to support the indications and to find clues for new therapeutic associations (e.g., ECT and chemo-, immuno-, radiation-therapy). The preliminary report from the University of Padua on the possible impact of ECT on patients' QoL deserves further and methodologically correct investigations. QoL is one of the main advantages to be explored in patients who undergo ECT because superficial tumor nodules can have an impact on patients' everyday life and activity of daily living. The undergoing validation process of the QoL questionnaire will become a dedicated tool in order to better assess the clinical benefit of ECT. By means of this questionnaire, a standardized evaluation of the QoL will be possible at any center that applies ECT. Moreover, this process will lead to a more accurate patient selection in order to treat only patients who could be successfully palliated, sparing the others from unnecessary, and often multiple treatment applications. Ultimately, one of the current limitations of the ECT application is represented by the size and depth of the tumor; the study with the new device and more deeply penetrating electrodes should extend the indications and reduce the number of ECT cycles currently necessary to manage bigger and deep lesions. Such patients could be spared from surgical procedure that are generally demanding and associated with high morbidity, especially in advanced and highly pretreated patients. In case of positive results from the undergoing trial, ECT could be offered as a minimally invasive and safe palliating procedure, also in case of large and deep symptomatic soft tissue metastases. In conclusion, at the present time, researchers are trying to establish the precise role of ECT within the frame of multidisciplinary management of cancer patients; the preliminary impressive results in terms of local response and disease control, taken together with the limited toxicity of this treatment approach, have prompted the beginning of new multicentered clinical trials that will allow/help to clarify some of the open issues (i.e., indications, benefits, combination with different therapies) raised since the introduction of ECT in the clinical practice. The presence of a multidisciplinary team, with professionals from different disciplines, may be useful to provide a comprehensive assessment and consultation when evaluating a patient for ECT treatment. Such a team can promote the coordination among members (especially surgical, medical, and radiation oncologists), thus providing a "check and balance" mechanism to ensure the best treatment to the patient with advanced stage disease. An optimal coordination between individual team members will also provide a forum for learning more about the strategies, resources, and possible combined anticancer treatment approaches in which ECT may play a promising role.

References

Belehradek M, Domenge C, Luboinski B et al. Electrochemotherapy, a new antitumor treatment. First clinical phase I–II trial. *Cancer* 1993; 72: 3694–3700.

Campana LG, Mocellin S, Basso M et al. Bleomycin-based electrochemotherapy: Clinical outcome from a single institution's experience with 52 patients. *Ann Surg Oncol* 2009; 16(1): 191–199.

Damron TA, Heiner J. Distant soft tissue metastases: A series of 30 new patients and 91 cases from the literature. *Ann Surg Oncol* 2000; 7: 526–534.

Gothelf A, Mir LM, Gehl J. Electrochemotherapy: Results of cancer treatment using enhanced delivery of bleomycin by electroporation. *Cancer Treat Rev* 2003; 29: 371–387.

Heller L, Pottinger C, Jaroszeski MJ, Gilbert R, Heller R. In vivo electroporation of plasmids encoding GM-CSF or interleukin-2 into existing B16 melanomas combined with electrochemotherapy induces long-term antitumor immunità. *Melanoma Res* 2000; 10(6): 577–583.

Huth JF, Eilber FR. Patterns of metastatic spread following resection of extremity soft tissue sarcomas and strategies for treatment. *Semin Oncol* 1998; 4: 20–26.

Marty M, Serša G, Garbay JR et al. Electrochemotherapy—An easy, highly effective and safe treatment of cutaneous and subcutaneous metastases: Results of ESOPE study. *EJC Suppl* 2006; 4: 3–13.

Miller AB, Hoogstraten B, Staquet M et al. Reporting results of cancer treatment. *Cancer* 1981; 47: 207–214.

Mir LM, Belehradek M, Domenge C et al. Electrochemotherapy, a new antitumor treatment: First clinical trial. *CR Acad Sci III* 1991; 313(13): 613–618.

Mir LM, Gehl J, Sersa G et al. Standard operating procedures of the Electrochemotherapy: Instructions for the use of bleomycin or cisplatin administered either systemically or locally and electric pulses delivered by the Cliniporator by means of invasive or non-invasive electrodes. *EJC Suppl* 2006; 4: 14–25.

Park JO, Lee SI, Song SY et al. Measuring response in solid tumors: Comparison of RECIST and WHO response criteria. *Jpn J Clin Oncol* 2003; 33: 533–537.

Plaza JA, Perez-Montiel D, Mayerson J et al. Metastases to soft tissue: A review of 118 cases over a 30-year period. *Cancer* 2008; 112(1): 193–203.

Quaglino P, Mortera C, Osella-Abate S et al. Electrochemotherapy with intravenous bleomycin in the local treatment of skin melanoma metastases. *Ann Surg Oncol* 2008; 15(8): 2215–2222.

Rolz-Cruz G, Kim CC. Tumor invasion of the skin. *Dermatol Clin* 2008; 26(1): 89–102.

Roux S, Bernat C, Al-Sakere B et al. Tumor destruction using electrochemotherapy followed by CpG oligodeoxynucleotide injection induces distant tumor responses. *Cancer Immunol Immunother* 2008; 57: 1291–1300.

Serša G, Kranjc S, Čemažar M et al. Improvement of combined modality therapy with cisplatin and radiation using electroporation of tumors. *Int J Radiat Oncol Biol Phys* 2000; 46: 1037–1041.

Therasse P, Arbuck SG, Eisenauer SA et al. New guidelines to evaluate the response to treatment in solid tumors. European Organization for Research and treatment of Cancer, National Cancer Institute of the United States, National Cancer Institute of Canada. *J Natl Cancer Inst* 2000; 92: 205–216.

Trotti A, Byhardt R, Stetz J et al. Common toxicity criteria: Version 2.0. an improved reference for grading the acute effects of cancer treatment: Impact on radiotherapy. *Int J Radiat Oncol Biol Phys* 2000; 47(1): 13–47.

Vezeridis MP, Moore R, Karakousis CP. Metastatic patterns in soft tissue sarcomas. *Arch Surg* 1983; 118: 915–918.

Wagner LI, Wenzel L, Shaw E et al. Patient-reported outcomes in Phase II cancer clinical trials: Lesson learned and future directions. *J Clin Oncol* 2007; 25: 5058–5062.

21

Tumor Blood Flow–Modifying Effects of Electroporation and Electrochemotherapy— Experimental Evidence and Implications for the Therapy

Tomaz Jarm

Maja Cemazar

Gregor Sersa

21.1 Introduction

Reversible *electroporation* or *electropermeabilization* (EP) is a highly effective method for nondestructive and transient modification of cell membrane permeability by means of a series of electric pulses characterized by high voltage (e.g., 1 kV) and short duration (e.g., 100 μs) (Orlowski et al. 1988). EP has gained considerable attention over the last decade, largely due to applications in which extracellular molecules that do not cross cell membrane easily, such as hydrophilic molecules, can be introduced into the cells in vastly increased quantities by means of EP. The combined use of EP with chemotherapeutic drugs has been named *electrochemotherapy* (ECT) (Mir et al. 1991). Over the last decade, numerous studies have been reported, that involve the treatment of experimental and clinical tumors by means of ECT (Sersa 2006, Sersa et al. 2006, Quaglino et al. 2008). The combined efforts of many research groups have recently culminated in the Cliniporator (IGEA, Carpi, Italy), the first CE-certified clinical instrument for EP, and in the Standard Operating Procedures for the use of the method in clinical oncology (Mir et al. 2006). ECT with chemotherapeutics bleomycin and cisplatin is now being used successfully in a steadily increasing number of hospitals and veterinary clinics for the treatment of solid tumors of various etiologies including melanoma skin metastases, head and neck tumors, basal cell carcinomas, and adenocarcinomas in humans (Heller et al. 1999, Marty et al. 2006, Sersa 2006, Cemazar et al. 2008). Recently ECT has been proposed as a treatment modality for skin melanoma metastases (Testori et al. 2009). In another procedure closely related to ECT and named *electrogenetherapy* (EGT), electric pulses different from those used in ECT are used not only to electropermeabilize the cell membranes but also to produce electrophoretic forces, which are both needed to transfer the extracellular genetic material into the cells (Satkauskas et al. 2002, 2005, Kanduser et al. 2009). Due to a more complex situation, the development of procedures for effective EGT is lagging behind that for ECT, but the potential of EGT for local gene therapy in tissues including muscle, skin, liver, and tumor without the use of viral vectors has already been demonstrated *in vivo* and in a clinical study (Neumann et al. 1982, Heller and Heller 2006, Cemazar et al. 2006, Prud'homme et al. 2006, Daud et al. 2008).

The main mechanism of antitumor effectiveness of ECT is the direct cytotoxic effect of the drug introduced into the cells by EP and entrapped there after resealing of the cell membrane (Mir et al. 2006). However, since the early studies the investigators have also been noticing transient but severe local blood flow–reducing effects of EP and ECT (Sersa et al. 1998, 1999b, Ramirez et al. 1998). Even though this indirect effect on tumors was recognized, its contribution was initially regarded as a side effect with minor significance for antitumor effectiveness of ECT, especially in the light of nonsignificant antitumor effect exerted by EP alone in the absence of chemotherapeutic drug. With mounting new evidence, however, this general view has been changing constantly. It is now evident that not only can the blood flow–modifying effect of ECT contribute significantly to its antitumor effectiveness, but that there are actually several distinct physiological pathways by which it does so.

There are several reasons why blood flow–modifying effects of EP alone and of ECT are important and must be studied. Understanding these effects contributes to better understanding of how EP and ECT work and can thus lead to further optimization of treatment protocols. Both direct and indirect effects of EP and ECT are of local nature (confined to the tumor itself and to minimal normal surrounding tissue), which by fine-tuning of the parameters can allow for truly individualized treatment. The blood flow–modifying effects of EP and ECT could be exploited in *vascular-targeted treatment*, a special class of antitumor treatments targeted at tumor blood supply (either on the level of existing tumor blood vessels or formation of new ones) rather than tumor cells directly (Chaplin et al. 1998, Tozer et al. 2005, Sersa et al. 2008). Exposure of tumor cells to the drug can be prolonged by impeding the drug washout from the tumor interstitium due to effective cutoff of tumor blood flow (the so-called *vascular lock* effect) and may thus improve the effectiveness of conventional chemotherapy by drug entrapment in the tumors. Because of the vascular lock, the *i.v.* administration of the drug must be performed before the application of EP. In addition, the increased tumor hypoxia and anoxia as a consequence of impeded blood flow can be used to increase effectiveness of hypoxia-dependent agents such as bioreductive drugs (Cemazar et al. 2001b). And finally, EP and ECT can be used effectively for alleviation and cosmetic effects in palliative treatment of hemorrhaging tumors and ulcers to stop the bleeding (Gehl and Geertsen 2000, Snoj et al. 2009).

In the following sections, we first outline the general characteristics of tumor vasculature and blood flow. This is followed by a brief description of some methods used for assessment of tumor blood flow and oxygenation and then by the demonstrated evidence of blood flow–modifying effects of EP and ECT with emphasis on experimental tumors. Finally, and based on this evidence, we suggest a model of physiological events taking place in a tumor within seconds, minutes, hours, and days following the application of EP and ECT.

21.2 Blood Flow in Solid Tumors

Solid tumors, in general, are physiologically different from normal tissues in many aspects. Even though there are also large histological differences between various tumor types, there are many common characteristics of tumor vasculature and blood flow (Jain 1987, Vaupel and Hockel 2000, Tozer et al. 2005, Folkman 2007).

In contrast to normal tissues, solid tumors in general are usually poorly perfused and oxygenated. This is largely due to inadequately developed tumor neovasculature, which is characterized by irregular and chaotic networks and branching patterns, irregular shape and length of microvessels, shunts and dead ends, which all contribute to increased resistance to blood flow and to presence of stagnant blood pools. All this leads to low blood flow and together with large distances between capillaries to development of hypoxic and acidic tissue environment. There may be large local variations in both the density of capillaries and oxygenation levels within the same tumor. But generally, the peripheral rim of center of most solid tumors is better vascularized and oxygenated than the center (Vaupel and Hockel 2000, Tozer et al. 2005).

Tumor blood vessel walls are structurally and functionally abnormal. Relative lack or paucity of smooth muscle cells and sympathetic enervation is common and contributes to inferior local regulation of microcirculatory blood flow. Abnormal basal membrane and endothelial lining with loose intercellular connections or even openings between endothelial cells result in high vascular permeability to large molecules, including plasma proteins. This leads to increased extravasation of liquids; tumor capillaries are said to be "leaky," which together with inadequate lymphatic drainage system contributes to elevated interstitial fluid pressure (IFP) in tumors (Pusenjak and Miklavcic 1997, Tozer et al. 2005).

These characteristics of tumor blood flow can represent obstacles for some antitumor treatments. For example, insufficient blood flow and elevated IFP in tumors impede effective delivery of chemotherapeutics into the tumor. Hypoxic environment makes tumors more resistant to radiotherapy. But on the other hand, the abnormal and inadequate tumor vasculature is also an attractive target for *vascular-disrupting therapies* and other new treatments specifically designed to take advantage of tumor vasculature's weaknesses (Tozer et al. 2005, Siemann and Horsman 2009).

21.3 Measurement of Tumor Blood Flow and Related Parameters

In the following section, we present measurement methods which are frequently used in experimental studies for assessment of various parameters of blood perfusion and oxygenation in solid tumors.

21.3.1 Contrast-Enhanced Magnetic Resonance Imaging

The use of contrast-enhancing agents in conjunction with MRI can provide insight into physiology and pathophysiology of tissues in addition to anatomical information offered by nonenhanced MRI. For measurement of blood flow–related parameters, the macromolecular agents are best suited because they produce relatively stable enhancement of blood pool in tissues for a prolonged time and remain largely confined to vascular space, thus enabling differentiation between the vascular and extravascular compartments. This is in contrast to low-molecular-weight agents such as paramagnetic Gd-DTPA (gadolinium diethylenetriaminepentaacetic acid), which equilibrates rapidly between the intra- and extravascular spaces after an intravenous injection, and is also rapidly cleared from the plasma by the renal function, so it is not suitable for blood pool analysis (Schmiedl et al. 1987). Macromolecular agents include substances such as albumin-(Gd-DTPA)$_{30}$, gadomer-17, and polylysine-DTPA, which can be used for the measurement of fractional tumor blood volume and microvascular permeability surface area product and have been used successfully in studies of effects of EP on tumor blood flow (Sersa et al. 1998, Ivanusa et al. 2001, Sersa et al. 2002). The measurements are based on a pixel-by-pixel comparison of a pre-contrast MRI image with a sequence of enhanced images taken at different intervals after intravenous injection of the contrast agent (see Demsar et al. 1997 for details).

21.3.2 Rubidium (^{86}Rb) Extraction Technique

Relative tissue perfusion as a proportion of cardiac output can be measured using extraction of the radioactive rubidium isotope ^{86}Rb (Sapirstein 1958, Sersa et al. 1999b). Briefly, the mouse is injected via tail vein with saline solution of ^{86}RbCl of known radioactivity. One minute later (the time needed to saturate tissues with rubidium) the mouse is sacrificed, the tail removed, and tissue samples of interest are extracted and weighed. Radioactivity of tissue samples is measured by a gamma counter and used to calculate the relative perfusion expressed as a percent of injected dose normalized per sample weight while applying corrections for the activity retained in the tail (injection site) and the background radioactivity. Such measurements can be used for comparison of perfusion levels between different tissues or organs. The method is simple but time consuming and requires work with radioactive material. Its major drawback is that only one measurement point in time per animal can be obtained, so this method requires a large number of experimental animals if a time course of perfusion changes is required.

21.3.3 Patent Blue-Violet Staining Technique

This *in vivo* staining technique uses patent blue-violet biological dye (otherwise commonly used in lymphography) as a contrast agent for visualization of perfused versus non-perfused tissue areas. Even though the results are of qualitative nature, it has been established experimentally that this method provides essentially the same results in solid tumors as the rubidium extraction technique with a benefit of being simpler, less time-consuming and involving no hazardous materials (Sersa et al. 1999a). The mouse is injected via tail vein with saline solution of patent blue-violet (PBV) dye. After 1 min (the time needed by the dye to distribute around the tissues) the mouse is sacrificed and tumor removed and cut along its largest diameter. Both cross sections are photographed and images can be analyzed using image processing software. The percentage of stained (perfused) and non-stained (non-perfused) cross-section area is estimated and used as a relative indicator of perfusion level (Sersa et al. 2008). In addition, the heterogeneous macroscopic distribution of perfused areas in the tumor can be visualized. However, only one measurement per time point per animal can be obtained; so this method requires a large number of experimental animals if a time course of perfusion changes is required.

21.3.4 Laser Doppler Flowmetry

Laser Doppler flowmetry (LDF) is a well-documented optical method, which allows for *in vivo* monitoring of relative blood flow on the level of microcirculation (Shepherd and Öberg 1990). It is based on a physical phenomenon of Doppler shift in wavelength of light. Lasers and fiber-optic probes are used to introduce a beam of coherent monochromatic light in the near-infrared part of the spectrum into the tissue. Most photons are absorbed or otherwise lost in the tissue, but some of them scatter on stationary and moving structures in the tissue and can be collected by the receiving fiber-optic probe and measured. Due to scattering of photons on moving structures (red blood cells in capillaries), the wavelength of these photons is changed (Doppler shift) and it is the spread of the wavelengths of reemerging photons that is used to calculate the relative microcirculatory blood flow. LDF is a very sensitive method with excellent time resolution and can be used to measure rapid changes in local blood flow. Even though LDF is in principle a noninvasive method, thin invasive probes should be used for measurement in tissues below the skin (such as subcutaneous tumors) due to relatively small sampling volume of the method. Anesthesia is usually required for animals to minimize all movements, preferably by an inhalation anesthetic, in which case animals must be kept at physiological temperature at all times to ensure stable physiological conditions (Jarm et al. 2002).

21.3.5 Power Doppler Ultrasonographic Imaging

Power Doppler ultrasonographic (US) imaging can be used to asses noninvasively the gross tumor blood flow changes and to visualize the distribution of areas with different perfusion rates within the same tumor. It requires use of a contrast agent such as SonoVue, a sulfur hexafluoride-filled micro-bubble contrast agent (SonoVue, Bracco, Milan, Italy) to produce sufficient contrast enhancement of perfused areas. In anesthetized mice, an *i.v.* injection of a solution containing tiny bubbles of this contrast agent gas produces strong enhancement of perfused tumor areas in US images within 5–10 s. Contrast enhancement is monitored just before and at regular intervals of 15 s for 5 min after the injection of the contrast agent. The procedure can be repeated at desired intervals after the treatment of tumors with blood flow–modifying treatments in order to follow the long-term effects (Sersa et al. 2008). The contrast agent SonoVue in particular is licensed for abdominal and vascular imaging in many countries and its use for tumor measurements has been tested and can be performed in clinical environment (Delorme and Krix 2006).

21.3.6 Electron Paramagnetic Resonance Oximetry

Electron paramagnetic resonance (EPR) oximetry is a method related to nuclear magnetic resonance method. It is based on paramagnetic properties of oxygen molecules. Its spin exchange interaction with a paramagnetic probe sensitive to oxygen allows essentially noninvasive monitoring of oxygen partial pressure (pO_2) at the same tissue location over long period of time (O'Hara et al. 2001, Sentjurc et al. 2004). A solid particle of paramagnetic probe such as a very small (e.g., 0.5 mm^3) amount of charcoal made from Bubinga tree is implanted into tissue under observation (Sentjurc et al. 2004). In case of tumor oxygenation studies, the probes are usually implanted in tumor periphery and center (to distinguish between the two regions) and in normal control tissue such as subcutis or muscle (Sersa et al. 2001, 2002, Sentjurc et al. 2004). When tissue is exposed to appropriate external magnetic excitation, the resulting EPR spectra of the paramagnetic material can be measured by a surface coil resonator. In the presence of oxygen, the spectral linewidth is broadened and the level of broadening depends on pO_2 (Swartz and Clarkson 1998). EPR linewidth of the implanted probe can be measured repeatedly at different time points of interest. Immobilization of experimental animals by anesthesia is required during measurements.

21.3.7 Polarographic Oximetry

Polarographic oximetry is a well-known and documented method for measurement of oxygen partial pressure (pO_2) in tissue. The principle and use of this method are extensively documented (see, for example, Vaupel et al., 1991). Briefly, a thin active electrode is inserted into the tissue of interest and the reference Ag/AgCl electrode is placed subcutaneously somewhere else. Due to a negative voltage the reduction and oxidation take place at the active and reference electrode, respectively, resulting in a very small electric current, which can be measured. A semipermeable membrane covers the active electrode to ensure that only oxygen contributes to the reduction process. When equilibrium in diffusion of O_2 molecules toward the cathode is reached, the resulting electric current is proportional to pO_2 in the tissue. In spite of some well-known drawbacks of this method (invasiveness and consumption of oxygen which affects the measurement at low pO_2 values typically encountered in tumors), its implementation in the Eppendorf Histograph instrument has been long considered a "golden standard" method for pO_2 measurements in tumors, largely due to a unique procedure that allows for sampling of tumor pO_2 in multitude of measurement points within the same tumor. It is however not well-suited for continuous monitoring of pO_2 changes or for repeated measurements due to its invasiveness. The method has also been used in clinical studies due to the recognized prognostic value of tumor oxygenation as an independent prognostic indicator in cancer, which influences tumor progression and treatment outcome (Menon and Fraker 2005).

21.3.8 Luminescence-Based Fiber-Optic Oximetry

This method implemented in the OxyLite instruments (Oxford Optronix, Oxford, United Kingdom) is considered an alternative to polarographic oximetry. The details about the method can be found in Collingridge et al. (1997). Briefly, pulses of light emitted by a blue LED diode are delivered via optical fiber to ruthenium chloride luminophore incorporated in a tip of a thin probe inserted into tissue of interest. The incident light induces fluorescence of ruthenium molecules at a longer wavelength. The lifetime of the fluorescence is inversely proportional to oxygen pO_2 in the surrounding tissue and pO_2 can be calculated using the Stern–Volmer relation. Tissue temperature at the probe tip must also be monitored and its fluctuations must be compensated for. In contrast to polarographic oximetry, the sensor of the luminescence-based oximetry does not consume oxygen and can therefore also be used to monitor pO_2 changes over longer periods. The same limitations due to invasiveness however apply and the method is currently still not approved for use in humans.

21.3.9 Histological Assessment of Hypoxia by Means of Hypoxia Marker Pimonidazole

Pimonidazole is a hypoxia marker known to bind preferentially to hypoxic tumor cells. It is reductively activated in an oxygen-dependent manner and is covalently bound to thiol-containing proteins in hypoxic cells, forming intracellular protein adducts. Detection of protein adducts of reductively activated pimonidazole using monoclonal antibodies can be used to detect hypoxic regions or even individual hypoxic cells in histological preparations. The experimental animal (such as a mouse) is injected intraperitoneally with pimonidazole approximately 16 h prior to tumor excision. Tumors are then formalin-fixed, paraffin-embedded, cut, and immunohistochemical staining is performed to detect pimonidazole (Raleigh et al. 2001, Cör et al. 2009).

21.4 Effects of Electroporation and Electrochemotherapy on Tumor Blood Flow

All effects of EP alone or ECT described in this section are strictly of local nature. All studies report that macroscopic changes in blood flow and related parameters as well as histological evidence of these effects were confined only to the tissue directly subjected to the treatment. No significant systemic effects on blood flow have ever been found.

21.4.1 Electroporation and Electrochemotherapy Protocols

A typical single application of EP commonly used in ECT consists of a sequence of eight square monopolar electric pulses delivered to tissue either noninvasively via parallel-plate electrodes or invasively using various arrangements of several needle electrodes (Mir et al. 2006). Duration of pulses and repetition frequency are most frequently set to 100 μs and 1 Hz, respectively. The amplitude of pulses is dependent on the interelectrode distance and is frequently reported as the voltage divided by the interelectrode distance. For invasive EP with needle electrodes inserted into tumors, lower voltages are used than for noninvasive EP with plate electrodes, due to absence of the skin barrier between the target tissue and the electrodes (Miklavcic et al. 2006). In a typical application of ECT, the electric pulses are delivered at the time when the maximum extracellular concentration of the chemotherapeutic drug is achieved (Sersa et al. 1995). Therefore the EP procedure is applied 3 min in mice or 8 min in humans after intravenous or immediately after intratumoral injection of bleomycin or cisplatin (Sersa et al. 1995, Domenge et al. 1996, Heller et al. 1997, Cemazar et al. 1998). The drugs are used at very low doses, which have no systemic side effects and only insignificant antitumor effect when used without EP. Even though the exact protocols for EP, drug dosage and administration, and ECT in various studies are optimized for particular applications (animal species, electrode type, tumor site and size, chemotherapeutic used, aim of the study) and can thus vary from the one outlined here, the protocol details will not be mentioned in this chapter unless necessary for the discussion, but can be found in the cited references.

21.4.2 Electroporation and Blood Flow

Application of EP causes an immediate and profound reduction in local blood flow in affected tissue as confirmed by different authors in solid tumors and in normal tissue (Ramirez et al. 1998, Sersa et al. 1998, 1999a, 2002, 2008, Gehl et al. 2002). The method with sufficient time resolution and sensitivity to show this rapid tissue response to EP in real time is LDF. In Figure 21.1, it can be observed that as soon as the first pulse in the series was delivered, blood flow started to decrease and reached a close-to-zero level within seconds. After a short period of approximately 1–2 min of almost complete absence of perfusion, the microcirculation started to recover and partial perfusion was restored within 10–15 min

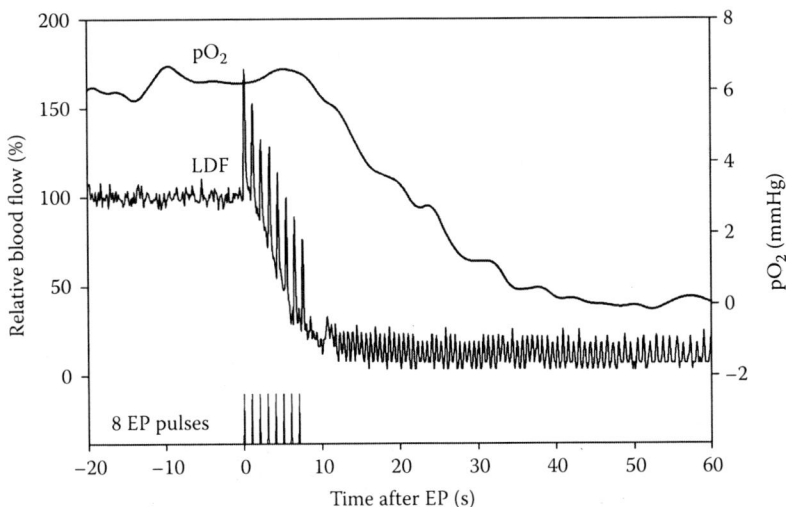

FIGURE 21.1 Decrease in blood flow and oxygenation of a tumor after application of EP. Rapid and profound decrease in blood flow (LDF; left axis) was followed closely by a decrease in oxygenation (pO₂; right axis) in a subcutaneous LPB tumor in an anesthetized C57Bl/6 mouse (previously unpublished data). Zero values on vertical axes are offset for better separation of the signals. Movement artifacts caused by individual electric pulses (muscle twitching) and respiration can be observed in the LDF data after time 0. EP protocol: 8 pulses, amplitude 1300 V/cm, duration 100 μs, 1 Hz, plate electrodes. OxyFlo Laser Doppler flowmeter and OxyLite luminescence-based oximeter with a combined LDF/pO₂ bare fiber probe were used for this measurement (Oxford Optronix, Oxford, United Kingdom).

after EP (Sersa et al. 2008). After this point, the gradual reperfusion continued but at a much slower rate with no significant further recovery for at least 2 h (see also Figure 21.4). According to the studies using rubidium extraction or PBV staining techniques, the maximum reduction by 70% in blood flow of Sa-1 tumors in A/J mice was reached 30 min after EP (Sersa et al. 1999b, 2002), but these methods are not suitable for assessment of rapid changes such as those observed with LDF within the first few minutes after ECT. In one of the first studies on effects of EP on blood flow using contrast-enhanced magnetic resonance imaging (CE-MRI) with albumin-(Gd-DTPA)$_{30}$ as the contrast agent, it was also shown in the same tumor model that 30 min after EP blood flow was severely impeded in comparison to nontreated tumors on contralateral side of the same animals (Sersa et al. 1998). Good agreement between the results of different methods used in several studies (LDF, PBV staining, rubidium extraction, CE-MRI, power Doppler US imaging) indicated that 1 h after EP blood flow in tumors was on average less than 50% of the pretreatment value or compared to the control tumors (Sersa et al. 1999a,b, 2003, 2008). But blood flow was almost completely restored after 24 h and became indistinguishable from that in the control tumors within 48 h after EP (Sersa et al. 1999a, 2002, 2008). In addition to supporting the results obtained by other methods, CE-MRI and PBV staining also provided clear evidence of heterogeneous distribution of blood flow in tumors with the periphery being considerably better perfused than the center (Ivanusa et al. 2001, Sersa et al. 2002, 2008).

If EP treatment was reapplied 24 h after the first application (when blood flow had not fully recovered yet), the time course of newly induced blood flow decrease was very similar to the original one but with some indication of slower recovery by 24 h after the second application (Sersa et al. 1999b). The rapid and complete initial decrease in blood flow and its duration observed in subcutaneous tumors was consistent with observations in various normal tissues such as liver, spleen, and pancreas as well as internal liver tumors in rabbits where the effect lasted for 10–15 min before gradual reperfusion (Ramirez et al. 1998). A study on leg muscles in mice showed the same rapid initial decrease in blood flow to zero level as seen in tumors but with considerably faster recovery in the later phase (Gehl et al. 2002).

Changes in blood flow in tumors after EP have been studied with respect to duration, amplitude, and number of pulses delivered. The level of blood flow reduction observed 30 min after EP in Sa-1 tumors in A/J mice was positively correlated with the number of pulses between 1 and 10 (at amplitude and duration of pulses 1300 V/cm and 100 µs, respectively) (Sersa et al. 1999a,b). It should be noted however that the rapid and extreme initial decrease in blood flow observed with eight pulses can also be observed by LDF after a single pulse, but this effect is always short lived in comparison to eight pulses (see Figure 21.2 for an example) and thus could not be detected by rubidium extraction technique 30 min after EP (Sersa et al. 1999b). Reduction in tumor blood flow was also dependent on the amplitude of pulses. With the number and duration of pulses fixed at 8 and 100 µs, respectively, it was found that 30 min after EP the decrease in blood flow could be detected only for amplitudes at and above 800 V/cm (value consistent with the threshold for the onset of reversible EP in tumors (Pavselj et al. 2005)), the level of decrease 30 min after treatment appeared positively dose-dependent up to 1300 V/cm with no further decrease at 1500 V/cm. It was interesting however that at amplitudes below 800 V/cm a small but statistically insignificant increase in tumor blood flow was observed (Sersa et al. 1999b).

In summary: the response of tumor blood flow to EP typically used in ECT is rapid (within seconds) and profound. It exhibits a short-lived (lasting for few minutes) and a relatively long-lived phase (lasting for at least 24 h). The reduction in blood flow is therefore reversible and according to its time course seems to involve at least two physiological mechanisms. The level and duration of blood flow reduction (the long-term phase) is positively dose-dependent on number, amplitude, and duration of pulses.

The results of a study on characterization of the effects of EP on blood flow in leg muscles of C57Bl/6 mice (Gehl et al. 2002) can help explain the observed phenomena in tumors. In this extensive study, the significance of vascular effects of EP was assessed by combining the data on transgene expression after EP, membrane electropermeabilization and resealing dynamics, induced tissue perfusion changes after

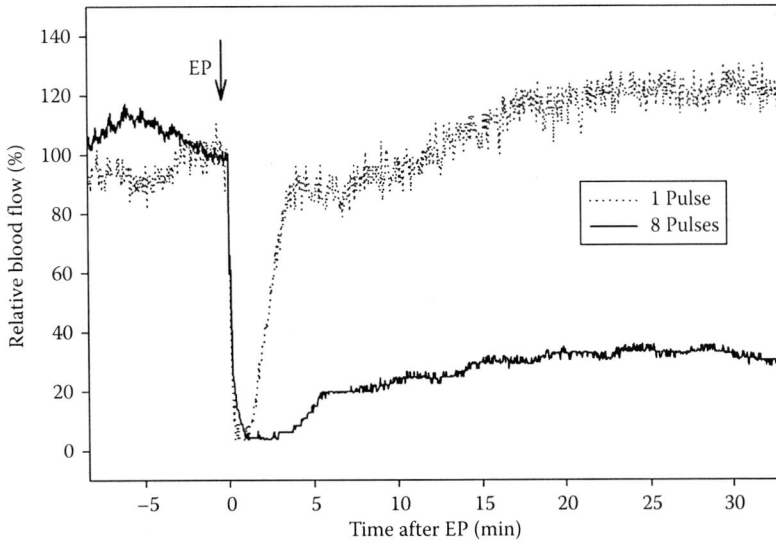

FIGURE 21.2 Difference in tumor blood flow decrease after EP using one or eight pulses. Data for LPB tumors in anesthetized C57Bl/6 mice are shown (previously unpublished data). EP protocol: amplitude 1040 V, duration of pulses 100 μs, 1 Hz, plate electrodes, interelectrode distance 8 mm. OxyFlo Laser Doppler flowmeter with a bare fiber probe was used (Oxford Optronix, Oxford, United Kingdom). Signals were filtered to remove movement artifacts caused by electric pulses (muscle twitching) and respiration.

more than 100 combinations of pulse durations and amplitudes, and the role of sympathetic nervous system. The effects on perfusion were measured qualitatively by means of PBV staining technique and expressed as a delay in staining observed in EP-treated muscle in comparison to nontreated contralateral muscle (Gehl et al. 2002).

As in tumors, the vascular response of muscles to EP was found to be a two-phase phenomenon. The first phase was attributed to sympathetically mediated vasoconstriction of afferent arterioles leading to a rapid and profound but short-lived (for 1–3 min) initial decrease in blood flow. Amplitude and duration of EP pulses had to be above the threshold for reversible permeabilization for this effect to take place. It was suggested that this phase was a Raynaud phenomenon–like reflexory reaction to permeabilization of muscle and/or vascular endothelial cells. The direct effect of pulses on smooth vascular musculature on the other hand was apparently not significantly involved in this response (Gehl et al. 2002).

The second phase of vascular response to EP was slower to develop and of much longer duration than the first phase (lasting up to 30 min). It depended on amplitude and duration of pulses and became progressively more significant for combinations of amplitudes and durations that either exceeded the threshold for irreversible permeabilization or caused very slow membrane resealing kinetics of muscle and vascular endothelial cells. It was suggested that this would lead to increased interstitial pressure (interstitial edema) and decreased intravascular pressure. The resolution of this long-lived phase resembles the kinetics of observed membrane resealing after EP.

Further information regarding this second phase of vascular response was provided by an *in vitro* study investigating the effects of EP on the cytoskeleton of cultured primary endothelial cells and on endothelial monolayer permeability (Kanthou et al. 2006). Human umbilical vein endothelial cells (HUVECs) were exposed to EP *in situ*. Immunofluorescence staining for F-actin, β-tubulin, vimentin, and VE-cadherin and western blotting for levels of phosphorylated myosin light chain and cytoskeletal proteins were performed. Endothelial monolayer permeability was determined by monitoring the passage of FITC-coupled dextran through endothelial monolayer. This study demonstrated that exposure of endothelial cells to electric pulses resulted in a profound disruption of microfilament and microtubule cytoskeletal networks, loss of contractility, and loss of VE-cadherin from cell-to-cell junctions within 5 min after EP. These effects were voltage-dependent and reversible, since cytoskeletal structures recovered within 60 min of EP, without any significant loss of cell viability. The cytoskeletal effects of EP were paralleled by a rapid rise in endothelial monolayer permeability, demonstrating that besides changes in cell membrane permeability, as suggested in (Gehl et al. 2002), changes in vascular endothelial monolayer permeability may contribute to the observed second phase effects of vascular response to EP of normal vessels (Kanthou et al. 2006).

In summary: permeabilization of muscle and endothelial cells induced two distinct vascular effects in muscle tissue: (1) a rapid and short-lived Raynaud-like reflexory vasoconstriction of afferent arterioles mediated by sympathetic nervous system, the resolution of which depended on the release of vasoconstricting spasm and (2) slower and longer lived increased endothelial monolayer permeability, that resulted in leakage of fluids and molecules from blood vessels as well as from permeabilized cells into extracellular space, leading to increased IFP and decreased intravascular pressure. The resolution of this phase depended on membrane resealing kinetics and recovery of endothelial cytoskeletal structures (Gehl et al. 2002, Kanthou et al. 2006).

Perfusion changes observed in tumor tissue after EP are therefore a combination of two phases schematically presented in Figure 21.3. A similar model with different kinetics was proposed for muscle tissue in Gehl et al. (2002). The two-phase effect of EP can also be recognized in Figure 21.4. The first rapid and short-lived phase in tumors closely resembled the one reported in the study on muscles both in the amplitude and kinetics of blood flow changes. On the other hand, there were important differences in the kinetics of resolution of the second phase between muscle and tumor tissues. In muscles (normal tissue), the second phase was resolved over a period of tens of minutes (Gehl et al. 2002) but in tumors (malignant tissue) it took at least 24 h to completely restore the blood flow (Sersa et al. 1999a,b, 2002). The reason for this apparent discrepancy lies most likely in qualitative and quantitative differences between normal and tumor vasculatures. It is reasonable to expect that a normal vasculature would recover faster from EP than tumor neovasculature.

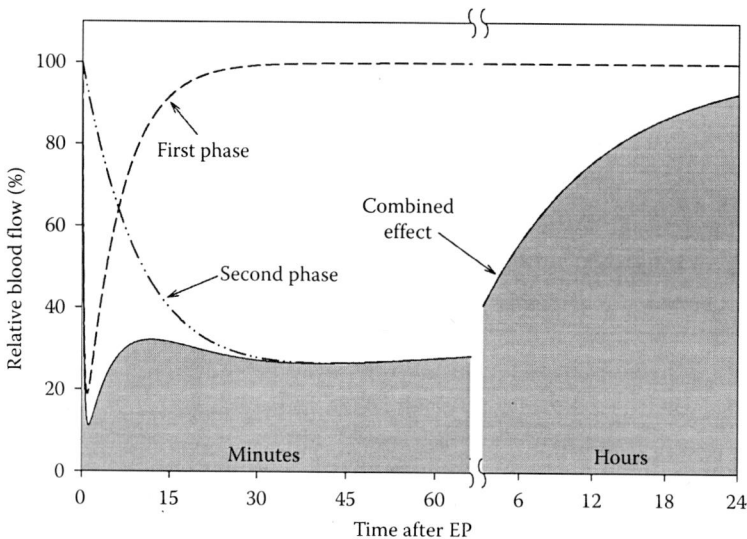

FIGURE 21.3 Model of the two-phase perfusion changes in subcutaneous tumors in mice. The model is based on the kinetics of blood flow changes observed in Sa-1 tumors in A/J mice after application of EP. The rapid first phase is followed by a much slower second phase of blood flow decrease. The sum of both phases represents total blood flow decrease observed by means of LDF and PBV staining technique. Compare also with the experimental data for tumors treated with EP and ECT in Figure 21.4.

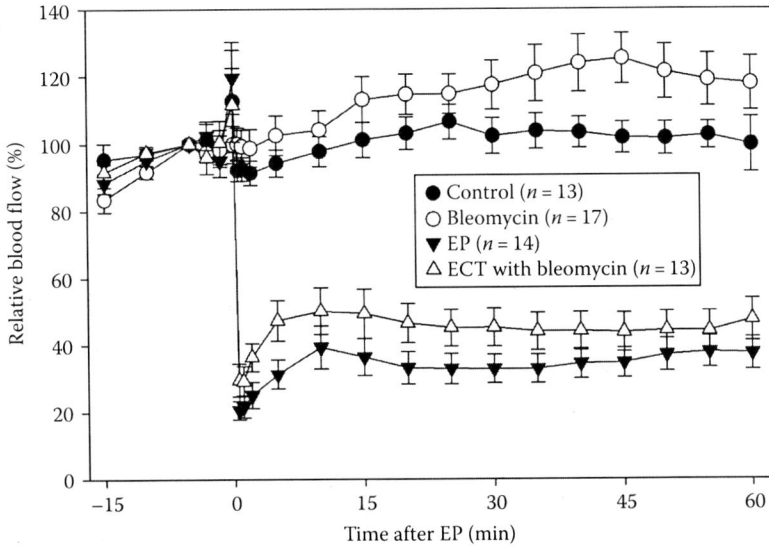

FIGURE 21.4 Blood flow changes after different treatments. Data for subcutaneous Sa-1 tumors in anesthetized A/J mice belonging to four experimental groups are shown. Data points represent the mean average blood flow and standard error of the mean based on filtered LDF signals. EP (eight pulses, amplitude 1300 V/cm, duration 100 μs, 1 Hz, plate electrodes) was delivered at time 0. Bleomycin (20 μg/mouse) was injected *i.v.* 3 min prior to application of EP. Values are expressed relatively as % of the value measured 5 min before time 0. The scatter of values just before time 0 was a result of manipulation of the animals due to injection and attachment of the electrodes. OxyFlo laser Doppler flowmeter with bare fiber probes was used (Oxford Optronix, Oxford, United Kingdom). (Reproduced from Sersa, G. et al., *Br. J. Cancer*, 98, 388, 2008, doi: 10.1038/sj.bjc.6604168, www.bjcancer.com.)

21.4.3 Electrochemotherapy and Blood Flow

Within the first hour (or so) after treatment, the effects of EP alone and ECT on blood flow are on average indistinguishable both in the extent and dynamics of changes (Sersa et al. 1999a, 2002, 2008). However, 8 h after treatment the differences are apparent: specifically, perfusion of EP-treated tumors is already partially recovered, while the perfusion in tumors treated with ECT with bleomycin is steadily decreasing up to 12 h after ECT when practically a complete shutdown of blood flow is reached as demonstrated by PBV and rubidium extraction data and supported further by power Doppler US imaging. Even 5 days after treatment blood flow is not restored (Sersa et al. 1999a). ECT with cisplatin in the same tumor model resulted in a similar time course of blood flow changes but with less extreme reduction in blood flow than observed with bleomycin over the course of 5 days (Sersa et al. 2002). In contrast, blood flow in tumors treated with either bleomycin or cisplatin in absence of EP was the same as that in control tumors over a period of 5 days after the treatment. Figure 21.4 presents the response of subcutaneous Sa-1 tumors on treatment with EP, bleomycin and ECT with bleomycin within the first hour after the treatment. Figure 21.5 summarizes the observed effects of EP and ECT with bleomycin and cisplatin on blood flow in Sa-1 tumors over a period of 5 days (based on combined data from (Sersa et al. 1999a, 2002, 2003)). Table 21.1 presents the antitumor effectiveness of these treatments.

In summary: Decrease of blood flow after ECT follows the same pattern as after EP alone within first hour after the treatment. But in contrast to EP alone, blood flow after ECT is further compromised at later time. The difference becomes significant by 8 h after the treatment and blood flow is essentially completely stopped 12 h after ECT with bleomycin and reduced to a lesser extent by ECT with cisplatin with no additional recovery at least 5 days after the treatment with either drug. The observed reduction in blood flow after ECT correlates well with the antitumor effectiveness of ECT with both drugs and depends on the drug dosage.

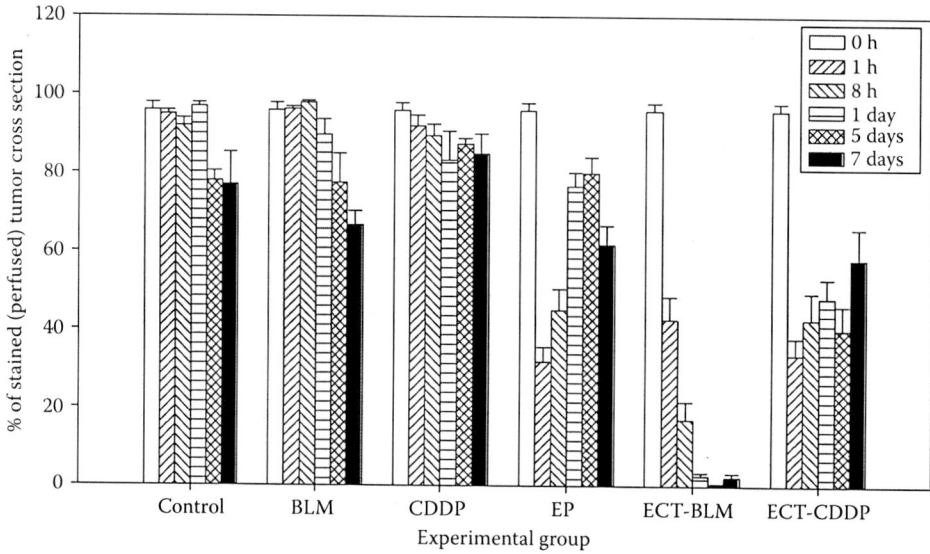

FIGURE 21.5 Blood flow at different times after the treatment. PBV staining technique was used. Mean and standard error of the mean values for at least six mice per bar are shown. Time interval 0 h represents the pretreatment value, which is the same for all experimental groups. The experimental groups shown are: control, bleomycin only (BLM, 5 mg/kg), cisplatin only (CDDP, 4 mg/kg), electroporation only (EP), ECT with BLM, and ECT with CDDP. EP protocol: 8 pulses, amplitude 1300 V/cm, duration 100 μs, 1 Hz, plate electrodes.

TABLE 21.1   Antitumor Effectiveness of ECT with Two Drugs on Sa-1 Tumors in A/J Mice

Treatment	N	Tumor Doubling Time (Days)	Tumor Growth Delay (Days)	Cure Rate (%)
Control	20	1.8 ± 0.1	—	0
EP	17	$3.1 \pm 0.2^*$	1.3 ± 0.2	0
Bleomycin (5 mg/kg)	20	1.9 ± 0.1	0.1 ± 0.1	0
Cisplatin (4 mg/kg)	10	$3.7 \pm 0.4^*$	1.9 ± 0.4	0
ECT with bleomycin	17	$34.5 \pm 2.9^*$	32.7 ± 2.9	70
ECT with cisplatin	10	$12.1 \pm 1.6^*$	10.3 ± 1.6	0

Reproduced from (Sersa et al. 2003) with permission.

Tumor growth delay was calculated with respect to the control group. In case of ECT with bleomycin, cured tumors were excluded from growth delay calculation. Mean ± SE of the mean values are given ($^*p < 0.05$).

21.4.4 Oxygenation in Tumors after EP and ECT

An important physiological parameter closely related to blood flow is tissue oxygenation, commonly expressed as partial pressure of oxygen (pO_2). It has been known for a long time that tumors in general are characterized by lower pO_2 values than the surrounding normal tissue and this was also confirmed for the Sa-1 tumor model in A/J mice by both EPR oximetry (Sentjurc et al. 2004) and the luminescence-based fiber-optic oximetry (Jarm et al. 2002). For an example of how closely related the microcirculatory blood flow and pO_2 in tumors are, see Figure 21.1, which shows that the rapid initial decrease in blood flow was followed with only a short delay by a decrease in pO_2 (measured by the luminescence-based oximetry at the same location as blood flow).

In a study on effects of ECT with cisplatin and bleomycin, it was shown that the time course and extent of pO_2 decrease after EP alone or after ECT was similar to that of the blood flow. The maximum decrease in pO_2 to about 30% of the pretreatment value was observed 2 h after the treatment and with steady recovery thereafter for EP alone and ECT with both drugs. Recovery of pO_2 to pretreatment values was faster in tumors treated with EP alone (8 h) than in tumors treated with ECT (2 days) (Sersa et al. 2002, 2008, Sentjurc et al. 2004). It is however interesting to note that while the perfusion in ECT-treated tumors did not recover even within 5 days (Figure 21.5), the oxygenation did. The reason for this may be the fact that the demand for oxygen in tumors was reduced due to a significant proportion of tumor cells being killed. And therefore the damaged but not completely destroyed tumor vasculature was able to supply sufficient blood flow for oxygenation recovery. In these EPR studies (Sentjurc et al. 2004, Sersa et al. 2008), the dose of cisplatin 4 mg/kg was the same as commonly used in previous studies on effects of ECT (Sersa et al. 2002), but the dose of bleomycin 1 mg/kg was five times smaller than in previous studies (Sersa et al. 1999a, 2003). This can explain why oxygenation could recover within 2 days after ECT with bleomycin given at a reduced dose. It is reasonable to expect that if bleomycin was given at the higher dose, oxygenation would not recover due to a complete shutdown and no further recovery of blood flow, as shown in earlier studies with a higher dose of bleomycin (Sersa et al. 1999a).

Development of hypoxia in tumors treated with EP and ECT was also observed in histological preparations using an exogenous hypoxia marker pimonidazole (Sersa et al. 2008). Control tumors and tumors treated with bleomycin alone contained only moderate pimonidazole-positive areas (ca. 10%). Tumors treated with EP alone or ECT developed larger hypoxic regions (up to 40%) immediately after the treatment, reaching the maximum extent of hypoxia in 2 h, which was retained for about 8 h with subsequent recovery to pretreatment values within 24 h. The recovery was faster in tumors treated with EP only (Sersa et al. 2008). These results were confirmed by an endogenous hypoxia marker glucose transporter 1 (Glut-1) (Cör et al. 2009). The increase in hypoxic regions was not as extreme as the observed blood flow changes and the recovery was also in contrast to no recovery observed in blood flow by several methods. Several possible explanations exist. One of them is that hypoxic regions may be underestimated by pimonidazole staining due to inadequate blood flow in at least some regions of tumors at the time of injection (large temporal and spatial heterogeneity in blood flow is typical for tumors). Pimonidazole also could not reach necrotic regions due to lack of blood flow thus resulting in falsely non-stained regions.

21.5 Synthesis—Models of Physiological Effects of EP and ECT on Tumor Blood Flow

21.5.1 The Proposed Series of Events in Blood Flow Changes after EP

The initial reaction to delivery of EP is most likely an immediate reflexory vasoconstriction of afferent arterioles and maybe even some larger vessels leading into tumor. The onset of this first short-lived phase of blood flow reduction is extremely rapid and profound; within a few seconds blood flow is practically arrested. The vascular spasm is gradually released after 1–2 min, when blood flow starts to recover. In a study on muscles it was shown that rapid vasoconstriction was sympathetically mediated in skeletal muscles treated with EP. The direct effect of EP on smooth vascular musculature however was discarded as a plausible mechanism of the observed vasoconstriction (Gehl et al. 2002). There is some doubt whether this initial phase could be a result of vasoconstriction due to paucity of smooth vascular musculature and sympathetic enervation in tumor vessels (Tozer et al. 2005, Sersa et al. 2008). However, the kinetics of this phase observed in tumors and muscles are the same. Furthermore, it has been shown that direct vasoconstrictive effect on tumor-supplying arterioles is actually the main mechanism of action of some chemical *vascular-disrupting agents* (VDAs) (Tozer et al. 2005). All this leads to an assumption that sympathetically mediated and/or directly stimulated constriction of smooth vascular musculature plays the principal role in the first phase of blood flow reduction in tumors treated with EP.

EP also causes a rapid and profound disruption of endothelial cytoskeleton and intercellular junctions, which consequently leads to swelling and rounding of endothelial cells and to compromised barrier function of the endothelial lining (Kanthou et al. 2006). Change in shape of endothelial cells increases vascular resistance for blood flow. The increased permeability of vascular walls to macromolecules is followed by protein leakage and leads to a decrease in oncotic pressure between the intra- and extravascular compartments and consequently to extravasation of liquids. The results are a development of interstitial edema, a buildup of IFP, a decreased intravascular pressure, and ultimately a compromised blood flow, not unlike what is seen after application of chemical VDAs (Tozer et al. 2005). The effect on endothelial cells *in vitro* was reported to be rapid, it developed within 5 min (Kanthou et al. 2006)), but not as rapid as the initial blood flow reduction observed in tumors or muscles after EP. So it is proposed that this second longer lived phase lags behind the first phase and becomes the predominant blood flow–reducing mechanism in tumors treated with EP 5–10 min after EP. It can be seen in LDF records of blood flow in tumors that blood flow stops increasing approximately 10–15 min after EP (Figure 21.4), which indicates a fully developed second phase of blood flow reduction. The resolution of this second phase is gradual and slow so that blood flow remains decreased up to about 24 h after the treatment of tumors (Sersa et al. 1999a,b, 2002). It was speculated that the resolution of the second phase follows the dynamics of endothelial cell membrane resealing after EP (Gehl et al. 2002). We suggest that the kinetics of reestablishment of endothelial cytoskeleton and the barrier function of endothelial cells may be an even more important determinant of resolution of the second phase in tumors. Blood flow reduction in tumors after EP is therefore reversible. But even this relatively short period of blood flow reduction (24 h) may contribute to the observed but practically insignificant growth delay in tumors treated by EP alone.

The proposed sequence of events after EP on the level of tumor microvasculature is summarized in the upper part of Figure 21.6.

21.5.2 The Proposed Series of Events in Blood Flow Changes after ECT

Up to few hours after ECT, the blood flow changes in tumors treated by ECT are identical to those observed in tumors treated by EP alone; so it is safe to assume that the main mechanisms of the observed effect are the same. However, 8 h after treatment, the differences between the effects of EP and ECT become obvious. Blood flow in EP-treated tumors continues to recover while blood flow in ECT-treated tumors continues to deteriorate or at least shows no further improvement up to 5 days after the treatment. The main reason for this is that the endothelial cells are affected by ECT in a similar way if not even more profoundly than the tumor cells. Affected endothelial cells cannot recover their structure and function and are ultimately pushed toward cell death. As a consequence, tumor vasculature is damaged beyond repair, blood flow cannot be restored, and the remaining tumor cells supplied by the affected vessels enter the so-called secondary cascade of induced cell death brought about by the lack of oxygen and nutrients and accumulation of waste products. The proposed sequence of events after ECT on the level of tumor microvasculature is summarized in the lower part of Figure 21.6.

21.5.3 The Role of EP and ECT in Antitumor Effectiveness of the Treatment

The increased cytotoxicity of chemotherapeutic drugs bleomycin and cisplatin induced by successful EP of tumor cells is undoubtedly the main mechanism involved in antitumor effectiveness of *in vivo* ECT. This is supported by (a) *in vitro* studies showing that exposure of tumor cells to EP increased cytotoxicity of bleomycin several 100-fold and that of cisplatin 10-fold (Orlowski et al. 1988, Sersa et al. 1995, Jaroszeski et al. 2000); (b) clearly increased accumulation of both drugs in tumors treated with electric pulses (Belehradek et al. 1994, Cemazar et al. 1999); and (c) the fact that both drugs being hydrophilic in nature become highly cytotoxic once they gain access to cytosol by means of EP (Sersa et al. 1995, Mir and Orlowski 1999). However, a discrepancy between increased cytotoxicity of cisplatin by EP *in vitro* (10-fold) and that observed in the same tumor cells *in vivo* (20-fold) led researchers to a conclusion that additional mechanisms must be involved (Cemazar et al. 1999). The following ones had been proposed:

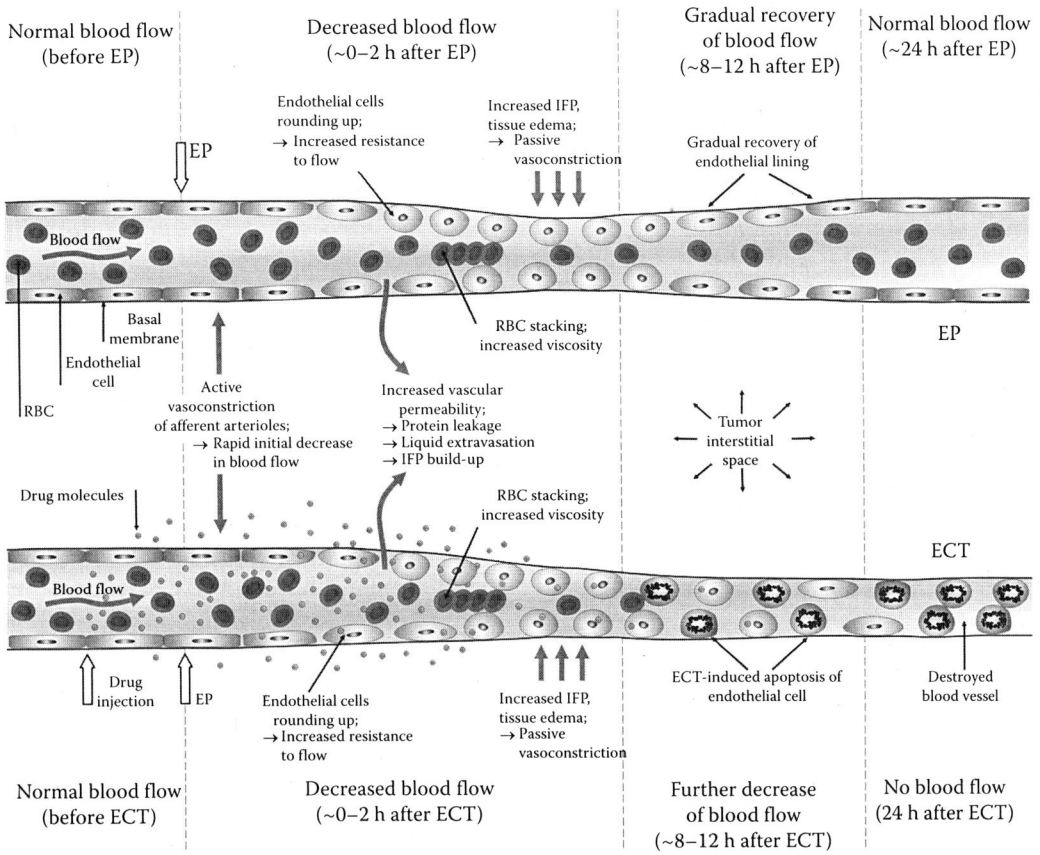

FIGURE 21.6 Model of blood flow changes in tumors after EP and ECT. The effects of EP alone (top) and of ECT (bottom) are presented on the level of a microcirculatory blood vessel. Application of the drug and electric pulses (EP) are indicated by block arrows.

the immune response (Mir et al. 1992, Sersa et al. 1997a,b), vascular-disrupting effects due to ECT of endothelial cells (Cemazar et al. 2001a, Sersa et al. 2008), and blood flow–modifying effects of ECT (Sersa et al. 1999a), the last two being closely related.

The importance of the host immune response for effective ECT was demonstrated in studies on immunocompetent and immunodeficient nude mice for ECT with cisplatin (Sersa et al. 1997b) and for ECT with bleomycin (Mir et al. 1991). In both studies, it was shown that growth delay induced by ECT was significantly greater in immunocompetent than in immunodeficient mice, and even more importantly, that complete tumor curability was only achievable in animals with normal immune response. Some other studies also showed potentiation of ECT effectiveness by combined use of ECT with various types of immunotherapy (Mir et al. 1992, 1995, Sersa et al. 1997a, Heller et al. 2000).

There is no doubt that EP alone and ECT in particular induce severe reduction in blood flow of treated tumors. What is less clear is to what extent this effect contributes to the overall antitumor effectiveness of ECT. While the onset of blood flow reduction follows the same pattern for EP alone and ECT, the long-term effects are hugely different: EP-induced changes are largely transient with full recovery within 24–48 h. ECT-induced reduction on the other hand is long-lived and may even be potentiated in the following days (depending on the drug dosage) to the point of complete shutdown of tumor blood flow in case of bleomycin and to a lesser degree in case of cisplatin with no recovery observed with either drug within 5 days after ECT.

EP is not a selective process and all cells in tumor treated with EP can be expected to be affected similarly (Sersa et al. 2003). Endothelial cells lining the tumor vasculature are thus also a likely target for EP and ECT. First, theoretical considerations, which took into account electrical properties of different tissue constituents and realistic microcirculatory blood vessel dimensions, have revealed that the electric field strength at the boundary between the blood and blood vessel wall (the location of endothelial cells) may be up to 40% higher than in the surrounding tissue further away from the vessel (Sersa et al. 2008). The reason for this is higher conductivity of blood in comparison to surrounding tissue. This indicates that the endothelial cells may be even more susceptible to the effects of EP than the tumor cells. Second, the endothelial cells are initially exposed to the highest concentration of the chemotherapeutic drug reached in blood vessels after an *i.v.* injection.

It is well known that the cell cytoskeleton that is composed of three major types of structures (microtubules, actin microfilaments, and intermediate filaments) is essential for maintaining cell shape and function. In case of vascular endothelial cells, the contractility and the integrity of intercellular junctions are responsible for regulation of vascular permeability (Kanthou et al. 2006). A study on human umbilical vein endothelial cells (HUVECs) provided conclusive evidence that the cytoskeleton and intercellular junctions are affected by EP. Exposure of confluent monolayers of HUVECs to electric pulses resulted in profound disruption of microfilament and microtubule cytoskeletal networks, loss of contractility, disruption of cell-to-cell junctions, and a rapid increase in endothelial monolayer permeability immediately after EP. The effects were voltage-dependent and almost fully reversible within 1–2 h after EP without a loss of endothelial cell viability. For voltages high enough to induce significant drop in cell viability, the restoration of perturbed cytoskeleton was severely impeded (Kanthou et al. 2006). All these results are in perfect agreement with the resolution of blood flow changes observed in tumors and suggest that changes in endothelial barrier function leading to increased permeability may be the main mechanism of the second long-term phase of blood flow decrease observed after EP *in vivo*. Due to inherent differences between normal and tumor vasculature, it can be expected that the disruption of cytoskeletal networks and intercellular junctions disrupted by EP would need more time to heal in tumor vasculature, which can explain why it takes longer to restore blood flow after EP in tumors (Sersa et al. 1999a,b, 2002) than in muscles (Gehl et al. 2002).

Damaged and killed endothelial cells lead to obstructed blood flow and to ischemic death of all cells supplied by the affected vessels. This effect brings ECT into the realm of the so-called vascular-targeted therapies. It was confirmed in *in vitro* studies using human dermal microvascular endothelial cells HMEC-1 that these cells are highly sensitive to ECT. While the cells were only moderately sensitive to cisplatin and bleomycin in absence of EP, the EP increased the cytotoxicity of cisplatin 10-fold and that of bleomycin even 5000-fold (compare these figures to previously mentioned 10-fold and several 100-fold increase *in vitro* cytotoxicity of cisplatin and bleomycin for tumor cells) (Cemazar et al. 2001a). Endothelial cells exhibited even greater sensitivity to ECT with bleomycin than some tumor cell lines (including Sa-1) but at the same time slightly better resistance to EP itself (Cemazar et al. 2001a). According to all this, ECT of endothelial cells should play a more important role in treatment of tumors with bleomycin than with cisplatin. The differences between the responses of endothelial cells treated with ECT with the two drugs *in vitro* are completely consistent with the overall blood flow reduction characteristics observed in tumors treated with the same treatments.

The profound and long-lasting reduction in blood flow after ECT with cisplatin was closely followed by similar extensive decrease in oxygenation, which can be expected to induce further cell death due to hypoxia. Indeed, it was found that the progression of necrosis in tumors to 90% of total tumor area correlated with oxygenation changes and also with the observed effects on overall tumor growth (Sersa et al. 2002). All these effects were more pronounced with bleomycin than with cisplatin, which was consistent with *in vitro* data on sensitivity of endothelial cells to these two drugs (Sersa et al. 1999a, 2003, Cemazar et al. 2001a).

Histological examination of tumors treated with ECT with bleomycin or cisplatin revealed development of massive necrosis (up to 100% of tumor mass in some cases with bleomycin) which reached maximum values 1–3 days after the treatment and never recovered thereafter (Sersa et al. 1999a, 2002, 2008). Necrotic areas were characterized by swollen tumor cells with picnotic nuclei and eosinophilic cytoplasm and sometimes by complete nuclear disappearance. Significant ECT-induced apoptosis of cells was observed even in non-necrotic areas of tumors treated with ECT with cisplatin (Sersa et al. 2002). In contrast to ECT-treated tumors, tumors treated with EP alone developed only transiently and significantly increased tissue necrosis within first 24 h after EP. The level of necrosis usually decreased to levels comparable to control tumors within a day or two. This was well correlated with the transient blood flow changes and with only moderate growth delay induced in tumors treated with EP alone. On the other hand, tumors treated with bleomycin alone were indistinguishable from control tumors in both growth delay, extent of necrosis, and blood flow changes (Sersa et al. 2002, 2008). Tumors treated with cisplatin alone developed slightly more necrosis and a small growth delay in comparison to control tumors (Sersa et al. 1999a).

The histological examination also revealed changes induced in blood vessels by EP or ECT with bleomycin (Sersa et al. 2008). One hour after EP alone or ECT the endothelial cells appeared swollen and rounded with narrowed lumen of blood vessels which is consistent with both the observed blood flow changes and demonstrated effects of EP on endothelial cytoskeleton and endothelial monolayer (Kanthou et al. 2006). Eight hours after ECT apoptotic morphological changes in endothelial cells were observed with some extravasation of erythrocytes, a clear indication of damaged blood vessels. These effects were absent in tumors treated with EP or bleomycin only or in control tumors.

The effects of EP and ECT on blood flow can contribute to antitumor effectiveness of ECT in two ways. The first one is the entrapment of the drug accumulated within tumor due to reduced blood flow—the so-called vascular lock effect—induced transiently by EP alone and permanently by ECT, which leads to extended exposure of tumor cells to the drug. The second one is the severely damaged tumor vasculature due to ECT of endothelial cells, which consequently leads to additional cascade of tumor cell death as a result of lack of oxygen and nutrients and accumulation of waste products in the tumor. Such a phenomenon, termed vascular-targeted therapy, has already been exploited in several studies using different approaches to affect tumor vasculature (Tozer et al. 2005). But this effect is also of extreme importance for complete elimination of the tumor in case of ECT because it helps to kill the cells that were not destroyed directly by the drug. In vitro studies confirmed that sensitivity of vascular endothelial cells to bleomycin (and to a lesser degree to cisplatin) was highly increased by ECT (Cemazar et al. 2001a).

In summary, we can therefore identify at least three pathways for how tumor cells in a tumor treated by ECT are killed.

1. Cells killed directly due to the increased uptake of the drug by electropermeabilized cells. This is the primary mechanism of ECT and these cells would die even in the absence of any antivascular effects of EP and ECT. Most cells in a successfully treated tumor treated with cisplatin or bleomycin probably belong to this class. But some unknown proportion of tumor cells would almost certainly escape this fate and remain viable either because some tumor regions were not exposed to high enough electric field or because extracellular drug concentration was not sufficient in some areas or due to other intrinsic characteristics of some tumor cells. In time, these remaining cells could thus give raise to tumor relapse.

2. Cells killed thanks to prolonged exposure of tumor cells to extracellular chemotherapeutic. Severely reduced blood flow after EP alone even without any long-term antivascular effects would effectively prevent or at least impede the washout of the drug from tumor tissue thus prolonging the exposure of tumor cells to high extracellular concentration of chemotherapeutic and increasing the chance of killing them. This class of cells is probably not significant in case of poorly permeant chemotherapeutics such as bleomycin and cisplatin, which require successful electropermeabilization in order to be effective at very low doses commonly used in ECT. Prolonged exposure of cells after membrane resealing to such low extracellular concentrations of poorly permeant drug molecules would not increase their cytotoxicity significantly. But this pathway of cell kill opens up

new possibilities for tumor treatment. It is reasonable to expect that this "vascular lock" or "drug entrapment" effect could increase effectiveness in case of some other drugs which cross the cell membrane more readily even after the membranes are resealed but whose effectiveness is limited due to rapid clearance from the extracellular space once the peak concentration is reached.

3. Cells killed due to lack of oxygen and nutrients and buildup of catabolic products. The long-term abrogation of blood flow in tumors due to destruction of tumor vasculature by ECT would wipe out this class of "resistant" cells. The importance of this mechanism of cell kill is clear in the case of presently used ECT but it also opens new possibilities for future vascular-targeted therapies.

Small but typically insignificant delay in growth of tumors treated by EP alone can be attributed to two distinct mechanisms. The first one is irreversible EP of a small proportion of tumor cells which leads to death or at least a temporary arrest in growth of the affected cells. In addition to this, the relatively long-lived decrease of blood flow in tumors due to reversible and irreversible EP of endothelial cells (duration of which is in agreement with the observed tumor growth delay after EP) leads to temporarily reduced supply of nutrients and removal of wastes from the tumor which impedes the growth of viable tumor cells and possibly also induces some minor additional cell death. In addition, the effectiveness of drugs that are activated in hypoxic environment (e.g., bioreductive drugs) is also increased due to this "vascular lock" caused by EP (Cemazar et al. 2001b).

21.5.4 Clinical Significance of Vascular-Disrupting Action of Electrochemotherapy

Bleeding metastases pose a treatment problem in many clinical situations, especially in melanoma metastases. There are only few treatment options available, isolated limb perfusion and intraarterial embolization (Sasso et al. 1995, Grunhagen et al. 2006). Based on the vascular-disrupting action of ECT with bleomycin (Sersa et al. 2008), several clinical cases of successful treatment of bleeding melanoma metastases have already been described. Gehl et al. have described two cases of melanoma metastases located on the skull and on the chest. The others reported immediate cessation of bleeding of the treated metastases, gradual formation of the scab, and eventually regression of the tumor nodules (Gehl and Geertsen 2000, 2006). Snoj et al. also reported successful treatment of bleeding metastasis on the thigh with the same treatment response (Snoj et al. 2009). As a further illustration, we report here a case of bleeding squamous cell carcinoma metastasis on the skull of a patient (see Figure 21.7). The tumor

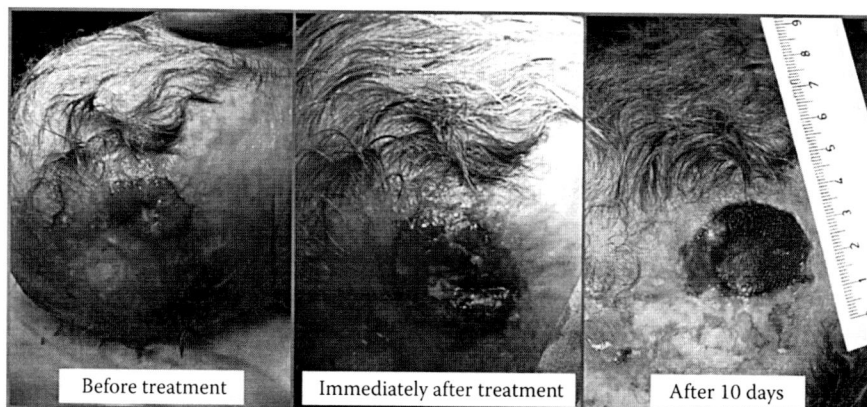

| Before treatment | Immediately after treatment | After 10 days |

FIGURE 21.7 (See color insert following page 268.) Use of ECT for palliative treatment of bleeding tumors. Bleeding squamous cell carcinoma metastasis was treated by ECT with bleomycin. Immediately after the treatment the bleeding was stopped. Within 10 days a scab formed and good antitumor effectiveness was evident.

stopped bleeding immediately (within seconds) after the application of electric pulses in ECT with bleomycin given intravenously. Within 10 days a scab gradually formed and good antitumor effect was recorded. All these cases support the proposed models of tumor blood flow–modifying action of EP and vascular-disrupting effect of ECT.

21.6 Conclusions

Tumor vasculature is now recognized as a valid target for anticancer treatments. In this chapter, we summarized the known effects of EP and ECT on blood flow in tumors and proposed the underlying mechanisms of action. In the case of EP, the duration and extent of blood flow reduction can be manipulated by using different voltages, number of pulses and pulse durations, and by repeating the application. This offers potential new application areas for combined treatment modalities with EP, such as hyperthermia and hypoxia-activated drugs. In addition, clinical applicability of EP for management of bleeding metastases has already been proven. In ECT, the effect of EP on tumor vasculature is combined with chemotherapy, which additionally damages tumor endothelial cells and can cause complete cessation of tumor blood flow. Due to this effect, ECT can also be included in the group of vascular-targeted therapies. As such it offers new opportunities for treatment modalities in combination with drugs targeted at tumor vasculature.

Acknowledgments

The authors would like to thank Dr. Simona Kranjc for her help in the preparation of the manuscript. The authors acknowledge the financial support of several research projects by the Slovenian Research Agency and by the European Commission.

Abbreviations

CE-MRI	contrast-enhanced magnetic resonance imaging
ECT	electrochemotherapy. Combined use of a chemotherapeutic and electric pulses to enhance cytotoxicity of the drug
EGT	electrogenetherapy. Combined use of genetic material and electric pulses to enhance the transfection of the cells
EP	electroporation or electropermeabilization. Use of electric pulses to increase the permeability of cell membrane for the purpose of ECT or EGT
EPR	electron paramagnetic resonance
IFP	interstitial fluid pressure
LDF	laser Doppler flowmetry
MRI	magnetic resonance imaging
PBV	patent blue-violet (biological dye)
pO_2	partial pressure of oxygen
US	ultrasound

References

Belehradek J, Orlowski S, Ramirez LH, Pron G, Poddevin B, Mir LM. 1994. Electropermeabilization of cells in tissues assessed by the qualitative and quantitative electroloading of bleomycin. *Biochimica et Biophysica Acta-Biomembranes* 1190(1):155–163.

Cemazar M, Milacic R, Miklavcic D, Dolzan V, Sersa G. 1998. Intratumoral cisplatin administration in electrochemotherapy: Antitumor effectiveness, sequence dependence and platinum content. *Anti-Cancer Drugs* 9(6):525–530.

Cemazar M, Miklavcic D, Scancar J, Dolzan V, Golouh R, Sersa G. 1999. Increased platinum accumulation in SA-1 tumour cells after in vivo electrochemotherapy with cisplatin. *British Journal of Cancer* 79(9–10):1386–1391.

Cemazar M, Parkins CS, Holder AL, Chaplin DJ, Tozer GM, Sersa G. 2001a. Electroporation of human microvascular endothelial cells: Evidence for an anti-vascular mechanism of electrochemotherapy. *British Journal of Cancer* 84(4):565–570.

Cemazar M, Parkins CS, Holder AL, Kranjc S, Chaplin DJ, Sersa G. 2001b. Cytotoxicity of bioreductive drug tirapazamine is increased by application of electric pulses in SA-1 tumours in mice. *Anticancer Research* 21(2A):1151–1156.

Cemazar M, Golzio M, Sersa G, Rols MP, Teissie J. 2006. Electrically-assisted nucleic acids delivery to tissues in vivo: Where do we stand? *Current Pharmaceutical Design* 12(29):3817–3825.

Cemazar M, Tamzali Y, Sersa G, Tozon N, Mir LM, Miklavcic D, Lowe R, Teissie J. 2008. Electrochemotherapy in veterinary oncology. *Journal of Veterinary Internal Medicine* 22(4):826–831.

Chaplin DJ, Hill SA, Bell KM, Tozer GM. 1998. Modification of tumor blood flow: Current status and future directions. *Seminars in Radiation Oncology* 8(3):151–163.

Collingridge DR, Young WK, Vojnovic B, Wardman P, Lynch EM, Hill SA, Chaplin DJ. 1997. Measurement of tumor oxygenation: A comparison between polarographic needle electrodes and a time-resolved luminescence-based optical sensor. *Radiation Research* 147(3):329–334.

Cör A, Cemazar M, Plazar N, Sersa G. 2009. Comparison between hypoxic markers pimonidazole and glucose transporter 1 (Glut-1) in murine fibrosarcoma tumours after electrochemotherapy. *Radiology and Oncology* 43(3):195–202.

Daud AI, DeConti RC, Andrews S, Urbas P, Riker AI, Sondak VK, Munster PN et al. 2008. Phase I trial of interleukin-12 plasmid electroporation in patients with metastatic melanoma. *Journal of Clinical Oncology* 26(36):5896–5903.

Delorme S, Krix M. 2006. Contrast-enhanced ultrasound for examining tumor biology. *Cancer Imaging* 6(1):148–152.

Demsar F, Roberts TPL, Schwickert HC, Shames DM, vanDijke CF, Mann JS, Saeed M, Brasch RC. 1997. A MRI spatial mapping technique for microvascular permeability and tissue blood volume based on macromolecular contrast agent distribution. *Magnetic Resonance in Medicine* 37(2):236–242.

Domenge C, Orlowski S, Luboinski B, DeBaere T, Schwaab G, Belehradek J, Mir LM. 1996. Antitumor electrochemotherapy—New advances in the clinical protocol. *Cancer* 77(5):956–963.

Folkman J. 2007. Opinion—Angiogenesis: An organizing principle for drug discovery? *Nature Reviews Drug Discovery* 6(4):273–286.

Gehl J, Geertsen PF. 2000. Efficient palliation of haemorrhaging malignant melanoma skin metastases by electrochemotherapy. *Melanoma Research* 10(6):585–589.

Gehl J, Geertsen PF. 2006. Palliation of haemorrhaging and ulcerated cutaneous tumours using electrochemotherapy. *EJC Supplements* 4(11):35–37.

Gehl J, Skovsgaard T, Mir LM. 2002. Vascular reactions to in vivo electroporation: Characterization and consequences for drug and gene delivery. *Biochimica et Biophysica Acta-General Subjects* 1569(1–3):51–58.

Grunhagen DJ, de Wilt JHW, Graveland WJ, van Geel AN, Eggermont AMM. 2006. The palliative value of tumor necrosis factor alpha-based isolated limb perfusion in patients with metastatic sarcoma and melanoma. *Cancer* 106(1):156–162.

Heller LC, Heller R. 2006. In vivo electroporation for gene therapy. *Human Gene Therapy* 17(9):890–897.

Heller R, Jaroszeski M, Perrott R, Messina J, Gilbert R. 1997. Effective treatment of B16 melanoma by direct delivery of bleomycin using electrochemotherapy. *Melanoma Research* 7(1):10–18.

Heller R, Gilbert R, Jaroszeski MJ. 1999. Clinical applications of electrochemotherapy. *Advanced Drug Delivery Reviews* 35(1):119–129.

Heller L, Pottinger C, Jaroszeski M-L, Gilbert R, Heller R. 2000. In vivo electroporation of plasmids encoding GM-CFS or interleukin-2 into existing B16 melanomas combined with electrochemotherapy induces long-term antitumour immunity. *Melanoma Research* 10(6):577–583.

Ivanusa T, Beravs K, Cemazar M, Jevtic V, Demsar F, Sersa G. 2001. MRI macromolecular contrast agents as indicators of changed tumor blood flow. *Radiology and Oncology* 35(2):139–147.

Jain RK. 1987. Transport of molecules across tumor vasculature. *Cancer and Metastasis Reviews* 6(4):559–593.

Jarm T, Sersa G, Miklavcic D. 2002. Oxygenation and blood flow in tumors treated with hydralazine: Evaluation with a novel luminescence-based optic sensor. *Technology and Health Care* 10:363–380.

Jaroszeski MJ, Dang V, Pottinger C, Hickey J, Gilbert R, Heller R. 2000. Toxicity of anticancer agents mediated by electroporation in vitro. *Anti-Cancer Drugs* 11(3):201–208.

Kanduser M, Miklavcic D, Pavlin M. 2009. Mechanisms involved in gene electrotransfer using high- and low-voltage pulses—An in vitro study. *Bioelectrochemistry* 74(2):265–271.

Kanthou C, Kranjc S, Sersa G, Tozer G, Zupanic A, Cemazar M. 2006. The endothelial cytoskeleton as a target of electroporation-based therapies. *Molecular Cancer Therapeutics* 5(12):3145–3152.

Marty M, Sersa G, Garbay JR, Gehl J, Collins CG, Snoj M, Billard V et al. 2006. Electrochemotherapy—An easy, highly effective and safe treatment of cutaneous and subcutaneous metastases: Results of ESOPE (European Standard Operating Procedures of Electrochemotherapy) study. *EJC Supplements* 4(11):3–13.

Menon C, Fraker DL. 2005. Tumor oxygenation status as a prognostic marker. *Cancer Letters* 221(2):225–235.

Miklavcic D, Corovic S, Pucihar G, Pavselj N. 2006. Importance of tumour coverage by sufficiently high local electric field for effective electrochemotherapy. *EJC Supplements* 4(11):45–51.

Mir LM, Orlowski S. 1999. Mechanisms of electrochemotherapy. *Advanced Drug Delivery Reviews* 35(1):107–118.

Mir LM, Orlowski S, Belehradek J, Paoletti C. 1991. Electrochemotherapy potentiation of antitumor effect of bleomycin by local electric pulses. *European Journal of Cancer* 27(1):68–72.

Mir LM, Orlowski S, Poddevin B, Belehradek J. 1992. Electrochemotherapy tumor treatment is improved by interleukin-2 stimulation of the host's defenses. *European Cytokine Network* 3(3):331–334.

Mir LM, Roth C, Orlowski S, Quintincolonna F, Fradelizi D, Belehradek J, Kourilsky P. 1995. Systemic antitumor effects of electrochemotherapy combined with histoincompatible cells secreting interleukin-2. *Journal of Immunotherapy* 17(1):30–38.

Mir LM, Gehl J, Sersa G, Collins CG, Garbay JR, Billard V, Geertsen PF, Rudolf Z, O'Sullivan GC, Marty M. 2006. Standard operating procedures of the electrochemotherapy: Instructions for the use of bleomycin or cisplatin administered either systemically or locally and electric pulses delivered by the Cliniporator (TM) by means of invasive or non-invasive electrodes. *EJC Supplements* 4(11):14–25.

Neumann E, Schaeferridder M, Wang Y, Hofschneider PH. 1982. Gene-transfer into mouse lyoma cells by electroporation in high electric-fields. *EMBO Journal* 1(7):841–845.

O'Hara JA, Blumenthal RD, Grinberg OY, Demidenko E, Grinberg S, Wilmot CM, Taylor AM, Goldenberg DM, Swartz HM. 2001. Response to radioimmunotherapy correlates with tumor pO(2) measured by EPR oximetry in human tumor xenografts. *Radiation Research* 155(3):466–473.

Orlowski S, Belehradek J, Paoletti C, Mir LM. 1988. Transient electropermeabilization of cells in culture - increase of the cyto-toxicity of anticancer drugs. *Biochemical Pharmacology* 37(24):4727–4733.

Pavselj N, Bregar Z, Cukjati D, Batiuskaite D, Mir LM, Miklavcic D. 2005. The course of tissue permeabilization studied on a mathematical model of a subcutaneous tumor in small animals. *IEEE Transactions on Biomedical Engineering* 52(8):1373–1381.

Prud'homme GJ, Glinka Y, Khan AS, Draghia-Akli R. 2006. Electroporation-enhanced nonviral gene transfer for the prevention or treatment of immunological, endocrine and neoplastic diseases. *Current Gene Therapy* 6(2):243–273.

Pusenjak J, Miklavcic D. 1997. Interstitial fluid pressure as an obstacle in treatment of solid tumors. *Radiology and Oncology* 31:291–297.

Quaglino P, Mortera C, Osella-Abate S, Barberis M, Illengo M, Rissone M, Savoia P, Bernengo MG. 2008. Electrochemotherapy with intravenous bleomycin in the local treatment of skin melanoma metastases. *Annals of Surgical Oncology* 15(8):2215–2222.

Raleigh JA, Chou SC, Bono EL, Thrall DE, Varia MA. 2001. Semiquantitative immunohistochemical analysis for hypoxia in human tumors. *International Journal of Radiation Oncology Biology Physics* 49:569–574.

Ramirez LH, Orlowski S, An D, Bindoula G, Dzodic R, Ardouin P, Bognel C, Belehradek J, Munck JN, Mir LM. 1998. Electrochemotherapy on liver tumours in rabbits. *British Journal of Cancer* 77(12):2104–2111.

Sapirstein LA. 1958. Regional blood flow by fractional distribution of indicators. *American Journal of Physiology* 193:161–168.

Sasso CM, Hubner C, Wall S. 1995. Intraarterial embolization of bleeding melanoma. *Journal of Vascular Nursing* 13(1):27–28.

Satkauskas S, Bureau MF, Puc M, Mahfoudi A, Scherman D, Miklavcic D, Mir LM. 2002. Mechanisms of in vivo DNA electrotransfer: Respective contributions of cell electropermeabilization and DNA electrophoresis. *Molecular Therapy* 5(2):133–140.

Satkauskas S, Andre F, Bureau MF, Scherman D, Miklavcic D, Mir LM. 2005. Electrophoretic component of electric pulses determines the efficacy of in vivo DNA electrotransfer. *Human Gene Therapy* 16(10):1194–1201.

Schmiedl U, Ogan M, Paajanen H, Marotti M, Crooks LE, Brito AC, Brasch RC. 1987. Albumin labeled with GD-DTPA as an intravascular, blood pool enhancing agent for MR imaging—Biodistribution and imaging studies. *Radiology* 162(1):205–210.

Sentjurc M, Cemazar M, Sersa G. 2004. EPR oximetry of tumors in vivo in cancer therapy. *Spectrochimica Acta Part A* 60:1379–1385.

Sersa G. 2006. The state-of-the-art of electrochemotherapy before the ESOPE study; advantages and clinical uses. *EJC Supplements* 4(11):52–59.

Sersa G, Cemazar M, Miklavcic D. 1995. Antitumor effectiveness of electrochemotherapy with cis-diamminedichloroplatinum(II) in mice. *Cancer Research* 55(15):3450–3455.

Sersa G, Cemazar M, Menart V, GabercPorekar V, Miklavcic D. 1997a. Anti-tumor effectiveness of electrochemotherapy with bleomycin is increased by TNF-alpha on SA-1 tumors in mice. *Cancer Letters* 116(1):85–92.

Sersa G, Miklavcic D, Cemazar M, Belehradek J, Jarm T, Mir LM. 1997b. Electrochemotherapy with CDDP on LPB sarcoma: Comparison of the anti-tumor effectiveness in immunocompetent and immunodeficient mice. *Bioelectrochemistry and Bioenergetics* 43:279–283.

Sersa G, Beravs K, Cemazar M, Miklavcic D, Demsar F. 1998. Contrast enhanced MRI assessment of tumor blood volume after application of electric pulses. *Electro- and Magnetobiology* 17:299–306.

Sersa G, Cemazar M, Miklavcic D, Chaplin DJ. 1999a. Tumor blood flow modifying effect of electrochemotherapy with bleomycin. *Anticancer Research* 19(5B):4017–4022.

Sersa G, Cemazar M, Parkins CS, Chaplin DJ. 1999b. Tumour blood flow changes induced by application of electric pulses. *European Journal of Cancer* 35(4):672–677.

Sersa G, Krzic M, Sentjurc M, Ivanusa T, Beravs K, Cemazar M, Auersperg M, Swartz HM. 2001. Reduced tumor oxygenation by treatment with vinblastine. *Cancer Research* 61(10):4266–4271.

Sersa G, Krzic M, Sentjurc M, Ivanusa T, Beravs K, Kotnik V, Coer A, Swartz HM, Cemazar M. 2002. Reduced blood flow and oxygenation in SA-1 tumours after electrochemotherapy with cisplatin. *British Journal of Cancer* 87(9):1047–1054.

Sersa G, Cemazar M, Miklavcic D. 2003. Tumor blood flow modifying effects of electrochemotherapy: A potential vascular targeted mechanism. *Radiology and Oncology* 37(1):43–48.

Sersa G, Cemazar M, Miklavcic D, Rudolf Z. 2006. Electrochemotherapy of tumours. *Radiology and Oncology* 40(3):163–174.

Sersa G, Jarm T, Kotnik T, Coer A, Podkrajsek M, Sentjurc M, Miklavcic D et al. 2008. Vascular disrupting action of electroporation and electrochemotherapy with bleomycin in murine sarcoma. *British Journal of Cancer* 98(2):388–398 (doi: 10.1038/sj.bjc.6604168, www.bjcancer.com).

Shepherd AP, Öberg PÅ. 1990. *Laser-Doppler Blood Flowmetry.* Norwell, MA: Kluwer Academic Publishers.

Siemann DW, Horsman MR. 2009. Vascular targeted therapies in oncology. *Cell and Tissue Research* 335(1):241–248.

Snoj M, Cemazar M, Srnovrsnik T, Kosir SP, Sersa G. 2009. Limb sparing treatment of bleeding melanoma recurrence by electrochemotherapy. *Tumori* 95(3):398–402.

Swartz HM, Clarkson RB. 1998. The measurement of oxygen in vivo using EPR techniques. *Physics in Medicine and Biology* 43(7):1957–1975.

Testori A, Rutkowski P, Marsden J, Bastholt L, Chiarion-Sileni V, Hauschild A, Eggermont AMM. 2009. Surgery and radiotherapy in the treatment of cutaneous melanoma. *Annals of Oncology* 20:22–29.

Tozer GM, Kanthou C, Baguley BC. 2005. Disrupting tumour blood vessels. *Nature Reviews Cancer* 5(6):423–435.

Vaupel P, Hockel M. 2000. Blood supply, oxygenation status and metabolic micromilieu of breast cancers: Characterization and therapeutic relevance (Review). *International Journal of Oncology* 17(5):869–879.

Vaupel P, Schlenger K, Knoop C, Hockel M. 1991. Oxygenation of human tumors—Evaluation of tissue oxygen distribution in breast cancers by computerized O2 tension measurements. *Cancer Research* 51(12):3316–3322.

22

Electrochemotherapy as Part of an Immunotherapy Strategy in the Treatment of Cancer

Julie Gehl

22.1 Cell Death after Electrochemotherapy

The key to understanding the use of electrochemotherapy (ECT) with immunotherapy is the unique way in which ECT leads to cell kill. The drug most frequently used in ECT is bleomycin, which acts as an enzyme causing single- or double-strand breaks in DNA, with 10–15 strand breaks per molecule (Tounekti et al. 2001). Bleomycin is a *uniquely* toxic molecule—in fact, all other DNA-targeting chemotherapeutic molecules target only one or few points of attack in the DNA strand. The reason that bleomycin is not used more in the clinic is that the cell membrane forms a formidable barrier to this hydrophilic, charged, and large molecule for which no efficient uptake mechanism exists (Gothelf et al. 2003).

It has been shown that the toxicity of bleomycin is increased a minimum of 300-fold when electroporation is performed (Gehl et al. 1998; Jaroszeski et al. 2000; Orlowski et al. 1988). One of the elements necessary for the activation of the immune system is a "danger signal" (Fuchs and Matzinger 1996). As ECT leads to the immediate cell death of tumor cells in the treated area, ECT could be regarded as a way to obtain such a danger signal. Other studies have used tumor tissue obtained by excision, exposed to toxic or lytic agents *ex vivo*, whereafter reintroduction into the patient has been performed. An example is described in Jocham et al. (2004), where a homogenate of tumor cells has been reintroduced into the patient after treatment of tumor cells *ex vivo*. In the case of ECT, cells are lysed *in situ* and exposed to immunogenic cells in their natural environment, with a danger signal to enhance interaction between the immune system and the treated tumor volume.

Interestingly, it has been shown both *in vitro* (Tounekti et al. 1993) and *in vivo* (Mekid et al. 2003) that, depending on the dose, bleomycin toxicity may lead to either apoptotic or necrotic cell death, or a mixture of both. A vast body of literature has examined the type of cell death induced by cancer therapy in *ex vivo* tumor samples compared with immunogenicity (Kepp et al. 2009). Interestingly, ECT may elicit a combined type of apoptotic and necrotic cell death, as variations in the drug dose through the tumor tissue may lead to different types of cell death in the same tumor (Mekid et al. 2003). Whereas many immunotherapy protocols using, e.g., dendritic cells are dependent on the patient's HLA-type (Human Leukocyte Antigen) in order for the chosen antigen to be correctly expressed by the antigen-presenting cells, ECT will work for patients with any HLA-type, leaving the tumor lysate *in situ*, and allowing an immune response in its natural environment.

22.2 Preclinical Data Supporting a Role for Electrochemoimmunotherapy

Several studies have shown that there is an immunogenic effect of electroporation. The weak immunogenic effect of electroporation alone can be strengthened by the addition of bleomycin and there is a synergistic effect when combined with interleukin-2 (IL-2) or other cytokines. The nature of this immunogenic effect is not yet established, but the release of antigens from the cytosol, possibly combined with changes in the cell membrane, are likely candidates. Animal studies have shown that the combination of ECT and IL-2 acts in a synergistic way by

- Enhancing the local antitumor effect in the treated area (Mir et al. 1995)
- Affecting tumors implanted in other sites of the same animal (Mir et al. 1995)
- Reducing the frequency of metastases, leading to longer survival (Orlowski et al. 1998; Ramirez et al. 1998)
- Inducing a memory effect so that treated animals did not develop tumors of a given type when rechallenged at a later time (Mir et al. 1995)
- Furthermore, blocking the immune system, can decrease the local and the systemic effect of ECT (Mir et al. 1992; Sersa et al. 1997)

Interestingly, a significant response and survival difference has been found in animal studies of different species (rats, mice, rabbits), as well as with various cell lines (representing melanoma, liver tumors, colorectal cancer) (Kuriyama et al. 2000; Miyazaki et al. 2003; Ramirez et al. 1998; Sersa et al. 1992).

22.3 Data from Clinical Studies

Two tumor histologies in particular are viewed as being immunogenic: renal cell carcinoma and malignant melanoma. Indeed it is with these cancer histologies that a major part of immunotherapy trials has taken place. Results generally show responses in a subset of patients (Rosenberg 2001), and at this time it is not possible to identify the potentially benefiting patient group ahead of treatment. High-dose cytokine treatment with, e.g., IL-2 has considerable toxicity, and therefore less toxic yet effective regimens have been sought.

As ECT of cutaneous malignant melanoma tumors is a straightforward procedure (Gehl and Geertsen 2000; Marty et al. 2006; Mir et al. 2006), the combined use of ECT and immunotherapy would obviously target malignant melanoma. The regression of lesions distant to the ECT treatment site has *not* been observed in previous clinical trials (Heller et al. 1998; Marty et al. 2006), and thus immunostimulatory compounds must be used in order to stimulate the immune system concomitantly with the ECT procedure.

At the time of conclusion of this chapter, two clinical trials involving immunotherapy in connection with electroporation-based therapies in cancer have been concluded. A trial using gene electrotransfer of IL-12 to patients with disseminated malignant melanoma was carried out at the University of South Florida, and is described in Chapter 19 by Heller and Heller, as well as in Daud et al. (2008).

22.3.1 ECT and Low-Dose IL-2 Study in Patients with Disseminated Malignant Melanoma

In a phase II study carried out at our institution (Gehl et al. 2001), 36 patients with disseminated malignant melanoma received ECT of up to five tumor nodules, and end point was response in metastatic lesions other than the ones treated with ECT. The treatment regimen consisted of ECT of up to five metastatic skin lesions, followed by injection of IL-2 (proleukin, Chiron, Holland), two MIU daily as flat dose for 21 days. The 3-week cycle could be given up to three times, but patients exited the trial in case of progression between treatments. The toxicity of the treatment was mild, indicating that this treatment may also be administered to patients in performance status or age groups where high-dose IL-2 therapy may be discouraged due to toxicity.

A total of 4 patients (4 of 36 or 11%) responded in distant lesions, 3 of whom remained alive with no evidence of disease at 5+ years after treatment. Patients were followed using ELISPOT (enzyme-linked immunosorbent spot analysis) to monitor immune responses to therapy. Several patients exhibited response to known melanoma antigens, and in one case tumor-infiltrating lymphocytes could be isolated from a responding lesion. In fact, it was shown that one particular T-cell clone was isolated from peripheral blood in response to the beginning of therapy, whereafter it disappeared, and subsequently T-cells with the same clonal characteristics could be isolated from a metastatic lesion in regression (Andersen et al. 2003).

Of the responders, all had skin or lymph node disease only. A subset of T-cells has been identified that circulate from skin to lymph nodes and back (Schrama et al. 2004), and this may explain that the cytotoxic attack is seen in skin and lymph nodes only, since the site of origin of the initial immunological response is in the treatment area in the skin.

22.4 Conclusion and Perspectives

As it has been stipulated already by other authors, a subset of cancer patients may benefit from immunotherapeutic approaches. ECT may fit into such a strategy by providing a type of autologous tumor cell vaccination, where acute tumor cell death in the ECT area may lead to exposure of cancer antigens to the immune system, and subsequent systemic immune response. To date, two clinical trials have used ECT or gene electrotransfer in an immunotherapeutic strategy, both with resultant systemic and long-term responses. Paraclinical evidence has shown activation of specific T-cell clones being able to home to areas of metastasis.

Although immunotherapy in general, and also electrochemoimmunotherapy or gene electrotransfer with immunostimulatory molecules, gives rise to few complete responses, it still may prove of value as a low-toxicity regimen in a patient category where efficient cancer therapy has been very hard to find.

References

Andersen, M.H., J. Gehl, S. Reker, L.O. Pedersen, J.C. Becker, P. Geertsen, and P.T. Straten. 2003. Dynamic changes of specific T cell responses to melanoma correlate with IL-2 administration. *Seminars in Cancer Biology.* 13:449–459.

Daud, A.I., R.C. DeConti, S. Andrews, P. Urbas, A.I. Riker, V.K. Sondak, P.N. Munster et al. 2008. Phase I trial of interleukin-12 plasmid electroporation in patients with metastatic melanoma. *Journal of Clinical Oncology.* 26:5896–5903.

Fuchs, J.E. and P. Matzinger. 1996. Is cancer dangerous to the immune system? *Seminars in Immunology.* 8:271–280.

Gehl, J., M.H. Andersen, P.T. Straten, and P. Geertsen. 2001. Tumor autovaccination by electrochemotherapy followed by low-dose Il-2 in advanced malignant melanoma: Efficient with low toxicity. *Journal of Clinical Oncology.* 20 [*Proc, Am Soc Clin Oncol*].

Gehl, J. and P. Geertsen. 2000. Efficient palliation of hemorrhaging malignant melanoma skin metastases by electrochemotherapy. *Melanoma Research*. 10:585–589.

Gehl, J., T. Skovsgaard, and L.M. Mir. 1998. Enhancement of cytotoxicity by electropermeabilization: An improved method for screening drugs. *Anti-Cancer Drugs*. 9:319–325.

Gothelf, A., L.M. Mir, and J. Gehl. 2003. Electrochemotherapy: Results of cancer treatment using enhanced delivery of bleomycin by electroporation. *Cancer Treatment Reviews*. 29:371–387.

Heller, R., M.J. Jaroszeski, D.S. Reintgen, C.A. Puleo, R.C. DeConti, R.A. Gilbert, and L.F. Glass. 1998. Treatment of cutaneous and subcutaneous tumors with electrochemotherapy using intralesional bleomycin. *Cancer*. 83:148–157.

Jaroszeski, M.J., V. Dang, C. Pottinger, J. Hickey, R. Gilbert, and R. Heller. 2000. Toxicity of anticancer agents mediated by electroporation in vitro. *Anti-Cancer Drugs*. 11:201–208.

Jocham, D., A. Richter, L. Hoffmann, K. Iwig, D. Fahlenkamp, G. Zakrzewski, E. Schmitt et al. 2004. Adjuvant autologous renal tumour cell vaccine and risk of tumour progression in patients with renal-cell carcinoma after radical nephrectomy: Phase III, randomised controlled trial. *Lancet*. 363:594–599.

Kepp, O., A. Tesniere, F. Schlemmer, M. Michaud, L. Senovilla, L. Zitvogel, and G. Kroemer. 2009. Immunogenic cell death modalities and their impact on cancer treatment. *Apoptosis*. 14:364–375.

Kuriyama, S., A. Mitoro, H. Tsujinoue, Y. Toyokawa, H. Nakatani, H. Yoshiji, T. Tsujimoto, H. Okuda, S. Nagao, and H. Fukui. 2000. Electrochemotherapy can eradicate established colorectal carcinoma and leaves a systemic protective memory in mice. *International Journal of Oncology*. 16:979–985.

Marty, M., G. Sersa, J.R. Garbay, J. Gehl, C.G. Collins, M. Snoj, V. Billard et al. 2006. Electrochemotherapy— An easy, highly effective and safe treatment of cutaneous and subcutaneous metastases: Results of ESOPE (European Standard Operating Procedures of Electrochemotherapy) study. *EJC Supplements*. 4:3–13.

Mekid, H., O. Tounekti, A. Spatz, M. Cemazar, F.Z. El Kebir, and L.M. Mir. 2003. In vivo evolution of tumour cells after the generation of double-strand DNA breaks. *British Journal of Cancer*. 88:1763–1771.

Mir, L.M., J. Gehl, G. Sersa, C.G. Collins, J.R. Garbay, V. Billard, P.F. Geertsen, Z. Rudolf, G.C. O'Sullivan, and M. Marty. 2006. Standard operating procedures of the electrochemotherapy: Instructions for the use of bleomycin or cisplatin administered either systemically or locally and electric pulses delivered by the Cliniporator (TM) by means of invasive or non-invasive electrodes. *EJC Supplements*. 4:14–25.

Mir, L.M., S. Orlowski, B. Poddevin, and J. Belehradek Jr. 1992. Electrochemotherapy tumor treatment is improved by interleukin-2 stimulation of the host's defenses. *European Cytokine Network*. 3:331–334.

Mir, L.M., C. Roth, S. Orlowski, F. Quintin-Colonna, D. Fradelizi, J. Belehradek Jr., and P. Kourilsky. 1995. Systemic antitumor effects of electrochemotherapy combined with histoincompatible cells secreting interleukin-2. *Journal of Immunotherapy with Emphasis on Tumor Immunology*. 17:30–38.

Miyazaki, S., Y. Gunji, H. Matsubara, H. Shimada, M. Uesato, T. Suzuki, T. Kouzu, and T. Ochiai. 2003. Possible involvement of antitumor immunity in the eradication of colon 26 induced by low-voltage electrochemotherapy with bleomycin. *Surgery Today*. 33:39–44.

Orlowski, S., D. An, J. Belehradek Jr., and L.M. Mir. 1998. Antimetastatic effects of electrochemotherapy and of histoincompatible interleukin-2-secreting cells in the murine Lewis lung tumor. *Anti-Cancer Drugs*. 9:551–556.

Orlowski, S., J. Belehradek Jr., C. Paoletti, and L.M. Mir. 1988. Transient electropermeabilization of cells in culture. Increase of the cytotoxicity of anticancer drugs. *Biochemical Pharmacology*. 37:4727–4733.

Ramirez, L.H., S. Orlowski, D. An, G. Bindoula, R. Dzodic, P. Ardouin, C. Bognel, J. Belehradek Jr., J. Munck, and L.M. Mir. 1998. Electrochemotherapy on liver tumors in rabbits. *British Journal of Cancer*. 77:2104–2111.

Rosenberg, S.A. 2001. Progress in human tumour immunology and immunotherapy. *Nature*. 411:380–384.

Schrama, D., X.A. Rong, A.O. Eggert, M.H. Andersen, L.O. Pedersen, E. Kampgen, T.N. Schumacher, R.R. Reisfeld, and J.C. Becker. 2004. Shift from systemic to site-specific memory by tumor-targeted IL-2. *Journal of Immunology*. 172:5843–5850.

Sersa, G., D. Miklavcic, U. Batista, S. Novakovic, F. Bobanovic, and L. Vodovnik. 1992. Anti-tumor effect of electrotherapy alone or in combination with interleukin-2 in mice with sarcoma and melanoma tumors. *Anti-Cancer Drugs.* 3:253–260.

Sersa, G., D. Miklavcic, M. Cemazar, J. Belehradek, T. Jarm, and L.M. Mir. 1997. Electrochemotherapy with CDDP on LPB sarcoma: Comparison of the anti-tumor effectiveness in immunocompetent and immunodeficient mice. *Bioelectrochemistry and Bioenergetics.* 43:279–283.

Tounekti, O., A. Kenani, N. Foray, S. Orlowski, and L. M. Mir. 2001. The ratio of single- to double-strand DNA breaks and their absolute values determine cell death pathway. *British Journal of Cancer.* 84:1272–1279.

Tounekti, O., G. Pron, J. Belehradek Jr., and L.M. Mir. 1993. Bleomycin, an apoptosis-mimetic drug that induces two types of cell death depending on the number of molecules internalized. *Cancer Research.* 53:5462–5469.

23

Combined Electrical Field and Ultrasound: A Nondrug-Based Method for Tumor Ablation*

Patrick F. Forde

Ciara Twomey

Gerald C. O'Sullivan

Declan M. Soden

23.1 Introduction

The successful and complete ablation of tumors continues to be a challenge in cancer therapy. Past research has focused much on development and validation of various systems for the delivery of both chemotherapeutic drugs and plasmid constructs to tumor tissue. These challenges lead to the exploitation of both ultrasound and electric field technologies to permeabilize the tissue in order to increase infiltration of DNA and drugs into tumors. However, as phenomena such as cellular adaptation and drug-resistance remain a concern whenever chemotherapeutic drugs are used, a physical means of tumor ablation may be preferable to enhancing delivery of cytotoxic compounds. Recent research demonstrated that sequential exposure of tumors to electric fields and ultrasound resulted in a significant decrease in tumor volume and may provide viable physical alternative means of clinical tumor ablation (Larkin et al., 2005; Rollan Haro et al., 2005).

---

* First two authors contributed equally to this work.

23.2 Electroporation

23.2.1 Definition and Background

In the late 1960s, electrical pulses were shown to have a lytic effect on a cell by irreversible disruption of the cell membrane (Sale and Hamilton, 1968). In later decades, it was seen that the modification of the electrical pulse parameters resulted in transient and reversible permeabilization of cell membranes. This could be used to introduce plasmid DNA to both eukaryotes and prokaryotes *in vitro* (Auer et al., 1976; Neumann et al., 1982); later, its use for *in vivo* applications was described. Cell or tissue electroporation can allow the transfer of biomolecules and chemicals alike from extracellular to intracellular locations by electronically disrupting the lipid bilayer to create short-lived pores in the cell membrane (Kinosita and Tsong, 1977). These pores effectively act as channels through which these molecules can passively travel into the cell.

This technique has been further exploited and refined for the delivery of both genes (electro-gene therapy) and otherwise impermeable drugs (electro-chemotherapy) in a therapeutic setting for the treatment of a wide variety of tumors (Goto et al., 2000; Sadadcharam et al., 2008). In both cases, the therapeutic agent, be it chemical or nucleic acid based, is first injected into the tumor. Subsequently, an electrical field is applied across the tissue, either by plate electrodes outside the tumor or needle electrodes inserted into it (Figure 23.1). This results in selective uptake of the agent by cells located within the electrical field while cells outside the electrical field are unaffected.

23.2.2 Use of Electroporation

Electroporation, alone or in combination with other components, has been shown over time to have many applications, both *in vitro* and *in vivo*, and is particularly suited to the treatment of solid tumors. Alone, irreversible electroporation has been shown to significantly decrease tumor volumes without incurring unwanted thermal effects (Al-Sakere et al., 2007a) (Figure 23.2). As well as being a noninvasive therapy, the therapeutic effect has been shown to be independent of contribution from the host immune system, meaning that it may be a useful therapy for immuno-compromised cancer patients (Al-Sakere et al., 2007b).

As mentioned, electro-chemotherapy significantly increases the efficacy of agents such as bleomycin and cisplatin (Figure 23.2). In clinical trials, it has been unequivocally successful in treating a variety of histologically tumors, ranging from malignant melanomas to head, neck, and locally recurrent breast cancers, as well as Mycosis fungoides (Peycheva and Daskalov, 2004; Tijink et al., 2006; Larkin et al., 2007; Quaglino et al., 2008; Sersa et al., 2008; Campana et al., 2009).

Electroporation also radically boosts the delivery of DNA, mRNA, siRNA, as well as viruses to both cultured cells and tumors alike (Ponsaerts et al., 2002; Mazda and Kishida, 2009; Radkevich-Brown

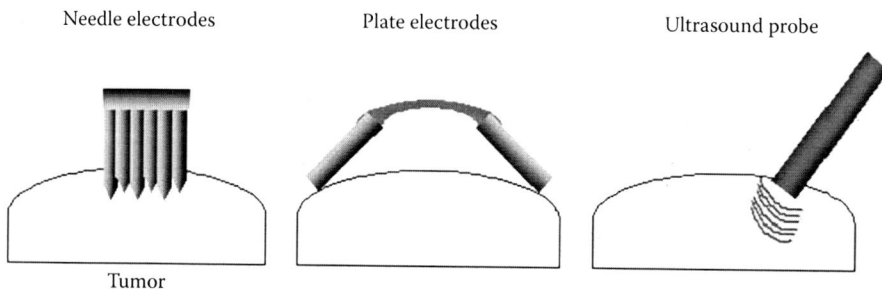

FIGURE 23.1 Schematic representation of the apparatus described in this chapter. Indicated and needle and plate electrodes and an ultrasound probe.

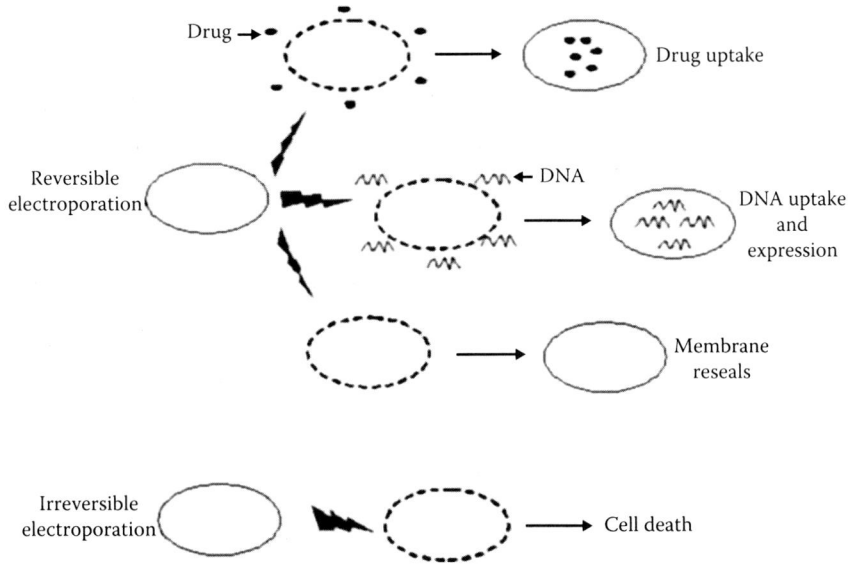

FIGURE 23.2 Overview of different forms and applications of reversible and irreversible electroporation. Reversible electroporation results in the transient formation of transmembrane pores, which reseal some time after cessation of application of the electric field. While the pores are open molecules such as DNA and drugs can diffuse into the cell and remain thereafter the pores seal. However, with irreversible electroporation, the disruption to the membrane is so great that the membrane cannot reseal, resulting in cell death.

et al., 2009) (Figure 23.2). Importantly, when constructs bearing sequences for immunological proteins such as B7-1 and GM-CSF were delivered to murine tumors via electroporation, considerably strong antitumor responses were seen and a Phase 1 trial in which plasmid DNA coding for IL-12 was delivered, via electroporation, to patients with metastatic melanoma was significantly successful (Daud et al., 2008). This method has also been successfully used to boost host immune system responses to antigens (Brave et al., 2009) and to obtain vaccine preparations from leukemic patients (Van Driessche et al., 2009).

Thus, as electro-gene and chemotherapy have proven to be viable, minimally invasive methods for tumor ablations and early studies have shown that electroporation alone has the potential to be further refined and developed as a therapeutic avenue.

23.2.3 Advantages and Disadvantages of Electroporation

As mentioned, electroporation increases the porosity of cell membranes at a cellular level. When applied to tumors, however, it can induce hypoperfusion in the tumor due to local vasoconstriction (Gehl et al., 2002). As a result, there is less diffusion of the drug to surrounding tissues than would otherwise occur, an advantage when treating tumor tissue as the surrounding healthy tissues are undamaged by the presence of the chemotherapeutic drug. As well as providing localized, noninvasive therapy, another advantage of electroporation is its efficacy: this technique can deliver DNA to almost 80% of cells in suspension and increases the cytotoxicity of chemotherapeutic drugs such as bleomycin up to 5000-fold (Cemazar et al., 2001). Similar trends have been seen for other agents such as cisplatin, pointing to a general application of electrochemotherapy in heightening the efficacy of hydrophilic chemotherapeutic drugs.

Electroporation does however require extensive optimization to maximize transfection efficacy and minimize irreversible cell damage. Depending on the cell shape and size to be electroporated, the voltage, the number of pulses, the length, and the frequency need to be altered in order to ensure a high degree of transfection with optimal parameters varying extensively between cell types. From a clinical perspective, the management of discomfort from the electrical pulses is of significant importance for the patient. Primarily local anesthetic and/or sedation have been used to negate the discomfort and an increase in the pulse frequency from 1 to 5000 Hz has resulted in a train of eight pulses being delivered in approximately 1 ms and thus reducing the sensation for the patients from eight individual pulses over 8 s to a single pulse.

23.2.4 Mechanism of Electroporation

Reversible electroporation induces reorganization in the phospholipid bilayer of the cell membrane. This is made possible by the ability of the membrane to be transiently permeabilized and the charges present on the membrane itself (Tsong, 1991). Pores of approximately 1 nm in width are formed when the membrane potential rises to 0.2–0.3 V; in a non-electroporated state cellular membrane potential is generally very low. This allows bidirectional passage of macromolecules, proteins, drugs, and other molecules between the intracellular and extracellular environments and indeed, the membranes of intracellular organelles can also become permeabilized during electroporation. As the nucleus is already porous, it does not become increasingly permeable (Somiari et al., 2000). Importantly, the cell membranes can reseal following cessation of electroporation, meaning that nucleic acids and compounds that had diffused into the cell in the presence of the electric field remain and exert their effects there.

More severe pulse parameters are employed in irreversible electroporation, rendering the cells incapable of recovery. In earlier studies irreversible electroporation induced a significant thermal effect and so was considered unsuitable for therapeutic application. However, mathematical modeling has proved invaluable in the therapeutic development of technology (Al-Sakere et al., 2007a). It has been used to define sets of parameters which affect only the target tissue and which minimize thermal side effects. Though thermal damage to tumor tissue can reduce tumor volume, the effects can be unpredictable as efficacy can be compromised by blood flow in the area and these methods can be uncomfortable for patients (Nikfarjam et al., 2006). Like reversible electroporation, only tissues within the electric field are affected, meaning that undiseased tissue can be easily spared. However, the tumor mass is depleted through the induction of necrosis and by vascular coagulation in the treated area (Al-Sakere et al., 2007a). Though this effect is independent of the immune system (Al-Sakere et al., 2007b), it is conceivable that unwanted side effects may occur when applied to large regions, which would correspondingly produce large areas of necrotic tissue. As necrotic tissue is not well recognized by phagocytes and so would not be as well cleared by the immune system as apoptotic tissue, a modified paradigm may therefore be preferable to a necrotic approach.

Therefore, a combined physical approach, first using parameters similar to those in reversible electroporation, and subsequently applying a second physical stress may yield a preferable alternative therapy for treatment of solid tumors. As cells are depleted of ATP during electroporation, cell death may be induced by application of a second stress: electroporation in this setting primes the tissue for apoptosis. Confirming this theory, we and others have proved that applying ultrasound to pre-electroporated tissues has significant potential as a viable physical cancer therapy (Larkin et al., 2005; Rollan Haro et al., 2005).

23.3 Ultrasound

23.3.1 Definition and Background

Ultrasound has a long history in medicine though it has traditionally been used for sonograms. However, like electric fields, this may be used to increase the delivery of biomolecules and therapeutic compounds to various tissues, including tumors. Ultrasound energy may be transformed into several forms of energy, which may exist at the same time within any given medium. The mechanisms of transformation into these other forms of energy are conventionally subdivided into two major categories comprising a nonthermal mechanism (mechanical and cavitational) and thermal mechanisms (O'brien, 2007).

23.3.2 Nonthermal Effects of Ultrasound

The primary effects of ultrasound are due to nonthermal mechanisms mediated by a mechanical means (acoustic streaming) and a process called cavitation. Acoustic streaming describes the physical force of sound waves which provide a driving force capable of displacing ions and small molecules (O'brien, 2007). This may act to modify the local environment of a cell. At the cellular level, organelles and molecules of different molecular weight exist; many are free floating and may be driven to move around more stationary structures (Paliwal and Mitragotri, 2006). This mechanical pressure applied by the ultrasound wave produces unidirectional movement of fluid along and around cell membranes which affects concentration gradients or various substances in the vicinity of an extracellular membrane (Miller et al., 2000). This causes a destabilization of the cellular membrane, causing increased cellular permeability and allows diffusion.

Cavitation essentially describes the biophysical interaction of gaseous inclusions (bubbles) within tissues when exposed to a waveform (e.g., ultrasound) (Wu and Nyborg, 2008). Cavitation causes expansion and compression of these small gas bubbles in tissue fluids, with a resulting increase in flow of surrounding fluid oscillating in response to the high- and low-pressure portions of the ultrasound wave (Hill, 1982). Two forms of acoustic cavitation exist: non-inertial (stable), which is described as a repetitive oscillation of a bubble over many acoustic cycles within the ultrasound field; and inertial (unstable), the process where the bubble rapidly collapses with the production of very high local temperature and/or pressures resulting in the production of toxic-free radicals and eventual tissue damage (Ayme and Carstensen, 1989). Cavitation is rare in biological systems and the additional use of gas based contrast agents increases the potential for cavitation.

23.3.3 Thermal Effects of Ultrasound

Tissue absorbs part of the mechanical energy from ultrasound through relaxation and thermo-viscous processes and the absorbed energy is converted into heat. As a result, the temperature rises in a localized area of the ultrasound source. This temperature rise is local as opposed to systemic (Miller and Brayman, 1997; Feril and Kondo, 2004). The acoustic intensities are much lower outside the focal point where heat generated is insufficient to damage cell; the intervening tissue is therefore unharmed, which is one of the main advantages of ultrasound (Feril and Kondo, 2004). The nondestructive ultrasound bioeffects on cells include the permeabilization of cell membranes for proteins transfer and the enhancement of reporter gene expression in benign and malignant cell types both in culture and *in vivo*.

Therapeutic ultrasound has been shown to cause cell lysis, inducing both necrosis and apoptosis *in vitro* depending on the parameters used. *In vivo*, regression of tumors, ablation of cancer tissue, coagulation of bleeding tissue and enhanced gene delivery is observed due to ultrasound induced destabilization and permeabilization of the cell membrane (Baker et al., 2001). Low-intensity ultrasound has been shown to induce cell lysis even without significant temperature rise and even at very low intensities (less than $0.5\,W/cm^2$) (Ward et al., 1999). Although lysis is commonly involved in ultrasound-induced cell death *in vitro*, *in vivo* this is less likely due to structural configurations of cells within the body (Ward et al., 1999).

The exact mechanism how ultrasound induced apoptosis occurs remains to be elucidated. Membrane damage by the physical effect of the ultrasound energy is one of the proposed means that brings about cellular damage and cell death (O'brien, 2007). The degree of membrane damage and the ability of the cells to repair the damage determine the mode of cell death. A possible mode is through generation of free radicals (reactive oxygen species) which may potentiate apoptotic cell death (Miller and Brayman, 1997). Certain ultrasound conditions results in optimal apoptosis, and ultrasound can also be combined with other apoptosis inducing methods to attain an enhanced level of apoptosis (Larkin et al., 2008).

23.4 Combined Electric Field and Ultrasound

23.4.1 Background

As mentioned, exposure of mammalian cells to high-intensity electric fields induces a permeabilizing effect which has been exploited to facilitate transport of drugs across the cell membrane. This permeabilization of the cell membrane suggests that the physical stability of that membrane is compromised following treatment with electric fields and may sensitize the membranes to a subsequent ultrasound stimulus. In this way, relatively low-intensity ultrasound could be used in the treatment of solid tumors, thereby reducing the negative aspects associated with High Intensity Focused Ultrasound (HIFU) and preclude the use of a drug entity as used in electrochemotherapeutic and ultrasound regimes alone (Larkin et al., 2008). This combined electric field and ultrasound (CEFUS) therapy has been applied to solid tumors to exert an antitumor effect (Larkin et al., 2005).

Application of low voltage electric pulses prior to the treatment of low frequency ultrasound effects has been shown to render cells apoptotic. The synergistic effect of the combined electric pulses and ultrasound, using optimal conditions for both, allows for a lesser amount of energy to be imparted from the individual components than when either is used alone in current clinical therapies. The development of this strategy has the potential to yield a widely clinically applicable anticancer therapy for the treatment of recurrent or inoperable tumors. The means by which this combinational treatment works is not fully understood, but the short application of low-intensity ultrasound suggests that delivery of electric pulses to cancers temporarily sensitizes the tumor tissue to the application of low-intensity ultrasound (Larkin et al., 2005).

23.4.2 CEFUS Protocol, Application, and Advantages

CEFUS has been used successfully to treat variety solid tumors in preclinical studies. A generalized CEFUS protocol has been established and is described below and outlined in Figure 23.3:

Step 1 *Electric field*: Two pulses of 1000 V/cm, lasting 1 ms were given 1 s apart.
Step 2 *Ultrasound*: A 1 MHz ultrasound transducer with a surface area of 5 cm^2 Ultrasound was irradiated for 2 min at an intensity of 3.5 W/cm^2 at a 35% duty cycle.

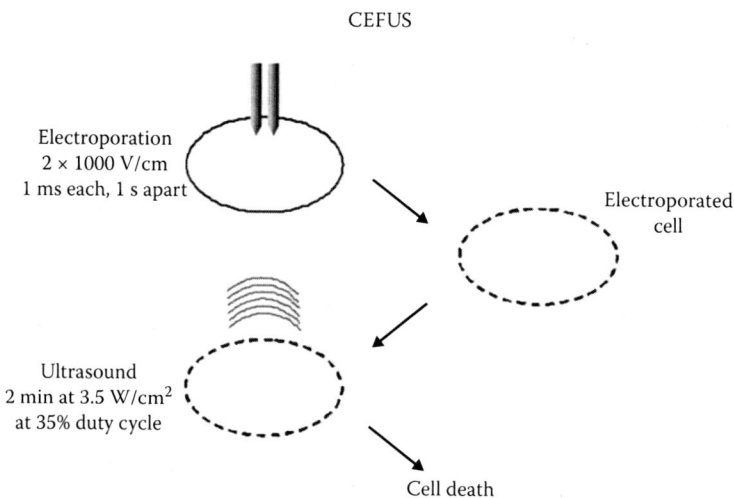

FIGURE 23.3 An overview of the CEFUS approach. First, an electric pulse is applied to the tumor. Subsequently, ultrasound is applied. The combined stress of both these phenomena results in ablation of the tumor.

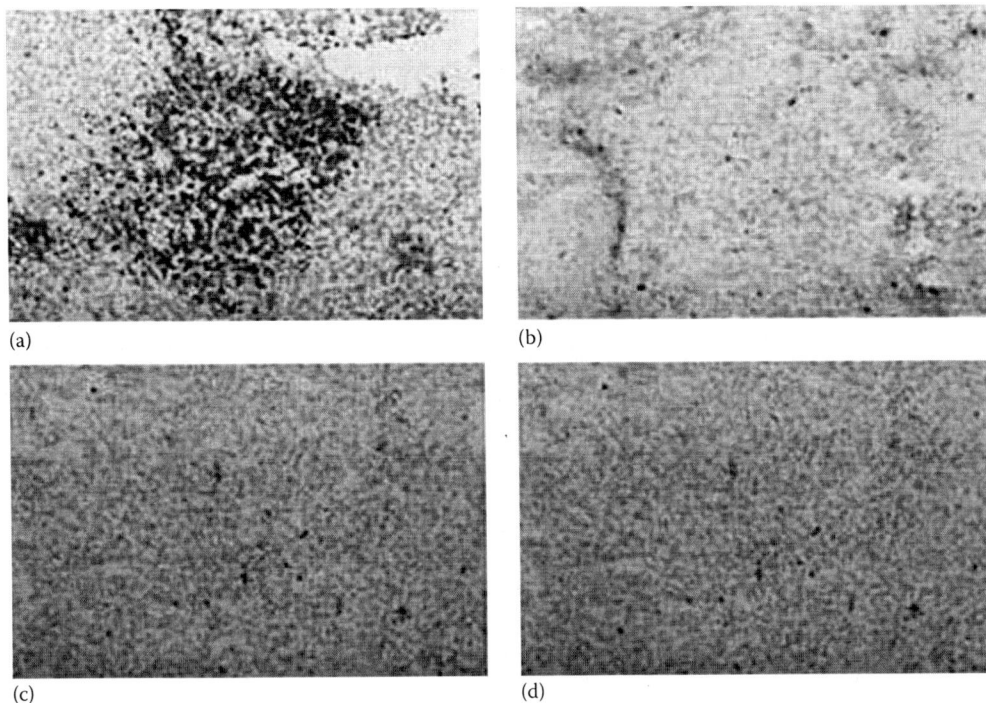

FIGURE 23.4 Apoptosis induced by CEFUS. Focal areas of apoptosis induced in C26 tumor tissue after (a) 24 h after CEFUS treatment, (b) no evidence of apoptosis is seen in the untreated control, (c) or in tumors treated by ultrasound, (d) or by electric field alone after 24 h. (From Larkin, J. et al., *Eur. J. Cancer*, 41, 1339, 2005. With permission.)

The CEFUS approach, using the outlined protocol, demonstrated encouraging antitumor effects (Larkin et al., 2005). The combination of ultrasound and electric pulses is essential to the successful ablation of tumors as it is not observed when the modalities are used individually. Furthermore, the order of application affects the degree and consistency of tumor regression observed, with the optimal response occurring when ultrasound is applied after the electric pulses.

23.4.3 Preclinical Application and Advantages of CEFUS

CEFUS treatment has been applied to mice bearing tumors of human esophageal adenocarcinoma (OE19) and a murine colonic adenocarcinoma (CT26). In both tumor types the treatment demonstrated apoptosis and regression of tumor (Larkin et al., 2005) (Figure 23.4).

When applied to normal tissue CEFUS demonstrated no apoptotic cell death although a mild inflammatory response was indicated but nothing significant (Larkin et al., 2005). This suggests that this treatment is a safe and effective treatment with in a clinical basis. Another advantage is that the area of treatment is limited to the area of the ultrasound probe, meaning that healthy tissue is unaffected and unharmed during this therapy (Larkin et al., 2005). Furthermore, as this therapy is not reliant on a chemical input, there are no concerns about phenomena such as cellular adaptation and drug-resistance as there would be in other therapies.

23.5 Conclusion

The application of CEFUS treatment suggests that this novel technology could potentially be of wide application in clinical practice for the treatment of solid tumors. It is a safe, economical, noninvasive treatment and does not require the use of chemotherapeutic drugs (Larkin et al., 2005). The treatment of localized cancer ideally should produce complete, irreversible tumor cell death without damage to the surrounding normal tissue and as such, would provide an extremely useful tool in the treatment of recurrent or inoperable tumors. Should tumors reoccur locally, or be incompletely ablated on first application of CEFUS, a second treatment could be safely applied as there are minimal side effects from this treatment. We anticipate that further research will uncover further benefits of CEFUS and its clinical potential will soon be realized.

References

Al-Sakere, B., Andre, F., Bernat, C., Connault, E., Opolon, P., Davalos, R. V., Rubinsky, B., and Mir, L. M. (2007a). Tumor ablation with irreversible electroporation. *PLoS One*, 2, e1135.

Al-Sakere, B., Bernat, C., Andre, F., Connault, E., Opolon, P., Davalos, R. V., and Mir, L. M. (2007b). A study of the immunological response to tumor ablation with irreversible electroporation. *Technol Cancer Res Treat*, 6, 301–306.

Auer, D., Brandner, G., and Bodemer, W. (1976). Dielectric breakdown of the red blood cell membrane and uptake of SV 40 DNA and mammalian cell RNA. *Naturwissenschaften*, 63, 391.

Ayme, E. J. and Carstensen, E. L. (1989). Cavitation induced by asymmetric distorted pulses of ultrasound: Theoretical predictions. *IEEE Trans Ultrason Ferroelectr Freq Control*, 36, 32–40.

Baker, K. G., Robertson, V. J., and Duck, F. A. (2001). A review of therapeutic ultrasound: Biophysical effects. *Phys Ther*, 81, 1351–1358.

Brave, A., Hallengard, D., Gudmundsdotter, L., Stout, R., Walters, R., Wahren, B., and Hallermalm, K. (2009). Late administration of plasmid DNA by intradermal electroporation efficiently boosts DNA-primed T and B cell responses to carcinoembryonic antigen. *Vaccine*, 27, 3692–3696.

Campana, L. G., Mocellin, S., Basso, M., Puccetti, O., De salvo, G. L., Chiarion-Sileni, V., Vecchiato, A., Corti, L., Rossi, C. R., and Nitti, D. (2009). Bleomycin-based electrochemotherapy: Clinical outcome from a single institution's experience with 52 patients. *Ann Surg Oncol*, 16, 191–199.

Canatella, P. J. and Prausnitz, M. R. (2001). Prediction and optimization of gene transfection and drug delivery by electroporation. *Gene Ther*, 8, 1464–1469.

Cemazar, M., Parkins, C. S., Holder, A. L., Chaplin, D. J., Tozer, G. M., and Sersa, G. (2001). Electroporation of human microvascular endothelial cells: Evidence for an anti-vascular mechanism of electrochemotherapy. *Br J Cancer*, 84, 565–570.

Daud, A. I., Deconti, R. C., Andrews, S., Urbas, P., Riker, A. I., Sondak, V. K., Munster, P. N. et al. (2008). Phase I trial of interleukin-12 plasmid electroporation in patients with metastatic melanoma. *J Clin Oncol*, 26, 5896–5903.

Feril, L. B. Jr. and Kondo, T. (2004). Biological effects of low intensity ultrasound: The mechanism involved, and its implications on therapy and on biosafety of ultrasound. *J Radiat Res (Tokyo)*, 45, 479–489.

Gehl, J., Skovsgaard, T., and Mir, L. M. (2002). Vascular reactions to in vivo electroporation: Characterization and consequences for drug and gene delivery. *Biochim Biophys Acta*, 1569, 51–58.

Goto, T., Nishi, T., Tamura, T., Dev, S. B., Takeshima, H., Kochi, M., Yoshizato, K. et al. (2000). Highly efficient electro-gene therapy of solid tumor by using an expression plasmid for the herpes simplex virus thymidine kinase gene. *Proc Natl Acad Sci USA*, 97, 354–359.

Hill, C. R. (1982). Ultrasound biophysics: A perspective. *Br J Cancer Suppl*, 5, 46–51.

Kinosita, K. Jr. and Tsong, T. Y. (1977). Formation and resealing of pores of controlled sizes in human erythrocyte membrane. *Nature*, 268, 438–441.

Larkin, J., Soden, D., Collins, C., Tangney, M., Preston, J. M., Russell, L. J., Mchale, A. P., Dunne, C., and O'sullivan, G. C. (2005). Combined electric field and ultrasound therapy as a novel anti-tumour treatment. *Eur J Cancer*, 41, 1339–1348.

Larkin, J. O., Casey, G. D., Tangney, M., Cashman, J., Collins, C. G., Soden, D. M., and O'sullivan, G. C. (2008). Effective tumor treatment using optimized ultrasound-mediated delivery of bleomycin. *Ultrasound Med Biol*, 34, 406–413.

Larkin, J. O., Collins, C. G., Aarons, S., Tangney, M., Whelan, M., O'reily, S., Breathnach, O., Soden, D. M., and O'sullivan, G. C. (2007). Electrochemotherapy: Aspects of preclinical development and early clinical experience. *Ann Surg*, 245, 469–479.

Mazda, O. and Kishida, T. (2009). Molecular therapeutics of cancer by means of electroporation-based transfer of siRNAs and EBV-based expression vectors. *Front Biosci (Elite Ed)*, 1, 316–331.

Miller, D. L., Kripfgans, O. D., Fowlkes, J. B., and Carson, P. L. (2000). Cavitation nucleation agents for nonthermal ultrasound therapy. *J Acoust Soc Am*, 107, 3480–3486.

Miller, M. W. and Brayman, A. A. (1997). Biological effects of ultrasound. The perceived safety of diagnostic ultrasound within the context of ultrasound biophysics: A personal perspective. *Echocardiography*, 14, 615–628.

Neumann, E., Schaefer-Ridder, M., Wang, Y., and Hofschneider, P. H. (1982). Gene transfer into mouse lyoma cells by electroporation in high electric fields. *Embo J*, 1, 841–845.

Nikfarjam, M., Muralidharan, V., Malcontenti-Wilson, C., Mclaren, W., and Christophi, C. (2006). Impact of blood flow occlusion on liver necrosis following thermal ablation. *ANZ J Surg*, 76, 84–91.

O'brien, W. D. Jr. (2007). Ultrasound-biophysics mechanisms. *Prog Biophys Mol Biol*, 93, 212–255.

Paliwal, S. and Mitragotri, S. (2006). Ultrasound-induced cavitation: Applications in drug and gene delivery. *Expert Opin Drug Deliv*, 3, 713–726.

Peycheva, E. and Daskalov, I. (2004). Electrochemotherapy of skin tumours: Comparison of two electroporation protocols. *J Buon*, 9, 47–50.

Ponsaerts, P., Van tendeloo, V. F., Cools, N., Van driessche, A., Lardon, F., NIJS, G., Lenjou, M. et al. (2002). mRNA-electroporated mature dendritic cells retain transgene expression, phenotypical properties and stimulatory capacity after cryopreservation. *Leukemia*, 16, 1324–1330.

Quaglino, P., Mortera, C., Osella-Abate, S., Barberis, M., Illengo, M., Rissone, M., Savoia, P., and Bernengo, M. G. (2008). Electrochemotherapy with intravenous bleomycin in the local treatment of skin melanoma metastases. *Ann Surg Oncol*, 15, 2215–2222.

Radkevich-Brown, O., Piechocki, M. P., Back, J. B., Weise, A. M., Pilon-Thomas, S., and Wei, W. Z. (2009). Intratumoral DNA electroporation induces anti-tumor immunity and tumor regression. *Cancer Immunol Immunother*, 59, 409–417.

Rollan Haro, A. M., Smyth, A., Hughes, P., Reid, C. N., and Mchale, A. P. (2005). Electro-sensitisation of mammalian cells and tissues to ultrasound: A novel tumour treatment modality. *Cancer Lett*, 222, 49–55.

Sadadcharam, M., Soden, D. M., and O'sullivan G. C. (2008). Electrochemotherapy: An emerging cancer treatment. *Int J Hyperthermia*, 24, 263–273.

Sale, A. J. and Hamilton, W. A. (1968). Effects of high electric fields on micro-organisms. 3. Lysis of erythrocytes and protoplasts. *Biochim Biophys Acta*, 163, 37–43.

Sersa, G., Miklavcic, D., Cemazar, M., Rudolf, Z., Pucihar, G., and Snoj, M. (2008). Electrochemotherapy in treatment of tumors. *Eur J Surg Oncol*, 34, 232–240.

Somiari, S., Glasspool-Malone, J., Drabick, J. J., Gilbert, R. A., Heller, R., Jaroszeski, M. J., and Malone, R. W. (2000). Theory and in vivo application of electroporative gene delivery. *Mol Ther*, 2, 178–187.

Tijink, B. M., De bree, R., Van dongen, G. A., and Leemans, C. R. (2006). How we do it: Chemo-electroporation in the head and neck for otherwise untreatable patients. *Clin Otolaryngol*, 31, 447–451.

Tsong, T. Y. (1991). Electroporation of cell membranes. *Biophys J*, 60, 297–306.

Van driessche, A., Van de velde, A. L., Nijs, G., Braeckman, T., Stein, B., De vries, J. M., Berneman, Z. N., and Van tendeloo, V. F. (2009). Clinical-grade manufacturing of autologous mature mRNA-electroporated dendritic cells and safety testing in acute myeloid leukemia patients in a phase I dose-escalation clinical trial. *Cytotherapy*, 11, 653–668.

Ward, M., Wu, J., and Chiu, J. F. (1999). Ultrasound-induced cell lysis and sonoporation enhanced by contrast agents. *J Acoust Soc Am*, 105, 2951–2957.

Wu, J. and Nyborg, W. L. (2008). Ultrasound, cavitation bubbles and their interaction with cells. *Adv Drug Deliv Rev*, 60, 1103–1116.

24

Combined Modality Therapy: Electrochemotherapy with Tumor Irradiation

Gregor Sersa

Simona Kranjc

Maja Cemazar

24.1 Introduction

Radiotherapy has an important role in the treatment of cancer; about one-half of all cancer patients receive radiotherapy during management of their disease. It is very useful and successful in controlling loco-regional diseases.

Radiotherapy as a single treatment modality has an important role in many clinical situations; however, its main role is as a component in multimodal management of cancer, together with surgery and chemotherapy. Both good local tumor control and control of metastatic spread of the tumors is of utmost importance for successful management of the disease. Based on this concept, systemic therapy has contributed greatly to the overall survival of cancer patients. However, good local tumor control by radiotherapy still remains a challenge. While almost any dose distribution can be created due to technological improvements, this still does not solve the problem of normal tissue present within the irradiated target volumes. Progress in solving this problem can only be expected by widening the therapeutic window between the tumor and normal tissue response to radiotherapy. This problem was approached by several different strategies, such as combination of radiotherapy with chemotherapy, hyperthermia, vascular-targeted drugs, or other molecular-targeted drugs (Baumann et al. 2004, 2008, Bernier 2009, Connell and Hellman 2009, Dewhirst et al. 2008, Mundt et al. 2006, Myerson et al. 2004, Teicher et al. 1988).

The combined treatment of radiotherapy and some common chemotherapeutic drugs has been shown to have an additive or even synergistic effect in local tumor control. The underlying mechanisms of potentiation of the radiation effect by chemotherapeutic drugs are numerous and include hypoxic sensitization, increased DNA injury, decreased DNA repair, increased apoptotic cell death, tumor vascular damage, cell cycle interference, and others. Besides enhancing the efficacy of radiotherapy, these agents add to local and systemic toxicity, so that the therapeutic index can vary (Boeckman et al. 2005, Dewhirst et al. 2008, Dolling et al. 1998, Dritschilo et al. 1979, Hall and Giaccia 2006, Lo Nigro et al. 2007, Mundt et al. 2006, Vernole et al. 2006). For a clinically successful approach (radiosensitization) of combined therapy, chemotherapeutic drugs should be administered shortly before or concomitantly with tumor irradiation (Kleinberg et al. 2007, Mundt et al. 2006). Platinum-based chemotherapeutic drugs are probably the most commonly used agents in these combined schedules. Several clinical trials have been conducted, which demonstrated that the addition of cisplatin significantly improved local tumor control and survival rates. Cisplatin induces different DNA injuries (single and double DNA strand crosslinks) from that caused by radiation (single- and double-strand breaks); consequently the efficacy of the repair process (sublethal and potentially lethal damage) is decreased in response to combined treatment and, therefore, radioresistant hypoxic cells may be better targeted (Boeckman et al. 2005, Dolling et al. 1998, Dritschilo et al. 1979, Hall and Giaccia 2006, Mundt et al. 2006). In addition to cisplatin, another chemotherapeutic drug often used in combination with radiotherapy is bleomycin. A radiosensitizing effect was observed, and radiosensitization of tumors by bleomycin is now used in the treatment of head and neck tumors (Fu et al. 1987, Hall and Giaccia 2006, Kleinberg et al. 2007, Smid et al. 2003, Strojan et al. 2005, Suntharalingam et al. 2001, Zakotnik et al. 2007). Bleomycin primarily acts on well-oxygenated cells since the major mechanism of bleomycin cytotoxicity is the formation of oxygen-free radicals that cause single and double DNA breaks. These DNA breaks increase the DNA damage caused by irradiation resulting in increased cell death (Chen et al. 2008, Hall and Giaccia 2006, Povirk 1996, Undevia et al. 2006, Vernole et al. 2006).

However, despite the fact that a combination of radiotherapy and chemotherapy resulted in an increased antitumor effect, an unacceptable proportion of patients still die of loco-regional disease progression. Therefore, additional improvements are sorely needed. One of the approaches that can be exploited is to increase uptake of the chemotherapeutic drugs into the tumors and cells in tumors by drug delivery systems such as electroporation. This increased drug uptake in the tumor cells would increase their intracellular accumulation and thus the radiosensitizing effect.

24.2 What Is Electrochemotherapy

Electrochemotherapy (ECT) combines administration of non-permeant chemotherapeutic drugs with the application of electric pulses to the tumors in order to facilitate drug delivery into cells (Mir 2006, Sersa et al. 2008). Thus, the enhanced drug delivery locally potentiates the chemotherapeutic drug effectiveness at the site of electric pulse application. So far, two chemotherapeutic drugs have proven to be effective in electrochemotherapy, bleomycin, and cisplatin. A severalfold increase in bleomycin and cisplatin cytotoxicity and in antitumor effectiveness has been shown in many preclinical electrochemotherapy studies. The increased drug delivery, both in tumor cells *in vitro* and tumors *in vivo*, was shown to be the predominant underlying mechanism (Mir 2006). Clinical studies conducted on patients with cutaneous and subcutaneous tumor nodules of different histology have demonstrated approximately 80% objective responses and approximately 70% long-lasting complete responses. Melanoma metastases were predominantly treated, as well as carcinomas and sarcomas, with the aim of palliative therapy (Byrne and Thompson 2006, Gothelf et al. 2003, Marty et al. 2006, Sersa 2006).

24.2.1 Radiosensitizing Effect of Electrochemotherapy with Cisplatin

Cisplatin is widely used in the treatment of cancer and has a cytotoxic as well as radiosensitizing effect. Its presence in tumors during irradiation presumably radiosensitizes tumor cells in addition to the combined cytotoxic action of the two modalities. *In vitro* and *in vivo* experimental systems have revealed a complex interaction of cisplatin and radiation. Some studies demonstrated a clear cisplatin-induced radiosensitization, especially in hypoxic conditions, whereas others showed only an additive effect (Candelaria et al. 2006, Marcu et al. 2006, Reboul 2004).

In order to increase the intracellular accumulation of drugs, several drug delivery systems for tumor targeting were developed, such as targeting by binding of the drugs to tumor-specific antibodies, magnetic drug targeting, incorporation of drugs into liposomes or other vehicles, or selectively increasing permeability of the plasma membrane of tumor cells by chemical or physical methods. However, reports utilizing drug delivery systems as a means of radiosensitization are scarce. Some attempts were made by intratumoral injection of cisplatin, and by a slow-release formulation in combined treatment with radiation, demonstrating a positive effect (Begg et al. 1994, Celikoglu et al. 2006, Ning et al. 1999, Yapp et al. 1998).

24.2.1.1 *In Vitro* Studies

Electroporation proved to be an effective drug delivery system for cisplatin; it increased cisplatin accumulation in the cells. Consequently, an up to 80-fold increase of cisplatin cytotoxicity was observed on different tumor cell lines *in vitro* (Kranjc et al. 2003a). The difference in sensitivity of cells to cisplatin may vary depending on the type of tumor cells, being either melanoma, carcinoma, or sarcoma cells. Melanoma cells have proved to be some of the most sensitive to cisplatin (Sersa et al. 1995). However, variability may also exist among the different carcinoma cells, as was demonstrated for SCK cells (IC_{50} = 14.8 ± 1.0 µg/mL; IC_{50}-concentration of cisplatin that caused 50% reduction of cell survival), which were more sensitive to cisplatin than EAT-E carcinoma cells (IC_{50} = 48.5 ± 1.5 µg/mL). In these cells, electroporation increased cisplatin cytotoxicity significantly, reducing the IC_{50} values to 3.4 ± 0.7 µg/mL in SCK cells and to 2.2 ± 0.9 µg/mL in EAT-E cells, demonstrating a 4.3- and 22-fold increase in cytotoxicity, respectively (Kranjc et al. 2003a). The results demonstrated that electroporation can increase the cytotoxic effect of cisplatin and renders more cisplatin-resistant cells, EAT-E cells as sensitive as SCK cisplatin-sensitive cells.

Furthermore, the tumor cells differ also in their radiosensitivity; therefore, more radioresistant tumors require different treatment strategies that aim to combine radiotherapy with other treatment modalities, including chemotherapy. In our studies, we approached this topic, by evaluation of electrochemotherapy with radiation on different tumor cell lines with different radiosensitivity. Specifically, the SCK and EAT-E cells that differ in radiosensitivity were used in the study. Their D_0 values, as a measure of cell radiosensitivity, was 1.2 Gy for SCK cells and 2.0 Gy for EAT-E cells. When irradiation with a single irradiation dose of 4 Gy was combined with exposure of cells to different concentrations of cisplatin, the radiation response of cells was significantly increased. Furthermore, if these cells (treated by irradiation and cisplatin) were exposed to electric pulses, the survival of cells was further decreased to the level of IC_{50} = 0.9 ± 0.2 µg/mL for SCK cells and to IC_{50} = 0.9 ± 0.3 µg/mL for EAT-E cells. This data demonstrated that electrochemotherapy of cells *in vitro* radiosensitizes the cells to approximately the same level, as measured by the IC_{50} values, regardless of the intrinsic sensitivity of the cells to irradiation (Kranjc et al. 2003a). Similar results were obtained on LPB sarcoma cells, where cells were treated with a single cisplatin concentration combined with electroporation and graded doses of irradiation (from 2 to 8 Gy). The results demonstrated that for the same level of cell kill, a 1.36-fold lower dose of irradiation is needed if the cells are pretreated with a combination of cisplatin and electroporation (Figure 24.1) (Kranjc et al. 2003b).

Altogether, these data indicate that electroporation of different tumor cells having different chemosensitivity and different radiosensitivity increases the radiosensitizing effect of cisplatin.

24.2.1.2 *In Vivo* Studies

Preclinical studies of electrochemotherapy with cisplatin and local tumor irradiation were done with single-dose irradiation on two different experimental tumor models, i.e., Ehrlich-Lettre ascites tumors (EATs) and LPB sarcoma tumors, in mice (Sersa et al. 2000, Kranjc et al. 2003a).

The goal of the first study (Sersa et al. 2000) was to evaluate whether electroporation of tumors can increase the radiosensitizing effect of cisplatin in EAT subcutaneous tumors, as a proof of principle. Cisplatin was injected intravenously at a dose of 4 mg/kg of body weight of the animals, and had no antitumor effect. Tumor irradiation in a single dose of 15 Gy resulted in 27% of tumor cures.

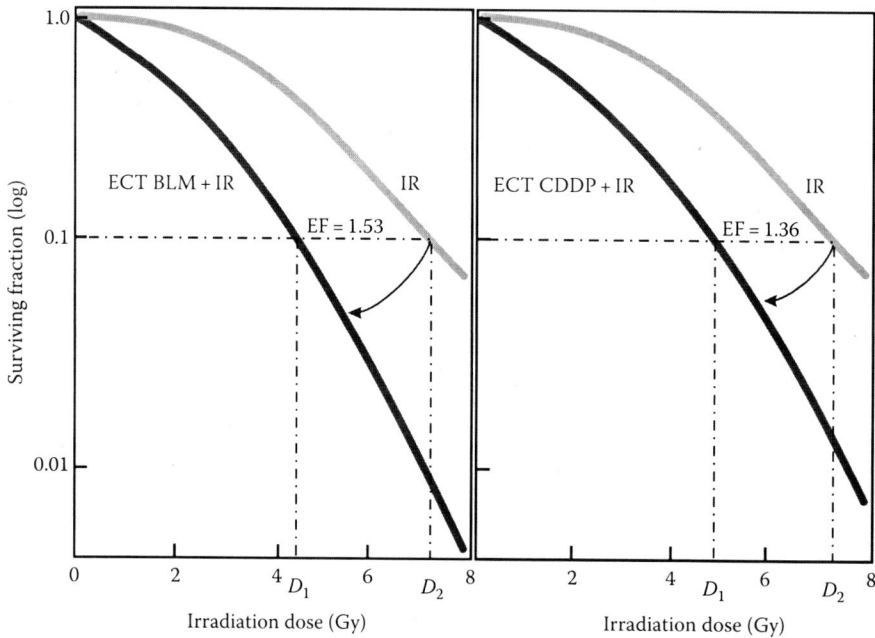

FIGURE 24.1 Radiation survival curves of LPB sarcoma cells exposed to different irradiation (IR) doses only and to a combination with bleomycin (BLM) or cisplatin (CDDP) and electroporation (EP) (ECT BLM + IR, ECT CDDP + IR). For further details see references Kranjc et al. (2003b, 2005). (Adapted from Kranjc, S. et al., *Radiat. Res.*, 172, 677, 2009.)

Electrochemotherapy was also an effective treatment, resulting in 12% tumor cures. As expected, injection of cisplatin 20 min prior to irradiation increased the radioresponse of tumors from 27% to 73% tumor cures. Electroporation of tumors alone, without the drug, also increased the radiation response of tumors from 27% to 54% tumor cures. However, electrochemotherapy given 20 min prior to tumor irradiation further increased the radioresponsiveness of tumors, resulting in 92% tumor cures (Sersa et al. 2000). The study demonstrated that delivery of cisplatin into cells by electroporation of tumors increases the radiosensitizing effect of cisplatin. The observed radiosensitization of electroporation was hypothesized to be due, on one hand, to the effect of electroporation on intracellular targets or to the perturbation of the plasma membrane, and, on the other hand, the possible mechanisms could also involve the effects of electric pulses on modification of tumor blood flow and oxygenation.

The second study (Kranjc et al. 2003a) was done on LPB sarcoma tumors with graded doses of irradiation in order to determine the tumor curability dose (TCD_{50}) of combined electrochemotherapy treatment and tumor irradiation in relation to tumor irradiation alone and irradiation of tumors combined with cisplatin or electroporation (Kranjc et al. 2003b). In addition, in order to elucidate the radiosensitization of electroporation, changes in tumor blood flow were measured. Treatment of animals with cisplatin (4 mg/kg) 20 min before local tumor irradiation significantly increased the radiation response of tumors. TCD_{50} values were lowered from 22.1 Gy with irradiation only to 19.6 Gy in combined treatment. However, the combination of application of electric pulses to the tumors and intravenous injection of cisplatin (electrochemotherapy) resulted in an even greater enhancement of the radiation response of the tumors (TCD_{50} = 14.2 Gy) (Figure 24.2). This study demonstrated that the radiocurability of tumors was significantly enhanced compared to only irradiated tumors (enhancement factor, EF = 1.6) and to tumors treated with cisplatin and irradiation (EF = 1.4) (Kranjc et al. 2003a).

In this study, it was also elucidated whether increased platinum accumulation in the tumors due to electroporation could be the underlying mechanism of increased radioresponsiveness of tumors (Kranjc et al. 2003b). Already a few minutes after the treatment, platinum accumulation in tumors treated with

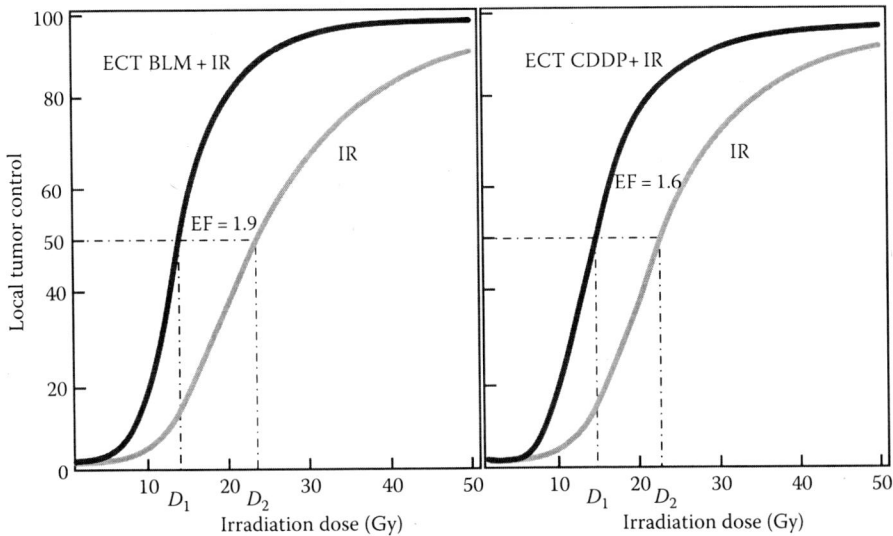

FIGURE 24.2 Local tumor control curves of LPB sarcoma tumors treated with electrochemotherapy (ECT) either with bleomycin (BLM) or cisplatin (CDDP) 20 min prior to irradiation (IR) (ECT BLM + IR or ECT CDDP + IR). For further details, see references Kranjc et al. (2003b, 2005). (Adapted from Kranjc, S. et al., *Radiat. Res.*, 172, 677, 2009.)

electrochemotherapy, with or without tumor irradiation, was approximately two times higher compared to platinum accumulation in tumors treated with cisplatin or cisplatin and irradiation. The difference remained evident up to 24 h after application of electric pulses to the tumors. Therefore, these results indicated that increased delivery of cisplatin to tumors by application of electric pulses might be the principal reason for the observed increase in cisplatin radiosensitization of tumors. Furthermore, it was demonstrated that the electroporation-mediated increase in platinum accumulation and the kinetics of platinum washout were not affected by an acute dose of radiation. Tumor blood flow changes were assessed at different time points up to 24 h after single and combined treatments (Kranjc et al. 2003b). No perfusion changes were observed in tumors after irradiation or cisplatin treatment, as well as after combination of both, compared to control untreated tumors. In contrast, tumor perfusion was significantly reduced immediately after application of electric pulses to the tumors with a recovery to 50% of the pretreatment value at 24 h. All combined modality treatment groups comprising electroporation responded in the same manner as the group treated with electroporation alone. The reduced blood flow could lead to radiobiologically relevant hypoxia. Therefore, the increased radiation response of tumors treated with electrochemotherapy as a result of higher platinum accumulation and cisplatin radiosensitization of hypoxic cells was counteracted by reduced tumor blood flow and oxygenation. Furthermore, compared to our previous study on EAT carcinoma (Sersa et al. 2000), the radiation effect in LPB sarcoma was not potentiated by electroporation alone (i.e., without the chemotherapeutic). In the case of EAT carcinoma tumors, one can presume that they are more sensitive and/or that electroporation generates more reactive oxygen species (as demonstrated *in vitro* by Bonnafous et al. 1999), which outweighs the effect of reduced tumor blood flow. However, in the case of LPB sarcoma tumors, one can presume that reduced oxygenation dominates other effects.

In the clinical setting, the radiosensitizing effect of electrochemotherapy with cisplatin was demonstrated in a patient with tubal dedifferentiated papillary adenocarcinoma skin metastases (Sersa et al. 1999). Skin metastases were palliatively treated by electrochemotherapy with cisplatin combined with single-dose irradiation. Due to a large number of metastases, which were up to 20 mm in diameter, it was possible to select groups of metastases that were treated by each modality or the combined treatment. Several metastases could not be treated and, therefore, served as control nodules.

Electrochemotherapy with cisplatin injected intratumorally had a good antitumor effect, with formation of a scab and no underlying viable tumor tissue. Irradiation of tumor nodules as a single treatment had no effect on the growth of the tumor nodules. Additionally, no effect of the treatment was observed when irradiation was combined with either of the modalities that were used in electrochemotherapy, cisplatin, or application of electric pulses, whereas when electrochemotherapy was combined with irradiation, quick and complete reduction in the size of the treated nodules was observed. Although in both electrochemotherapy alone and electrochemotherapy combined with irradiation the response to the treatment was complete, the effect was quicker after electrochemotherapy combined with radiotherapy (Sersa et al. 1999).

24.2.2 Radiosensitizing Effect of Electrochemotherapy with Bleomycin

The radiosensitizing effect of bleomycin was demonstrated in preclinical and clinical studies (Jiang et al. 1989, Kleinberg et al. 2007, Kranjc et al. 2005, Molin et al. 1981, Smid et al. 2003, Strojan et al. 2005, Suntharalingam et al. 2001, Teicher et al. 1988, Zakotnik et al. 2007). The use of bleomycin 5–20 min prior to a single-dose or fractionated irradiation schedule resulted in sensitization of murine mammary carcinoma C3H and fibrosarcoma FSaIIC tumors; an enhancement factor of up to 1.4 was obtained with the single-dose or fractionated regime (Jiang et al. 1989, Molin et al. 1981, Teicher et al. 1988). Similarly, when bleomycin was used in monotherapy or in combination with other chemotherapeutic drugs concurrently with a fractionated irradiation regime, loco-regional control and survival in patients was improved by up to 30%. Radiosensitization of tumors by bleomycin is an established treatment approach in the treatment of head and neck tumors (Smid et al. 2003, Strojan et al. 2005, Zakotnik et al. 2007).

24.2.2.1 *In Vitro* Studies

Electroporation increases cytotoxicity of bleomycin several 1000-fold due to cell membrane permeabilization allowing increased transport and accumulation of bleomycin into the cells (Belehradek et al. 1994). This potentiation is due to the limited transport of bleomycin through the cell membrane without electroporation. The usual transmembrane transport of BLM is receptor-mediated, and, because of the limited number of carriers, bleomycin does not exert its cytotoxic effect to full extent without electroporation of the cell membrane (Pron et al. 1993). Only a few hundred molecules in the cell are needed for its cytotoxic effect. Due to bleomycin's radiosensitizing effect, electroporation can potentiate this effect. An *in vitro* study has demonstrated that very low doses of bleomycin can already have a good radiosensitizing effect (Kranjc et al. 2005). Exposure of cells to either bleomycin, electroporation, or electrochemotherapy statistically significantly increased the radiation response of LPB sarcoma cells. The increase in radiation response was the lowest when the cells were exposed to bleomycin (EF = 1.19), probably due to a low bleomycin concentration. On the other hand, the radiosensitivity of cells was greatly enhanced when the cells were pretreated with electrochemotherapy (EF = 1.53) (Figure 24.1). Compared to electrochemotherapy with cisplatin, radiosensitization was more pronounced in the case of electrochemotherapy with bleomycin (EF = 1.36 for cisplatin and 1.53 for bleomycin), demonstrating high cytotoxicity of bleomycin within the cells (Figure 24.1). When the cells were exposed to electric pulses prior to irradiation, the EF was 1.25 (Kranjc et al. 2003b). Therefore, the radiosensitizing effect of electrochemotherapy *in vitro* might not just be due to increased bleomycin delivery into the cells by electroporation, but also to radiosensitization by electroporation of cells (Kranjc et al. 2005), presumably by the formation of free radicals (Bonnafous et al. 1999).

24.2.2.2 *In Vivo* Studies

Preclinical studies on electrochemotherapy with bleomycin and local tumor irradiation were done with single and fractionated irradiation on three experimental tumor models: LPB and SA-1 sarcoma tumors and CaNT carcinoma tumors in mice.

The first study was done with the same goal as the study with electrochemotherapy with cisplatin, i.e., to provide proof of principle that electrochemotherapy with bleomycin has a radiosensitizing effect on LPB sarcoma tumors (Kranjc et al. 2005). To further elucidate the underlying mechanisms for *in vivo* radiosensitization with electroporation, tumor oxygenation was measured after treatment. Solid subcutaneous tumors were treated by electrochemotherapy with bleomycin (0.5 mg/kg of body weight per mouse; intravenous injection) 20 min prior to single-dose tumor irradiation. Tumors on the back of the animals were irradiated with graded doses of x-rays, ranging from 5 to 50 Gy. As the endpoint for evaluation of antitumor effectiveness, the local tumor control (TCD_{50} assay) was used (Kranjc et al. 2005). The results demonstrated that neither the treatment of animals with bleomycin alone nor application of electric pulses to the tumors prior to irradiation of tumors had any effect on local tumor control. These results, which differ from the *in vitro* results, suggest that additional mechanisms are involved in antitumor effectiveness of electroporation combined with radiation *in vivo*. Electrochemotherapy of tumors increased the radiation response of tumors. The TCD_{50} value was reduced from 23.1 Gy in mice that were treated with irradiation alone, to 12.4 Gy in mice that were treated with electrochemotherapy prior to irradiation (Figure 24.2). Since the enhancement factor was 1.9, it is evident that electroporation of tumors significantly contributed to radiosensitization of tumors with bleomycin, specifically because combined treatment of electrochemotherapy with irradiation was statistically significantly more effective than treatment of tumors that were irradiated in combination with electroporation or bleomycin alone (Kranjc et al. 2005).

As already mentioned, electroporation of tumors induces changes in the tumor blood flow (see also Chapter 21). These changes occur immediately after electroporation; the tumor blood flow is almost completely abrogated (Kranjc et al. 2003b). However, thereafter, the restitution of tumor perfusion starts to occur, with 50% restoration of tumor blood flow within 24 h after electroporation of tumors. These changes in tumor perfusion correlated with a decreased partial tumor oxygen tension (pO_2) in the LPB tumors (Kranjc et al. 2005). It was demonstrated that 5 min after electroporation of the tumors, pO_2 was reduced in the center of the tumor by 75% and in the periphery of the tumors by 50%. Immediately thereafter the tumors started to reoxygenate. However, 6 h after electroporation, the oxygen level was still at 70% of the pretreatment level and, even after 24 h, was not completely restored. Specifically, 20 min after electroporation of tumors, i.e., at the time of tumor irradiation, there was still reduced tumor oxygenation in the tumor center as well as in the tumor periphery (4.6 ± 0.6 mmHg; 6.1 ± 0.3 mmHg, respectively) at the level of radiobiologically relevant hypoxia. In normal tissue, the skeletal muscle and subcutaneous tissue pO_2 levels were not affected by application of electric pulses (Kranjc et al. 2005).

Since electroporation-induced tumor hypoxia was radiobiologically relevant, it could have an effect on the tumor response to irradiation. However, the curability of the tumors treated with electroporation combined with tumor irradiation was the same as the curability of tumors that were only irradiated. In contrast, our *in vitro* results demonstrated that electroporation of cells enhanced the radiation response. Therefore, some other mechanisms affect the radiation response in tumors. It is already known that reactive oxygen species are formed after electroporation of cells *in vitro* (Bonnafous et al. 1999). Reactive oxygen species are known to contribute to radiation damage of the cells. Therefore, predisposition of cells to irradiation by electroporation is also expected, and was previously demonstrated *in vitro* (West 1992). *In vivo* on solid tumors, the radiation response depended on the tumor type. Electroporation enhanced the radiation response in EAT carcinoma, but not in LPB sarcoma. These differences in response are not yet clear, but could be tumor type-dependent (Sersa et al. 2000, Kranjc et al. 2003b).

In the overall response of tumors to radiation after electroporation of tumors, a radiobiologically relevant hypoxia is formed, which could counteract the effect of ionizing radiation, but its effect is not exerted since it is counteracted by the formation of reactive oxygen species, also caused by electroporation.

In order to bring this combined treatment closer to clinical practice, the interaction between electrochemotherapy with bleomycin and single and fractionated irradiation was studied (Kranjc et al. 2009). The interaction was evaluated on two tumor models with different histology and radiosensitivity; radiosensitive SA-1 sarcoma tumors, and radioresistant CaNT adenocarcinoma. Again, the tumor irradiation with a single x-ray dose or fractionated regime split into 2 Gy per fraction, 5 days per week, were used (Figure 24.3).

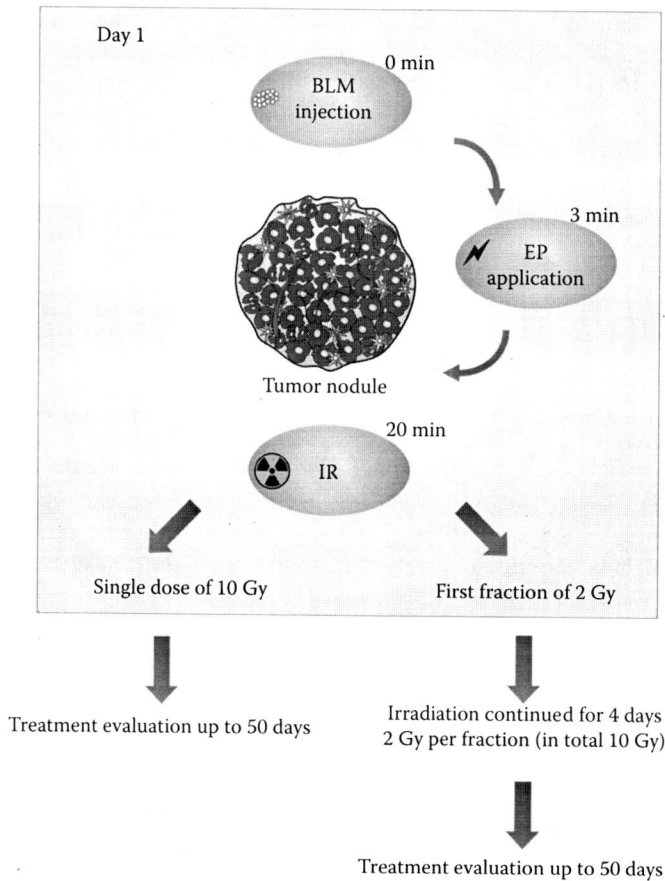

FIGURE 24.3   Treatment protocol of combined electrochemotherapy (ECT) and tumor irradiation (IR). SA-1 tumors were treated by a single-dose irradiation 20 min after electrochemotherapy (an i.v. bleomycin [BLM] injection and application of electric pulses [EP]). In a fractionated irradiation regime SA-1 tumors were irradiated with fractionated irradiation five times with 2 Gy per fraction after ECT, which was performed only on the first day. For further details, see reference Kranjc et al. (2009). (Adapted from Kranjc, S. et al., *Radiat. Res.*, 172, 677, 2009.)

Electrochemotherapy was performed once, 20 min prior to the first fraction in the fractionated regime or tumor irradiation with a single dose. The bleomycin dose (0.5 mg bleomycin/kg of body weight per mouse) with a minor antitumor effect in electrochemotherapy was chosen. This dose does not induce complete responses of the tumors; it enables the evaluation of the interaction with tumor irradiation on the level of tumor growth delay. For this reason, low-irradiation doses for tumor irradiation were chosen. SA-1 tumors were irradiated with a cumulative dose of 10 Gy in a single or fractionated regime, since the tumors are radiosensitive and their TCD_{50} value is 25 Gy, whilst the more radioresistant CaNT tumors were irradiated with a cumulative dose of 20 Gy since their TCD_{50} value is 70 Gy (Kranjc et al. 2009).

Radiosensitization of tumors by electrochemotherapy with bleomycin in combination with single-dose irradiation was effective on both radiosensitive SA-1 and radioresistant adenocarcinoma CaNT tumors. The resulting growth delay was more than the sum of the effects of bleomycin or application of electric pulses to the tumors combined with tumor irradiation. As with the single-dose irradiation, tumor irradiation in the fractionated regime was not significantly potentiated when combined with systemic administration of bleomycin or with electroporation of tumors. Radiosensitization of the tumors by electrochemotherapy in combination with fractionated irradiation was effective on both radiosensitive SA-1 and radioresistant CaNT tumors. Again, the resulting growth delay was more than the sum

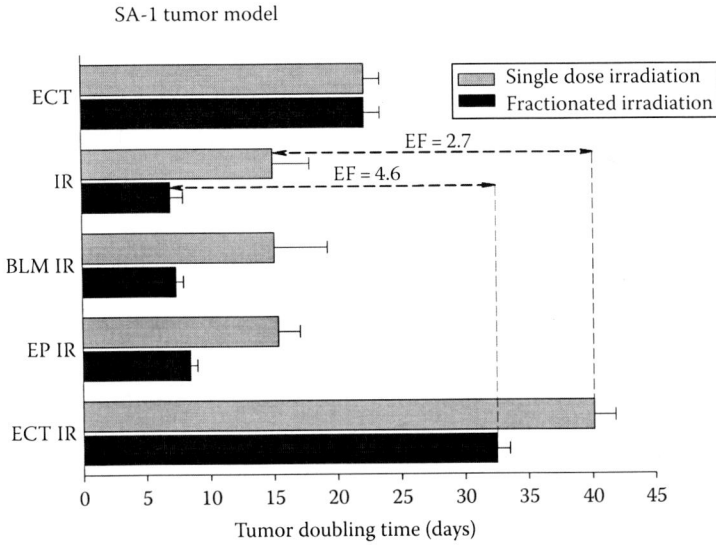

FIGURE 24.4 Antitumor effectiveness of electrochemotherapy (ECT) with bleomycin (BLM) combined with tumor irradiation (IR) with either the single-dose or fractionated radiation regime in a SA-1 sarcoma tumor model. Data represent the mean and standard error of the mean, which were pooled from at least eight animals per treatment group. Enhancement factor (EF) of radioresponse was calculated based on the tumor doubling time of the compared experimental groups.

of the effects of bleomycin or the application of electric pulses to the tumors combined with tumor irradiation. The potentiating effect of electrochemotherapy was more pronounced in combination with fractionated tumor irradiation than with single-dose irradiation. The radiation response was potentiated in combination with electrochemotherapy by a factor of 2.7 in single-dose irradiation and by a factor of 4.6 in the fractionated regime of SA-1 tumors (Figure 24.4). In CaNT tumors, the potentiation was by a factor of 2.3 in combination with single-dose irradiation and 3.1 in the fractionated regime. Although the effect was evident in both tumor models, the data indicated that potentiation of the radiation response was more pronounced in the radiosensitive sarcoma SA-1 tumor model than in the radioresistant adenocarcinoma CaNT tumor model (Kranjc et al. 2009).

24.2.3 Side Effects of the Treatment

To provide the therapeutic index (normal tissue damage versus local tumor control at the tumor dose that results in 50% tumor cures) of the combined electrochemotherapy and tumor irradiation, evaluation of the side effects is a prerequisite. All of the studies performed so far reported minimal or no, general or local side effects in the irradiation field. The most extensive data is reported in our latest study, which evaluated animal body weight changes during and after the treatment, as well as the skin reaction around the tumor in the irradiation field (Kranjc et al. 2009). Single or combined treatment had a minimal effect on animal weight loss. Animals lost weight at the beginning of the treatment and during the fractionated radiation regime, but their weight stabilized thereafter, and the animals were in good physical condition assessed by monitoring the animal's appetite, locomotion, coat, and general appearance. The skin reaction around the tumors in the irradiation field was arbitrarily scored. Single-dose irradiation alone or in combination with electrochemotherapy provoked much more serious normal tissue damage a week earlier compared to the fractionated radiotherapy regime, inducing edema, erythema, and moist desquamation with moderate ulceration. Skin damage after the fractionated radiation regime alone or in combination with electrochemotherapy was observed only after 30–40 days after the first tumor irradiation. A comparison of the level of skin damage demonstrated that single-dose irradiation resulted in around four times higher skin damage compared to the fractionated regime, regardless of the treatment combinations. These side effects however proved to be acceptable and provide a sufficient therapeutic index for further development of this combined treatment approach.

24.3 Conclusions

Several preclinical *in vitro* and *in vivo* studies have demonstrated that electroporation can increase the radiosensitizing effect of the chemotherapeutic drugs cisplatin and bleomycin. It was demonstrated that already a single electrochemotherapy before tumor irradiation significantly increases the radiation response of tumors without any increase in skin damage in the irradiation field. This effect was demonstrated by single-dose irradiation as well as in the fractionated regime. The potentiation of the radioresponse was demonstrated on radiosensitive as well as on radioresistant tumors. In addition, one patient was already treated by combination of electrochemotherapy and irradiation, demonstrating promising results. All these data indicate that this approach of combining electrochemotherapy with radiotherapy can be further translated into the clinic, predominantly in the treatment of accessible tumor masses that cannot be successfully treated by either of the treatments as a single modality.

References

Baumann M, Krause M, Zips D et al. 2004. Molecular targeting in radiotherapy of lung cancer. *Lung Cancer* 45(Suppl 2):S187–S197.

Baumann M, Krause M, Hill R. 2008. Exploring the role of cancer stem cells in radioresistance. *Nat Rev Cancer* 8(7):545–554.

Begg AC, Deurloo MJ, Kop W, Bartelink H. 1994. Improvement of combined modality therapy with cisplatin and radiation using intratumoral drug administration in murine tumors. *Radiother Oncol* 31(2):129–137.

Belehradek J Jr, Orlowski S, Ramirez LH, Pron G, Poddevin B, Mir LM. 1994. Electropermeabilization of cells in tissues assessed by the qualitative and quantitative electroloading of bleomycin. *Biochim Biophys Acta* 1190(1):155–163.

Bernier J. 2009. Current state-of-the-art for concurrent chemoradiation. *Semin Radiat Oncol* 19(1):3–10.

Boeckman HJ, Trego KS, Turchi JJ. 2005. Cisplatin sensitizes cancer cells to ionizing radiation via inhibition of nonhomologous end joining. *Mol Cancer Res* 3(5):277–285.

Bonnafous P, Vernhes MC, Teissie J, Gabriel B. 1999. The generation of reactive-oxygen species associated with long-lasting pulse-induced electropermeabilization of mammalian cells is based on a nondestructive alteration of the plasma membrane. *Biochim Biophys Acta* 1461:123–134.

Byrne CM, Thompson JF. 2006. Role of electrochemotherapy in the treatment of metastatic melanoma and other metastatic and primary skin tumors. *Expert Rev Anticancer Ther* 6(5):671–678.

Candelaria M, Garcia-Arias A, Cetina L, Duenas-Gonzalez A. 2006. Radiosensitizers in cervical cancer. Cisplatin and beyond. *Radiat Oncol* 1:15. doi: 10.1186/1748-717X-1-15.

Celikoglu F, Celikoglu SI, York AM, Goldberg EP. 2006. Intratumoral administration of cisplatin through a bronchoscope followed by irradiation for treatment of inoperable non-small cell obstructive lung cancer. *Lung Cancer* 51(2):225–236.

Chen J, Ghorai MK, Kenney G, Stubbe J. 2008. Mechanistic studies on bleomycin-mediated DNA damage: Multiple binding modes can result in double-stranded DNA cleavage. *Nucleic Acids Res* 36(11):3781–3790.

Connell PP, Hellman S. 2009. Advances in radiotherapy and implications for the next century: A historical perspective. *Cancer Res* 69(2):383–392.

Dewhirst MW, Cao Y, Moeller B. 2008. Cycling hypoxia and free radicals regulate angiogenesis and radiotherapy response. *Nat Rev Cancer* 8(6):425–437.

Dolling JA, Boreham DR, Brown DL, Mitchel RE, Raaphorst GP. 1998. Modulation of radiation-induced strand break repair by cisplatin in mammalian cells. *Int J Radiat Biol* 74(1):61–69.

Dritschilo A, Piro AJ, Kelman AD. 1979. Effect of cis-platinum on the repair of radiation-damage in plateau phase Chinese-hamster (V-79) cells. *Int J Radiat Oncol Biol Phys* 5(8):1345–1349.

Fu KK, Phillips TL, Silverberg IJ et al. 1987. Combined radiotherapy and chemotherapy with bleomycin and methotrexate for advanced inoperable head and neck-cancer—Update of a Northern California Oncology Group Randomized Trial. *J Clin Oncol* 5(9):1410–1418.

Gothelf A, Mir LM, Gehl J. 2003. Electrochemotherapy: Results of cancer treatment using enhanced delivery of bleomycin by electroporation. *Cancer Treat Rev* 29(5):371–387.

Hall EJ, Giaccia AJ. 2006. Chemotherapeutic agents from the perspective of the radiation biologist. In *Radiobiology for the Radiologist*, 6th edn. Lippincot Williams & Wilkins, Philadelphia, PA, pp. 440–468.

Jiang GL, Ang KK, Thames HD, Wong CS, Wendt CD. 1989. Response of plateau-phase C3H 10T1/2 cells to radiation and concurrent administration of bleomycin. *Radiat Res* 120(2):306–312.

Kleinberg L, Gibson MK, Forastiere AA. 2007. Chemoradiotherapy for localized esophageal cancer: Regimen selection and molecular mechanisms of radiosensitization. *Nat Clin Pract Oncol* 4(5):282–294.

Kranjc S, Cemazar M, Grosel A, Pipan Z, Sersa G. 2003a. Effect of electroporation on radiosensitization with cisplatin in two cell lines with different chemo- and radiosensitivity. *Radiol Oncol* 37:101–107.

Kranjc S, Cemazar M, Grosel A, Scancar J, Sersa G. 2003b. Electroporation of LPB sarcoma cells in vitro and tumors in vivo increases the radiosensitizing effect of cisplatin. *Anticancer Res* 23(1A):275–281.

Kranjc S, Cemazar M, Grosel A, Sentjurc M, Sersa G. 2005. Radiosensitizing effect of electrochemotherapy with bleomycin in LPB sarcoma cells and tumors in mice. *BMC Cancer* 5:115.

Kranjc S, Tevz G, Kamensek U, Vidic S, Cemazar M, Sersa G. 2009. Radiosensitizing effect of electrochemotherapy in a fractionated radiation regime in radiosensitive murine sarcoma and radioresistant adenocarcinoma tumor model. *Radiat Res* 172:677–685.

Lo Nigro C, Arnolfo E, Taricco E. et al. 2007. The cisplatin-irradiation combination suggests that apoptosis is not a major determinant of clonogenic death. *Anti-Cancer Drugs* 18(6):659–667.

Marcu L, Bezak E, Olver I. 2006. Scheduling cisplatin and radiotherapy in the treatment of squamous cell carcinomas of the head and neck: A modelling approach. *Phys Med Biol* 51(15):3625–3637.

Marty M, Sersa G, Garbay JR et al. 2006. Electrochemotherapy - An easy, highly effective and safe treatment of cutaneous and subcutaneous metastases: Results of ESOPE (European Standard Operating Procedures of Electrochemotherapy) study. *EJC Suppl* 4(11):3–13.

Mir LM. 2006. Bases and rationale of the electrochemotherapy. *EJC Suppl* 4(11):38–44.

Molin J, Sogaard PE, Overgaard J. 1981. Experimental studies on the radiation-modifying effect of bleomycin in malignant and normal mouse tissue in vivo. *Cancer Treat Rep* 65(7–8):583–589.

Mundt AJ, Roeske JC, Chung TD, Weichselbaum RR. 2006. Principles of radiation oncology. In Kufe DW, Bast RC, Hait WN, Hong WK, Pollock RE, Weichselbaum RR, Holland JF, and Frei III E (eds.), *Cancer Medicine*, 7th edn. BC Decker Inc., Ontario, Canada, pp. 517–536.

Myerson RJ, Roti Roti JL, Moros EG, Straube WL, Xu M. 2004. Modelling heat-induced radiosensitization: Clinical implications. *Int J Hyperther* 20(2):201–212.

Ning SC, Yu N, Brown DM, Kanekal S, Knox SJ. 1999. Radiosensitization by intratumoral administration of cisplatin in a sustained-release drug delivery system. *Radiother Oncol* 50(2):215–223.

Povirk LF. 1996. DNA damage and mutagenesis by radiomimetic DNA-cleaving agents: Bleomycin, neocarzinostatin and other enediynes. *Mutat Res* 355(1–2):71–89.

Pron G, Belehradek J Jr, Mir LM. 1993. Identification of a plasma membrane protein that specifically binds bleomycin. *Biochem Biophys Res Commun* 194(1):333–337.

Reboul FL. 2004. Radiotherapy and chemotherapy in locally advanced non-small cell lung cancer: Preclinical and early clinical data. *Hematol-Oncol Clin N* 18(1):41–53.

Sersa G. 2006. The state-of-the-art of electrochemotherapy before the ESOPE study: Advantages and clinical uses. *EJC Suppl* 4(11):52–59.

Sersa G, Cemazar M, Miklavcic D. 1995. Antitumor effectiveness of electrochemotherapy with cis-diamminedichloroplatinum(I) in mice. *Cancer Res* 55:3450–3455.

Sersa G, Cemazar M, Rudolf Z, Fras P. 1999. Adenocarcinoma skin metastases treated by electrochemotherapy with cisplatin combined with radiation. *Radiol Oncol* 33:291–296.

Sersa G, Kranjc S, Cemazar M. 2000. Improvement of combined modality therapy with cisplatin and radiation using electroporation of tumors. *Int J Radiat Oncol Biol Phys* 46(4):1037–1041.

Sersa G, Miklavcic D, Cemazar M, Rudolf Z, Pucihar G, Snoj M. 2008. Electrochemotherapy in treatment of tumours. *EJSO* 34(2):232–240.

Smid L, Budihna M, Zakotnik B et al. 2003. Postoperative concomitant irradiation and chemotherapy with mitomycin C and bleomycin for advanced head-and-neck carcinoma. *Int J Radiat Oncol Biol Phys* 56(4):1055–1062.

Strojan P, Soba E, Budihna M, Auersperg M. 2005. Radiochemotherapy with vinblastine, methotrexate, and bleomycin in the treatment of verrucous carcinoma of the head and neck. *J Surg Oncol* 92(4):278–283.

Suntharalingam N, Haas ML, Van Echo DA et al. 2001. Predictors of response and survival after concurrent chemotherapy and radiation for locally advanced squamous cell carcinomas of the head and neck. *Cancer* 91(3):548–554.

Teicher BA, Herman TS, Holden SA. 1988. Combined modality therapy with bleomycin, hyperthermia, and radiation. *Cancer Res* 48(22):6291–6297.

Undevia SD, Ratain MJ, Plunkett W. 2006. Pharmacology. In Kufe DW, Bast RC, Hait WN, Hong WK, Pollock RE, Weichselbaum RR, Holland JF, Frei E III (eds.), *Cancer Medicine*, 7th edn. BC Decker Inc., Ontario, Canada, pp. 617–629.

Vernole P, Tedeschi B, Tentori L et al. 2006. Role of the mismatch repair system and p53 in the clastogenicity and cytotoxicity induced by bleomycin. *Mutat Res* 594(1–2):63–77.

West CML. 1992. A potential pitfall in the use of electroporation-cellular radiosensitization by pulsed high-voltage electric-fields. *Int J Radiat Biol* 61(3):329–334.

Yapp DTT, Lloyd DK, Zhu JL, Lehnert SM. 1998. The potentiation of the effect of radiation treatment by intratumoral delivery of cisplatin. *Int J Radiat Oncol Biol Phys* 42(2):413–420.

Zakotnik B, Budihna M, Smid L et al. 2007. Patterns of failure in patients with locally advanced head and neck cancer treated postoperatively with irradiation or concomitant irradiation with mitomycin C and bleomycin. *Int J Radiat Oncol Biol Phys* 67(3):685–690.

25

Irreversible Electroporation in Medicine

Boris Rubinsky

25.1 Introduction

The phenomenon of irreversible electroporation (IRE)—damage to cells due to the applications of short electrical pulses—has been recognized, in various forms, for centuries (Rubinsky, 2007). It is probable that Fuller (1898) reported the first practical application of IRE for the sterilization of river water. Until recently, the main application of IRE was in the food industry where it is used for breaking the cell membrane to release its contents and for energy efficient sterilization (Toepfl et al., 2006). The biophysical phenomenon referred to as IRE in life sciences is known in food technology as *pulsed electric field* processing or *electroplasmolysis* in reference to the lysis of cell membranes in tissue, for extracting their contents, or the bactericidal effect in the treatment of fluids. The "pulsed electrical field" concept is broader than just IRE. It has been recently shown to also include the effects of nanoscale pulses on intracellular components (Beebe et al., 2002).

The first commercial use of what may be IRE was described in 1961 (Doevenspeck, 1961). He describes commercial installations using electrical pulses to break apart cellular components for industrial food–related processing of animal meat through means that involve the electrical discharge of electrical pulses from carbon electrodes through the treated material. It should be emphasized that the paper does not specifically refer to the breakdown of the cell membrane. Neither does it provide specific values for the electrical pulses used. Furthermore, Doevenspeck also reported results showing that these electrical pulses can inactivate microorganisms with what he considered to be a nonthermal effect producing a small increase in temperature of at most 30°C; which in fact is high enough to cause thermal damage.

The interest of the food industry in the so-called bactericidal action of electrical fields motivated a series of three seminal papers by Sale and Hamilton (1967, 1968) and Hamilton and Sale (1967). These papers are extraordinary in that they set the basis for the field of IRE and contain the ingredients of many of the future studies in what subsequently was called the field of electroporation in general.

In the subsequent decades, the discipline of "pulsed electrical fields" has split into applications for the food industry (including some commercial devices that have eventually failed commercially) and the use of reversible electroporation in the life sciences field (Neumann et al., 1982; Zimmermann, 1982; Okino and Mohri, 1987; Orlowski et al., 1988). The discipline has separated because the food industry

453

was interested in destroying the cell membrane and microorganisms while the life sciences discipline was interested in causing a reversible breach of the cell membrane. Therefore, the electric field strength at which a transition occurs between the reversible and irreversible electroporation phenomena was the upper boundary of the electric field strengths studied for the life sciences applications and the lower boundary of the electric field strengths studied for the food industry applications.

Because medical technology resides in the field of life sciences, the potential of IRE to be used alone, as a nonchemical method for tissue ablation, was mostly ignored. It is quite possible that the idea for using IRE for tissue ablation in medicine has crossed the mind of many researchers (e.g., the last line in the abstract of Piñero et al. (1997), which was, however, not developed any further in the text).

Our research in the field of nonthermal IRE in medicine was inspired by the early studies of R.C. Lee and his colleagues on the cellular damages caused by accidental electrical pulses (Lee and Kolodney, 1987; Lee et al., 1988). They have recognized that high electrical fields simultaneously produce a thermal effect due to Joule heating as well as the cell membrane electroporation effect, and that it is difficult to determine what mode of damage has actually destroyed the cells. They write, "In this case, cells would be vulnerable to both thermal and nonthermal injury mechanisms. Under these conditions the primary cause of cell rupture would have to be determined since both electroporation and heating may lead to disruption and both may be cytotoxic."

Having worked in the field of thermal ablation for decades, including the microscale (Rubinsky, 1997), we have recognized that while the final outcome of IRE and thermal damage on cell viability is the same, cell death, there are many very important aspects at the microscale that are different. Unlike electroporation that affects the treated volume, only one type of molecule, the cell membrane lipid bilayer, and thermal, radiation, as well as chemical (alcohol) ablation indiscriminately affects every molecule in the treated volume. Therefore, when a volume of tissue is ablated by IRE only, the surgery is molecular, i.e., targeted at one type of molecule only. However, when thermal effects ablate a volume of tissue, the ablation is macroscopic and volumetric. It occurred to us that having a minimally invasive surgical technique that can ablate only one certain type of molecule in a volume of tissue, the cell membrane, is advantageous because it spares important other tissue components, such as the extracellular matrix and electrically nonconductive molecular structures. However, the Joule heating effects are inseparable from the electroporation effects. In the food industry, irreversible electroporation is used to either break the cell membrane of the ex vivo food or to destroy microorganisms. While there is some concern about thermal effects changing the quality of the food, the food industry does not try to maintain a biological functional extracellular matrix in the treated product. In contrast, there is great importance to maintaining a functional extracellular matrix and therefore to completely avoid potential damaging thermal effects when using IRE for treatment of live tissue in a molecular surgery mode. When we began to develop the idea of using IRE for tissue ablation in a nonthermal molecular surgery mode, it was not obvious that there are any electric pulsed fields that can affect substantial volumes of tissue to be of practical value while ablating in that volume only the cell membrane without producing any thermal damaging effects to the organic molecules in that volume. To this end, we have developed a methodology for studying the electrical and thermal fields that occur during the application of electrical pulses to a volume of tissue. Using this methodology, we have found that while limited, such a domain of electrical pulses exists (Davalos et al., 2004, 2005); thereby showing that the field of nonthermal IRE for medical applications is a viable mode of molecular surgery for ablation of tissue volumes.

The goal of this brief review is to first introduce the mathematical model required for the design of nonthermal irreversible treatment protocols and than to conduct a series of recent experimental studies that demonstrate the value of applying IRE as molecular surgery in the nonthermal mode.

25.2 Mathematical Models for Nonthermal Irreversible Electroporation

Mathematical modeling for treatment planning and experiment analysis has been shown to be an important aspect of the research in electroporation (Miklavcic et al., 1998). The mathematical tools involved in research on electrochemotherapy are the solution of the field equation subject to voltage boundary conditions on the electrodes (Miklavcic et al., 1998) and mass transfer equations (Granot and Rubinsky, 2008). Treatment planning and experimental analysis for nonthermal IRE, in addition to the field equation, employs the bio-heat equation coupled with a kinetic equation for thermal damage (Davalos et al., 2004, 2005; Edd et al., 2006; Becker and Kuznetsov, 2007; Edd and Davalos, 2007; Davalos and Rubinsky, 2008; Daniels and Rubinsky, 2009). The mathematical modeling required to assess and analyze the thermal damage during nonthermal IRE are listed and discussed below.

First, the field equation is solved

$$\nabla \cdot (\sigma \nabla \phi) = 0 \tag{25.1}$$

subject to voltage boundary conditions on the electrodes

$$\phi(\text{electrodes}) = \text{prescribed} \tag{25.2}$$

where
  σ is the electrical conductivity of the tissue
  ϕ is the local electrical potential

It should be emphasized that the electrical conductivity of the tissue can change during the process of electroporation and, therefore, the problem may become nonlinear as the local conductivity may become related to the local electrical field. In addition, the boundary conditions prescribed here are of the first kind. It is possible that contact impedance between the electrodes and the tissue may be a more appropriate boundary, e.g., Somersalo et al. (1992). The boundaries of the domain that are not in contact with the electrodes are treated as either infinite or insulated.

The local electrical field, E, can be calculated from the electrical potential:

$$E = \nabla \phi \tag{25.3}$$

This electrical field affects the local conductivity through electroporation effects and makes the problem nonlinear. In addition, it yields a local heat source, P, through Joule heating. The local electrical power dissipation is given by

$$P = \sigma \left(|\nabla \phi| \right)^2 \tag{25.4}$$

The equation most commonly used to calculate the temperature during electroporation in biological tissues is the bio-heat equation (Pennes 1984) to which the local power dissipation is added as a heat source. It should be noted that the validity of this equation is sometimes questioned. Nevertheless, it is the most widely used bio-heat equation and no practical alternative is available.

$$\nabla \cdot (k \nabla T) + w_b c_b (T_a - T) + q'' + P = \rho c_p \frac{\partial T}{\partial t} \tag{25.5}$$

where
  k is the thermal conductivity of the tissue
  T is temperature
  $w_b c_b$ are the product of the volumetric mass flow of blood and blood heat capacity
  q'' is the volumetric metabolic heat
  t is time
  ρc_p is the product of tissue density and heat capacity of tissue

This equation is solved subject to initial temperature conditions, which in the case of the living body is often taken as 37°C. The boundary conditions are taken as either 37°C or adiabatic. A review of various models of the bio-heat equation can be found in Rubinsky (2006).

Once the time-dependent temperature is calculated, it can be introduced into an Arrhenius type of chemical reaction kinetics equation that correlates tissue damage, Ω, to temperature, T, and time, t (Henriques and Moritz, 1947)

$$\Omega = \int \zeta e^{-E/RT} dt \qquad (25.6)$$

where
   ζ is a frequency factor
   E is the activation energy
   R is the universal gas constant

It should be emphasized that the thermal damage is a function of both time and temperature, and long-term exposure to temperatures as low as 42°C can cause thermal damage. Nevertheless, when the exposure is in seconds, 50°C is sometimes taken as a target temperature (Diller, 1992).

25.3 Experimental Results of Nonthermal Irreversible Electroporation

In this section, I will try to highlight the aspects of nonthermal IRE that can be directly attributed to the unique "molecular surgery" mode of this procedure. The first experimental study in which nonthermal IRE was used on an animal model is (Edd et al., 2006). In that study, a rat liver was exposed to a 20 ms IRE pulse and the rat was sacrificed a couple of hours later. The liver was flushed through its vasculature prior to embedding and H&E staining. Figure 25.1 illustrates the unique outcome of NTIRE, which has emerged through the use of flushing in the preparation of the samples.

Figure 25.1 shows the margin of an NRIRE-treated liver. The left-hand side is the normal liver and the right-hand side is treated. Flushing the liver has confirmed the unique aspects of NTIRE. The lights spaces between cells on the left side are the intact sinusoids that were flushed of red blood cells. The hepatocytes around the intact sinusoids are also intact. On the other hand, on the right-hand side, the sinusoids are filled with what is destroyed red blood cells and other cellular debris. Flushing did not remove the debris from the sinusoids. However, as predicted for the NTIRE mode, the procedure does

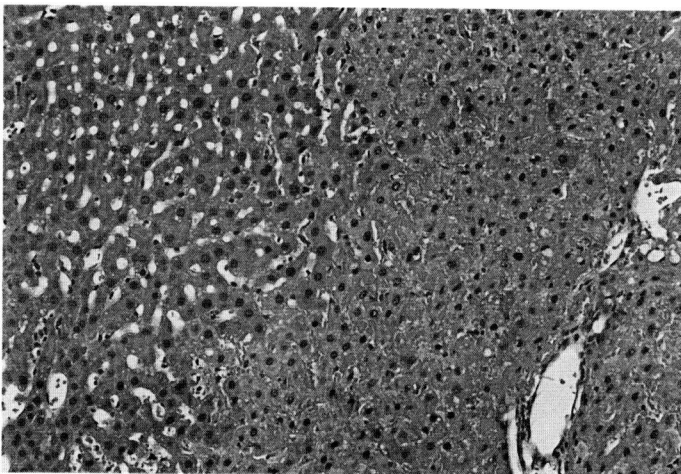

FIGURE 25.1 **(See color insert following page 268.)** H&E stained liver that has undergone NTIRE. The left-hand side is the normal liver and the right-hand side is the electroporated liver. Note the sharp line of distinction between the treated and untreated areas.

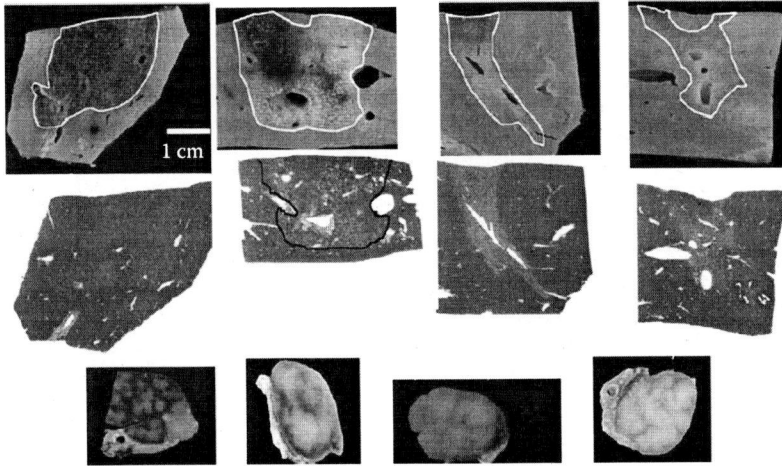

FIGURE 25.2 (See color insert following page 268.) Cross section through NTIRE treated pig liver after: 24 h (first column from left), 3 days (second column), 7 days (third column) 14 days (fourth column). Top row—macroscopic cross section, middle row—H&E stained section, and bottom row—lymph nodes. (From Rubinsky, B. et al., *Technol. Cancer Res. Treat.*, 6, 37, 2007. With permission.)

not affect the extracellular matrix and other tissue molecules, except the cell membrane. Therefore, the mechanical integrity of large blood vessels remains intact. The patency of the large blood vessels and ducts is evident from the bottom right-hand side images of clear large blood vessels in the midst of the NTIRE-treated region. Because of the unique aspects of NTIRE, the mechanical structure of these blood vessels remained intact and accessible to flushing.

The value of the unique mode of molecular surgery associated with NTIRE is even better illustrated by Figure 25.2, which has come from the first long-term study of NTIRE (Rubinsky et al., 2007).

Figure 25.2 shows a long-term follow-up of pig liver treated with NTIRE. Here also the liver was flushed prior to embedding in formalin. There are several aspects here that are unique. The top and bottom rows show the macroscopic and H&E stained cross section of the liver. The treated areas are outlined. When performing the analysis, the most amazing aspect of NTIRE was the speed with which the treated liver regenerated. As an anecdotal aside, this speed almost caused the entire project to be abandoned. From our experience with other modes of tissue ablation, such as cryosurgery (Onik et al., 1991), in all thermal ablations of the liver, the healing process takes months and very often scar tissue is left behind. Therefore, we began our experiments by analyzing the outcome of the NTIRE procedure after 2 weeks. To our surprise, we were unable to find the treated site, as seen in the last column of Figure 25.2. Only when we went back and re-did these very difficult and costly large animal experiments with histology taken at 24 h, 3 days, and 7 days did we recognize that NTIRE does ablate tissue in a unique way, thanks to the molecular surgery aspect of the procedure. It is particularly interesting to see in the second and third columns that NTIRE can ablate tissue all the way to the margin of the larger blood vessels—an aspect of tissue ablation of great importance in the treatment of unresectable tumors near large blood vessels. Also, it is interesting to notice that, as in the case of the flushed rat liver, the flushed pig liver also had patent and open large blood vessels within the treated area. Such open and mechanically intact large blood vessels do not occur during thermal treatment. The availability of such open conduits allow the immune system excellent access to all the parts of the treated body, through the blood flow. In other modes of ablation, the immune system needs to remove the dead cells by diffusion through the volume of treated tissue from the outer margin of the treated tissue. In NTIRE, the entire volume of the treated tissue is accessible through the large blood vessels and the immune system can remove the dead cells through the entire treated system. This effect is evident from the bottom row of Figure 25.2, which shows the lymph nodes. It is

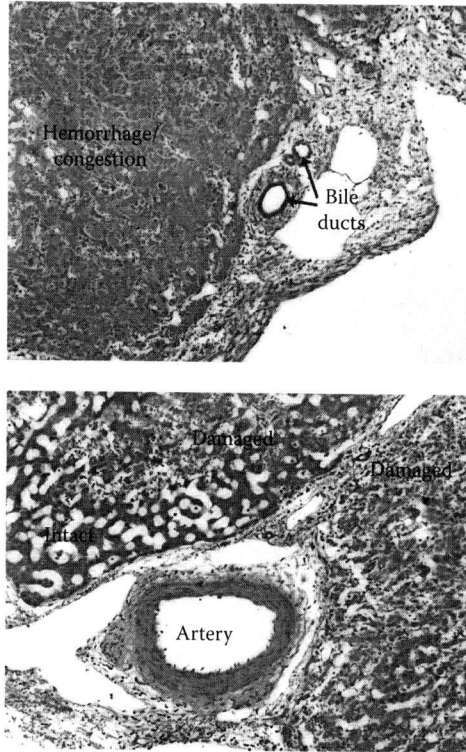

FIGURE 25.3   **(See color insert following page 268.)** Intact bile ducts and arteries within NTIRE treated tissue. (From Rubinsky, B. et al., *Technol. Cancer Res. Treat.*, 6, 37, 2007. With permission.)

evident that after 24 h the lymph nodes are inflamed and active; however, within a week the inflammation disappeared.

Figure 25.3 shows that bile ducts as well as arteries remain intact in the NTIRE treated volume of tissue. Figure 25.4 deals with the long-term treatment of the prostate with NTIRE (Onik et al., 2007). The study of Onik et al. (2007) deals with the dog prostate. It again illustrates the value of a molecular surgery procedure that targets only the cell membrane and leaves the rest of the tissue molecular components intact. For instance, harming the urethra during thermal, radiation, or chemical (alcohol) ablation of the prostate is a major source of clinical complications. In this study, we have applied the NTIRE pulses across the urethra and because of the selective molecular nature of the procedure, the mechanical scaffold has remained intact; while all the cells around it (including the endothelial cells) became ablated. The figure also shows that the nerve has remained intact. In the thermal treatment of the prostate, the entire treated volume is affected, including the axons. In NTIRE, the myelin (unless it is breached) serves as insulation from the electrical fields (Daniels and Rubinsky, 2009). Therefore, as shown in Figure 25.4, the nerve is intact while the surrounding tissue is ablated by NTIRE.

Another exciting outcome of the molecular selective nature of NTIRE becomes evident when it is applied directly on blood vessels, with possible applications for the treatment of restenosis (Maor et al., 2007, 2008, 2009). The significance of the molecular surgery nature of NTIRE emerges from Figure 25.5.

Figure 25.5 compares a cross section through a normal rat artery (top panel) to a cross section through a NTIRE-treated artery a week after the treatment. It is evident that the smooth muscles have completely

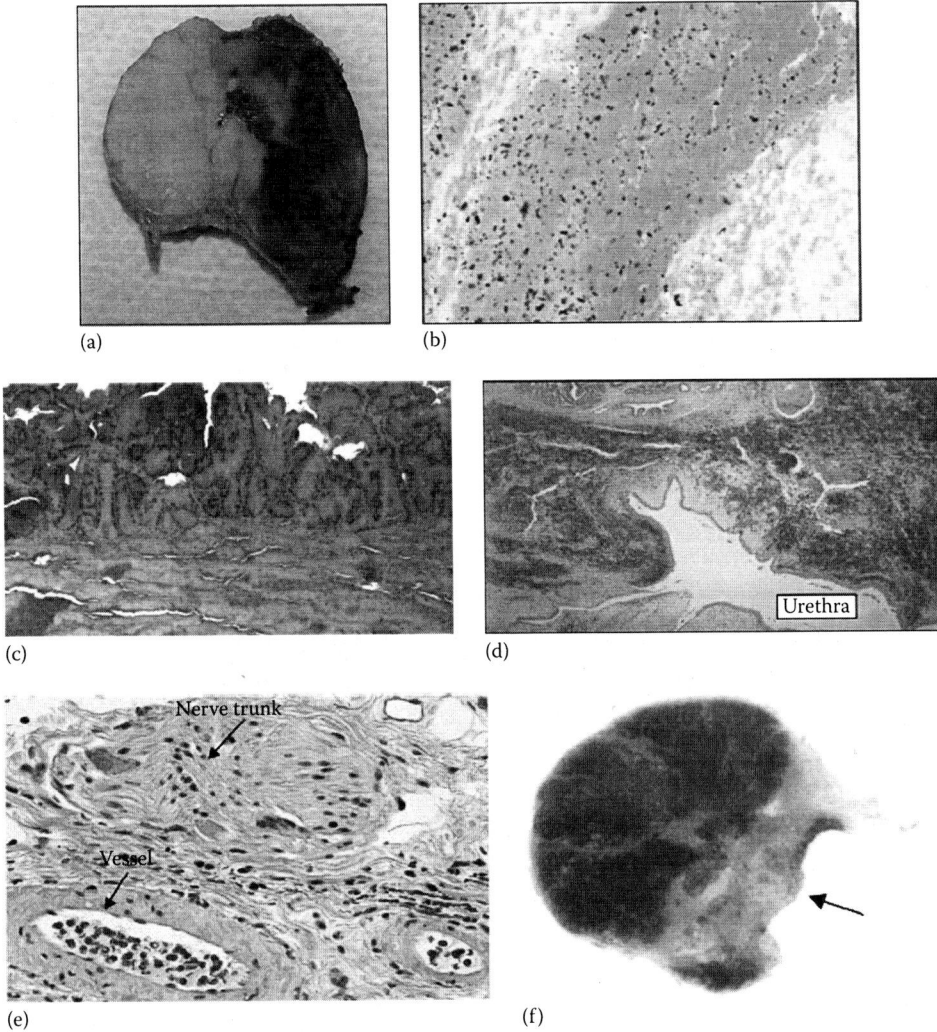

FIGURE 25.4 **(See color insert following page 268.)** Dog prostate treated with NTIRE. (From Onik, G. et al., *Technol. Cancer Res. Treat.* 6, 295, 2007. With permission.) (a) Gross pathology of the IRE lesion at 24 h. The right side of the gland is hemorrhagic (pulses = 8, Kv = 1). (b) Photomicrograph of prostate tissue that has been electroporated at 24 h. No glandular elements are visible (pulses = 8, Kv = 1). (c) Photomicrograph at the margin of the IRE lesion. A very narrow zone of transition between normal and necrotic tissue is noted at the margin (pulses = 8, Kv = 1). (d) The urethra is noted at the center of the micrograph as the open space at 24 h. Sub-mucosal hemorrhage is noted but the integrity of the urethra is still intact (pulses = 8, Kv = 1). (e) Photomicrograph of the neurovascular bundle after electroporation at two weeks (pulses = 80, Kv = 1.5). It can be seen that both the vessel and the nerve trunk show no evidence for necrosis. (f) Whole mount slide of a prostate where the right side of the gland was electroporated 2 weeks prior. There is marked shrinkage of the lobe with replacement by fibrous tissue (pulses = 80, Kv = 1.5).

disappeared after 1 week. However, the mechanical and structural integrity of the blood vessels extracellular matrix has remained intact, due to the molecular selective nature of NTIRE. Furthermore, because the extracellular scaffold has remained intact, it can serve as a matrix on which the embryonic endothelial cells in the blood stream can attach and form a new endothelial layer. It should be emphasized that a major reason for the formation of restenosis is because the blood vessel endothelial layer is

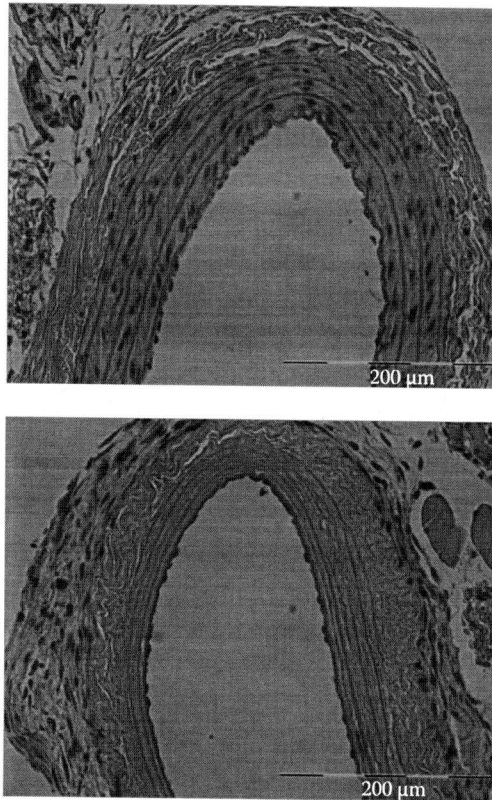

FIGURE 25.5 **(See color insert following page 268.)** Cross section through a rat carotid artery. Top panel shows a normal artery. The vascular smooth muscles as well as the endothelial cells around the lumen are evident. An NTIRE treated artery a week after the procedure. Note the complete absence of the smooth muscles as well as the beginning of the formation of the endothelial cells around the lumen.

destroyed in angioplasty and that layer is needed for inhibiting wild smooth muscle growth. The unique aspect of NTIRE holds promise in the treatment of restenosis.

25.4 Summary

Nonthermal IRE is emerging as an important new modality in the armamentarium of surgeons. It has some unique aspects in relation to other nonchemical surgical procedures. It is probably the first procedure that can treat a volume of tissue in a molecular selective surgery mode. This brief review has discussed the methodology and some first promising results of NTIRE. When this review was written, there were already over 20 centers that were in various stages of clinical implementation of the technique. Close to 50 patients with prostate, kidney, lymph, and lung cancer were already treated with this procedure with good results.

Acknowledgment

The support of the Arnold and Barbara Silverman Distinguished Professor of Bioengineering Chair Funds for the development of the field of nonthermal irreversible electroporation is gratefully acknowledged.

References

Beebe SJ, Fox PM, Rec LJ, Somers K, Stark RH, Schoenbach KH. 2002. Nanosecond pulsed electric field (nsPEF) effects on cells and tissues: Apoptosis induction and tumor growth inhibition. *IEEE Transactions on Plasma Science*. 30:286–292.

Becker SM, Kuznetsov AV. 2007. Thermal damage reduction associated with in vivo skin electroporation: A numerical investigation justifying aggressive pre-cooling. *International Journal of Heat and Mass Transfer*. 50:105–116.

Diller KR. 1992. Modeling of bioheat transfer processes at high and low temperatures, in: Choi YI (ed.), *Bioengineering Heat Transfer*, Academic Press, Inc., Boston, MA, pp. 157–357.

Daniels C, Rubinsky B. 2009. Electrical field and temperature model of nonthermal irreversible electroporation in heterogeneous tissues. *Journal of Biomechanical Engineering-ASME Transactions*. 131(7):071006.

Davalos RV, Rubinsky B. 2008. Temperature considerations during irreversible electroporation. *International Journal of Heat and Mass Transfer*. 51:5617–5622.

Davalos RV, Rubinsky B. 2004. Tissue ablation with irreversible electroporation. US patent application 0043345 A1, USPTO. The Regents of the University of California, Oakland, CA.

Davalos RV, Mir LM, Rubinsky B. 2005. Tissue ablation with irreversible electroporation. *Annals of Biomedical Engineering*. 33:223.

Doevenspeck H. 1961. Influencing cells and cell walls by electrostatic impulses. *Fleishwirtshaft* 13:986–987.

Edd JF, Davalos RV. 2007. Mathematical modeling of irreversible electroporation for treatment planning. *Technology in Cancer Research and Treatment*. 6:275–286.

Edd, JF, Horowitz L, Davalos RV, Mir LM, Rubinsky B. 2006. In-vivo results of a new focal tissue ablation technique: Irreversible electroporation. *IEEE Transactions on Biomedical Engineering*. 53(4):1409–1415.

Fuller, GV. 1898. Report on the investigations into the purification of the Ohio river water at Louisville Kentucky. D. Van Nostrand Company, New York.

Granot Y, Rubinsky, B. 2008. Mass transfer model for drug delivery in tissue cells with reversible electroporation. *International Journal of Heat and Mass Transfer*. 51:5610–5616.

Henriques FC, Moritz AR. 1947. Studies in thermal injuries: The predictability and the significance of thermally induced rate processes leading to irreversible epidermal damage. *Archives of Pathology*. 43:489–502.

Hamilton WA, Sale AJH. 1967. Effects of high electric fields on microorganisms. 2. Mechanism of action of the lethal effect. *Biochimica et Biophysica Acta*. 148:789–800.

Lee RC, Kolodney MS. 1987. Electrical injury mechanisms: Electrical breakdown of cell membranes. *Plastic & Reconstructive Surgery*. 80:672–679.

Lee RC, Gaylor DC, Bhatt D, Israel DA. 1988. Role of cell membrane rupture in the pathogenesis of electrical trauma. *The Journal of Surgical Research*. 44:709–719.

Miklavcic D, Beravs K, Semrov D, Cemazar M, Demsar F, Sersa G. 1998 The importance of electric field distribution for effective in vivo electroporation of tissues. *Biophysical Journal*. 74:2152–2158.

Maor E, Ivorra A, Leor J, Rubinsky B. 2007. The effect of irreversible electroporation on blood vessels. *Technology in Cancer Research and Treatment*. 6(4):307–312.

Maor E, Ivorra A, Leor J, Rubinsky B. 2008. Irreversible electroporation attenuates neointimal formation after angioplasty. *IEEE Transactions on Biomedical Engineering*. 55(9):2268–2274.

Maor E, Ivorra A, Rubinsky B. 2009. Non thermal irreversible electroporation: Novel technology for vascular smooth muscle cells ablation. *PLoS ONE* 4(3):e4757. doi:10.1371/journal.pone.0004757.

Neumann E, Schaeffer-Ridder M, Wang Y, Hofschneider PH. 1982. Gene transfer into mouse lymphoma cells by electroporation in high electric fields. *EMBO Journal*. 1:841–845.

Okino M, Mohri H. 1987. Effects of a high-voltage electrical impulse and an anticancer drug on in vivo growing tumors. *Japanese Journal of Cancer Research*. 78:1319–1321.

Onik G, Rubinsky B, Zemel R, Weaver L, Diamond D, Cobb C, Porterfield B. 1991. Ultrasound-guided hepatic cryosurgery in the treatment of metastatic colon carcinoma. Preliminary results. *Cancer*. 67:901–907.

Onik G, Mikus P, Rubinsky B. 2007. Irreversible electroporation: Implications for prostate ablation. *Technology in Cancer Research and Treatment*. 6(4):295–300.

Orlowski S, Belehradek JJ, Paoletti C, Mir LM. 1988 Transient electropermeabilization of cells in culture. Increase of the cytotoxicity of anticancer drugs. *Biochemical Pharmacology*. 34:4727–4733.

Pennes HH. 1984. Analysis of tissue and arterial blood temperatures in the resting forearm. *Journal of Applied Physiology*. 1:93–122.

Piñero J, Lopez-Baena M, Ortiz T, Cortes F. 1997. Apoptotic and necrotic cell death are both induced by electroporation in HL60 human promyeloid leukaemia cells. *Apoptosis*. 2:330–336.

Rubinsky B. 1997. Microscale heat transfer in biological systems. *Experimental Heat Transfer*. 10(1):1–29.

Rubinsky B. 2006. Chapter 26. Numerical bio-heat transfer, in: Minkowycz WJ, Sparrow EM, Murthy JY. (eds.), *Handbook of Numerical Heat Transfer*, John Wiley & Sons, Inc., New York.

Rubinsky B. 2007. Irreversible electroporation in medicine. *Technology in Cancer Research and Treatment*. 6(4):255–260.

Rubinsky B, Onik G, Mikus P. 2007. Irreversible electroporation: A new ablation modality—Clinical implications. *Technology in Cancer Research and Treatment*. 6(1):37–48.

Sale AJH, Hamilton WA. 1967. Effects of high electric fields on microorganisms. 1. Killing of bacteria and yeasts. *Biochimica et Biophysica Acta*. 148:781–788.

Sale AJH, Hamilton WA. 1968. Effects of high electric fields on microorganisms. 3. Lysis of erythrocytes and protopasts. *Biochimica et Biophysica Acta*. 163:37–43.

Somersalo E, Cheney M, Isaacson D. 1992. Existence and uniqueness for electrode models for electric current computed tomography. *SIAM Journal of Applied Mathematics*. 52:1023–1040.

Toepfl S, Mathys A, Heinz V, Knorr D. 2006. Review: Potential of high hydrostatic pressure and pulsed electric fields for energy efficient and environmentally friendly food processing. *Food Reviews International*. 22:405–423.

Zimmermann U. 1982. Electric field-mediated fusion and related electrical phenomena. *Biochimica et Biophysica Acta*. 694:227–277.

26

Food and Biomaterials Processing Assisted by Electroporation

Nikolai Lebovka

Eugene Vorobiev

26.1 Introduction

Electric field application in biomaterials and food processing has a long history. The efficiency of direct and alternating electrical current treatments for microbial killing was first demonstrated at the end of the nineteenth century (Prochownick and Spaeth 1890). Then, different practical applications of this technique were tested (Stone 1909, Beattie 1914, Anderson and Finkelstein 1919, Prescott 1927, Fettermann 1928, Getchell 1935) and electric apparatuses for the treatment of fluid foods were patented (Jones 1897, Anglim 1923, Ball 1937). Later on, Flaumenbaum (1949) reported an increase of the juice yield (up to 10%) from prunes, apples, and grapes under their electrical treatment using the voltage of 220 V and industrial frequency of 50 Hz. Zagorulko (1958) reported the application of DC and AC electric fields for the enhancement of the diffusion extraction of juice from sugar beets. An increase in extraction yield was explained by the electrical breakage of cellular membranes and was named as electroplasmolysis.

The first successful application of pulsed electric field (PEF) for the disintegration of biomaterials was reported in 1960 (Doevenspeck 1961). Later on, several industrial AC setups were implemented at canning and wine production enterprises in Bendery (Moldova) and different treatment cells were tested in practice (Flaumenbaum 1968). Many examples of low-gradient alternating electric field applications for acceleration of extractions from grapes, apples, and sugar beet were reported, and these data were explained by both the thermal heating and electrical destruction of cellular tissue (Kogan 1968, Matov and Reshetko 1968, Rogov and Gorbatov 1974, Bazhal et al. 1983, Rogov 1988). However, earlier industrial AC and DC applications were restricted by technical difficulties related with the erosion of electrodes, excessive overheating of products, and high power consumption.

At the same time, the academic studies of the biotechnological applications of the PEF treatment were continuously developing, starting from 1960. The membrane electroporation concept was theoretically worked out and different practically useful examples of enhanced electropermeabilization, electrofusion of cells, and electroinsertion of proteins and nanoparticles into the cell membranes were demonstrated (Weaver and Chizmadzhev 1996, Zimmermann 1996).

Starting from 1990, many new practical applications of the PEF method involving higher voltage PEF generators were proposed. It was shown that PEF treatment can be used as an efficient tool for microbial killing; food preservation; the acceleration of drying, pressing, and diffusion; and selective extraction (Gulyi et al. 1994, Barbosa-Cánovas et al. 1998, Eshtiaghi and Knorr 1999, Vorobiev and Lebovka 2008). The PEF-assisted processing of food and biomaterials allows nonthermal impact on membrane components through the mechanism of electroporation, and it is more preferable, as far as thermal heating may result in deterioration of material quality, color, and denaturation of intracellular components (Ravishankar et al. 2008). However, the mechanisms of the PEF-induced effect in real food and biomaterials are rather complex and still poorly understood.

This chapter reviews the experimental and theoretical data dealing with the electroporation of suspensions and plant tissues, and, also, the current state of practical applications of PEF for the acceleration of drying, freezing, pressing, diffusion, and selective extraction processes.

26.2 Fundamentals of Electroporation in Relation to Biomaterial and Food Processing

26.2.1 Phenomenon of Electropermeabilization

The biological membranes possess rather strong barrier functions and their permeabilization facilitating the transport of molecular components is restricted. The molecular transport can be accelerated by the introduction of thermal defects (or pores) in the membrane structure. The fraction of the membrane surface, covered by thermally induced pores (or damage degree), may be estimated as $P \approx \exp(-W/kT)$ where W is the activation energy necessary for pore formation, $k = 1.380 \times 10^{-23}$ J/K is the Boltzmann constant, and T is the absolute temperature. At ambient conditions ($T = 298$ K), the activation energy W is rather high as compared with the thermal energy kT, and the thermal factor may be insignificant. For example, for a typical value of $W/kT \approx 100$ (Lebedeva 1987), $P \approx 10^{-44} \ll 1$. However, external electric fields may increase permeabilization and provoke the damage of membranes.

In general, the phenomenon of electropermeabilization and damage of the membranes is still far from being understood in full. Different approaches, accounting for pore formation (electroporation), electric field induced instabilities (mechanical, hydrodynamical, viscous-elastic, thermal, and osmotic), chemical imbalances, and Joule overheating of the membrane surface, were proposed for its description (Ho and Mittal 1996, Weaver and Chizmadzhev 1996, Lebovka et al. 2000b, Chen et al. 2006, Dimitrov and Sowers 1990).

The electroporation theory predicts a drastic increase of pore concentration in the membrane P if a transmembrane voltage u_m is applied to it, which can be estimated as $P \approx \exp(-W^*/(1 + (u_m/u_c)^2))$. Here, u_c is the material parameter that characterizes the electroporation response of the membrane. High voltage application normally results in a noticeable increase of the concentration of pores. For example, at $u_m = 10u_c$, the above equation gives a rather high concentration of pores $P \approx 0.37$, which can be the result of membrane damage.

It is widely accepted that membrane damage is of a threshold character and the enhanced permeabilization or damage of membrane occurs when transmembrane potential u_m (or applied electric field strength) exceeds the certain threshold value (typically, about 0.2–1.0 V) (El-Mashak and Tsong 1985, Weaver and Chizmadzhev 1996, Pavlin et al. 2008).

Note that the field amplification factor k of the electric field strength E on membranes ($k = E_m/E$) is roughly proportional to the ratio of the cell size R and membrane thickness h. Owing to a huge difference between h (≈ 5 nm) and cell size ($R \approx 1$–$100\,\mu$m), the field amplification factor can be rather large, $k \sim 10^2$–10^4. Moreover, for plant tissues with larger cells (with $R \approx 100\,\mu$m) as compared with microbial cells ($R \approx 1$–$10\,\mu$m), the induced transmembrane potentials and charging phenomena can be greatly influenced by electrical conductivities of the membrane, extracellular medium, and cytoplasm (Vorobiev and Lebovka 2008).

However, the extent of electroporation and its stability also depends upon the treatment protocol, shape of pulses, pulse duration, and intervals between pulses (Canatella et al. 2001).

26.2.2 Electrically Induced Damage of Membranes, Cells, and Tissues

The degree of membrane permeabilization/damage mainly depends upon the value of the transmembrane potential, its distribution across the membrane surface, and the duration of electric field application. In turn, the electrically induced damage of cells may be determined by many factors related to the structure of the biological object.

When the voltage u_m is applied to the individual membrane (Figure 26.1a), its lifetime τ_m may be estimated from the electroporation theory as (Weaver and Chizmadzhev 1996)

$$\tau_m = \frac{\tau_\infty}{P} = \tau_\infty \exp(W/kT(1 + (u_m/u_o)^2)), \qquad (26.1)$$

where τ_∞ is a parameter corresponding to the limit of membrane lifetime at very high temperatures, $\tau = \tau_\infty$ at $kT \to \infty$, when $P \to 1$.

The W and τ_∞ parameters of the material depend on the membrane structure and composition. For example, their experimentally estimated values are $W \approx 270$ kJ/mol and $\tau_\infty \approx 3.7 \times 10^{-7}$ s for general

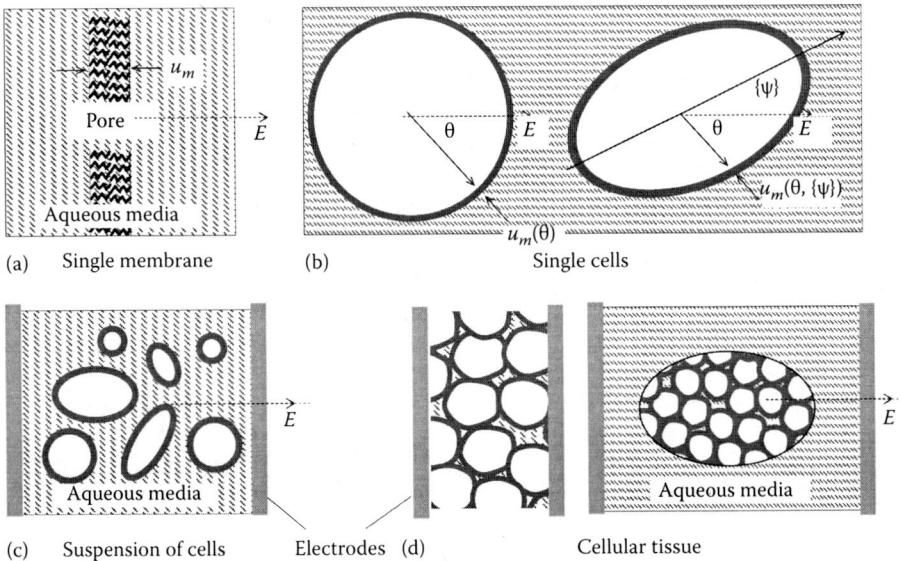

FIGURE 26.1 Effects of the external electric field E, applied to the individual membrane (a), individual cells having spherical or elongated shape (b), suspensions of cells (c), and bio-tissues (d).

lipid membranes (Lebedeva 1987) or $W \approx 166\,\text{kJ/mol}$ and $\tau_\infty \approx 10^{-23}\,\text{s}$ for membranes in sugarbeet cells (Lebovka et al. 2007a).

The transmembrane potential u_m is proportional to the cell size and the larger cells get damaged earlier than the smaller ones (Schwan 1957). Moreover, the transmembrane potential u_m is a complex function of the cell geometry, direction of the external field, and its location on the membrane surface (Fricke 1953, Bernhardt and Pauly 1973, Zimmermann et al. 1974, Kotnik and Miklavčič 2000, Gimsa and Wachner 2001). For example, for spherical cells, the value of u_m may be estimated using Schwan's equation (Schwan 1957) $u_m = 1.5\,fRE\cos\theta$, where $f \approx 1$ is a parameter depending on the electrophysical and dimensional properties of the membrane, cell, and surrounding medium (Figure 26.1b). The transmembrane potential u_m is maximal at cell poles.

For elongated cells, the maximum of the transmembrane potential u_m and electropermeabilization efficiency were observed (Valic et al. 2003, Agarwal et al. 2007, Toepfl et al. 2007) when the cells were oriented by their longest axes in parallel to the external electric field (i.e., $\psi = 0$ in Figure 26.1b). The lifetime of a cell τ_m in an external field may be estimated by averaging the damage probability over the cell surface (Lebovka et al. 2002, Lebovka and Vorobiev 2004). It can be a complex function of the shape of cells and their orientation. For suspended cells (Figure 26.1c) and cells in a tissue (Figure 26.1d), the electroporation efficiency may be influenced by the aggregation of cells, their arrangement, the local density of the cells, solute concentration, and local electric field distribution (Canatella et al. 2004, Pavlin et al. 2007, Pucihar et al. 2007). The external fields can affect the orientation (Lebovka and Vorobiev 2007) and aggregation of cells (Toepfl 2006) in suspensions. Redistribution of the local fields inside a tissue is possible also during the PEF treatment (Lebovka et al. 2000a, 2001).

If the field is applied to the samples of the whole tissue, placed in an aqueous medium between the plane electrodes (Figure 26.1d), the electric field inside the tissue E_i differs from the external field E (Grimi et al. 2010). In a general case, E_i depends on the shape of the sample and mean conductivities of the tissue sample σ and aqueous medium σ_w, their dielectric characteristics, and frequency f; so the relation between E_i and E may be rather complex. For a spherical sample, the field inside it is homogeneous, and in the limit of small frequencies f it can be calculated as $E_i = 3E/(2 + \sigma/\sigma_w)$ (Hart and Marino 1986). This equation gives $E_i = 0$ for conductive inclusion in a nonconductive medium, $\sigma_w \ll \sigma$, and $E_i = (3/2)E$ for the opposite case, when $\sigma_w \gg \sigma$.

It is useful to study the electric stability of membrane systems by monitoring their damage degree P. For a single membrane, the damage degree P can be defined as a fraction of the membrane surface covered by pores. For suspension of cells or plant tissue, the value of P can be defined as a ratio of the number of damaged cells and the total number of cells. The value of P can be estimated directly through microscopic observations (Fincan and Dejmek 2002), or using any indirect technique involving the measurements of electrical conductivity (Rogov and Gorbatov 1974, Lebovka et al. 2000a, Angersbach et al. 2002, Vorobiev and Lebovka 2006), diffusion coefficients (Jemai and Vorobiev 2001, Lebovka et al. 2007b), or acoustic response (Grimi et al. 2010).

However, the application of all these procedures for the estimation of P in suspensions or plant tissues is not simple and is still controversial. Diffusion techniques, based on solute extraction or convective drying experiments, are indirect and invasive for biological objects, and they may impact the structure of the tissue (Vorobiev et al. 2005, Lebovka et al. 2007b). The most widely accepted method is based on electrical conductivity measurements (Lebovka et al. 2000a, Vorobiev and Lebovka 2006, El Zakhem et al. 2006a,b). It is very useful, as far as the electrical conductivity σ grows during the electrical treatment of the cellular systems (Lebovka et al. 2000a, Angersbach et al. 2002, Vorobiev and Lebovka 2006). The electrical conductivity disintegration index Z_c defined as (Rogov and Gorbatov 1974)

$$Z_c = \frac{\sigma - \sigma_i}{\sigma_d - \sigma_i} \tag{26.2}$$

may be assumed to be a good approximation of the damage degree P. Here, σ is the electrical conductivity of the cellular system measured at a low frequency (1–5 kHz), and indices "i" and "d" correspond to the intact and totally damaged objects, respectively.

This method has a limitation, as far as it requires supplementary measurements for the determination of σ_d in a maximally damaged material. The maximally damaged material may be obtained by freeze-thawing the material or by its strong PEF treatment using a high strength electric field and long duration of PEF treatment (Lebovka et al. 2007a). In principle, the procedure of the material destruction can influence the value of σ_d, and this question has not been investigated in detail yet. Another alternative of Z estimation (Angersbach et al. 2002) excludes the necessity of σ_d determination. However, this method requires the measurement of electrical conductivity at rather high electric field frequencies (3–50 MHz), and it is based on a theoretical model simulating the dielectric behavior of a cellular material.

Equation 26.2 gives reasonable values of the electrical conductivity disintegration index Z_c; $Z_c = 0$ for the intact tissue and $Z_c = 1$ for the totally disintegrated material. However, though Z_c depends on P, no simple relation between Z_c and P exists and it is assumed to be strongly nonlinear (Vorobiev and Lebovka 2006). The simple empirical Archie's power equation (Archie 1942) $Z_c \approx P^m$, where experimentally estimated exponents m fall within the range of 1.8–2.5 for biological tissues, such as tissues of apples, carrots, or potatoes (Lebovka et al. 2002), can be a reasonable approximation of this relation.

The theoretical dependence between Z_c and P can be approximated more strictly using the general effective medium (GEM) equation (McLachlan et al. 2000):

$$\frac{c(\sigma_i^{1/s} - \sigma^{1/s})}{\sigma_i^{1/s} + c_c \sigma^{1/s}} + \frac{\sigma_d^{1/t} - \sigma^{1/t}}{\sigma_d^{1/t} + c_c \sigma^{1/t}} = 0. \qquad (26.3)$$

Here, s and t are the exponents of the random percolation theory, $s = 0.73$, $t = 2.0$ for 3D materials, $c = 1/P - 1$, $c_c = 1/P_c - 1$, and P_c is the percolation threshold corresponding to transitions from low ($\sigma = \sigma_i$) to high ($\sigma = \sigma_d$) conductivity (Stauffer and Aharony 1994).

Equation 26.3 allows for the estimation of σ versus P, which can be recalculated as Z_c versus P dependence. Figure 26.2 presents the Z_c versus P dependences (dashed lines), calculated using the typical values for apple tissue: $\sigma_i = 0.006$ S/m, $\sigma_d = 0.1$ S/m at $P_c = 0.1$ (upper line), and $P_c = 0.6$ (lower line).

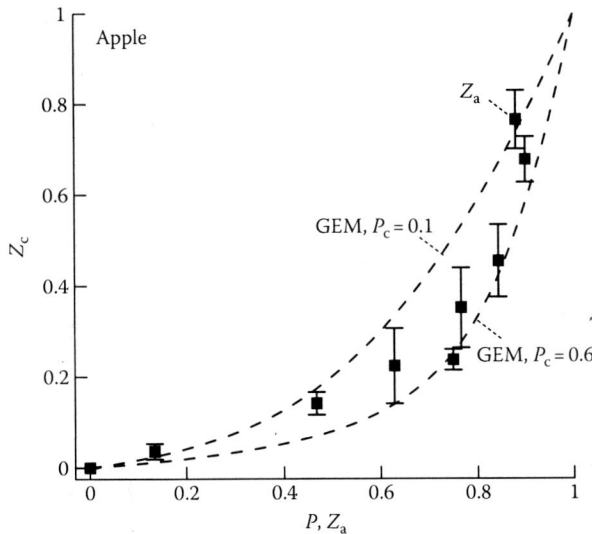

FIGURE 26.2 Nonlinear dependences between electrical conductivity disintegration index Z_c and damage degree P (dashed lines) or acoustical disintegration index Z_a (filled squares are experimental data for the apple tissue.

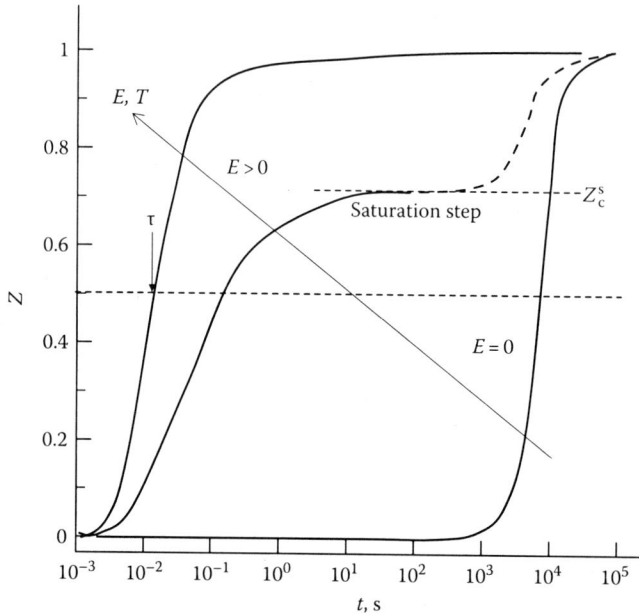

FIGURE 26.3 The temporal evolution of the disintegration index Z. Increase of electric field strength E and/or temperature T results in acceleration of biomaterial damage.

In a general case, the percolation threshold parameter P_c depends upon the arrangement of cells inside the tissue, and limits of $P_c = 0.1$–0.6 are quite reasonable (Stauffer and Aharony 1994). Using Z_c instead of P results in a systematic underestimation of the damage degree.

Recently, the acoustic disintegration index Z_a, defined similarly to the electrical conductivity disintegration index $Z_a = (F - F_i)/(F_d - F_i)$ (here F is the measured index of firmness), was proposed for the characterization of the damage degree P (Grimi et al. 2010). The square symbols in Figure 26.2 correspond to the Z_c versus Z_a dependence experimentally determined for apples. The experimental data fall within the theoretically predicted curves of Z_c versus Z_a. So, the acoustic disintegration index Z_a seems to be better adopted for the characterization of the damage degree P (Grimi et al. 2010).

The cell damage kinetics is determined by many factors related to the composition and structure of biomaterials and usually it cannot be described by the simple first order kinetic law. So, strict determination of the lifetime of a cell τ_m from the experimental data is impracticable, and it is useful to introduce the characteristic damage time τ, defined as the time needed for attaining half of the maximal damage ($Z \approx 1/2$) (Bazhal et al. 2003). The characteristic damage time τ characterizes the rate of the tissue damage. The electrically induced damage depends on the treatment protocol, and its rate grows with the increase of the electric field strength E and temperature T (Figure 26.3) (Lebovka et al. 2005a,b).

The characteristic damage time, τ, values were strongly dependent on the tissue type (Lebovka et al. 2002). It was experimentally observed that the characteristic damage time τ (reflecting resistance to damage under the PEF treatment) decreased in the following order of sequence: onion → orange → apple → tomato (Grimi et al. 2010) and banana → apple → carrot (Bazhal 2001, Bazhal et al. 2003). The differences in the damage resistance can be explained by differences in the cell sizes, nature of membranes and their composition, and in tissue porosity.

26.2.3 Synergetics of Electrical and Thermal Treatments

The temporal evolution of the disintegration index Z (either conductivity or acoustical) is schematically presented in Figure 26.3. Electro-processing at a moderate electric field strength ($E < 100\,\text{V/cm}$) and ambient conditions, or thermal processing alone at a moderate temperature ($T < 50°C$), requires a long time of treatment and high energy consumption as a consequence.

For example, the thermal damage of a biomaterial at ambient conditions was noticeable only if the time of treatment exceeded 10^5 s and could be accelerated only at elevated temperatures above 50°C. Moreover, a rather complex kinetics with an intermediate saturation step (when the disintegration index Z reaches a plateau, $Z = Z_s$) was often observed at long PEF treatment (Figure 26.3) at moderate electric fields ($E < 300$ V/cm) and at moderate temperatures ($T < 50°C$) (Lebovka et al. 2001, 2007a). For example, the maximal disintegration index Z_s was of the order of 0.75 at $E = 100$ V/cm for sugar beet tissue (Lebovka et al. 2007a). The steplike behavior of $Z(t)$ is typical for inhomogeneous tissues, such as the red beetroots (Shynkaryk 2007, Shynkaryk et al. 2008). Such saturation at the level of $Z = Z_s$, possibly, reflects the presence of a wide distribution of cell survivability, related with spreading in cell geometries and sizes. It was experimentally observed that the saturation level Z_s increased with an increase of both E (Lebovka et al. 2001) and T (Lebovka et al. 2007a). For tissues with relatively homogeneous structures (potatoes, apples, chicory, etc.), this saturation behavior was less pronounced and was not practically observed at higher fields $E > 500$ V/cm. If PEF stops at the saturation level (Figure 26.3), the scenario of the further evolution can be different. The cells can partially reseal at a very small level of disintegration (Knorr et al. 2001). But higher levels of disintegration usually results in a further increase of Z after a relatively long time (Lebovka et al. 2001, Angersbach et al. 2002).

Usually, the simultaneous PEF and thermal treatment exerts a synergetic effect on the damage of tissues (Lebovka et al. 2005a,b, 2007a) and an increase of temperature T or electric field strength E results in a drastic drop of the characteristic damage time by many orders of magnitude. The example of such behavior for chicory tissue is demonstrated in Figure 26.4. Moreover, the energy of tissue damage activation W was a decreasing function of electric field strength E (Insert in Figure 26.4). The electro-thermal synergetism of tissue damage, possibly, reflects existing softening transitions in membranes at temperatures within 20°C–55°C (Exerova and Nikolova 1992, Mouritsen and Jørgensen 1997).

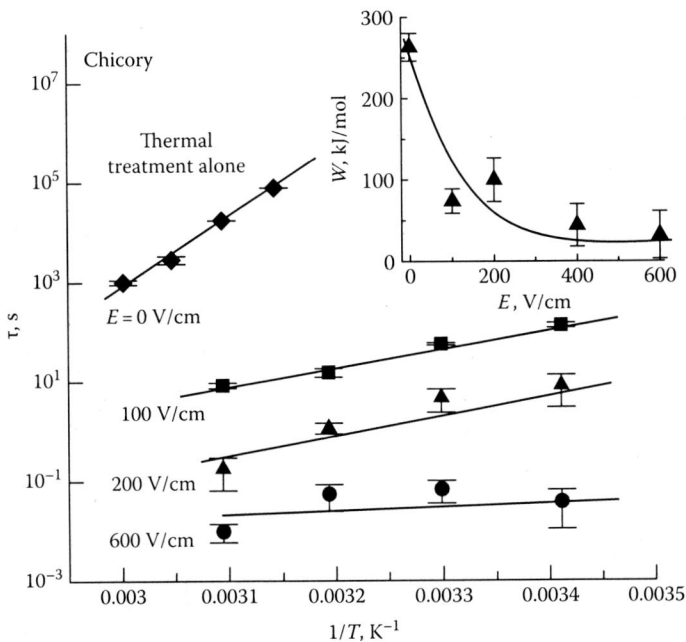

FIGURE 26.4 Arrhenius plots of characteristic damage time τ for chicory tissue at different values of electric field strength E. Inset shows damage activation energy W versus E. The PEF treatment was done at pulse duration $t_i = 10$ ms and distance between pulses $\Delta t = 10$ ms.

A noticeable drop in the breakdown transmembrane voltage u_m of a single membrane was experimentally observed near the region of thermal softening ($\approx$50°C) (Zimmermann 1986). The cell membrane fluidity and its domain structure exert a noticeable influence on the electropermeabilization of cells (Kandušer et al. 2008).

Relations between the characteristic damage time τ and electric field strength E or temperature T may be rather complex. The experimental data for potato tissue were fitted successfully by the following equation (Lebovka et al. 2005a)

$$\tau_m = \tau_\infty \exp(W/kT(1+(E/E_0)^2)), \qquad (26.4)$$

where τ_∞, W, and E_0 are adjustable empirical parameters. Note that that this equation resembles Equation 26.1 by its form. However, it is fully empirical and has no mathematical justification based on the mechanisms of electroporation processes in tissues.

The very interesting effects of simultaneous electro-thermal treatment were observed in ohmic heating experiments. The ohmic heating naturally develops during electric treatment and causes a rise in temperature. Changes in the tissue structure, electrically induced during its ohmic heating, can be essential (Wang and Sastry 2002). A direct method was proposed based on the experimental observations of electrical conductivity changes during the ohmic heating, for monitoring of electroporation changes, and it was shown that ohmic heating at an electric field strength E of the order of 20–80 V/cm induced the structural changes inside tissues, related to losses of membrane barrier functions (Lebovka et al. 2005a,b, 2007a).

26.2.4 Energy Consumption

The energy consumption Q (volume density of the energy input) during PEF treatment is an important factor for the estimation of the industrial attractiveness of the electro-thermal technology of disintegration. High damage of the tissue requires sufficient power consumption, which can be estimated using the following expression

$$Q = \int_0^t \sigma(t)E^2 \, dt, \qquad (26.5)$$

where electrical conductivity $\sigma(t)$ is the function of time owing to the developing damage of material and owing to an increase in temperature caused by ohmic heating.

The energy consumption Q is roughly proportional to the product $\bar{\sigma}\tau E^2$, where $\bar{\sigma}$ is the time-averaged conductivity. The value of $\tau(E)$ decreases with an increase of the electric field strength E, however, the product τE^2 goes through a minimum (Lebovka et al. 2002). There exists some optimum value of electric field strength $E \approx E_0$, which corresponds to minimum power consumption (Bazhal et al. 2003). The further increase of E results in a progressive increase of the energy consumption, but gives no additional increase in the conductivity disintegration index Z. For vegetable and fruit tissues, the typical values of E_0 lies in the interval of $E = 200–700$ V/cm (Bazhal et al. 2003). Specific energy consumption, required for effective electropermeabilization of cells in tissues, typically lies in the interval of 1–10 kJ/kg.

Note that energetically the PEF method is practically ideal for producing damaged plant tissues as compared with other methods of treatment like mechanical ($Q = 20–40$ kJ/kg), enzymatic ($Q = 60–100$ kJ/kg), and heating or freezing/thawing ($Q > 100$ kJ/kg) (Toepfl et al. 2006). However, for bacterial inactivation and food preservation, high optimal electric field strength is required ($E_0 = 15–40$ kV/cm) and that naturally results in noticeably higher specific energy consumption on the order of 40–1000 kJ/kg (Toepfl et al. 2006).

26.2.5 Electroporation Efficiency as a Function of Pulse Protocol

Experiments show that electroporation efficiency depends on pulse parameters, such as amplitude of pulse (or electric field strength E), its shape, duration t_i, number of repeats n, and intervals between pulses Δt (Canatella et al. 2001). The pulse shapes commonly used in PEF generators are exponential decay or near-rectangular shape and they may be either monopolar or bipolar. The square-wave generators are more expensive and require more complex equipment than the exponential decay generators. But square-wave generators have better energy performance and demonstrate higher disintegrating efficiency in experiments with micro-cell inactivation (Zhang et al. 1994). Bipolar pulses seem to be more advantageous, as they cause additional stress in the membrane structure and result in more effective electroporation response. Moreover, the application of bipolar pulses offers minimum energy consumption with reduced deposition of solids on the electrodes and decreased food electrolysis (Chang 1989, Qin et al. 1994, Wouters and Smelt 1997).

The typical PEF protocol for bipolar pulses of a near-rectangular shape is presented in Figure 26.5. Usually, a series of N pulses (train) are used in laboratory experiments, and each separate train consists of n pulses with pulse duration t_i; distances between pulses, Δt; and pause Δt_t after each train. The total time of PEF treatment is regulated by the variation of the number of series N and is calculated as $t_{PEF} = nNt_i$. Such protocols with adjustable pauses between trains allows for fine regulation of the disintegration index Z without noticeable temperature elevation during the PEF treatment.

The main relevant parameters, determining efficiency of the PEF damage, are the applied electric fiend strength E and the total treatment time t_{PEF} (Sale and Hámilton 1967). The higher electric field strength leads to better damage efficiency (Canatella et al. 2001, Toepfl et al. 2007), however, electrical power consumption and ohmic heating become essential at high fields.

The distance between pulses, Δt, was shown to be an essential parameter affecting PEF disintegration efficiency (Lebovka et al. 2001). For example, the protocol with longer distances between pauses at fixed values of E and t_{PEF} displayed accelerated kinetics of disintegration of the apple tissue. The obtained results were explained accounting for the moisture diffusion processes inside the cellular structure. The inactivation of *Escherichia coli* was also affected by pulse distance Δt (Evrendilek and Zhang 2005), however, the observed results are still ambiguous.

The effects of pulse duration t_i on PEF inactivation and tissue damage efficiency at fixed values of E and t_{PEF} were previously discussed (Martín-Belloso et al. 1997, Wouters et al. 1999, Mañas et al. 2000, Raso et al. 2000, Aronsson et al. 2001, Abram et al. 2003, Sampedro et al. 2007, De Vito et al. 2008). Some authors have demonstrated that inactivation was more efficient at higher pulse widths at invariable quantities of the applied energy (Martín-Belloso et al. 1997, Abram et al. 2003), but others observed little effect of the pulse width on inactivation at equal energy inputs (Mañas et al. 2000, Raso et al. 2000, Sampedro et al. 2007, Fox et al. 2008). The effect of pulse width on microbial inactivation seems to vary

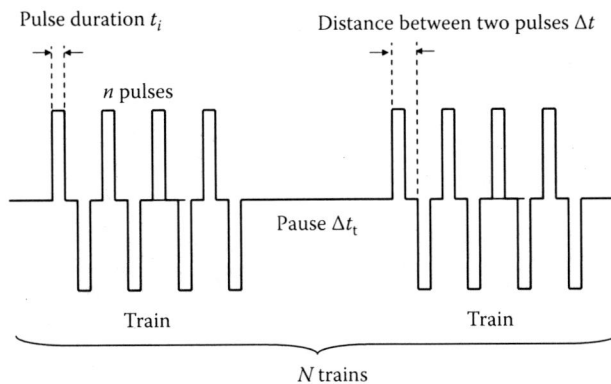

FIGURE 26.5 Typical pulsed electric field protocol.

depending on electric field strength; still, the obtained results are controversial (Wouters et al. 1999, Aronsson et al. 2001). The critical review of the effect of pulse duration on electroporation efficiency in relation to therapeutic applications was recently published (Teissié et al. 2008). It was shown that the response of a biosystem on PEF treatment includes multistep events with well-defined kinetics and its successful practical applications require a careful choice of the pulse duration t_i in nanosecond and microsecond diapasons.

The distinct correlations between pulse duration and damage efficiency were recently demonstrated in tissue cellular materials with large cells (De Vito et al. 2008). The theory predicts the deceleration of the membrane charging processes for large cells, when the membrane charging time t_c may be rather large, $t_c \approx 10^{-5}$–10^{-4}s (Kotnik et al. 1998). An efficient PEF treatment requires long pulse duration t_i as compared with the membrane charging time t_c in order to reach the maximum transmembrane voltage. So, at larger values of t_c, longer pulses will be required for attaining the desired voltage amplitude and we can expect higher PEF efficiency for longer pulse width t_i. Existing experiments support this conclusion and clearly demonstrate the influence of the pulse duration t_i (10–1000 μs) on the efficiency of the PEF-treatment of grapes, apples, and potatoes (De Vito et al. 2008, Grimi et al. 2010). Longer pulses were more effective and their effect was particularly pronounced at room temperature and moderate electric fields ($E = 100$–300 V/cm) (De Vito et al. 2008).

26.3 Electro-Processing of Liquid Foods and Suspensions

In liquid foods and suspensions, the cells are randomly distributed in the surrounding liquid medium (Figure 26.1c). The efficiency of electroporation in such systems is mainly determined by electric field distribution on the surface of the cell membranes. In turn, this distribution may be governed by the concentration, shape, and orientation of cells. So, the efficiency of electro-processing (PEFs or high voltage electric discharge, HVED) and optimal protocol (electric field strength, pulse shape, pulse length, total time of treatment, temperature) may be dependent on many details characterizing the structure of fluid suspensions and their response to the external electric fields. Moreover, the physiological state of cells, cell wall construction, state of cell aggregation (Wouters and Smelt 1997) as well as their properties, such as the electrical conductivity of liquid medium, pH, the presence of surfactants, cell density, etc. (Barbosa-Cánovas et al. 1998, Susil et al. 1998, Pavlin et al. 2002) are very important in the electro-processing of fluid foods and suspensions.

26.3.1 Bacterial Inactivation of Liquid Foods

The bacterial inactivation requires rather high critical electric fields and presumably reflects the relatively small size of bacterial cells. Note that the cells of microorganisms display a variety of shapes and dimensions (Bergey 1986) depending on the culture condition and age. Usually, the near-spherical, ellipsoidal or ovoid (cocci), cylindroidal (bacilli), and spiral or comma-like (spirilli) shapes occur. *Escherichia coli* and Salmonella cells are rodlike and are 0.4–0.6 μm in diameter and 2–4 μm in length, cells of *Leptospira* spp. are like very long rods with 0.1 μm in diameter and 20 μm in length, cells of Staphylococcus spp are spherical with diameters of 0.5–1.5 μm, *Saccharomyces cerevisiae* have ellipsoid or near spherical forms of cells with principal dimensions between 2 and 15 μm, cells of *Klebsiella pneumoniae* are ovoid with a mean dimension of 0.4 μm, and *Vibrio cholerae* have comma-like cells with the principal dimensions of 0.5 and 1.5–3 μm, respectively (Bergey 1986).

For example, the commonly reported field strength required for high disintegration of anisotropic *E. coli* cells lies within $E = 10$–35 kV/cm (Grahl and Märkl 1996, Aronsson et al. 2001, 2005, Amiali 2006, Bazhal et al. 2006), though smaller fields, $E = 1.25$–3.75 kV/cm, also can affect permeabilization in the membranes of *E. coli* cells (Eynard et al. 1998). The PEF-induced orientation of the rodlike *E. coli* cells can even facilitate their electropermeabilization (Eynard et al. 1998). The effective disintegration of *E. coli* at small electric fields below 5 kV/cm requires a longer time of treatment (El Zakhem et al. 2006b, 2007), yet very little is known about the mechanisms of PEF-induced disintegration, cell survivability, and the extent of extraction of the intracellular components at small electric fields.

The physical mechanisms of the PEF-induced effects in liquid foods and suspensions are rather complex and are not well known yet. Numerous studies have investigated the effect of PEF application on microbial inactivation (Barbosa-Cánovas et al. 1998), cells electrofusion (Teissié et al. 2005), and transport of nanoparticles or biopolymers across the cell wall into the recipient cells (Van Wert and Saunders 1992, Jen et al. 2004). Usually, the electric pulses of high intensity ($E \approx 10$–$50\,$kV/cm) and of small duration (typically, 1–100 μs) cause a temporary loss of the cell membrane permeability (electroporation) and ion leakage (Weaver and Chizmadzhev 1996) without a significant increase in temperature or undesirable effects on cell components.

Leakage of cytoplasmic ions during the PEF application influences the ionic concentration of the medium and its electrical conductivity (Kinosita and Tsong 1977, Eynard et al. 1992, El Zakhem et al. 2006a,b). The damage of *S. cerevisiae* and *E. coli* cells was accompanied by a decrease of the cells' size, which reflects the PEF-induced leakage of the intracellular components. In principle, it allows the application of the electrical conductivity method for an estimation of the degree of damage to the microorganisms during the PEF treatment of an aqueous suspension. The conductometric approach was used for the continuous monitoring of the degree of damage to the cell (*S. cerevisiae* and *E. coli* cells) and it was applied for studying the effects of temperature and surfactant on inactivation efficiency (El Zakhem et al. 2006a,b). Figure 26.6 shows examples of electrical conductivity disintegration index Z evolution under the thermal or simultaneous PEF + thermal treatments (El Zakhem et al. 2007). The experimental data support the existence of synergy between the PEF and thermal treatments in *E. coli* inactivation.

The PEF inactivation efficiency at a high concentration of cells may be affected by the formation of inter-cell aggregates (Figure 26.7a and b) (Calleja 1984, Zhang et al. 1994, El Zakhem et al. 2006b). The aggregation of cells upon electric field application may reflect the enhancement of attraction between the near-neighbor cells due to forces produced by the induced dipoles, accelerated electrophoretic, or dielectrophoretic movement of cells, changes in electro-surface properties of cells, etc. (Zhang et al. 1994, Molinari et al. 2004, El Zakhem et al. 2006a,b, Wosik et al. 2006). The possibility

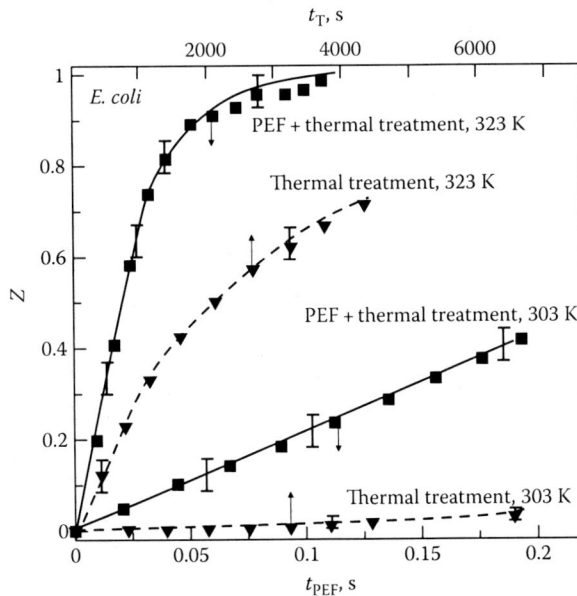

FIGURE 26.6 The electrical conductivity disintegration index Z versus effective PEF treatment time (t_{PEF}) and thermal treatment time (t_T) for 1% *E. coli* aqueous suspensions. PEF treatment was done at $E = 5\,$kV/cm, $t_i = 10^{-3}\,$s, $\Delta t = 5 \times 10^{-2}\,$s and $\Delta t_t = 1\,$s.

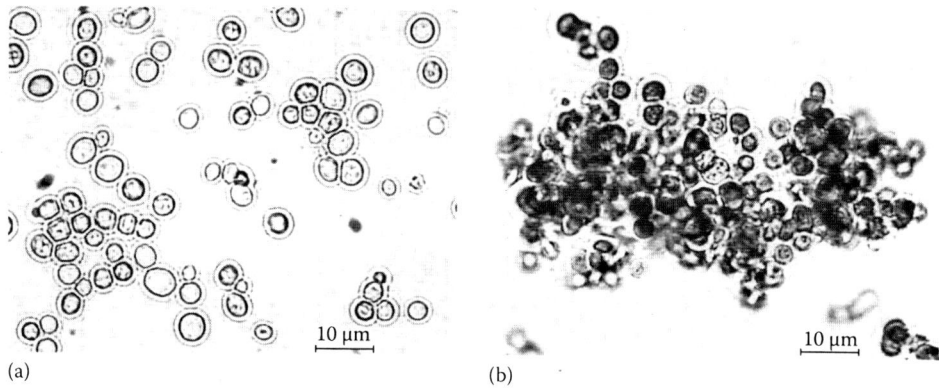

FIGURE 26.7 Microphotos of untreated (a) and PEF treated (b) 0.5% aqueous suspension of *S. cerevisiae*. The PEF treatment was done at $T = 328\,K$, $E = 7.5\,kV/cm$, and $t_{PEF} = 0.04\,s$.

of the formation of a "pearl chain,", in which the cells are in very close contact with each other, was described by Zimmermann (1986, 1992). Moreover, this effect was greatly enhanced by the presence of nonuniform fields, originated by the inhomogeneous distribution of cells (Zimmermann 1992). The PEF treatment resulted in the formation of rather large cell flocs, and this effect was significant. One of the possible mechanisms of cell aggregation enhancement in suspensions may be related to changes in the net charge of cells, provoked by PEF application. For example, it was demonstrated (El Zakhem et al. 2006b) that the "alive" *S. cerevisiae* cells have negative ζ-potential as compared with PEF-treated "died" cells (Figure 26.8).

So, PEF treatment can provoke an electrostatic attraction between "alive" and "dead" cells. PEF inactivation of *S. cerevisiae* at a high yeast concentration in suspension may be affected by the formation of clusters or "pearl chain" cells (Zhang et al. 1994, Zimmermann 1986, 1992). In principle, this effect may facilitate the PEF-induced damage due to the formation of "equivalent cells" of larger volume or may protect cells against PEF-induced damage (Zhang et al. 1994). However, the incomplete damage of cells

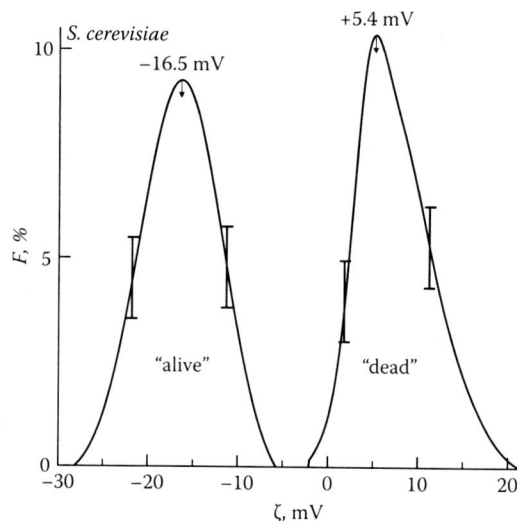

FIGURE 26.8 The ζ-potential distribution function F for "alive" cells in untreated suspensions and "died" cells in PEF-treated suspensions of *S. cerevisiae*. The PEF treatment of 0.5% yeast suspension was done at $T = 293\,K$, $E = 5\,kV/cm$, and $t_{PEF} = 0.1\,s$.

inside the clusters is also possible; it can occur because of the formation of the low-conductive cores (consisting of damaged cells) enveloping the surface of intact cells inside the floc of cells. Moreover, the theoretical calculations predict dependence of induced transmembrane potential on cell density and arrangement (Susil et al. 1998, Pavlin et al. 2002) and that higher voltage amplitude or longer pulse duration is required to cause the same effects of poration if the cells are in a cluster (Joshi et al. 2008).

The efficiency of electroporation of the PEF-treated cells may be increased by the addition of the supplementary chemical reagents and nanoparticles. The data regarding the improvement of the damage efficiency in suspensions by the addition of surfactants, peptides, dimethylsulfoxide (DMSO), or polylysine were reported (Melkonyan et al. 1996, Diederich et al. 1998, Tung et al. 1999, El Zakhem et al. 2007). The use of nanotubes for the enhancement of cell electroporation was recently discussed (Rojas-Chapana et al. 2004, Yantzi and Yeow 2005, Raffa et al. 2009). Owing to the so-called lightening rod effect, the nanotubes have the ability to strongly enhance the electric field at their ends, thus making them ideal for localized electroporation. It was demonstrated that nanotubes can be used as nanotools enabling electropermeabilization of cells at rather low electric fields (40–60 V/cm) (Raffa et al. 2009).

26.3.2 Accelerated Extraction of Products from Cells

The disruption of some microorganisms (*E. coli*, *S. cerevisiae*) is a very important step in the industrial extraction of products (valuable proteins, cytoplasmic enzymes, polysaccharides, etc.), which are present inside the cells (Engler 1985). The existing techniques of cell disruption are based on the application of different treatments: mechanical (high-pressure homogenization [HPH], wet milling), chemical (organic solvents, enzymes, detergents), and physical (sonification, freeze–thaw, electrically assisted treatment). Successful recovery of the intracellular products assumes the preservation of their content and the removal of the cell debris. Mechanical methods are most appropriate for the large-scale disruption of cells and allow for the attainment of high recovery of the intracellular material. However, they are restricted by temperature elevation, require multiple passes with supplementary cooling, and their final products contain large quantities of cell debris (Brookman 1974, Engler 1985, Lovitt et al. 2000). A thermal treatment at $T > 50°C$ results in the damage of the yeast membranes, but also causes denaturation and degradation of many valuable intracellular components. The chemical methods are rather expensive, usually result in low recovery of the intracellular material, and require additional purification in the downstream processes (Harrison et al. 1991).

Nowadays, electroporation-assisted extraction is expected to be promising for the recovery of the homogeneous and heterogeneous intracellular proteins with wide biotechnological applications (Ohshima et al. 1995, Ganeva et al. 2003, Suga et al. 2006, 2007). The electroporation-assisted extraction efficiency can depend on the cell strain, age of culture, temperature of cultivation, and medium; however, general relations between the extraction efficiency and mode of PEF treatment are not yet well understood. The PEF treatment removes the membrane barriers and accelerates the release of the ionic content, but it has practically no influence on the cell walls. So, the high level of conductivity disintegration index Z can reflect the high level of the cytoplasmic ions leakage, while the protein extraction may still be low. The efficiency of releasing the high molecular weight content of cells is questionable and may depend on the cell strain, age of culture, time of extraction, temperature, and many other factors. The electroextraction of proteins from the suspensions of *S. cerevisiae* was observed at 3.2 kV/cm (Ganeva et al. 2003), but high efficiency ($\approx 85\%$ yield) required long extraction after PEF treatment (>4 h at 30°C). PEF application provided the selective release of the intracellular protein from the yeast cells (Ohshima et al. 1995). The PEF treatment of *S. cerevisiae* at 5 kV/cm allowed for the attainment of the high level of conductivity disintegration index, $Z \approx 1$, with more intensive release of peptides and proteins than nucleic bases (El Zakhem et al. 2006b). However, in PEF-treated suspensions of wine yeast cells (*S. cerevisiae bayanus*, strain DV10), the relatively small release of proteins was observed even at high levels of $Z \approx 0.8$ (Shynkaryk et al. 2009). More effective extraction of the high molecular weight content (e.g., proteins) requires more powerful mechanical disintegration of cell walls, which is provided by high voltage electrical discharges (HVED) and HPH techniques (Lovitt et al. 2000, Vorobiev and Lebovka 2008). The synergistic enhancement of protein release from yeasts can be attained using combined disruption techniques. It was shown that a combined HVED+HPH technique helped in reaching a high level of protein extraction at smaller pressure or number of passes through the homogenizer (Shynkaryk et al. 2009).

26.4 Processing of Plant Foods Assisted by Electroporation

26.4.1 Convective Drying

The drying processes consume an appreciable part of the total energy used in the food and bioprocessing industry (Chou and Chua 2001). Moreover, drying at elevated temperatures can produce undesirable changes in pigments, vitamins, and flavoring agents (Aguilera et al. 2003). Different drying-assisting techniques were discussed, such as microwave (Beaudry et al. 2003), electrohydrodynamic (Bajgai and Hashinaga 2001, Cao et al. 2004, Li et al. 2005) and osmotic pre-treatment (Chua et al. 2004), pre-treatment by chemical reagents, and ohmic heating (Zhong and Lima 2003, Salengke and Sastry 2005).

Electroporation was shown to be a promising tool for the enhancement of drying thermally sensitive foods (Ade-Omowaye et al. 2003, Lebovka et al. 2006, 2007b; Toepfl 2006, Shynkaryk 2007). The main idea is based on the fact that electrically induced disintegration and electroporation may facilitate the diffusion of the moisture inside a porous plant material and, thus, enhance its drying. The effect of both PEF treatment and ohmic preheating in electrically assisted drying on the processing time and energy consumption were investigated for different soft cellular tissues at the drying temperatures T_d within 30°C–70°C (Lebovka et al. 2006, 2007b; Shynkaryk 2007). The drying rate was shown to be accelerated for highly PEF disintegrated tissues.

In general, PEF pretreatment of apples, potatoes, and red beets at a moderate electric field $E = 400$ V/cm allowed for a reduction in the drying temperature by 20°C–25°C. PEF pretreatment resulted in a greater degree of tissue shrinkage and, hence, an increase in the rehydration time. However, the textural properties of the rehydrated samples with and without PEF treatment were seen to be similar. For potato tissue, the thermal pre-treatment of samples had practically no beneficial effect on the drying rate at high temperatures ($T = 70$°C), though the same thermal pre-treatment at mild temperatures ($T = 50$°C) increased the moisture effective diffusion coefficient D_e and gave an effect comparable with that for the PEF pre-treated samples (Lebovka et al. 2007b). Additionally, the spectral data obtained for the red beet tissue have demonstrated the benefits of electric-pretreatment for drying regimes in terms of colorant preservation, which seems to be promising for future industrial applications of this technique.

Figure 26.9 presents the effective diffusion coefficient D_e and PEF treatment time t_{PEF} versus conductivity disintegration index Z for the potato. The conductivity disintegration index Z was a growing function of PEF treatment time t_{PEF}: the high was t_{PEF}, the faster was the drying process. In fact, PEF treatment releases moisture from the damaged cells and enhances the transport processes, which results in an increase of the effective diffusion coefficient D_e and the drying rate. The observed nonlinear dependence between the diffusion coefficient D_e and Z was satisfactory approximated using a parallel model of diffusion, where the PEF-damaged tissue was represented as a mixture of intact and damaged cells with different diffusion coefficients (Lebovka et al. 2007b).

PEF pretreatment at high electric fields ($E = 0.5$–1.5 kV/cm) also allowed for significant acceleration of the drying rate of carrots at 70°C, and the most prominent parameter was the input energy per pulse (Gachovska et al. 2008). A similar behavior was observed for PEF-treated samples of okra at 4 kV/cm. Here, significant increases in the diffusivity coefficient and drying rate were observed (Adedeji et al. 2008). A comparative study on the effect of chemical, microwave, and PEF pretreatments on the drying process (at 65°C) and quality of raisins has shown that though the drying rate of the chemically pre-treated samples was the highest when compared with other methods of pretreatment, the PEF-treatment allows for obtaining of the samples with high total content of soluble solids along with their good market quality (Dev et al. 2008).

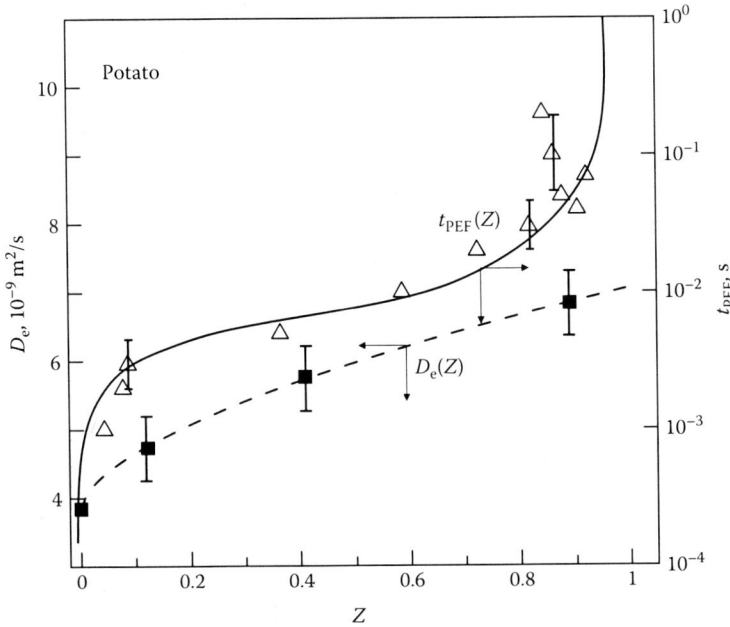

FIGURE 26.9 Effective diffusion coefficient D_e (closed squares) and PEF treatment time t_{PEF} (open triangles) versus conductivity disintegration index Z for potato. PEF treatment was done at the temperature of 298 K, electric field strength $E = 400$ V/cm, pulse duration $t_i = 10^{-3}$ s, pulse repetition time $\Delta t = 10^{-2}$ s. The D_e values were evaluated from the drying data obtained at drying temperature $T_d = 50°C$.

The efficiency of tissue drying can also be greatly influenced by temperature conditions (Lebovka et al. 2008). Figure 26.10 presents the evolutions of temperature inside apple tissue T_s versus dimensionless moisture content $\omega = (M - M_e)/(M_0 - M_e)$ for different drying temperature operational conditions. Here, M is the moisture content and the subscripts "0" and "e" refer to the initial and equilibrium (final) moisture contents, respectively. In mode I (Figure 26.10a), the inlet air temperature T_a was constant and temperature inside the sample T_s was an increasing function of time. In mode II (Figure 26.10b), the inlet air temperature T_a was adjusted in order to maintain the constancy of the temperature inside the sample T_s. Though the highest drying rate was always observed for the freeze-thawing pretreatment, this process is rather energy consuming and requires ≈280 kJ/kg (Toepfl and Knorr 2006). Better drying characteristics (drying rates and power consumptions) were observed for PEF pretreated samples and the drying mode provided a constant temperature inside the sample (mode II). For example, at $T_d = 50°C$, the effective diffusion coefficients were 5.6×10^{-9} m²/s (mode I) and 8.5×10^{-9} m²/s (mode II) for PEF treated apples (high disintegration index, $Z \approx 1$) as compared with 3.9×10^{-9} m²/s (mode I) and 4.4×10^{-9} m²/s (mode II) for untreated samples ($Z = 0$) (Lebovka et al. 2008).

26.4.2 Freezing and Freeze-Drying Behavior

Freezing is widely used for the preservation of food products because it results in minimal deterioration of the original color, flavor, texture, or nutritional values of the products. The quality of frozen food is considered to be inversely related to the extent of freezing-induced cellular dehydration, the size of the ice crystals, and their location inside the food (Delgado and Sun 2001, Li and Sun 2002). Moreover, freezing is often used as an intermediate stage of the freeze-drying technique for receiving high-quality dried materials. However, wide application of freeze-drying is still limited due to high energy consumption and long times of drying.

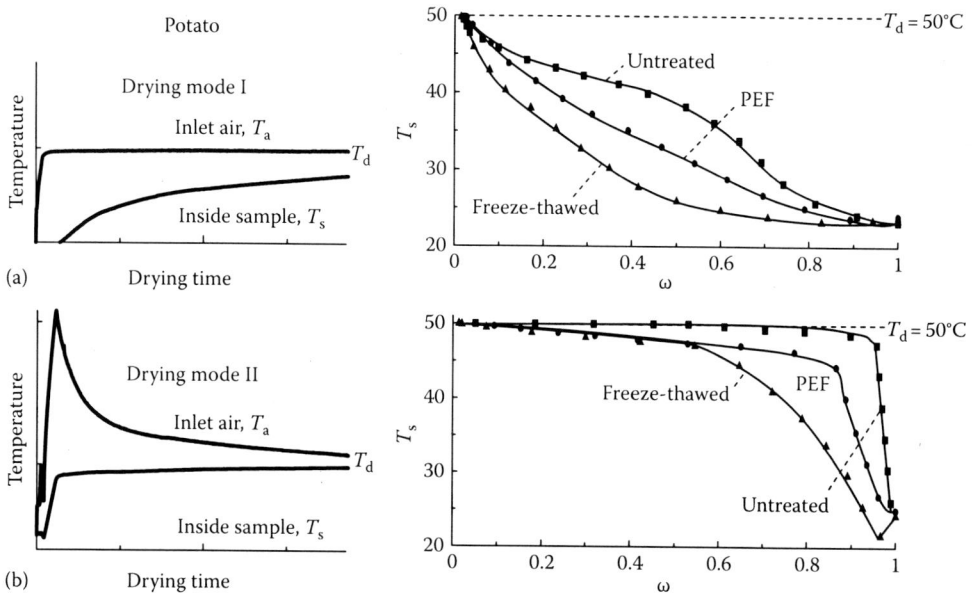

FIGURE 26.10 Effects of the drying modes on temperature inside the apple tissue T_s versus dimensionless moisture content ω for untreated, PEF, and freeze-thawed treated apple tissues. In mode I (a), the inlet air temperature T_a was constant and in mode II (b), the inlet air temperature T_a was adjusted in order to maintain the constancy of the temperature inside the sample T_s. PEF treatment was done to a high level of $Z \approx 0.9$ at the temperature of 298 K, electric field strength $E = 400$ V/cm, pulse duration $t_i = 10^{-3}$ s, pulse repetition time $\Delta t = 10^{-2}$ s. The final drying temperature of inlet air was $T_d = 50°C$.

PEF-induced electroporation may positively affect the freezing processes and its application seems to be promising for assisting in sub-zero processing of biomaterials. The release of cytoplasm in the PEF pre-treated tissues is expected to influence the freezing process, but may also result in a change of the spatial gradients (concentration and/or temperature) inside the sample. Moreover, removal of the membrane components makes the osmotic pre-treatment of tissues more efficient. For example, it was recently demonstrated that PEF application in combination with vacuum impregnation by trehalose drastically improved the freezing tolerance of the spinach leaves (Phoon et al. 2008).

The effects of PEF on the freezing, freeze-drying, and rehydration behavior of different vegetable tissues (potato, spinach leaves, green beans) were recently studied in detail Jalté et al. 2009, Ben Ammar et al. 2009. It was shown that electroporation results in a noticeable reduction of the freezing times, increased deformation of cells, and larger intercellular spaces in the PEF pre-treated samples. PEF pre-treatment also improved the rate of freeze-drying and the quality and rehydration of the samples (Jalté et al. 2009).

26.4.3 Pressing

Extraction by pressing, also known as solid-liquid expression, is widely used in the production of sugar, wine, fruit, and vegetable juices (Schwartzberg 1997). Pressing at a moderate pressure is usually insufficient for the effective rupture of cells and requires a long time of pressing.

Different pre-treatment techniques (fine grinding, thermal, and enzyme maceration) are widely used for the facilitation of pressing. However, their application can result in plant tissue degradation and juice pollution. In the last decade, many examples of electroporation-assisted pressing were demonstrated (Eshtiaghi and Knorr 1999, Bazhal 2001, Bazhal et al. 2001, Bouzrara and Vorobiev 2000, 2001, 2003, Bouzrara 2001, Jemai and Vorobiev 2002, 2006, Lebovka et al. 2003, Chalermchatand and Dejmek 2005, Praporscic 2005, Praporscic et al. 2005, Toepfl 2006).

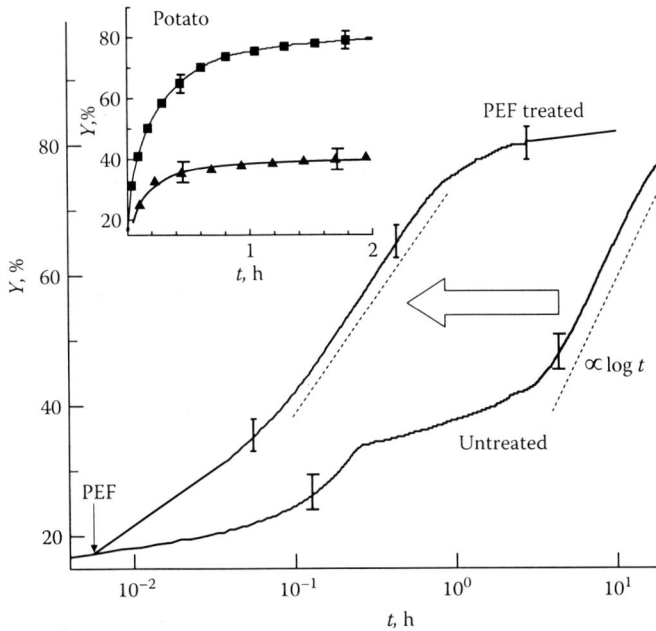

FIGURE 26.11 Typical consolidation curves $Y(t)$ at constant pressure of 5 bars for untreated and PEF treated potato slices. PEF pretreatment was done at 20 s of pressing using the following parameters $E = 200\,V/cm$, $t_i = 100\,\mu s$, $\Delta t = 10\,ms$, and $t_{PEF} = 3\,s$.

Figure 26.11 presents typical juice yield Y versus pressing time t for untreated and PEF-treated potato slices (Lebovka et al. 2003). Normally, different consolidation periods were observed for untreated tissues and consolidation became logarithmically long after a certain transition time on the order of 2–3 h, i.e., $Y \propto \ln t$. The time of industrial pressing is usually restricted by 1–2 h and juice yield Y, obtained at such time of pressing, is rather limited; thus, for potato it is below 40% (see inset in Figure 26.11). PEF application allows for the shifting of the logarithmically long stage toward shorter times; as a result, the juice yield during the first 1–2 h of pressing becomes noticeably larger; it reaches the order of 80% for potato (Lebovka et al. 2003). The relation between compressing behavior and the textural properties of PEF-treated potatoes in the pressure range within 0.5 to 4 MPa were recently discussed (Grimi et al. 2009b). The fracture pressure and dependencies of deformation versus time and electrical conductivity versus deformation were analyzed for different levels of PEF-induced disintegration.

The methodology of the PEF-assisted pressing was successfully tested for different plant tissues, such as apple (Bazhal and Vorobiev 2000); carrot (Bouzrara 2001); sugar beet (Eshtiaghi and Knorr 1999, Bouzrara and Vorobiev 2000, 2001, 2003, Bouzrara 2001, Jemai and Vorobiev 2002, 2006, Praporscic et al. 2005); spinach (Bouzrara 2001); haricots, topinambours, and red cabbages (Vorobiev et al. 2002); potatoes and onions (Vorobiev et al. 2004); artichokes (Marchal et al. 2004); and grapes (Praporscic et al. 2007a, Grimi et al. 2009a).

The different regimes of pressing involving PEF pretreatment before pressing, initial pressurization, and intermediate PEF treatment, with both constant and variable pressures (studied within the range of 0.5–10 bars) and slices of different sizes, were carefully tested in order to optimize juice yield and quality as well as energy consumption (Bouzrara and Vorobiev 2000, 2001, Praporscic et al. 2005). The energy input, required for efficient intermediate PEF treatment of the pressurized slices, was rather insignificant on the order of 2–5 kJ/kg of raw material (Bazhal and Vorobiev 2000, Bouzrara 2001). The plant tissue conditioning by mild heating at 45°C–50°C resulted in its softening and influenced the textural properties of foods (apples, carrots, potatoes). Lebovka et al. (2004a,b) have demonstrated that such conditioning enhances expression kinetics. A combination of thermal and PEF pre-treatments clearly demonstrates the synergetic effect on tissue damage efficiency (Vorobiev and Lebovka 2006). Praporscic et al. (2005, 2006) have compared the effects of ohmic heating and PEF treatments on kinetics of juice expression from the sugar beet gratings. A remarkable synergetic effect was observed when the sugar beet tissue was conditioned by ohmic heating with further PEF treatment (Praporscic et al. 2005).

The PEF-assisted pressing also allowed for the improvement of qualitative characteristics of the expressed juices (Bouzrara and Vorobiev 2001, 2003, Vorobiev et al. 2005, Toepfl 2006, Praporscic et al. 2007b). Systematic investigations of the PEF-assisted pressing of sugar beet gratings have demonstrated an enhancement in the purity of the extracted juices and sugar concentration (Bouzrara and Vorobiev 2001), moreover, sugar crystals, obtained from PEF-extracted juice, were less colored that sugar crystals obtained from the factory juice (Jemai and Vorobiev 2006).

An improvement in the juice pressing from the grapes (Muscadelle, Sauvignon, and Semillon) was reported (Praporscic et al. 2007a), and it was shown that in this case an intermediate PEF application with field strength of $E = 750$ V/cm and the total treatment duration $t_{PEF} = 0.3$ s resulted in a substantial increase of the juice yield and a decrease of the juice absorbance and turbidity. Two regimes of PEF-assisted juice extraction from Chardonnay white grapes were compared: one with constant (1.0 bar) pressure and the other with a progressive pressure increase (from 0 to 1 bar) (Grimi et al. 2009a). The PEF treatment at $E = 400$ V/cm and $t_{PEF} = 0.1$ s resulted in a juice yield increase from 67% to 75% (1 h of pressing). It was shown that no significant effect of PEF treatment on turbidity and content of poly-phenols was observed for the constant pressure regime, while the progressive pressure increase regime (Grimi et al. 2009a) resulted in an elevation of the content of polyphenols (by more than 15%).

Different combinations of the PEF-assisted pressing and washing operations allowed for the regulation of the selective extraction of the valuable components from plant tissues (Grimi et al. 2007). It was shown that the application of the PEF-assisted washing-pressing or pressing-washing-pressing procedures to large slices of carrot allowed for the retention of high concentrations of carotenoids inside the press cake, though the second procedure released the water-soluble components (including sugars) into the juice more efficiently. The obtained "sugar-free" concentrate of carrot was rich in vitamins and carotenoids and could be used as an additive in diet foods.

26.4.4 Accelerated Diffusion and Solute Extraction

The solute extraction is a very important unit operation in such industrial applications as the extraction of sugar, vegetable oils, natural colorants, aromas, and other valuable cell components (Schwartzberg and Chao 1982). Nonthermal membrane breakdown under the PEF treatment allows for the optimization of the process and a decrease in the diffusion temperature and power consumption (Jemai and Vorobiev 2002, 2003, El-Belghiti and Vorobiev 2004, 2005a,b, Fincan et al. 2004, El-Belghiti et al. 2005, Corrales et al. 2008).

For example, the thermal treatment of the beet tissue at 70°C–74°C is used in industrial sugar processing. Such an elevated temperature causes overheating of the cell walls, the release of polluting substances, and the development of hydrolytic degradation reactions affecting the purity of the juice. As a result, the amount of pectin passing into the juice increases sharply with a rise in temperature, which complicates considerably the sugar juice purification (Van der Poel et al. 1998). PEF processing allows us to overcome these difficult circumstances accompanying high temperature diffusion. Lebovka et al. (2007a) have compared the kinetics of thermal diffusion and diffusion coefficients D_e for untreated and PEF-treated at $E = 100–400$ V/cm sugar beet slices. At 70°C, the values of D_e were nearly the same for both untreated and PEF-pretreated slices ($1–1.5 \times 10^{-9}$ m²/s). The activation energy for the PEF-treated slices was $W = 21 \pm 2$ kJ/mol, which is close to the activation energy of sugar in the aqueous solutions, $W \approx 22$ kJ/mol (Lysjanskii 1973). However, the activation energy for the untreated slices was noticeably higher, $W = 75 \pm 5$ kJ/mol, which possibly reflects the interrelations of the restricted diffusion and thermally induced damage effects in the untreated sugar beet tissue. The sugar diffusivity inside the PEF-treated tissue remained rather high even at moderate temperatures, and the value of D_e was nearly the same for the PEF-treated tissue at 50°C and for the tissue not treated by PEF at 70°C. Moreover, cold diffusion applied to the PEF-treated tissue resulted in higher juice purity (Lebovka et al. 2007a). Comparison of the processes of static and centrifugal aqueous extraction (at different moderate temperatures $T = 25°C–50°C$) from PEF-treated coarse sugar beet slices (1.5 mm in thickness) was done (El-Belghiti 2005, El-Belghiti and Vorobiev 2005b, El-Belghiti et al. 2005). It was shown that extraction kinetics was much faster above the acceleration threshold (150 g). At such centrifugal acceleration of a highly disintegrated tissue, $Z \approx 1$, the solute concentration of 97% was reached after 25 min of aqueous extraction at 25°C and just after 15 min of aqueous extraction at 35°C. The impact of PEF treatment at higher electric fields ($E = 1–7$ kV/cm, pulse width of 2–5 µs) on enhancement of the solid-liquid extraction of sucrose from the sugar beet ($T = 20°C–70°C$) was also studied recently (López et al. 2009a). It was demonstrated that the effect of the field strength was higher at lower temperatures and the sucrose yield increased with field strength, time of extraction, and temperature. The PEF treatment of red beetroots by 2 µs pulses at 7 kV/cm resulted in the highest betanine yield and rate of its extraction using extracting media with pH 3.5 at 30°C (López et al. 2009b).

Qualitatively similar accelerated diffusion results were also reported for other plant materials, such as apple (Jemai and Vorobiev 2002), carrot (El-Belghiti and Vorobiev 2005b), fennel (El-Belghiti et al. 2008), chicory (Loginova et al. 2010), alfalfa (Gachovska et al. 2006, 2009), red beetroot (Fincan et al. 2004, López et al. 2009b), rapeseed (Guderjan et al. 2007), grape (López et al. 2008a,b, Corrales et al. 2008, Boussetta et al. 2009), and red raspberry (Luo et al. 2008, 2009).

It was demonstrated that fine (0.5 mm thick) and coarse (1.5 mm thick) carrot slices had almost the same extraction kinetics after the PEF treatment (El-Belghiti and Vorobiev 2005b). Such results confirm the attractiveness of PEF treatment, especially for the coarse particles. For very fine grinding of the fennel tissue, when all the cells were broken mechanically, PEF application did not give any additional acceleration of the extraction process, however, extraction from the coarse slices was considerably improved after the optimal PEF treatment (El-Belghiti et al. 2008). Moreover, the PEF treatment of coarse slices allowed for obtaining high purity extracts with smaller turbidity than extracts from thin slices.

PEF treatment of alfalfa mash, obtained by the blending of alfalfa leaves, increased the extraction of juice with the improved dry matter, protein, and mineral content (Gachovska et al. 2006, 2009). The PEF-assisted processing with an electric field of high intensity was shown to be effective for the extraction of betulin (non-polar compound) (at $E = 40$ kV/cm) (Yin et al. 2008a) and polysaccharide (at $E = 30$ kV/cm) (Yin et al. 2008b). PEF application in the solvent extraction processing of rapeseed (*Brassica napus*) allowed for an increase in the rapeseed oil yield and content of functional food ingredients in the rapeseed oil (Guderjan et al. 2007). The positive effects of PEF on the extraction of phenolic compounds from grapes (Corrales et al. 2008, López et al. 2008a,b, Boussetta et al. 2009) and the improvement of anthocyanins composition in extracts from grape (López et al. 2008b) and from red raspberry (Luo et al. 2008, 2009) were also recently reported.

26.5 Conclusions

Our days gave rise to many examples of electroporation applications for the processing of food and biomaterials (e.g., for acceleration of drying, freezing, pressing, diffusion, and selective extraction). The electroporation-assisted techniques demonstrated high efficiency and they were more advantageous as compared with other techniques that may result in deterioration of the product quality. However, the successful industrial implementation of electroporation still requires more fundamental knowledge of many complex phenomena related to the impact of electric fields on biomaterials.

Acknowledgments

The authors would like to thank the "Pole Regional Génie des Procédés" (Picardie, France) for the provision of financial support. The authors also thank Dr. N. S. Pivovarova for her help with the preparation of the manuscript.

References

Abram F, Smelt JPPM, Bos R, Wouters PC. 2003. Modelling and optimization of inactivation of *Lactobacillus plantarum* by pulsed electric field treatment. *J Appl Microbiol* 94: 571–579.

Adedeji AA, Gachovska TK, Ngadi MO, Raghavan GSV. 2008. Effect of pretreatments on drying characteristics of okra. *Drying Technol* 26: 1251–1256.

Ade-Omowaye BI, Rastogi NK, Angersbach A, Knorr D. 2003. Combined effects of pulsed electric field pre-treatment and partial osmotic dehydration on air drying behaviour of red bell pepper. *J Food Eng* 60: 89–98.

Agarwal A, Zudans I, Weber EA, Olofsson J, Orwar O, Stephen G, Weber SG. 2007. Effect of cell size and shape on single-cell electroporation. *Anal Chem* 79: 3589–3596.

Aguilera JM, Chiralt A, Fito P. 2003. Food dehydration and product structure. *Trends Food Sci Technol* 14: 432–437.

Amiali M. 2006. Inactivation of *Escherichia coli* O157:H7 and *Salmonella enteritidis* in liquid egg products using pulsed electric field. PhD thesis, Montreal, Quebec, Canada: McGill University.

Anderson AK, Finkelstein R. 1919. A study of electro-pure process of treating milk. *J Diary Sci* 2: 374–406.

Angersbach A, Heinz V, Knorr D. 2002. Evaluation of process-induced dimensional changes in the membrane structure of biological cells using impedance measurement. *Biotechnol Prog* 18: 597–603.

Anglim TH. 1923. Method and apparatus for pasteurization milk. U.S. Patent 1,468,871, United States Patent and Trademark Office, Washington, DC.

Archie GE. 1942. The electrical resistivity log as an aid in determining some reservoir characteristics. *Trans AIME* 146: 54–62.

Aronsson K, Lindgren M, Johansson BR, Rönner U. 2001. Inactivation of microorganisms using pulsed electric fields: The influence of process parameters on *Escherichia coli, Listeria innocua, Leuconostoc mesenteroides* and *Saccharomyces cerevisiae. Innov Food Sci Emerg Technol* 2: 41–54.

Aronsson K, Rönner U, Borch E. 2005. Inactivation of *Escherichia coli, Listeria innocua* and *Saccharomyces cerevisiae* in relation to membrane permeabilization and subsequent leakage of intracellular compounds due to pulsed electric field processing. *Int J Food Microbiol* 99: 19–32.

Ball CO. 1937. Apparatus for pasteurization milk. U.S. Patent 2,091,263, United States Patent and Trademark Office, Washington, DC.

Bajgai TR, Hashinaga F. 2001. High electric field drying of Japanese radish. *Drying Technol* 19: 2291–2302.

Barbosa-Cánovas GV, Pothakamury UR, Palou E, Swanson B. 1998. *Nonthermal Preservation of Foods.* New York: Marcel Dekker.

Bazhal IG, Gulyi IS, Bobrovnik LD. 1983. Extraction of sugar from sugar beet in a constant electric field. *Izvestija vuzov: Pischevye technologii* (*Proc. Inst. Higher Ed. USSR: Food Ind.*) N5: 49–51 (in Russian).

Bazhal M. 2001. Etude du mécanisme d'électropermeabilisation des tissus végétaux. Application à l'extraction du jus des pommes. PhD thesis, Compiègne, France: UTC.

Bazhal MI, Vorobiev EI. 2000. Electrical treatment of apple cossettes for intensifying juice pressing. *J Sci Food Agric* 80: 1668–1674.

Bazhal MI, Lebovka NI, Vorobiev EI. 2001. Pulsed electric field treatment of apple tissue during compression for juice extraction. *J Food Eng* 50: 129–139.

Bazhal MI, Lebovka NI, Vorobiev EI. 2003. Optimisation of pulsed electric field strength for electroplasmolysis of vegetable tissues. *Biosyst Eng* 86: 339–345.

Bazhal MI, Ngadi MO, Raghavan GSV, Smith JP. 2006. Inactivation of *Escherichia coli* O157:H7 in liquid whole egg using combined pulsed electric field and thermal treatments. *LWT—Food Sci Technol* 39: 420–426.

Beattie JM. 1914. Electrical treatment of milk for infant feeding. *Br J State Med* 24: 97–113.

Beaudry C, Raghavan GSV, Rennie TJ. 2003. Microwave finish drying of osmotically dehydrated cranberries. *Drying Technol* 21: 1797–1810.

Ben Ammar J, Van Hecke E, Lebovka N, Vorobiev E, Lanoisellé J-L. 2009. Pulsed electric fields improve freezing process. In: Vorobiev E, Lebovka N, Van Hecke E, Lanoisellé J-L (eds.), *Proceeding of BFE*, UTC, Compiègne, France, C16, pp. 1–6.

Bergey L. 1986–1989. *Manual of Systematic Bacteriology*, Vols. 1–4. Baltimore, MD: Williams & Wilkins.

Bernhardt J, Pauly H. 1973. On the generation of potential differences across the membranes of ellipsoidal cells in an alternating electrical field. *Biophysik* 10: 89–98.

Boussetta N, Lebovka NI, Vorobiev EI, Adenier H, Bedel-Cloutour C, Lanoisellé J-L. 2009. Electrically assisted extraction of soluble matter from chardonnay grape skins for polyphenol recovery. *J Agric Food Chem* 57: 1491–1497.

Bouzrara H, Vorobiev EI. 2000. Beet juice extraction by pressing and pulsed electric fields. *Int Sugar J* 102: 194–200.

Bouzrara H. 2001. Amélioration du pressage de produits végétaux par Champ Electrique Pulsé. Cas de la betterave à sucre. PhD thesis, Compiègne, France: UTC.

Bouzrara H, Vorobiev EI. 2001. Non-thermal pressing and washing of fresh sugarbeet cossettes combined with a pulsed electrical field. *Zucker* 126: 463–466.

Bouzrara H, Vorobiev EI. 2003. Solid/liquid expression of cellular materials enhanced by pulsed electric field. *Chem Eng Process* 42: 249–257.

Brookman JSG. 1974. Mechanism of cell disintegration in a high pressure homogenizer. *Biotechnol Bioeng* 16: 371–383.

Calleja GB. 1984. *Microbial Aggregation*. Boca Raton, FL: CRC Press.

Cao W, Nishiyama Y, Koide S., Lu ZH. 2004. Drying enhancement of rough rice by an electric field. *Biosyst Eng* 87: 445–451.

Canatella PJ, Karr JF, Petros JA, Prausnitz MR. 2001. Quantitative study of electroporation mediated uptake and cell viability. *Biophys J* 80: 755–764.

Canatella PJ, Black MM, Bonnichsen DM, McKenna C, Prausnitz MR. 2004. Tissue electroporation: Quantification and analysis of heterogeneous transport in multicellular environments. *Biophys J* 86: 3260–3268.

Chalermchatand Y, Dejmek P. 2005. Effect of pulsed electric field pretreatment on solid–liquid expression from potato tissue. *J Food Eng* 71: 164–169.

Chang DC. 1989. Cell poration and cell fusion using an oscillating electric field. *Biophys J* 56: 641–652.

Chen C, Smye SW, Robinson MP, Evans JA. 2006. Membrane electroporation theories: A review. *Med Biol Eng Comput* 44: 5–14.

Chou SK, Chua KJ. 2001. New hybrid drying technologies for heat sensitive foodstuffs. *Trends Food Sci Technol* 12: 359–369.

Chua KJ, Chou SK, Mujumdar AS, Ho JC, Hon CK. 2004. Radiant-convective drying of osmotic treated agro-products: effect on drying kinetics and product quality. *Food Control* 15: 145–158.

Corrales M, Toepfl S, Butz P, Knorr D, Tauscher B. 2008. Extraction of anthocyanins from grape by-products assisted by ultrasonic, high hydrostatic pressure or pulsed electric fields: A comparison. *Innov Food Sci Emerg Technol* 9: 85–91.

Delgado AE, Sun D-W. 2001. Heat and mass transfer models for predicting freezing process––A review. *J Food Eng* 47: 157–174.

De Vito F, Ferrari G, Lebovka NI, Shynkaryk NV, Vorobiev E. 2008. Pulse duration and efficiency of soft cellular tissue disintegration by pulsed electric fields. *Food Bioprocess Technol* 1: 307–313.

Dev SRS, Padmini T, Adedeji A, Gariépy Y, Raghavan GSV. 2008. A comparative study on the effect of chemical, microwave, and pulsed electric pretreatments on convective drying and quality of raisins. *Drying Technol* 26: 1238–1243.

Diederich A, Bhr G, Winterhalter M. 1998. Influence of polylysine on the rupture of negatively charged membranes. *Langmuir* 14: 4597–4605.

Dimitrov DS, Sowers AE. 1990. Membrane electroporation—Fast molecular exchange by electroosmosis. *Biochim Biophys Acta* 1022: 381–392.

Doevenspeck H. 1961. Influencing cells and cell walls by electrostatic impulses. *Fleischwirtschaft* 13: 968–987.

El-Belghiti K. 2005. Effets d'un champ électrique pulsé sur le transfert de matière et sur les caractéristiques végétales. PhD thesis, Compiègne, France: UTC.

El-Belghiti K, Vorobiev EI. 2004. Mass transfer of sugar from beets enhanced by pulsed electric field. *Trans IChemE* 82: 226–230.

El-Belghiti K, Vorobiev EI. 2005a. Kinetic model of sugar diffusion from sugar beet tissue treated by pulsed electric field. *J Sci Food Agric* 85: 213–218.

El-Belghiti K, Vorobiev EI. 2005b. Modelling of solute aqueous extraction from carrots subjected to a pulsed electric field pre-treatment. *Biosyst Eng* 90: 289–294.

El-Belghiti K, Rabhi Z, Vorobiev EI. 2005. Effect of the centrifugal force on the aqueous extraction of solute from sugar beet tissue pretreated by a pulsed electric field. *J Food Process Eng* 28: 346–358.

El-Belghiti K, Moubarik A, Vorobiev EI. 2008. Aqueous extraction of solutes from fennel (foeniculum vulgare) assisted by pulsed electric field. *J Food Process Eng* 31: 548–563.

El-Mashak EM, Tsong TY. 1985. Ion selectivity of temperature-induced and electric field induced pores in dipalmitoylphosphatidylcholine vesicles. *Biochem* 24: 2884–2888.

El Zakhem H, Lanoisellé J-L, Lebovka NI, Nonus M, Vorobiev EI. 2006a. Behavior of yeast cells in aqueous suspension affected by pulsed electric field. *J Colloid Interface Sci* 300: 553–563.

El Zakhem H, Lanoisellé J-L, Lebovka NI, Nonus M, Vorobiev EI. 2006b. The early stages of *Saccharomyces cerevisiae* yeast suspensions damage in moderate pulsed electric fields. *Colloids Surf* B47: 189–197.

El Zakhem H, Lanoisellé J-L, Lebovka NI, Nonus M, Vorobiev EI. 2007. Influence of temperature and surfactant on *Escherichia coli* inactivation in aqueous suspensions treated by moderate pulsed electric fields. *Int J Food Microbiol* 120: 259–265.

Engler CR. 1985. Disruption of microbial cells in comprehensive biotechnology. In: Moo-Young M (ed.), *Comprehensive Biotechnology*. Oxford, U.K.: Pergamon Press.

Eshtiaghi MN, Knorr D. 1999. Method for treating sugar beet. International Patent. Patent WO 99/6434.

Evrendilek GA, Zhang QH. 2005. Effects of pulse polarity and pulse delaying time on pulsed electric fields-induced pasteurization of *E. coli* O157:H7. *J Food Eng* 68: 271–276.

Exerova D, Nikolova A. 1992. Phase transitions in phospholipid foam bilayers. *Langmuir* 8: 3102–3108.

Eynard N, Sixou S, Duran N, Teissié J. 1992. Fast kinetics studies of *Escherichia coli* electrotransformation. *Eur J Biochem* 209: 431–436.

Eynard N, Rodriguez F, Trotard J, Teissié J. 1998. Electrooptics studies of *Escherichia coli* electropulsation: Orientation, permeabilization, and gene transfer. *Biophys J* 75: 2587–2596.

Fettermann JC. 1928. Electrical conductivity method of processing milk. *Agric Eng* 9: 107.

Fincan M, Dejmek P. 2002. In situ visualization of the effect of a pulsed electric field on plant tissue. *J Food Eng* 55: 223–230.

Fincan M, De Vito F, Dejmek P. 2004. Pulsed electric field treatment for solid–liquid extraction of red beetroot pigment. *J Food Eng* 64: 381–388.

Flaumenbaum BL. 1949. Electrical treatment of fruits and vegetables before extraction of juice. *Trudy OTIKP* 3: 15–20 (in Russian).

Flaumenbaum BL. 1968. Anwendung der Elektroplasmolyse bei der Herstellung von Fruchtsäften. *Flüssiges Obst* 35: 19–22.

Fricke H. 1953. The electric permittivity of a dilute suspension of membrane-covered ellipsoids. *J Appl Phys* 24: 644–646.

Fox MB, Esveld DC, Mastwijk H, Boom RM. 2008. Inactivation of *L. plantarum* in a PEF microreactor. The effect of pulse width and temperature on the inactivation. *Innov Food Sci Emerg Technol* 9: 101–108.

Gachovska TK, Ngadi MO, Raghavan GSV. 2006. Pulsed electric field assisted juice extraction from alfalfa. *Can Biosyst Eng/Le Genie des biosystems au Canada* 48: 3.33–3.37.

Gachovska TK, Adedeji AA, Ngadi M, Raghavan GVS. 2008. Drying characteristics of pulsed electric field-treated carrot. *Drying Technol* 26: 1244–1250.

Gachovska TK, Adedeji AA, Ngadi MO. 2009. Influence of pulsed electric field energy on the damage degree in alfalfa tissue. *J Food Eng* 95: 558–563.

Ganeva V, Galutzov B, Teissié J. 2003. High yield electroextraction of proteins from yeast by a flow process. *Anal Biochem* 315: 77–84.

Getchell BE. 1935. Electric pasteurization of milk. *Agric Eng* 16: 408–410.

Gimsa J, Wachner D. 2001. Analytical description of the transmembrane voltage induced on arbitrarily oriented ellipsoidal and cylindrical cells. *Biophys J* 81: 1888–1896.

Grahl T, Märkl H. 1996. Killing of microorganisms by pulsed electric fields. *Appl Microbiol Biotechnol* 45: 148–157.

Grimi N, Praporscic I, Lebovka N, Vorobiev E. 2007. Selective extraction from carrot slices by pressing and washing enhanced by pulsed electric fields. *Sep Purif Technol* 58: 267–273.

Grimi N, Lebovka NI, Vorobiev E, Vaxelaire J. 2009a. Effect of a pulsed electric field treatment on expression behavior and juice quality of Chardonnay grape. *Food Biophys* 4: 191–198.

Grimi N, Lebovka NI, Vorobiev E, Vaxelaire J. 2009b. Compressing behavior and texture evaluation for potatoes pretreated by pulsed electric field. *J Texture Stud* 40: 208–224.

Grimi N, Mamouni F, Lebovka NI, Vorobiev E, Vaxelaire J. 2010. Acoustic impulse response in plant tissues treated by pulsed electric field. *Biosyst Eng* 105: 266–272.

Guderjan M, Elez-Martínez P, Knorr D. 2007 Application of pulsed electric fields at oil yield and content of functional food ingredients at the production of rapeseed oil. *Innov Food Sci Emerg Technol* 8: 55–62.

Gulyi IS, Lebovka NI, Mank VV, Kupchik MP, Bazhal MI, Matvienko AB, Papchenko AY. 1994. *Scientific and Practical Principles of Electrical Treatment of Food Products and Materials.* Kiev, Ukraine: UkrINTEI (in Russian).

Harrison STL, Dennis JS, Chase HA. 1991. Combined chemical and mechanical processes for the disruption of bacteria. *Bioseparation* 2: 95–105.

Hart FX, Marino AA. 1986. Penetration of electric fields into a concentric-sphere model of biological tissue. *Med Biol Eng Comput* 24: 105–108.

Ho SY, Mittal GS. 1996. Electroporation of cell membranes: A review. *Crit Rev Biotechnol* 16: 349–362.

Jalté M, Lanoisellé J-L, Lebovka NI, Vorobiev E. 2009. Freezing of potato tissue pre-treated by pulsed electric fields. *LWT—Food Sci Technol* 42: 576–580.

Jemai AB, Vorobiev E. 2001. Enhancement of the diffusion characteristics of apple slices due to moderate electric field pulses (MEFP). In: Welti-Chanes J, Barbosa-Cánovas GV, Aguilera JM (eds.), *Proceedings of the 8th International Congress on Engineering and Food*, Puebla, Mexico. Lancaster, PA: Technomic Publishing Co., pp. 1504–1508.

Jemai AB, Vorobiev E. 2002. Effect of moderate electric field pulse (MEFP) on the diffusion coefficient of soluble substances from apple slices. *Int J Food Sci Technol* 37: 73–86.

Jemai AB, Vorobiev E. 2003. Enhancing leaching from sugar beet cossettes by pulsed electric field. *J Food Eng* 59: 405–412.

Jemai AB, Vorobiev E. 2006. Pulsed electric field assisted pressing of sugar beet slices: Towards a novel process of cold juice extraction. *Biosyst Eng* 93: 57–68.

Jen CP, Chen YH, Fan CS, Yeh CS, Lin YC, Shieh DB, Wu CL, Chen DH, Chou CH. 2004. A nonviral transfection approach in vitro: The design of a gold nanoparticle vector joint with microelectromechanical systems. *Langmuir* 20: 1369–1374.

Jones F. 1897. Apparatus for electrically treating liquids. U.S. Patent 592,735, United States Patent and Trademark Office, Washington, DC.

Joshi RP, Mishra A, Schoenbach KH. 2008. Model assessment of cell membrane breakdown in clusters and tissues under high-intensity electric pulsing. *IEEE Trans Plasma Sci* 36: 1680–1688.

Kandušer M, Šentjurc M, Miklavčič D. 2008. The temperature effect during pulse application on cell membrane fluidity and permeabilization. *Bioelectrochemistry* 74: 52–57.

Kinosita K Jr, Tsong TY. 1977. Hemolysis of human erythrocytes by transient electric field. *Proc Natl Acad Sci USA* 74: 1923–1927.

Knorr D, Angersbach A, Eshtiaghi MN, Heinz V, Lee D-U. 2001. Processing concepts based on high intensity electric field pulses. *Trends Food Sci Technol* 12: 129–135.

Kogan FI. 1968. Electrophysical methods in canning technologies of foodstuff. Kiev: Tehnika (in Russian).

Kotnik T, Miklavčič D. 2000. Analytical description of transmembrane voltage induced by electric fields on spheroidal cells. *Biophys J* 79: 670–679.

Kotnik T, Miklavčič D, Slivnik T. 1998. Time course of transmembrane voltage induced by time-varying electric fields: A method for theoretical analysis and its application. *Bioelectrochem Bioenerg* 45: 3–16.

Lebedeva NE. 1987. Electric breakdown of bilayer lipid membranes at short times of voltage effect. *Biol Membr* 4: 994–998 (in Russian).

Lebovka NI, Vorobiev EI. 2004. On the origin of the deviation from the first-order kinetics in inactivation of microbial cells by pulsed electric fields. *Int J Food Microbiol* 91: 83–89.

Lebovka NI, Vorobiev E. 2007. The kinetics of inactivation of spheroidal microbial cells by pulsed electric fields. E-print arXiv:0704.2750v1, 1–22.

Lebovka NI, Bazhal MI, Vorobiev EI. 2000a. Simulation and experimental investigation of food material breakage using pulsed electric field treatment. *J Food Eng* 44: 213–223.

Lebovka NI, Melnyk RM, Kupchik MP, Bazhal MI, Serebrjakov RA. 2000b. Local generation of ohmic heat on cellular membranes during the electrical treatment of biological tissues. *Sci Notes NaUKMA*, 4: 54–60 (in Ukrainian).

Lebovka NI, Bazhal MI, Vorobiev EI. 2001. Pulsed electric field breakage of cellular tissues: Visualization of percolative properties. *Innov Food Sci Emerg Technol* 2: 113–125.

Lebovka NI, Bazhal MI, Vorobiev EI. 2002. Estimation of characteristic damage time of food materials in pulsed-electric fields. *J Food Eng* 54: 337–346.

Lebovka NI, Praporscic I, Vorobiev EI. 2003. Enhanced expression of juice from soft vegetable tissues by pulsed electric fields: Consolidation stages analysis. *J Food Eng* 59: 309–317.

Lebovka NI, Praporscic I, Vorobiev EI. 2004a. Combined treatment of apples by pulsed electric fields and by heating at moderate temperature. *J Food Eng* 65: 211–217.

Lebovka NI, Praporscic I, Vorobiev EI. 2004b. Effect of moderate thermal and pulsed electric field treatments on textural properties of carrots, potatoes and apples. *Innov Food Sci Emerg Technol* 5: 9–16.

Lebovka NI, Praporscic I, Ghnimi S, Vorobiev E. 2005a. Temperature enhanced electroporation under the pulsed electric field treatment of food tissue. *J Food Eng* 69: 177–184.

Lebovka NI, Praporscic I, Ghnimi S, Vorobiev E. 2005b. Does electroporation occur during the ohmic heating of food. *J Food Sci* 70: 308–311.

Lebovka NI, Shynkaryk MV, Vorobiev E. 2006. Drying of potato tissue pretreated by ohmic heating. *Drying Technol* 24: 1–11.

Lebovka NI, Shynkaryk MV, El-Belghiti K, Benjelloun H, Vorobiev EI. 2007a. Plasmolysis of sugarbeet: Pulsed electric fields and thermal treatment. *J Food Eng* 80: 639–644.

Lebovka NI, Shynkaryk NV, Vorobiev EI. 2007b. Pulsed electric field enhanced drying of potato tissue. *J Food Eng* 78: 606–613.

Lebovka NI, Shynkaryk NV, Vorobiev EI. 2008. Electrically assisted drying of food plants at different temperature operational conditions. In: *ICEF 10: International Congress of Engineering and Food.* Viña del Mar, Chile. http://www.icef10.com/

Li B, Sun D-W. 2002. Novel methods for rapid freezing and thawing of foods—A review. *J Food Eng* 54: 175–182.

Li F-D, Li L-T, Sun J-F, Tatsumi E. 2005. Electrohydrodynamic (EHD) drying characteristic of okara cake. *Drying Technol* 23: 565–580.

Loginova KV, Shynkaryk MV, Lebovka NI, Vorobiev EI. 2010 Acceleration of soluble matter extraction from chicory with pulsed electric fields. *J Food Eng* 96: 374–379.

López N, Puértolas E, Condón S, Álvarez I, Raso J. 2008a. Effects of pulsed electric fields on the extraction of phenolic compounds during the fermentation of must of Tempranillo grapes. *Innov Food Sci Emerg Technol* 9: 477–482.

López N, Puértolas E, Condón S, Álvarez I, Raso J. 2008b. Application of pulsed electric fields for improving the maceration process during vinification of red wine: Influence of grape variety. *Eur Food Res Technol* 227: 1099–1107.

López N, Puértolas E, Condón S, Raso J, Álvarez I. 2009a. Enhancement of the solid–liquid extraction of sucrose from sugar beet (*Beta vulgaris*) by pulsed electric fields. *LWT—Food Sci Technol* 42: 1674–1680.

López N, Puértolas E, Condón S, Raso J, Alvarez I. 2009b. Enhancement of the extraction of betanine from red beetroot by pulsed electric fields. *J Food Eng* 90: 60–66.

Lovitt RW, Jones M, Collins SE, Coss GM, Yau CP, Attouch C. 2000. Disruption of baker's yeast using a disrupter of simple and novel geometry. *Process Biochem* 36: 415–421.

Luo W, Zhang R, Wang L, Chen J, Guan Z, Liao X, Mo M. 2008. Effects of PEF-assisted extraction of anthocyanin in red raspberry. *Annual Report—Conference on Electrical Insulation and Dielectric Phenomena*, CEIDP 4772923, pp. 630–632.

Luo W, Zhang R-B, Wang L-M, Chen J, Liao X-J, Mo M-B, Guan Z-C. 2009. Effect of PEF on extraction of anthocyanin. *Gaodianya Jishu/High Voltage Eng* 35: 1430–1433.

Lysjanskii VM. 1973. *The Extraction Process of Sugar from Sugarbeet: Theory and Calculations (Process ekstractzii sahara iz svekly: teorija i raschet)*. Moscow: Pischevaja Promyshlennost (in Russian).

Mañas P, Barsotti L, Cheftel JC. 2000. Microbial inactivation by pulsed electric fields in a batch treatment chamber: Effects of some electrical parameters and food constituents. *Innov Food Sci Emerg Technol* 2: 239–249.

Marchal L, Muravetchi V, Vorobiev EI, Bonhoure JP. 2004. Recovery of inulin from Jerusalem Artichoke Tubers: Development of a pressing method assisted by pulsed electric field. In: *International Congress on Engineering and Food*, Montpellier, France, CD-Rom, 6 p.

Martín-Belloso O, Vega-Mercado H, Qin BL, Chang FJ, Barbosa-Cánovas GV, Swanson BG. 1997. Inactivation of *Escherichia coli* suspended in liquid egg using pulsed electric fields. *J Food Process Preserv* 21: 193–208.

Matov BI, Reshetko EV. 1968. *Electrophysical Methods in Food Industry*. Kishinev: Kartja Moldavenjaske (in Russian).

McLachlan DS, Cai K, Chiteme C, Heiss WD. 2000. An analysis of dispersion measurements in percolative metal-insulator systems using analytic scaling functions. *Physica B* 279: 66–68.

Melkonyan H, Sorg C, Klempt M. 1996. Electroporation efficiency in mammalian cells is increased by dimethyl sulfoxide (DMSO). *Nucleic Acids Res* 24: 4356–4357.

Molinari P, Pilosof AMR, Jagus RJ. 2004. Effect of growth phase and inoculum size on the inactivation of *Saccharomyces cerevisiae* in fruit juices by pulsed electric fields. *Food Res Int* 37: 793–798.

Mouritsen OG, Jørgensen K. 1997. Small-scale lipid-membrane structure: Simulation versus experiment. *Curr Opin Struct Biol* 7: 518–527.

Ohshima T, Sato M, Saito M. 1995. Selective release of intracellular protein using pulsed electric field. *J Electrostat* 35: 103–112.

Pavlin M, Pavselj N, Miklavčič D. 2002. Dependence of induced transmembrane potential on cell density, arrangement, and cell position inside a cell system. *IEEE Trans Biomed Eng* 49: 605–612.

Pavlin M, Leben V, Miklavčič D. 2007. Electroporation in dense cell suspension—Theoretical and experimental analysis of ion diffusion and cell permeabilization. *Biochim Biophys Acta* 1770: 12–23.

Pavlin M, Kotnik T, Miklavčič D, Kramar P, Lebar AM. 2008. Electroporation of planar lipid bilayers and membranes. In: Leitmanova LA (ed.), *Advances in Planar Lipid Bilayers and Liposomes*. Amsterdam, the Netherlands: Elsevier, pp. 165–226.

Phoon PY, Galindo FG, Vicente A, Dejmek P. 2008. Pulsed electric field pulsed electric field in combination with vacuum impregnation with trehalose improves the freezing tolerance of spinach leaves. *J Food Eng* 88: 144–148.

Praporscic I. 2005. Influence du traitement combiné par champ électrique pulsé et chauffage modéré sur les propriétés physiques et sur le comportement au pressage de produits végétaux. PhD thesis, Compiègne, France: UTC.

Praporscic I, Ghnimi S, Vorobiev EI. 2005. Enhancement of pressing sugar beet cuts by combined ohmic heating and pulsed electric field treatment. *J Food Process Preserv* 29: 378–389.

Praporscic I, Lebovka NI, Ghnimi S, Vorobiev EI. 2006. Ohmically heated, enhanced expression of juice from apple and potato tissues. *Biosyst Eng* 93: 199–204.

Praporscic I, Lebovka NI, Vorobiev EI, Mietton-Peuchot M. 2007a. Pulsed electric field enhanced expression and juice quality of white grapes. *Sep Purif Technol* 52: 520–526.

Praporscic I, Shynkaryk MV, Lebovka NI, Vorobiev EI. 2007b. Analysis of juice colour and dry matter content during pulsed electric field enhanced expression of soft plant tissues. *J Food Eng* 79: 662–670.

Prescott SC. 1927. The treatment of milk by an electrical method. *Am J Public Health* 17: 221–223.

Prochownick L, Spaeth F. 1890. Über die keimtötende Wirkung des galvanischen Stroms. *DMW* 26: 564–565.

Pucihar G, Kotnik T, Teissié J, Miklavčič D. 2007. Electropermeabilization of dense cell suspensions. *Eur Biophys J* 36: 173–185.

Qin BL, Zhang Q, Swanson BG, Pedrow PD. 1994. Inactivation of microorganisms by different pulsed electric fields of different voltage waveforms. *IEEE Trans Dielectr Electr Insulator* 1: 1047–1057.

Raffa V, Ciofani G, Cuschieri A. 2009. Enhanced low voltage cell electropermeabilization by boron nitride nanotubes. *Nanotechnology* 20: 075104.

Ravishankar S, Zhang H, Kempkes ML. 2008. Pulsed electric fields. *Food Sci Technol Int* 14: 429–432.

Rojas-Chapana JA, Correa-Duarte MA, Ren Z, Kempa K, Giersig M. 2004. Enhanced introduction of gold nanoparticles into vital *Acidothiobacillus ferrooxidans* by carbon nanotube-based microwave electroporation. *Nano Lett* 4: 985–988.

Raso J, Álvarez I, Condón S, Sala-Trepat FJ. 2000. Predicting inactivation of *Salmonella senftenberg* by pulsed electric fields. *Innov Food Sci Emerg Technol* 1: 21–29.

Rogov IA. 1988. *Electrophysical Methods of Foods Product Processing*. Moscow: Agropromizdat (in Russian).

Rogov IA, Gorbatov AV. 1974. *Physical Methods of Foods Processing*. Moscow: Pischevaja Promyshlennost (in Russian).

Sale A, Hamilton W. 1967. Effect of high electric fields on microorganisms. I. Killing of bacteria and yeast. *Biochim Biophys Acta* 148: 781–788.

Salengke S, Sastry SK. 2005. Effect of ohmic pretreatment on the drying rate of grapes and ad-sorption isotherm of raisins. *Drying Technol* 23: 551–564.

Sampedro F, Rivas A, Rodrigo D, Martínez A, Rodrigo M. 2007. Pulsed electric fields inactivation of *Lactobacillus plantarum* in an orange juice–milk based beverage: Effect of process parameters. *J Food Eng* 80: 931–938.

Schwan HP. 1957. Electrical properties of tissue and cell suspensions. In: Lawrence JH, Tobias A (eds.), *Advances in Biological and Medical Physics*. New York: Academic Press, pp. 147–209.

Schwartzberg HG. 1997. Expression of fluid from biological solids. *Sep Purif Methods* 26: 1–213.

Schwartzberg HG, Chao RY. 1982. Solute diffusivities in leaching processes. *Food Technol* 36: 73–86.

Shynkaryk MV. 2007. Influence de la perméabilisation membranaire par champ électrique sur la performance de séchage des végétaux. PhD thesis, Compiègne, France: UTC.

Shynkaryk MV, Lebovka NI, Vorobiev EI. 2008. Pulsed electric fields and temperature effects on drying and rehydration of red beetroots. *Drying Technol* 26: 695–704.

Shynkaryk MV, Lebovka NI, Lanoisellé J-L, Nonus M, Bedel-Clotour C, Vorobiev EI. 2009. Electrically-assisted extraction of bio-products using high pressure disruption of yeast cells (*Saccharomyces cerevisiae*). *J Food Eng* 92: 189–195.

Stauffer D, Aharony A. 1994. *Introduction to Percolation Theory*. London: Taylor & Francis.

Stone GE. 1909. Influence of electricity on micro-organisms. *Bot Gazette* 48(5): 359–379.

Suga M, Goto A, Hatakeyama T. 2006. Control by osmolarity and electric field strength of electro-induced gene transfer and protein release in fission yeast cells. *J Electrostat* 64: 796–801.

Suga M, Goto A, Hatakeyama T. 2007. Electrically induced protein release from *Schizosaccharomyces pombe* cells in a hyperosmotic condition during and following a high electropulsation. *J Biosci Bioeng* 103: 298–302.

Susil R, Semrov D, Miklavčič D. 1998. Electric field induced transmembrane potential depends on cell density and organization. *Electromagn Biol Med* 17: 391–399.

Teissié J, Golzio M, Rols MP. 2005. Mechanisms of cell membrane electropermeabilisation: A minireview of our present (lack of?) knowledge. *Biochim Biophys Acta* 1724: 270–280.

Teissié J, Escoffre JM, Rols MP, Golzio M. 2008. Time dependence of electric field effects on cell membranes. A review for a critical selection of pulse duration for therapeutical applications. *Radiol Oncol* 42: 196–206.

Toepfl S. 2006. Pulsed electric fields (PEF) for permeabilization of cell membranes in food- and bioprocessing—Applications, process and equipment design and cost analysis. PhD thesis, Berlin, Germany: Institut für Lebensmitteltechnologie und Lebensmittelchemie.

Toepfl S, Knorr D. 2006. Pulsed electric fields as a pretreatment technique in drying processes. *Stewart Postharvest Rev* 4: 1–6.

Toepfl S, Mathys A, Heinz V, Knorr D. 2006. Potential of high hydrostatic pressure and pulsed electric fields for energy efficient and environmentally friendly food processing. *Food Rev Int* 22: 405–423.

Toepfl S, Heinz V, Knorr D. 2007. High intensity pulsed electric fields applied for food preservation. *Chem Eng Process* 46: 537–546.

Tung L, Troiano GC, Sharma V, Raphael RM, Stebe KJ. 1999. Changes in electroporation thresholds of lipid membranes by surfactants and peptides. *Ann NY Acad Sci* 888: 249–265.

Valic B, Golzio M, Pavlin M, Schatz A, Faurie C, Gabriel B, Teissié J, Rols M-P, Miklavčič D. 2003. Effect of electric field induced transmembrane potential on spheroidal cells: Theory and experiment. *Eur Biophys J* 32: 519–528.

Van der Poel PW, Schiweck H, Schwartz T. 1998. *Sugar Technology Beet and Cane Sugar Manufacture*. Denver, CO: Beet Sugar Development Foundation.

Van Wert SL, Saunders JA. 1992. Electrofusion and electroporation of plants. *Plant Physiol* 99: 365–367.

Vorobiev EI, Lebovka NI. 2006. Extraction of intercellular components by pulsed electric fields. In: Raso J, Heinz V (eds.), *Pulsed Electric Field Technology for the Food Industry. Fundamentals and Applications*. New York: Springer, pp. 153–194.

Vorobiev EI, Lebovka, NI (eds.). 2008. *Electrotechnologies for Extraction from Food Plants and Biomaterials*. New York: Springer.

Vorobiev EI, Bazhal MI, Bouzrara H. 2002. Solid-liquid expression of biological materials enhanced by electroosmosis and pulsed electric field. In: *Symposium on Emerging Technologies for the Food Industry*, Madrid, Spain.

Vorobiev EI, Lebovka NI, Praporscic I, Muravetchi V. 2004. Stages of constant rate and constant pressure solid-liquid expression enhanced by pulsed electric field. In: *Proceedings of 9 World Filtration Congress*, New Orleans, LA. CD-Rom, 9 p.

Vorobiev EI, Jemai AB, Bouzrara H, Lebovka NI, Bazhal MI. 2005. Pulsed electric field assisted extraction of juice from food plants. In: Barbosa-Cánovas G, Tapia MS, Cano MP (eds.), *Novel Food Processing Technologies*. New York: CRC Press, pp. 105–130.

Wang WC, Sastry SK. 2002. Effects of moderate electrothermal treatments on juice yield from cellular tissue. *Innov Food Sci Emerg Technol* 3: 371–377.

Weaver JC, Chizmadzhev YA. 1996. Theory of electroporation: A review. *Bioelectrochem Bioenerg* 41: 135–160.

Wosik J, Padmaraj D, Darne C, Zagozdozon-Wosik W. 2006. Dielectrophoresis of biological cells and single-walled carbon nanotubes. ISSO Annual Report, pp. 108–110.

Wouters PC, Smelt, JPPM. 1997. Inactivation of microorganisms with pulsed electric fields: Potential for food preservation. *Food Biotechnol* 11: 193–229.

Wouters PC, Dutreux N, Smelt JPPM, Lelieveld HLM. 1999. Effects of pulsed electric fields on inactivation kinetics of *Listeria innocua*. *Appl Environ Microbiol* 65: 5364–5371.

Yantzi JD, Yeow JTW. 2005. Carbon nanotube enhanced pulsed electric field electroporation for biomedical applications. In: *IEEE International Conference on Mechatronics Automation*, Vol. 4, Niagara Falls, Canada, pp. 1872–1877.

Yin Y, Cui Y, Ding H. 2008a. Optimization of betulin extraction process from *Inonotus obliquus* with pulsed electric fields. *Innov Food Sci Emerg Technol* 9: 306–310.

Yin Y, Cui Y, Wang T. 2008b. Study on extraction of polysaccharide from *Inonotus obliquus* by high intensity pulsed electric fields. *Nongye Jixie Xuebao/Trans Chinese Soc Agric Machinery* 39: 89–92.

Zagorulko AJa. 1958. Technological parameters of beet desugaring process by the selective electroplasmolysis. In: *New Physical Methods of Foods Processing*. Moscow: Izdatelstvo GosINTI (in Russian), pp. 21–27.

Zhang Q, Monsalve-Gonzalez A, Qin BL, Barbosa-Cánovas GV and Swanson BG. 1994. Inactivation of *Saccharomyces cerevisiae* in apple juice by square-wave and exponential—Decay pulsed electric fields. *J Food Process Eng* 17: 469–478.

Zhong T, Lima M. 2003. The effect of ohmic heating on vacuum drying rate of sweet potato tissue. *Bioresour Technol* 87: 215–220.

Zimmermann U. 1986. Electrical breakdown, electropermeabilization and electrofusion. *Rev Physiol Biochem Pharmacol* 105: 175–256.

Zimmermann U. 1992. Electric field-mediated fusion and related electrical phenomena. *Biochim Biophys Acta* 694: 227–277.

Zimmermann U. 1996. The effect of high intensity electric field pulses on eucaryotic cell membrane: Fundamentals and applications. In: Zimmermann U, Neil GA (eds.), *Electromanipulation of Cells*. Boca Raton, FL: CRC Press, pp. 1–106.

Zimmermann U, Pilwat G, Riemann F. 1974. Dielectric breakdown of cell membranes. *Biophys J* 14: 881–899.

27

In Vivo Electroporation: An Important Injury Mechanism in Electrical Shock Trauma

Iskandar Barakat

Jill Gallaher

Hongfeng Chen

Raphael C. Lee

27.1 Introduction

Most people have experienced some form of electrical shock considering our increasing exposure to electrical power. Public health records suggest that electrical injuries result in more than 3000 admissions to specialized burn units in the United States, accounting for 3%–4% of all burn-related injuries. In addition, it is estimated that there are far more electrical shock accidents that do not reach medical attention than the ones that do (Tkachenko et al. 1999). Up to 40% of serious electrical injuries are fatal, resulting in an estimated 1000 deaths annually. There is a bimodal distribution of environmental electrical injuries with respect to age; work-related injuries affect mostly young adults and domestic accidents occur more in children (Spies and Trohman 2006). In the American workplace, it is the fifth most common occupational hazard.

For years, electrical injury was considered to be mediated only by Joule heating. However, it is now well established that electrical field exposure can affect living tissues through Joule heating, electropore formation in cellular membranes, electroconformational changes in membrane proteins, and mechanical trauma resulting from the stimulation of muscle contraction. All these effects are dependent on the magnitude of the tissue field strength and the duration of field exposure. Clinical pathology may manifest early or late and any organ can be affected.

In trained hands, by controlling the field strength, the frequency, and exposure duration, electrical shock trauma can serve as a very useful therapeutic tool. The cardiac defibrillator is the most commonly recognized example. But, in addition, electrical current can be applied to the brain to arrest intractable seizures or to induce seizures to manage major psychiatric illnesses. These methods have been used and accepted for years. There are several other important examples. When used as therapy, the exposure is calibrated and controlled to limit the extent of injury to an amount that can be rapidly repaired by natural cellular healing mechanisms.

The subject of this chapter is the explanation of the clinically observed consequences of electrical shock injury, which exceeds natural repair and thus results in permanent tissue damage. Of course, it is the cell membrane that has the greatest vulnerability to damage by passage of electrical current through the body. Because the human anatomy is nonuniform and the tissues are anisotropic in their electrical properties, current passage through the body establishes tissue electrical fields that vary widely in magnitude from one anatomical location to another. The strongest electric fields are induced across the lipid bilayer of cell membranes. Therefore, the primary focus in understanding electrical shock pathogenesis focuses on the consequences of cell membrane alteration (Lee et al. 1988).

27.2 Injury Biophysics

Injury resulting from electrical shock may be mediated by effects of electric forces or elevated temperatures acting on the molecular scale (Lee and Astumian 1996). Indeed both mechanisms may ultimately lead to molecular conformational changes that alter protein function or permeabilize the cell membrane. It is this cell membrane damage, characterized by both breakdown of structural integrity and loss of function, that is the primary mechanism for the loss of tissue viability following electrical shock (Lee 2005). In this chapter, we will discuss the general concepts of what is known about the pathophysiological significance of electroporation in electrical injuries.

27.3 Electropore Formation

Electroporation is primarily mediated from the action of dielectric stresses acting on the interface between the high electrically polar water molecules and the transient defects in the lipid packing order within a bilayer (Lee et al. 1995, Ho and Mittal 1996, Weaver and Chizmadzhev 1996). The applied stress induces rearrangements of the ionic polar head-groups of bilayer lipids leading to localized hydrophilic pores (Böckmann et al. 2008). By the early 1980s, electroporation techniques were being used to introduce a variety of molecules including drugs and foreign DNA into cells and tissue (Chang et al. 1992, Aihara and Miyazaki 1998). Electroporation has been used on isolated cells to (a) introduce enzymes, antibodies, viruses, and other agents or particles for intracellular assays (Weaver 1993, Neumann et al. 1999, Teissié et al. 1999); (b) induce cell fusion (Neumann et al. 1989, Chang et al. 1992); and (c) insert or embed macromolecules into the cell membrane (Lynch and Davey 1996). Successful defibrillation of the heart also relies on electrically induced pores to reduce electrical nonuniformities and synchronize individual cells within the cardiac tissue. Electric shock is the only effective treatment for ventricular fibrillation and the success or failure of defibrillation has been attributed to one or both of these two mechanisms: (a) success in extinguishing ongoing fibrillatory activity (Zipes et al. 1975) and (b) failure to reinitiate a new arrhythmia (Frazier et al. 1989). Electroporation may produce both of these effects (Al-Khadra et al. 2000).

A major factor in determining the pore's kinetics is the physical state of the lipid bilayer, whether liquid crystal or fluidic, which is mostly governed by the temperature of the membrane. This is an important variable in tissue electroporation because tissue temperature varies considerably from one anatomic location to another. For example, skin temperature is typically lower than the temperature in cardiac muscle. Cell membranes are viscoelastic fluids. As such, the force required to deform the membrane under electrical stress is not only dependent on the amplitude of the applied voltage, but the duration of the application. The threshold transmembrane potential required for electroporation has been found to be in the range of 250–350 mV for field exposure greater than several milliseconds (Gowrishankar et al. 1998, Bier et al. 1999). Because of the viscous mechanical properties of the cell membrane, the threshold to create a pore is higher for shorter duration pulses.

Lifetimes of electropores may be transient or stable, depending on the magnitude of the induced transmembrane potential, its duration, membrane composition, and temperature. Transient pores have been observed to occur at lower applied voltages (340–480 mV) with sealing times averaged around 9 min whereas stable pores dominate at higher applied voltages (>540 mV) in rat skeletal muscle (Bier et al. 1999). In another study using liposomes, after an applied 1 V field was turned off, a stable micrometer-sized hole was seen to close within milliseconds (Chang et al. 1992). After application of a short electroporating field pulse, the sealing of transiently electroporated cells is spontaneous (Bier et al. 1999). However, stable defects require active modes of repair. Cells have evolved specialized proteins and molecular transport processes for sealing damaged membranes (McNeil and Steinhardt 2003, Togo 2004). The genetic expression of proteins involved in membrane sealing is increased to respond and adapt to repetitive membrane disruption. The sealing of electropores requires reordering of membrane lipids and removal of water molecules from the pore, which are both time- and energy-consuming processes (Gabriel and Teissié 1997, 1998). Sealing kinetics is often orders of magnitude slower than electropore formation. When the rate of molecular sealing events is overwhelmed by the rate of electroporation, as in the case of larger membrane defects, active transport of cytoplasmic transport vesicles is required to seal or patch the defect.

27.4 Current Path Parameters

The passage of current through the body exposes any cell or tissue in its path to electric fields. The risk of electroporation for any particular cell in the current path scales with the magnitude of the induced transmembrane potential (V_m), pulse duration, and membrane properties. The peak transmembrane potential induced by an externally applied electric field scales with the dimensions of the cell in the direction of the applied field. Generally, for media-suspended isolated cells with a characteristic diameter of 10–20 µm, the threshold field strength in the suspension media for electroporation is ~100 kV/m.

The fields required to electroporate adult skeletal muscle or peripheral nerve tissue are also much less due to their large lengths. As skeletal muscle cells in the extremities and peripheral nerve cells of humans and larger animals exceed 8 cm and 2 m, respectively, the electroporation thresholds begin to lower significantly (Lee et al. 2000). Electric fields as small as 0.2 kV/m may damage the membranes of these cells rendering them hundreds of fold more vulnerable to electroporation than small blood cells.

There is no set path in which electric current must pass through the body. Most frequently, the upper extremity is part of the current path found in major electrical shocks. The current passing along the long axis of the arm induces large muscle and nerve membrane potentials in cells with lengthwise orientation approximately parallel to the direction of the field lines (Danielson et al. 2000, Lee et al. 2000). The transmembrane potentials induced on the membranes of these cells are significantly larger than those experienced by skeletal muscle cells in any other orientation (Gaylor et al. 1988).

The cell membrane can be described as an "insulating shell" containing a conducting medium surrounded by a conducting buffer (Ramos and Teissié 2000). Normally, the intracellular and extracellular fluids have nearly equal ionic osmolarities, thus, their conductivities are similar. The lipid membrane conductivity, however, is typically 10^6–10^8 folds less than that of the surrounding fluid. Consequently, electric current established in the extracellular space by low-frequency fields are to a variable degree shielded from the cytoplasm by the electrically insulating cell membrane (Lee and Astumian 1996). This shielding leads to an induced transmembrane potential, which is sensitive to the geometry of the cell and its orientation in the field. A simple electrical cable model can provide quantitative insight into the importance of these parameters when considering the risk of electroporation injury.

27.5 Cable Model Theories

A distributed circuit model of a cell can be made with a series of parallel resistors and capacitors (Jack et al. 1975, Adrian 1983) as illustrated in Figure 27.1a. This circuit model of the membrane is combined with the specific resistivities of the intracellular and extracellular media to yield the cable

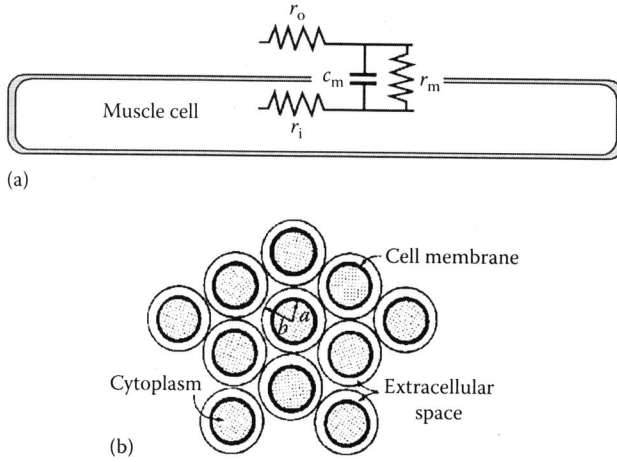

FIGURE 27.1   Circuit model of the cell. (a) Lumped element cable circuit model of a typical muscle cell. Cell length is equal to 2L. (b) A cross-sectional view of the arrangement of cells in a bundle which, in our model, is a hexagonal array of elongated cells with cell radius *a*, and extracellular fluid radius *b*.

circuit representation. In the presence of an applied uniform field $E(t)$ to the cell's axis, a transmembrane potential will be superimposed on the natural resting potential of the membrane.

The induced transmembrane potential distribution can be estimated using an approach proposed by Gaylor et al. for both isolated muscle cells and for cells within intact tissue (Gaylor et al. 1988). Analyzing the circuit model leads to a differential equation for the induced transmembrane potential, $V_m(z, t)$, along the long axis of the cell, z:

$$\lambda_m^2 \frac{\partial^2 V_m(z,t)}{\partial z^2} = V_m(z,t) + \tau_m \frac{\partial V_m(z,t)}{\partial t} \tag{27.1}$$

The parameters λ_m and τ_m are the membrane electrical space constant and the membrane charging time constant, respectively. They can be defined as

$$\lambda_m = \sqrt{\frac{1}{(r_i + r_o)g_m}}, \quad \tau_m = \frac{c_m}{g_m} \tag{27.2}$$

where

   r_i and r_o are the resistivities (Ω/cm) of the intracellular and extracellular fluids, respectively
   c_m and g_m are the capacitance per unit length (F/cm) and the conductance per unit length (S/cm) of the membrane, respectively

For the case of a single cell in an infinitely large bath of extracellular fluid, r_o is negligible compared with r_i since extracellular space is much greater than the intracellular space. Thus, for a constant applied electric field E_o, an approximate solution arises for the induced transmembrane potential of the cell

$$V_m(z) = \frac{\lambda_m E_o}{\cosh(L/\lambda_m)} \frac{\sinh(z/\lambda_m)}{1 + \lambda_m r_i G_e \tanh(L/\lambda_m)} \tag{27.3}$$

where

   G_e is the conductance of the membrane at the end-caps of the cylindrical cell
   L is the cell half-length

Equation 27.3 suggests that the induced transmembrane potential will increase according to the length of the cell up to the point where the size of the cell exceeds the length of the electrical space constant. For the largest skeletal muscle cells, the electrical space constant λ_m reaches a maximum value of about 2 cm, which means that the induced electrical potential reaches a maximum value of $E_o \lambda_m$ even when the physical length of the cell greatly exceeds the electrical space constant. For human skeletal muscle cells typically 2–8 cm in length, Equation 27.3 gives a membrane potential at the ends of the cells of four to five orders of magnitude larger than the applied field (Gaylor et al. 1988).

There is a second effect due to cell crowding. For cells within intact tissue subjected to an electric field, the previous analysis can be used to include the effects of neighboring cells on the induced transmembrane potential. The cells are assumed to be ordered parallel to each other in a hexagonal array as illustrated in Figure 27.1b. To facilitate the comparison of induced transmembrane potential in tissue with the case of isolated cells, the quantity $V_c/2L$ is used as the "source" term, where V_c is the voltage drop across the full length of a cell. For the isolated cell case, $V_c/2L$ is equal to the applied field amplitude E_o. The boundaries are modeled as cylinders of radius a, and the cross-sectional area of the extracellular fluid surrounding each cell with radial thickness $b - a$ will determine the extracellular resistivity r_o. Generally, for cells that are not on the muscle surface, r_o is not negligible compared with r_i, which significantly affects the value of the space constant, λ_m, in Equation 27.3.

Moreover, the extracellular electric field amplitude between muscle cells is not constant in the z axis. The induced transmembrane potential for a cell embedded in other cells is approximately

$$V_m(z) = \frac{V_c}{2L} \frac{\lambda_m (r_o + r_i)/r_i}{\cosh(L/\lambda_m)} \frac{\sinh(z/\lambda_m)}{1 + \lambda_m ((r_i + r_o)G_e + r_o/(Lr_i)) \tanh(L/\lambda_m)} \tag{27.4}$$

A comparison of the result predicted for cells within intact tissue (Equation 27.4) with the result predicted for isolated cells (Equation 27.3) suggests that cells surrounded by neighboring cells experience a higher induced maximum transmembrane potential than that of isolated cells. The decreased extracellular space increases resistance outside the cells and allows more current to flow through the ends of the cells producing the larger potential. Thus, the more closely packed the cells, the higher the induced potential.

The above analysis relates to the effects of an applied static electric field. Is this relevant to the contact with alternating electric currents that is used in standard commercial power supplies that operate near 60 Hz? The transmembrane potential response will depend on the charging time of the cell. Cooper gives a modal transient solution to the cable equation (Cooper 1986)

$$V_m(z,t) = \sum_{n=1}^{\infty} A_n \sin\left(\frac{\alpha_n z}{\lambda_m}\right) e^{(1+\alpha_n^2) t_n/\tau_m} \tag{27.5}$$

with

$$\alpha_n = \frac{n\pi\lambda_m}{2L} \tag{27.6}$$

Thus, the time-dependent terms decay with a time constant:

$$t_n = \frac{\tau_m}{1 + \left(n\pi \dfrac{\lambda_m}{2L}\right)^2} \tag{27.7}$$

The time τ_l is the maximum time constant and represents the time required for the cell to attain the transmembrane potential distribution predicted by the cable model analysis. For long skeletal muscle and peripheral nerve cells, this time approaches τ_m (typically between 1 and 3 ms) but it decreases for cells of shorter length.

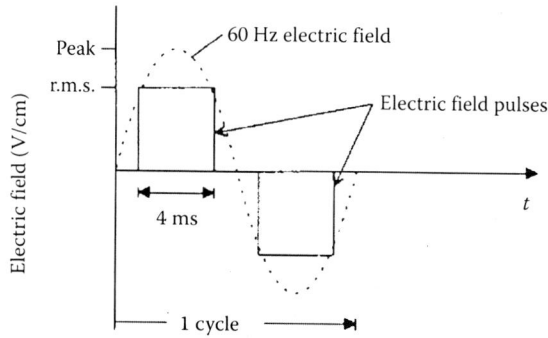

FIGURE 27.2 Approximation for a commercial sinusoidal electric current. Alternating 4 ms discontinuous square wave field pulses can roughly approximate a standard 60 Hz commercial sinusoidal electric current. The pulse magnitude will equal the average value of the field within the 4 ms timeframe.

For a 60 Hz applied field, large muscle and nerve cells should reach a steady-state level during each excursion of the sine wave because the root-mean-square duration of the 60 Hz sine wave is about 4 ms (Figure 27.2).

The model predicts a strong nonlinear dependence of induced V_m along the direction of the field. Moreover, cells that are surrounded by other cells can be expected to be more vulnerable to injury than cells on the muscle surface because they experience a larger imposed transmembrane potential. The increased potential for electroporation in electrical shock injuries of longer and more densely packed muscle and nerve cells in the extremities must be taken into account in assessing tissue damage. These crowding and size discrepancies must also be considered when predicting *in vivo* outcomes with *in vitro* studies.

27.6 Thermal Injury

A discussion about electrical shock injury would be incomplete and misleading if thermal injury aspects were not mentioned. Current passage in a conducting media will increase the temperature through Joule heating, dielectric heating, or both. Because the components of cell membranes are held together only by forces of hydration, the lipid bilayer is the most vulnerable to thermal insult (Gershfeld and Murayama 1988, Lee et al. 2000). Alterations in the membrane molecular structure disturbs the membrane transport processes. Temperatures of only 6°C above normal (i.e., 43°C) may disrupt the structural integrity of the lipid bilayer (Moussa et al. 1979). As melting events of the membrane components is approached, the permeability of cell membrane increases. This may be due to increased area fluctuations causing local maximums in lateral compressibility and/or larger domain boundaries between solid and liquid phases where these pores are thought to occur (Heimburg 2007, Blicher et al. 2009). Experiments on fibroblasts have demonstrated that heat-induced membrane permeabilization also begins to appear at temperatures greater than 45°C (Merchant et al. 1998). Though not as susceptible as the bilayer components, there is also thermal denaturation of membrane-bound proteins, which may lead to irreversible unfolded conformations and aggregation at supraphysiological temperatures (Despa et al. 2005).

27.7 Current Path through the Body

The pathway that a current takes through the body during electrical shock determines the tissues at risk and the clinical manifestations. When the primary cause of tissue injury is electroporation and not heat, clinical diagnosis is challenging. Electroporated tissue appears to be edematous without other visible changes. A histological analysis of tissue biopsies has characteristic changes (Figure 27.3b). When electroporation damage exceeds cell membrane resealing capability, this injury will progress to necrosis.

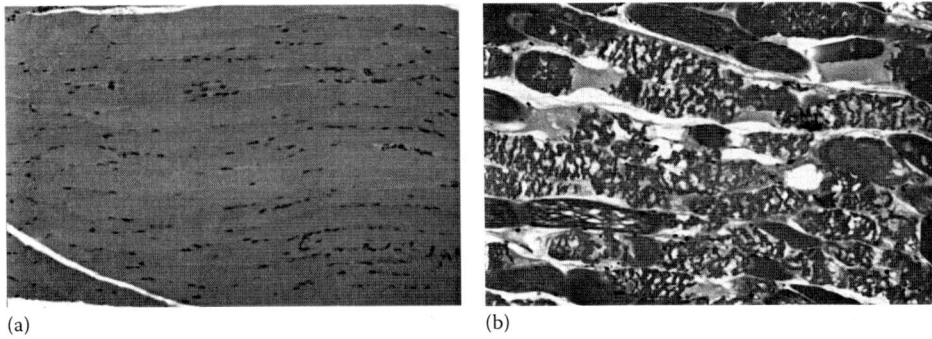

(a)   (b)

FIGURE 27.3   Human muscle cells as they appear in a normal and an electroporated patient. (a) Hematoxylin and Eosin stained skeletal muscle biopsy taken from a normal subject and seen under 40 times power magnification. Notice the regularly aligned sarcomeres as compared with figure (b) which shows edematous cells with loss of normal architecture and contraction band necrosis.

At frequencies less than 10 kHz, the skin is the primary resistive barrier to the flow of current into the body. Deeply calloused skin can have 20–70 times greater resistance (Childbert et al. 1985). This high resistance may result in a large fractional voltage drop across the interface and a significant amount of energy will be dissipated at the skin surface. If the body makes contact with a voltage drop greater than 200 V and the power source generates several amperes of current, the epidermis is destroyed almost instantaneously. Eliminating the epidermal barrier lowers total body resistance and leads to accelerated tissue damage (Püschel et al. 1985). Respiratory arrest may follow electrical shock from damage to the respiratory musculature or associated nervous system. This may be caused by tetanic contraction or paralysis of respiratory muscles in the lungs (Spies and Trohman 2006). Current passing through the heart or thorax can also cause cardiac dysrhythmias (Jensen et al. 1987) and/or direct myocardial damage (Ku et al. 1989, Walton et al. 1988). Electrical current passage directly through the brain will result in different outcomes than for current passage limited to one extremity. When the current path travels through brain tissue, it can lead to respiratory arrest, seizures, and paralysis (Hooshmand et al. 1989).

The current path through the body is a major determinant of the outcome, yet damage cannot always be predicted solely by pathway. Neurological manifestations of electrical shock can be especially perplexing because cognitive and emotional consequences often follow electrical shock (Cooper et al. 1992, Kelley et al. 1994, Pliskin et al. 1998, 2006, Chico et al. 1999). This is common even when no current passes directly through the brain. While the exact mechanism of this effect is not understood, it is generally recognized that the nervous system is tightly connected. Damage to peripheral nerves in the extremities can manifest in the brain.

27.8 Challenges in Clinical Diagnosis

To illustrate the clinical consequences of the most common modes of electrical shock injury, the following clinical cases of an electrician on the job and a young boy at home are presented. Mr. V is a patient who was recently evaluated by the physicians of the Chicago Electrical Trauma Research Institute. He is a 40-year-old patient suffering from multiple post-electric injury complaints. The injury occurred while performing his job as an electrician in May 2000. His right (dominant) hand came in electrical contact with what he thought was a "de-energized" 12,500 V transformer circuit breaker. The contact was mediated by the ignition of an arc between his hand and the high-voltage side of the transformer. A snapping loud arc was heard, and then the transformer faulted, which generated a high-energy acoustic shock wave that knocked the patient to the ground in front of the transformer housing. On contact, Mr. V experienced what he described as a "burning" sensation and severe pain in the back of his head as if being shot. This was followed by a brief

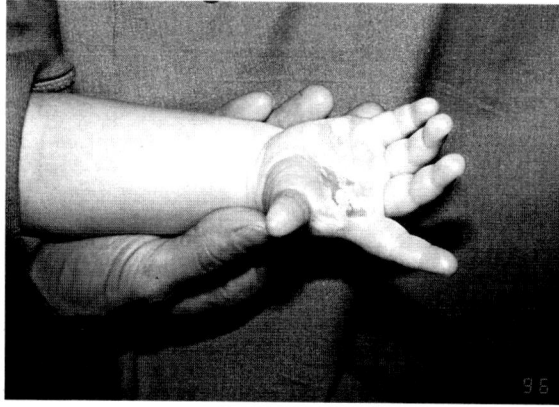

FIGURE 27.4 Right hand of a 1-year-old child after brief contact with an open fluorescent bulb power supply. Besides the superficial burn, this child presented with temporary muscle paralysis of the lumbrical muscle of his right index finger and the inability to flex his index following a brief contact with an open fluorescent bulb supply. Patient fully recovered within a couple weeks.

period of loss of consciousness after which he woke up confused on the way to the hospital with a burning sensation in his hands and feet.

In the emergency ward, Mr. V was complaining of chest pain, headache, dizziness, numbness of his hands and feet, and diffuse upper body myalgias. No major skin burns were noted except for a small arc wound. Mr. V's electrocardiogram was normal. Neuro-imaging results revealed no evidence of abnormalities. No evidence of skull fractures was found at that time. He was discharged the following day.

Mr. V followed up immediately with his primary care physician and was referred to a neurologist because of "black-out" episodes with subsequent disorientation. A brain MRI was completed but did not show any abnormalities. Mr. V tried several anticonvulsants, which did not help these episodes. A 24 h cardiac rhythm evaluation revealed episodic supra ventricular tachycardias.

Over the past 5 years, Mr. V has undergone multiple evaluations from neurologists and neuropsychologists. Various diagnostic imaging techniques have failed to show a clear pathology yet neuropsychological studies show considerable disabilities. He has been receiving treatment for post-traumatic stress disorder and depression since 2001. Mr. V returned to light-duty work 9 months following the injury but finds simple tasks such as using as screw driver or walking down a flight a stairs to require his full concentration. He remains unable to return to the dangerous power line work he was engaged in pre-trauma.

Another example is the case of the young boy in Figure 27.4 who got electrocuted while briefly holding on to an open fluorescent bulb supply with his bare wet hand. He presented with a superficial burn to his hand and temporary loss of function of the intrinsic muscles of his hand. His symptoms resolved completely after a couple of weeks and the wound healed without grafting.

A multidisciplinary medical team approach is often needed to manage electrical shock patients due to the various symptoms that it may present (Capelli-Schellpfeffer et al. 1995, Chico et al. 1999). There is considerable need to increase the research efforts to understand the relationship between electrical contact parameters such as voltage, current capacity, and duration of current passage, frequency, anatomical current path, and pre-existing medical conditions. These many variables increase the complexity of diagnosing and treating electrical shock.

27.9 Summary

The significance of nonthermal modes of tissue damage such as electroporation in the pathogenesis of electrical shock injury has become increasingly recognized over the past 20 years. Peripheral neurons and skeletal muscle cells are particularly vulnerable to electroporation injury because excitable tissue must detect and respond to action potentials, which in the context of this discussion have relatively small associated electric fields. When the electrical shock is very short, there is often very little visible burn injury. However, clinical manifestations of such injury can be severe. Understanding underlying mechanisms of injury and developing an effective therapy to treat electroporated membranes is a high priority.

References

Adrian RH. 1983. Electrical properties of striated muscle. In *Handbook of Physiology.* Bethesda, MD: American Physiological Society.

Aihara H, Miyazaki JI. 1998. Gene transfer into muscle by electroporation in vivo. *Nature Biotechnology* 16:867–870.

Al-Khadra A, Nikolski V, Efimov IR. 2000. The role of electroporation in defibrillation. *Circulation Research* 87(9):797–804.

Bier M, Hammer SM, Canaday DJ, Lee R. 1999. Kinetics of sealing for transient electropores in isolated mammalian skeletal muscle cells. *Bioelectromagnetics* 20:194–201.

Blicher A, Wodzinska K, Fidorra M, Winterhalter M, Heimburg T. 2009. The temperature dependence of lipid membrane permeability, its quantized nature, and the influence of anesthetics. *Biophysical Journal* 96(11):4581–4591.

Böckmann RA, de Groot BL, Kakorin S, Neumann E, Grubmüller H. 2008. Kinetics, statistics, and energetics of lipid membrane electroporation studied by molecular dynamics simulations. *Biophysical Journal* 95(4):1837–1850.

Capelli-Schellpfeffer M, Toner M, Lee RC, Astumian RD. 1995. Advances in the evaluation and treatment of electrical and thermal injury emergencies. *IEEE Transactions on Industrial Applications* 31(5):1147–1152.

Chang DC, Chassy BM, Saunders JA, Sowers AE. 1992. *Guide to Electroporation and Electrofusion.* New York: Academic Press.

Chico M, Capelli-Schellpfeffer M, Kelley KM, Lee RC. 1999. Management and coordination of post-acute medical care for electrical trauma survivors. *Annals of the New York Academy of Science* 888:334–342.

Childbert M, Maiman D, Sances A Jr., Myklebust J, Prieto TE, Swiontek T, Heckman M, Pintar K. 1985. Measure of tissue resistivity in experimental electric burns. *Journal of Trauma* 25:209.

Cooper MS. 1986. Electrical cable theory, transmembrane ion fluxes, and the motile responses of tissue cells to external electrical fields. In *Bioelectric Interactions Symposium IEEE/Engineering in Medicine and Biology Society, 7th Annual Conference*, Chicago, IL.

Cooper MA, Andrews CJ, ten Duis HJ. 1992. Neuropsychological aspects of lightning injury. In *Proceedings of the 9th International Conference on Atmospheric Physics*, St. Petersburg, Russia.

Danielson J, Capelli M, Lee RC. 2000. Electrical injury of the upper extremity. *Hand Clinics* 16:225–234.

Despa F, Orgill DP, Neuwalder J, Lee RC. 2005. The relative thermal stability of tissue macromolecules and cellular structure in burn injury. *Burns* 31(5):568–577.

Frazier DW, Wolf PD, Wharton JM, Tang AS, Smith WM, Ideker RE. 1989. Stimulus-induced critical point. Mechanism for electrical initiation of reentry in normal canine myocardium. *The Journal of Clinical Investigation* 83(3):1039–1052.

Gabriel B, Teissié J. 1997. Direct observation in the millisecond time range of fluorescent molecule asymmetrical interaction with the electropermeabilized cell membrane. *Biophysical Journal* 73:2630–2637.

Gabriel B, Teissié J. 1998. Mammalian cell electropermeabilization as revealed by millisecond imaging of fluorescence changes of ethidium bromide in interaction with the membrane. *Bioelectrochemistry and Bioenergetics* 47:113–118.

Gaylor DG, Prakah-Asante K, Lee RC. 1988. Significance of cell size and tissue structure in electrical trauma. *Journal of Theoretical Biology* 133(2):223–237.

Gershfeld NL, Murayama M. 1988. Thermal instability of red blood cell membrane bilayers: Temperature dependence of hemolysis. *Journal of Membrane Biology* 101:62–72.

Gowrishankar TR, Chen W, Lee RC. 1998. Non-linear microscale alterations in membrane transport by electropermeabilization. *Annals of the New York Academy of Science* 858:205–216.

Heimburg T. 2007. *Thermal Biophysics of Membranes*. Berlin, Germany: Wiley-VCH.

Ho SY, Mittal GS. 1996. Electroporation of cell membranes: A review. *Critical Reviews in Biotechnology* 16(4):349–362.

Hooshmand H, Radfar F, Beckner E. 1989. The neurophysiologic aspects of electrical injuries. *Clinical EEG* 20(2):111–120.

Jack JJB, Noble D, Tsien RW. 1975. *Electric Current Flow in Excitable Cells*. Oxford, U.K.: Oxford University Press.

Jensen PJ, Thomsen PE, Bagger JP, Norgaard A, Baandrup U. 1987. Electrical injury causing ventricular arrhythmias. *British Heart Journal* 57:279–283.

Kelley KM, Pliskin NH, Meyer G, Lee RC. 1994. Neuropsychiatric aspects of electrical injury. The nature of psychiatric disturbance. *Annals of the New York Academy of Sciences* 720:213–218.

Ku CS, Lin SL, Hsu TL, Wang SP, Chang MS. 1989. Myocardial damage associated with electrical injury. *The American Heart Journal* 118:621–624.

Lee RC. 2005. Cell injury by electric forces. *Annals New York Academy of Sciences* 1066:85–91.

Lee RC, Astumian RD. 1996. Physiochemical basis for thermal and non-thermal 'burn' injuries. *Burns* 22(7):509–519.

Lee RC, Aarsvold JN, Chen W, Astumian RD, Capelli-Schellpfeffer M, Kelley KM, Pliskin NH. 1995. Biophysical mechanisms of cell membrane damage in electric shock. *Seminars in Neurology* 15(4):367–374.

Lee RC, Gaylor DC, Bhatt D, Israel DA. 1988. Role of cell membrane rupture in the pathogenesis of electrical trauma. *The Journal of Surgical Research* 44(6):709–719.

Lee RC, Zhang D, Hannig J. 2000. Biophysical injury mechanisms in electrical shock trauma. *Annual Review of Biomedical Engineering* 2:477–509.

Lynch PT, Davey MR. 1996. *Electrical Manipulation of Cells*. New York: Chapman & Hall, p. 292.

McNeil PL, Steinhardt RA. 2003. Plasma membrane disruption: Repair, prevention, adaptation. *Annual Review of Cell and Developmental Biology* 19:697–731.

Merchant FA, Holmes WH, Capelli-Schellpfeffer M, Lee RC, Toner M. 1998. Poloxamer 188 enhances functional recovery of lethally heat-shocked fibroblasts. *The Journal of Surgical Research* 74:131–140.

Moussa NA, Tell EN, Cravalho EG. 1979. Time progression of hemolysis of erythrocyte populations exposed to supraphysiologic temperatures. *Journal of Biomechanical Engineering* 101:213–217.

Neumann E, Sowers AE, Jordan CA. 1989. *Electroporation and Electrofusion in Cell Biology*. New York: Plenum Press, p. 436.

Neumann E, Kakorin S, Toensing K. 1999. Fundamentals of electroporative delivery of drugs and genes. *Bioelectrochemistry and Bioenergetics* 48:3–16.

Pliskin NH, Capelli-Schellpfeffer M, Law RT, Malina AC, Kelley KM, Lee RC. 1998. Neuropsychological symptom presentation following electrical injury. *The Journal of Trauma* 44(4):709–715.

Pliskin NH, Ammar AM, Fink JW, Hill K, Malina AC, Ramati A, Kelley KM, Lee RC. 2006. Neuropsychological changes following electrical injury. *Journal of the International Neuropsychological Society* 12:17–23.

Püschel K, Brinkmann B, Lieske K. 1985. Ultrastructural alteration of skeletal muscles after electrical shock. *The American Journal of Forensic Medicine Pathology* 6(4):296–300.

Ramos C, Teissié J. 2000. Electrofusion: A biophysical modification of cell membrane and a mechanism in exocytosis. *Biochemie* 82:511–518.

Spies C, Trohman R. 2006. Narrative review: Electrocution and life-threatening electrical injuries. *Annals of Internal Medicine* 145:531–537.

Teissié J, Eynard N, Gabriel B, Rols MP. 1999. Electropermeabilization of cell membranes. *Advanced Drug Delivery Reviews* 35:3–19.

Tkachenko TA, Kelley KM, Pliskin NH, Fink JW. 1999. Electrical injury through the eyes of professional electricians. *Annals of the New York Academy of Sciences* 888:42–59.

Togo T. 2004. Long-term potentiation of wound-induced exocytosis and plasma membrane repair is dependent on cAMP-response element-mediated transcription via a protein kinase C- and p38 MAPK-dependent pathway. *The Journal of Biological Chemistry* 279(43):44996–45003.

Walton AS, Harper RW, Coggins GL. 1988. Myocardial infarction after electrocution. *The Medical Journal of Australia* 148(7):365–367.

Weaver JC. 1993. Electroporation: A general phenomenon for manipulating cells and tissue. *Journal of Cellular Biochemistry* 51(4):426–435.

Weaver JC, Chizmadzhev YA. 1996. Theory of electroporation: A review. *Bioelectrochemistry and Bioenergetics* 41(2):135–160.

Zipes DP, Fischer J, King RM, Nicoll AD, Jolly WW. 1975. Termination of ventricular fibrillation in dogs by depolarizing a critical amount of myocardium. *The American Journal of Cardiology* 36(1):37–44.

Index

DATE DUE